beginning algebra

seventh edition

M000289849

The *Streeter/Hutchison* Series in Mathematics

Baratto | Bergman

beginning algebra

seventh edition

Stefan Baratto

Clackamas Community College

Barry Bergman

Clackamas Community College

 Higher Education

Boston   Burr Ridge, IL   Dubuque, IA   New York   San Francisco   St. Louis
Bangkok   Bogotá   Caracas   Kuala Lumpur   Lisbon   London   Madrid   Mexico City
Milan   Montreal   New Delhi   Santiago   Seoul   Singapore   Sydney   Taipei   Toronto

Higher Education

BEGINNING ALGEBRA, SEVENTH EDITION

Published by McGraw-Hill, a business unit of The McGraw-Hill Companies, Inc., 1221 Avenue of
the Americas, New York, NY 10020. Copyright © 2008 by The McGraw-Hill Companies, Inc. All rights
reserved. No part of this publication may be reproduced or distributed in any form or by any means, or
stored in a database or retrieval system, without the prior written consent of The McGraw-Hill
Companies, Inc., including, but not limited to, in any network or other electronic storage or
transmission, or broadcast for distance learning.

Some ancillaries, including electronic and print components, may not be available to customers outside
the United States.

This book is printed on recycled, acid-free paper containing 10% postconsumer waste.

4 5 6 7 8 9 0 QPD/QPD 0 9 8
1 2 3 4 5 6 7 8 9 0 QPD/QPD 0 9 8 7 6

ISBN 978–0–07–304831–4
MHID 0–07–304831–3

ISBN 978–0–07–325783–9
MHID 0–07–325783–4 (Annotated Instructor's Edition)

Publisher: *Elizabeth J. Haefele*
Sponsoring Editor: *Richard Kolasa*
Senior Developmental Editor: *Michelle L. Flomenhoft*
Marketing Manager: *Barbara Owca*
Project Manager: *Lora Kalb*
Senior Production Supervisor: *Kara Kudronowicz*
Senior Media Project Manager: *Sandra M. Schnee*
Media Producer: *Amber M. Huebner*
Senior Designer: *David W. Hash*

Cover/Interior Designer: *Asylum Studios; Mona
Grigaliunas,Tom J. Nemoda*
Senior Photo Research Coordinator: *Lori Hancock*
Photo Research: *Connie Mueller*
Supplement Producer: *Melissa M. Leick*
Compositor: *Interactive Composition Corporation*
Typeface: *10/12 New Times Roman*
Printer: *Quebecor World Dubuque, IA*

www.mhhe.com

Don Hutchison has literally changed our lives. Don provided us with the opportunity of a lifetime by inviting us to write with him on the Streeter/Hutchison texts. Just as Don learned a great deal from James Streeter, we have learned and grown in many ways thanks to Don. Not only does Don have the wonderful ability to present mathematical ideas in simple, clear terms, he is also generous, has a wonderful family, and is a pleasure to spend time with. Don serves as an excellent role model for both of us, but most importantly, we are proud to call Don a friend.

Stefan Baratto and Barry Bergman

about the authors

Stefan Baratto

Stefan currently enjoys teaching math at Clackamas Community College in Oregon. He has also taught at the University of Oregon, Southeast Missouri State University, York County Technical College (Maine), and in New York City middle schools.

Stefan is involved in many professional organizations, including AMATYC and ORMATYC, where he is a frequent presenter. He has also used statistics, technology, and Web design to apply math to various fields.

Stefan, and his wife Peggy, make their home in Portland and spend their time enjoying the wonders of Oregon and the Pacific Northwest.

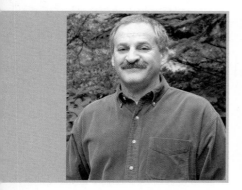

Barry Bergman

Barry has enjoyed teaching mathematics to a wide variety of students over the years. He began in the field of adult basic education and moved into the teaching of high school mathematics in 1977. He taught at that level for 11 years, at which point he served as a K–12 mathematics specialist for his county. This work allowed him the opportunity to help promote the emerging NCTM Standards in his region.

In 1990 Barry began the present portion of his career, having been hired to teach at Clackamas Community College. He maintains a strong interest in the appropriate use of technology and visual models in the learning of mathematics.

Throughout the past 28 years, Barry has played an active role in professional organizations. As a member of OCTM, he contributed several articles and activities to the group's journal. He has made presentations at OCTM, NCTM, ORMATYC, and ICTCM conferences. Barry also served as an officer of ORMATYC for four years and participated on an AMATYC committee to provide feedback to revisions of NCTM's Standards.

Donald Hutchison

Don began teaching in a preschool while he was an undergraduate. He subsequently taught children with disabilities, adults with disabilities, high school mathematics, and college mathematics. Although all of these positions were challenging and satisfying, it was breaking a challenging lesson into teachable components that he most enjoyed.

At Clackamas Community College he found his professional niche. The community college allowed him to focus on teaching within a department that constantly challenged faculty and students to expect more. Under the guidance of Jim Streeter, Don learned to present his approach to teaching in the form of a textbook.

Don has also been an active member of many professional organizations. He has been president of ORMATYC, AMATYC committee chair, and ACM curriculum committee member. He has presented at AMATYC, ORMATYC, AACC, MAA, ICTCM, and numerous other conferences.

Beginning Algebra   The Streeter/Hutchison Series in Mathematics

contents

Chapter 8

Systems of Linear Equations

Chapter 9

Exponents and Radicals

Chapter 10

Quadratic Equations

You are about to begin a course in mathematics. We made every attempt to provide a text that will help you understand what mathematics is about and how to use it effectively. We made no assumptions about your previous experience with mathematics. Your progress through the course will depend on the amount of time and effort you devote to the course and your previous background in math. There are some specific features in this book that will aid you in your studies. Here are some suggestions about how to use this book. (Keep in mind that a review of *all* the chapter and summary material will further enhance your ability to grasp later topics and to move more effectively through the text.)

1. If you are in a lecture class, make sure that you take the time to read the appropriate text section *before* your instructor's lecture on the subject. Then take careful notes on the examples that your instructor presents during class.

2. After class, work through similar examples in the text, making sure that you understand each of the steps shown. Examples are followed in the text by *Check Yourself* exercises. You can best learn mathematics by being involved in the process, and that is the purpose of these exercises. Always have a pencil and paper at hand, work out the problems presented, and check your results immediately. If you have difficulty, go back and carefully review the previous exercises. Make sure you understand what you are doing and why. The best test of whether you do understand a concept lies in your ability to explain that concept to one of your classmates. Try working together.

3. At the end of each chapter section you will find a set of exercises. Work these carefully to check your progress on the section you have just finished. You will find the solutions for the odd-numbered exercises following the problem set. If you have difficulties with any of the exercises, review the appropriate parts of the chapter section. If your questions are not completely cleared up, by all means do not become discouraged. Ask your instructor or an available tutor for further assistance. A word of caution: Work the exercises on a regular (preferably daily) basis. Again, learning mathematics requires becoming involved. As is the case with learning any skill, the main ingredient is practice.

4. When you complete a chapter, review by using the *Summary.* You will find all the important terms and definitions in this section, along with examples illustrating all the techniques developed in the chapter. Following the *Summary* are *Summary Exercises* for further practice. The exercises are keyed to chapter sections, so you will know where to turn if you are still having problems.

5. When you finish with the *Summary Exercises,* try the *Self-Test* that appears at the end of each chapter. It is an actual practice test you can work on as you review for in-class testing. Again, answers with section references are provided.

6. Finally, an important element of success in studying mathematics is the process of regular review. We provide a series of *Cumulative Reviews* throughout the textbook, beginning at the end of Chapter 2. These tests will help you review not only the concepts of the chapter that you have just completed but also those of previous chapters. Use these tests in preparation for any midterm or final exams. If it appears that you have forgotten some concepts that are being tested, don't worry. Go back and review the sections where the idea was initially explained or the appropriate chapter *Summary.* That is the purpose of the *Cumulative Review.*

We hope that you will find our suggestions helpful as you work through this material, and we wish you the best of luck in the course.

Stefan Baratto

Barry Bergman

Donald Hutchison

Beginning Algebra   The Streeter/Hutchison Series in Mathematics

preface

Message from the Authors

We believe the key to learning mathematics, at any level, is active participation—
MASTERING MATH THROUGH PRACTICE. Students who are active participants in the
learning process have the opportunity to construct their own mathematical ideas
and make connections to previously studied material. Such participation leads to
understanding, retention, success, and confidence. We developed this text with that
philosophy in mind and integrated many features throughout the book to reflect that
philosophy. Our goal is to provide *content in context*. The opening vignette for each
chapter, one activity in that chapter, and several section exercises all relate to the
same topic in order to engage students and allow them to see the relevance of
mathematics. The *Check Yourself* exercises are designed to keep the students active
and involved with every page of exposition. The optional calculator references involve
students actively in the development of mathematical ideas.

The exercise sets are organized to showcase the different types of exercises
available to the students: Basic Skills, Advanced Skills, Vocational-Technical
Applications, Calculator/Computer, Above and Beyond (which encompasses
challenging exercises, writing exercises, and collaborative exercises), and Getting
Ready. Answer blanks for these exercises appear in the margin, allowing the student
to actively use the text, making it more than just a reference tool. Many of these
exercises are designed to awaken interest and insight in students; all are meant to
provide continual practice and reinforcement of the topics being learned. For those
seeking additional understanding, some exercises include icons next to them to
indicate that a video stepping through that exercise is available. Not all of the
exercises will be appropriate for every student, but each one gives another opportunity
for both the instructor and the student. Our hope is that every student who uses this
text will become a better mathematical thinker as a result.

Overview

When preparing to write this text, the authors solicited feedback from the market about what types of pedagogical tools would better help students to understand and retain the key beginning algebra concepts, in addition to what content should be updated or expanded. Increasing the number of applications and integrating them throughout the sections was one of the most prevalent responses. Based on this feedback, the authors focused on these themes as they revised the text.

In terms of pedagogical tools, this worktext seeks to provide carefully detailed explanations and accessible pedagogy to introduce beginning algebra concepts to the students. The authors use a three-pronged approach to present the material and encourage critical thinking skills. The areas used to create the framework are communication, pattern recognition, and problem solving. Items such as *Math Anxiety* boxes, *Check Yourself* exercises, and *Activities* represent this approach and the underlying philosophy of mastering math through practice. A new feature is *Reading Your Text*—these quick exercises presented at the end of each section quiz students' vocabulary knowledge and help strengthen their communication skills.

Market research has reinforced the importance of the exercise sets to the students' ability to process the content. To that end, the exercise sets in this edition have been expanded, organized, and clearly labeled (as previously outlined). Vocational and professional-technical exercises have been added throughout. In addition, exercises with fractions, decimals, and negative numbers have been added as appropriate. Repeated exposure to this consistent structure should help advance the students' skills in relating to mathematics.

Features

A number of features have contributed to the previous success of the Hutchison texts. All of these features are also included in this text, and help reinforce the authors' philosophy of *mastering math through practice*. Each feature was thoroughly discussed by the authors and review panels during the development of the text. More than ever, we are confident that the entire learning package is of value to your students and to you as an instructor. We will describe each of the key features of our package.

> ## Make the Connection— Chapter-Opening Vignettes

The chapter-opening vignettes were substantially revised to provide students interesting, relevant scenarios that will capture their attention and engage them in the upcoming material. Furthermore, exercises and *Activities* related to the *Opening Vignette* were added or updated in each chapter. These exercises are marked with a special icon.

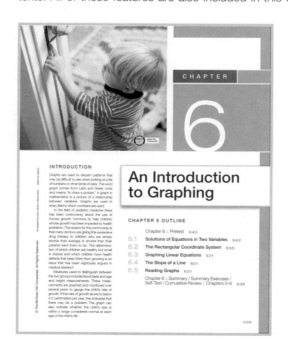

preface

> Activities

Activities are included in each chapter. The *Activities* promote active learning by requiring students to find, interpret, and manipulate real-world data. Each *Activity* relates to the chapter-opening vignette, providing cohesiveness to the chapter. Students can complete the *Activities* on their own, but these are best solved in small groups.

> Check Yourself Exercises

Check Yourself exercises have been the hallmark of the Hutchison series; they are designed to actively involve students throughout the learning process. Every example is followed by an exercise that encourages students to solve a problem similar to the one just presented and check/practice what they have just learned. Answers are provided at the end of the section for immediate feedback.

Example 1 — Identifying the Commutative Properties

< Objective 1 >

(a) $5 + 9 = 9 + 5$

This is an application of the commutative property of addition.

(b) $5 \cdot 9 = 9 \cdot 5$

This is an application of the commutative property of multiplication.

Check Yourself 1

Identify the property being applied.

(a) $7 + 3 = 3 + 7$ (b) $7 \cdot 3 = 3 \cdot 7$

> Overcoming Math Anxiety Boxes

Overcoming Math Anxiety boxes are located within the first few chapters. These suggestions are designed to be timely and useful. They are similar to the suggestions most instructors make in class, but having them in the text provides another opportunity to impact the student.

Overcoming Math Anxiety

Hint #5
Working Together
How many of your classmates do you know? Whether you are by nature gregarious or shy, you have much to gain by getting to know your classmates.

1. It is important to have someone to call when you miss class or are unclear on an assignment.

2. Working with another person is almost always beneficial to both people. If you don't understand something, it helps to have someone to ask about it. If you do understand something, nothing cements that understanding more than explaining the idea to another person.

3. Sometimes we need to commiserate. If an assignment is particularly frustrating, it is reassuring to find that it is also frustrating for other students.

4. Have you ever thought you had the right answer, but it doesn't match the answer in the text? Frequently the answers are equivalent, but that's not always easy to see. A different perspective can help you see that. Occasionally there is an error in a textbook (here we are talking about *other* textbooks). In such cases it is wonderfully reassuring to find that someone else has the same answer you do.

preface

> Reading Your Text

The new *Reading Your Text* feature is a set of quick exercises presented at the end of each section to quiz students' vocabulary knowledge. These exercises are designed to encourage careful reading of the text. If students do not understand the vocabulary, they cannot communicate effectively. The *Reading Your Text* exercises address the vocabulary issue with which many students struggle in learning and understanding mathematics. Answers to these exercises are provided at the end of the book.

> End-of-Section Exercises

The comprehensive *End-of-Section Exercises* have been reorganized to more clearly identify the different types of exercises being presented. This structure highlights the progression in level and type of exercise for each section. This will not only provide clarity for the student, but will also make it easier for the instructor to determine the exercises for their assignments. The *application exercises* that are now integrated into every section are a crucial component of this organization.

> Getting Ready Exercises

Starting with Chapter 1, each section ends with a set of *Getting Ready Exercises*. These short exercises are designed to help students prepare for material in the next section of the text. If the students have any difficulty with these exercises, a section for review is referenced in brackets.

> Summary and Summary Exercises

The comprehensive *Summary* at the end of each chapter enables students to review important concepts. The *Summary Exercises* provide an opportunity for the student to practice these important concepts. The answers to odd-numbered exercises are provided in the answers appendix.

Reading Your Text

The following fill-in-the-blank exercises are designed to ensure that you understand some of the key vocabulary used in this section.

SECTION 1.2

(a) When two negative numbers are added, the sign of the sum is _____.

(b) The sum of two numbers with different signs is given the sign of the number with the larger _____ value.

(c) _____ is called the additive identity.

(d) When subtracting negative numbers, change the minus to a plus and replace the second number with its _____.

XIV

preface

> Icon Key

Make the Connection—exercise, activity, and chapter opener relate to each other

Check Yourself—exercise tied to preceding example

Calculator—example can be done using a calculator

Video—exercise has a video that walks through the solution

Caution—points out potential trouble spots

> Margin Notes and Recall Notes

Margin notes throughout the text are designed to help students focus on important topics and techniques, while *Recall notes* give references to previously learned material.

NOTE

This definition tells us that a composite number has factors other than 1 and itself.

RECALL

A product is the result of multiplication.

> Learning Objectives

Learning Objectives are clearly identified for each section. Annotations for the *Objectives* appear next to examples, showing when a particular objective is about to be developed. References are also included within the exercise sets to help students quickly identify examples related to topics where they need more practice.

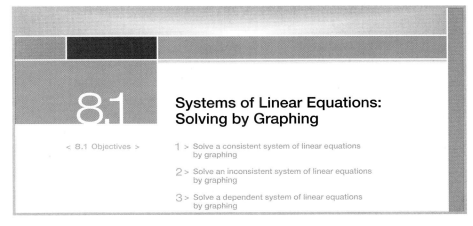

8.1

Systems of Linear Equations: Solving by Graphing

< 8.1 Objectives >

1 > Solve a consistent system of linear equations by graphing

2 > Solve an inconsistent system of linear equations by graphing

3 > Solve a dependent system of linear equations by graphing

> Self-Tests

Self-Tests appear in each chapter to provide students with an opportunity to check their progress and to review important concepts, as well as provide confidence and guidance in preparing for in-class tests or exams. The answers to the *Self-Test* exercises are given at the end of the book. Section references are given with the answers to help the student.

preface

> Cumulative Reviews

Cumulative Reviews are included starting with Chapter 2, following the *Self-Tests*. These reviews help students build on previously covered material and give them an opportunity to reinforce the skills necessary in preparing for midterm and final exams. These reviews assist students with the retention of knowledge throughout the course. The answers to these exercises are also given at the end of the book, along with section references.

> Graph Paper Card

A *Graph Paper* card is bound into the back of the book. This perforated card can be torn out and copied as needed by the students. An electronic version of the card is available through the book's website in the Information Center.

Content Changes

- In *Chapter 0,* more complex order-of-operations examples and exercises are now included. The coverage of mixed numbers has also expanded. In addition, the review of material on percent has been revised, and the summary table has expanded to include a more complete list of words signifying arithmetic operations.

- In *Chapter 1,* more complex order-of-operations content is included and there is now more material involving negative numbers. Coverage of the median and summation notation has been removed.

- In *Chapter 2,* the algorithm for solving linear equations has been clarified and is used consistently throughout. The discussion of the mean has been removed.

- In *Chapter 3,* the properties of exponents are now named. Examples with negative coefficients are included, and material is included to provide a more gradual development of evaluating expressions. The exposition for scientific notation with regard to small numbers is improved, and the section on special products has been incorporated into the section on multiplying polynomials.

- In *Chapter 4,* factoring by grouping is incorporated into the first section. The ac-method is now part of Section 4.3, while factoring cubes is now in Section 4.5.

- *Chapter 5* contains examples simplifying expressions using the GCF. Complex radical expressions are now covered before equations.

- In *Chapter 6,* some of the exercises have been revised to increase the number of non-integer results.

- *Chapter 7* has additional examples on the discussion of slope. It includes more on graphing of boundary lines.

- In *Chapter 8,* a discussion demonstrating that either equation may be used in the substitution method is included. The discussion of the system of linear inequalities now includes new examples and some revised examples.

XVI

Beginning Algebra   The Streeter/Hutchison Series in Mathematics

- *Chapter 9* now includes higher-order roots, arithmetic with cube-root expressions, rationalizing denominators with more depth, examples modeling solving application formulas for specified variables, and applications of the Pythagorean theorem.

- In *Chapter 10,* material on the discriminant and on finding the vertex of a parabola is now included.

Supplements
Multimedia Supplements

www.mathzone.com

MathZone

McGraw-Hill's **MathZone** is a complete, online tutorial and course management system for mathematics and statistics, designed for greater ease of use than any other system available. Accessible with selected McGraw-Hill texts, the system allows instructors to **create and share courses and assignments** with colleagues and adjuncts with only a few clicks of the mouse. All assignments, questions, e-Professors, online tutoring, and video lectures are directly tied to **text-specific** materials.

MathZone courses are customized to your textbook, but you can **edit** questions and algorithms, **import** your own content, and **create** announcements and due dates for assignments.

MathZone has **automatic grading** and reporting of easy-to-assign, algorithmically generated homework, quizzing, and testing. All student activity within MathZone is automatically recorded and available to you through a **fully integrated grade book** that can be downloaded to Excel.

MathZone offers:

- **Practice exercises** based on the text and generated in an unlimited number for as much practice as needed to master any topic you study.

- **Videos** of classroom instructors giving lectures and showing you how to solve exercises from the text.

- **e-Professors** to take you through animated, step-by-step instructions (delivered via on-screen text and synchronized audio) for solving exercises in the book, allowing you to digest each step at your own pace.

- **NetTutor,** which offers live, personalized tutoring via the Internet.

Instructor's Testing and Resource CD (Instructors Only)

This cross-platform CD-ROM includes a **computerized test bank** utilizing Brownstone Diploma® algorithm-based testing software that enables users to create customized exams quickly. This user-friendly program enables instructors to search for questions by topic, format, or difficulty level; to edit existing questions or to add new ones; and to scramble questions and answer keys for multiple versions of the same test. Hundreds of text-specific open-ended and multiple-choice questions are included in the question bank. Sample chapter tests and final exams in Microsoft Word® and PDF formats are also provided.

NetTutor

Available through MathZone, NetTutor is a revolutionary system that enables students to interact with a live tutor over the World Wide Web. NetTutor's web-based, graphical chat capabilities enable students and tutors to use mathematical notation and even to draw graphs as they work through a problem together. Students can also submit questions and receive answers, browse previously answered questions, and view previous live-chat sessions. Tutors are familiar with the textbook's objectives and problem-solving styles.

Video Lectures on Digital Video Disk (DVD)

In the videos, qualified teachers work through selected exercises from the textbook, following the solution methodology employed in the text. The video series is available on DVD or online as an assignable element of MathZone. The DVDs are closed-captioned for the hearing impaired, subtitled in Spanish, and meet the Americans with Disabilities Act Standards for Accessible Design. Instructors may use them as resources in a learning center, for online courses, and/or to provide extra help for students who require extra practice.

ALEKS

www.aleks.com

ALEKS (**A**ssessment and **LE**arning in **K**nowledge **S**paces) is an artificial intelligence-based system for individualized mathematics learning, available over the Web 24/7. ALEKS's unique adaptive questioning continually assesses each student's math knowledge. ALEKS then provides an individualized learning path, guiding the student in the selection of appropriate new study material. ALEKS 3.0 now links to text-specific video, multimedia tutorials, and textbook pages in PDF format. The system records each student's progress toward mastery of curricular goals in a robust classroom management system.

ALEKS improves students' performance by assessing what they know, guiding them to what they are ready to learn, and helping them master key mathematical concepts.

Printed Supplements

Annotated Instructor's Edition (Instructors Only)

This ancillary contains answers to exercises in the text, including answers to all *End-of-Section Exercises,* all *Summary Exercises, Self-Tests,* and *Cumulative Reviews.* These answers are printed in a special color for ease of use by the instructor and are located on the appropriate pages throughout the text. *Exercises, Self-Tests, Summary Exercises,* and *Cumulative Reviews* are annotated with section references to aid the instructor who may have omitted certain sections from study.

Instructor's Solutions Manual (Instructors Only)

The Instructor's Solutions Manual provides comprehensive, **worked-out solutions** to all exercises in the text. The methods used to solve the problems in the manual are the same as those used to solve the examples in the textbook.

Student's Solutions Manual

The Student's Solutions Manual provides comprehensive, **worked-out solutions** to all of the odd-numbered exercises. The steps shown in the solutions match the style of solved examples in the textbook.

Acknowledgments

Putting together a text such as this one requires much more than an author team. There are three primary groups that make it possible to write and assemble the material. The most important of those groups is the set of end-users, the students. Thanks to the many students who have contributed to the development of this text by telling us what worked (and didn't work) in helping them to learn the material.

At the other end of the process are the many people who participated in the production of the text. The McGraw-Hill team includes our publisher, Elizabeth Haefele, who assembled this fine team; our editor, Rich Kolasa, who made certain that we had the resources and support that we needed to put the project together; and the project manager, Lora Kalb. Other important contributors at or through McGraw-Hill include Barb Owca, David Hash, Kara Kudronowicz, Amber Huebner, Lori Hancock, Sandra Schnee, Harold (Hal) Whipple, Jennifer Smith, and Melissa Leick. Special thanks go to our developmental editor, Michelle Flomenhoft. Michelle kept us on track and on time. She also ensured that we would never suffer the loneliness of waking up to an empty e-mail in-box.

The third group that made this edition possible was the set of reviewers who took the time to make observations and suggestions that helped improve this text. We specifically want to thank the Board of Advisors who contributed feedback throughout the process. We also thank the 6th-edition users for their willingness to respond to surveys as we refined our revision plan.

Board of Advisors

Sequetta Desgraves, *North Lake College*

Kathy Griffin, *The Art Institute of Pittsburgh*

Suzanne Harris-Smith, *Albuquerque Technical Vocational Institute*

Frank Pizza, *Heald College—Lead Member*

Suzanne Trabucco, *Nassau Community College*

Manuscript Reviewers

Sheila Anderson, *Housatonic Community College*

Beverly R. Broomell, *Suffolk County Community College*

Gail G. Burkett, *Palm Beach Community College*

Jon Davidson, *Southern State Community College*

Laurie Delitsky, *Nassau Community College*

Matt Flacche, *Camden County College*

Robin Fulton, *East Mississippi Community College*

Frank J. Gelbwasser, *New York City College of Technology*

Dorris Holland, *Tarrant County College District–South Campus*

Linda Horner, *Columbia State Community College*

Paula Kreinbihl, *Nassau Community College*

Anna P. Leder, *Gloucester County College*

Maureen Loiacano, *Montgomery College*

Jessica Lowenfeld-Bosworth, *Nassau Community College*

Denise Lujan, *University of Texas–El Paso*

Louise Matoax, *Miami Dade College*

Carlea McAvoy, *South Puget Sound Community College*

Michael Montano, *Riverside Community College*

Stacey Moore, *Wallace State Community College*

Linda J. Murphy, *Northern Essex Community College*

Dale Nauta, *Orange Coast College*

Janet Orwick, *Dutchess Community College*

Euguenia Peterson, *Richard J. Daley College*

Marilyn G. Platt, *Gaston College*

Renee Quick, *Wallace State Community College*

R. Dwayne Ramey, *Cumberland College*

Shubhada Sagdeo, *Davenport University–Kalamazoo*

John Shirey, *Triangle Tech–Pittsburgh*

Brandon Smith, *Wallace State Community College*

Sharon Steuer, *Nassau Community College*

Abolhassan Taghavy, *Richard J. Daley College*

Sandi Tannen, *Camden County College*

Randolph J. Taylor, *Las Positas College*

Sheela L. Whelan, *Westchester Community College*

Kenneth Williams, *Davenport University–Dearborn*

Susan Working, *Grossmont College*

A commitment to accuracy

You have a right to expect an accurate textbook, and McGraw-Hill invests considerable time and effort to make sure that we deliver one. Listed below are the many steps we take to make sure this happens.

Our accuracy verification process

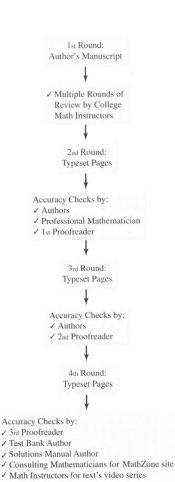

1st Round:
Author's Manuscript

↓

✓ Multiple Rounds of
Review by College
Math Instructors

↓

2nd Round:
Typeset Pages

↓

Accuracy Checks by:
✓ Authors
✓ Professional Mathematician
✓ 1st Proofreader

↓

3rd Round:
Typeset Pages

↓

Accuracy Checks by:
✓ Authors
✓ 2nd Proofreader

↓

4th Round:
Typeset Pages

↓

Accuracy Checks by:
✓ 3rd Proofreader
✓ Test Bank Author
✓ Solutions Manual Author
✓ Consulting Mathematicians for MathZone site
✓ Math Instructors for text's video series

↓

Final Round:
Printing

↓

✓ Accuracy Check by
4th Proofreader

First Round

Step 1: Numerous **college math instructors** review the manuscript and report on any errors that they may find, and the authors make these corrections in their final manuscript.

Second Round

Step 2: Once the manuscript has been typeset, the **authors** check their manuscript against the first page proofs to ensure that all illustrations, graphs, examples, exercises, solutions, and answers have been correctly laid out on the pages, and that all notation is correctly used.

Step 3: An outside, **professional mathematician** works through every example and exercise in the page proofs to verify the accuracy of the answers.

Step 4: A **proofreader** adds a triple layer of accuracy assurance in the first pages by hunting for errors; then a second, corrected round of page proofs is produced.

Third Round

Step 5: The **author team** reviews the second round of page proofs for two reasons: (1) to make certain that any previous corrections were properly made, and (2) to look for any errors they might have missed on the first round.

Step 6: A **second proofreader** is added to the project to examine the new round of page proofs to double check the author team's work and to lend a fresh, critical eye to the book before the third round of paging.

Fourth Round

Step 7: A **third proofreader** inspects the third round of page proofs to verify that all previous corrections have been properly made and that there are no new or remaining errors.

Step 8: Meanwhile, in partnership with **independent mathematicians,** the text accuracy is verified from a variety of fresh perspectives:

- The **test bank author** checks for consistency and accuracy while preparing the computerized test item file.

- The **solutions manual author** works every single exercise and verifies their answers, reporting any errors to the publisher.

- A **consulting group of mathematicians,** who write material for the text's MathZone site, notifies the publisher of any errors encountered in the page proofs.

- A video production company employing **expert math instructors** for the text's videos will alert the publisher of any errors found in the page proofs.

Final Round

Step 9: The **project manager,** who has overseen the book from the beginning, performs a **fourth proofread** of the textbook during the printing process, providing a final accuracy review.

⇒ What results is a mathematics textbook that is as accurate and error-free as is humanly possible, and our authors and publishing staff are confident that our many layers of quality assurance have produced textbooks that are the leaders of the industry for their integrity and correctness.

CHAPTER

0

chapter 0 > Make the Connection

INTRODUCTION

Number systems and methods to record patterns in their surroundings were developed by every culture in history. The Mayans in Central America had one of the most sophisticated number systems in the world in the twelfth century A.D. The Chinese number system dates from around 1200 B.C.E.

The oldest evidence of a number system is from Africa near modern-day Swaziland. Archeologists found a bone that was notched in a numerical pattern and dates from about 35,000 B.C.E.

The roots of algebra first appear in the 4,000-year-old Babylonian culture, in what is now Iraq. The Babylonians developed ways to record useful numerical relationships so that they would be easy to remember, easy to record, and helpful in solving problems. Some of the formulas developed by the Babylonians are still in use today.

You are about to embark on an exciting and useful endeavor: learning to use algebra to help you solve problems. It will take some time and effort, but do not be discouraged. Everyone can master this topic—people just like you have used it for many centuries!

An Arithmetic Review

CHAPTER 0 OUTLINE

1

This pretest provides a preview of the types of exercises you will encounter in each section of this chapter. The answers for these exercises can be found in the back of the text. If you are working on your own, or ahead of the class, this pretest can help you identify the sections in which you should focus more of your time.

Name _____

Section _____ Date _____

Answers

1. _____
2. _____
3. _____
4. _____
5. _____
6. _____ 7. _____
8. _____ 9. _____
10. _____ 11. _____
12. _____ 13. _____
14. _____ 15. _____
16. _____ 17. _____
18. _____
19. _____
20. _____
21. _____
22. _____
23. _____
24. _____ 25. _____
26. _____ 27. _____
28. _____ 29. _____
30. _____

0.1 **1.** List all the factors of 42.

2. For the group of numbers 2, 3, 6, 7, 9, 17, 18, 21, and 23, list the prime numbers and the composite numbers.

3. Find the prime factorization for each of the following numbers.
(a) 60　　(b) 350

4. Find the greatest common factor (GCF) for each of the following groups of numbers.
(a) 12 and 32　　(b) 24, 36, and 42

5. Find the least common multiple (LCM) for each of the following groups of numbers.
(a) 4, 5, and 10　　(b) 36, 20, and 30

Perform the indicated operations.

0.2 **6.** $\dfrac{3}{5} \cdot \dfrac{25}{12}$　　**7.** $\dfrac{6}{7} \div \dfrac{12}{21}$　　**8.** $\dfrac{5}{6} + \dfrac{3}{4}$

9. $\dfrac{17}{18} - \dfrac{5}{9}$　　0.3 **10.** $8.123 - 4.356$　　**11.** $7.16 \cdot 3.19$

0.2 **12.** $2\dfrac{2}{5} \cdot 5\dfrac{1}{4}$　　**13.** $1\dfrac{5}{6} \div 2\dfrac{4}{9}$　　0.3 **14.** $3.896 \div 1.6$

Evaluate the following expressions.

0.4 **15.** $21 - 3 \cdot 5$　　**16.** $3 \cdot 4 - 2^2$　　**17.** $(18 \div 9) \cdot 2 + 3^2$

18. $(15 - 12 + 5) \div 2^2$

0.3 **19.** Write 23% as (a) a fraction and (b) a decimal.

20. Write 0.035 as a percent.

0.5 **21.** Represent the following integers on the number line shown: $6, -8, 4, -2, 10$.

⟵+-+-+-+-+-+-+-+-+-+-+-+-+-+-+-+-+-+-+-+⟶
　　　　　　　　　　0

22. Place the following data set in ascending order: $5, -2, -4, 0, -1, 1$.

23. Determine the maximum and minimum of the following data set: $-4, 1, -5, 7, 3, 2$.

Evaluate the following expressions.

0.5 **24.** $|-5|$　　**25.** $|6|$　　**26.** $|11 - 5|$

27. $|-11| - |5|$　　**28.** $|4 + 5| - |6 - 3|$

Find the opposite of each of the following.

29. -16　　**30.** 23

0.1

Prime Factorization and Least Common Multiples

Beginning Algebra

The Streeter/Hutchison Series in Mathematics

< 0.1 Objectives >

1 > Find the factors of a whole number

2 > Determine whether a number is prime, composite, or neither

3 > Find the prime factorization for a number

4 > Find the GCF for two or more numbers

5 > Find the LCM for two or more numbers

▶ **Overcoming Math Anxiety**

Over the first few chapters, we present you with a series of class-tested techniques that are designed to improve your performance in this math class.

Hint #1   Become familiar with your textbook.

Perform each of the following tasks.

1. Use the Table of Contents to find the title of Section 5.1.

2. Use the index to find the earliest reference to the term *mean.* (By the way, this term has nothing to do with the personality of either your instructor or the textbook author!)

3. Find the answer to the first Check Yourself exercise in Section 0.1.

4. Find the answers to the Self-Test for Chapter 1.

5. Find the answers to the odd-numbered exercises in Section 0.1.

6. In the margin notes for Section 0.1, find the definition for the term *relatively prime.*

Now you know where some of the most important features of the text are. When you have a moment of confusion, think about using one of these features to help you clear up that confusion.

How would you organize the following list of objects: cow, dog, daisy, fox, lily, sunflower, cat, tulip?

Although there are many ways to organize the objects, most people would break them into two groups, the animals and the flowers. In mathematics, we call a group of things that have something in common a **set.**

Definition	
Set	A **set** is a collection of distinct objects that are grouped together into a single unit. Each member of a set is called an *element*.

We generally describe a set in one of two ways:

List the elements of the set.

Describe the rule(s) used to determine whether a given object is a member of the set.

We usually use braces to enclose the elements of a set when we are listing them:

{cow, dog, fox, cat}   or   {daisy, lily, sunflower, tulip}

Of course, in mathematics many (but not all) of the sets we are interested in are sets of numbers.

The numbers used to count things—1, 2, 3, 4, 5, and so on—are called the **natural** (or **counting**) **numbers.** The **whole numbers** consist of the natural numbers and zero—0, 1, 2, 3, 4, 5, and so on. They can be represented on a number line like the one shown. Zero (0) is considered the origin.

Beginning Algebra   The Streeter/Hutchison Series in Mathematics   © The McGraw-Hill Companies. All Rights Reserved.

> **NOTE**
>
> The centered dot indicates multiplication.

The origin

The number line continues forever in both directions.

Any whole number can be written as a product of two whole numbers. For example, we say that $12 = 3 \cdot 4$. We call 3 and 4 **factors** of 12.

Definition	
Factor	A **factor** of a whole number is another whole number that *divides exactly* into that number. This means that the division has a remainder of 0.

 Example 1   **Finding Factors**

< Objective 1 >

> **NOTES**
>
> 3 and 6 can also be called *divisors* of 18. They divide 18 exactly.
>
> This is a complete list of the factors. There are no other whole numbers that divide 18 exactly. Note that the factors of 18, except for 18 itself, are all *smaller* than 18.

List all factors of 18.

$3 \cdot 6 = 18$ Because $3 \cdot 6 = 18$, 3 and 6 are factors (or divisors) of 18.

$2 \cdot 9 = 18$ 2 and 9 are also factors of 18.

$1 \cdot 18 = 18$ 1 and 18 are factors of 18.

1, 2, 3, 6, 9, and 18 are all the factors of 18.

 Check Yourself 1*

List all factors of 24.

---

* Check Yourself answers appear at the end of each section in this book.

Listing factors leads us to an important classification of whole numbers. Any whole number larger than 1 is either a *prime* or a *composite* number.

Definition

Prime Number

A **prime number** is any whole number greater than 1 that has only 1 and itself as factors.

NOTE

A whole number greater than 1 always has itself and 1 as factors. Sometimes these are the *only* factors. For instance, 1 and 3 are the only factors of 3.

As examples, 2, 3, 5, and 7 are prime numbers. Their only factors are 1 and themselves.

To check whether a number is prime, one approach is simply to divide the smaller primes, 2, 3, 5, 7, and so on, into the given number. If no factors other than 1 and the given number are found, the number is prime.

The Sieve of Eratosthenes is an easy method for identifying prime numbers.

NOTE

How large can a prime number be? There is no largest prime number!

As of December 2005, the largest known prime number is $2^{30,402,457} - 1$. If you are curious, this is a number with 9,152,052 *digits*. Of course, a computer was used to find this number and verify that it is prime.

By the time you read this, someone may very well have found an even larger prime number.

Step by Step

The Sieve of Eratosthenes

Step 1 Write down a sequence of counting numbers, beginning with the number 2. In the example below, we stop at 50.

Step 2 Start at the number 2. Delete every *second* number after the 2. Each of the deleted numbers has 2 as a factor. This means that each deleted number is a composite number.

Step 3 Move to the number 3. Delete every *third* number after 3 (some numbers will already have been deleted). Each deleted number is divisible by 3, so each deleted number is not prime.

Step 4 Move to the next undeleted number, which is 5 (you should already have deleted 4). Delete every fifth number after 5.

Step 5 Continue this process, deleting every seventh number after 7, and so on.

Step 6 When you have finished, the numbers that remain are prime.

NOTE

The mathematician, geographer, and astronomer Eratosthenes was born in Cyrene (now Shahhat, Libya) in 276 B.C.E. He worked in Athens, Greece, and Alexandria, Egypt. He died in 194 B.C.E.

Among other contributions, Eratosthenes computed the size of Earth and devised our system of latitude and longitude.

	2	3	4	5	6	7	8	9	10
11	12	13	14	15	16	17	18	19	20
21	22	23	24	25	26	27	28	29	30
31	32	33	34	35	36	37	38	39	40
41	42	43	44	45	46	47	48	49	50

The prime numbers less than 50 are 2, 3, 5, 7, 11, 13, 17, 19, 23, 29, 31, 37, 41, 43, and 47.

The Streeter/Hutchison Series in Mathematics   Beginning Algebra

Example 2	Identifying Prime Numbers

< Objective 2 >

Which of the numbers 17, 29, and 33 are prime?

17 is a prime number.   1 and 17 are the only factors.

29 is a prime number.   1 and 29 are the only factors.

33 is *not* prime.   1, 3, 11, and 33 are all factors of 33.

Note: For two-digit numbers, if the number is *not* a prime, it will have one or more of the numbers 2, 3, 5, or 7 as factors.

Check Yourself 2

Which of the following numbers are prime numbers?

2, 6, 9, 11, 15, 19, 23, 35, 41

NOTE

This definition tells us that a composite number has factors other than 1 and itself.

We can now define a second class of whole numbers.

Definition	
Composite Number	A **composite number** is any whole number greater than 1 that is not prime.

Example 3	Identifying Composite Numbers

Which of the following numbers are composite: 18, 23, 25, and 38?

18 is a composite number.   1, 2, 3, 6, 9, and 18 are all factors of 18.

23 is not a composite number.   1 and 23 are the only factors. This means that 23 is a *prime number.*

25 is a composite number.   1, 5, and 25 are factors.

38 is a composite number.   1, 2, 19, and 38 are factors.

Check Yourself 3

Which of the following numbers are composite numbers?

2, 6, 10, 13, 16, 17, 22, 27, 31, 35

By the definitions of prime and composite numbers:

Property	
0 and 1	The whole numbers 0 and 1 are neither prime nor composite.

To **factor a number** means to write the number as a product of whole-number factors.

Example 4	**Factoring a Composite Number**

< Objective 3 >

Factor the number 10.

$10 = 2 \cdot 5$

The order in which you write the factors does not matter, so $10 = 5 \cdot 2$ would also be correct.

Of course, $10 = 10 \cdot 1$ is also a correct statement. However, in this section we are interested in products that use factors other than 1 and the given number.

Factor the number 21.

$21 = 3 \cdot 7$

 Check Yourself 4

Factor 35.

In writing composite numbers as a product of factors, there may be several different possible factorizations.

Example 5	**Factoring a Composite Number**

NOTE

There have to be at least two different factorizations, because a composite number has factors other than 1 and itself.

Find three ways to factor 72.

$72 = 8 \cdot 9$

$\quad = 6 \cdot 12$

$\quad = 3 \cdot 24$

 Check Yourself 5

Find three ways to factor 42.

We now want to write composite numbers as a product of their **prime factors.** Look again at the first factored line of Example 5. The process of factoring can be continued until all the factors are prime numbers.

Example 6	**Factoring a Composite Number into Prime Factors**

NOTE

This is often called a **factor tree.**

$72 = \quad 8 \quad \cdot \quad 9$

$\quad = 2 \cdot 4 \quad \cdot 3 \cdot 3$

$\quad = 2 \cdot 2 \cdot 2 \cdot 3 \cdot 3$

4 is still not prime, and so we continue by factoring 4.

72 is now written as a product of prime factors.

Beginning Algebra   The Streeter/Hutchison Series in Mathematics   © The McGraw-Hill Companies. All Rights Reserved.

NOTES

Finding the prime factorization of a number is a skill that is used when adding fractions.

Because $2 \cdot 3 = 6 = 3 \cdot 2$, the order in which we write the factors does not matter. As a matter of convention, we usually write the factors in size order.

When we write 72 as $2 \cdot 2 \cdot 2 \cdot 3 \cdot 3$, no further factorization is possible. This is called the **prime factorization** of 72.

Now, what if we start with the second factored line from the same example, $72 = 6 \cdot 12$?

$72 = \quad 6 \quad \cdot \quad 12$ Continue to factor 6 and 12.

$= 2 \cdot 3 \cdot \quad 3 \cdot 4$ Continue again to factor 4. Other choices for the factors of 12 are possible. The end result is

$= 2 \cdot 3 \cdot 3 \cdot 2 \cdot 2$ always the same.

No matter which factor pair you begin with, you will always finish with the same set of prime factors. In this case, the factor 2 appears three times and the factor 3 appears twice. The order in which we write the factors does not matter.

Check Yourself 6

We could also write

$72 = 2 \cdot 36$

Continue the factorization.

Property

The Fundamental Theorem of Arithmetic

There is exactly one prime factorization for any composite number.

The method shown in Example 6 always works. However, an easier method for factoring composite numbers exists. This method is particularly useful when factoring large numbers, in which case factoring with a number tree becomes unwieldy.

Property

Factoring by Division

To find the prime factorization of a number, divide the number by a series of primes until the final quotient is a prime number.

 Example 7   **Finding Prime Factors**

NOTE

The prime factorization is the product of all the prime divisors and the final quotient.

To write 60 as a product of prime factors, divide 2 into 60 for a quotient of 30. Continue to divide by 2 again for the quotient 15. Because 2 does not divide exactly into 15, we try 3. Because the quotient 5 is prime, we are done.

$$\dfrac{30}{2)\overline{60}} \searrow \dfrac{15}{2)\overline{30}} \searrow \dfrac{5}{3)\overline{15}} \quad \text{Prime}$$

Our factors are the prime divisors and the final quotient. We have

$60 = 2 \cdot 2 \cdot 3 \cdot 5$

Check Yourself 7

Complete the process to find the prime factorization of 90.

$$\begin{array}{c} \overset{45}{2)90} \end{array} \longrightarrow \begin{array}{c} \overset{?}{?)45} \end{array}$$

Remember to continue until the final quotient is prime.

Writing composite numbers in their completely factored form can be simplified if we use a format called **continued division.**

Example 8	Finding Prime Factors Using Continued Division

NOTE

In each short division, we write the quotient *below* rather than above the dividend. This is just a convenience for the next division.

Use the continued-division method to divide 60 by a series of prime numbers.

$$\text{Primes} \begin{cases} 2)60 \\ 2)30 \\ 3)15 \\ 5 \end{cases}$$

Stop when the final quotient is prime.

To write the factorization of 60, we include each divisor used and the final prime quotient. In our example, we have

$$60 = 2 \cdot 2 \cdot 3 \cdot 5$$

Check Yourself 8

Find the prime factorization of 234.

We know that a factor or a divisor of a whole number divides that number exactly. The factors or divisors of 20 are

1, 2, 4, 5, 10, and 20

NOTE

Again the factors of 20, other than 20 itself, are less than 20.

Each of these numbers divides 20 exactly, that is, with no remainder.

Next, we look at common factors or divisors. A **common factor** or **divisor** of two numbers is any factor that divides both the numbers exactly.

Example 9	Finding Common Factors

< Objective 4 >

Look at the numbers 20 and 30. Is there a common factor for the two numbers?
First, we list the factors. Then, we circle the ones that appear in both lists.

Factors

20:    ①, ②, 4, ⑤, ⑩, 20

30:    ①, ②, 3, ⑤, 6, ⑩, 15, 30

We see that 1, 2, 5, and 10 are common factors of 20 and 30. Each of these numbers divides both 20 and 30 exactly.

Our later work with fractions will require that we find the greatest common factor of a group of numbers.

Definition

Greatest Common Factor

The **greatest common factor (GCF)** of a group of numbers is the *largest* number that divides each of the given numbers exactly.

In the first part of Example 9, the common factors of the numbers 20 and 30 were listed as

1, 2, 5, 10   Common factors of 20 and 30

The GCF of the two numbers is 10, because 10 is the *largest* of the four common factors.

Check Yourself 9

List the factors of 30 and 36, and then find the GCF.

The method of Example 9 also works in finding the GCF of a group of more than two numbers.

Example 10   **Finding the GCF by Listing Factors**

Find the GCF of 24, 30, and 36. We list the factors of each of the three numbers.

24:    ①, ②, ③, 4, ⑥, 8, 12, 24
30:    ①, ②, ③, 5, ⑥, 10, 15, 30
36:    ①, ②, ③, 4, ⑥, 9, 12, 18, 36

6 is the GCF of 24, 30, and 36.

NOTE

Looking at the three lists, we see that 1, 2, 3, and 6 are common factors.

Check Yourself 10

Find the GCF of 16, 24, and 32.

NOTE

If there are no common prime factors, the GCF is 1.

The process shown in Example 10 is very time-consuming when larger numbers are involved. A better approach to the problem of finding the GCF of a group of numbers uses the prime factorization of each number.

Step by Step

Finding the GCF

Step 1   Write the prime factorization for each of the numbers in the group.
Step 2   Locate the prime factors that appear in every prime factorization.
Step 3   The GCF is the *product* of all the common prime factors.

Beginning Algebra   The Streeter/Hutchison Series in Mathematics

> **Example 11**   **Finding the GCF**

Find the GCF of 20 and 30.

Step 1   Write the prime factorizations of 20 and 30.

$20 = 2 \cdot 2 \cdot 5$
$30 = 2 \cdot 3 \cdot 5$

Step 2   Find the prime factors common to each number.

$20 = ② \cdot 2 \cdot ⑤$
$30 = ② \cdot 3 \cdot ⑤$ 2 and 5 are the common prime factors.

Step 3   Form the product of the common prime factors.

$2 \cdot 5 = 10$

10 is the GCF.

Check Yourself 11

Find the GCF of 30 and 36.

To find the GCF of a group of more than two numbers, we use the same process.

> **Example 12**   **Finding the GCF**

Find the GCF of 24, 30, and 36.

$24 = ② \cdot 2 \cdot 2 \cdot ③$
$30 = ② \cdot ③ \cdot 5$
$36 = ② \cdot 2 \cdot ③ \cdot 3$

2 and 3 are the prime factors common to *all three numbers.*

$2 \cdot 3 = 6$ is the GCF.

Check Yourself 12

Find the GCF of 15, 30, and 45.

Sometimes, two numbers have no common factors other than 1.

Example 13 | Finding the GCF

NOTE

If two numbers, such as 15 and 28, have no common factor other than 1, they are called **relatively prime**.

Find the GCF of 15 and 28.

$15 = 3 \cdot 5$
$28 = 2 \cdot 2 \cdot 7$

There are no common prime factors listed. But remember that 1 is a factor of every whole number.

1 is the GCF.

 Check Yourself 13

Find the GCF of 30 and 49.

Another idea that will be important in our work with fractions is the concept of *multiples*. Every whole number has an associated group of multiples.

Definition

Multiples   The **multiples** of a number are the product of that number with the natural numbers 1, 2, 3, 4, 5, ...

 Example 14 | Listing Multiples

< Objective 5 >

NOTE

Other than 3 itself, the multiples of 3 are all larger than 3.

List the multiples of 3.
    The multiples of 3 are

$3 \cdot 1, 3 \cdot 2, 3 \cdot 3, 3 \cdot 4, \ldots$

The three dots indicate that the list continues without stopping.

or

$3, 6, 9, 12, \ldots$

An easy way of listing the multiples of 3 is to think of *counting by threes*.

 Check Yourself 14

List the first seven multiples of 4.

You may see the relationship between factors and multiples. Saying "12 is a multiple of 3" is the same as saying "3 is a factor of 12."
    Sometimes we need to find common multiples of two or more numbers.

Definition

Common Multiples   If a number is a multiple of each of a group of numbers, it is called a **common multiple** of the numbers; that is, it is a number that is evenly divisible by all the numbers in the group.

Beginning Algebra   The Streeter/Hutchison Series in Mathematics

Example 15 | **Finding Common Multiples**

NOTE

15, 30, 45, and 60 are multiples of *both* 3 and 5.

NOTE

We can say that both 3 and 5 are common factors of 15, 30, 45, and 60.

Find four common multiples of 3 and 5.
  Some common multiples of 3 and 5 are

15, 30, 45, 60

Check Yourself 15

List the first six multiples of 6. Then look at your list from Check Yourself 14 and list some common multiples of 4 and 6.

In our later work with fractions, we will use the *least common multiple* of a group of numbers.

Definition

Least Common Multiple

The **least common multiple** (LCM) of a group of numbers is the *smallest* number that is a multiple of each number in the group.

It is possible to simply list the multiples of each number and then find the LCM by inspection.

Example 16 | **Finding the LCM**

NOTE

48 is also a common multiple of 6 and 8, but we are looking for the *smallest* such number.

Find the LCM of 6 and 8.

Multiples

6:   6, 12, 18, ⟨24⟩, 30, 36, 42, 48, . . .

8:   8, 16, ⟨24⟩, 32, 40, 48, . . .

We see that 24 is the smallest number common to both lists. So 24 is the LCM of 6 and 8.

NOTE

For instance, if a number appears three times in the factorization of a number, it will be included at least three times in forming the least common multiple.

Check Yourself 16

Find the LCM of 20 and 30 by listing the multiples of each number.

The technique of Example 16 will work for any group of numbers. However, it becomes tedious for larger numbers. Here is an easier approach.

Step by Step

Finding the LCM	**Step 1**	Write the prime factorization for each of the numbers in the group.
	Step 2	List the prime factors that occur the greatest number of times in any one prime factorization.
	Step 3	Form the product of those prime factors, using each factor the greatest number of times it occurs in any one factorization.

Some students prefer to line up the factors to help remember the process of finding the LCM of a group of numbers.

 Example 17   **Finding the LCM**

NOTE

Line up the *like* factors vertically.

To find the LCM of 10 and 18, we factor:

$$10 = 2 \qquad \cdot 5$$
$$18 = 2 \cdot 3 \cdot 3$$
$$\overline{2 \cdot 3 \cdot 3 \cdot 5}\qquad \text{Bring down the factors.}$$

2 and 5 appear, at most, one time in any one factorization. 3 appears twice in one factorization.

$$2 \cdot 3 \cdot 3 \cdot 5 = 90$$

So 90 is the LCM of 10 and 18.

 Check Yourself 17

Use the method of Example 17 to find the LCM of 24 and 36.

The procedure is the same for a group of more than two numbers.

 Example 18   **Finding the LCM**

NOTE

The different factors that appear are 2, 3, and 5.

To find the LCM of 12, 18, and 20, we factor:

$$12 = 2 \cdot 2 \cdot 3$$
$$18 = 2 \quad \cdot 3 \cdot 3$$
$$20 = 2 \cdot 2 \qquad \cdot 5$$
$$\overline{2 \cdot 2 \cdot 3 \cdot 3 \cdot 5}$$

2 and 3 appear twice in one factorization; 5 appears just once.

$$2 \cdot 2 \cdot 3 \cdot 3 \cdot 5 = 180$$

So 180 is the LCM of 12, 18, and 20.

 Check Yourself 18

Find the LCM of 3, 4, and 6.

Check Yourself ANSWERS

1. 1, 2, 3, 4, 6, 8, 12, and 24   **2.** 2, 11, 19, 23, and 41 are prime numbers.

3. 6, 10, 16, 22, 27, and 35 are composite numbers.   **4.** 5 · 7

5. 2 · 21, 3 · 14, 6 · 7   **6.** 2 · 2 · 2 · 3 · 3

7.
$$
\begin{array}{ccc}
45 & 15 & 5 \\
2\overline{)90} & 3\overline{)45} & 3\overline{)15}
\end{array}
$$

$90 = 2 \cdot 3 \cdot 3 \cdot 5$

8. 2 · 3 · 3 · 13

9. 30: ①, ②, ③, 5, ⑥, 10, 15, 30

36: ①, ②, ③, 4, ⑥, 9, 12, 18, 36

The GCF is 6.

10. 16: ①, ②, ④, ⑧, 16

24: ①, ②, 3, ④, 6, ⑧, 12, 24

32: ①, ②, ④, ⑧, 16, 32

The GCF is 8.

11. 30 = ② · ③ · 5

36 = ② · 2 · ③ · 3

The GCF is 2 · 3 = 6.

12. 15   **13.** The GCF is 1; 30 and 49 are relatively prime.

14. The first seven multiples of 4 are 4, 8, 12, 16, 20, 24, and 28.

15. 6, 12, 18, 24, 30, 36; some common multiples of 4 and 6 are 12, 24, and 36.

16. The multiples of 20 are 20, 40, 60, 80, 100, 120, . . . ; the multiples of 30 are 30, 60, 90, 120, 150, . . . ; the LCM of 20 and 30 is 60, the smallest number common to both lists.

17. 24 = 2 · 2 · 2 · 3

$$
\frac{36 = 2 \cdot 2 \quad \cdot 3 \cdot 3}{2 \cdot 2 \cdot 2 \cdot 3 \cdot 3 = 72}
$$

18. 12

Reading Your Text

The following fill-in-the-blank exercises are designed to ensure that you understand some of the key vocabulary used in this section.

SECTION 0.1

(a) The centered dot in the expression 3 · 4 indicates _____.

(b) A composite number is any whole number greater than 1 that is not _____.

(c) A pair of numbers that have no common factor other than 1 are called _____ prime.

(d) Saying "12 is a _____ of 3" is the same as saying "3 is a factor of 12."

0.1 exercises

MathZone

Boost your grade at mathzone.com!
> Practice Problems
> NetTutor
> Self-Tests
> e-Professors
> Videos

Name _____

Section _____ Date _____

Answers

1. _____   2. _____

3. _____

4. _____

5. _____   6. _____

7. _____

8. _____

9. _____

10. _____

11. _____   12. _____

13. _____

14. _____

15. _____

16. _____

17. _____   18. _____

19. _____   20. _____

21. _____   22. _____

< Objective 1 >

List the factors of each of the following numbers.

1. 8

2. 6

3. 10

4. 12

5. 15

6. 21

7. 24

8. 32

9. 64 > Videos

10. 66

11. 13

12. 37

< Objective 2 >

Use the following list of numbers for exercises 13 and 14.

0, 1, 15, 19, 23, 31, 49, 55, 59, 87, 91, 97, 103, 105

13. Which of the given numbers are prime? > Videos

14. Which of the given numbers are composite?

15. List all the prime numbers between 30 and 50.

16. List all the prime numbers between 55 and 75.

< Objective 3 >

Find the prime factorization of each number.

17. 20

18. 22

19. 30

20. 35

21. 51

22. 24

23. 63

24. 94

25. 70

26. 90

27. 88

28. 100

29. 130

30. 66

31. 315

32. 400

33. 225

34. 132

35. 189   > Videos

36. 330

In later mathematics courses, you often will want to find factors of a number with a given sum or difference. The following problems use this technique.

37. Find two factors of 24 with a sum of 10.

38. Find two factors of 15 with a difference of 2.

39. Find two factors of 30 with a difference of 1.

40. Find two factors of 28 with a sum of 11.

< Objective 4 >

Find the GCF for each of the following groups of numbers.

41. 4 and 6

42. 6 and 9

43. 10 and 15

44. 12 and 14

45. 21 and 24

46. 22 and 33

47. 20 and 21

48. 28 and 42

49. 18 and 24   > Videos

50. 35 and 36

51. 45, 60, and 75

52. 36, 54, and 180

53. 12, 36, and 60

54. 15, 45, and 90

55. 105, 140, and 175

56. 32, 80, and 112

57. 25, 75, and 150

58. 36, 72, and 144

Answers

23.	24.
25.	26.
27.	28.
29.	30.

31.

32.

33.

34.

35.

36.

37.	38.
39.	40.
41.	42.
43.	44.
45.	46.
47.	48.
49.	50.
51.	52.
53.	54.
55.	56.
57.	58.

Answers

59. _____

60. _____

61. _____

62. _____

63. _____

64. _____

65. _____

66. _____

67. _____

68. _____

69. _____

70. _____

71. _____

72. _____

73. _____

74. _____

75. _____

76. _____

< Objective 5 >

Find the LCM for each of the following groups of numbers. Use whichever method you wish.

59. 12 and 15   **60.** 12 and 21   **61.** 18 and 36

62. 25 and 50   **63.** 25 and 40   **64.** 10 and 14

65. 3, 5, and 6   **66.** 2, 8, and 10   **67.** 18, 21, and 28

 > Videos

68. 8, 15, and 20   **69.** 20, 30, and 40   **70.** 12, 20, and 35

Basic Skills | Advanced Skills | Vocational-Technical Applications | Calculator/Computer | **Above and Beyond** | Getting Ready

71. Prime numbers that differ by two are called *twin primes.* Examples are 3 and 5, 5 and 7, and so on. Find one pair of twin primes between 85 and 105.

chapter 0 > Make the Connection

72. The following questions refer to twin primes (see exercise 71).

  (a) Search for, and make a list of several pairs of twin primes in which the primes are greater than 3.

chapter 0 > Make the Connection

  (b) What do you notice about each number that lies *between* a pair of twin primes?

  (c) Write an explanation for your observation in part **(b).**

73. Obtain (or imagine that you have) a quantity of square tiles. Six tiles can be arranged in the shape of a rectangle in two different ways:

  (a) Record the dimensions of the rectangles shown.

  (b) If you use seven tiles, how many different rectangles can you form?

  (c) If you use ten tiles, how many different rectangles can you form?

  (d) What kind of number (of tiles) permits *only one* arrangement into a rectangle? *More than* one arrangement?

74. The number 10 has four factors: 1, 2, 5, and 10. We can say that 10 has an even number of factors. Investigate several numbers to determine which numbers have an *even number* of factors and which numbers have an *odd number* of factors.

75. A natural number is said to be *perfect* if it is equal to the sum of its divisors, except itself.

  (a) Show that 28 is a perfect number.

chapter 0 > Make the Connection

  (b) Identify another perfect number less than 28.

76. Find the smallest natural number that is divisible by all of the following: 2, 3, 4, 6, 8, 9.

77. Suppose that a school has 1,000 lockers and that they are all closed. A person passes through, opening every other locker, beginning with locker 2. Then another person passes through, changing every third locker (closing it if it is open, opening it if it is closed), starting with locker 3. Yet another person passes through, changing every fourth locker, beginning with locker 4. This process continues until 1,000 people pass through.

(a) At the end of this process, which locker numbers are closed?

(b) Write an explanation for your answer to part **(a)**.
(Hint: It may help to attempt exercise 74 first.)

Answers

We provide the answers for the odd-numbered problems at the end of each exercise set.

1. 1, 2, 4, 8   **3.** 1, 2, 5, 10   **5.** 1, 3, 5, 15

7. 1, 2, 3, 4, 6, 8, 12, 24   **9.** 1, 2, 4, 8, 16, 32, 64   **11.** 1, 13

13. 19, 23, 31, 59, 97, 103   **15.** 31, 37, 41, 43, 47   **17.** $2 \cdot 2 \cdot 5$

19. $2 \cdot 3 \cdot 5$ **21.** $3 \cdot 17$ **23.** $3 \cdot 3 \cdot 7$ **25.** $2 \cdot 5 \cdot 7$ **27.** $2 \cdot 2 \cdot 2 \cdot 11$

29. $2\overline{)130}$
       $5\overline{)65}$
         $\overline{13}$
       $130 = 2 \cdot 5 \cdot 13$

31. $3 \cdot 3 \cdot 5 \cdot 7$ **33.** $3 \cdot 3 \cdot 5 \cdot 5$

35. $3\overline{)189}$ **37.** 4, 6   **39.** 5, 6   **41.** 2   **43.** 5   **45.** 3
       $3\overline{)63}$
       $3\overline{)21}$
         $\overline{7}$
       $189 = 3 \cdot 3 \cdot 3 \cdot 7$

47. 1   **49.** 6   **51.** 15   **53.** 12   **55.** 35   **57.** 25   **59.** 60

61. 36   **63.** 200   **65.** 30   **67.** 252   **69.** 120

71. 101, 103   **73.** Above and Beyond   **75.** Above and Beyond

77. Above and Beyond

0.2

Fractions and Mixed Numbers

< 0.2 Objectives >

1 > Simplify a fraction

2 > Multiply and divide fractions

3 > Add and subtract fractions

4 > Write fractions as mixed numbers

5 > Multiply and divide mixed numbers

6 > Add and subtract mixed numbers

This section provides a review of the basic arithmetic operations—addition, subtraction, multiplication, and division—with fractions and mixed numbers.

In Section 0.1, we identified the set of whole numbers as the set consisting of the numbers 0, 1, 2, 3, and so on. In this section, we look at the set of positive numbers that can be written as fractions.

There are two types of fractions that we examine here: proper fractions and improper fractions. **Proper fractions** are those fractions that are less than 1, such as $\frac{1}{2}$ and $\frac{2}{3}$ (the numerator is less than the denominator). **Improper fractions** are those fractions that are greater than or equal to 1, such as $\frac{7}{2}$ and $\frac{19}{5}$ (the numerator is greater than the denominator).

> **NOTE**
>
> The set of numbers that can be written as fractions is called the set of *rational numbers*.

Every whole number can be written in fraction form, $\frac{a}{b}$, in which the denominator $b \neq 0$. In fact, there are many fraction forms for each number. This is because the fraction bar can be interpreted as division. For example, we can write $2 \div 2$ as $\frac{2}{2}$. Of course, this is another way of writing the whole number 1. Any fraction in which the numerator and the denominator are the same is a representation of the number 1 because any nonzero number divided by itself is 1.

$$1 = \frac{2}{2} \qquad 1 = \frac{12}{12} \qquad 1 = \frac{257}{257}$$

These fractions are called **equivalent fractions** because they all represent the same number.

To determine whether two fractions are equivalent or to find equivalent fractions, we use the **Fundamental Principle of Fractions.** The Fundamental Principle of Fractions arises from the idea that multiplying any number by 1 does not change the number.

Property

The Fundamental Principle of Fractions	$\dfrac{a}{b} = \dfrac{a \cdot c}{b \cdot c}$ or $\dfrac{a \cdot c}{b \cdot c} = \dfrac{a}{b}$, in which neither b nor c is zero.

 Example 1   **Rewriting Fractions**

< Objective 1 >

NOTE

Each representation is a numeral, or name for the number. Each number has many names.

NOTE

In each case, we use the Fundamental Principle of Fractions with c equal to a different number.

Write three fractional representations for each number.

(a) $\dfrac{2}{3}$

We use the Fundamental Principle of Fractions to multiply the numerator and denominator by the same number.

$$\frac{2}{3} = \frac{2 \cdot 2}{3 \cdot 2} = \frac{4}{6}$$

 $\dfrac{4}{6}$

$$\frac{2}{3} = \frac{2 \cdot 3}{3 \cdot 3} = \frac{6}{9}$$

 $\dfrac{6}{9}$

$$\frac{2}{3} = \frac{2 \cdot 10}{3 \cdot 10} = \frac{20}{30}$$

 $\dfrac{20}{30}$

(b) 5

$$5 = \frac{5 \cdot 2}{1 \cdot 2} = \frac{10}{2}$$

$$5 = \frac{5 \cdot 3}{1 \cdot 3} = \frac{15}{3}$$

$$5 = \frac{5 \cdot 100}{1 \cdot 100} = \frac{500}{100}$$

Check Yourself 1

Write three fractional representations for each number.

(a) $\dfrac{5}{8}$ **(b)** $\dfrac{4}{3}$ **(c)** 3

The simplest fractional representation for a number has the smallest numerator and denominator. Fractions written in this form are said to be **simplified.**

| Example 2 | Simplifying Fractions |

Simplify each fraction.

(a) $\dfrac{22}{55}$ (b) $\dfrac{35}{45}$ (c) $\dfrac{24}{36}$

We first find the prime factors for the numerator and for the denominator.

(a) $\dfrac{22}{55} = \dfrac{2 \cdot 11}{5 \cdot 11}$

NOTE

Often, we use the convention of "canceling" a factor that appears in both the numerator and denominator to prevent careless errors.

$$\dfrac{7 \cdot 5}{3 \cdot 3 \cdot 5} = \dfrac{7 \cdot \cancel{5}}{3 \cdot 3 \cdot \cancel{5}}$$
$$= \dfrac{7}{3 \cdot 3}$$
$$= \dfrac{7}{9}$$

We then use the Fundamental Principle of Fractions.

$$\dfrac{22}{55} = \dfrac{2 \cdot 11}{5 \cdot 11} = \dfrac{2}{5}$$

(b) $\dfrac{35}{45} = \dfrac{7 \cdot 5}{3 \cdot 3 \cdot 5} = \dfrac{7 \cdot 5}{9 \cdot 5}$

Using the fundamental principle to remove the common factor of 5 yields

$$\dfrac{35}{45} = \dfrac{7}{9}$$

(c) $\dfrac{24}{36} = \dfrac{2 \cdot 2 \cdot 2 \cdot 3}{2 \cdot 2 \cdot 3 \cdot 3}$

NOTE

With practice, you will be able to simplify fractions mentally.

$$\dfrac{\cancel{2} \cdot \cancel{2} \cdot 2 \cdot \cancel{3}}{\cancel{2} \cdot \cancel{2} \cdot 3 \cdot \cancel{3}} = \dfrac{2}{3}$$

Removing the common factor $2 \cdot 2 \cdot 3$ yields

$$\dfrac{24}{36} = \dfrac{2}{3}$$

 Check Yourself 2

Simplify each fraction.

(a) $\dfrac{21}{33}$ (b) $\dfrac{15}{30}$ (c) $\dfrac{12}{54}$

The Fundamental Principle of Fractions is really based on the way in which we multiply fractions. To multiply a pair of fractions, we multiply the numerators—the result becomes the numerator of their product. Then, we multiply the denominators—the result becomes the denominator of the product.

| Property |

Multiplying Fractions

$$\dfrac{a}{b} \cdot \dfrac{c}{d} = \dfrac{a \cdot c}{b \cdot d}$$

When multiplying two fractions, rewrite them in factored form, and then simplify before multiplying. To multiply a fraction by a whole number, we rewrite the whole number as a fraction in which the denominator is 1.

The Streeter/Hutchison Series in Mathematics   Beginning Algebra

Example 3 | **Multiplying Fractions**

< Objective 2 >

Find the product of the two fractions.

RECALL

A product is the result of multiplication.

$$\frac{9}{2} \cdot \frac{4}{3}$$

$$\frac{9}{2} \cdot \frac{4}{3} = \frac{9 \cdot 4}{2 \cdot 3}$$

$$= \frac{\cancel{3} \cdot 3 \cdot \cancel{2} \cdot 2}{\cancel{2} \cdot \cancel{3}}$$

$$= \frac{3 \cdot 2}{1}$$

$$= \frac{6}{1} \qquad \text{The denominator of 1 is not necessary.}$$

$$= 6$$

 Check Yourself 3

Multiply and simplify each pair of fractions.

(a) $\dfrac{3}{5} \cdot \dfrac{10}{7}$ \qquad\qquad (b) $\dfrac{12}{5} \cdot \dfrac{10}{6}$

Property

Dividing Fractions

$$\frac{a}{b} \div \frac{c}{d} = \frac{a}{b} \cdot \frac{d}{c} = \frac{a \cdot d}{b \cdot c}$$

This rule says that to divide two fractions, invert the divisor (flip the second fraction) and multiply.

 Example 4 | **Dividing Fractions**

Find the quotient.

$$\frac{7}{3} \div \frac{5}{6}$$

$$\frac{7}{3} \div \frac{5}{6} = \frac{7}{3} \cdot \frac{6}{5} = \frac{7 \cdot 6}{3 \cdot 5}$$

$$= \frac{7 \cdot 2 \cdot 3}{3 \cdot 5} = \frac{7 \cdot 2}{5}$$

$$= \frac{14}{5}$$

Beginning Algebra   The Streeter/Hutchison Series in Mathematics   © The McGraw-Hill Companies. All Rights Reserved.

Check Yourself 4

Find the result from dividing the two fractions.

$$\frac{9}{2} \div \frac{3}{5}$$

NOTE

Find the LCM of the set of denominators (as described in Section 0.1). The result is the LCD.

The next property tells us how to add fractions when they have the same denominator.

Property

Adding Fractions

$$\frac{a}{b} + \frac{c}{b} = \frac{a + c}{b}$$

When adding two fractions with different denominators, find the **least common denominator (LCD)** first. The LCD is the smallest number that both denominators evenly divide. After rewriting the fractions with this denominator, add the numerators, and then simplify the result.

 Example 5   Adding Fractions

< Objective 3 >

Find the sum of the two fractions.

RECALL

A sum is the result of addition.

$$\frac{5}{8} + \frac{7}{12}$$

The fractions $\frac{5}{8}$ and $\frac{7}{12}$ have different denominators. In order to add them, we need to find equivalent fractions that have the same denominator. To find the least common denominator (LCD) of these fractions, we find the LCM of their denominators 8 and 12.

From Section 0.1, we know that the LCM of 8 and 12 is 24, so we rewrite each fraction as an equivalent fraction with a denominator of 24.

RECALL

To find equivalent fractions, multiply each fraction by 1.

$$\frac{5}{8} \cdot \frac{3}{3} = \frac{15}{24} \qquad 24 \div 8 = 3$$

$$\frac{7}{12} \cdot \frac{2}{2} = \frac{14}{24} \qquad 24 \div 12 = 2$$

$$\frac{5}{8} + \frac{7}{12} = \frac{15}{24} + \frac{14}{24} = \frac{29}{24} \qquad \text{This fraction cannot be simplified.}$$

Check Yourself 5

Find the sum for each pair of fractions.

(a) $\dfrac{4}{5} + \dfrac{7}{9}$

(b) $\dfrac{5}{6} + \dfrac{4}{15}$

Property

Subtracting Fractions

$$\frac{a}{b} - \frac{c}{b} = \frac{a-c}{b}$$

Subtracting fractions is treated exactly like adding them, except the numerator becomes the difference of the two numerators.

▶ **Example 6** | **Subtracting Fractions**

Find the difference.

$$\frac{7}{9} - \frac{1}{6}$$

The LCD is 18. We rewrite the fractions with that denominator.

$$\frac{7}{9} = \frac{14}{18}$$

$$\frac{1}{6} = \frac{3}{18}$$

$$\frac{7}{9} - \frac{1}{6} = \frac{14}{18} - \frac{3}{18} = \frac{11}{18}$$ This fraction cannot be simplified.

RECALLS

The difference is the result of subtraction.

We cannot simplify $\dfrac{11}{18}$

because 11 and 18 are relatively prime.

Check Yourself 6

Find the difference $\dfrac{11}{12} - \dfrac{5}{8}$.

Another way to write an improper fraction is as a *mixed number.*

Definition

Mixed Number | A **mixed number** is the sum of a whole number and a proper fraction.

For our later work it will be important to be able to change back and forth between improper fractions and mixed numbers. Because an improper fraction represents a number that is greater than or equal to 1, we have the following rule:

Writing Improper Fractions as Mixed Numbers

An improper fraction can always be written as either a mixed number or a whole number.

To do this, remember that you can think of a fraction as indicating division. The numerator is divided by the denominator. This leads us to the following rule:

Writing an Improper Fraction as a Mixed Number

Step 1   Divide the numerator by the denominator.
Step 2   If there is a remainder, write the remainder over the original denominator.

 Example 7   **Writing a Fraction as a Mixed Number**

< Objective 4 >

Write $\dfrac{17}{5}$ as a mixed number.

Divide 17 by 5.

NOTE

Both forms are correct, but in subsequent courses you will find that improper fractions are preferred to mixed numbers.

$$5\overline{)17} \quad \begin{array}{r} 3 \\ \underline{15} \\ 2 \end{array}$$

$$\frac{17}{5} = 3\frac{2}{5}$$

Remainder
Original denominator
Quotient

    Check Yourself 7

Write $\dfrac{32}{5}$ as a mixed number.

In order to write a mixed number as an equivalent improper fraction, we write the whole-number part as an equivalent fraction with the same denominator as the fraction part. We then add the two fractions. This is illustrated in Example 8.

 Example 8   **Writing a Mixed Number as an Improper Fraction**

(a) Write $3\dfrac{2}{5}$ as an equivalent improper fraction.

Because the fraction part has a denominator of 5, we write the whole-number part as a fraction with 5 as its denominator.

NOTE

With practice, you should be able to do this mentally.

$$3\frac{2}{5} = 3 + \frac{2}{5} = 3 \cdot \frac{5}{5} + \frac{2}{5} \qquad 3 = \frac{3}{1}$$

$$= \frac{15}{5} + \frac{2}{5} = \frac{15 + 2}{5}$$

$$= \frac{17}{5}$$

NOTE

Multiply the denominator, 7, by the whole number, 4, and add the numerator, 5.

(b) Write $4\frac{5}{7}$ as an improper fraction.

$$4\frac{5}{7} = \frac{(7 \cdot 4) + 5}{7} = \frac{33}{7}$$

 Check Yourself 8

Write $5\frac{3}{8}$ as an improper fraction.

When multiplying two mixed numbers, it is usually easier to change the mixed numbers to improper fractions and then perform the multiplication. Example 9 illustrates this method.

Example 9   **Multiplying Two Mixed Numbers**

< Objective 5 >

Multiply.

$$3\frac{2}{3} \cdot 2\frac{1}{2} = \frac{11}{3} \cdot \frac{5}{2} \qquad \text{Change the mixed numbers to improper fractions.}$$

$$= \frac{11 \cdot 5}{3 \cdot 2} = \frac{55}{6} = 9\frac{1}{6}$$

 > CAUTION

Be careful! Students sometimes think of

$$3\frac{2}{3} \cdot 2\frac{1}{2} \quad \text{as} \quad (3 \cdot 2) + \left(\frac{2}{3} \cdot \frac{1}{2}\right)$$

This is *not* the correct multiplication pattern. You must first change the mixed numbers to improper fractions.

 Check Yourself 9

Multiply.

$$2\frac{1}{3} \cdot 3\frac{1}{2}$$

When dividing mixed numbers, simply write the mixed or whole numbers as improper fractions as the first step. Then proceed with the division. Example 10 illustrates this approach.

Example 10   Dividing Two Mixed Numbers

Divide.

$$2\frac{3}{8} \div 1\frac{3}{4} = \frac{19}{8} \div \frac{7}{4}$$ Write the mixed numbers as improper fractions.

$$= \frac{19}{\overset{}{\underset{2}{8}}} \times \frac{\overset{1}{4}}{7}$$ Invert the divisor and multiply as before.

$$= \frac{19}{14} = 1\frac{5}{14}$$

Check Yourself 10

Divide $3\frac{1}{5} \div 2\frac{2}{5}$.

When adding or subtracting mixed numbers, first write the mixed numbers as improper fractions and then proceed as you would when adding or subtracting fractions. Example 11 illustrates these concepts.

Example 11   Adding and Subtracting Mixed Numbers

< Objective 6 >

(a) Add, and write the result as a mixed number.

$$3\frac{1}{6} + 2\frac{3}{8} = \frac{19}{6} + \frac{19}{8}$$ The LCD of the fractions is 24. Rename them with that denominator.

$$= \frac{76}{24} + \frac{57}{24}$$ Then add as before.

$$= \frac{133}{24}$$

$$= 5\frac{13}{24}$$

NOTE

$$\begin{array}{r} 5 \\ 24\overline{)133} \\ \underline{120} \\ 13 \end{array}$$

(b) Subtract.

$$8\frac{7}{10} - 3\frac{3}{8} = \frac{87}{10} - \frac{27}{8}$$ Write the fractions with denominator 40.

$$= \frac{348}{40} - \frac{135}{40}$$ Subtract as before.

$$= \frac{213}{40}$$

This can be written as $5\frac{13}{40}$.

Check Yourself 11

Perform the indicated operation. Write your result as a mixed number.

(a) $5\dfrac{7}{10} + 3\dfrac{5}{6}$ 　　　　　　　　　　(b) $7\dfrac{11}{12} - 3\dfrac{5}{8}$

To subtract a mixed number from a whole number, we use the same techniques.

 Example 12 　　　**Subtracting Mixed Numbers**

Subtract.

$$6 - 2\dfrac{3}{4}$$

NOTE

$$6 = \dfrac{6}{1} = \dfrac{24}{4}$$

Multiply the numerator and denominator by 4 to form a common denominator.

$$6 - 2\dfrac{3}{4} = \dfrac{24}{4} - \dfrac{11}{4}$$

Write both the whole number and the mixed number as improper fractions with a common denominator.

$$= \dfrac{13}{4}$$

This can be written as $3\dfrac{1}{4}$.

Check Yourself 12

Subtract $7 - 3\dfrac{2}{5}$.

When adding mixed numbers, some students prefer to take advantage of the fact that a mixed number is the sum of a whole number and a fraction. To do this, add the whole-number parts and add the fraction parts separately, and then combine the two. You may need to simplify the fraction. Example 13 illustrates this.

 Example 13 　　　**Adding Mixed Numbers**

Add $6\dfrac{2}{5} + 4\dfrac{4}{5}$.

$$6\dfrac{2}{5} + 4\dfrac{4}{5} = 6 + 4 + \dfrac{2}{5} + \dfrac{4}{5}$$

 > CAUTION

Do not use this method to subtract mixed numbers.

$$= 10 + \dfrac{6}{5}$$

Add the whole-number and fraction parts separately.

$$= 10 + 1\dfrac{1}{5}$$

After simplifying the fraction, you may need to add more to the whole-number part.

$$= 11\dfrac{1}{5}$$

Check Yourself 13

Add $4\frac{3}{4} + 3\frac{2}{3}$.

In algebra, we usually use improper fractions rather than mixed numbers, so when we need to add mixed numbers, we will generally write them as improper fractions and then add them following the procedure shown in Example 11(a).

There are many applications in which fractions and mixed numbers are used. Examples 14 and 15 illustrate some of these.

Example 14   An Application of Fractions and Mixed Numbers

NOTE

The word *of* usually indicates multiplication.

Chair rail molding 136 inches (in.) long must be cut into pieces of $31\frac{1}{3}$ in. each. How many pieces can be cut from the molding?

$$136 \div 31\frac{1}{3} = 136 \div \frac{94}{3} = \frac{136}{1} \div \frac{94}{3} = \frac{136}{1} \cdot \frac{3}{94} = \frac{408}{94} = \frac{204}{47} = 4\frac{16}{47}$$

So four full-length pieces can be cut from the molding.

 Check Yourself 14

After a family party, $10\frac{2}{3}$ cupcakes were left. If Amanda took $\frac{3}{8}$ of these, how many did she take?

Example 15   An Application of Fractions and Mixed Numbers

José must trim $2\frac{5}{16}$ feet (ft) from a board 8 ft long. How long will the board be after it is cut?

$$8 - 2\frac{5}{16} = 7\frac{16}{16} - 2\frac{5}{16} = 5\frac{11}{16}$$

The board will be $5\frac{11}{16}$ ft long after it is cut.

 Check Yourself 15

Three pieces of lumber measure $5\frac{3}{8}$ ft, $7\frac{1}{2}$ ft, and $9\frac{3}{4}$ ft. What is the total length of the lumber?

Check Yourself ANSWERS

1. Answers will vary.   2. (a) $\dfrac{7}{11}$; (b) $\dfrac{1}{2}$; (c) $\dfrac{2}{9}$ 3. (a) $\dfrac{6}{7}$; (b) 4

4. $\dfrac{15}{2}$ 5. (a) $\dfrac{71}{45}$; (b) $\dfrac{11}{10}$ 6. $\dfrac{7}{24}$ 7. $6\dfrac{2}{5}$ 8. $\dfrac{43}{8}$

9. $\dfrac{49}{6}$ or $8\dfrac{1}{6}$ 10. $\dfrac{4}{3}$ or $1\dfrac{1}{3}$ 11. (a) $\dfrac{143}{15}$ or $9\dfrac{8}{15}$; (b) $\dfrac{103}{24}$ or $4\dfrac{7}{24}$

12. $\dfrac{18}{5}$ or $3\dfrac{3}{5}$ 13. $8\dfrac{5}{12}$ 14. 4   15. $22\dfrac{5}{8}$ ft

Reading Your Text

The following fill-in-the-blank exercises are designed to ensure that you understand some of the key vocabulary used in this section.

SECTION 0.2

(a) The numerator of a _____ fraction is less than the denominator.

(b) Two fractions that represent the same quantity are called _____ fractions.

(c) The _____ of a set of fractions is the same as the LCM of their denominators.

(d) A mixed number is the sum of a whole number and a proper _____.

The Streeter/Hutchison Series in Mathematics   Beginning Algebra

Name _____

Section _____ Date _____

Answers

1. _____

2. _____

3. _____

4. _____

5. _____

6. _____

7. _____

8. _____

9. _____ 10. _____

11. _____ 12. _____

13. _____ 14. _____

15. _____ 16. _____

17. _____ 18. _____

19. _____ 20. _____

21. _____ 22. _____

Basic Skills | Advanced Skills | Vocational-Technical Applications | Calculator/Computer | Above and Beyond | Getting Ready

< Objective 1 >

Give three equivalent fractions for each given fraction.

1. $\dfrac{3}{7}$ 2. $\dfrac{4}{9}$

3. $\dfrac{7}{8}$ 4. $\dfrac{11}{13}$

5. $\dfrac{10}{17}$ 6. $\dfrac{9}{16}$

7. $\dfrac{6}{11}$ 8. $\dfrac{15}{16}$

Write each fraction in simplest form.

9. $\dfrac{8}{12}$ 10. $\dfrac{12}{15}$

11. $\dfrac{10}{14}$ 12. $\dfrac{15}{50}$

13. $\dfrac{12}{18}$ 14. $\dfrac{28}{35}$

15. $\dfrac{35}{40}$ 16. $\dfrac{21}{24}$

17. $\dfrac{11}{44}$ 18. $\dfrac{10}{25}$

19. $\dfrac{12}{36}$ 20. $\dfrac{18}{48}$

21. $\dfrac{48}{60}$ 22. $\dfrac{48}{66}$

23. $\dfrac{105}{135}$ > Videos

24. $\dfrac{54}{126}$

25. $\dfrac{15}{44}$

26. $\dfrac{10}{63}$

< Objective 2 >

Multiply. Be sure to simplify each product.

27. $\dfrac{3}{4} \cdot \dfrac{7}{5}$

28. $\dfrac{2}{3} \cdot \dfrac{8}{5}$

29. $\dfrac{3}{5} \cdot \dfrac{5}{7}$

30. $\dfrac{6}{11} \cdot \dfrac{8}{6}$

31. $\dfrac{6}{13} \cdot \dfrac{4}{9}$

32. $\dfrac{5}{9} \cdot \dfrac{6}{11}$

33. $\dfrac{3}{11} \cdot \dfrac{7}{9}$

34. $\dfrac{3}{10} \cdot \dfrac{5}{9}$

Divide. Write each result in simplest form.

35. $\dfrac{5}{21} \div \dfrac{25}{14}$

36. $\dfrac{1}{5} \div \dfrac{3}{4}$

37. $\dfrac{2}{5} \div \dfrac{1}{3}$

38. $\dfrac{8}{9} \div \dfrac{4}{3}$

39. $\dfrac{8}{9} \div \dfrac{11}{15}$

40. $\dfrac{8}{15} \div \dfrac{2}{5}$

41. $\dfrac{5}{27} \div \dfrac{15}{54}$ > Videos

42. $\dfrac{5}{27} \div \dfrac{25}{36}$

< Objective 3 >

Add.

43. $\dfrac{2}{5} + \dfrac{1}{4}$

44. $\dfrac{2}{3} + \dfrac{3}{10}$

45. $\dfrac{2}{5} + \dfrac{7}{15}$

46. $\dfrac{3}{10} + \dfrac{7}{12}$

Answers

23. _____   24. _____

25. _____   26. _____

27. _____   28. _____

29. _____   30. _____

31. _____

32. _____

33. _____

34. _____

35. _____

36. _____

37. _____

38. _____

39. _____

40. _____

41. _____

42. _____

43. _____

44. _____

45. _____

46. _____

Answers

47. _____ 48. _____

49. _____ 50. _____

51. _____ 52. _____

53. _____ 54. _____

55. _____ 56. _____

57. _____

58. _____

59. _____

60. _____

61. _____

62. _____

63. _____

64. _____

65. _____

66. _____

67. _____

68. _____

69. _____

70. _____

47. $\dfrac{3}{8} + \dfrac{5}{12}$

48. $\dfrac{5}{36} + \dfrac{7}{24}$

49. $\dfrac{2}{15} + \dfrac{9}{20}$

50. $\dfrac{9}{14} + \dfrac{10}{21}$

51. $\dfrac{7}{15} + \dfrac{13}{18}$

52. $\dfrac{12}{25} + \dfrac{19}{30}$

53. $\dfrac{1}{2} + \dfrac{1}{4} + \dfrac{1}{8}$

54. $\dfrac{1}{3} + \dfrac{1}{5} + \dfrac{1}{10}$

Subtract.

55. $\dfrac{8}{9} - \dfrac{3}{9}$

56. $\dfrac{9}{10} - \dfrac{6}{10}$

57. $\dfrac{5}{8} - \dfrac{1}{8}$

58. $\dfrac{11}{12} - \dfrac{7}{12}$

59. $\dfrac{7}{8} - \dfrac{2}{3}$ > Videos

60. $\dfrac{5}{6} - \dfrac{3}{5}$

61. $\dfrac{11}{18} - \dfrac{2}{9}$

62. $\dfrac{5}{6} - \dfrac{1}{4}$

< Objective 4 >

Write the following fractions as mixed numbers.

63. $\dfrac{17}{4}$

64. $\dfrac{200}{11}$

Write the following mixed numbers as fractions.

65. $3\dfrac{1}{4}$

66. $6\dfrac{3}{4}$

< Objectives 5–6 >

Perform the indicated operations.

67. $2\dfrac{2}{9} + 3\dfrac{5}{9}$

68. $5\dfrac{2}{9} + 6\dfrac{4}{9}$

69. $1\dfrac{1}{3} + 2\dfrac{1}{5}$

70. $2\dfrac{1}{4} + 1\dfrac{1}{6}$

71. $3\dfrac{2}{5} - 1\dfrac{4}{5}$

72. $5\dfrac{3}{7} - 2\dfrac{1}{7}$

73. $3\dfrac{2}{3} - 2\dfrac{1}{4}$

74. $5\dfrac{4}{5} - 1\dfrac{1}{6}$

75. $2\dfrac{2}{5} \cdot 3\dfrac{3}{4}$

76. $2\dfrac{2}{7} \cdot 2\dfrac{1}{3}$

77. $3\dfrac{1}{2} \div 2\dfrac{4}{5}$

78. $3\dfrac{3}{4} \div 1\dfrac{3}{8}$

Solve the following applications.

79. **CRAFTS** Roseann is making shirts for her three children. One shirt requires $\dfrac{1}{2}$ yard (yd) of material, a second shirt requires $\dfrac{1}{3}$ yd of material, and the third shirt requires $\dfrac{1}{4}$ yd of material. How much material is required for all three shirts?

80. **SCIENCE** José rode his trail bike for 10 miles. Two-thirds of the distance was over a mountain trail. How long is the mountain trail?

81. **BUSINESS AND FINANCE** You make $240 a day on a job. What will you receive for working $\dfrac{2}{3}$ of a day?

82. **STATISTICS** A survey has found that $\dfrac{3}{4}$ of the people in a city own pets. Of those who own pets, $\dfrac{2}{3}$ have cats. What fraction of those surveyed own cats?

83. **SOCIAL SCIENCE** The scale on a map is 1 in. = 200 miles (mi). What actual distance, in miles, does $\dfrac{3}{8}$ in. represent?

84. **BUSINESS AND FINANCE** A family uses $\dfrac{2}{5}$ of its monthly income for housing and utilities on average. If the family's monthly income is $1,750, what is spent for housing and utilities? What amount remains?

Answers

71. _____

72. _____

73. _____

74. _____

75. _____

76. _____

77. _____

78. _____

79. _____

80. _____

81. _____

82. _____

83. _____

84. _____

85. SOCIAL SCIENCE Of the eligible voters in an election, $\frac{3}{4}$ were registered. Of those registered, $\frac{5}{9}$ actually voted. What fraction of those people who were eligible voted?

86. STATISTICS A survey has found that $\frac{7}{10}$ of the people in a city own pets. Of those who own pets, $\frac{2}{3}$ have dogs. What fraction of those surveyed own dogs?

87. SCIENCE A jet flew at an average speed of 540 mi/h on a $4\frac{2}{3}$-h flight. What was the distance flown?

88. GEOMETRY A piece of land that has $11\frac{2}{3}$ acres is being subdivided for home lots. It is estimated that $\frac{2}{7}$ of the area will be used for roads. What amount remains to be used for lots?

89. GEOMETRY To find the approximate circumference or distance around a circle, we multiply its diameter by $\frac{22}{7}$. What is the circumference of a circle with a diameter of 21 in.?

90. GEOMETRY The length of a rectangle is $\frac{6}{7}$ yd, and its width is $\frac{21}{26}$ yd. What is its area in square yards? (The area of a rectangle is the product of its length and its width.)

Basic Skills | Advanced Skills | Vocational-Technical Applications | Calculator/Computer | **Above and Beyond** | Getting Ready

91. Every fraction (rational number) has a corresponding decimal form that either terminates or repeats. For example, $\frac{5}{16} = 0.3125$ (the decimal form terminates), and $\frac{4}{11} = 0.363636\ldots$ (the decimal form repeats). Investigate a number of fractions to determine which ones terminate and which ones repeat. (Hint: You can focus on the denominator; study the prime factorizations of several denominators.)

92. Complete the following sums:

$$\frac{1}{2} + \frac{1}{4} =$$

$$\frac{1}{2} + \frac{1}{4} + \frac{1}{8} =$$

$$\frac{1}{2} + \frac{1}{4} + \frac{1}{8} + \frac{1}{16} =$$

Based on these, predict the sum:

$$\frac{1}{2} + \frac{1}{4} + \frac{1}{8} + \frac{1}{16} + \frac{1}{32} + \frac{1}{64} + \frac{1}{128}$$

 chapter 0 > Make the Connection

Answers

92. _____

Answers

For exercises 1–7, answers will vary.

1. $\dfrac{6}{14}, \dfrac{9}{21}, \dfrac{12}{28}$ **3.** $\dfrac{14}{16}, \dfrac{35}{40}, \dfrac{70}{80}$ **5.** $\dfrac{20}{34}, \dfrac{30}{51}, \dfrac{100}{170}$ **7.** $\dfrac{12}{22}, \dfrac{18}{33}, \dfrac{24}{44}$

9. $\dfrac{2}{3}$ **11.** $\dfrac{5}{7}$ **13.** $\dfrac{2}{3}$ **15.** $\dfrac{7}{8}$ **17.** $\dfrac{1}{4}$ **19.** $\dfrac{1}{3}$ **21.** $\dfrac{4}{5}$

23. $\dfrac{7}{9}$ **25.** $\dfrac{15}{44}$ **27.** $\dfrac{21}{20}$ **29.** $\dfrac{3}{7}$ **31.** $\dfrac{8}{39}$ **33.** $\dfrac{7}{33}$ **35.** $\dfrac{2}{15}$

37. $\dfrac{6}{5}$ **39.** $\dfrac{40}{33}$ **41.** $\dfrac{2}{3}$ **43.** $\dfrac{13}{20}$ **45.** $\dfrac{13}{15}$ **47.** $\dfrac{19}{24}$ **49.** $\dfrac{7}{12}$

51. $\dfrac{107}{90}$ **53.** $\dfrac{7}{8}$ **55.** $\dfrac{5}{9}$ **57.** $\dfrac{1}{2}$ **59.** $\dfrac{5}{24}$ **61.** $\dfrac{7}{18}$ **63.** $4\dfrac{1}{4}$

65. $\dfrac{13}{4}$ **67.** $5\dfrac{7}{9}$ **69.** $3\dfrac{8}{15}$ **71.** $1\dfrac{3}{5}$ **73.** $1\dfrac{5}{12}$ **75.** 9 **77.** $1\dfrac{1}{4}$

79. $1\dfrac{1}{12}$ yd   **81.** \$160   **83.** 75 mi   **85.** $\dfrac{5}{12}$ **87.** 2,520 mi

89. 66 in.   **91.** Above and Beyond

0.3

Decimals and Percents

< 0.3 Objectives >

1 > Write a fraction as a decimal

2 > Write a decimal as a fraction

3 > Add and subtract decimals

4 > Multiply and divide decimals

5 > Write a percent as a fraction or decimal

6 > Write a decimal or fraction as a percent

Because a fraction can be interpreted as division, we can divide the numerator of a fraction by its denominator to write the fraction as an equivalent decimal. The result is called its **decimal equivalent.**

 Example 1   Writing a Fraction as a Decimal

< Objective 1 >

RECALL

5 can be written as 5.0, 5.00, 5.000, and so on. In this case, we continue the division by adding zeros to the dividend until a 0 remainder is reached.

Write $\dfrac{5}{8}$ as a decimal.

$$
\begin{array}{r}
0.625 \\
8\overline{)5.000} \\
\underline{4\,8} \\
20 \\
\underline{16} \\
40 \\
\underline{40} \\
0
\end{array}
$$

Because $\dfrac{5}{8}$ means $5 \div 8$, divide 8 into 5.

We see that $\dfrac{5}{8} = 0.625$; 0.625 is the decimal equivalent of $\dfrac{5}{8}$.

 Check Yourself 1

Find the decimal equivalent of $\dfrac{7}{8}$.

You should recall that the decimal 0.625 in Example 1 means $\dfrac{625}{1,000}$. This is equivalent to $\dfrac{625}{1,000} \cdot \dfrac{10}{10} = \dfrac{6,250}{10,000} = 0.6250$.

The Streeter/Hutchison Series in Mathematics   Beginning Algebra

We could choose to round our answer to some specific decimal place. If we round to the nearest tenth, then we are rounding to one decimal place $0.625 = 0.6$ (to the nearest tenth), whereas if we rounded to the nearest hundredth, we are rounding to two decimal places, $0.625 = 0.63$ (to the nearest hundredth).

When a decimal does not terminate, we usually round it to a specific place, as in Example 2.

| Example 2 | Writing a Fraction as a Decimal |

Write $\dfrac{3}{7}$ as a decimal. Round the answer to the nearest thousandth.

$$\begin{array}{r} 0.4285 \\ 7\overline{)3.0000} \\ \underline{2\,8} \\ 20 \\ \underline{14} \\ 60 \\ \underline{56} \\ 40 \\ \underline{35} \\ 5 \end{array}$$

In this example, we are choosing to round to three decimal places, so we must add enough zeros to carry the division to four decimal places.

So $\dfrac{3}{7} = 0.429$ (to the nearest thousandth).

Check Yourself 2

Find the decimal equivalent of $\dfrac{5}{12}$ to the nearest thousandth.

If the decimal equivalent of a fraction does *not* terminate, it will *repeat* a sequence of digits. These decimals are called **repeating decimals.**

| Example 3 | Writing a Fraction as a Repeating Decimal |

Write $\dfrac{5}{11}$ as a decimal.

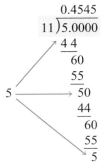

$$\begin{array}{r} 0.4545 \\ 11\overline{)5.0000} \\ \underline{4\,4} \\ 60 \\ \underline{55} \\ 50 \\ \underline{44} \\ 60 \\ \underline{55} \\ 5 \end{array}$$

As soon as a remainder repeats itself, as 5 does here, the pattern of digits will repeat in the quotient.

$$\dfrac{5}{11} = 0.4545\ldots$$

$$= 0.\overline{45}$$

Check Yourself 3

Use the bar notation to write the decimal equivalent of $\dfrac{5}{7}$. (Be patient. You have to divide for a while to find the repeating pattern.)

To write a decimal as a fraction, write the decimal without the decimal point. This is the **numerator** of the fraction. The **denominator** of the fraction is a 1 followed by as many zeros as there are places in the decimal. The next two examples illustrate this process.

Example 4   Writing a Decimal as a Fraction

< Objective 2 >

$$0.7 = \frac{7}{10} \qquad 0.09 = \frac{9}{100} \qquad 0.257 = \frac{257}{1,000}$$

One place   One zero   Two places   Two zeros   Three places   Three zeros

Check Yourself 4

Write as fractions.

(a) 0.3   **(b)** 0.311

When a decimal is written as an equivalent fraction, the common fraction that results should be simplified.

Example 5   Converting a Decimal to a Fraction

NOTE

Divide the numerator and denominator of $\dfrac{395}{1,000}$ by 5.

Convert 0.395 to a fraction and simplify the result.

$$0.395 = \frac{395}{1,000} = \frac{79}{200}$$

Check Yourself 5

Write 0.275 as a fraction.

We add and subtract decimals using place value, just as we do with whole numbers. You must be sure to align the decimal points, as illustrated in Example 6.

Example 6   Adding or Subtracting Two Decimals

< Objective 3 >

Perform the indicated operation.

(a) Add 2.356 and 15.6.

Aligning the decimal points, we get

$$\begin{array}{r} 2.356 \\ +\,15.600 \\ \hline 17.956 \end{array}$$ Although the zeros are not necessary, they ensure proper alignment.

RECALL

We borrowed from the tenths place, just as we would with whole numbers.

$$\overset{7\ \ 0\ 1}{8.\,\cancel{1}\,0}$$
$$-3.\,8\ 4$$
$$\overline{4.\,2\ 6}$$

(b) Subtract 3.84 from 8.1.

Again, we align the decimal points.

$$\begin{array}{r} 8.10 \\ -3.84 \\ \hline 4.26 \end{array}$$ When subtracting, always add zeros so that the right columns line up.

Check Yourself 6

Perform the indicated operation.

(a) 34.76 + 2.419   **(b)** 71.82 − 8.197

Example 7 illustrates the multiplication of two decimal fractions.

 | **Example 7** | **Multiplying Two Decimals**

< Objective 4 >

RECALL

Multiply as you would with whole numbers. The final result is given the same number of decimal places as the total number of decimal places in both factors.

Multiply 4.6 and 3.27.

$$\begin{array}{r} 4.6 \\ \times\ 3.27 \\ \hline 322 \\ 920 \\ 13800 \\ \hline 15.042 \end{array}$$

It is not necessary to align decimals being multiplied. Note that the two factors have a total of three digits to the right of the decimal point.

The decimal point of the product is moved three digits to the left.

Check Yourself 7

Multiply 5.8 and 9.62.

Dividing decimals is a bit trickier. In order to divide when the divisor is a decimal, we multiply both the dividend and divisor by a large enough power of 10 that the divisor becomes a whole number. We show you how to set this up in Example 8.

 | **Example 8** | **Rewriting a Problem That Requires Dividing by a Decimal**

NOTE

It is always easier to rewrite a division problem so that you're dividing by a whole number. Dividing by a whole number makes it easy to place the decimal point in the quotient.

Rewrite the division so that the divisor is a whole number.

$$2.57 \div 3.4 = \frac{2.57}{3.4}$$ Write the division as a fraction.

$$= \frac{2.57 \times 10}{3.4 \times 10}$$ We multiply the numerator and denominator by 10 so that the divisor is a whole number. This *does not change* the value of the fraction.

$$= \frac{25.7}{34}$$ Multiplying by 10, shift the decimal point in the numerator and denominator *one place to the right*.

$$= 25.7 \div 34$$ Our division problem is rewritten so that the divisor is a whole number.

So

$$2.57 \div 3.4 = 25.7 \div 34$$ After we multiply the numerator and denominator by 10, we see that $2.57 \div 3.4$ is the same as $25.7 \div 34$.

 Check Yourself 8

Rewrite the division problem so that the divisor is a whole number.

3.42 ÷ 2.5

NOTE

Of course, multiplying by any whole-number power of 10 greater than 1 is just a matter of shifting the decimal point to the right.

Do you see the rule suggested by Example 8? We multiplied the numerator and the denominator (the dividend and the divisor) by 10. We made the divisor a whole number without altering the actual digits involved. All we did was shift the decimal point in the divisor and dividend the same number of places. This leads us to the following procedure.

Step by Step

To Divide by a Decimal		
	Step 1	Move the decimal point in the divisor *to the right,* making the divisor a whole number.
	Step 2	Move the decimal point in the dividend to the right *the same number of places.* Add zeros if necessary.
	Step 3	Place the decimal point in the quotient directly above the decimal point of the dividend.
	Step 4	Divide as you would with whole numbers.

Here is an example using the division rule.

▶   **Example 9**   Rounding the Result of Dividing by a Decimal

Divide 1.573 by 0.48 and give the quotient to the nearest tenth.

Write

$$0.48\,\overline{)1.57\,3}$$ Shift the decimal points two places to the right to make the divisor a whole number.

Now divide:

$$
\begin{array}{r}
3.27 \\
48\overline{)157.30} \\
\underline{144} \\
13\,3 \\
\underline{9\,6} \\
3\,70 \\
\underline{3\,36} \\
34
\end{array}
$$

Add a 0 to carry the division to the hundredths place. In this case, we want to find the quotient to the nearest tenth.

NOTE

Once the division statement is rewritten, place the decimal point in the quotient above that in the dividend.

Round 3.27 to 3.3. So

$$1.573 \div 0.48 = 3.3 \quad \text{(to the nearest tenth)}$$

Check Yourself 9

Divide, rounding the quotient to the nearest tenth.

3.4 ÷ 1.24

We have used fractions and decimals to name parts of a whole. **Percents** can also be used to accomplish this. The word *percent* means "for each hundred." We can think of percents as fractions whose denominators are 100. So 25% can be written as $\frac{25}{100}$ or, in simplified form, $\frac{1}{4}$.

Because there are different ways of naming the parts of a whole, you need to know how to change from one of these ways to another. First, we look at writing a percent as a fraction. Because a percent is a fraction or a ratio with denominator 100, we can use the following rule.

Property

Writing a Percent as a Fraction

To write a percent as a fraction, replace the percent symbol with $\frac{1}{100}$.

We use this rule in Example 10.

Example 10   **Writing a Percent as a Fraction**

< Objective 5 >

Write each percent as a fraction.

(a) $7\% = 7\left(\frac{1}{100}\right) = \frac{7}{100}$

(b) $25\% = 25\left(\frac{1}{100}\right) = \frac{25}{100} = \frac{1}{4}$ Always simplify fractions.

 Check Yourself 10

Write 12% as a fraction.

RECALL

$\frac{1}{100} = 0.01$

In Example 10, we wrote percents as fractions by replacing the percent sign with $\frac{1}{100}$ and multiplying. How do we convert percents when we are working with decimals? Just move the decimal point two places to the left. This gives us a second rule for rewriting percents.

Property

Writing a Percent as a Decimal

To write a percent as a decimal, replace the percent symbol with $\frac{1}{100}$. As a result of multiplying by $\frac{1}{100}$, the decimal point will move two places to the left.

Example 11   **Writing a Percent as a Decimal**

RECALL

Multiplying by $\frac{1}{100}$ is the same as dividing by 100.

NOTE

A percent greater than 100 gives a decimal greater than 1.

Write each percent as a decimal.

(a) $25\% = 25\left(\frac{1}{100}\right) = 0.25$ The decimal point in 25% is understood to be after the 5.

(b) $4.5\% = 4.5\left(\frac{1}{100}\right) = 0.045$ We must add a zero to move the decimal point.

(c) $130\% = 130\left(\frac{1}{100}\right) = 1.30$

 Check Yourself 11

Write as decimals.

(a) 5%   **(b)** 3.9%   **(c)** 115%

Writing a decimal as a percent is the opposite of writing a percent as a decimal. We simply reverse the process. Here is the rule:

Property

Writing a Decimal as a Percent

To write a decimal as a percent, move the decimal point *two* places to the *right* and attach the percent symbol.

 Example 12   **Writing a Decimal as a Percent**

< Objective 6 >

Write each decimal as a percent.

(a) $0.18 = 18\%$
(b) $0.03 = 3\%$
(c) $1.25 = 125\%$

 Check Yourself 12

Write each decimal as a percent.

(a) 0.27   **(b)** 0.045   **(c)** 1.3

The following rule allows us to write fractions as percents.

Property

Writing a Fraction as a Percent

To write a fraction as a percent, write the decimal equivalent of the fraction by dividing. Then, move the decimal point two places to the right and attach the percent symbol.

Example 13 — **Writing a Fraction as a Percent**

Write each fraction as a percent.

(a) $\dfrac{3}{5} = 0.60$ To find the decimal equivalent, just divide the denominator into the numerator.

Now write the percent.

$\dfrac{3}{5} = 0.60 = 60\%$

(b) $\dfrac{1}{8} = 0.125 = 12.5\%$ or $12\dfrac{1}{2}\%$

(c) $\dfrac{1}{3} = 0.\overline{3} = 0.33\overline{3} = 33.\overline{3}\%$ or $33\dfrac{1}{3}\%$

Check Yourself 13

Change each fraction to a percent equivalent.

(a) $\dfrac{3}{4}$ (b) $\dfrac{3}{8}$ (c) $\dfrac{2}{3}$

Example 14 illustrates one of the many applications using decimals.

Example 14 — **An Application of Decimals**

NOTE

In most applications involving money, you should round to either the nearest dollar or the nearest cent.

Lucretia's car gets approximately 20 miles per gallon (mi/gal) of fuel. If 1 gal of fuel costs $1.93, how much does it cost her to drive 125 mi?

$125 \div 20 = 6.25$ gal

$6.25 \cdot 1.93 = \$12.06$ (rounded)

Check Yourself 14

The art department has a budget of $195.75 to purchase art supplies. After purchasing 35 paintbrushes for $1.92 each, six jars of paint remover for $0.93 each, and four cans of blue paint for $2.95 each, how much money was left in the budget?

Check Yourself ANSWERS

1. 0.875   **2.** 0.417   **3.** $0.\overline{714285}$ **4.** (a) $\dfrac{3}{10}$; (b) $\dfrac{311}{1,000}$ **5.** $\dfrac{11}{40}$

6. (a) 37.179; (b) 63.623   **7.** 55.796   **8.** 34.2 ÷ 25   **9.** 2.7

10. $\dfrac{12}{100}$ or $\dfrac{3}{25}$ **11.** (a) 0.05; (b) 0.039; (c) 1.15   **12.** (a) $\dfrac{27}{100} = 27\%$;

(b) 4.5%; (c) 130%   **13.** (a) 75%; (b) 37.5%; (c) $66.\overline{6}\%$ or $66\dfrac{2}{3}\%$

14. $111.17

Reading Your Text

The following fill-in-the-blank exercises are designed to ensure that you understand some of the key vocabulary used in this section.

SECTION 0.3

(a) To write a fraction as a _____, divide the numerator by the denominator.

(b) We use bar notation to indicate a _____ decimal.

(c) Fractions, decimals, and _____ are all ways of naming parts of a whole.

(d) To write a percent as a decimal, move the decimal two places to the _____, and remove the percent symbol.

Basic Skills | Advanced Skills | Vocational-Technical Applications | Calculator/Computer | Above and Beyond | Getting Ready

< Objective 1 >

Find the decimal equivalents for each of the following fractions.

1. $\dfrac{3}{4}$

2. $\dfrac{4}{5}$

3. $\dfrac{9}{20}$

4. $\dfrac{3}{10}$

5. $\dfrac{1}{5}$

6. $\dfrac{1}{8}$

7. $\dfrac{5}{16}$

8. $\dfrac{11}{20}$

9. $\dfrac{7}{10}$

10. $\dfrac{7}{16}$

11. $\dfrac{27}{40}$

12. $\dfrac{17}{32}$

Find the decimal equivalents rounded to the indicated place.

13. $\dfrac{5}{6}$;   thousandth

14. $\dfrac{7}{12}$;   hundredth

15. $\dfrac{4}{15}$;   thousandth

Write the decimal equivalents using the bar notation.

16. $\dfrac{1}{18}$

17. $\dfrac{4}{9}$

18. $\dfrac{3}{11}$ > Videos

< Objective 2 >

Write each of the following as a fraction and simplify.

19. 0.9

20. 0.3

MathZone

Boost your grade at mathzone.com!
> Practice Problems
> NetTutor
> Self-Tests
> e-Professors
> Videos

Name _____

Section _____ Date _____

Answers

1. _____   2. _____

3. _____   4. _____

5. _____   6. _____

7. _____   8. _____

9. _____   10. _____

11. _____   12. _____

13. _____

14. _____

15. _____

16. _____

17. _____

18. _____

19. _____

20. _____

Answers

21. _____

22. _____

23. _____

24. _____

25. _____

26. _____

27. _____

28. _____

29. _____

30. _____

31. _____ 32. _____

33. _____ 34. _____

35. _____ 36. _____

37. _____ 38. _____

39. _____ 40. _____

41. _____ 42. _____

43. _____ 44. _____

45. _____ 46. _____

47. _____ 48. _____

49. _____ 50. _____

21. 0.8

22. 0.6

23. 0.37

24. 0.97

25. 0.587

26. 0.379

27. 0.48 > Videos

28. 0.75

29. 0.58

30. 0.65

< Objectives 3–4 >

Perform the indicated operations.

31. $7.1562 + 14.78$

32. $6.2358 + 3.14$

33. $11.12 + 8.3792$

34. $6.924 + 5.2$

35. $9.20 - 2.85$

36. $17.345 - 11.12$

37. $18.234 - 13.64$

38. $21.983 - 9.395$

39. $3.21 \cdot 2.1$

40. $15.6 \cdot 7.123$

41. $6.29 \cdot 9.13$ > Videos

42. $8.245 \cdot 3.1$

Divide.

43. $16.68 \div 6$

44. $43.92 \div 8$

45. $1.92 \div 4$

46. $5.52 \div 6$

47. $5.48 \div 8$

48. $2.76 \div 8$

49. $13.89 \div 6$

50. $21.92 \div 5$

51. $185.6 \div 32$

52. $165.6 \div 36$

53. $79.9 \div 34$

54. $179.3 \div 55$

55. $52\overline{)13.78}$

56. $76\overline{)26.22}$

57. $0.6\overline{)11.07}$ > Videos

58. $0.8\overline{)10.84}$

59. $3.8\overline{)7.22}$

60. $2.9\overline{)13.34}$

61. $5.2\overline{)11.622}$

62. $6.4\overline{)3.616}$

< Objective 5 >

Write as fractions.

63. 6%

64. 17%

65. 75%

66. 20%

67. 65%

68. 48%

69. 50%

70. 52%

71. 46%

72. 35%

73. 66% > Videos

74. 4%

Write as decimals.

75. 20%

76. 70%

77. 35%

78. 75%

Answers

51. _____ 52. _____

53. _____ 54. _____

55. _____ 56. _____

57. _____ 58. _____

59. _____ 60. _____

61. _____ 62. _____

63. _____ 64. _____

65. _____

66. _____

67. _____

68. _____

69. _____

70. _____

71. _____

72. _____

73. _____

74. _____

75. _____

76. _____

77. _____

78. _____

Answers

79. _____ 80. _____

81. _____ 82. _____

83. _____ 84. _____

85. _____ 86. _____

87. _____

88. _____

89. _____

90. _____

91. _____

92. _____

93. _____

94. _____

95. _____

96. _____

97. _____

98. _____

99. _____

100. _____

101. _____

102. _____

79. 39% **80.** 27%

81. 5% **82.** 7%

83. 135% **84.** 250%

85. 240% **86.** 160%

< Objective 6 >

Write each decimal as a percent.

87. 4.40 **88.** 5.13

89. 0.065 **90.** 0.095

91. 0.025 **92.** 0.085

93. 0.002 **94.** 0.008

Write each fraction as a percent.

95. $\dfrac{1}{4}$ **96.** $\dfrac{4}{5}$

97. $\dfrac{2}{5}$ **98.** $\dfrac{1}{2}$

99. $\dfrac{1}{5}$ **100.** $\dfrac{3}{4}$

101. $\dfrac{5}{8}$ **102.** $\dfrac{7}{8}$

Basic Skills | Advanced Skills | Vocational-Technical Applications | Calculator/Computer | **Above and Beyond** | Getting Ready

Answers

103. **STATISTICS** On a math quiz, Adam answered 18 of 20 questions correctly, or $\frac{18}{20}$ of the quiz. Write the decimal equivalent of this fraction.

```
18/20    Name: Adam

2 x 3 = 6        5 x 4 = 20
1 + 5 = 6        3 x 4 = 12
2 x 5 = 10       5 x 2 = 10
4 + 5 = 9        5 + 4 = 9
15 - 2 = 13      15 - 4 = 11
4 x 3 = 12    ✓ 8 x 3 = 22
3 + 6 = 9        6 + 3 = 9
9 + 4 = 13       5 + 6 = 11
✓ 3 + 9 = 11     6 + 9 = 15
1 x 2 = 2        2 x 1 = 2
```

104. **STATISTICS** In a weekend baseball tournament, Joel had 4 hits in 13 times at bat. That is, he hit safely $\frac{4}{13}$ of the time. Write the decimal equivalent for Joel's hitting, rounding to three decimal places. (That number is Joel's batting average.)

105. **BUSINESS AND FINANCE** A restaurant bought 50 glasses at a cost of $39.90. What was the cost per glass, to the nearest cent?

106. **BUSINESS AND FINANCE** The cost of a case of 48 items is $28.20. What is the cost of an individual item, to the nearest cent?

107. **BUSINESS AND FINANCE** An office bought 18 handheld calculators for $284. What was the cost per calculator, to the nearest cent?

108. **BUSINESS AND FINANCE** Al purchased a new refrigerator that cost $736.12 with interest included. He paid $100 as a down payment and agreed to pay the remainder in 18 monthly payments. What amount will he be paying per month?

109. **BUSINESS AND FINANCE** The cost of a television set with interest is $490.64. If you make a down payment of $50 and agree to pay the balance in 12 monthly payments, what will be the amount of each monthly payment?

Answers

103. _____

104. _____

105. _____

106. _____

107. _____

108. _____

109. _____

Answers

1. 0.75   **3.** 0.45   **5.** 0.2   **7.** 0.3125   **9.** 0.7   **11.** 0.675

13. 0.833   **15.** 0.267   **17.** $0.\overline{4}$ **19.** $\frac{9}{10}$ **21.** $\frac{4}{5}$ **23.** $\frac{37}{100}$

25. $\dfrac{587}{1,000}$ **27.** $\dfrac{12}{25}$ **29.** $\dfrac{29}{50}$ **31.** 21.9362   **33.** 19.4992

35. 6.35   **37.** 4.594   **39.** 6.741   **41.** 57.4277   **43.** 2.78

45. 0.48   **47.** 0.685   **49.** 2.315   **51.** 5.8   **53.** 2.35

55. 0.265   **57.** 18.45   **59.** 1.9   **61.** 2.235   **63.** $\dfrac{3}{50}$ **65.** $\dfrac{3}{4}$

67. $\dfrac{13}{20}$ **69.** $\dfrac{1}{2}$ **71.** $\dfrac{23}{50}$ **73.** $\dfrac{33}{50}$ **75.** 0.2   **77.** 0.35

79. 0.39   **81.** 0.05   **83.** 1.35   **85.** 2.4   **87.** 440%   **89.** 6.5%

91. 2.5%   **93.** 0.2%   **95.** 25%   **97.** 40%   **99.** 20%

101. 62.5%   **103.** 0.9   **105.** $0.80 or 80¢   **107.** $15.78

109. $36.72

0.4

Exponents and the Order of Operations

< 0.4 Objectives >

1 > Write a product of factors in exponential form

2 > Evaluate an expression involving several operations

Often in mathematics we define symbols that allow us to write a mathematical statement in a more compact or "shorthand" form. This is an idea that you have encountered before. For example, the repeated addition

$$5 + 5 + 5$$

can be rewritten as

$$3 \cdot 5$$

Thus, multiplication is shorthand for repeated addition.

In algebra, we frequently have a number or variable that is repeated as a factor in an expression several times. For instance, we might have

$$5 \cdot 5 \cdot 5$$

To abbreviate this product, we write

$$5 \cdot 5 \cdot 5 = 5^3$$

This is called **exponential notation** or **exponential form.** The exponent or power, here 3, indicates the number of times that the factor or base, here 5, appears in a product.

Exponent or power

$$5 \cdot 5 \cdot 5 = 5^3$$

Factor or base

This is read "5 to the third power" or "5 cubed."

NOTE

5 + 5 + 5 = 15
and
3 · 5 = 15

RECALL

A factor is a number or a variable that is being multiplied by another number or variable.

> CAUTION

5^3 is *not* the same as 5 · 3.
$5^3 = 5 \cdot 5 \cdot 5 = 125$ and
$5 \cdot 3 = 15$.

Example 1	Writing Products in Exponential Form

< Objective 1 >

Write $3 \cdot 3 \cdot 3 \cdot 3$ using exponential form.
    The number 3 appears four times in the product, so

Four factors of 3

$$3 \cdot 3 \cdot 3 \cdot 3 = 3^4$$

This is read "3 to the fourth power."

53

Check Yourself 1

Rewrite each expression using exponential form.

(a) $4 \cdot 4 \cdot 4 \cdot 4 \cdot 4 \cdot 4$ **(b)** $7 \cdot 7 \cdot 7 \cdot 7$

> CAUTION

Only one of these results can be correct.

NOTE

Parentheses, brackets, and fraction bars are all examples of grouping symbols. You will learn other grouping symbols in later chapters.

To evaluate an arithmetic expression, you need to know the order in which the operations are done. To see why, simplify the expression $5 + 2 \cdot 3$.

Method 1   or   *Method 2*

$\underline{5 + 2} \cdot 3$ $5 + \underline{2 \cdot 3}$

Add first.   Multiply first.

$= 7 \cdot 3$ $= 5 + 6$

$= 21$ $= 11$

Because we get different answers depending on how we do the problem, the language of mathematics would not be clear if there were no agreement on which method is correct. The following rules tell us the order in which operations should be done.

Step by Step

The Order of Operations

Step 1   Evaluate all expressions inside grouping symbols first.
Step 2   Evaluate all expressions involving exponents.
Step 3   Do any multiplication or division in order, working from left to right.
Step 4   Do any addition or subtraction in order, working from left to right.

Example 2   Evaluating Expressions

< Objective 2 >

NOTE

Method 2 shown above is the correct one.

Evaluate $5 + 2 \cdot 3$.

There are no parentheses or exponents, so start with step 3: First multiply and then add.

$5 + 2 \cdot 3$
           Multiply first.
$= 5 + 6$
           Then add.
$= 11$

Check Yourself 2

Evaluate the following expressions.

(a) $20 - 3 \cdot 4$ **(b)** $9 + 6 \div 3$

When there are no parentheses, evaluate the exponents first.

Example 3	Evaluating Expressions

Evaluate $5 \cdot 3^2$.

$5 \cdot 3^2 = 5 \cdot 9$

        Evaluate the power first.

$\quad\quad = 45$

Check Yourself 3

Evaluate $4 \cdot 2^4$.

Both scientific and graphing calculators correctly interpret the order of operations, as demonstrated in Example 4.

Example 4	Using a Calculator to Evaluate Expressions

Use your scientific or graphing calculator to evaluate each expression. Round the answer to the nearest tenth.

(a) $24.3 + 6.2 \cdot 3.53$

When evaluating expressions by hand, you must consider the order of operations. In this case, the multiplication must be done first, and then the addition. With a calculator, you need only enter the expression correctly. The calculator is programmed to follow the order of operations.

 › Calculator

Entering   24.3 $\boxed{+}$ 6.2 $\boxed{\times}$ 3.53 $\boxed{\text{ENTER}}$

yields the evaluation 46.186. Rounding to the nearest tenth, we have 46.2.

(b) $2.45^3 - 49 \div 8{,}000 + 12.2 \cdot 1.3$

Some calculators use the caret ($\wedge$) to designate powers. Others use the symbol x^y (or y^x).

Entering 2.45 $\boxed{\wedge}$ 3 $\boxed{-}$ 49 $\boxed{\div}$ 8000 $\boxed{+}$ 12.2 $\boxed{\times}$ 1.3 $\boxed{\text{ENTER}}$

or   2.45 $\boxed{y^x}$ 3 $\boxed{-}$ 49 $\boxed{\div}$ 8000 $\boxed{+}$ 12.2 $\boxed{\times}$ 1.3 $\boxed{=}$

yields the evaluation 30.56. Rounding to the nearest tenth, we have 30.6.

NOTE

With most graphing calculators, the final command is $\boxed{\text{ENTER}}$. With most scientific calculators, the key is marked $\boxed{=}$.

Check Yourself 4

Use your scientific or graphing calculator to evaluate each expression.

(a) $67.89 - 4.7 \cdot 12.7$ **(b)** $4.3 \cdot 55.5 - 3.75^3 + 8{,}007 \div 1{,}600$

Operations inside grouping symbols are always done first.

Example 5   **Evaluating Expressions**

Evaluate $(5 + 2) \cdot 3$.

Do the operation inside the parentheses as the first step.

$(5 + 2) \cdot 3 = 7 \cdot 3 = 21$

— Add.

> ✓ **Check Yourself 5**
>
> Evaluate $4 \cdot (9 - 3)$.

The principle is the same when more than two "levels" of operations are involved.

Example 6   **Using Order of Operations**

(a) Evaluate $4 \cdot (2 + 3)^3$.

— Add inside the parentheses first.

$4 \cdot (2 + 3)^3 = 4 \cdot (5)^3$

Evaluate the power.

$= 4 \cdot 125$

Multiply.

$= 500$

(b) Evaluate $5 \cdot (7 - 3)^2 - 10$.

Evaluate the expression inside the parentheses.

$5 \cdot (7 - 3)^2 - 10 = 5(4)^2 - 10$

Evaluate the power.

$= 5 \cdot 16 - 10$

Multiply.

$= 80 - 10 = 70$

Subtract.

> ✓ **Check Yourself 6**
>
> Evaluate.
>
> **(a)** $4 \cdot 3^3 - 8 \cdot 11$ **(b)** $12 + 4 \cdot (2 + 3)^2$

The correct order of operations must be followed within a set of grouping symbols, as shown in Example 7.

Example 7   **Using Order of Operations**

Evaluate $3 \cdot [(1 + 2)^2 - 5] + 8$.

We evaluate the expression in the parentheses within the brackets first. Next, we evaluate the exponent before proceeding to the subtraction. After evaluating everything within the brackets, we follow the correct order of operations by multiplying first, and then adding.

$$
\begin{aligned}
3 \cdot [(1 + 2)^2 - 5] + 8 &= 3 \cdot [(3)^2 - 5] + 8 \\
&= 3 \cdot [9 - 5] + 8 \\
&= 3 \cdot (4) + 8 \\
&= 12 + 8 \\
&= 20
\end{aligned}
$$

Check Yourself 7

Evaluate $8 - 2 \cdot [(5 - 3)^2 - 1]$.

We stated that parentheses and brackets are not the only types of grouping symbols. Example 8 demonstrates the fraction bar as a grouping symbol.

Example 8   **Using the Order of Operations with Grouping Symbols**

> C A U T I O N

You may not "cancel" the 2's, because the numerator is being added, not multiplied.
$\dfrac{2 + 14}{2}$ is incorrect!

Evaluate $3 + \dfrac{2 + 14}{2} \cdot 5$.

$$
\begin{aligned}
3 + \dfrac{2 + 14}{2} \cdot 5 &= 3 + \dfrac{16}{2} \cdot 5 \qquad \text{The fraction bar acts as a grouping symbol.}\\
&= 3 + 8 \cdot 5 \qquad \text{We perform the division first because}\\
&= 3 + 40 \qquad\quad\ \ \text{it precedes the multiplication.}\\
&= 43
\end{aligned}
$$

Check Yourself 8

Evaluate $4 \cdot \dfrac{3^2 + 2 \cdot 3}{5} + 6$.

Check Yourself ANSWERS

1. (a) 4^6; (b) 7^4 2. (a) 8; (b) 11   3. 64   4. (a) 8.2; (b) 190.92
5. 24   6. (a) 20; (b) 112   7. 2   8. 18

Reading Your Text

The following fill-in-the-blank exercises are designed to ensure that you understand some of the key vocabulary used in this section.

SECTION 0.4

(a) Multiplication is shorthand for repeated _____.

(b) The _____ or power indicates the number of times the base appears in a product.

(c) Operations inside _____ symbols are done first when evaluating an expression.

(d) _____, brackets, and fraction bars are all examples of grouping symbols.

< Objective 1 >

Write each expression in exponential form.

1. $7 \cdot 7 \cdot 7 \cdot 7$

2. $2 \cdot 2 \cdot 2 \cdot 2 \cdot 2 \cdot 2$

3. $6 \cdot 6 \cdot 6 \cdot 6 \cdot 6$

4. $4 \cdot 4 \cdot 4 \cdot 4 \cdot 4 \cdot 4 \cdot 4$

5. $8 \cdot 8 \cdot 8 \cdot 8 \cdot 8 \cdot 8 \cdot 8 \cdot 8 \cdot 8 \cdot 8$

6. $10 \cdot 10 \cdot 10$

7. $15 \cdot 15 \cdot 15 \cdot 15 \cdot 15 \cdot 15$

8. $31 \cdot 31 \cdot 31 \cdot 31 \cdot 31 \cdot 31 \cdot 31 \cdot 31 \cdot 31 \cdot 31$

< Objective 2 >

Evaluate each of the following expressions.

9. $5 + 3 \cdot 4$

10. $10 - 4 \cdot 2$

11. $(7 + 2) \cdot 6$

12. $(10 - 4) \cdot 2$

13. $12 - 8 \div 4$

14. $20 + 10 \div 2$ > Videos

15. $(24 - 12) \div 6$

16. $(10 + 20) \div 5$

17. $8 \cdot 7 + 2 \cdot 2$

18. $56 \div 7 - 8 \div 4$

19. $7 \cdot (8 + 3) \cdot 3$

20. $48 \div (8 - 4) \div 2$

21. $3 \cdot 5^2$

22. $5 \cdot 2^3$

23. $(2 \cdot 4)^2$

24. $(5 \cdot 2)^3$

25. $4 \cdot 3^2 - 2$ > Videos

26. $3 \cdot 2^4 - 8$

MathZone

Boost your grade at mathzone.com!
> Practice Problems
> NetTutor
> Self-Tests
> e-Professors
> Videos

Name _____

Section _____ Date _____

Answers

1. _____   2. _____

3. _____   4. _____

5. _____   6. _____

7. _____   8. _____

9. _____   10. _____

11. _____   12. _____

13. _____   14. _____

15. _____   16. _____

17. _____   18. _____

19. _____   20. _____

21. _____   22. _____

23. _____   24. _____

25. _____   26. _____

Answers

27. _____

28. _____

29. _____

30. _____

31. _____

32. _____

33. _____

34. _____

35. _____

36. _____

37. _____

38. _____

39. _____

40. _____

41. _____

42. _____

43. _____ 44. _____

45. _____ 46. _____

47. _____ 48. _____

49. _____ 50. _____

27. $5 - [3 \cdot (4 - 2)^2] + (3 \cdot 5)$

28. $14 \div 7 \cdot [12 \div (4 - 2)^2 \cdot 5] - 3^3$

> Videos

29. $3 \cdot 2^4 - 6 \cdot 2$

30. $4 \cdot 2^3 - 5 \cdot 6$

31. $4 \cdot (2 + 6)^2$

32. $3 \cdot (8 - 4)^2$

33. $(4 \cdot 2 + 6)^2$

34. $[25 \div (2^3 - 3)] \cdot 2$

35. $64 \div [(16 \div 2 \cdot 4) - 16]$

36. $5 \cdot (4 - 2)^3$

37. $12 - \dfrac{2 + 3}{2} \cdot 3$

38. $10 \cdot \dfrac{3^2 - 4}{4} + 6$

39. $3 \cdot \dfrac{2 \cdot 4 + 1 \cdot 5^2}{2 + 2} - 6$

40. $\dfrac{16 - (2 + 1)^2 + 1}{2^3}$

41. $(4 \cdot 2 + 3)^2 - 25$ > Videos

42. $8 + (2 \cdot 3 + 3)^2$

43. $2 \cdot [16 - (1 + 3)^2]$

44. $[(2 + 3)^2 - 4 \cdot 5] + 7$

45. Social Science Over the last 2,000 years, Earth's population has doubled approximately five times. Use exponential notation to write an expression that indicates doubling five times.

46. Geometry The volume of a cube with each edge of length 9 in. is given by $9 \cdot 9 \cdot 9$. Write the volume using exponential notation.

| Basic Skills | Advanced Skills | Vocational-Technical Applications | **Calculator/Computer** | Above and Beyond | Getting Ready |

Use a calculator to evaluate each expression. Round your results to the nearest tenth.

47. $(1.2)^3 \div 2.0736 \cdot 2.4 + 1.6935 - 2.4896$

48. $(5.21 \cdot 3.14 - 6.2154) \div 5.12 - 0.45625$

49. $1.23 \cdot 3.169 - 2.05194 + (5.128 \cdot 3.15 - 10.1742)$

50. $4.56 + (2.34)^4 \div 4.7896 \cdot 6.93 \div 27.5625 - 3.1269 + (1.56)^2$

Basic Skills | Advanced Skills | Vocational-Technical Applications | Calculator/Computer | **Above and Beyond** | Getting Ready
▲

Insert grouping symbols in the proper place(s) so that the given value of the expression is obtained.

51. $36 \div 4 + 2 - 4; \quad 2$

52. $48 \div 3 \cdot 2 - 2 \cdot 3; \quad 2$

53. $6 + 9 \div 3 + 16 - 4 \cdot 2; \quad 29$

54. $5 - 3 \cdot 2 + 8 \cdot 5 - 2; \quad 28$

51. _____

52. _____

53. _____

54. _____

Answers

1. 7^4 **3.** 6^5 **5.** 8^{10} **7.** 15^6 **9.** 17   **11.** 54   **13.** 10

15. 2   **17.** 60   **19.** 231   **21.** 75   **23.** 64   **25.** 34

27. 8   **29.** 36   **31.** 256   **33.** 196   **35.** 4   **37.** $\dfrac{9}{2}$

39. $\dfrac{75}{4}$ **41.** 96   **43.** 0   **45.** 2^5 **47.** 1.2   **49.** 7.8

51. $36 \div (4 + 2) - 4$ **53.** $(6 + 9) \div 3 + (16 - 4) \cdot 2$

0.5

Positive and Negative Numbers

< 0.5 Objectives >

1 > Represent integers on a number line

2 > Order real numbers

3 > Find the opposite of a number

4 > Evaluate numerical expressions involving absolute value

When numbers are used to represent physical quantities (altitudes, temperatures, and money are examples), it is often necessary to distinguish between *positive* and *negative* amounts. It is convenient to represent these quantities with plus (+) or minus (−) signs. Some instances of this are shown here.

The altitude of Mount Whitney is 14,495 ft *above* sea level (+14,495).

14,495 ft

Mount Whitney

The altitude of Death Valley is 282 ft *below* sea level (−282).

−282 ft

Death Valley

The temperature in Chicago is 10° *below* zero (−10).

An account could show a *gain* of $100 (+100) or a *loss* of $100 (−100).

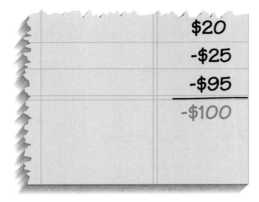

These numbers suggest the need to extend the number system to include both positive numbers (like +100) and negative numbers (like −282).

To represent the negative numbers, we extend the number line to the *left* of zero and name equally spaced points.

Numbers used to name points to the right of zero are positive numbers. They can be written with a positive (+) sign, but are usually written with no sign at all.

+6 and 9 are positive numbers

Numbers used to name points to the left of zero are negative numbers. They are always written with a negative (−) sign.

−3 and −20 are negative numbers.

Read "negative 3."

Positive and negative numbers are both examples of **integers.**

Here, the number line is extended to include both positive and negative numbers.

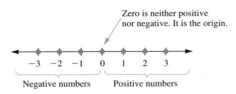

Zero is neither positive nor negative. It is the origin.

Negative numbers   Positive numbers

Definition

Integers

The **integers** consist of the natural numbers, their negatives, and zero. We can represent the set of integers by

$$\{\ldots, -3, -2, -1, 0, 1, 2, 3, \ldots\}.$$

The sets of three dots are called *ellipses* and indicate that a pattern continues.

| **Example 1** | Representing Integers on the Number Line

< Objective 1 >

Represent the following integers on the number line shown.

−3, −12, 8, 15, −7

Check Yourself 1

Represent the following integers on a number line.

−1, −9, 4, −11, 8, 20

The set of numbers on the number line is *ordered*. The numbers get smaller moving to the left on the number line and larger moving to the right.

When a set of numbers is written from smallest to largest, the numbers are said to be in *ascending order.*

 | **Example 2** | Ordering Integers

< Objective 2 >

Place each set of numbers in ascending order.

(a) 9, −5, −8, 3, 7

From smallest to largest, the numbers are

−8, −5, 3, 7, 9   Note that this is the order in which the numbers appear on a number line as we move from left to right.

(b) 3, −2, 18, −20, −13

From smallest to largest, the numbers are

−20, −13, −2, 3, 18

Check Yourself 2

Place each set of numbers in ascending order.

(a) 12, −13, 15, 2, −8, −3   **(b)** 3, 6, −9, −3, 8

The least and greatest numbers in a set are called the **extreme values.** The least element is called the **minimum,** and the greatest element is called the **maximum.**

Example 3   **Labeling Extreme Values**

For each set of numbers, determine the minimum and maximum values.

(a) 9, −5, −8, 3, 7

From our previous ordering of these numbers, we see that −8, the least element, is the minimum, and 9, the greatest element, is the maximum.

(b) 3, −2, 18, −20, −13

−20 is the minimum, and 18 is the maximum.

> **Check Yourself 3**
>
> **For each set of numbers, determine the minimum and maximum values.**
>
> **(a)** 12, −13, 15, 2, −8, −3   **(b)** 3, 6, −9, −3, 8

Integers are not the only kind of signed numbers. Decimals and fractions can also be thought of as signed numbers.

 Example 4   **Identifying Numbers That Are Integers**

Which of the numbers 145, −28, 0.35, and $-\dfrac{2}{3}$ are integers?

(a) 145 is an integer.

(b) −28 is an integer.

(c) 0.35 is not an integer.

(d) $-\dfrac{2}{3}$ is not an integer.

> **Check Yourself 4**
>
> **Which of the following numbers are integers?**
>
> −23   1,054   −0.23   0   −500 $-\dfrac{4}{5}$

NOTE

0 is the opposite of 0.

We refer to the negative of a number as its *opposite*. But what is the opposite of the opposite of a number? It is the number itself. Example 5 illustrates this concept.

Example 5   **Finding Opposites**

< Objective 3 >

Find the opposite for each number.

(a) 5   The opposite of 5 is −5.

(b) −9   The opposite of −9 is 9.

> **Check Yourself 5**
>
> **Find the opposite for each number.**
>
> (a) 17   (b) −12

An important idea for our work in this chapter is the **absolute value** of a number. This represents the distance of the point named by the number from the origin on the number line.

The absolute value of 5 is 5. The absolute value of −5 is also 5.

The absolute value of a positive number or zero is itself. The absolute value of a negative number is its opposite.

In symbols we write

$$|5| = 5 \qquad \text{and} \qquad |-5| = 5$$

Read "the absolute   Read "the absolute
value of 5."   value of negative 5."

The absolute value of a number does *not* depend on whether the number is to the right or to the left of the origin, but on its *distance* from the origin.

Example 6   **Simplifying Absolute Value Expressions**

< Objective 4 >

(a) $|7| = 7$

(b) $|-7| = 7$

(c) $-|-7| = -7$

This is the *negative,* or opposite, of the absolute value of negative 7.

(d) $|-10| + |10| = 10 + 10 = 20$

Absolute value bars serve as another set of grouping symbols, so do the operation *inside* first.

(e) $|8 - 3| = |5| = 5$

(f) $|8| - |3| = 8 - 3 = 5$

Here, evaluate the absolute values and then subtract.

Check Yourself 6

Evaluate.

(a) $|8|$ (b) $|-8|$ (c) $-|-8|$

(d) $|-9| + |4|$ (e) $|9 - 4|$ (f) $|9| - |4|$

Check Yourself ANSWERS

1.

2. (a) $-13, -8, -3, 2, 12, 15$; (b) $-9, -3, 3, 6, 8$
3. (a) Minimum is -13, maximum is 15; (b) Minimum is -9, maximum is 8
4. $-23, 1{,}054, 0$, and -500 5. (a) -17; (b) 12
6. (a) 8; (b) 8; (c) -8; (d) 13; (e) 5; (f) 5

Reading Your Text

The following fill-in-the-blank exercises are designed to ensure that you understand some of the key vocabulary used in this section.

SECTION 0.5

(a) When numbers are used to represent physical quantities, it is often necessary to distinguish between positive and _____ quantities.

(b) It is convenient to represent negative quantities with a _____ sign.

(c) The _____ consist of the natural numbers, their negatives, and zero.

(d) When a set of numbers is written from smallest to largest, the numbers are said to be in _____ order.

0.5 exercises

Name _____

Section _____ Date _____

Answers

1. _____
2. _____
3. _____
4. _____
5. _____
6. _____
7. _____
8. _____
9. _____
10. _____
11. _____
12. _____
13. _____
14. _____
15. _____
16. _____

Basic Skills | Advanced Skills | Vocational-Technical Applications | Calculator/Computer | Above and Beyond | Getting Ready

< Objective 1 >

Represent each quantity with an integer.

1. An altitude of 400 ft above sea level

2. An altitude of 80 ft below sea level

3. A loss of $200

4. A profit of $400

5. A decrease in population of 25,000

6. An increase in population of 12,500

Represent the integers on the number lines shown.

7. 5, −15, 18, −8, 3

8. −18, 4, −5, 13, 9

Which numbers in the following sets are integers?

9. $\left\{ 5, -\dfrac{2}{9}, 175, -234, -0.64 \right\}$ > Videos

10. $\left\{ -45, 0.35, \dfrac{3}{5}, 700, -26 \right\}$

< Objective 2 >

Place each of the following sets of numbers in ascending order.

11. 3, −5, 2, 0, −7, −1, 8

12. −2, 7, 1, −8, 6, −1, 0

13. 9, −2, −11, 4, −6, 1, 5   > Videos

14. 23, −18, −5, −11, −15, 14, 20

15. −6, 7, −7, 6, −3, 3

16. 12, −13, 14, −14, 15, −15

For each set, determine the maximum and minimum values.

17. $5, -6, 0, 10, -3, 15, 1, 8$

18. $9, -1, 3, 11, -4, 2, 5, -2$

19. $21, -15, 0, 7, -9, 16, -3, 11$

20. $-22, 0, 22, -31, 18, -5, 3$

21. $3, 0, 1, -2, 5, 4, -1$

22. $2, 7, -3, 5, -10, -5$

< Objective 3 >
Find the opposite of each number.

23. 15 **24.** 18

25. 15 **26.** 34

27. -19 **28.** -6

29. -7 **30.** -54

< Objective 4 >
Evaluate.

31. $|17|$ **32.** $|28|$

33. $|-19|$ **34.** $|-7|$

35. $-|21|$ **36.** $-|3|$

37. $-|-8|$ **38.** $-|-13|$

39. $|-2|+|3|$ **40.** $|4|+|-3|$

41. $|-9|+|9|$ **42.** $|11|+|-11|$

Answers

17. ___
18. ___
19. ___
20. ___
21. ___
22. ___
23. ___
24. ___
25. ___
26. ___
27. ___
28. ___
29. ___
30. ___
31. ___ 32. ___
33. ___ 34. ___
35. ___ 36. ___
37. ___ 38. ___
39. ___ 40. ___
41. ___ 42. ___

Answers

43. _____

44. _____

45. _____

46. _____

47. _____

48. _____

49. _____

50. _____

51. _____

52. _____

53. _____

54. _____

55. _____

56. _____

57. _____

58. _____

59. _____

60. _____

43. $\left| -6 \right| - \left| 6 \right|$

44. $\left| 5 \right| - \left| -5 \right|$

45. $\left| 15 \right| - \left| 8 \right|$

46. $\left| 11 \right| - \left| 3 \right|$

47. $\left| 15 - 8 \right|$

48. $\left| 11 - 3 \right|$

49. $\left| -9 \right| + \left| 2 \right|$

50. $\left| -7 \right| + \left| 4 \right|$

51. $\left| 7 \right| - \left| -6 \right|$

52. $\left| -9 \right| - \left| -4 \right|$

Represent each quantity with a real number.

53. **SCIENCE AND TECHNOLOGY** The erosion of 5 centimeters (cm) of topsoil from an Iowa corn field.

> Videos

54. **SCIENCE AND TECHNOLOGY** The formation of 2.5 cm of new topsoil on the African savanna.

55. **BUSINESS AND FINANCE** The withdrawal of $50 from a checking account.

56. **BUSINESS AND FINANCE** The deposit of $200 into a savings account.

57. **SCIENCE AND TECHNOLOGY** The temperature change pictured.

58. **BUSINESS AND FINANCE** An increase of 75 points in the Dow-Jones average.

59. **STATISTICS** An eight-game losing streak by the local baseball team.

60. **SOCIAL SCIENCE** An increase of 25,000 in the population of the city.

61. **BUSINESS AND FINANCE** A country exported $90,000,000 more than it imported, creating a positive trade balance.

62. **BUSINESS AND FINANCE** A country exported $60,000,000 less than it imported, creating a negative trade balance.

Determine whether each of the following statements is **true** *or* **false**.

63. All natural numbers are integers.

64. Zero is an integer.

65. All integers are whole numbers.

66. All real numbers are integers.

67. All negative integers are whole numbers.

68. Zero is neither positive nor negative.

Basic Skills | Advanced Skills | Vocational-Technical Applications | Calculator/Computer | **Above and Beyond** | Getting Ready

---

For each collection of numbers given in exercises 69 to 72, answer the following:

   (a) Which number is smallest?
   (b) Which number lies farthest from the origin?
   (c) Which number has the largest absolute value?
   (d) Which number has the smallest absolute value?

69. $-6, 3, 8, 7, -2$

70. $-8, 3, -5, 4, 9$

71. $-2, 6, -1, 0, 2, 5$

72. $-9, 0, -2, 3, 6$

Place absolute value bars in the proper location(s) on the left side of the expression so that the equation is true.

73. $6 + (-2) = 4$

74. $8 + (-3) = 5$

75. $6 + (-2) = 8$ > Videos

76. $8 + (-3) = 11$

Answers

61. _____

62. _____

63. _____

64. _____

65. _____

66. _____

67. _____

68. _____

69. _____

70. _____

71. _____

72. _____

73. _____

74. _____

75. _____

76. _____

77. Simplify each of the following:

$$-(-7) \qquad -(-(-7)) \qquad -(-(-(-7)))$$

Based on your answers, generalize your results.

Answers

1. 400 or (+400)   **3.** -200 **5.** $-25,000$

7.

9. 5, 175, -234

11. $-7, -5, -1, 0, 2, 3, 8$ **13.** $-11, -6, -2, 1, 4, 5, 9$

15. $-7, -6, -3, 3, 6, 7$ **17.** Max: 15; min: -6 **19.** Max: 21, min: -15

21. Max: 5; min: -2 **23.** -15 **25.** -15 **27.** 19   **29.** 7

31. 17   **33.** 19   **35.** -21 **37.** -8 **39.** 5   **41.** 18   **43.** 0

45. 7   **47.** 7   **49.** 11   **51.** 1   **53.** -5 **55.** -50

57. $-10°F$ **59.** -8 **61.** $+90,000,000$ **63.** True   **65.** False

67. False   **69.** **(a)** -6; **(b)** 8; **(c)** 8; **(d)** -2

71. **(a)** -2; **(b)** 6; **(c)** 6; **(d)** 0   **73.** $|6 + (-2)| = 4$ **75.** $|6| + |-2| = 8$

77. Above and Beyond

Definition/Procedure	Example	Reference
Prime Factorization and Least Common Multiples		Section 0.1
Factor A **factor** of a whole number is another whole number that divides exactly into that number, leaving a remainder of zero.	The factors of 12 are 1, 2, 3, 4, 6, and 12.	*p. 4*
Prime Number Any whole number greater than 1 that has only 1 and itself as factors.	7, 13, 29, and 73 are prime numbers.	*p. 5*
Composite Number Any whole number greater than 1 that is not prime.	8, 15, 42, and 65 are composite numbers.	*p. 6*
Zero and One 0 and 1 are classified as neither prime nor composite numbers.		*p. 6*
Greatest Common Factor (GCF) The GCF is the *largest* number that is a factor of each of a group of numbers.	The GCF of 21 and 24 is 3.	*p. 10*
To Find the GCF 1. Write the prime factorization for each of the numbers in the group. 2. Locate the prime factors that appear in every prime factorization. 3. The GCF is the product of all the common prime factors. If there are no common prime factors, the GCF is 1.	To find the GCF of 24, 30, and 36: $24 = ② \cdot 2 \cdot 2 \cdot ③$ $30 = ② \cdot ③ \cdot 5$ $36 = ② \cdot 2 \cdot ③ \cdot 3$ The GCF is $2 \cdot 3 = 6$.	*p. 10*
Least Common Multiple (LCM) The LCM is the *smallest* number that is a multiple of each of a group of numbers.	The LCM of 21 and 24 is 168.	*p. 13*
To Find the LCM 1. Write the prime factorization for each of the numbers in the group. 2. Find all the prime factors that appear in any one of the prime factorizations. 3. Form the product of those prime factors, using each factor the greatest number of times it occurs in any one factorization.	To find the LCM of 12, 15, and 18: $12 = 2 \cdot 2 \cdot 3$ $15 = 3 \cdot 5$ $\underline{18 = 2 \cdot 3 \cdot 3}$ $ 2 \cdot 2 \cdot 3 \cdot 3 \cdot 5$ The LCM is $2 \cdot 2 \cdot 3 \cdot 3 \cdot 5$, or 180.	
Fractions and Mixed Numbers		Section 0.2
The Fundamental Principle of Fractions $\dfrac{a}{b} = \dfrac{a \cdot c}{b \cdot c}$ in which neither b nor c is zero.	$\dfrac{2}{3} = \dfrac{2 \cdot 3}{3 \cdot 3} = \dfrac{6}{9}$	*p. 21*

Continued

Definition/Procedure	Example	Reference

Multiplying Fractions
1. Multiply numerator by numerator. This gives the numerator of the product.
2. Multiply denominator by denominator. This gives the denominator of the product.
3. Simplify the resulting fraction if possible.

In multiplying fractions, it is usually easiest to factor and simplify the numerator and denominator *before* multiplying.

$\dfrac{5}{8} \cdot \dfrac{3}{7} = \dfrac{5 \cdot 3}{8 \cdot 7} = \dfrac{15}{56}$

$\dfrac{5}{9} \cdot \dfrac{3}{10} = \dfrac{\overset{1}{\cancel{5}} \cdot \overset{1}{\cancel{3}}}{\underset{3}{\cancel{9}} \cdot \underset{2}{\cancel{10}}} = \dfrac{1}{6}$

p. 22

Dividing Fractions
Invert the divisor and multiply.

$\dfrac{3}{7} \div \dfrac{4}{5} = \dfrac{3}{7} \cdot \dfrac{5}{4} = \dfrac{15}{28}$

p. 23

To Add or Subtract Fractions with Different Denominators
1. Find the LCD of the fractions.
2. Change each fraction to an equivalent fraction with the LCD as a common denominator.
3. Add or subtract the resulting like fractions as before.

$\dfrac{3}{4} + \dfrac{7}{10} = \dfrac{15}{20} + \dfrac{14}{20} = \dfrac{29}{20}$

$\dfrac{8}{9} - \dfrac{5}{6} = \dfrac{16}{18} - \dfrac{15}{18} = \dfrac{1}{18}$

p. 24

Mixed Number
The sum of a whole number and a proper fraction.

$2\dfrac{1}{3}$ and $5\dfrac{7}{8}$ are mixed numbers.

p. 25

To Write an Improper Fraction as a Mixed Number
1. Divide the numerator by the denominator. The quotient is the whole-number portion of the mixed number.
2. If there is a remainder, write the remainder over the original denominator. This gives the fractional portion of the mixed number.

To write $\dfrac{22}{5}$ as a mixed number:

$\begin{array}{r} 4 \\ 5)\overline{22} \\ \underline{20} \\ 2 \end{array}$ ⟵ Quotient ⟵ Remainder

$\dfrac{22}{5} = 4\dfrac{2}{5}$

p. 26

To Write a Mixed Number as an Improper Fraction
1. Multiply the denominator of the fraction by the whole-number portion of the mixed number.
2. Add the numerator of the fraction to that product.
3. Write that sum over the original denominator to form the improper fraction.

Whole number

Denominator

Numerator

$5\dfrac{3}{4} = \dfrac{(4 \cdot 5) + 3}{4} = \dfrac{23}{4}$

Denominator

p. 26

To Add or Subtract Mixed Numbers
1. Rewrite as improper fractions.
2. Add or subtract the fractions.
3. Rewrite the results as a mixed number if required.

$5\dfrac{1}{2} - 3\dfrac{3}{4} = \dfrac{11}{2} - \dfrac{15}{4} = \dfrac{22}{4} - \dfrac{15}{4}$

$= \dfrac{7}{4} = 1\dfrac{3}{4}$

p. 28

Decimals and Percents

Section 0.3

To Write a Fraction as a Decimal
1. Divide the numerator of the fraction by its denominator.
2. The quotient is the decimal equivalent of the common fraction.

To write $\dfrac{1}{2}$ as a decimal:

$\begin{array}{r} 0.5 \\ 2)\overline{1.0} \\ \underline{1\ 0} \\ 0 \end{array}$

p. 38

Definition/Procedure	Example	Reference

To Write a Terminating Decimal Less Than 1 as a Fraction
1. Write the digits of the decimal without the decimal point. This is the numerator of the fraction.
2. The denominator of the fraction is a 1 followed by as many zeros as there are places in the decimal.

To write 0.275 as a fraction:
$$0.275 = \frac{275}{1,000} = \frac{11}{40}$$

p. 40

To Add Decimals
1. Write the numbers being added in column form with their decimal points in a vertical line.
2. Add just as you would with whole numbers.
3. Place the decimal point of the sum in line with the decimal points of the addends.

To add 2.7, 3.15, and 0.48:

$$\begin{array}{r} 2.7 \\ 3.15 \\ +\ 0.48 \\ \hline 6.33 \end{array}$$

p. 40

To Subtract Decimals
1. Write the numbers being subtracted in column form with their decimal points in a vertical line. You may have to place zeros to the right of the existing digits.
2. Subtract just as you would with whole numbers.
3. Place the decimal point of the difference in line with the decimal points of the numbers being subtracted.

To subtract 5.875 from 8.5:

$$\begin{array}{r} 8.500 \\ -\ 5.875 \\ \hline 2.625 \end{array}$$

p. 41

To Multiply Decimals
1. Multiply the decimals as though they were whole numbers.
2. Add the number of decimal places in the factors.
3. Place the decimal point in the product so that the number of decimal places in the product is the sum of the number of decimal places in the factors.

To multiply 2.85×0.045:

$$\begin{array}{r} 2.85 \longleftarrow \text{Two places} \\ \times\ 0.045 \longleftarrow \text{Three places} \\ \hline 1425 \\ 11400 \\ \hline 0.12825 \longleftarrow \text{Five places} \end{array}$$

p. 41

To Divide by a Decimal
1. Move the decimal point in the divisor to the right, making the divisor a whole number.
2. Move the decimal point in the dividend to the right the same number of places. Add zeros if necessary.
3. Place the decimal point in the quotient directly above the decimal point of the dividend.
4. Divide as you would with whole numbers.

To divide 16.5 by 5.5, move the decimal points:

$$\begin{array}{r} 3 \\ 5.5\overline{)16.5} \\ \underline{16\ 5} \\ 0 \end{array}$$

p. 42

Percent
Along with fractions and decimals another way of naming parts of a whole. Percent means per hundred.

$$21\% = 21\left(\frac{1}{100}\right) = \frac{21}{100} = 0.21$$

p. 43

To Write a Percent as a Fraction or Decimal
1. *To write a percent as a fraction,* replace the percent symbol with $\frac{1}{100}$ and multiply.
2. *To write a percent as a decimal,* remove the percent symbol, and move the decimal point two places to the left.

$$37\% = 37\left(\frac{1}{100}\right) = \frac{37}{100}$$

$$37\% = 0.37$$

p. 43

Continued

Definition/Procedure	Example	Reference

To Write a Decimal or Fraction as a Percent

1. *To write a decimal as a percent,* move the decimal point two places to the right, and attach the percent symbol.

2. *To write a fraction as a percent,* write the decimal equivalent of the fraction, and then change that decimal to a percent.

$0.58 = 58\%$

$\dfrac{3}{5} = 0.60 = 60\%$

p. 44

Exponents and the Order of Operations

Section 0.4

Using Exponents

Base The number that is raised to a power.

Exponent The exponent is written to the right and above the base. The exponent tells the number of times the base is to be used as a factor.

$$\underset{\text{Base}}{5}\overset{\text{Exponent}}{^3} = \underset{\text{Three factors}}{5 \cdot 5 \cdot 5} = 125$$

p. 53

The Order of Operations

Mixed operations in an expression should be done in the following order:

1. Do any operations inside grouping symbols.
2. Evaluate any powers.
3. Do all multiplication and division in order, from left to right.
4. Do all addition and subtraction in order, from left to right.

$4 \cdot (2 + 3)^2 - 7$
$= 4 \cdot 5^2 - 7$
$= 4 \cdot 25 - 7$
$= 100 - 7$
$= 93$

p. 54

Positive and Negative Numbers

Section 0.5

Positive Numbers

Numbers used to name points to the right of the origin on the number line.

p. 63

Negative Numbers

Numbers used to name points to the left of the origin on the number line.

Integers

The natural (or counting) numbers, their negatives, and zero. The integers are $\{\ldots, -3, -2, -1, 0, 1, 2, 3, \ldots\}$

Opposite of a Number

The opposite of a number is the negative of that number.

The opposite of 2 is -2.
The opposite of -9 is 9.

p. 65

Absolute Value

The distance (on the number line) between the point named by an integer and the origin.

$|7| = 7$
$|-10| = 10$

p. 66

This summary exercise set is provided to give you practice with each of the objectives of this chapter. Each exercise is keyed to the appropriate chapter section. When you are finished, you can check your answers to the odd-numbered exercises in the back of the text. If you have difficulty with any of these questions, go back and reread the examples from that section. The answers to the even-numbered exercises appear in the *Instructor's Solutions Manual.* Your instructor will give you guidelines on how best to use these exercises in your instructional setting.

0.1 *List all the factors of the given numbers.*

1. 52

2. 41

3. 76

4. 315

Use the group of numbers 2, 5, 7, 11, 14, 17, 21, 23, 27, 39, and 43.

5. List the prime numbers; then list the composite numbers.

Find the prime factorization for the given numbers.

6. 48

7. 420

8. 60

9. 180

Find the greatest common factor (GCF).

10. 15 and 20

11. 30 and 31

12. 72 and 180

13. 240 and 900

Find the least common multiple (LCM).

14. 4 and 12

15. 8 and 16

16. 18 and 24

17. 12 and 18

0.2 *Write three fractional representations for each number.*

18. $\dfrac{5}{7}$

19. $\dfrac{3}{11}$

20. $\dfrac{4}{9}$

21. Write the fraction $\dfrac{24}{64}$ in simplest form.

22. The Pennsylvania Turnpike, from the Ohio border to the New Jersey border is 360 miles long. Miranda and Carl agree to hike along the turnpike in order to raise money for their favorite charity. On the first day, they hike $23\frac{3}{4}$ miles. The second day, they hike $24\frac{2}{3}$ miles, and on the third day they hike another $17\frac{7}{10}$ miles. How many miles did they walk over the first three days? How much farther do they have to hike in order to complete the entire distance?

0.3 *Perform the indicated operations.*

23. $\dfrac{7}{15} \cdot \dfrac{5}{21}$

24. $\dfrac{10}{27} \cdot \dfrac{9}{20}$

25. $\dfrac{5}{12} \div \dfrac{5}{8}$

26. $\dfrac{7}{15} \div \dfrac{14}{25}$

27. $\dfrac{5}{6} + \dfrac{11}{18}$

28. $\dfrac{5}{18} + \dfrac{7}{12}$

29. $\dfrac{11}{18} - \dfrac{2}{9}$

30. $\dfrac{11}{27} - \dfrac{5}{18}$

31. $5.123 + 6.4$

32. $10.127 - 5.49$

33. $5.26 \cdot 3.796$

34. $6\dfrac{5}{7} + 3\dfrac{4}{7}$

35. $5\dfrac{7}{10} + 3\dfrac{11}{12}$

36. $7\dfrac{7}{9} - 3\dfrac{4}{9}$

37. $6\dfrac{5}{12} - 3\dfrac{5}{8}$

38. $5\dfrac{1}{3} \cdot 1\dfrac{4}{5}$

39. $3\dfrac{2}{5} \cdot \dfrac{5}{8}$

40. $3\dfrac{3}{8} \div 2\dfrac{1}{4}$

Divide and round the quotient to the nearest thousandth.

41. $3.042 \div 0.37$

42. $0.2549 \div 2.87$

Write the percent as a fraction or a mixed number.

43. 2%

44. 20%

45. 37.5%

46. 150%

47. $233\dfrac{1}{3}$%

48. 300%

Write the percents as decimals.

49. 75%

50. 4%

51. 6.25%

52. 13.5%

53. 0.6%

54. 225%

Write as percents.

55. 0.06

56. 2.4

57. 7

58. 0.035

59. 0.005

60. $\dfrac{7}{10}$

61. $\dfrac{2}{5}$

62. $1\dfrac{1}{4}$

63. $2\dfrac{2}{3}$

64. Pierce's monthly electric bill comes to \$84.52 under his equal-payment program. How much will he pay for electricity over a full year?

0.4 *Evaluate each of the following expressions.*

65. $18 - 3 \cdot 5$

66. $(18 - 3) \cdot 5$

67. $5 \cdot 4^2$

68. $(5 \cdot 4)^2$

69. $5 \cdot 3^2 - 4$

70. $5 \cdot (3^2 - 4)$

71. $5 \cdot (4 - 2)^2$

72. $5 \cdot 4 - 2^2$

73. $(5 \cdot 4 - 2)^2$

74. $3 \cdot (5 - 2)^2$

75. $3 \cdot 5 - 2^2$

76. $(3 \cdot 5 - 2)^2$

77. $8 \div 4 \cdot 2$

78. $36 + 4 \cdot 2 - 7 \cdot 6$

79. $4^2 - 2 \cdot \dfrac{10 - 3^2}{4 - (1 + 1)}$

80. $3 \cdot 2^2 + \dfrac{18 - (12 + 2^2)}{2^2}$

0.5

81. Represent the following integers on the number line shown: $6, -18, -3, 2, 15, -9$.

Place each of the following sets in ascending order.

82. $4, -3, 6, -7, 0, 1, -2$

83. $-7, 8, -8, 1, 2, -3, 3, 0, 7$

For each data set determine the maximum and minimum.

84. $4, -2, 5, 1, -6, 3, -4$

85. $-4, 2, 5, -9, 8, 1, -6$

Find the opposite of each number.

86. 17

87. -63

Evaluate.

88. $|9|$

89. $|-9|$

90. $-|9|$

91. $-|-9|$

92. $|12-8|$

93. $|8|+|-12|$

94. $-|8+12|$

95. $|-18|-|-12|$

96. $|-7|-|-3|$

97. $|-9|+|-5|$

98. At the beginning of the month, Tyler had $33.15 in his checking account. He deposited his $425.87 paycheck and paid his $314.89 student loan bill. What is the balance in his checking account?

This purpose of this chapter test is to help you check your progress so that you can find sections and concepts that you need to review before the next exam. Allow yourself about an hour to take this test. At the end of that hour, check your answers against those given in the back of this text. If you missed any, note the section reference that accompanies the answer. Go back to that section and reread the examples until you have mastered that particular concept.

Name _____

Section _____ Date _____

Answers

1. Which of the numbers 5, 9, 13, 17, 22, 27, 31, and 45 are prime numbers? Which are composite numbers?

2. Find the prime factorization for 264.

Find the greatest common factor (GCF) for the given numbers.

3. 36 and 84

4. 16, 24, and 72

Find the least common multiple (LCM) for the given numbers.

5. 12 and 27

6. 3, 4, and 18

Perform the indicated operations.

7. $\dfrac{8}{21} \cdot \dfrac{3}{4}$

8. $\dfrac{7}{12} \div \dfrac{28}{36}$

9. $\dfrac{3}{4} + \dfrac{5}{6}$

10. $\dfrac{8}{21} - \dfrac{2}{7}$

11. $3.25 + 4.125$

12. $16.234 - 12.35$

13. $7.29 \cdot 3.15$

14. $6.10 \cdot 13.1$

15. $2\dfrac{2}{3} \cdot 1\dfrac{2}{7}$

16. $4\dfrac{1}{6} + 3\dfrac{3}{4}$

17. $3\dfrac{5}{6} - 2\dfrac{2}{9}$

18. $3.969 \div 0.54$

Write as fractions.

19. 7%

20. 72%

Answers

1. _____

2. _____

3. _____   4. _____

5. _____   6. _____

7. _____

8. _____

9. _____

10. _____

11. _____

12. _____

13. _____

14. _____

15. _____

16. _____

17. _____

18. _____

19. _____

20. _____

Answers

21.	22.
23.	24.
25.	26.
27.	28.
29.	30.
31.	
32.	
33.	
34.	
35.	
36.	
37.	
38.	
39.	
40.	
41.	
42.	
43.	
44.	
45.	

Write as decimals.

21. 42%   **22.** 6%   **23.** 160%

Write as percents.

24. 0.03   **25.** 0.042

26. $\dfrac{2}{5}$ **27.** $\dfrac{5}{8}$

Write using exponents.

28. $4 \cdot 4 \cdot 4 \cdot 4$ **29.** $9 \cdot 9 \cdot 9 \cdot 9 \cdot 9$

Evaluate the following expressions.

30. $23 - 4 \cdot 5$ **31.** $4 \cdot 5^2 - 35$

32. $4 \cdot (2 + 4)^2$ **33.** $16 \cdot 2 - 5^2$

34. $(3 \cdot 2 - 4)^3$ **35.** $8 - 3 \cdot 2 + 5$

36. Represent the following integers on the number line shown: $5, -12, 4, -7, 18, -17$.

37. Place the following data set in ascending order: $4, -3, -6, 5, 0, 2, -2$.

38. Determine the maximum and minimum of the following data set: $3, 2, -5, 6, 1, -2$.

Evaluate.

39. $|7|$ **40.** $|-7|$

41. $|18 - 7|$ **42.** $|18| - |-7|$

43. $-|24 - 5|$

Find the opposite of each number.

44. 40   **45.** -19

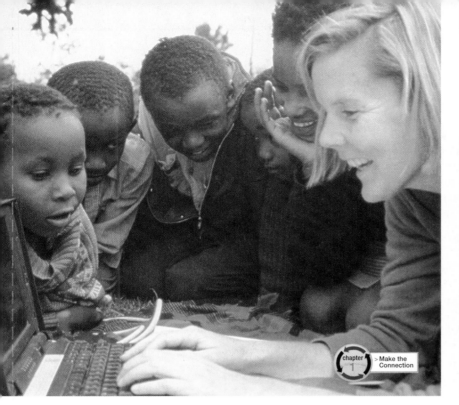

CHAPTER

1

> Make the
Connection

The Language of Algebra

INTRODUCTION

Anthropologists and archeologists investigate modern human cultures and societies as well as cultures that existed so long ago that their characteristics must be inferred from objects found buried in lost cities or villages. With methods such as carbon dating, it has been established that large, organized cultures existed around 3000 B.C.E. in Egypt, 2800 B.C.E. in India, no later than 1500 B.C.E. in China, and around 1000 B.C.E. in the Americas.

Which is older, an object from 3000 B.C.E. or an object from 500 A.D.? An object from 500 A.D. is about $2,000 - 500$ years old, or about 1,500 years old. But an object from 3000 B.C.E. is about $2,000 + 3,000$ years old, or about 5,000 years old. Why subtract in the first case but add in the other? Because of the way years are counted before the common era (B.C.E.) and after the birth of Christ (A.D.), the B.C.E. dates must be considered as *negative* numbers.

Very early on, the Chinese accepted the idea that a number could be negative; they used red calculating rods for positive numbers and black for negative numbers. Hindu mathematicians in India worked out the arithmetic of negative numbers as long ago as 400 A.D., but western mathematicians did not recognize this idea until the sixteenth century. It would be difficult today to think of measuring things such as temperature, altitude, and money without using negative numbers.

CHAPTER 1 OUTLINE

1000 B.C.E. = −1000   1000 A.D. = +1000

← Count   Count →

Name _____

Section _____ Date _____

This pretest provides a preview of the types of exercises you will encounter in each section of this chapter. The answers for these exercises can be found in the back of the text. If you are working on your own, or ahead of the class, this pretest can help you identify the sections in which you should focus more of your time.

Answers

Identify the property that is illustrated by the following statements.

1.1 **1.** $8 \cdot 9 = 9 \cdot 8$

1. _____

2. $3(4 + 2) = 3 \cdot 4 + 3 \cdot 2$

2. _____

3. $9 + (1 + 7) = (9 + 1) + 7$

3. _____

Perform the indicated operations.

1.2–1.3 **4.** $-7 + (-3)$ **5.** $8 + (-9)$ **6.** $(-3) + (-2)$

4. _____ **5.** _____

7. $-\dfrac{7}{4} + \dfrac{3}{4}$ **8.** $8 - 11$ **9.** $-8 - 11$

6. _____ **7.** _____

10. $9 - (-3)$ **11.** $6 + (-6)$ **12.** $(-7)(-3)$

8. _____ **9.** _____

13. $(3.5)(4)$ **14.** $(3)\left(\dfrac{1}{6}\right)$ **15.** $\dfrac{-27 + 6}{-3}$

10. _____ **11.** _____

Write each of the following using symbols.

1.4 **16.** 8 less than x

12. _____ **13.** _____

17. The quotient when w is divided by the product of x and 17

14. _____ **15.** _____

Identify which are expressions and which are not.

16. _____ **17.** _____

18. $7x - 5 = 11$

19. $3x - 2(x + 1)$

18. _____

Evaluate the following expressions.

19. _____

1.5 **20.** $5 - 4^2 \cdot 3 \div 6$ **21.** $(45 - 3 \cdot 5) + 5^2$

20. _____ **21.** _____

22. If $x = -2$, $y = 7$, and $w = -4$, evaluate the expression $\dfrac{x^2 y}{w}$.

Combine like terms.

22. _____ **23.** _____

1.6 **23.** $5w^2 t + 3w^2 t$ **24.** $4a^2 - 3a + 5 + 7a - 2 - 5a^2$

24. _____

Divide.

25. _____

1.7 **25.** $\dfrac{96x^3 y^5}{8x^2 y^3}$

1.1
Properties of Real Numbers

< 1.1 Objectives >

1 > Recognize applications of the commutative properties

2 > Recognize applications of the associative properties

3 > Recognize applications of the distributive property

> **Overcoming Math Anxiety**

RECALL

Section 0.1 begins with the first Overcoming Math Anxiety hint.

Over the first few chapters, we present you with a series of class-tested techniques that are designed to improve your performance in your math class. Hint #2   Become familiar with your syllabus.

In your first class meeting, your instructor probably gave you a class syllabus. If you have not already done so, incorporate important information into a calendar and address book.

1. Write all important dates in your calendar. This includes the date and time of the final exam, test dates, quiz dates, and homework due dates. Never allow yourself to be surprised by a deadline!

2. Write your instructor's name, contact information, and office number in your address book. Also include your instructor's office hours. Make it a point to see your instructor early in the term. Although not the only person who can help you, your instructor is an important resource to help clear up any confusion you may have.

3. Make note of other resources that are available to you. This includes tutoring, CDs and DVDs, and Web pages.

Given all of these resources, it is important that you never let confusion or frustration mount. If you "can't get it" from the text, try another resource. All of these resources are there specifically for you, so take advantage of them!

NOTE

We work only with real numbers in this text.

Everything that we do in *algebra* is based on the **properties of real numbers.** Before being introduced to algebra, you should understand these properties.

The **commutative property** tells us that we can add or multiply in any order.

Property

The Commutative Properties

If *a* and *b* are any numbers,

1. $a + b = b + a$ Commutative property of addition
2. $a \cdot b = b \cdot a$ Commutative property of multiplication

You may notice that we used the letters a and b rather than numbers in the **Property** box. We use these letters to indicate that these properties are true for any choice of real numbers.

 Example 1   **Identifying the Commutative Properties**

< Objective 1 >

(a) $5 + 9 = 9 + 5$

This is an application of the commutative property of addition.

(b) $5 \cdot 9 = 9 \cdot 5$

This is an application of the commutative property of multiplication.

Check Yourself 1

Identify the property being applied.

(a) $7 + 3 = 3 + 7$ **(b)** $7 \cdot 3 = 3 \cdot 7$

We also want to be able to change the grouping when simplifying expressions. Regrouping is possible because of the **associative properties.** Numbers can be grouped in any manner to find a sum or a product.

Property

The Associative Properties

If a, b, and c are any numbers,

1. $a + (b + c) = (a + b) + c$ Associative property of addition

2. $a \cdot (b \cdot c) = (a \cdot b) \cdot c$ Associative property of multiplication

---

 Example 2   **Demonstrating the Associative Properties**

< Objective 2 >

RECALL

As we saw in Section 0.4, we always do the operation in the parentheses first.

(a) Show that $2 + (3 + 8) = (2 + 3) + 8$.

$2 + \underbrace{(3 + 8)}_{\text{Add first.}}$ $\underbrace{(2 + 3)}_{\text{Add first.}} + 8$

$= 2 + 11$ $= 5 + 8$

$= 13$ $= 13$

So

$2 + (3 + 8) = (2 + 3) + 8$

(b) Show that $\frac{1}{3} \cdot (6 \cdot 5) = \left(\frac{1}{3} \cdot 6\right) \cdot 5$.

$\frac{1}{3} \cdot \underbrace{(6 \cdot 5)}$
Multiply first.

$\left(\underbrace{\frac{1}{3} \cdot 6}\right) \cdot 5$
Multiply first.

$= \frac{1}{3} \cdot (30)$

$= 10$

$= (2) \cdot 5$

$= 10$

So

$\frac{1}{3} \cdot (6 \cdot 5) = \left(\frac{1}{3} \cdot 6\right) \cdot 5$

Check Yourself 2

Show that the following statements are true.

(a) $3 + (4 + 7) = (3 + 4) + 7$

(b) $3 \cdot (4 \cdot 7) = (3 \cdot 4) \cdot 7$

(c) $\left(\frac{1}{5} \cdot 10\right) \cdot 4 = \frac{1}{5} \cdot (10 \cdot 4)$

NOTE

The area of a rectangle is the product of its length and width:
$A = L \cdot W$

The **distributive property** involves addition and multiplication together. We can illustrate this property with an application.

Suppose that we want to find the total of the two areas shown in the following figure.

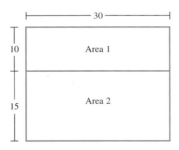

We can find the total area by multiplying the length by the overall width, which is found by adding the two widths.   [or]   We can find the total area as a sum of the two areas.

Length	Overall width		(Area 1) Length · Width		(Area 2) Length · Width
$\overbrace{30}$	$\cdot$	$\overbrace{(10 + 15)}$	$\overbrace{30 \cdot 10}$	$+$	$\overbrace{30 \cdot 15}$

$= 30 \cdot 25$

$= 750$

$= 300 + 450$

$= 750$

So

$30 \cdot (10 + 15) = 30 \cdot 10 + 30 \cdot 15$

This leads us to the following property.

Property

The Distributive Property	If a, b, and c are any numbers,
	$a(b + c) = a \cdot b + a \cdot c$ and $(b + c)a = b \cdot a + c \cdot a$

Example 3   Using the Distributive Property

< Objective 3 >

NOTE

Notice the pattern.

$a(b + c) = a \cdot b + a \cdot c$

We "distributed" the multiplication "over" the addition.

Use the distributive property to remove the parentheses in the following.

(a) $5(3 + 4)$

$5(3 + 4) = 5 \cdot 3 + 5 \cdot 4$ We could also say $5(3 + 4) = 5 \cdot 7 = 35$
$= 15 + 20 = 35$

(b) $\dfrac{1}{3}(9 + 12) = \dfrac{1}{3} \cdot 9 + \dfrac{1}{3} \cdot 12$

$= 3 + 4 = 7$

NOTE

It is also true that

$\dfrac{1}{3}(9 + 12) = \dfrac{1}{3}(21) = 7$

Check Yourself 3

Use the distributive property to remove the parentheses.

(a) $4(6 + 7)$ **(b)** $\dfrac{1}{5}(10 + 15)$

Example 4 requires that you identify which property is being demonstrated. Look for patterns that help you to remember each of the properties.

Example 4   Identifying Properties

Name the property demonstrated.

(a) $3(8 + 2) = 3 \cdot 8 + 3 \cdot 2$ demonstrates the distributive property.

(b) $2 + (3 + 5) = (2 + 3) + 5$ demonstrates the associative property of addition.

(c) $3 \cdot 5 = 5 \cdot 3$ demonstrates the commutative property of multiplication.

Check Yourself 4

Name the property demonstrated.

(a) $2 \cdot (3 \cdot 5) = (2 \cdot 3) \cdot 5$

(b) $4(-2 + 4) = 4(-2) + 4 \cdot 4$

(c) $\dfrac{1}{2} + 8 = 8 + \dfrac{1}{2}$

Check Yourself ANSWERS

1. **(a)** Commutative property of addition;
   (b) commutative property of multiplication
2. **(a)** $3 + (4 + 7) = 3 + 11 = 14$ **(b)** $3 \cdot (4 \cdot 7) = 3 \cdot 28 = 84$
   $(3 + 4) + 7 = 7 + 7 = 14$ $(3 \cdot 4) \cdot 7 = 12 \cdot 7 = 84$
   (c) $\left(\dfrac{1}{5} \cdot 10\right) \cdot 4 = 2 \cdot 4 = 8$

   $\dfrac{1}{5} \cdot (10 \cdot 4) = \dfrac{1}{5} \cdot 40 = 8$
3. **(a)** $4 \cdot 6 + 4 \cdot 7 = 24 + 28 = 52$; **(b)** $\dfrac{1}{5} \cdot 10 + \dfrac{1}{5} \cdot 15 = 2 + 3 = 5$
4. **(a)** Associative property of multiplication; **(b)** distributive property;
   (c) commutative property of addition

Reading Your Text

The following fill-in-the-blank exercises are designed to ensure that you understand some of the key vocabulary used in this section.

SECTION 1.1

(a) You should write all important class dates in your _____.

(b) Everything we do in algebra is based on the _____ of real numbers.

(c) The _____ property of multiplication states that $a \cdot (b \cdot c) = (a \cdot b) \cdot c$.

(d) The _____ of a rectangle is the product of its length and width.

Name _____

Section _____ Date _____

Answers

1. _____
2. _____
3. _____
4. _____
5. _____
6. _____
7. _____
8. _____
9. _____
10. _____
11. _____
12. _____
13. _____
14. _____
15. _____
16. _____
17. _____
18. _____
19. _____
20. _____

Basic Skills | Advanced Skills | Vocational-Technical Applications | Calculator/Computer | Above and Beyond | Getting Ready

< Objectives 1–3 >

Identify the property that is illustrated by each of the following statements.

1. $5 + 9 = 9 + 5$

2. $6 + 3 = 3 + 6$

3. $2 \cdot (3 \cdot 5) = (2 \cdot 3) \cdot 5$

4. $3 \cdot (5 \cdot 6) = (3 \cdot 5) \cdot 6$

5. $\dfrac{1}{4} \cdot \dfrac{1}{5} = \dfrac{1}{5} \cdot \dfrac{1}{4}$

6. $7 \cdot 9 = 9 \cdot 7$

7. $8 + 12 = 12 + 8$

8. $6 + 2 = 2 + 6$

9. $(5 \cdot 7) \cdot 2 = 5 \cdot (7 \cdot 2)$

10. $(8 \cdot 9) \cdot 2 = 8 \cdot (9 \cdot 2)$

11. $7 \cdot (2 \cdot 5) = (7 \cdot 2) \cdot 5$

12. $\dfrac{1}{2} \cdot 6 = 6 \cdot \dfrac{1}{2}$

13. $2(3 + 5) = 2 \cdot 3 + 2 \cdot 5$

14. $5 \cdot (4 + 6) = 5 \cdot 4 + 5 \cdot 6$

> Videos

15. $5 + (7 + 8) = (5 + 7) + 8$

16. $8 + (2 + 9) = (8 + 2) + 9$

17. $\left(\dfrac{1}{3} + 4\right) + \dfrac{1}{5} = \dfrac{1}{3} + \left(4 + \dfrac{1}{5}\right)$

18. $(5 + 5) + 3 = 5 + (5 + 3)$

19. $7 \cdot (3 + 8) = 7 \cdot 3 + 7 \cdot 8$

20. $5 \cdot (6 + 8) = 5 \cdot 6 + 5 \cdot 8$

Verify that each of the following statements is true by evaluating each side of the equation separately and comparing the results.

21. $7 \cdot (3 + 4) = 7 \cdot 3 + 7 \cdot 4$

22. $4 \cdot (5 + 1) = 4 \cdot 5 + 4 \cdot 1$

23. $2 + (9 + 8) = (2 + 9) + 8$

24. $6 + (15 + 3) = (6 + 15) + 3$

25. $\frac{1}{3} \cdot (6 \cdot 3) = \left(\frac{1}{3} \cdot 6\right) \cdot 3$

26. $2 \cdot (9 \cdot 10) = (2 \cdot 9) \cdot 10$

27. $5 \cdot (2 + 8) = 5 \cdot 2 + 5 \cdot 8$

28. $\frac{1}{4} \cdot (10 + 2) = \frac{1}{4} \cdot 10 + \frac{1}{4} \cdot 2$

29. $(3 + 12) + 8 = 3 + (12 + 8)$

30. $(8 + 12) + 7 = 8 + (12 + 7)$

31. $(4 \cdot 7) \cdot 2 = 4 \cdot (7 \cdot 2)$

32. $(6 \cdot 5) \cdot 3 = 6 \cdot (5 \cdot 3)$

33. $\frac{1}{2} \cdot (2 + 6) = \frac{1}{2} \cdot 2 + \frac{1}{2} \cdot 6$

34. $\frac{1}{3} \cdot (6 + 9) = \frac{1}{3} \cdot 6 + \frac{1}{3} \cdot 9$

35. $\left(\frac{2}{3} + \frac{1}{6}\right) + \frac{1}{3} = \frac{2}{3} + \left(\frac{1}{6} + \frac{1}{3}\right)$

36. $\frac{3}{4} + \left(\frac{5}{8} + \frac{1}{2}\right) = \left(\frac{3}{4} + \frac{5}{8}\right) + \frac{1}{2}$

37. $(2.3 + 3.9) + 4.1 = 2.3 + (3.9 + 4.1)$ > Videos

38. $(1.7 + 4.1) + 7.6 = 1.7 + (4.1 + 7.6)$

39. $\frac{1}{2} \cdot (2 \cdot 8) = \left(\frac{1}{2} \cdot 2\right) \cdot 8$

40. $\frac{1}{5} \cdot (5 \cdot 3) = \left(\frac{1}{5} \cdot 5\right) \cdot 3$

41. $\left(\frac{3}{5} \cdot \frac{5}{6}\right) \cdot \frac{4}{3} = \frac{3}{5} \cdot \left(\frac{5}{6} \cdot \frac{4}{3}\right)$

42. $\frac{4}{7} \cdot \left(\frac{21}{16} \cdot \frac{8}{3}\right) = \left(\frac{4}{7} \cdot \frac{21}{16}\right) \cdot \frac{8}{3}$

43. $2.5 \cdot (4 \cdot 5) = (2.5 \cdot 4) \cdot 5$

44. $4.2 \cdot (5 \cdot 2) = (4.2 \cdot 5) \cdot 2$

Answers

21. _____
22. _____
23. _____
24. _____
25. _____
26. _____
27. _____
28. _____
29. _____
30. _____
31. _____
32. _____
33. _____
34. _____
35. _____
36. _____
37. _____
38. _____
39. _____
40. _____
41. _____
42. _____
43. _____
44. _____

The Streeter/Hutchison Series in Mathematics   Beginning Algebra

Answers

45. _____

46. _____

47. _____

48. _____

49. _____

50. _____

51. _____

52. _____

53. _____

54. _____

55. _____

56. _____

57. _____

58. _____

59. _____

60. _____

61. _____

62. _____

63. _____

64. _____

65. _____

66. _____

Use the distributive property to remove the parentheses in each of the following expressions. Then simplify your result where possible.

45. $3(2 + 6)$

46. $5(4 + 6)$

47. $2(12 + 10)$

48. $9(1 + 8)$

49. $0.1(2 + 10)$

50. $1.2(3 + 8)$

51. $\dfrac{2}{3}(6 + 9)$

52. $\dfrac{1}{2}\left(4 + \dfrac{1}{3}\right)$

53. $\dfrac{1}{3} \cdot (15 + 9)$

54. $\dfrac{1}{6} \cdot (36 + 24)$

Basic Skills | **Advanced Skills** | Vocational-Technical Applications | Calculator/Computer | Above and Beyond | Getting Ready
▲

Use the properties of addition and multiplication to complete each of the following statements.

55. $5 + 7 = \quad + 5$

56. $(5 + 3) + 4 = 5 + (\quad + 4)$

57. $(8)(3) = (3)(\quad)$

58. $8(3 + 4) = 8 \cdot 3 + \quad \cdot 4$

59. $7(2 + 5) = 7 \cdot \quad + 7 \cdot 5$

60. $4 \cdot (2 \cdot 4) = (\quad \cdot 2) \cdot 4$

Use the indicated property to write an expression that is equivalent to each of the following expressions.

61. $3 + 7$ (commutative property of addition)

62. $2(3 + 4)$ (distributive property)

63. $5 \cdot (3 \cdot 2)$ (associative property of multiplication)

64. $(3 + 5) + 2$ (associative property of addition)

65. $2 \cdot 4 + 2 \cdot 5$ (distributive property)

66. $7 \cdot 9$ (commutative property of multiplication)

Basic Skills | Advanced Skills | Vocational-Technical Applications | Calculator/Computer | **Above and Beyond** | Getting Ready
▲

Evaluate each of the following pairs of expressions. Then answer the given question.

67. $8 - 5$ and $5 - 8$
Do you think subtraction is commutative?

68. $12 \div 3$ and $3 \div 12$
Do you think division is commutative?

69. $(12 - 8) - 4$ and $12 - (8 - 4)$
Do you think subtraction is associative?

70. $(48 \div 16) \div 4$ and $48 \div (16 \div 4)$
Do you think division is associative?

71. $3(6 - 2)$ and $3 \cdot 6 - 3 \cdot 2$
Do you think multiplication is distributive over subtraction?

72. $\dfrac{1}{2}(16 - 10)$ and $\dfrac{1}{2} \cdot 16 - \dfrac{1}{2} \cdot 10$

Do you think multiplication is distributive over subtraction?

Complete the statement using the

    (a) Distributive property
    (b) Commutative property of addition
    (c) Commutative property of multiplication

73. $5 \cdot (3 + 4) =$ **74.** $6 \cdot (5 + 4) =$

Identify the property that is used.

75. $5 + (6 + 7) = (5 + 6) + 7$ **76.** $5 + (6 + 7) = 5 + (7 + 6)$

> Videos

77. $4 \cdot (3 + 2) = 4 \cdot (2 + 3)$ **78.** $4 \cdot (3 + 2) = (3 + 2) \cdot 4$

Answers

67. _____

68. _____

69. _____

70. _____

71. _____

72. _____

73. _____

74. _____

75. _____

76. _____

77. _____

78. _____

Answers

(a) _____

(b) _____

(c) _____

(d) _____

(e) _____

(f) _____

Find each sum or difference.

(a) $3 + (8 + 9)$ **(b)** $6 + (12 + 3)$

(c) $(3 + 8) + (9 + 4)$ **(d)** $15 - 11 - (2 + 1)$

(e) $\dfrac{3}{5} + \dfrac{4}{15}$ **(f)** $\dfrac{12}{27} - \dfrac{2}{9}$

Answers

1. Commutative property of addition **3.** Associative property of multiplication **5.** Commutative property of multiplication
7. Commutative property of addition **9.** Associative property of multiplication **11.** Associative property of multiplication
13. Distributive property **15.** Associative property of addition
17. Associative property of addition **19.** Distributive property
21. $49 = 49$ **23.** $19 = 19$ **25.** $6 = 6$ **27.** $50 = 50$

29. $23 = 23$ **31.** $56 = 56$ **33.** $4 = 4$ **35.** $\dfrac{7}{6} = \dfrac{7}{6}$

37. $10.3 = 10.3$ **39.** $8 = 8$ **41.** $\dfrac{2}{3} = \dfrac{2}{3}$ **43.** $50 = 50$ **45.** 24

47. 44 **49.** 1.2 **51.** 10 **53.** 8 **55.** 7 **57.** 8 **59.** 2
61. $7 + 3$ **63.** $(5 \cdot 3) \cdot 2$ **65.** $2 \cdot (4 + 5)$ **67.** No **69.** No
71. Yes **73. (a)** $5 \cdot 3 + 5 \cdot 4$; **(b)** $5 \cdot (4 + 3)$; **(c)** $(3 + 4) \cdot 5$
75. Associative property of addition **77.** Commutative property of addition

(a) 20 **(b)** 21 **(c)** 24 **(d)** 1 **(e)** $\dfrac{13}{15}$ **(f)** $\dfrac{2}{9}$

1.2

Adding and Subtracting Real Numbers

< 1.2 Objectives >

1 > Find the sum of two real numbers

2 > Find the difference of two real numbers

In Section 0.5 we introduced the idea of numbers with different signs. Now we will examine the four arithmetic operations (addition, subtraction, multiplication, and division) to see how those operations are performed when these numbers are involved. We start by considering addition.

An application may help. As before, we represent a gain of money as a positive number and a loss as a negative number.

If you gain $3 and then gain $4, the result is a gain of $7:

$$3 + 4 = 7$$

If you lose $3 and then lose $4, the result is a loss of $7:

$$-3 + (-4) = -7$$

If you gain $3 and then lose $4, the result is a loss of $1:

$$3 + (-4) = -1$$

If you lose $3 and then gain $4, the result is a gain of $1:

$$-3 + 4 = 1$$

The number line can be used to illustrate the addition of these numbers. Starting at the origin, we move to the *right* for positive numbers and to the *left* for negative numbers.

| Example 1 | Adding Two Negative Numbers |

< Objective 1 >

(a) Add $-3 + (-4)$.

Start at the origin and move 3 units to the left. Then move 4 more units to the left to find the sum. From the number line we see that the sum is

$$-3 + (-4) = -7$$

95

(b) Add $-\dfrac{3}{2} + \left(-\dfrac{1}{2}\right)$.

As before, we start at the origin. From that point move $\dfrac{3}{2}$ units left. Then move another $\dfrac{1}{2}$ unit left to find the sum. In this case

$$-\dfrac{3}{2} + \left(-\dfrac{1}{2}\right) = -2$$

Check Yourself 1

Add.

(a) $-4 + (-5)$ **(b)** $-3 + (-7)$

(c) $-5 + (-15)$ **(d)** $-\dfrac{5}{2} + \left(-\dfrac{3}{2}\right)$

You have probably noticed some helpful patterns in the previous examples. These patterns will allow you to do the work mentally without having to use the number line. Look at the following rule.

Property

Adding Real Numbers with the Same Sign	If two numbers have the same sign, add their absolute values. Give the sum the sign of the original numbers. In other words, the sum of two positive numbers is positive and the sum of two negative numbers is negative.

Again, we use the number line to illustrate the addition of two numbers. This time the numbers have *different* signs.

Example 2   **Adding Two Numbers with Different Signs**

(a) Add $3 + (-6)$.

First move 3 units to the right of the origin. Then move 6 units to the left.

$3 + (-6) = -3$

(b) Add $-4 + 7$.

This time move 4 units to the left of the origin as the first step. Then move 7 units to the right.

$$-4 + 7 = 3$$

 Check Yourself 2

Add.

(a) $7 + (-5)$ **(b)** $4 + (-8)$ **(c)** $-\dfrac{1}{3} + \dfrac{16}{3}$ **(d)** $-7 + 3$

RECALL

We first encountered absolute values in Section 0.5.

You have no doubt noticed that, in adding a positive number and a negative number, sometimes the sum is positive and sometimes it is negative. This depends on which of the numbers has the larger absolute value. This leads us to the second part of our addition rule.

Property

Adding Real Numbers with Different Signs

If two numbers have different signs, subtract their absolute values, the smaller from the larger. Give the result the sign of the number with the larger absolute value.

 Example 3   **Adding Positive and Negative Numbers**

(a) $7 + (-19) = -12$

Because the two numbers have different signs, subtract the absolute values $(19 - 7 = 12)$. The sum has the sign $(-)$ of the number with the larger absolute value.

(b) $-\dfrac{13}{2} + \dfrac{7}{2} = -3$

Subtract the absolute values $\left(\dfrac{13}{2} - \dfrac{7}{2} = \dfrac{6}{2} = 3\right)$. The sum has the sign $(-)$ of the number with the larger absolute value, $-\dfrac{13}{2}$.

(c) $-8.2 + 4.5 = -3.7$

Subtract the absolute values $(8.2 - 4.5 = 3.7)$. The sum has the sign $(-)$ of the number with the larger absolute value, -8.2.

Check Yourself 3

Add mentally.

(a) $5 + (-14)$ **(b)** $-7 + (-8)$ **(c)** $-8 + 15$

(d) $7 + (-8)$ **(e)** $-\dfrac{2}{3} + \left(-\dfrac{7}{3}\right)$ **(f)** $5.3 + (-2.3)$

In Section 1.1 we discussed the commutative, associative, and distributive properties. There are two other properties of addition that we should mention. First, the sum of any number and 0 is always that number. In symbols,

---

Property

Additive Identity Property	For any number a, $a + 0 = 0 + a = a$ In words, no number loses its identity after addition with 0. Zero is called the **additive identity**.

▶ **Example 4**   **Adding the Identity**

Add.

(a) $9 + 0 = 9$

(b) $0 + \left(-\dfrac{5}{4}\right) = -\dfrac{5}{4}$

(c) $(-25) + 0 = -25$

Check Yourself 4

Add.

(a) $8 + 0$ **(b)** $0 + \left(-\dfrac{8}{3}\right)$ **(c)** $(-36) + 0$

NOTES

The opposite of a number is also called the **additive inverse** of that number.

3 and -3 are opposites.

Recall that every number has an *opposite*. It corresponds to a point the same distance from the origin as the given number, but in the opposite direction.

The opposite of 9 is -9.

The opposite of -15 is 15.

Our second property states that the sum of any number and its opposite is 0.

The Streeter/Hutchison Series in Mathematics   Beginning Algebra

Property

Additive Inverse Property

For any number *a*, there exists a number −*a* such that

$$a + (-a) = (-a) + a = 0$$

We could also say that −*a* represents the opposite of the number *a*. The sum of any number and its opposite, or additive inverse, is 0.

Example 5   **Adding Inverses**

(a) $9 + (-9) = 0$

(b) $-15 + 15 = 0$

(c) $(-2.3) + 2.3 = 0$

(d) $\dfrac{4}{5} + \left(-\dfrac{4}{5}\right) = 0$

Check Yourself 5

Add.

(a) $(-17) + 17$ (b) $12 + (-12)$

(c) $\dfrac{1}{3} + \left(-\dfrac{1}{3}\right)$ (d) $-1.6 + 1.6$

To begin our discussion of subtraction when signed numbers are involved, we can look back at a problem using natural numbers. Of course, we know that

$$8 - 5 = 3$$

From our work in adding real numbers, we know that it is also true that

$$8 + (-5) = 3$$

NOTE

This is the *definition* of subtraction.

Comparing these equations, we see that the results are the same. This leads us to an important pattern. Any subtraction problem can be written as a problem in addition. Subtracting 5 is the same as adding the opposite of 5, or −5. We can write this fact as follows:

$$8 - 5 = 8 + (-5) = 3$$

This leads us to the following rule for subtracting real numbers.

Property

Subtracting Real Numbers

1. Rewrite the subtraction problem as an addition problem.
   a. Change the minus sign to a plus sign.
   b. Replace the number being subtracted with its opposite.
2. Add the resulting signed numbers as before.
   In symbols,
   $$a - b = a + (-b)$$

Example 6 illustrates the use of this definition while subtracting.

| **▷** | **Example 6** | **Subtracting Real Numbers** |

< Objective 2 >

Simplify each expression.

Change the subtraction symbol (−)
to an addition symbol (+).

(a) $15 - 7 = 15 + (-7)$

Replace 7 with its opposite, −7.

$\qquad = 8$

(b) $9 - 12 = 9 + (-12) = -3$

(c) $-6 - 7 = -6 + (-7) = -13$

(d) $-\dfrac{3}{5} - \dfrac{7}{5} = -\dfrac{3}{5} + \left(-\dfrac{7}{5}\right) = -\dfrac{10}{5} = -2$

(e) $2.1 - 3.4 = 2.1 + (-3.4) = -1.3$

(f) Subtract 5 from −2. We write the statement as $-2 - 5$ and proceed as before:

$-2 - 5 = -2 + (-5) = -7$

 Check Yourself 6

Subtract.

| **(a)** $18 - 7$ | **(b)** $5 - 13$ | **(c)** $-7 - 9$ |
| **(d)** $-\dfrac{5}{6} - \dfrac{7}{6}$ | **(e)** $-2 - 7$ | **(f)** $5.6 - 7.8$ |

The subtraction rule is used in the same way when the number being subtracted is negative. Change the subtraction to addition. Replace the negative number being subtracted with its opposite, which is positive. Example 7 illustrates this principle.

| **▷** | **Example 7** | **Subtracting Real Numbers** |

Simplify each expression.

Change the subtraction
to an addition.

(a) $5 - (-2) = 5 + (+2) = 5 + 2 = 7$

Replace −2 with its opposite, +2 or 2.

(b) $7 - (-8) = 7 + (+8) = 7 + 8 = 15$

(c) $-9 - (-5) = -9 + 5 = -4$

(d) $-12.7 - (-3.7) = -12.7 + 3.7 = -9$

(e) $-\dfrac{3}{4} - \left(-\dfrac{7}{4}\right) = -\dfrac{3}{4} + \left(+\dfrac{7}{4}\right) = \dfrac{4}{4} = 1$

(f) Subtract -4 from -5. We write

$$-5 - (-4) = -5 + 4 = -1$$

Check Yourself 7

Subtract.

(a) $8 - (-2)$　　　　**(b)** $3 - (-10)$　　　　**(c)** $-7 - (-2)$
(d) $-9.8 - (-5.8)$　　**(e)** $7 - (-7)$

NOTE

If your calculator is different from the ones we describe, refer to your manual, or ask your instructor for assistance.

In order to use a calculator to do arithmetic with real numbers, there are some keys you should become familiar with.

The first key is the subtraction key, $\boxed{-}$. This key is usually found in the right column of calculator keys along with the other "operation" keys such as addition, multiplication, and division.

The second key to find is the one for negative numbers. On graphing calculators, it usually looks like $\boxed{(-)}$, whereas on scientific calculators, the key usually looks like $\boxed{+/-}$. In either case, the negative number key is usually found in the bottom row.

One very important difference between the two types of calculators is that when using a graphing calculator, you input the negative sign before keying in the number (as it is written). When using a scientific calculator, you input the negative number button after keying in the number.

In Example 8, we illustrate this difference, while showing that subtraction remains the same.

　Example 8　　**Subtracting with a Calculator**

NOTE

Graphing calculators usually use an $\boxed{\text{ENTER}}$ key while scientific calculators have an $\boxed{=}$ key.

Use a calculator to find each difference.

(a) $-12.43 - 3.516$

Graphing Calculator

$\boxed{(-)}$ 12.43 $\boxed{-}$ 3.516 $\boxed{\text{ENTER}}$　　　The negative number sign comes before the number.

The display should read -15.946.

Scientific Calculator

12.43 $\boxed{+/-}$ $\boxed{-}$ 3.516 $\boxed{=}$　　　The negative number sign comes after the number.

The display should read -15.946.

```
-12.43 – 3.516
                    -15.946
23.56 – -4.7
                     28.26
■
```

(b) $23.56 - (-4.7)$

Graphing Calculator

23.56 ⊟ (−) 4.7 ENTER   The negative number sign comes before the number.

The display should read 28.26.

Scientific Calculator

23.56 ⊟ 4.7 +/− =   The negative number sign comes after the number.

The display should read 28.26.

Check Yourself 8

Use your calculator to find the difference.

(a) $-13.46 - 5.71$ **(b)** $-3.575 - (-6.825)$

| Example 9 | An Application Involving Real Numbers |

Oscar owned four stocks. This year his holdings in Cisco went up $2,250, in AT&T they went down $1,345, in Texaco they went down $5,215, and in IBM they went down $1,525. How much less are his holdings worth at the end of the year compared to the beginning of the year?

To find the change in Oscar's holdings, we add the amounts that went up and subtract the amounts that went down.

$\$2,250 - \$1,345 - \$5,215 - \$1,525 = -\$5,835$

Oscar's holdings are worth $5,835 less at the end of the year.

Check Yourself 9

A bus with fifteen people stopped at Avenue A. Nine people got off and five people got on. At Avenue B six people got off and eight people got on. At Avenue C four people got off the bus and six people got on. How many people were now on the bus?

Check Yourself ANSWERS

1. **(a)** -9; **(b)** -10; **(c)** -20; **(d)** -4 2. **(a)** 2; **(b)** -4; **(c)** 5; **(d)** -4
3. **(a)** -9; **(b)** -15; **(c)** 7; **(d)** -1; **(e)** -3; **(f)** 3
4. **(a)** 8; **(b)** $-\dfrac{8}{3}$; **(c)** -36 5. **(a)** 0; **(b)** 0; **(c)** 0; **(d)** 0
6. **(a)** 11; **(b)** -8; **(c)** -16; **(d)** -2; **(e)** -9; **(f)** -2.2 7. **(a)** 10; **(b)** 13;
(c) -5; **(d)** -4; **(e)** 14   8. **(a)** -19.17; **(b)** 3.25   9. 15 people

Reading Your Text

The following fill-in-the-blank exercises are designed to ensure that you understand some of the key vocabulary used in this section.

SECTION 1.2

(a) When two negative numbers are added, the sign of the sum is _____.

(b) The sum of two numbers with different signs is given the sign of the number with the larger _____ value.

(c) _____ is called the additive identity.

(d) When subtracting negative numbers, change the minus to a plus and replace the second number with its _____.

1.2 exercises

Boost your grade at mathzone.com!

> Practice Problems
> NetTutor
> Self-Tests
> e-Professors
> Videos

Name _____

Section _____ Date _____

Answers

1. _____

2. _____

3. _____

4. _____

5. _____

6. _____

7. _____ 8. _____

9. _____ 10. _____

11. _____ 12. _____

13. _____ 14. _____

15. _____ 16. _____

17. _____ 18. _____

19. _____ 20. _____

21. _____ 22. _____

23. _____ 24. _____

Basic Skills | Advanced Skills | Vocational-Technical Applications | Calculator/Computer | Above and Beyond | Getting Ready

< Objective 1 >

Add.

1. $3 + 6$

2. $8 + 7$

3. $\dfrac{4}{5} + \dfrac{6}{5}$

4. $\dfrac{7}{3} + \dfrac{8}{3}$

5. $\dfrac{1}{2} + \dfrac{4}{5}$

6. $\dfrac{2}{3} + \dfrac{5}{9}$

7. $-4 + (-1)$

8. $-1 + (-9)$

9. $-\dfrac{1}{2} + \left(-\dfrac{3}{8}\right)$ > Videos

10. $-\dfrac{4}{7} + \left(-\dfrac{3}{14}\right)$

11. $-1.6 + (-2.3)$

12. $-3.5 + (-2.6)$

13. $3 + (-9)$

14. $11 + (-7)$

15. $\dfrac{3}{4} + \left(-\dfrac{1}{2}\right)$

16. $\dfrac{2}{3} + \left(-\dfrac{1}{6}\right)$

17. $13.4 + (-11.4)$

18. $5.2 + (-9.2)$

19. $-5 + 3$

20. $-12 + 17$

21. $-\dfrac{4}{5} + \dfrac{9}{20}$

22. $-\dfrac{11}{6} + \dfrac{5}{12}$ > Videos

23. $-8.6 + 4.9$

24. $-3.6 + 7.6$

25. $0 + (-8)$

26. $-15 + 0$

27. $7 + (-7)$

28. $-12 + 12$

29. $-4.5 + 4.5$

30. $\dfrac{2}{3} + \left(-\dfrac{2}{3}\right)$

< Objective 2 >

Subtract.

31. $82 - 45$

32. $45 - 82$

33. $18 - 20$

34. $136 - 352$

35. $\dfrac{8}{7} - \dfrac{15}{7}$

36. $\dfrac{17}{8} - \dfrac{9}{8}$

37. $5.4 - 7.9$

38. $11.7 - 4.5$

39. $-3 - 1$

40. $-15 - 8$

41. $-14 - 9$

42. $-8 - 12$

43. $-\dfrac{2}{5} - \dfrac{7}{10}$

44. $-\dfrac{7}{18} - \dfrac{5}{9}$

45. $-3.4 - 4.7$

46. $-8.1 - 7.6$

47. $5 - (-11)$

48. $8 - (-4)$

49. $12 - (-7)$

50. $3 - (-10)$

Answers

25. _____

26. _____

27. _____

28. _____

29. _____

30. _____

31. _____

32. _____

33. _____

34. _____

35. _____

36. _____

37. _____

38. _____

39. _____ 40. _____

41. _____ 42. _____

43. _____ 44. _____

45. _____ 46. _____

47. _____ 48. _____

49. _____ 50. _____

Answers

51. _____

52. _____

53. _____

54. _____

55. _____

56. _____

57. _____

58. _____

59. _____

60. _____

61. _____

62. _____

63. _____

64. _____

65. _____

51. $\dfrac{3}{4} - \left(-\dfrac{3}{2}\right)$ > Videos

52. $\dfrac{11}{16} - \left(-\dfrac{5}{8}\right)$

53. $8.3 - (-5.7)$

54. $14.5 - (-54.6)$

55. $-28 - (-11)$

56. $-11 - (-16)$

57. $-19 - (-27)$

58. $-13 - (-4)$

59. $-\dfrac{3}{4} - \left(-\dfrac{11}{4}\right)$

60. $-\dfrac{5}{8} - \left(-\dfrac{1}{2}\right)$

Solve each of the following problems.

61. **BUSINESS AND FINANCE** Amir has $100 in his checking account. He writes a check for $23 and makes a deposit of $51. What is his new balance?

62. **BUSINESS AND FINANCE** Olga has $250 in her checking account. She deposits $52 and then writes a check for $77. What is her new balance?

Bal: _____ 250
Dep: _____ 52
CK # 1111: _____ 77

63. **STATISTICS** On four consecutive running plays, Duce Staley of the Philadelphia Eagles gained 23 yards, lost 5 yards, gained 15 yards, and lost 10 yards. What was his net yardage change for the series of plays?

64. **BUSINESS AND FINANCE** Ramon owes $780 on his VISA account. He returns three items costing $43.10, $36.80, and $125.00 and receives credit on his account. Next, he makes a payment of $400. He then makes a purchase of $82.75. How much does Ramon still owe?

65. **SCIENCE AND MEDICINE** The temperature at noon on a June day was 82°. It fell by 12° over the next 4 h. What was the temperature at 4:00 P.M.?

66. STATISTICS Chia is standing at a point 6,000 ft above sea level. She descends to a point 725 ft lower. What is her distance above sea level?

67. BUSINESS AND FINANCE Omar's checking account was overdrawn by $72. He wrote another check for $23.50. How much was his checking account overdrawn after writing the check?

68. BUSINESS AND FINANCE Angelo owed his sister $15. He later borrowed another $10. What integer represents his current financial condition?

69. STATISTICS A local community college had a decrease in enrollment of 750 students in the fall of 2005. In the spring of 2006, there was another decrease of 425 students. What was the total decrease in enrollment for both semesters?

70. SCIENCE AND MEDICINE At 7 A.M., the temperature was $-15°F$. By 1 P.M., the temperature had increased by $18°F$. What was the temperature at 1 P.M.?

Answers

66. _____

67. _____

68. _____

69. _____

70. _____

71. _____

72. _____

73. _____

74. _____

75. _____

76. _____

77. _____

78. _____

79. _____

80. _____

81. _____

82. _____

| Basic Skills | **Advanced Skills** | Vocational-Technical Applications | Calculator/Computer | Above and Beyond | Getting Ready |

Evaluate each expression.

71. $9 + (-7) + 6 + (-5)$

72. $(-4) + 6 + (-3) + 0$

73. $-8 - 4 - 1 - (-2) - (-5)$

74. $6 - (-9) - 7 - (-5)$

75. $3 - 7 + (-12) - (-2) - 9$

76. $-12 + (-5) - 7 - (-13) + 4$

77. $-\dfrac{3}{2} + \left(-\dfrac{7}{4}\right) + \dfrac{1}{4}$

78. $-\dfrac{1}{2} + \dfrac{1}{3} + \left(-\dfrac{5}{6}\right)$

> Videos

79. $2.3 + (-5.4) - (-2.9)$

80. $-5.4 - (-2.1) + (-3.5)$

81. $-\dfrac{1}{2} - \left(-\dfrac{3}{4}\right) + (-2) - 3\dfrac{1}{2} + \dfrac{3}{2}$

82. $0.25 + 0.7 - 1.5 - (-2.95) + (-3.1)$ > Videos

Answers

83. _____

84. _____

85. _____

86. _____

87. _____

88. _____

83. MECHANICAL ENGINEERING A pneumatic actuator is operated by a pressurized air reservoir. At the beginning of the operator's shift, the pressure in the reservoir was 126 pounds per square inch (psi). At the end of each hour, the operator records the change in pressure of the reservoir. The values recorded for this shift were a drop of 12 psi, a drop of 7 psi, a rise of 32 psi, a drop of 17 psi, a drop of 15 psi, a rise of 31 psi, a drop of 4 psi, and a drop of 14 psi. What is the pressure in the tank at the end of the shift?

84. MECHANICAL ENGINEERING A diesel engine for an industrial shredder has an 18-quart oil capacity. When the maintenance technician checked the oil, it was 7 quarts low. Later that day, she added 4 quarts to the engine. What was the oil level after the 4 quarts were added?

ELECTRICAL ENGINEERING Dry cells or batteries have a positive terminal and a negative terminal. When the cells are correctly connected in series (positive to negative), the voltages of the cells can be added together. If a cell is connected and its terminals are reversed, the current will flow in the opposite direction.

For example, if three 3-volt cells are supposedly connected in series but one cell is inserted backwards, the resulting voltage is 3 volts.

3 volts + 3 volts + (−3) volts = 3 volts

The voltages are added together because the cells are in series, but you must pay attention to the current flow.

Now solve exercises 85 and 86.

85. Assume you have a 24-volt cell and a 12-volt cell with their negative terminals connected. What would the resulting voltage be if measured from the positive terminals?

| + 24 V − | − 12 V + |

86. If a 24-volt cell, an 18-volt cell, and 12-volt cell are supposed to be connected in series and the 18-volt cell is accidentally reversed, what would the total voltage be?

+ 24 V −   − 18 V +   + 12 V −

MANUFACTURING TECHNOLOGY At the beginning of the week, there were 2,489 lb of steel in inventory. Report the change in steel inventory for the week if the end-of-week inventory is:

87. 2.581 lb   **88.** 2.111 lb

Basic Skills | Advanced Skills | Vocational-Technical Applications | **Calculator/Computer** | Above and Beyond | Getting Ready
▲

Use your calculator to evaluate each expression.

89. $-4.1967 - 5.2943$

90. $5.3297 - (-4.1897)$

91. $-4.1623 - (-3.1468)$

92. $-3.6829 - 4.5687$

93. $-6.3267 + 8.6789 - (-6.6712) + (-5.3245)$

94. $32.456 + (-67.004) - (-21.6059) - 13.4569$

Basic Skills | Advanced Skills | Vocational-Technical Applications | Calculator/Computer | **Above and Beyond** | Getting Ready
▲

95. En route to their 2006 Super Bowl victory, the game-by-game rushing leaders for the Pittsburgh Steelers playoff run are shown below, along with yardage gained.

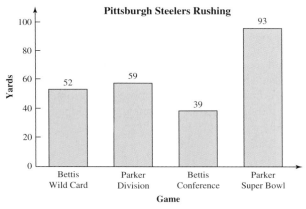

Source: ESPN. com

Use a real number to represent the change in the rushing yardage given from one game to the next.

(a) From the wild card game to the division game

(b) From the division game to the conference championship

(c) From the conference championship to the Super Bowl

96. In this chapter, it is stated that "Every number has an opposite." The opposite of 9 is -9. This corresponds to the idea of an opposite in English. In English an opposite is often expressed by a prefix, for example, *un-* or *ir-*.

(a) Write the opposite of these words: unmentionable, uninteresting, irredeemable, irregular, uncomfortable.

Answers

89. _____

90. _____

91. _____

92. _____

93. _____

94. _____

95. _____

96. _____

Answers

97. _____

98. _____

(b) What is the meaning of these expressions: not uninteresting, not irredeemable, not irregular, not unmentionable?

(c) Think of other prefixes that *negate* or change the meaning of a word to its *opposite*. Make a list of words formed with these prefixes, and write a sentence with three of the words you found. Make a sentence with two words and phrases from each of the lists. Look up the meaning of the word *irregardless*.

What is the value of $-[-(-5)]$? What is the value of $-(-6)$? How does this relate to the previous examples? Write a short description about this relationship.

97. The temperature on the plains of North Dakota can change rapidly, falling or rising many degrees in the course of an hour. Here are some temperature changes during each day over a week.

Day	Mon.	Tues.	Wed.	Thurs.	Fri.	Sat.	Sun.
Temp. change from 10 A.M. to 3 P.M.	$+13°$	$+20°$	$-18°$	$+10°$	$-25°$	$-5°$	$+15°$

Write a short speech for the TV weather reporter that summarizes the daily temperature change.

98. How long ago was the year 1250 B.C.E.? What year was 3,300 years ago? Make a number line and locate the following events, cultures, and objects on it. How long ago was each item in the list? Which two events are the closest to each other? You may want to learn more about some of the cultures in the list and the mathematics and science developed by that culture.

chapter 1 >Make the Connection

Inca culture in Peru—1400 A.D.

The *Ahmes Papyrus,* a mathematical text from Egypt—1650 B.C.E.

Babylonian arithmetic develops the use of a zero symbol—300 B.C.E.

First Olympic Games—776 B.C.E.

Pythagoras of Greece is born—580 B.C.E.

Mayans in Central America independently develop use of zero—500 A.D.

The *Chou Pei,* a mathematics classic from China—1000 B.C.E.

The *Aryabhatiya,* a mathematics work from India—499 A.D.

Trigonometry arrives in Europe via the Arabs and India—1464 A.D.

Arabs receive algebra from Greek, Hindu, and Babylonian sources and develop it into a new systematic form—850 A.D.

Development of calculus in Europe—1670 A.D.

Rise of abstract algebra—1860 A.D.

Growing importance of probability and development of statistics—1902 A.D.

99. Complete the following statement: "$3 - (-7)$ is the same as _____ because . . ." Write a problem that might be answered by doing this subtraction.

100. Explain the difference between the two phrases: "a number subtracted from 5" and "a number less than 5." Use algebra and English to explain the meaning of these phrases. Write other ways to express subtraction in English. Which ones are confusing?

| Basic Skills | Advanced Skills | Vocational-Technical Applications | Calculator/Computer | Above and Beyond | **Getting Ready** ▲ |

Add.

    (a) $(-1) + (-1) + (-1) + (-1)$
    (b) $3 + 3 + 3 + 3 + 3$
    (c) $9 + 9 + 9$
    (d) $(-10) + (-10) + (-10)$
    (e) $(-5) + (-5) + (-5) + (-5) + (-5)$
    (f) $(-8) + (-8) + (-8) + (-8)$

Answers

1. 9   **3.** 2   **5.** $\dfrac{13}{10}$ **7.** -5 **9.** $-\dfrac{7}{8}$ **11.** -3.9 **13.** -6

15. $\dfrac{1}{4}$ **17.** 2   **19.** -2 **21.** $-\dfrac{7}{20}$ **23.** -3.7 **25.** -8

27. 0   **29.** 0   **31.** 37   **33.** -2 **35.** -1 **37.** -2.5

39. -4 **41.** -23 **43.** $-\dfrac{11}{10}$ **45.** -8.1 **47.** 16   **49.** 19

51. $\dfrac{9}{4}$ **53.** 14   **55.** -17 **57.** 8   **59.** 2   **61.** $128

63. 23 yd   **65.** 70°   **67.** $95.50   **69.** 1,175   **71.** 3   **73.** -6

75. -23 **77.** -3 **79.** -0.2 **81.** $-\dfrac{15}{4}$ **83.** 120 psi

85. 12 V   **87.** $+92$ lb   **89.** -9.491 **91.** -1.0155 **93.** 3.6989
95. (a) $+7$; **(b)** -20; **(c)** $+54$ **97.** Above and Beyond
99. Above and Beyond   **(a)** -4 **(b)** 15   **(c)** 27   **(d)** -30
(e) -25 **(f)** -32

Answers

99. _____

100. _____

(a) _____

(b) _____

(c) _____

(d) _____

(e) _____

(f) _____

1.3

Multiplying and Dividing Real Numbers

< 1.3 Objectives >

1 > Find the product of two real numbers

2 > Find the quotient of two real numbers

3 > Use the order of operations to evaluate expressions involving real numbers

When you first considered multiplication in arithmetic, it was thought of as repeated addition. What does our work with the addition of numbers with different signs tell us about multiplication when real numbers are involved?

$$3 \cdot 4 = \underline{4 + 4 + 4} = 12$$

We interpret multiplication as repeated addition to find the product, 12.

Now, consider the product $(3)(-4)$:

$$(3)(-4) = (-4) + (-4) + (-4) = -12$$

Looking at this product suggests the first portion of our rule for multiplying numbers with different signs. The product of a positive number and a negative number is negative.

Property

Multiplying Real Numbers with Different Signs

The product of two numbers with different signs is negative.

To use this rule in multiplying two numbers with different signs, multiply their absolute values and attach a negative sign.

Example 1

Multiplying Numbers with Different Signs

< Objective 1 >

Multiply.

(a) $(5)(-6) = -30$

The product is negative.

(b) $(-10)(10) = -100$

NOTE

Multiply numerators together and then denominators and simplify.

(c) $(8)(-12) = -96$

(d) $\left(-\dfrac{3}{4}\right)\left(\dfrac{2}{5}\right) = -\dfrac{3}{10}$

Check Yourself 1

Multiply.

(a) $(-7)(5)$ **(b)** $(-12)(9)$ **(c)** $(-15)(8)$ **(d)** $\left(-\dfrac{4}{7}\right)\left(\dfrac{14}{5}\right)$

The product of two negative numbers is harder to visualize. The following pattern may help you see how we can determine the sign of the product.

NOTES

This number is decreasing by 1.

$(-1)(-2)$ is the opposite of -2.

$(3)(-2) = -6$

$(2)(-2) = -4$

$(1)(-2) = -2$ Do you see that the product is *increasing* by 2 each time?

$(0)(-2) = 0$

$(-1)(-2) = 2$

What should the product $(-2)(-2)$ be? Continuing the pattern shown, we see that

$(-2)(-2) = 4$

This suggests that the product of two negative numbers is positive. We can extend our multiplication rule.

Property

Multiplying Real Numbers with the Same Sign

The product of two numbers with the same sign is positive.

    Example 2   **Multiplying Real Numbers with the Same Sign**

Multiply.

(a) $9 \cdot 7 = 63$ The product of two positive numbers (same sign, $+$) is positive.

(b) $(-8)(-5) = 40$ The product of two negative numbers (same sign, $-$) is positive.

(c) $\left(-\dfrac{1}{2}\right)\left(-\dfrac{1}{3}\right) = \dfrac{1}{6}$

Check Yourself 2

Multiply.

(a) $10 \cdot 12$ **(b)** $(-8)(-9)$ **(c)** $\left(-\dfrac{2}{3}\right)\left(-\dfrac{6}{7}\right)$

Two numbers, 0 and 1, have special properties in multiplication.

Property

Multiplicative Identity Property	The product of 1 and any number is that number. In symbols, $a \cdot 1 = 1 \cdot a = a$ The number 1 is called the **multiplicative identity** for this reason.

Property

Multiplicative Property of Zero	The product of 0 and any number is 0. In symbols, $a \cdot 0 = 0 \cdot a = 0$

Example 3   **Multiplying Real Numbers Involving 0 and 1**

Find each product.

(a) $(1)(-7) = -7$

(b) $(15)(1) = 15$

(c) $(-7)(0) = 0$

(d) $0 \cdot 12 = 0$

(e) $\left(-\dfrac{4}{5}\right)(0) = 0$

Check Yourself 3

Multiply.

(a) $(-10)(1)$ **(b)** $(0)(-17)$ **(c)** $\left(\dfrac{5}{7}\right)(1)$ **(d)** $(0)\left(\dfrac{3}{4}\right)$

Before we continue, consider the following equivalent fractions:

$$-\frac{1}{a} = \frac{-1}{a} = \frac{1}{-a}$$

Any of these forms can occur in the course of simplifying an expression. The first form is generally preferred.

To complete our discussion of the properties of multiplication, we state the following.

Property

Multiplicative Inverse Property

For any number a, where $a \neq 0$, there is a number $\dfrac{1}{a}$ such that

$$a \cdot \frac{1}{a} = 1$$

$\dfrac{1}{a}$ is called the **multiplicative inverse**, or the **reciprocal**, of a.
The product of any nonzero number and its reciprocal is 1.

Example 4 illustrates this property.

 Example 4 | **Multiplying Reciprocals**

(a) $3 \cdot \dfrac{1}{3} = 1$ The reciprocal of 3 is $\dfrac{1}{3}$.

(b) $-5\left(-\dfrac{1}{5}\right) = 1$ The reciprocal of -5 is $\dfrac{1}{-5}$ or $-\dfrac{1}{5}$.

(c) $\dfrac{2}{3} \cdot \dfrac{3}{2} = 1$ The reciprocal of $\dfrac{2}{3}$ is $\dfrac{1}{\frac{2}{3}}$, or $\dfrac{3}{2}$.

Check Yourself 4

Find the multiplicative inverse (or the reciprocal) of each of the following numbers.

(a) 6   **(b)** -4 **(c)** $\dfrac{1}{4}$ **(d)** $-\dfrac{3}{5}$

You know from your work in arithmetic that multiplication and division are related operations. We can use that fact, and our work in the earlier part of this section, to determine rules for the division of numbers with different signs. Every equation involving division can be stated as an equivalent equation involving multiplication. For instance,

$\dfrac{15}{5} = 3$ can be restated as $15 = 5 \cdot 3$

$\dfrac{-24}{6} = -4$ can be restated as $-24 = (6)(-4)$

$\dfrac{-30}{-5} = 6$ can be restated as $-30 = (-5)(6)$

These examples illustrate that because the two operations are related, the rules of signs that we stated in the earlier part of this section for multiplication are also true for division.

Property

Dividing Real Numbers

1. The quotient of two numbers with different signs is negative.
2. The quotient of two numbers with the same sign is positive.

Again, the rules are easy to use. To divide two numbers with different signs, divide their absolute values. Then attach the proper sign according to the rules stated in the box.

Example 5   **Dividing Real Numbers**

< Objective 2 >

Divide.

(a) Positive $\longrightarrow \dfrac{28}{7} = 4 \longleftarrow$ Positive
Positive $\longrightarrow$

(b) Negative $\longrightarrow \dfrac{-36}{-4} = 9 \longleftarrow$ Positive
Negative $\longrightarrow$

(c) Negative $\longrightarrow \dfrac{-42}{7} = -6 \longleftarrow$ Negative
Positive $\longrightarrow$

(d) Positive $\longrightarrow \dfrac{75}{-3} = -25 \longleftarrow$ Negative
Negative $\longrightarrow$

(e) Positive $\longrightarrow \dfrac{15.2}{-3.8} = -4 \longleftarrow$ Negative
Negative $\longrightarrow$

Check Yourself 5

Divide.

(a) $\dfrac{-55}{11}$ **(b)** $\dfrac{80}{20}$ **(c)** $\dfrac{-48}{-8}$ **(d)** $\dfrac{144}{-12}$ **(e)** $\dfrac{-13.5}{-2.7}$

You should be very careful when 0 is involved in a division problem. Remember that 0 divided by any nonzero number is just 0. Recall that

$$\dfrac{0}{-7} = 0 \qquad \text{because} \qquad 0 = (-7)(0)$$

However, if zero is the *divisor,* we have a special problem. Consider

$$\dfrac{9}{0} = ?$$

This means that $9 = 0 \cdot ?$.

Can 0 times a number ever be 9? No, so there is no solution.

Because $\dfrac{9}{0}$ cannot be replaced by any number, we agree that *division by 0 is not allowed.* We say that

Property

Division by Zero   Division by 0 is undefined.

 Example 6   **Dividing Numbers Involving Zero**

NOTE

The expression $\dfrac{0}{0}$ is called an **indeterminate form.** You will learn more about these in later math courses.

Divide, if possible.

(a) $\dfrac{7}{0}$ is undefined.

(b) $\dfrac{-9}{0}$ is undefined.

(c) $\dfrac{0}{5} = 0$

(d) $\dfrac{0}{-8} = 0$

 Check Yourself 6

Divide if possible.

(a) $\dfrac{0}{3}$ **(b)** $\dfrac{5}{0}$ **(c)** $\dfrac{-7}{0}$ **(d)** $\dfrac{0}{-9}$

As we stated in Section 0.4, the fraction bar serves as a *grouping symbol*. This means that all operations in the numerator and denominator should be performed separately. Then the division is done as the last step. Example 7 illustrates this procedure.

 Example 7   **Operations with Grouping Symbols**

< Objective 3 >

Evaluate each expression.

(a) $\dfrac{(-6)(-7)}{3} = \dfrac{42}{3} = 14$ Multiply in the numerator, and then divide.

(b) $\dfrac{3 + (-12)}{3} = \dfrac{-9}{3} = -3$ Add in the numerator, and then divide.

(c) $\dfrac{-4 + (2)(-6)}{-6 - 2} = \dfrac{-4 + (-12)}{-6 - 2}$ Multiply in the numerator. Then add in the numerator and subtract in the denominator.

$\qquad\qquad = \dfrac{-16}{-8} = 2$ Divide as the last step.

Check Yourself 7

Evaluate each expression.

(a) $\dfrac{-4 + (-8)}{6}$ (b) $\dfrac{3 - (2)(-6)}{-5}$ (c) $\dfrac{(-2)(-4) - (-6)(-5)}{(-4)(11)}$

Evaluating fractions with a calculator poses a special problem. Example 8 illustrates this problem.

Example 8   **Using a Calculator to Divide**

 > Calculator

Use your scientific calculator to evaluate each fraction.

(a) $\dfrac{4}{2 - 3}$

As you can see, the correct answer should be -4. To get this answer with your calculator, you must place the denominator in parentheses. The keystroke sequence is

$4\ \boxed{\div}\ \boxed{(}\ 2\ \boxed{-}\ 3\ \boxed{)}\ \boxed{=}$

(b) $\dfrac{-7 - 7}{3 - 10}$

In this problem, the correct answer is 2. This result can be found on your calculator by placing the numerator in parentheses and then placing the denominator in parentheses. The keystroke sequence is

$\boxed{(}\ 7\ \boxed{+/-}\ \boxed{-}\ 7\ \boxed{)}\ \boxed{\div}\ \boxed{(}\ 3\ \boxed{-}\ 10\ \boxed{)}\ \boxed{=}$

When evaluating a fraction with a calculator, it is safest to use parentheses in both the numerator and the denominator.

NOTE

The keystroke sequence for a graphing calculator is

$\boxed{(}\ \boxed{(-)}\ 7\ \boxed{-}\ 7\ \boxed{)}\ \boxed{\div}$
$\boxed{(}\ 3\ \boxed{-}\ 10\ \boxed{)}\ \boxed{ENTER}$

(-7 – 7)/(3 – 10)
                    2

Check Yourself 8

Evaluate using your calculator.

(a) $\dfrac{-8}{5 - 7}$ (b) $\dfrac{-3 - 2}{-13 + 23}$

The order of operations that you learned in Section 0.4 remains the same when performing computations involving negative numbers. You must remain vigilant, though, with any negative signs.

Example 9   **Order of Operations**

Evaluate each expression.

(a) $7(-9 + 12)$ Evaluate inside the parentheses first.
    $= 7(3) = 21$

The Streeter/Hutchison Series in Mathematics   Beginning Algebra

NOTE

Note that
$(-5)^2 = (-5)(-5) = 25$ but
that $-5^2 = -25$. The power
applies *only* to the 5 in the
latter expression.

(b) $(-8)(-7) - 40$ Multiply first, then subtract.

$\quad = 56 - 40 = 16$

(c) $(-5)^2 - 3$ Evaluate the power first.

$\quad = (-5)(-5) - 3$

$\quad = 25 - 3 = 22$

(d) $-5^2 - 3$

$\quad = -25 - 3$

$\quad = -28$

Check Yourself 9

Evaluate each expression.

(a) $8(-9 + 7)$ **(b)** $(-3)(-5) + 7$

(c) $(-4)^2 - (-4)$ **(d)** $-4^2 - (-4)$

Many students have difficulty applying the distributive property when negative numbers are involved. Just remember that the sign of a number "travels" with that number.

Example 10	Applying the Distributive Property with Negative Numbers

RECALL

We usually enclose negative numbers in parentheses in the middle of an expression to avoid careless errors.

RECALL

We use brackets rather than nesting parentheses to avoid careless errors.

Evaluate each expression.

(a) $-7(3 + 6) = -7 \cdot 3 + (-7) \cdot 6$ Apply the distributive property.

$\qquad = -21 + (-42)$ Multiply first, then add.

$\qquad = -63$

(b) $-3(5 - 6) = -3[5 + (-6)]$ First, change the subtraction to addition.

$\qquad = -3 \cdot 5 + (-3)(-6)$ Distribute the -3.

$\qquad = -15 + 18$ Multiply first, then add.

$\qquad = 3$

(c) $5(-2 - 6) = 5[-2 + (-6)]$

$\qquad = 5 \cdot (-2) + 5 \cdot (-6)$

$\qquad = -10 + (-30)$

$\qquad = -40$ The sum of two negative numbers is negative.

Check Yourself 10

Evaluate each expression.

(a) $-2(-3 + 5)$ **(b)** $4(-3 + 6)$ **(c)** $-7(-3 - 8)$

Another thing to keep in mind when working with negative signs is the way in which you should evaluate multiple negative signs. Our approach takes into account two ways of looking at positive and negative numbers.

First, a negative sign indicates the opposite of the number that follows. For instance, we have already said that the opposite of 5 is -5, whereas the opposite of -5 is 5. This last instance can be translated as $-(-5) = 5$.

Second, any number must correlate to some point on the number line. That is, any nonzero number is either positive or negative. No matter how many negative signs a quantity has, you can always simplify it so that it is represented by a positive or a negative number (with zero or one negative sign).

Example 11 | **Simplifying Negative Signs**

NOTES

You should see a pattern emerge. An even number of negative signs gives a positive number, whereas an odd number of negative signs produces a negative number.

In this text, we generally choose to write negative fractions with the negative sign outside the fraction, such as $-\dfrac{1}{2}$.

Simplify each expression.

(a) $-(-(-(-4)))$

The opposite of -4 is 4, so $-(-4) = 4$.

The opposite of 4 is -4, so $-(-(-4)) = -4$. The opposite of this last number, -4, is 4, so

$$-(-(-(-4))) = 4$$

(b) $-\dfrac{-3}{4}$

This is the opposite of $\dfrac{-3}{4}$, which is $\dfrac{3}{4}$, a positive number.

Check Yourself 11

Simplify each expression.

(a) $-(-(-(-(-(-12))))))$ **(b)** $-\dfrac{-2}{-3}$

Example 12 | **An Application of Multiplying and Dividing Real Numbers**

Three partners own stock worth $4,680. One partner sells it for $3,678. How much did each partner lose?

First find the total loss: $4,680 - $3,678 = $1,002

Then divide the total loss by 3: $\dfrac{\$1,002}{3} = \$334$

Each person lost $334.

Check Yourself 12

Sal and Vinnie invested $8,500 in a business. Ten years later they sold the business for $22,000. How much profit did each make?

Property

The Product of Two Negative Numbers

NOTE

Here is a more detailed explanation of why the product of two negative numbers is positive.

From our earlier work, we know that the sum of a number and its opposite is 0:

$$5 + (-5) = 0$$

Multiply both sides of the equation by -3:

$$(-3)[5 + (-5)] = (-3)(0)$$

Because the product of 0 and any number is 0, on the right we have 0.

$$(-3)[5 + (-5)] = 0$$

We use the distributive property on the left.

$$(-3)(5) + (-3)(-5) = 0$$

We know that $(-3)(5) = -15$, so the equation becomes

$$-15 + (-3)(-5) = 0$$

We now have a statement of the form

$$-15 + \boxed{} = 0$$

in which $\boxed{}$ is the value of $(-3)(-5)$. We also know that $\boxed{}$ is the number that must be added to -15 to get 0, so $\boxed{}$ is the opposite of -15, or 15. This means that

$$(-3)(-5) = 15 \qquad \text{The product is positive!}$$

It doesn't matter what numbers we use in this argument. The resulting product of two negative numbers will always be positive.

Check Yourself ANSWERS

1. (a) -35; (b) -108; (c) -120; (d) $-\dfrac{8}{5}$ 2. (a) 120; (b) 72; (c) $\dfrac{4}{7}$

3. (a) -10; (b) 0; (c) $\dfrac{5}{7}$; (d) 0   4. (a) $\dfrac{1}{6}$; (b) $-\dfrac{1}{4}$; (c) 4; (d) $-\dfrac{5}{3}$

5. (a) -5; (b) 4; (c) 6; (d) -12; (e) 5   6. (a) 0; (b) undefined;

(c) undefined; (d) 0   7. (a) -2; (b) -3; (c) $\dfrac{1}{2}$ 8. (a) 4; (b) -0.5

9. (a) -16; (b) 22; (c) 20; (d) -12 10. (a) -4; (b) 12; (c) 77

11. (a) -12; (b) $-\dfrac{2}{3}$ 12. $6,750

Reading Your Text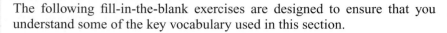

The following fill-in-the-blank exercises are designed to ensure that you understand some of the key vocabulary used in this section.

SECTION 1.3

(a) The product of two numbers with different signs is always _____.

(b) The product of two numbers with the same sign is always _____.

(c) The number _____ is called the multiplicative identity.

(d) Division by _____ is undefined.

< Objective 1 >

Multiply.

1. $4 \cdot 10$

2. $3 \cdot 14$

3. $(-4)(10)$

4. $(3)(-14)$

5. $(-4)(-10)$

6. $(-3)(-14)$

7. $(-13)(5)$

8. $(11)(-9)$

9. $(-4)(-17)$

10. $(-23)(-8)$

11. $4 \cdot \left(-\dfrac{3}{2}\right)$ > Videos

12. $(-9)\left(-\dfrac{2}{3}\right)$

13. $\left(-\dfrac{1}{4}\right) \cdot (-8)$

14. $\left(-\dfrac{5}{3}\right) \cdot (6)$

15. $\left(-\dfrac{2}{3}\right)\left(\dfrac{3}{5}\right)$

16. $\left(\dfrac{5}{8}\right)\left(-\dfrac{2}{3}\right)$

17. $\left(-\dfrac{1}{2}\right)\left(-\dfrac{10}{3}\right)$ > Videos

18. $\left(-\dfrac{7}{10}\right)\left(-\dfrac{5}{8}\right)$

19. $3.25 \cdot (-4)$

20. $(5.4)(-5)$

21. $(-1.1)(-1.2)$

22. $(0.8)(-3.5)$

23. $0 \cdot (-18)$

24. $(-5)(0)$

MathZone

Boost your grade at mathzone.com!

> Practice Problems
> NetTutor

> Self-Tests
> e-Professors
> Videos

Name _____

Section _____ Date _____

Answers

1. _____

2. _____

3. _____

4. _____

5. _____   6. _____

7. _____   8. _____

9. _____   10. _____

11. _____   12. _____

13. _____   14. _____

15. _____   16. _____

17. _____   18. _____

19. _____   20. _____

21. _____   22. _____

23. _____   24. _____

SECTION 1.3   123

Answers

25. _____

26. _____

27. _____

28. _____

29. _____

30. _____

31. _____

32. _____

33. _____

34. _____

35. _____

36. _____

37. _____

38. _____

39. _____

40. _____

41. _____

42. _____

43. _____

44. _____

45. _____ 46. _____

25. $\left(-\dfrac{11}{12}\right)(0)$

26. $(0)(-2.37)$

27. $\left(-\dfrac{1}{2}\right)(2)$

28. $\left(-\dfrac{1}{3}\right)(-3)$

29. $\left(-\dfrac{3}{2}\right)\left(-\dfrac{2}{3}\right)$

30. $\left(-\dfrac{4}{7}\right)\left(\dfrac{7}{4}\right)$

< Objective 2 >
Divide.

31. $\dfrac{70}{14}$

32. $48 \div 6$

33. $(-35) \div (-7)$

34. $\dfrac{-48}{-12}$

35. $\dfrac{50}{-5}$

36. $\dfrac{-60}{15}$ > Videos

37. $\dfrac{-125}{5}$

38. $\dfrac{24}{-8}$

39. $\dfrac{-11}{-1}$

40. $\dfrac{-13}{1}$

41. $\dfrac{32}{-1}$

42. $\dfrac{-1}{-8}$

43. $\dfrac{0}{-8}$

44. $\dfrac{-10}{0}$

45. $\dfrac{-14}{0}$

46. $\dfrac{0}{-2}$

< Objective 3 >

Evaluate each expression.

47. $\dfrac{(-6)(-3)}{2}$

48. $\dfrac{(-9)(5)}{-3}$

49. $\dfrac{(-8)(2)}{-4}$

50. $\dfrac{(7)(-8)}{-14}$

51. $\dfrac{24}{-4-8}$

52. $\dfrac{-36}{-7+3}$

53. $\dfrac{55-19}{-12-6}$

54. $\dfrac{-11-7}{-14+8}$

55. $\dfrac{5-7}{4-4}$

56. $\dfrac{-3-(-3)}{6-10}$

57. $5(7-2)$

58. $5(2-7)$

59. $-3(-2-5)$

60. $-2[-7-(-3)]$

61. $(-2)(3)-5$

62. $(8)(-6)-27$

63. $(-5)(-2)-12$

64. $(-7)(-3)-25$

65. $-3+(-2)(4)$

66. $-5-(-5)(4)$

67. $12-(-3)(-4)$

68. $20+(-4)(-5)$

69. $(-8)^2-5^2$

70. $(-8)^2-(-4)^2$

Answers

47. _____

48. _____

49. _____

50. _____

51. _____

52. _____

53. _____

54. _____

55. _____

56. _____

57. _____

58. _____

59. _____

60. _____

61. _____

62. _____

63. _____

64. _____

65. _____   66. _____

67. _____   68. _____

69. _____   70. _____

Answers

71. _____

72. _____

73. _____

74. _____

75. _____

76. _____

77. _____

78. _____

79. _____

80. _____

81. _____

82. _____

71. $-8^2 - (-5)^2$

72. $-8^2 - 4^2$ > Videos

73. $-(-(-(-(-3))))$

74. $-(-(-(-3.45)))$

75. $\dfrac{-(-2)}{-(-8)}$

76. $\dfrac{-3}{-(-(-4))}$

77. **SCIENCE AND MEDICINE** The temperature is $-6°F$ at 5:00 in the evening. If the temperature drops $2°F$ every hour, what is the temperature at 1:00 A.M.?

78. **SCIENCE AND MEDICINE** A woman lost 42 pounds (lb) while dieting. If she lost 3 lb each week, how long has she been dieting?

79. **BUSINESS AND FINANCE** Patrick worked all day mowing lawns and was paid $9 per hour. If he had $125 at the end of a 9-h day, how much did he have before he started working?

80. **BUSINESS AND FINANCE** Suppose that you and your two brothers bought equal shares of an investment for a total of $20,000 and sold it later for $16,232. How much did each person lose?

81. **SCIENCE AND MEDICINE** Suppose that the temperature outside is dropping at a constant rate. At noon, the temperature is $70°F$ and it drops to $58°F$ at 5:00 P.M. How much did the temperature change each hour?

82. **SCIENCE AND MEDICINE** A chemist has 84 ounces (oz) of a solution. He pours the solution into test tubes. Each test tube holds $\dfrac{2}{3}$ oz. How many test tubes can he fill?

Beginning Algebra   The Streeter/Hutchison Series in Mathematics

*Complete each of the following statements with **never, sometimes,** or **always.***

83. A product made up of an odd number of negative factors is _____ negative.

84. A product of an even number of negative factors is _____ negative.

85. The quotient $\dfrac{x}{y}$ is _____ positive.

86. The quotient $\dfrac{x}{y}$ is _____ negative.

Basic Skills | **Advanced Skills** | Vocational-Technical Applications | Calculator/Computer | Above and Beyond | Getting Ready

87. $4 \cdot 8 \div 2 - 5^2$

88. $36 \div 4 \cdot 3 - (-25)$

89. $-8 + 14 - 2 \cdot 4 - 3$

90. $(-3)^3 - (-8)(-2)$

91. $8 + [2(-3) + 3]^2$

92. $-8^2 - 5^2 + 8 \div (4 \cdot 2)$

93. $\dfrac{-\dfrac{3}{8}}{\dfrac{3}{4}}$

94. $\left(-\dfrac{5}{12}\right) \div \left(-\dfrac{3}{16}\right)$

95. $\left(\dfrac{7}{4}\right) \div \left(-\dfrac{3}{2}\right)$

96. $\dfrac{-\dfrac{1}{2}}{-\dfrac{3}{4}}$

97. $\left(-1\dfrac{1}{2}\right)\left(3\dfrac{1}{3}\right)$

98. $\left(-2\dfrac{1}{2}\right)\left(-3\dfrac{3}{4}\right)$

99. $\left(-5\dfrac{1}{4}\right) \div \left(-2\dfrac{1}{2}\right)$ > Videos

100. $\left(1\dfrac{1}{3}\right) \div \left(-6\dfrac{2}{3}\right)$

Answers

83. _____

84. _____

85. _____

86. _____

87. _____

88. _____

89. _____

90. _____

91. _____

92. _____

93. _____

94. _____

95. _____

96. _____

97. _____

98. _____

99. _____

100. _____

Answers

101. _____

102. _____

103. _____

104. _____

105. _____

106. _____

107. _____

108. _____

109. _____

101. MANUFACTURING TECHNOLOGY Companies occasionally sell products at a loss in order to draw in customers or as a reward to good customers. The theory is that customers will buy other products along with the discounted product and the net result will be a profit.

Beguhn Industries sells five different products. On product A, they make $18 each; on product B, they lose $4 each; product C makes $11 each; product D makes $38 each; and product E loses $15 each. During the previous month, Beguhn Industries sold 127 units of product A, 273 units of product B, 201 units of product C, 377 units of product D, and 43 units of product E.

Calculate the profit or loss for the month.

102. MECHANICAL ENGINEERING The bending moment created by a center support on a steel beam is approximated by the formula $-\dfrac{1}{4}PL^3$, in which P is the load on each side of the center support and L is the length of the beam on each side of the center support (assuming a symmetrical beam and load).

If the total length of the beam is 24 ft (12 ft on each side of the center) and the total load is 4,124 lb (2,062 lb on each side of the center), what is the bending moment (in ft-lb) at the center support?

Use a calculator to evaluate each expression to the nearest thousandth.

103. $\dfrac{7}{4-5}$

104. $\dfrac{-8}{-4+2}$

105. $\dfrac{-6-9}{-4+1}$

106. $\dfrac{-10+4}{-7+10}$

107. $(-1.23) \cdot (3.4)$

108. $\dfrac{(3.55)(-12.12)}{(-6.4)}$

109. $3.4 - 5.1^2 + (-1.02)^2 \div 22 \cdot (-4.8)$

Basic Skills | Advanced Skills | Vocational-Technical Applications | Calculator/Computer | **Above and Beyond** | Getting Ready

110. Some animal ecologists in Minnesota are planning to reintroduce a group of animals into a wilderness area. The animals, mammals on the endangered species list, will be released into an area where they once prospered and where there is an abundant food supply. But, the animals will face predators. The ecologists expect that the number of mammals will grow about 25 percent each year but that 30 of the animals will die from attacks by predators and hunters.

    The ecologists need to decide how many animals they should release to establish a stable population. Work with other students to try several beginning populations and follow the numbers through 8 years. Is there a number of animals that will lead to a stable population? Write a letter to the editor of your local newspaper explaining how to decide what number of animals to release. Include a formula for the number of animals next year based on the number this year. Begin by filling out this table to track the number of animals living each year after the release:

110. _____

(a) _____

(b) _____

(c) _____

(d) _____

No. Initially Released	Year							
	1	2	3	4	5	6	7	8
20	+ ___ − ___ = _____							
100	+ ___ − ___ = _____							
200	+ ___ − ___ = _____							

Basic Skills | Advanced Skills | Vocational-Technical Applications | Calculator/Computer | Above and Beyond | **Getting Ready**

The area of a rectangle is given by the product of its length and width $(A_{\text{Rectangle}} = L \cdot W)$. Find the area of each rectangle described.

(a)

6 ft, 4 ft

(b) $L = 6$ in.; $W = \dfrac{2}{3}$ in.

The area of a triangle is found by multiplying the length of its base and its height, and then halving the result $\left(A_{\text{Triangle}} = \dfrac{1}{2} b \cdot h\right)$. Find the area of each triangle.

(c) $b = \dfrac{4}{5}$ ft; $h = 2$ ft

(d)

0.5 in., 0.9 in.

Answers

1. 40 **3.** -40 **5.** 40 **7.** -65 **9.** 68 **11.** -6 **13.** 2

15. $-\dfrac{2}{5}$ **17.** $\dfrac{5}{3}$ **19.** -13 **21.** 1.32 **23.** 0 **25.** 0

27. -1 **29.** 1 **31.** 5 **33.** 5 **35.** -10 **37.** -25 **39.** 11

41. -32 **43.** 0 **45.** Undefined **47.** 9 **49.** 4 **51.** -2

53. -2 **55.** Undefined **57.** 25 **59.** 21 **61.** -11 **63.** -2

65. -11 **67.** 0 **69.** 39 **71.** -89 **73.** -3 **75.** $\dfrac{1}{4}$

77. $-22°$F **79.** \$44 **81.** 2.4°F **83.** always **85.** sometimes

87. -9 **89.** -5 **91.** 17 **93.** $-\dfrac{1}{2}$ **95.** $-\dfrac{7}{6}$ **97.** -5

99. $2\dfrac{1}{10}$ **101.** $+\$17,086$ **103.** -7 **105.** 5 **107.** -4.182

109. -24.88 **(a)** 24 ft^2 **(b)** 4 in.^{2} **(c)** $\dfrac{4}{5}$ ft^2 **(d)** 0.225 in.^{2}

1.4 From Arithmetic to Algebra

< 1.4 Objectives >

1 > Use the symbols and language of algebra

2 > Identify algebraic expressions

In arithmetic, you learned how to do calculations with numbers using the basic operations of addition, subtraction, multiplication, and division.

In algebra, we still use numbers and the same four operations. However, we also use letters to represent numbers. Letters such as x, y, L, and W are called **variables** when they represent numerical values.

Here we see two rectangles whose lengths and widths are labeled with numbers.

If we need to represent the length and width of *any* rectangle, we can use the variables L and W.

NOTE

In arithmetic:
+ denotes addition;
− denotes subtraction;
× denotes multiplication;
÷ denotes division.

You are familiar with the four symbols ($+$, $−$, $\times$, $\div$) used to indicate the fundamental operations of arithmetic.

To see how these operations are indicated in algebra, we begin with addition.

Definition

Addition	$x + y$ means the *sum* of x and y or x *plus* y.

Example 1 — Writing Expressions That Indicate Addition

< Objective 1 >

(a) *The sum of a and* 3 is written as $a + 3$.

(b) *L plus W* is written as $L + W$.

131

(c) 5 *more than m* is written as $m + 5$.

(d) *x increased by* 7 is written as $x + 7$.

(e) 15 *added to x* is written as $x + 15$.

Check Yourself 1

Write, using symbols.

(a) The sum of *y* and 4   **(b)** *a* plus *b*

(c) 3 more than *x*   **(d)** *n* increased by 6

Similarly, we use a minus sign to indicate subtraction.

Definition

Subtraction $x - y$ means the *difference* of x and y or x minus y.

 Example 2   **Writing Expressions That Indicate Subtraction**

(a) *r minus s* is written as $r - s$.

(b) *The difference of m and* 5 is written as $m - 5$.

(c) *x decreased by* 8 is written as $x - 8$.

(d) 4 *less than a* is written as $a - 4$.

(e) 12 *subtracted from y* is written as $y - 12$.

Check Yourself 2

Write, using symbols.

(a) *w* minus *z*   **(b)** The difference of *a* and 7

(c) *y* decreased by 3   **(d)** 5 less than *b*

(e) *m* subtracted from 6

You have seen that the operations of addition and subtraction are written exactly the same way in algebra as in arithmetic. This is not true in multiplication because the sign $\times$ looks like the letter *x*. So in algebra we use other symbols to show multiplication to avoid any confusion. Here are some ways to write multiplication.

Definition

Multiplication

A centered dot	$x \cdot y$
Parentheses	$(x)(y)$
Writing the letters next to each other	xy

All these expressions indicate the *product* of x and y or x times y. x and y are called the **factors** of the product xy.

Example 3 | **Writing Expressions That Indicate Multiplication**

(a) The product of 5 and a is written as $5 \cdot a$, $(5)(a)$, or $5a$. The last expression, $5a$, is the shortest and the most common way of writing the product.

(b) 3 times 7 can be written as $3 \cdot 7$ or $(3)(7)$.

(c) Twice z is written as $2z$.

(d) The product of 2, s, and t is written as $2st$.

(e) 4 more than the product of 6 and x is written as $6x + 4$.

 Check Yourself 3

Write, using symbols.

(a) m times n (b) The product of h and b
(c) The product of 8 and 9   (d) The product of 5, w, and y
(e) 3 more than the product
    of 8 and a

Before we move on to division, we define the ways that we can combine the symbols we have learned so far.

Definition

Expression | An **expression** is a meaningful collection of numbers, variables, and symbols of operation.

 Example 4 | **Identifying Expressions**

< Objective 2 >

NOTE

Not every collection of symbols is an expression.

(a) $2m + 3$ is an expression. It means that we multiply 2 and m, and then add 3.

(b) $x + \cdot + 3$ is not an expression. The three operations in a row have no meaning.

(c) $y = 2x - 1$ is not an expression. The equal sign is not an operation sign.

(d) $3a + 5b - 4c$ is an expression. Its meaning is clear.

 Check Yourself 4

Identify which are expressions and which are not.

(a) $7 - \cdot x$ (b) $6 + y = 9$
(c) $a + b - c$ (d) $3x - 5yz$

To write more complicated products in algebra, we need some "punctuation marks." Parentheses ( ) mean that an expression is to be thought of as a single quantity. Brackets [ ] are used in exactly the same way as parentheses in algebra. Example 5 shows the use of these signs of grouping.

Example 5   Expressions with More Than One Operation

(a) 3 times the sum of a and b is written as

$$3\underbrace{(a + b)}$$

The sum of a and b is a single quantity, so it is enclosed in parentheses.

(b) The sum of 3 times a and b is written as $3a + b$.

(c) 2 times the difference of m and n is written as $2(m - n)$.

(d) The product of s plus t and s minus t is written as $(s + t)(s - t)$.

(e) The product of b and 3 less than b is written as $b(b - 3)$.

Check Yourself 5

Write, using symbols.

(a) Twice the sum of p and q
(b) The sum of twice p and q
(c) The product of a and the quantity $b - c$
(d) The product of x plus 2 and x minus 2
(e) The product of x and 4 more than x

Now we look at the operation of division. In arithmetic, we use the division sign $\div$, the long division symbol $\overline{)}$, and fraction notation. For example, to indicate the quotient when 9 is divided by 3, we may write

$$9 \div 3 \quad \text{or} \quad 3\overline{)9} \quad \text{or} \quad \frac{9}{3}$$

Definition

Division $\dfrac{x}{y}$ means x *divided by* y or *the quotient of* x *and* y.

Example 6   Writing Expressions That Indicate Division

(a) m divided by 3 is written as $\dfrac{m}{3}$.

(b) The quotient when a plus b is divided by 5 is written as $\dfrac{a + b}{5}$.

(c) The sum p plus q divided by the difference p minus q is written as $\dfrac{p + q}{p - q}$.

Check Yourself 6

Write, using symbols.

(a) *r* divided by *s*
(b) The quotient when *x* minus *y* is divided by 7
(c) The difference *a* minus 2 divided by the sum *a* plus 2

Notice that we can use many different letters to represent variables. In Example 6, the letters *m*, *a*, *b*, *p*, and *q* represented different variables. We often choose a letter that reminds us of what it represents, for example, *L* for *length* and *W* for *width*.

Example 7 | Writing Geometric Expressions

(a) *Length* times *width* is written $L \cdot W$.

(b) One-half of the *base* times the *height* is written $\frac{1}{2} b \cdot h$.

(c) *Length* times *width* times *height* is written $L \cdot W \cdot H$.

(d) Pi (π) times *diameter* is written πd.

Check Yourself 7

Write each geometric expression, using symbols.

(a) Two times *length* plus two times *width*
(b) Two times pi (π) times *radius*

Algebra can be used to model a variety of applications, such as the one shown in Example 8.

Example 8 | Modeling Applications with Algebra

Carla earns $10.25 per hour in her job. Write an equation that sets her weekly gross pay equal to an expression involving the number of hours she works per week.

We need to name the variables we are using. Let *P* be Carla's weekly gross pay and *h* be the number of hours she works in a week.

Then the equation $P = 10.25h$ gives Carla's weekly gross pay based on the number of hours she works.

Check Yourself 8

The specifications for an engine cylinder call for the stroke length to be 2 more than twice the diameter of the cylinder. Write an equation relating stroke length to cylinder diameter.

We close this section by listing many of the common words used to indicate arithmetic operations.

NOTE

The word *twice* indicates multiplication by 2.

Summary: Words Indicating Operations

The operations listed are usually indicated by the words shown.

Addition (+)   Plus, and, more than, increased by, sum
Subtraction (−)   Minus, from, less than, decreased by, difference
Multiplication (·)   Times, of, by, product
Division (÷)   Divided, into, per, quotient

Check Yourself ANSWERS

1. (a) $y + 4$; **(b)** $a + b$; **(c)** $x + 3$; **(d)** $n + 6$ **2. (a)** $w - z$; **(b)** $a - 7$;
(c) $y - 3$; **(d)** $b - 5$; **(e)** $6 - m$ **3. (a)** mn; **(b)** hb; **(c)** $8 \cdot 9$ or $(8)(9)$;
(d) $5wy$; **(e)** $8a + 3$ **4. (a)** Not an expression; **(b)** not an expression;
(c) an expression; **(d)** an expression   **5. (a)** $2(p + q)$; **(b)** $2p + q$;
(c) $a(b - c)$; **(d)** $(x + 2)(x - 2)$; **(e)** $x(x + 4)$ **6. (a)** $\dfrac{r}{s}$; **(b)** $\dfrac{x - y}{7}$;
(c) $\dfrac{a - 2}{a + 2}$ **7. (a)** $2L + 2W$; **(b)** $2\pi r$ **8.** $s = 2d + 2$

Reading Your Text

The following fill-in-the-blank exercises are designed to ensure that you understand some of the key vocabulary used in this section.

SECTION 1.4

(a) In algebra, we often use letters, called _____, to represent numerical values that can vary depending on the application.

(b) $x - y$ means the _____ of x and y.

(c) $x \cdot y$, $(x)(y)$, and xy are all ways of indicating _____ in algebra.

(d) When choosing a _____ to use as a variable, it is often a good idea to choose one that reminds us of what the variable represents.

1.4 exercises

MathZone

Boost your grade at mathzone.com!
- > Practice Problems
- > NetTutor
- > Self-Tests
- > e-Professors
- > Videos

Name _____

Section _____ Date _____

< Objective 1 >

Write each of the following phrases, using symbols.

1. The sum of c and d

2. a plus 7

3. w plus z

4. The sum of m and n

5. x increased by 5

6. 4 more than c

7. 10 more than y

8. m increased by 4

9. b minus a

10. 5 less than w

11. b decreased by 4

12. r minus 3

13. 6 less than r

14. x decreased by 3

15. w times z

16. The product of 3 and c

17. The product of 5 and t

18. 8 times a

19. The product of 8, m, and n

20. The product of 7, r, and s

21. The product of 3 and the quantity p plus q

22. The product of 5 and the sum of a and b

23. Twice the sum of x and y

24. 7 times the sum of m and n

25. The sum of twice x and y

26. The sum of 3 times m and n

27. Twice the difference of x and y

28. 3 times the difference of a and c

Answers

1. _____   2. _____

3. _____   4. _____

5. _____   6. _____

7. _____   8. _____

9. _____   10. _____

11. _____   12. _____

13. _____   14. _____

15. _____   16. _____

17. _____   18. _____

19. _____   20. _____

21. _____   22. _____

23. _____   24. _____

25. _____   26. _____

27. _____   28. _____

Answers

29. _____

30. _____

31. _____

32. _____

33. _____

34. _____

35. _____

36. _____

37. _____

38. _____

39. _____

40. _____

41. _____

42. _____

43. _____

44. _____

45. _____

46. _____

47. _____

48. _____

49. _____   50. _____

29. The quantity a plus b times the quantity a minus b

30. The product of x plus y and x minus y

31. The product of m and 3 more than m

32. The product of a and 7 less than a > Videos

33. x divided by 5

34. The quotient when b is divided by 8

35. The result of a minus b, divided by 9

36. The difference x minus y, divided by 9

37. The sum of p and q, divided by 4

38. The sum of a and 5, divided by 9

39. The sum of a and 3, divided by the difference of a and 3

40. The sum of m and n, divided by the difference of m and n

< Objective 2 >

Identify which are expressions and which are not.

41. $2(x + 5)$

42. $4 - (x + 3)$

43. $m \div + 4$

44. $6 + a = 7$ > Videos

45. $y(x + 3)$

46. $8 = 4b$

47. $2a + 5b$

48. $4x + \cdot 7$

49. **SOCIAL SCIENCE** The earth's population has doubled in the last 40 years. If we let x represent the earth's population 40 years ago, what is the population today?

50. **SCIENCE AND MEDICINE** It is estimated that the earth is losing 4,000 species of plants and animals every year. If S represents the number of species living last year, how many species are on the earth this year?

51. BUSINESS AND FINANCE The simple interest (I) earned when a principal (P) is invested at a rate (r) for a time (t) is calculated by multiplying the principal times the rate times the time. Write an expression for the interest earned.

52. SCIENCE AND MEDICINE The kinetic energy of a particle of mass m is found by taking one-half the product of the mass and the square of the velocity v. Write an expression for the kinetic energy of a particle.

Match each phrase with the proper expression.

53. 8 decreased by x **(a)** $x - 8$

54. 8 less than x **(b)** $8 - x$

55. The difference between 8 and x

56. 8 from x

Basic Skills | **Advanced Skills** | Vocational-Technical Applications | Calculator/Computer | Above and Beyond | Getting Ready

Write each of the following phrases, using symbols. Use the variable x to represent the number in each case.

57. 5 more than a number   **58.** A number increased by 8

59. 7 less than a number   **60.** A number decreased by 10

61. 9 times a number   **62.** Twice a number

63. 6 more than 3 times a number

64. 5 times a number, decreased by 10

65. Twice the sum of a number and 5

66. 3 times the difference of a number and 4   > Videos

67. The product of 2 more than a number and 2 less than that same number

68. The product of 5 less than a number and 5 more than that same number

69. The quotient of a number and 7

70. A number divided by 3

Answers

51. _____

52. _____

53. _____

54. _____

55. _____

56. _____

57. _____

58. _____

59. _____

60. _____

61. _____

62. _____

63. _____

64. _____

65. _____

66. _____

67. _____

68. _____

69. _____

70. _____

Answers

71. _____

72. _____

73. _____

74. _____

75. _____

76. _____

77. _____

78. _____

79. _____

80. _____

81. _____

82. _____

83. _____

84. _____

71. The sum of a number and 5, divided by 8

72. The quotient when 7 less than a number is divided by 3

73. 6 more than a number divided by 6 less than that same number

74. The quotient when 3 more than a number is divided by 3 less than that same number

Write each of the following geometric expressions using the given symbols.

75. Four times the length of a side (s)

76. $\dfrac{4}{3}$ times π times the cube of the radius (r)

77. The radius (r) squared times the height (h) times π

78. Twice the length (L) plus twice the width (W)

79. One-half the product of the height (h) and the sum of two unequal sides $(b_1$ and $b_2)$

> Videos

80. Six times the length of a side (s) squared

Basic Skills | Advanced Skills | **Vocational-Technical Applications** | Calculator/Computer | Above and Beyond | Getting Ready

▲

---

81. ALLIED HEALTH The standard dosage given to a patient is equal to the product of the desired dose D and the available quantity Q divided by the available dose H. Write an expression for the standard dosage.

82. INFORMATION TECHNOLOGY Mindy is the manager of the help desk at a large cable company. She notices that, on average, her staff can handle 50 calls/hr. Last week, during a thunderstorm, the call volume increased from 65 calls/hr to 150 calls/hr. To figure out the average number of customers in the system, she needs to take the quotient of the average rate of customer arrivals (the call volume) a and the average rate at which customers are served h minus the average rate of customer arrivals a. Write an expression for the average number of customers in the system.

83. CONSTRUCTION TECHNOLOGY K Jones Manufacturing produces hex bolts and carriage bolts. They sold 284 more hex bolts than carriage bolts last month. Write an expression that describes the number of carriage bolts they sold last month.

84. ELECTRICAL ENGINEERING Electrical power P is the product of voltage V and current I. Express this relationship algebraically.

Basic Skills | Advanced Skills | Vocational-Technical Applications | Calculator/Computer | **Above and Beyond** | Getting Ready

▲

Answers

Translate each of the given algebraic expressions into words. Exchange papers with another student to edit each other's writing. Be sure the meaning in English is the same as in algebra. Note: Each expression is not a complete sentence, so your English does not have to be a complete sentence, either. As an example.

Algebra: $2(x - 1)$

English (some possible answers):

  One less than a number is doubled

  A number decreased by one, and then multiplied by two

85. $n + 3$

86. $\dfrac{x + 2}{5}$

87. $3(5 + a)$

88. $3 - 4n$

89. $\dfrac{x + 6}{x - 1}$

90. $\dfrac{x^2 - 1}{(x - 1)^2}$

85. _____

86. _____

87. _____

88. _____

89. _____

90. _____

(a) _____

(b) _____

(c) _____

(d) _____

(e) _____

(f) _____

Basic Skills | Advanced Skills | Vocational-Technical Applications | Calculator/Computer | Above and Beyond | **Getting Ready**

▲

Simplify.

  (a) $\dfrac{6 \cdot 2 + 8}{5 - 3}$

  (b) $\dfrac{4 \cdot 5 - 8}{8 - 4 \div 2}$

  (c) $\dfrac{-8 + 3 \cdot 2}{-12 \div 6}$

  (d) $\dfrac{-3^2 - (-4 - 1)}{-2 \cdot 2}$

  (e) $8 \div 4 - 3 \cdot 2$

  (f) $6^2 - 18 \div 2 \cdot 3$

Answers

1. $c + d$ **3.** $w + z$ **5.** $x + 5$ **7.** $y + 10$ **9.** $b - a$

11. $b - 4$ **13.** $r - 6$ **15.** wz **17.** $5t$ **19.** $8mn$

21. $3(p + q)$ **23.** $2(x + y)$ **25.** $2x + y$ **27.** $2(x - y)$

29. $(a + b)(a - b)$ **31.** $m(m + 3)$ **33.** $\dfrac{x}{5}$ **35.** $\dfrac{a - b}{9}$

37. $\dfrac{p + q}{4}$ **39.** $\dfrac{a + 3}{a - 3}$ **41.** Expression   **43.** Not an expression

45. Expression   **47.** Expression   **49.** $2x$ **51.** Prt **53.** (b)

55. (b)   **57.** $x + 5$ **59.** $x - 7$ **61.** $9x$ **63.** $3x + 6$

65. $2(x + 5)$ **67.** $(x + 2)(x - 2)$ **69.** $\dfrac{x}{7}$ **71.** $\dfrac{x + 5}{8}$

73. $\dfrac{x + 6}{x - 6}$ **75.** $4s$ **77.** $\pi r^2 h$ **79.** $\dfrac{1}{2}h(b_1 + b_2)$ **81.** $\dfrac{DQ}{H}$

83. $H - 284$ **85.** Above and Beyond   **87.** Above and Beyond

89. Above and Beyond   **(a)** 10   **(b)** 2   **(c)** 1   **(d)** 1

(e) -4 **(f)** 9

1.5

Evaluating Algebraic Expressions

< 1.5 Objectives >

1 > Evaluate algebraic expressions given any real-number value for the variables

2 > Use a calculator to evaluate algebraic expressions

When using algebra to solve problems, we often want to find the value of an algebraic expression, given particular values for the variables. Finding the value of an expression is called *evaluating the expression* and uses the following steps.

Step by Step

| To Evaluate an Algebraic Expression | Step 1 | Replace each variable by the given number value. |
| | Step 2 | Do the necessary arithmetic operations, following the rules for order of operations. |

 Example 1   Evaluating Algebraic Expressions

< Objective 1 >

NOTE

We use parentheses when we make the initial substitution. This helps us to avoid careless errors.

Suppose that $a = 5$ and $b = 7$.

(a) To evaluate $a + b$, we replace a with 5 and b with 7.

$a + b = (5) + (7) = 12$

(b) To evaluate $3ab$, we again replace a with 5 and b with 7.

$3ab = 3 \cdot (5) \cdot (7) = 105$

 Check Yourself 1

If $x = 6$ and $y = 7$, evaluate.

(a) $y - x$ $7 - 6 = 1$

(b) $5xy$ $5 \cdot 6 \cdot 7 = 210$

We are now ready to evaluate algebraic expressions that require following the rules for the order of operations.

 Example 2   Evaluating Algebraic Expressions

Evaluate the following expressions if $a = 2$, $b = 3$, $c = 4$, and $d = 5$.

(a) $5a + 7b = 5(2) + 7(3)$ Multiply first.

$\qquad\qquad = 10 + 21 = 31$ Then add.

> CAUTION

This is different from

$(3c)^2 = (3 \cdot 4)^2$
$\quad\quad = 12^2 = 144$

(b) $3c^2 = 3(4)^2$ Evaluate the power.

$\quad\quad = 3 \cdot 16 = 48$ Then multiply.

(c) $7(c + d) = 7[(4) + (5)]$ Add inside the brackets.

$\quad\quad\quad\quad = 7 \cdot 9 = 63$

(d) $5a^4 - 2d^2 = 5(2)^4 - 2(5)^2$ Evaluate the powers.

$\quad\quad\quad\quad\quad = 5 \cdot 16 - 2 \cdot 25$ Multiply.

$\quad\quad\quad\quad\quad = 80 - 50 = 30$ Subtract.

Check Yourself 2

If $x = 3$, $y = 2$, $z = 4$, and $w = 5$, evaluate the following expressions.

(a) $4x^2 + 2$ **(b)** $5(z + w)$ **(c)** $7(z^2 - y^2)$

To evaluate an algebraic expression when a fraction bar is used, do the following: Start by doing all the work in the numerator, then do all the work in the denominator. Divide the numerator by the denominator as the last step.

Example 3   **Evaluating Algebraic Expressions**

If $p = -2$, $q = 3$, and $r = 4$, evaluate:

(a) $\dfrac{8p}{r}$

Replace p with -2 and r with 4.

$\dfrac{8p}{r} = \dfrac{8(-2)}{(4)} = \dfrac{-16}{4} = -4$ Divide as the last step.

RECALL

Again, the fraction bar is a grouping symbol, like parentheses. Work first in the numerator and then in the denominator.

(b) $\dfrac{7q + r}{p + q} = \dfrac{7(3) + (4)}{(-2) + (3)}$ Now evaluate the top and bottom separately.

$\quad\quad\quad\quad = \dfrac{21 + 4}{(-2) + (3)}$

$\quad\quad\quad\quad = \dfrac{25}{1} = 25$

Check Yourself 3

Evaluate the following if $c = -5$, $d = 8$, and $e = 3$.

(a) $\dfrac{6c}{e}$ **(b)** $\dfrac{4d + e}{c}$ **(c)** $\dfrac{10d - e}{d + e}$

Often, you will use a calculator or computer to evaluate an algebraic expression. We demonstrate how to do this in Example 4.

| ▶ | **Example** 4 | **Using a Calculator to Evaluate an Expression** |

< Objective 2 >

Use a calculator to evaluate each expression.

(a) $\dfrac{4x + y}{z}$ if $x = 2$, $y = 1$, and $z = 3$.

Begin by making each of the substitutions.

$$\frac{4x + y}{z} = \frac{4(2) + (1)}{3}$$

Then, enter the numerical expression into a calculator.

$\boxed{(}\ 4\ \boxed{\times}\ 2\ \boxed{+}\ 1\ \boxed{)}\ \boxed{\div}\ 3\ \boxed{\text{ENTER}}$ Remember to enclose the entire numerator in parentheses.

The display should read 3.

(b) $\dfrac{7x - y}{3z - x}$ if $x = 2$, $y = 6$, and $z = -2$.

Again, we begin by substituting:

$$\frac{7x - y}{3z - x} = \frac{7(2) - (6)}{3(-2) - 2}$$

Then, we enter the expression into a calculator.

$\boxed{(}\ 7\ \boxed{\times}\ 2\ \boxed{-}\ 6\ \boxed{)}\ \boxed{\div}\ \boxed{(}\ 3\ \boxed{\times}\ \boxed{(-)}\ 2\ \boxed{-}\ 2\ \boxed{)}\ \boxed{\text{ENTER}}$

The display should read -1.

RECALL

If you are using a scientific calculator, the denominator should be entered as

$\boxed{(}\ 3\ \boxed{\times}\ 2\ \boxed{+/-}\ \boxed{-}\ 2\ \boxed{)}\ \boxed{=}$

Check Yourself 4

Use a calculator to evaluate each expression if $x = 2$, $y = -6$, and $z = 5$.

(a) $\dfrac{2x + y}{z}$ **(b)** $\dfrac{4y - 2z}{3x}$

It is important to remember that a calculator follows the correct order of operations when evaluating an expression. For example, if we omit the parentheses in Example 4(b) and enter

$7\ \boxed{\times}\ 2\ \boxed{-}\ 6\ \boxed{\div}\ 3\ \boxed{\times}\ \boxed{(-)}\ 2\ \boxed{-}\ 2\ \boxed{\text{ENTER}}$

the calculator will interpret our input as $7 \cdot 2 - \dfrac{6}{3} \cdot (-2) - 2$, which is not what we wanted.

Whether working with a calculator or pencil and paper, you must remember to take care both with signs and with the order of operations.

The Streeter/Hutchison Series in Mathematics   Beginning Algebra

▶ | Example 5 | **Evaluating Expressions**

RECALL

The rules for the order of operations call for us to multiply first, and then add.

Evaluate $5a + 4b$ if $a = -2$ and $b = 3$.

Replace a with -2 and b with 3.

$$5a + 4b = 5(-2) + 4(3)$$
$$= -10 + 12$$
$$= 2$$

Check Yourself 5

Evaluate $3x + 5y$ if $x = -2$ and $y = -5$.

We follow the same rules no matter how many variables are in the expression.

▶ | Example 6 | **Evaluating Expressions**

Evaluate the following expressions if $a = -4$, $b = 2$, $c = -5$, and $d = 6$.

This becomes $-(-20)$, or $+20$.

(a) $7a - 4c = 7(-4) - 4(-5)$
$$= -28 + 20$$
$$= -8$$

Evaluate the power first, and then multiply by 7.

 > C A U T I O N

When a squared variable is replaced by a negative number, square the negative.

$(-5)^2 = (-5)(-5) = 25$

↑

The exponent applies to -5!

$-5^2 = -(5 \cdot 5) = -25$

↑

The exponent applies only to 5!

(b) $7c^2 = 7(-5)^2 = 7 \cdot 25$
$$= 175$$

(c) $b^2 - 4ac = (2)^2 - 4(-4)(-5)$
$$= 4 - 4(-4)(-5)$$
$$= 4 - 80$$
$$= -76$$

Add inside the brackets first.

(d) $b(a + d) = (2)[(-4) + (6)]$
$$= 2(2)$$
$$= 4$$

Check Yourself 6

Evaluate if $p = -4$, $q = 3$, and $r = -2$.

(a) $5p - 3r$ **(b)** $2p^2 + q$ **(c)** $p(q + r)$
(d) $-q^2$ **(e)** $(-q)^2$

If an expression involves a fraction, remember that the fraction bar is a grouping symbol. This means that you should do the required operations first in the numerator and then in the denominator. Divide as the last step.

 Example 7   **Evaluating Expressions**

Evaluate the following expressions if $x = 4$, $y = -5$, $z = 2$, and $w = -3$.

(a) $\dfrac{z - 2y}{x} = \dfrac{(2) - 2(-5)}{(4)} = \dfrac{2 + 10}{4}$

$= \dfrac{12}{4} = 3$

(b) $\dfrac{3x - w}{2x + w} = \dfrac{3(4) - (-3)}{2(4) + (-3)} = \dfrac{12 + 3}{8 + (-3)}$

$= \dfrac{15}{5} = 3$

Check Yourself 7

Evaluate if $m = -6$, $n = 4$, and $p = -3$.

(a) $\dfrac{m + 3n}{p}$ **(b)** $\dfrac{4m + n}{m + 4n}$

Let us now look at a business and finance application of this section's content.

 Example 8   **A Business and Finance Application**

The simple interest earned on a principal P at a growth rate r for time t, in years, is given by the product Prt.

Find the simple interest earned on a \$6,000 investment if the growth rate is 0.03 and the principal is invested for 2 years.

We substitute the known variable values and compute.

$Prt = (6,000)(0.03)(2) = 360$

The investment earns \$360 in simple interest over a 2-year period.

NOTE

The principal is the amount invested.

The growth rate is usually given as a percentage.

 Check Yourself 8

In most of the world, temperature is given using a Celsius scale. In the U.S., though, we generally use the Fahrenheit scale. The formula to convert temperatures from Fahrenheit to Celsius is

$\dfrac{5}{9}(F - 32)$

If the temperature is reported to be 41°F, what is the Celsius equivalent?

We provide the following chart as a reference guide for entering expressions into a calculator.

	Algebraic Notation	Calculator Notation
Addition	$6 + 2$	6 $\boxed{+}$ 2
Subtraction	$4 - 8$	4 $\boxed{-}$ 8
Multiplication	$(3)(-5)$	3 $\boxed{\times}$ $\boxed{(-)}$ 5 or 3 $\boxed{\times}$ 5 $\boxed{+/-}$
Division	$\dfrac{8}{6}$	8 $\boxed{\div}$ 6
Exponential	3^4	3 $\boxed{\wedge}$ 4 or 3 $\boxed{y^x}$ 4
	$(-3)^4$	$\boxed{(}$ $\boxed{(-)}$ 3 $\boxed{)}$ $\boxed{\wedge}$ 4 or $\boxed{(}$ 3 $\boxed{+/-}$ $\boxed{)}$ $\boxed{y^x}$ 4

 Check Yourself ANSWERS

1. **(a)** 1; **(b)** 210   2. **(a)** 38; **(b)** 45; **(c)** 84   3. **(a)** −10; **(b)** −7; **(c)** 7

4. **(a)** $-\dfrac{2}{5}$; **(b)** $-\dfrac{17}{3}$ 5. −31   6. **(a)** −14; **(b)** 35; **(c)** −4; **(d)** −9; **(e)** 9

7. **(a)** −2; **(b)** −2   8. 5°C

Reading Your Text

The following fill-in-the-blank exercises are designed to ensure that you understand some of the key vocabulary used in this section.

SECTION 1.5

(a) To evaluate an algebraic expression, first replace each _____ by the given numerical value.

(b) When evaluating an algebraic expression that includes a _____ bar, you should do all the work in the numerator and denominator separately.

(c) Scientific and graphing calculators follow the familiar order of _____.

(d) The amount borrowed or invested is known as the _____.

1.5 exercises

Name _____

Section _____ Date _____

Answers

1. _____	2. _____
3. _____	4. _____
5. _____	6. _____
7. _____	8. _____
9. _____	10. _____
11. _____	12. _____
13. _____	14. _____
15. _____	16. _____
17. _____	18. _____
19. _____	20. _____
21. _____	22. _____
23. _____	24. _____
25. _____	26. _____
27. _____	28. _____
29. _____	30. _____

Basic Skills | Advanced Skills | Vocational-Technical Applications | Calculator/Computer | Above and Beyond | Getting Ready

< Objective 1 >

Evaluate each of the expressions if $a = -2$, $b = 5$, $c = -4$, and $d = 6$.

1. $3c - 2b$

2. $4c - 2b$

3. $8b + 2c$

4. $7a - 2c$

5. $-b^2 + b$

6. $(-b)^2 + b$

7. $3a^2$

8. $6c^2$

9. $c^2 - 2d$

10. $3b^2 + 4c$

11. $2a^2 + 3b^2$

12. $4b^2 - 2c^2$

13. $2(a + b)$

14. $5(b - c)$

15. $-4(2c - a)$

16. $6(3c - d)$

17. $a(b + 3c)$

18. $c(3a - d)$

19. $\dfrac{6d}{c}$

20. $\dfrac{8c}{2a}$

21. $\dfrac{3d + 2c}{b}$

22. $\dfrac{2b + 3d}{2a}$

23. $\dfrac{2b - 3a}{c + 2d}$ > Videos

24. $\dfrac{3d - 2b}{5a + d}$

25. $d^2 - b^2$

26. $c^2 - a^2$

27. $(d - b)^2$

28. $(c - a)^2$

29. $(d - b)(d + b)$

30. $(c - a)(c + a)$

31. $d^3 - b^3$

32. $c^3 + a^3$

33. $(d - b)^3$

34. $(c + a)^3$

35. $(d - b)(d^2 + db + b^2)$

36. $(c + a)(c^2 - ac + a^2)$

> Videos

37. $-(b + a)^2$

38. $(d - a)^2$

39. $3a - 2b + \dfrac{2d}{c}$

40. $4b + 5d - \dfrac{c}{a}$

41. $a^2 + 2ad + d^2$

42. $b^2 - 2bc + c^2$

Evaluate each expression if $x = -3$, $y = 5$, and $z = \dfrac{2}{3}$.

43. $x^2 - y$

44. $\dfrac{y - x}{z}$

45. $z - y^2$

46. $z - \dfrac{z + x}{y - x}$

Evaluate each expression if $m = 4$, $n = -\dfrac{3}{2}$, and $p = \dfrac{2}{3}$.

47. $mn - np + m^2$

48. $n^2 + 2np + p^2$

> Videos

49. $\dfrac{mn}{np}$

50. $-\dfrac{np}{mn}$

51. **SCIENCE AND MEDICINE** The formula for the total resistance in a parallel circuit is given by the formula $R_T = \dfrac{R_1 R_2}{R_1 + R_2}$. Find the total resistance if $R_1 = 6$ ohms (Ω) and $R_2 = 10\ \Omega$.

52. **GEOMETRY** The formula for the area of a triangle is given by $A = \dfrac{1}{2}bh$. Find the area of a triangle if $b = 4$ cm and $h = 8$ cm.

Answers

31. _____ 32. _____

33. _____ 34. _____

35. _____

36. _____

37. _____

38. _____

39. _____

40. _____

41. _____

42. _____

43. _____

44. _____

45. _____

46. _____

47. _____

48. _____

49. _____

50. _____

51. _____

52. _____

Answers

53. _____

54. _____

55. _____

56. _____

57. _____

58. _____

53. **GEOMETRY** The perimeter of a rectangle of length L and width W is given by the formula $P = 2L + 2W$. Find the perimeter when $L = 10$ in. and $W = 5$ in.

54. **BUSINESS AND FINANCE** The simple interest I on a principal of P dollars at interest rate r for time t, in years, is given by $I = Prt$. Find the simple interest on a principal of $6,000 at 3% for 2 years. (Hint: 3% = 0.03)

> Videos

55. **BUSINESS AND FINANCE** Use the simple interest formula to find the principal if the total interest earned was $150 and the rate of interest was 4% for 2 years.

56. **BUSINESS AND FINANCE** Use the simple interest formula to find the rate of interest if $5,000 earns $300 interest in 3 years.

57. **SCIENCE AND MEDICINE** A formula that relates Celsius and Fahrenheit temperature is $F = \frac{9}{5}C + 32$. If the current temperature is $-10°C$, what is the Fahrenheit temperature?

58. **GEOMETRY** If the area of a circle whose radius is r is given by $A = \pi r^2$, in which $\pi \approx 3.14$, find the area when $r = 3$ meters (m).

Basic Skills | **Advanced Skills** | Vocational-Technical Applications | Calculator/Computer | Above and Beyond | Getting Ready

In each of the following problems, decide whether the given values for the variables make the statement **true** *or* **false.**

59. $x - 7 = 2y + 5$; $x = 22, y = 5$

60. $3(x - y) = 6$; $x = 5, y = -3$

61. $2(x + y) = 2x + y$; $x = -4, y = -2$

 > Videos

62. $x^2 - y^2 = x - y$; $x = 4, y = -3$

Basic Skills | Advanced Skills | **Vocational-Technical Applications** | Calculator/Computer | Above and Beyond | Getting Ready

63. ALLIED HEALTH The concentration, in micrograms per milliliter (mcg/mL), of an antihistamine in a patient's bloodstream can be approximated using the expression $-2t^2 + 13t + 1$, in which t is the number of hours since the drug was administered. Approximate the concentration of the antihistamine 1 hour after being administered.

64. ALLIED HEALTH Use the expression given in exercise 63 to approximate the concentration of the antihistamine 3 hours after being administered.

65. ELECTRICAL ENGINEERING Evaluate $\dfrac{rT}{5,252}$ for $r = 1,180$ and $T = 3$ (round to the nearest thousandth).

66. MECHANICAL ENGINEERING The kinetic energy (in joules) of a particle is given by $\dfrac{1}{2}mv^2$. Find the kinetic energy of a particle if its mass is 60 kg and its velocity is 6 m/s.

Basic Skills | Advanced Skills | Vocational-Technical Applications | **Calculator/Computer** | Above and Beyond | Getting Ready

< Objective 2 >

Use your calculator to evaluate each expression if $x = -2.34$, $y = -3.14$, *and* $z = 4.12$. *Round your results to the nearest tenth.*

67. $x + yz$

68. $y - 2z$

69. $x^2 - z^2$

70. $x^2 + y^2$

71. $\dfrac{xy}{z - x}$

72. $\dfrac{y^2}{zy}$

73. $\dfrac{2x + y}{2x + z}$

74. $\dfrac{x^2 y^2}{xz}$

Answers

59. _____

60. _____

61. _____

62. _____

63. _____

64. _____

65. _____

66. _____

67. _____

68. _____

69. _____

70. _____

71. _____

72. _____

73. _____

74. _____

Beginning Algebra   The Streeter/Hutchison Series in Mathematics   © The McGraw-Hill Companies. All Rights Reserved.

Answers

75. _____

76. _____

77. _____

78. _____

79. _____

80. _____

81. _____

82. _____

83. _____

Use your calculator to evaluate each expression if $m = 232$, $n = -487$, *and* $p = 58$. *Round your results to the nearest tenth.*

75. $m + np^2$

76. $p - (m + 2n)$

77. $(p + n)^2 - m^2$

78. $\dfrac{pm - 2n}{n - 2m}$

79. $\dfrac{n^2 - p^2}{p^2 - m^2}$

80. $m^2 + (-n)^2 + (-p^2)$

Basic Skills	Advanced Skills	Vocational-Technical Applications	Calculator/Computer	**Above and Beyond**	Getting Ready

81. Write an English interpretation of each of the following algebraic expressions.

(a) $(2x^2 - y)^3$

(b) $3n - \dfrac{n - 1}{2}$

(c) $(2n + 3)(n - 4)$

82. Is it true that $a^n + b^n = (a + b)^n$? Try a few numbers and decide whether this is true for all numbers, for some numbers, or never true. Write an explanation of your findings and give examples.

83. Enjoyment of patterns in art, music, and language is common to all cultures, and many cultures also delight in and draw spiritual significance from patterns in numbers. One such set of patterns is that of the "magic" square. One of these squares appears in a famous etching by Albrecht Dürer, who lived from 1471 to 1528 in Europe. He was one of the first artists in Europe to use geometry to give perspective, a feeling of three dimensions, in his work.

The magic square in his work is this one:

16	3	2	13
5	10	11	8
9	6	7	12
4	15	14	1

Why is this square "magic"? It is magic because every row, every column, and both diagonals add to the same number. In this square there are sixteen spaces for the numbers 1 through 16.

Part 1: What number does each row and column add to?

Write the square that you obtain by adding −17 to each number. Is this still a magic square? If so, what number does each column and row add to? If you add 5 to each number in the original magic square, do you still have a magic square? You have been studying the operations of addition, multiplication, subtraction, and division with integers and with rational numbers. What operations can you perform on this magic square and still have a magic square? Try to find something that will not work. Use algebra to help you decide what will work and what won't. Write a description of your work and explain your conclusions.

Part 2: Here is the oldest published magic square. It is from China, about 250 B.C.E. Legend has it that it was brought from the River Lo by a turtle to the Emperor Yii, who was a hydraulic engineer.

4	9	2
3	5	7
8	1	6

Check to make sure that this is a magic square. Work together to decide what operation might be done to every number in the magic square to make the sum of each row, column, and diagonal the *opposite* of what it is now. What would you do to every number to cause the sum of each row, column, and diagonal to equal zero?

| Basic Skills | Advanced Skills | Vocational-Technical Applications | Calculator/Computer | Above and Beyond | **Getting Ready** ▲ |

Simplify.

(a) $(8 + 9) - 5$

(b) $15 - 4 - 11$

(c) $5(4 + 3) - 9$

(d) $-3(5 - 7) + 11$

(e) $-6(-9 + 7) - 4$

(f) $8 - 7(-2 - 6)$

Answers

1. -22 **3.** 32 **5.** -20 **7.** 12 **9.** 4 **11.** 83 **13.** 6

15. 24 **17.** 14 **19.** -9 **21.** 2 **23.** 2 **25.** 11 **27.** 1

29. 11 **31.** 91 **33.** 1 **35.** 91 **37.** -9 **39.** -19

41. 16 **43.** 4 **45.** $-\dfrac{73}{3}$ **47.** 11 **49.** 6 **51.** $3.75\ \Omega$

53. 30 in. **55.** $\$1,875$ **57.** $14°F$ **59.** True **61.** False

63. 12 mcg/mL **65.** 0.674 **67.** -15.3 **69.** -11.5 **71.** 1.1

73. 14 **75.** $-1,638,036$ **77.** $130,217$ **79.** -4.6

81. Above and Beyond **83.** Above and Beyond **(a)** 12 **(b)** 0

(c) 26 **(d)** 17 **(e)** 8 **(f)** 64

1.6

Adding and Subtracting Terms

< 1.6 Objectives >

1 > Identify terms and like terms

2 > Combine like terms

To find the perimeter of (or the distance around) a rectangle, we add 2 times the length and 2 times the width. In the language of algebra, this can be written as

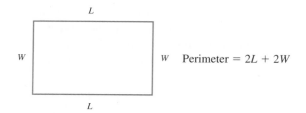

W Perimeter $= 2L + 2W$

We call $2L + 2W$ an **algebraic expression,** or more simply an **expression.** Recall from Section 1.5 that an expression allows us to write a mathematical idea in symbols. It can be thought of as a meaningful collection of letters, numbers, and operation signs.

Some expressions are

$$5x^2 \qquad 3a + 2b \qquad 4x^3 + (-2y) + 1$$

Addition and subtraction signs break expressions into smaller parts called *terms.*

Definition

Term

A **term** is a number, or the product of a number and one or more variables, raised to a whole-number power.

In an expression, each sign ($+$ or $-$) is a part of the term that follows the sign.

Example 1   **Identifying Terms**

< Objective 1 >

NOTE

This is usually written as $4x^3 - 2y + 1$.

(a) $5x^2$ has one term.

(b) $\underbrace{3a} + \underbrace{2b}$ has two terms: $3a$ and $2b$.
     Term Term

(c) $\underbrace{4x^3} + \underbrace{(-2y)} + \underbrace{1}$ has three terms: $4x^3$, $-2y$, and 1.
     Term   Term   Term

(d) $x - y$ has two terms: x and $-y$.

154

Check Yourself 1

List the terms of each expression.

(a) $2b^4$ **(b)** $5m + 3n$ **(c)** $2s^2 - 3t - 6$

NOTE

We usually shorten this to the "coefficient."

Note that a term in an expression may have any number of factors. For instance, $5xy$ is a term. It has factors of 5, x, and y. The number factor of a term is called the **numerical coefficient.** So for the term $5xy$, the numerical coefficient is 5.

Example 2   **Identifying the Numerical Coefficient**

(a) $4a$ has the numerical coefficient 4.

(b) $6a^3b^4c^2$ has the numerical coefficient 6.

(c) $-7m^2n^3$ has the numerical coefficient -7.

(d) Because $1 \cdot x = x$, the numerical coefficient of x is understood to be 1.

Check Yourself 2

Give the numerical coefficient for each of the following terms.

(a) $8a^2b$ **(b)** $-5m^3n^4$ **(c)** y

If terms contain exactly the *same letters* (or variables) raised to the *same powers,* they are called **like terms.**

Example 3   **Identifying Like Terms**

(a) The following are like terms.

$6a$ and $7a$

$5b^2$ and b^2 Each pair of terms has the same letters,

$10x^2y^3z$ and $-6x^2y^3z$ with each letter raised to the same power—the numerical coefficients can be

$-3m^2$ and m^2 any number.

(b) The following are *not* like terms.

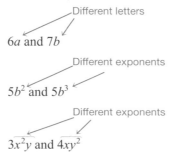

Different letters

$6a$ and $7b$

Different exponents

$5b^2$ and $5b^3$

Different exponents

$3x^2y$ and $4xy^2$

Check Yourself 3

Circle the like terms.

$$5a^2b \quad ab^2 \quad a^2b \quad -3a^2 \quad 4ab \quad 3b^2 \quad -7a^2b$$

Like terms of an expression can always be combined into a single term. Look at the following:

RECALL

We use the distributive property from Section 1.1.

$$\underbrace{2x}_{x + x} \; + \; \underbrace{5x}_{x + x + x + x + x} \; = \; \underbrace{7x}_{x + x + x + x + x + x + x}$$

Rather than having to write out all those x's, try

$$2x + 5x = (2 + 5)x = 7x$$

In the same way,

$$9b + 6b = (9 + 6)b = 15b$$

and $10a + (-4a) = (10 + (-4))a = 6a$

This leads us to the following rule.

Step by Step

To Combine Like Terms	To combine like terms, use the following steps.
	Step 1 Add or subtract the numerical coefficients.
	Step 2 Attach the common variables.

 Example 4   **Combining Like Terms**

< Objective 2 >

Combine like terms.

(a) $8m + 5m = (8 + 5)m = 13m$

(b) $5pq^3 - 4pq^3 = 5pq^3 + (-4pq^3) = 1pq^3 = pq^3$

> **CAUTION**

Do not change the exponents when combining like terms.

(c) $7a^3b^2 - 7a^3b^2 = 7a^3b^2 + (-7a^3b^2) = 0a^3b^2 = 0$

Check Yourself 4

Combine like terms.

(a) $6b + 8b$ **(b)** $12x^2 - 3x^2$

(c) $8xy^3 - 7xy^3$ **(d)** $9a^2b^4 - 9a^2b^4$

The idea is the same when expressions involve more than two terms.

| Example 5 | Combining Like Terms |

NOTE

The distributive property can be used over any number of like terms.

Combine like terms.

(a) $5ab - 2ab + 3ab$

$= 5ab + (-2ab) + 3ab$

$= (5 + (-2) + 3)ab = 6ab$

Only like terms can be combined.

(b) $\overline{8x - 2x} + 5y$

$= (8 + (-2))x + 5y$

$= 6x + 5y$

Like terms   Like terms

NOTE

With practice, you will do this mentally instead of writing out all of these steps.

(c) $5m + 8n \qquad + 4m - 3n$

$= (5m + 4m) + (8n + (-3n))$

$= \qquad 9m \quad + \qquad 5n$

Here we have used both the associative and commutative properties.

(d) $4x^2 + 2x - 3x^2 + x$

$= (4x^2 + (-3x^2)) + (2x + x)$

$= x^2 + 3x$

As these examples illustrate, combining like terms often means changing the grouping and the order in which the terms are written. Again, all this is possible because of the properties of addition that we introduced in Section 1.1.

Check Yourself 5

Combine like terms.

(a) $4m^2 - 3m^2 + 8m^2$ **(b)** $9ab + 3a - 5ab$ **(c)** $4p + 7q + 5p - 3q$

As you have seen in arithmetic, subtraction can be performed directly. As this is the form used for most of mathematics, we will use that form throughout this text. Just remember, by using negative numbers, you can always rewrite a subtraction problem as an addition problem.

| Example 6 | Combining Like Terms |

Combine the like terms.

(a) $2xy - 3xy + 5xy$

$= (2 - 3 + 5)xy$

$= 4xy$

(b) $5a - 2b + 7b - 8a$

$= (5a - 8a) + (-2b + 7b)$

$= -3a + 5b$

Check Yourself 6

Combine like terms.

(a) $4ab + 5ab - 3ab - 7ab$ **(b)** $2x - 7y - 8x - y$

Let us now look at a business and finance application of this section's content.

Example 7	A Business and Finance Application

NOTE

A business can compute the profit it earns on a product by subtracting the costs associated with the product from the revenue earned by the product.

S-Bar Electronics, Inc., sells a certain server for $1,410. It pays the manufacturer $849 for each server and there are fixed costs of $4,500 per week associated with the servers.

Let x be the number of servers bought and sold during the week.

Then, the revenue earned by S-Bar Electronics, Inc., from these servers can be modeled by the formula

$$R = 1,410x$$

The cost can be modeled with the formula

$$C = 849x + 4,500$$

Therefore, the profit can be modeled by the difference between the revenue and the cost.

$$P = 1,410x - 849x - 4,500$$

NOTE

A negative profit would mean the company suffered a loss.

Simplify the given profit formula.

The like terms are $1,410x$ and $-849x$. We combine these to give a simplified formula

$$P = 561x - 4,500$$

Check Yourself 7

S-Bar Electronics, Inc., also sells 19-in. flat-screen monitors for $799 each. The monitors cost them $489 each. Additionally, there are weekly fixed costs of $3,150 associated with the sale of the monitors. We can model the profit earned on the sale of y monitors with the formula

$$P = 799y - 489y - 3,150$$

Simplify the profit formula.

Check Yourself ANSWERS

1. **(a)** $2b^4$; **(b)** $5m, 3n$; **(c)** $2s^2, -3t, -6$ **2.** **(a)** 8; **(b)** -5; **(c)** 1
3. The like terms are $5a^2b, a^2b$, and $-7a^2b$ **4.** **(a)** $14b$; **(b)** $9x^2$; **(c)** xy^3; **(d)** 0
5. **(a)** $9m^2$; **(b)** $4ab + 3a$; **(c)** $9p + 4q$ **6.** **(a)** $-ab$; **(b)** $-6x - 8y$
7. $310y - 3,150$

Reading Your Text

The following fill-in-the-blank exercises are designed to ensure that you understand some of the key vocabulary used in this section.

SECTION 1.6

(a) The product of a number and a variable is called a _____.

(b) The _____ factor of a term is called a coefficient.

(c) Terms that contain exactly the same variables raised to the same powers are called _____ terms.

(d) The _____ property enables us to combine like terms into a single term.

1.6 exercises

Basic Skills | Advanced Skills | Vocational-Technical Applications | Calculator/Computer | Above and Beyond | Getting Ready

< Objective 1 >

List the terms of the following expressions.

1. $5a + 2$

2. $7a - 4b$

3. $4x^3$

4. $3x^2$

5. $3x^2 + 3x - 7$

6. $2a^3 - a^2 + a$

Circle the like terms in the following groups of terms.

7. $5ab, 3b, 3a, 4ab$

8. $9m^2, 8mn, 5m^2, 7m$

9. $4xy^2, 2x^2y, 5x^2, -3x^2y, 5y, 6x^2y$

> Videos

10. $8a^2b, 4a^2, 3ab^2, -5a^2b, 3ab, 5a^2b$

< Objective 2 >

Combine the like terms.

11. $4m + 6m$

12. $6a^2 + 8a^2$

13. $7b^3 + 10b^3$

14. $7rs + 13rs$

15. $21xyz + 7xyz$

16. $-3mn^2 + 9mn^2$

17. $9z^2 - 3z^2$

18. $7m - 6m$

19. $9a^5 - 9a^5$

20. $13xy - 9xy$

21. $19n^2 - 18n^2$

22. $7cd - 7cd$

23. $21p^2q - 6p^2q$

24. $17r^3s^2 - 8r^3s^2$

25. $5x^2 - 3x^2 + 9x^2$

26. $13uv + uv - 12uv$

27. $11b - 9a - 6b$

28. $5m^2 - 3m + 6m^2$

29. $7x + 5y - 4x - 4y$

30. $6a^2 + 11a + 7a^2 - 9a$

31. $4a + 7b + 3 - 2a + 3b - 2$

32. $5p^2 + 2p + 8 + 4p^2 + 5p - 6$

33. **GEOMETRY** Provide a simplified expression for the perimeter of the rectangle shown.

34. **GEOMETRY** Provide a simplified expression for the perimeter of the triangle shown.

35. **GEOMETRY** A rectangle has sides that measure $8x + 9$ in. and $6x - 7$ in. Provide a simplified expression for its perimeter.

36. **GEOMETRY** A triangle has sides measuring $3x + 7$ mm, $4x - 9$ mm, and $5x + 6$ mm. Find the simplified expression that represents its perimeter.

37. **BUSINESS AND FINANCE** The cost of producing x units of an item is $C = 150 + 25x$. The revenue from selling x units is $R = 90x - x^2$. The profit is given by the revenue minus the cost. Find the simplified expression that represents the profit.

38. **BUSINESS AND FINANCE** The revenue from selling y units is $R = 3y^2 - 2y + 5$ and the cost of producing y units is $C = y^2 + y - 3$. Find the simplified expression that represents the profit.

Basic Skills | **Advanced Skills** | Vocational-Technical Applications | Calculator/Computer | Above and Beyond | Getting Ready

Simplify each expression by combining like terms.

39. $\frac{2}{3}m + 3 + \frac{4}{3}m$

40. $\frac{a}{5} - 2 + \frac{4a}{5}$

41. $\frac{13x}{5} + 2 - \frac{3x}{5} + 5$

42. $\frac{17}{12}y + 7 + \frac{7}{12}y - 3$

43. $2.3a + 7 + 4.7a + 3$

44. $5.8m + 4 - 2.8m + 11$

Answers

29. _____
30. _____
31. _____
32. _____
33. _____
34. _____
35. _____
36. _____
37. _____
38. _____
39. _____
40. _____
41. _____
42. _____
43. _____
44. _____

Answers

45. _____

46. _____

47. _____

48. _____

49. _____

50. _____

51. _____

52. _____

53. _____

54. _____

55. _____

56. _____

57. _____

58. _____

59. _____

60. _____

Rewrite each statement as an algebraic expression. Simplify each expression, if possible.

45. Find the sum of $5a^4$ and $8a^4$.

46. Find the sum of $9p^2$ and $12p^2$.

47. Find the difference between $15a^3$ and $12a^3$.

48. Subtract $5m^3$ from $18m^3$.

49. Subtract $3mn^2$ from the sum of $9mn^2$ and $5mn^2$.

50. Find the difference between the sum of $6x^2y$ and $12x^2y$, and $4x^2y$.

Use the distributive property to remove the parentheses in each expression. Then, simplify each expression by combining like terms.

51. $2(3x + 2) + 4$ **52.** $3(4z + 5) - 9$

53. $5(6a - 2) + 12a$ **54.** $7(4w - 3) - 25w$

55. $4s + 2(s + 4) + 4$ **56.** $5p + 4(p + 3) - 8$

Basic Skills | Advanced Skills | **Vocational-Technical Applications** | Calculator/Computer | Above and Beyond | Getting Ready
▲

57. ALLIED HEALTH The ideal body weight, in pounds, for a woman can be approximated by substituting her height, in inches, into the formula $105 + 5(h - 60)$. Use the distributive property to simplify the expression.

58. ALLIED HEALTH Use exercise 57 to approximate the ideal body weight for a woman who stands 5 ft 4 in. tall.

59. MECHANICAL ENGINEERING A primary beam can support a load of $54p$. A second beam is added that can support a load of $32p$. What is the total load that the two beams can support?

60. MECHANICAL ENGINEERING Two objects are spinning on the same axis. The moment of inertia of the first object is $\dfrac{6^3}{12}b$. The moment of inertia of the second object is given by $\dfrac{30^3}{36}b$. The total moment of inertia is given by the sum of the moments of inertia of the two objects. Write a simplified expression for the total moment of inertia for the two objects described.

61. Write a paragraph explaining the difference between n^2 and $2n$.

62. Complete the explanation: "x^3 and $3x$ are not the same because . . ."

63. Complete the statement: "$x + 2$ and $2x$ are different because . . ."

64. Write an English phrase for each given algebraic expression:

   (a) $2x^3 + 5x$ **(b)** $(2x + 5)^3$ **(c)** $6(n + 4)^2$

65. Work with another student to complete this exercise. Place $>$, $<$, or $=$ in the blank in these statements.

1^2____2^1 What happens as the table of numbers is extended? Try more examples.

2^3____3^2

3^4____4^3 What sign seems to occur the most in your table? $>$, $<$,

4^5____5^4 or $=$?

Write an algebraic statement for the pattern of numbers in this table. Do you think this is a pattern that continues? Add more lines to the table and extend the pattern to the general case by writing the pattern in algebraic notation. Write a short paragraph stating your conjecture.

66. Work with other students on this exercise.

Part 1: Evaluate the three expressions $\dfrac{n^2 - 1}{2}, n, \dfrac{n^2 + 1}{2}$ using odd values of n: 1, 3, 5, 7, etc. Make a chart like the one below and complete it.

n	$a = \dfrac{n^2 - 1}{2}$	$b = n$	$c = \dfrac{n^2 + 1}{2}$	a^2	b^2	c^2
1						
3						
5						
7						
9						
11						
13						

Part 2: The numbers a, b, and c that you get in each row have a surprising relationship to each other. Complete the last three columns and work together to discover this relationship. You may want to find out more about the history of this famous number pattern.

Answers

61. _____

62. _____

63. _____

64. _____

65. _____

66. _____

Answers

(a) _____

(b) _____

(c) _____

(d) _____

(e) _____

(f) _____

Write the following using exponential notation.

(a) $4 \cdot 4 \cdot 4$ **(b)** $6 \cdot 6 \cdot 6 \cdot 6 \cdot 6 \cdot 6$

(c) $3 \cdot 3 \cdot 3 \cdot 3 \cdot 3$ **(d)** $(-2) \cdot (-2) \cdot (-2)$

(e) $(-8) \cdot (-8) \cdot (-8) \cdot (-8)$ **(f)** $9 \cdot 9 \cdot 9 \cdot 9 \cdot 9 \cdot 9 \cdot 9 \cdot 9$

Answers

1. $5a, 2$ **3.** $4x^3$ **5.** $3x^2, 3x, -7$ **7.** $5ab, 4ab$

9. $2x^2y, -3x^2y, 6x^2y$ **11.** $10m$ **13.** $17b^3$ **15.** $28xyz$ **17.** $6z^2$

19. 0 **21.** n^2 **23.** $15p^2q$ **25.** $11x^2$ **27.** $-9a + 5b$

29. $3x + y$ **31.** $2a + 10b + 1$ **33.** $4x^2 + 4x - 2$ cm   **35.** $28x + 4$ in.

37. $-x^2 + 65x - 150$ **39.** $2m + 3$ **41.** $2x + 7$ **43.** $7a + 10$

45. $13a^4$ **47.** $3a^3$ **49.** $11mn^2$ **51.** $6x + 8$ **53.** $42a - 10$

55. $6s + 12$ **57.** $5h - 195$ **59.** $86p$ **61.** Above and Beyond

63. Above and Beyond   **65.** Above and Beyond   **(a)** 4^3 **(b)** 6^6

(c) 3^5 **(d)** $(-2)^3$ **(e)** $(-8)^4$ **(f)** 9^8

The Streeter/Hutchison Series in Mathematics   Beginning Algebra

1.7

Multiplying and Dividing Terms

< 1.7 Objectives >

1 > Find the product of algebraic terms

2 > Find the quotient of algebraic terms

In Section 0.4, we introduced exponential notation. Remember that the exponent tells us how many times the base is to be used as a factor.

Exponent
↓

$$2^5 = 2 \cdot 2 \cdot 2 \cdot 2 \cdot 2 = 32$$

Base   The fifth power of 2

The notation can also be used when working with letters or variables.

$$x^4 = \underbrace{x \cdot x \cdot x \cdot x}_{4 \text{ factors}}$$

Now look at the product $x^2 \cdot x^3$.

$$x^2 \cdot x^3 = \underbrace{(x \cdot x)}\underbrace{(x \cdot x \cdot x)} = \underbrace{x \cdot x \cdot x \cdot x \cdot x} = x^5$$

2 factors + 3 factors = 5 factors

So

$$x^2 \cdot x^3 = x^{2+3} = x^5$$

You should recall from the previous section that in order to combine a pair of terms into a single term, we must have like terms. For instance, we cannot combine the sum $x^2 + x^3$ into a single term.

On the other hand, when we multiply a pair of unlike terms, as above, their product is a single term.

This leads us to the following property of exponents.

> **NOTE**
>
> In general,
>
> $$x^m = \underbrace{x \cdot x \cdot \cdots \cdot x}_{m \text{ factors}}$$
>
> in which m is a natural number. **Natural numbers** are the numbers we use for counting: 1, 2, 3, and so on.

> **NOTE**
>
> The exponent of x^5 is the *sum* of the exponents in x^2 and x^3.

Property

The Product Property of Exponents

For any integers m and n and any real number a,

$$a^m \cdot a^n = a^{m+n}$$

In words, to multiply expressions with the same base, keep the base and add the exponents.

 Example 1   Using the Product Property of Exponents

< Objective 1 >

> CAUTION

In part (c), the product is *not* 9^6.
The base does not change.

(a) $a^5 \cdot a^7 = a^{5+7} = a^{12}$

(b) $x \cdot x^8 = x^1 \cdot x^8 = x^{1+8} = x^9$ $x = x^1$

(c) $3^2 \cdot 3^4 = 3^{2+4} = 3^6$

(d) $y^2 \cdot y^3 \cdot y^5 = y^{2+3+5} = y^{10}$

(e) $x^3 \cdot y^4$ *cannot* be simplified. The bases are not the same.

 Check Yourself 1

Multiply.

(a) $b^6 \cdot b^8$ **(b)** $y^7 \cdot y$ **(c)** $2^3 \cdot 2^4$ **(d)** $a^2 \cdot a^4 \cdot a^3$

NOTE

Although we have several factors, this is still a single term.

Suppose that numerical coefficients are involved in a product. To find the product, multiply the coefficients and then use the product property of exponents to combine the variables.

$2x^3 \cdot 3x^5 = (2 \cdot 3)(x^3 \cdot x^5)$ Multiply the coefficients.

$\qquad\quad = 6x^{3+5}$ Add the exponents.

$\qquad\quad = 6x^8$

You may have noticed that we have again changed the order and grouping. This method uses the commutative and associative properties that we introduced in Section 1.1.

 Example 2   Using the Product Property of Exponents

NOTE

Again, we have written out all the steps. You can do the multiplication mentally with practice.

Multiply.

(a) $5a^4 \cdot 7a^6 = (5 \cdot 7)(a^4 \cdot a^6) = 35a^{10}$

(b) $y^2 \cdot 3y^3 \cdot 6y^4 = (1 \cdot 3 \cdot 6)(y^2 \cdot y^3 \cdot y^4) = 18y^9$

(c) $2x^2y^3 \cdot 3x^5y^2 = (2 \cdot 3)(x^2 \cdot x^5)(y^3 \cdot y^2) = 6x^7y^5$

 Check Yourself 2

Multiply.

(a) $4x^3 \cdot 7x^5$ **(b)** $3a^2 \cdot 2a^4 \cdot 2a^5$ **(c)** $3m^2n^4 \cdot 5m^3n$

What about dividing expressions when exponents are involved? For instance, what if we want to divide x^5 by x^2? We can use the following approach to division:

$$\frac{x^5}{x^2} = \frac{\overbrace{x \cdot x \cdot x \cdot x \cdot x}^{5\ factors}}{\underbrace{x \cdot x}_{2\ factors}} = \frac{x \cdot x \cdot x \cdot x \cdot x}{x \cdot x}$$

We can divide by
2 factors of x.

$$= \overbrace{x \cdot x \cdot x}^{3\ factors} = x^3$$

NOTE

The exponent of x^3 is the *difference* of the exponents in x^5 and x^2.

So

$$\frac{x^5}{x^2} = x^{5-2} = x^3$$

This leads us to a second property of exponents.

Property

The Quotient Property of Exponents

For any integers m and n, and any nonzero number a,

$$\frac{a^m}{a^n} = a^{m-n}$$

In words, to divide expressions with the same base, keep the base and subtract the exponents.

Example 3   **Using the Quotient Property of Exponents**

< Objective 2 >

Divide the following.

(a) $\dfrac{y^7}{y^3} = y^{7-3} = y^4$

RECALL

We can write $\dfrac{a^3 b^5}{a^2 b^2}$ as $\dfrac{a^3}{a^2} \cdot \dfrac{b^5}{b^2}$ because this is how we multiply fractions.

(b) $\dfrac{m^6}{m} = \dfrac{m^6}{m^1} = m^{6-1} = m^5$

Apply the quotient property to each variable separately.

(c) $\dfrac{a^3 b^5}{a^2 b^2} = a^{3-2} \cdot b^{5-2} = ab^3$

Check Yourself 3

Divide.

(a) $\dfrac{m^9}{m^6}$ **(b)** $\dfrac{a^8}{a}$ **(c)** $\dfrac{a^3 b^5}{a^2}$ **(d)** $\dfrac{r^5 s^6}{r^3 s^2}$

If numerical coefficients are involved, just divide the coefficients and then use the quotient property of exponents to divide the variables, as shown in Example 4.

Example 4	**Using the Quotient Property of Exponents**

Divide the following.

Subtract the exponents.

(a) $\dfrac{6x^5}{3x^2} = 2x^{5-2} = 2x^3$

6 divided by 3

20 divided by 5

(b) $\dfrac{20a^7b^5}{5a^3b^4} = 4a^{7-3} \cdot b^{5-4}$

Again apply the quotient property to each variable separately.

$= 4a^4b$

 Check Yourself 4

Divide.

(a) $\dfrac{4x^3}{2x}$ **(b)** $\dfrac{20a^6}{5a^2}$ **(c)** $\dfrac{24x^5y^3}{4x^2y^2}$

Check Yourself ANSWERS

1. (a) b^{14}; **(b)** y^8; **(c)** 2^7; **(d)** a^9 **2. (a)** $28x^8$; **(b)** $12a^{11}$; **(c)** $15m^5n^5$

3. (a) m^3; **(b)** a^7; **(c)** ab^5; **(d)** r^2s^4 **4. (a)** $2x^2$; **(b)** $4a^4$; **(c)** $6x^3y$

Reading Your Text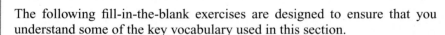

The following fill-in-the-blank exercises are designed to ensure that you understand some of the key vocabulary used in this section.

SECTION 1.7

(a) When multiplying expressions with the same base, _____ the exponents.

(b) When multiplying expressions with the same base, the _____ does not change.

(c) When multiplying expressions with the same base, _____ the coefficients.

(d) To divide expressions with the same base, keep the base and _____ the exponents.

Beginning Algebra   The Streeter/Hutchison Series in Mathematics

| Basic Skills | Advanced Skills | Vocational-Technical Applications | Calculator/Computer | Above and Beyond | Getting Ready |

1.7 exercises

< Objective 1 >

Multiply.

1. $x^5 \cdot x^7$

2. $b^2 \cdot b^4$

3. $3^2 \cdot 3^6$

4. $y^6 \cdot y^4$

5. $a^9 \cdot a$

6. $3^4 \cdot 3^5$

7. $z^{10} \cdot z^3$

8. $x^6 \cdot x^3$

9. $p^5 \cdot p^7$

10. $s^6 \cdot s^9$

11. $x^3y \cdot x^2y^4$

12. $m^2n^3 \cdot mn^4$

13. $w^3 \cdot w^4 \cdot w^2$

14. $x^5 \cdot x^4 \cdot x^6$

15. $m^3 \cdot m^2 \cdot m^4$

16. $r^3 \cdot r \cdot r^5$ > Videos

17. $a^3b \cdot a^2b^2 \cdot ab^3$

18. $w^2z^3 \cdot wz \cdot w^3z^4$

19. $p^2q \cdot p^3q^5 \cdot pq^4$

20. $c^3d \cdot c^4d^2 \cdot cd^5$

21. $2a^5 \cdot 3a^2$

22. $5x^3 \cdot 3x^2$

23. $x^2 \cdot 3x^5$

24. $2m^4 \cdot 6m^7$

25. $5m^3n^2 \cdot 4mn^3$

26. $7x^2y^5 \cdot 6xy^4$

27. $4x^5y \cdot 3xy^2$

28. $5a^3b \cdot 10ab^4$

MathZone

Boost your grade at mathzone.com!

> Practice Problems
> NetTutor
> Self-Tests
> e-Professors
> Videos

Name _____

Section _____ Date _____

Answers

1.	2.
3.	4.
5.	6.
7.	8.
9.	10.
11.	12.
13.	14.
15.	16.
17.	18.
19.	20.
21.	22.
23.	24.
25.	26.
27.	28.

Answers

29. _____

30. _____

31. _____

32. _____

33. _____

34. _____

35. _____

36. _____

37. _____

38. _____

39. _____

40. _____

41. _____

42. _____

43. _____

44. _____

45. _____

46. _____

47. _____

48. _____

49. _____

50. _____

29. $2a^2 \cdot a^3 \cdot 3a^7$

30. $2x^3 \cdot 3x^4 \cdot x^5$

31. $3c^2d \cdot 4cd^3 \cdot 2c^5d$

32. $5p^2q \cdot p^3q^2 \cdot 3pq^3$

33. $5m^2 \cdot m^3 \cdot 2m \cdot 3m^4$

34. $3a^3 \cdot 2a \cdot a^4 \cdot 2a^5$

35. $2r^3s \cdot rs^2 \cdot 3r^2s \cdot 5rs$

36. $6a^2b \cdot ab \cdot 3ab^3 \cdot 2a^2b$

< Objective 2 >

Divide.

37. $\dfrac{a^{10}}{a^7}$

38. $\dfrac{m^8}{m^2}$

39. $\dfrac{y^{10}}{y^4}$

40. $\dfrac{b^9}{b^4}$

41. $\dfrac{p^{15}}{p^{10}}$

42. $\dfrac{s^{15}}{s^9}$

43. $\dfrac{x^5y^3}{x^2y^2}$

44. $\dfrac{s^5t^4}{s^3t^2}$

45. $\dfrac{10m^6}{5m^4}$

46. $\dfrac{8x^5}{4x}$

47. $\dfrac{24a^7}{6a^4}$

48. $\dfrac{25x^9}{5x^8}$

49. $\dfrac{26m^8n}{13m^6}$

50. $\dfrac{30a^4b^5}{6b^4}$

51. $\dfrac{35w^4z^6}{5w^2z}$

52. $\dfrac{48p^6q^7}{8p^4q}$

53. $\dfrac{48x^4y^5z^9}{24x^2y^3z^6}$ > Videos

54. $\dfrac{25a^5b^4c^3}{5a^4bc^2}$

| Basic Skills | **Advanced Skills** | Vocational-Technical Applications | Calculator/Computer | Above and Beyond | Getting Ready |

Simplify each expression, if possible.

55. $3a^4b^3 \cdot 2a^2b^4$

56. $2xy^3 \cdot 3xy^2$

57. $2a^3b + 3a^2b$

58. $2xy^3 + 3xy^2$

59. $2x^2y^3 \cdot 3x^2y^3$

60. $5a^3b^2 \cdot 10a^3b^2$

61. $2x^3y^2 + 3x^3y^2$

62. $5a^3b^2 + 10a^3b^2$

63. $\dfrac{8a^2b \cdot 6a^2b}{2ab}$

64. $\dfrac{6x^2y^3 \cdot 9x^2y^3}{3x^2y^2}$ > Videos

65. $\dfrac{8a^2b + 6a^2b}{2ab}$

66. $\dfrac{6x^2y^3 + 9x^2y^3}{3x^2y^2}$

| Basic Skills | Advanced Skills | Vocational-Technical Applications | Calculator/Computer | **Above and Beyond** | Getting Ready |

67. Complete the following statements:

    (a) a^n is negative when _____ because _____ .

    (b) a^n is positive when _____ because _____ .

                (give all possibilities)

68. "Earn Big Bucks!" reads an ad for a job. "You will be paid 1 cent for the first day and 2 cents for the second day, 4 cents for the third day, 8 cents for the fourth day, and so on, doubling each day. Apply now!" What kind of deal is this—where is the big money offered in the headline? The fine print at the bottom of the ad says: "Highly qualified people may be paid $1,000,000 for

Answers

51. _____

52. _____

53. _____

54. _____

55. _____

56. _____

57. _____

58. _____

59. _____

60. _____

61. _____

62. _____

63. _____

64. _____

65. _____

66. _____

67. _____

68. _____

the first 28 working days if they choose." Well, *that* does sound like big bucks! Work with other students to decide which method of payment is better and how much better. You may want to make a table and try to find a formula for the first offer.

69. An oil spill from a tanker in pristine Prince William Sound in Alaska begins in a circular shape only 2 ft across. The area of the circle is $A = \pi r^2$. Make a table to decide what happens to the area if the diameter is doubling each hour. How large will the spill be in 24 h? (Hint: The radius is one-half the diameter.)

2 ft

Answers

1. x^{12} **3.** 3^8 **5.** a^{10} **7.** z^{13} **9.** p^{12} **11.** $x^5 y^5$ **13.** w^9
15. m^9 **17.** $a^6 b^6$ **19.** $p^6 q^{10}$ **21.** $6a^7$ **23.** $3x^7$ **25.** $20m^4 n^5$
27. $12x^6 y^3$ **29.** $6a^{12}$ **31.** $24c^8 d^5$ **33.** $30m^{10}$ **35.** $30r^7 s^5$
37. a^3 **39.** y^6 **41.** p^5 **43.** $x^3 y$ **45.** $2m^2$ **47.** $4a^3$
49. $2m^2 n$ **51.** $7w^2 z^5$ **53.** $2x^2 y^2 z^3$ **55.** $6a^6 b^7$
57. Cannot simplify   **59.** $6x^4 y^6$ **61.** $5x^3 y^2$ **63.** $24a^3 b$
65. $7a$ **67.** Above and Beyond   **69.** Above and Beyond

Definition/Procedure	Example	Reference
Properties of Real Numbers		Section 1.1
The Commutative Properties		
If a and b are any numbers, **1.** $a + b = b + a$ **2.** $a \cdot b = b \cdot a$	$3 + 8 = 8 + 3$ $2 \cdot 5 = 5 \cdot 2$	*p. 85*
The Associative Properties		
If a, b, and c are any numbers, **1.** $a + (b + c) = (a + b) + c$ **2.** $a \cdot (b \cdot c) = (a \cdot b) \cdot c$	$3 + (7 + 12) = (3 + 7) + 12$ $2 \cdot (5 \cdot 12) = (2 \cdot 5) \cdot 12$	*p. 86*
The Distributive Property		
If a, b, and c are any numbers, $a(b + c) = a \cdot b + a \cdot c$	$6 \cdot (8 + 15) = 6 \cdot 8 + 6 \cdot 15$	*p. 88*
Adding and Subtracting Real Numbers		Section 1.2
Addition		
1. If two numbers have the same sign, add their absolute values. Give the sum the sign of the original numbers.	$9 + 7 = 16$ $(-9) + (-7) = -16$	*p. 96*
2. If two numbers have different signs, subtract their absolute values, the smaller from the larger. Give the result the sign of the number with the larger absolute value.	$15 + (-10) = 5$ $(-12) + 9 = -3$	*p. 97*
Subtraction		

1. Rewrite the subtraction problem as an addition problem by:
  a. Changing the subtraction symbol to an addition symbol
  b. Replacing the number being subtracted with its opposite
2. Add the resulting signed numbers as before.
	$16 - 8 = 16 + (-8)$
	$\quad = 8$
	$8 - 15 = 8 + (-15)$
	$\quad = -7$
	$-9 - (-7) = -9 + 7$
	$\quad = -2$
		p. 99

Continued

Definition/Procedure	Example	Reference

Multiplying and Dividing Real Numbers — Section 1.3

Multiplication

Multiply the absolute values of the two numbers.
1. If the numbers have different signs, the product is negative.
2. If the numbers have the same sign, the product is positive.

$5(-7) = -35$
$(-10)(9) = -90$
$8 \cdot 7 = 56$
$(-9)(-8) = 72$

p. 112

p. 113

Division

Divide the absolute values of the two numbers.
1. If the numbers have different signs, the quotient is negative.
2. If the numbers have the same sign, the quotient is positive.

$\dfrac{-32}{4} = -8$

$\dfrac{75}{-5} = -15$

$\dfrac{20}{5} = 4$

$\dfrac{-18}{-9} = 2$

p. 116

From Arithmetic to Algebra — Section 1.4

Addition

$x + y$ means the *sum of x and y* or *x plus y*. Some other words indicating addition are "more than" and "increased by."

The sum of x and 5 is $x + 5$.
7 more than a is $a + 7$.
b increased by 3 is $b + 3$.

p. 131

Subtraction

$x - y$ means the *difference of x and y* or *x minus y*. Some other words indicating subtraction are "less than" and "decreased by."

The difference of x and 3 is $x - 3$.
5 less than p is $p - 5$.
a decreased by 4 is $a - 4$.

p. 132

Multiplication

$x \cdot y$
$(x)(y)$ All these mean the *product* of x and y or *x times y*.
xy

The product of m and n is mn.
The product of 2 and the sum of a and b is $2(a + b)$.

p. 132

Definition/Procedure	Example	Reference

Expressions

An expression is a meaningful collection of numbers, variables, and signs of operation.

$3x + y$ is an expression
$3x = y$ is not an expression

p. 133

Division

$\dfrac{x}{y}$ means x *divided by y* or the *quotient* when x is divided by y.

n divided by 5 is $\dfrac{n}{5}$.

The sum of a and b, divided by 3, is $\dfrac{a + b}{3}$.

p. 134

Evaluating Algebraic Expressions

Section 1.5

Evaluating Algebraic Expressions

To evaluate an algebraic expression:
1. Replace each variable or letter with its number value.
2. Do the necessary arithmetic, following the rules for the order of operations.

Evaluate $2x + 3y$ if $x = 5$ and $y = -2$.
$$2x + 3y$$
$$= 2(5) + (3)(-2)$$
$$= 10 - 6 = 4$$

p. 142

Adding and Subtracting Terms

Section 1.6

Term

A term is a number or the product of a number and one or more variables.

p. 154

Combining Like Terms

To combine like terms:
1. Add or subtract the numerical coefficients (the numbers multiplying the variables).
2. Attach the common variables.

$5x + 2x = 7x$
$5 + 2$
$8a - 5a = 3a$
$8 - 5$

p. 156

Continued

Definition/Procedure	Example	Reference
Multiplying and Dividing Terms		Section 1.7
The Product Property of Exponents		
$a^m \cdot a^n = a^{m+n}$	$x^7 \cdot x^3 = x^{7+3} = x^{10}$	p. 165
The Quotient Property of Exponents		
$\dfrac{a^m}{a^n} = a^{m-n}$	$\dfrac{y^7}{y^3} = y^{7-3} = y^4$	p. 167

This summary exercise set is provided to give you practice with each of the objectives of this chapter. Each exercise is keyed to the appropriate chapter section. When you are finished, you can check your answers to the odd-numbered exercises in the back of the text. If you have difficulty with any of these questions, go back and reread the examples from that section. The answers to the even-numbered exercises appear in the *Instructor's Solutions Manual.* Your instructor will give you guidelines on how best to use these exercises in your instructional setting.

1.1 *Identify the property that is illustrated by each of the following statements.*

1. $5 + (7 + 12) = (5 + 7) + 12$

2. $2(8 + 3) = 2 \cdot 8 + 2 \cdot 3$

3. $4 \cdot (5 \cdot 3) = (4 \cdot 5) \cdot 3$

4. $4 \cdot 7 = 7 \cdot 4$

Verify that each of the following statements is true by evaluating each side of the equation separately and comparing the results.

5. $8(5 + 4) = 8 \cdot 5 + 8 \cdot 4$

6. $2(3 + 7) = 2 \cdot 3 + 2 \cdot 7$

7. $(7 + 9) + 4 = 7 + (9 + 4)$

8. $(2 + 3) + 6 = 2 + (3 + 6)$

9. $(8 \cdot 2) \cdot 5 = 8(2 \cdot 5)$

10. $(3 \cdot 7) \cdot 2 = 3 \cdot (7 \cdot 2)$

Use the distributive law to remove parentheses.

11. $3(7 + 4)$

12. $4(2 + 6)$

13. $\dfrac{1}{2}(5 + 8)$

14. $0.05(1.35 + 8.1)$

1.2 *Add.*

15. $-3 + (-8)$

16. $10 + (-4)$

17. $6 + (-6)$

18. $-16 + (-16)$

19. $-18 + 0$

20. $\dfrac{3}{8} + \left(-\dfrac{11}{8}\right)$

21. $5.7 + (-9.7)$

22. $-18 + 7 + (-3)$

Subtract.

23. $8 - 13$

24. $-7 - 10$

25. $10 - (-7)$

26. $-5 - (-1)$

27. $-9 - (-9)$

28. $0 - (-2)$

29. $-\dfrac{5}{4} - \left(-\dfrac{17}{4}\right)$

30. $7.9 - (-8.1)$

Use a calculator to perform the indicated operations.

31. $489 + (-332)$

32. $1,024 - (-3,206)$

33. $-234 + (-321) - (-459)$

34. $981 - 1,854 - (-321)$

35. $4.56 + (-0.32)$

36. $-32.14 - 2.56$

37. $-3.112 - (-0.1) + 5.06$

38. $10.01 - 12.566 + 2$

39. $13 - (-12.5) + 4\frac{1}{4}$

40. $3\frac{1}{8} - 6.19 + (-8)$

1.3 *Multiply.*

41. $(10)(-7)$

42. $(-8)(-5)$

43. $(-3)(-15)$

44. $(1)(-15)$

45. $(0)(-8)$

46. $\left(\frac{2}{3}\right)\left(-\frac{3}{2}\right)$

47. $(-4)\left(\frac{3}{8}\right)$

48. $\left(-\frac{5}{4}\right)(-1)$

Divide.

49. $\frac{80}{16}$

50. $\frac{-63}{7}$

51. $\frac{-81}{-9}$

52. $\frac{0}{-5}$

53. $\frac{32}{-8}$

54. $\frac{-7}{0}$

Perform the indicated operations.

55. $\frac{-8 + 6}{-8 - (-10)}$

56. $\frac{-6 - 1}{5 - (-2)}$

57. $\frac{25 - 4}{-5 - (-2)}$

58. $\frac{3 - (-6)}{-4 + 2}$

1.4 *Write, using symbols.*

59. 5 more than y

60. c decreased by 10

61. The product of 8 and a

62. The quotient when y is divided by 3

63. 5 times the product of m and n

64. The product of a and 5 less than a

65. 3 more than the product of 17 and x

66. The quotient when a plus 2 is divided by a minus 2

Identify which are expressions and which are not.

67. $4(x + 3)$

68. $7 \div \cdot 8$

69. $y + 5 = 9$

70. $11 + 2(3x - 9)$

1.5 *Evaluate each of the following expressions.*

71. $18 - 3 \cdot 5$

72. $(18 - 3) \cdot 5$

73. $5 \cdot 4^2$

74. $(5 \cdot 4)^2$

75. $5 \cdot 3^2 - 4$

76. $5(3^2 - 4)$

77. $5(4 - 2)^2$

78. $5 \cdot 4 - 2^2$

79. $(5 \cdot 4 - 2)^2$

80. $3(5 - 2)^2$

81. $3 \cdot 5 - 2^2$

82. $(3 \cdot 5 - 2)^2$

Evaluate the expressions if $x = -3$, $y = 6$, $z = -4$, and $w = 2$.

83. $3x + w$

84. $5y - 4z$

85. $x + y - 3z$

86. $5z^2$

87. $3x^2 - 2w^2$

88. $3x^3$

89. $5(x^2 - w^2)$

90. $\dfrac{6z}{2w}$

91. $\dfrac{2x - 4z}{y - z}$

92. $\dfrac{3x - y}{w - x}$

93. $\dfrac{x(y^2 - z^2)}{(y + z)(y - z)}$

94. $\dfrac{y(x - w)^2}{x^2 - 2xw + w^2}$

1.6 *List the terms of the expressions.*

95. $4a^3 - 3a^2$

96. $5x^2 - 7x + 3$

Circle like terms.

97. $5m^2, -3m, -4m^2, 5m^3, m^2$

98. $4ab^2, 3b^2, -5a, ab^2, 7a^2, -3ab^2, 4a^2b$

Combine like terms.

99. $5c + 7c$

100. $2x + 5x$

101. $4a - 2a$

102. $6c - 3c$

103. $9xy - 6xy$

104. $5ab^2 + 2ab^2$

105. $7a + 3b + 12a - 2b$

106. $6x - 2x + 5y - 3x$

107. $5x^3 + 17x^2 - 2x^3 - 8x^2$

108. $3a^3 + 5a^2 + 4a - 2a^3 - 3a^2 - a$

109. Subtract $4a^3$ from the sum of $2a^3$ and $12a^3$.

110. Subtract the sum of $3x^2$ and $5x^2$ from $15x^2$.

1.7 *Simplify.*

111. $\dfrac{x^{10}}{x^3}$

112. $\dfrac{a^5}{a^4}$

113. $\dfrac{x^2 \cdot x^3}{x^4}$

114. $\dfrac{m^2 \cdot m^3 \cdot m^4}{m^5}$

115. $\dfrac{18p^7}{9p^5}$

116. $\dfrac{24x^{17}}{8x^{13}}$

117. $\dfrac{30m^7n^5}{6m^2n^3}$

118. $\dfrac{108x^9y^4}{9xy^4}$

119. $\dfrac{48p^5q^3}{6p^3q}$

120. $\dfrac{52a^5b^3c^5}{13a^4c}$

121. $(4x^3)(5x^4)$

122. $(3x)^2(4xy)$

123. $(8x^2y^3)(3x^3y^2)$

124. $(-2x^3y^3)(-5xy)$

125. $(6x^4)(2x^2y)$

Write an algebraic expression to model each application.

126. **CONSTRUCTION** If x ft are cut off the end of a board that is 23 ft long, how much is left?

127. **BUSINESS AND FINANCE** Joan has 25 nickels and dimes in her pocket. If x of these are dimes, how many of the coins are nickels?

128. **SOCIAL SCIENCE** Sam is 5 years older than Angela. If Angela is x years old now, how old is Sam?

129. **BUSINESS AND FINANCE** Margaret has $5 more than twice as much money as Gerry. Write an expression for the amount of money that Margaret has.

130. **GEOMETRY** The length of a rectangle is 4 m more than the width. Write an expression for the length of the rectangle.

131. **NUMBER PROBLEM** A number is 7 less than 6 times the number n. Write an expression for the number.

132. **CONSTRUCTION** A 25-ft plank is cut into two pieces. Write expressions for the length of each piece.

133. **BUSINESS AND FINANCE** Bernie has x dimes and q quarters in his pocket. Write an expression for the amount of money that Bernie has in his pocket.

This purpose of this chapter test is to help you check your progress so that you can find sections and concepts that you need to review before the next exam. Allow yourself about an hour to take this test. At the end of that hour, check your answers against those given in the back of this text. If you missed any, note the section reference that accompanies the answer. Go back to that section and reread the examples until you have mastered that particular concept.

Name _____

Section _____ Date _____

Answers

Identify the property that is illustrated by each of the following statements.

1. $6 \cdot 7 = 7 \cdot 6$

2. $2(6 + 7) = 2 \cdot 6 + 2 \cdot 7$

3. $4 + (3 + 7) = (4 + 3) + 7$

Use the distributive property to remove parentheses. Then simplify your result.

4. $3(5 + 2)$

5. $4(5x + 3)$

Identify which are expressions and which are not.

6. $5x + 6 = 4$

7. $4 + (6 + x)$

Add.

8. $-8 + (-5)$

9. $6 + (-9)$

10. $(-9) + (-12)$

11. $-\dfrac{5}{3} + \dfrac{8}{3}$

Subtract.

12. $9 - 15$

13. $-10 - 11$

14. $5 - (-4)$

15. $-7 - (-7)$

Multiply.

16. $(8)(-5)$

17. $(-9)(-7)$

18. $(4.5)(-6)$

19. $(6)(-4)$

Write, using symbols.

20. 5 less than a

21. The product of 6 and m

1. _____

2. _____

3. _____

4. _____

5. _____

6. _____

7. _____

8. _____

9. _____

10. _____

11. _____

12. _____

13. _____

14. _____

15. _____

16. _____ 17. _____

18. _____ 19. _____

20. _____ 21. _____

Answers

22. _____

23. _____

24. _____

25. _____

26. _____

27. _____

28. _____

29. _____

30. _____

31. _____

32. _____

33. _____

34. _____

35. _____

36. _____

37. _____

38. _____

39. _____ 40. _____

41. _____ 42. _____

22. 4 times the sum of m and n

23. The quotient when the sum of a and b is divided by 3

Evaluate each expression.

24. $\dfrac{-100}{4}$

25. $\dfrac{-36 + 9}{-9}$

26. $\dfrac{(-15)(-3)}{-9}$

27. $\dfrac{9}{0}$

Evaluate the following expressions.

28. $29 - 3 \cdot 4$

29. $4 \cdot 5^2 - 35$

30. $4(2 + 4)^2$

31. $16 \div (-4) + (-5)$

32. If $x = 2$, $y = -1$, and $z = 3$, evaluate the expression $\dfrac{9x^2y}{3z}$.

Combine like terms.

33. $9a + 4a$

34. $10x + 8y + 9x - 3y$

35. Subtract $9a^2$ from the sum of $12a^2$ and $5a^2$.

Multiply.

36. $a^5 \cdot a^9$

37. $2x^3y^2 \cdot 4x^4y$

Divide.

38. $\dfrac{9x^9}{3x^3}$

39. $\dfrac{20a^3b^5}{5a^2b^2}$

40. $\dfrac{x^{10} \cdot x^5}{x^6}$

41. **SOCIAL SCIENCE** Tom is 8 years younger than twice Moira's age. Write an expression for Tom's age. Let x represent Moira's age.

42. **GEOMETRY** The length of a rectangle is 4 more than twice the width. Write an expression for the length of the rectangle.

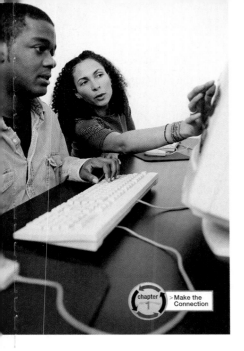

chapter 1 > Make the Connection

Activity 1 ::
An Introduction to Searching

Each activity in this text is designed to either enhance your understanding of the topics of the preceding chapter or provide you with a mathematical extension of those topics, or both. The activities can be undertaken by one student, but they are better suited for a small group project. Occasionally it is only through discussion that different facets of the activity become apparent. For material related to this activity, visit the text website at www.mhhe.com/streeter.

As you work through this text, you will encounter several activities. Many (though not all) of these activities will require you to use the Internet. This activity is designed to familiarize you with searching the Internet for information.

If you are new to computers or the Internet, your instructor (or perhaps a friend) can help you get started. You will need to access the Internet through one of the many Web browsers, such as Microsoft's Internet Explorer, Netscape Navigator, Mozilla Firefox, Opera, or AOL's browser.

We will ask that you "bookmark" various websites—some browsers refer to bookmarks as "favorites," or "preferences." Bookmarking a website allows you to return to the site at a later time by choosing it from a list of bookmarked sites. This way, you do not have to search for it every time you wish to visit that site.

Most Web browsers provide you with a way to bookmark a website. When you are looking at a website that you wish to bookmark, click on the menu that says Bookmarks (or Favorites, Preferences, or whatever name your browser uses), and click on the menu item Add Bookmark (or Add to Favorites, Create Preference, etc.). The next time you want to look at the website that you bookmarked, you need only click on the Bookmark menu and click on the title of the site saved.

Activity:

First, you will need to connect to the Internet. Next, you will need to access a page containing a *search engine* that searches the World Wide Web at large (rather than just doing an internal search). Many *default* home pages (such as www.msn.com and www.aol.com) contain a *search* field. If yours does not, several of the more popular sites that do are listed here:

> http://www.altavista.com
> http://www.google.com
> http://www.lycos.com
> http://www.yahoo.com

Access one of these websites or use some other search engine, as you see fit. Now we are ready to continue.

I.

1. Locate your state government home page. You may accomplish this by typing your state name into the search field and clicking on Go or Find. A (long) list of "matches" will appear. Read the blurbs to determine which belongs to the state government. Note: You should not need to go through more than the first page or two of listings. Bookmark this site (as previously described).

2. Find an official e-mail address for the governor of your state. Find the names of the senators for your state. See if you can locate e-mail addresses for the senators of your state. Note: If their e-mail addresses are not provided, then we will find them later.

3. Locate the home page of your college or university, or the flagship university in your state.

4. Repeat steps 1 to 3 for a different state.

II.

1. Search for the White House in Washington, D.C. Find an official e-mail address for the President of the United States.

2. Find the home page for the U.S. Congress. If you could not find e-mail addresses for senators on the state pages (step I.2), then find one through the U.S. Congress website.

3. Locate the home page of the U.S. Census Bureau and bookmark this site. Find the census data for your state. How many people live in your state, according to the most recent census? How many people live in your city or hometown?

4. Repeat step 3 for the second state that you used in part I (choose the state capital as the city).

III.

1. Search for weather data for your city or hometown (the National Weather Service is a good starting point). What is the predicted weather for tomorrow? What was the weather yesterday? What was the weather one month ago? What was the weather one year ago, today?

2. Repeat step 1 for the capital of the second state that you used in part I.

IV.

1. Choose a country on some other continent. Can you locate the name and e-mail address of that nation's leader? Census data? Weather?

2. Locate a home page for a university in the country found in step 1.

3. Locate the home page for the United Nations (UN) and bookmark this site. Does the UN site provide some of the information asked for in step 1? Do the UN data differ from the data provided by the nation itself?

V.

1. Locate two different images of the so-called Mandelbrot fractal. These can be the same image with different colors, or they can be two different views of the fractal.

2. Locate an image of a different fractal.

3. Locate information on fractals in general. (What are they?) Write a paragraph describing fractals in general and express your thoughts on the fractals you found in steps 1 and 2.

VI.

1. Search for a business or an organization that hires people in your field; if you are not already on a career path, determine a potential career for yourself. Bookmark this site.

2. Write a paragraph describing the organization as it has presented itself on its website.

3. Find the name and an e-mail address of a person whom you would contact if you were applying for a job.

4. Find information describing the application process for someone seeking employment with this business or organization.

5. Are there forms that the site provides for job seekers? Can you download them (using Save As or a link)? If forms exist, what format are these forms in (Adobe pdf files, Microsoft Word documents, or some other format)?

You have now searched through the World Wide Web for information. Later, you will be asked to use the skills gained here to complete some of the other activities in this text.

chapter 2 > Make the Connection

CHAPTER

2

Equations and Inequalities

INTRODUCTION

Every year, millions of people travel to other countries for business and pleasure. When traveling to another country, you need to consider many things, such as passports and visas, immunizations, local sights, restaurants and hotels, and language.

Another consideration when traveling internationally is currency. Nearly every country has its own money. For example, the Japanese currency is the yen (¥), the British use the pound (£), and the Canadians use Canadian dollars (CAN$), whereas the United States of America uses the US$.

When visiting another country, you need to acquire the local currency. Many sources publish exchange rates for currency on a daily basis. For instance, on May 20, 2006, *Yahoo!Finance* listed the US$ to CAN$ exchange rate as 1.1191. We can use this to construct an equation to determine the amount of Canadian dollars that one receives for U.S. dollars.

$$C = 1.1191U$$

in which U represents the amount of US$ to be exchanged and C represents the amount of CAN$ to be received.

The equation is an ancient tool used to solve problems and describe numerical relationships accurately and clearly. In this chapter, you will learn methods to solve linear equations and will practice writing equations that accurately describe problem situations.

CHAPTER 2 OUTLINE

Name _____

Section _____ Date _____

Answers

1. _____   2. _____

3. _____   4. _____

5. _____   6. _____

7. _____   8. _____

9. _____

10. _____

11. _____

12. _____

13. _____

14. _____

15. _____

16. _____

17. _____

18. _____

19. _____

20. _____

This pretest provides a preview of the types of exercises you will encounter in each section of this chapter. The answers for these exercises can be found in the back of the text. If you are working on your own, or ahead of the class, this pretest can help you identify the sections in which you should focus more of your time.

State whether the number shown in parentheses is a solution for the given equation.

2.1
1. $4x - 9 = 15$ (7)   **2.** $7x - 5 = 3x + 31$ (9)

Solve the following equations and check your results.

2.2–2.3 **3.** $8x - 23 = 7x - 15$ **4.** $6x + 15 - 2x = 18 + 8x - 5x - 15$

5. $7x - 15 = 4x + 6$ **6.** $\dfrac{5}{7}x = 25$

7. $9x - (x - 5) = -(6 - x) - 3$

8. $2(3 - x) + 15 = 8(4x - 5) - 7$

Solve for the indicated variable.

2.4
9. $P = 2L + 2W$ for W **10.** $5x - 3y = 14$ for y

Translate each statement into an algebraic equation.

11. 5 more than 4 times a number is 17.

12. 4 times the sum of a number and 6 is 6 more than 10 times the number.

2.4–2.5 Solve the following word problems.

13. NUMBER PROBLEM 7 times a number decreased by 5 is 37. Find the number.

14. NUMBER PROBLEM The sum of 2 consecutive odd integers is 32. Find the integers.

15. GEOMETRY The perimeter of a rectangle is 34 cm. If the length is 1 cm more than 3 times the width, what are the dimensions of the rectangle?

16. BUSINESS AND FINANCE A state sales tax rate is 2.5%. If the tax on a purchase is $13.50, what was the amount of the purchase?

17. BUSINESS AND FINANCE A house sells for $250,000 and the rate of commission is 5.25%. How much will the salesperson make for the sale?

18. BUSINESS AND FINANCE A stereo system is marked down from $790 to $671.50. What was the rate of discount?

Solve the following inequalities.

2.6
19. $x - 7 \leq 8$ **20.** $6 - 2x \geq 9 + x$

2.1

Solving Equations by the Addition Property

< 2.1 Objectives >

1 > Determine whether a given number is a solution for an equation

2 > Identify expressions and equations

3 > Use the addition property to solve an equation

4 > Use the distributive property in solving equations

▶ Overcoming Math Anxiety

Over the first few chapters, we present you with a series of class-tested techniques that are designed to improve your performance in this math class. Hint #3   Don't procrastinate!

1. Do your math homework while you are still fresh. If you wait until too late at night, your tired mind will have much more difficulty understanding the concepts.

2. Do your homework the day it is assigned. The more recent the explanation, the easier it is to recall.

3. When you finish your homework, try reading through the next section one time. This will give you a sense of direction when you next hear the material. This works in a lecture or lab setting.

Remember that, in a typical math class, you are expected to do two or three hours of homework for each weekly class hour. This means two or three hours per night. Schedule the time and stick to your schedule.

In this chapter we begin working with one of the most important tools of mathematics, the equation. The ability to recognize and solve various types of equations is probably the most useful algebraic skill you will learn. We will continue to build upon the methods of this chapter throughout the text. To begin, we define the word *equation*.

Definition

| **Equation** | An **equation** is a mathematical statement that two expressions are equal. |

Some examples are $3 + 4 = 7$, $x + 3 = 5$, $P = 2L + 2W$.

As you can see, an equal sign ($=$) separates the two expressions. These expressions are usually called the *left side* and the *right side* of the equation.

$$x + 3 = 5$$

Left side   Equals   Right side

NOTE

An equation such as

$x + 3 = 5$

is called a **conditional equation** because it can be either true or false, depending on the value given to the variable.

Just as the balance scale may be in balance or out of balance, an equation may be either true or false. For instance, $3 + 4 = 7$ is true because both sides name the same number. What about an equation such as $x + 3 = 5$ that has a letter or variable on one side? Any number can replace x in the equation. However, only one number will make this equation a true statement.

$$x + 3 = 5$$

$$\text{If } x = 2 \begin{cases} 1 & (1) + 3 = 5 \text{ is false} \\ 2 & (2) + 3 = 5 \text{ is true} \\ 3 & (3) + 3 = 5 \text{ is false} \end{cases}$$

The number 2 is called the **solution** (or *root*) of the equation $x + 3 = 5$ because substituting 2 for x gives a true statement.

Definition

Solution   A **solution** for an equation is any value for the variable that makes the equation a true statement.

Example 1   Verifying a Solution

< Objective 1 >

(a) Is 3 a solution for the equation $2x + 4 = 10$?

To find out, replace x with 3 and evaluate $2x + 4$ on the left.

NOTE

We could represent the solution using set notation as {3}.

Left side		*Right side*
$2(3) + 4$	$\overset{?}{=}$	10
$6 + 4$	$\overset{?}{=}$	10
10	$=$	10

Because $10 = 10$ is a true statement, 3 is a solution of the equation.

(b) Is 5 a solution of the equation $3x - 2 = 2x + 1$?

To find out, replace x with 5 and evaluate each side separately.

RECALL

The rules for order of operations require that we multiply first; then add or subtract.

Left side		*Right side*
$3(5) - 2$	$\overset{?}{=}$	$2(5) + 1$
$15 - 2$	$\overset{?}{=}$	$10 + 1$
13	$\neq$	11

Because the two sides do not name the same number, we do not have a true statement, and 5 is not a solution.

Check Yourself 1

For the equation

$$2x - 1 = x + 5$$

(a) Is 4 a solution?   **(b)** Is 6 a solution?

You may be wondering whether an equation can have more than one solution. It certainly can. For instance,

$$x^2 = 9$$

has two solutions. They are 3 and -3 because

$$3^2 = 9 \quad \text{and} \quad (-3)^2 = 9$$

In this chapter, however, we work with *linear equations in one variable.* These are equations that can be put into the form

$$ax + b = 0$$

in which the variable is x, a and b are any numbers, and a is not equal to 0. In a linear equation, the variable can appear only to the first power. No other power (x^2, x^3, and so on) can appear. Linear equations are also called **first-degree equations.** The degree of an equation in one variable is the highest power to which the variable appears.

NOTE

This is an example of a **quadratic equation.** We will consider such equations in Chapter 4 and then again in Chapter 10.

Property

Linear Equations

Linear equations in one variable are equations that can be written in the form

$$ax + b = 0 \qquad a \neq 0$$

Every such equation has exactly one solution.

 Example 2   Identifying Expressions and Equations

< Objective 2 >

NOTE

In part (e) we see that an equation that includes a variable in a denominator is not a linear equation.

Label each of the following as an expression, a linear equation, or an equation that is not linear.

(a) $4x + 5$ is an expression.

(b) $2x + 8 = 0$ is a linear equation.   *expression*

(c) $3x^2 - 9 = 0$ is an equation that is not linear.   *expression*

(d) $5x = 15$ is a linear equation.   *linear*

(e) $5 - \dfrac{7}{x} = 4x$ is an equation that is not linear.

Check Yourself 2

Label each as an expression, a linear equation, or an equation that is not linear.

(a) $2x^2 = 8$ (b) $2x - 3 = 0$ (c) $5x - 10$

(d) $2x + 1 = 7$ (e) $\dfrac{3}{x} - 4 = x$

It is not difficult to find the solution for an equation such as $x + 3 = 8$ by guessing the answer to the question "What plus 3 is 8?" Here the answer to the question is 5, which is also the solution for the equation. But for more complicated equations we need something more than guesswork. A better method is to transform the given equation to an *equivalent equation* whose solution can be found by inspection.

Definition

Equivalent Equations

Equations that have exactly the same solutions are called **equivalent equations**.

NOTE

In some cases we write the equation in the form

$\square = x$

The number will be our solution when the equation has the variable isolated on either side.

The following are all equivalent equations:

$2x + 3 = 5 \qquad 2x = 2 \qquad$ and $\qquad x = 1$

They all have the same solution, 1. We say that a linear equation is *solved* when it is transformed to an equivalent equation of the form

$x = \square$

↑ The variable is alone on the left side.   The right side is some number, the solution.

The *addition property of equality* is the first property you need to transform an equation to an equivalent form.

Property

The Addition Property of Equality

If $a = b$

then $a + c = b + c$

In words, adding the same quantity to both sides of an equation gives an equivalent equation.

RECALL

An equation is a statement that the two sides are equal. Adding the same quantity to both sides does not change the equality or "balance."

Recall that we said that a true equation was like a scale in balance.

The Streeter/Hutchison Series in Mathematics   Beginning Algebra

The addition property is equivalent to adding the same weight to both sides of the scale. It remains in balance.

NOTE

This scale represents

$a + c = b + c$

 Example 3   Using the Addition Property to Solve an Equation

< Objective 3 >

Solve

$x - 3 = 9$

Remember that our goal is to isolate x on one side of the equation. Because 3 is being subtracted from x, we can add 3 to remove it. We must use the addition property to add 3 to both sides of the equation.

NOTE

To check, replace x with 12 in the original equation:

$x - 3 \overset{?}{=} 9$

$(12) - 3 \overset{?}{=} 9$

$9 = 9$

Because we have a true statement, 12 is the solution.

$$
\begin{array}{rcl}
x - 3 &=& 9 \\
+\ 3 & & +3 \\
\hline
x & =& 12
\end{array}
$$

Adding 3 "undoes" the subtraction and leaves x alone on the left.

Because 12 is the solution for the equivalent equation $x = 12$, it is the solution for our original equation.

 Check Yourself 3

Solve and check.

x − 5 = 4

The addition property also allows us to add a negative number to both sides of an equation. This is really the same as subtracting the same quantity from both sides.

 Example 4   Using the Addition Property to Solve an Equation

Solve

$x + 5 = 9$

In this case, 5 is *added* to x on the left. We can use the addition property to add a -5 to both sides. Because $5 + (-5) = 0$, this "undoes" the addition and leaves the variable x alone on one side of the equation.

$$
\begin{array}{rcl}
x + 5 &=& 9 \\
-\ 5 & & -5 \\
\hline
x & =& 4
\end{array}
$$

RECALL

Earlier, we stated that we could write an equation in the equivalent forms $x = \square$ or $\square = x$, in which $\square$ represents some number. Suppose we have an equation like

$12 = x + 7$

Adding -7 isolates x on the right:

$$\begin{array}{r} 12 = x + 7 \\ -7 \quad\quad -7 \\ \hline 5 = x \end{array}$$

The solution is 5.

The solution is 4. To check, replace x with 4:

$(4) + 5 = 9$ (True)

Check Yourself 4

Solve and check.

$x + 6 = 13$

What if the equation has a variable term on both sides? We have to use the addition property to add or subtract a term involving the variable to get the desired result.

▶   **Example 5**   Using the Addition Property to Solve an Equation

Solve

$5x = 4x + 7$

We start by subtracting $4x$ from both sides of the equation. Do you see why? Remember that an equation is solved when we have an equivalent equation of the form $x = \square$.

RECALL

Subtracting $4x$ is the same as adding $-4x$.

$$\begin{array}{r} 5x = \quad 4x + 7 \\ -4x \quad -4x \\ \hline x = \quad\quad 7 \end{array}$$

⎰ Subtracting $4x$ from
⎱ both sides *removes*
  $4x$ from the right.

To check: Because 7 is a solution for the equivalent equation $x = 7$, it should be a solution for the original equation. To find out, replace x with 7.

$$5(7) \stackrel{?}{=} 4(7) + 7$$
$$35 \stackrel{?}{=} 28 + 7$$
$$35 = 35 \quad\quad \text{(True)}$$

Check Yourself 5

Solve and check.

$7x = 6x + 3$

You may have to apply the addition property more than once to solve an equation. Look at Example 6.

▶   **Example 6**   Using the Addition Property to Solve an Equation

Solve

$7x - 8 = 6x$

NOTE

We could add 8 to both sides, and then subtract $6x$. However, we find it easiest to bring the variable terms to one side first, and then work with the constant (or numerical) term.

We want all variables on *one* side of the equation. If we choose the left, we subtract $6x$ from both sides of the equation. This removes the $6x$ from the right:

$$\begin{array}{rcr} 7x - 8 = & & 6x \\ -6x & & -6x \\ \hline x - 8 = & & 0 \end{array}$$

We want the variable alone, so we add 8 to both sides. This isolates x on the left.

$$\begin{array}{rcr} x - 8 = & & 0 \\ +8 & & +8 \\ \hline x & = & 8 \end{array}$$

The solution is 8. We leave it to you to check this result.

Check Yourself 6

Solve and check.

$9x + 3 = 8x$

Often an equation has more than one variable term *and* more than one number. You have to apply the addition property twice to solve these equations.

Example 7   Using the Addition Property to Solve an Equation

Solve

$5x - 7 = 4x + 3$

We would like the variable terms on the left, so we start by subtracting $4x$ from both sides of the equation:

$$\begin{array}{rcr} 5x - 7 = & & 4x + 3 \\ -4x & & -4x \\ \hline x - 7 = & & 3 \end{array}$$

Now, to isolate the variable, we add 7 to both sides.

NOTE

You could just as easily have added 7 to both sides and *then* subtracted $4x$. The result would be the same. In fact, some students prefer to combine the two steps.

$$\begin{array}{rcr} x - 7 = & & 3 \\ +7 & & +7 \\ \hline x & = & 10 \end{array}$$

The solution is 10. To check, replace x with 10 in the original equation:

$$5(10) - 7 \overset{?}{=} 4(10) + 3$$
$$43 = 43 \qquad \text{(True)}$$

Check Yourself 7

Solve and check.

(a) $4x - 5 = 3x + 2$ (b) $6x + 2 = 5x - 4$

In solving an equation, you should always simplify each side as much as possible before using the addition property.

 Example 8   **Simplifying an Equation**

Solve

$$\underset{\text{Like terms}}{5 + 8x - 2} = \underset{\text{Like terms}}{2x - 3 + 5x}$$

Because like terms appear on each side of the equation, we start by combining the numbers on the left (5 and -2). Then we combine the like terms ($2x$ and $5x$) on the right. We have

$$3 + 8x = 7x - 3$$

Now we can apply the addition property, as before.

$$
\begin{array}{rcl}
3 + 8x = & 7x - 3 & \\
\underline{-7x = -7x} & & \text{Subtract } 7x. \\
3 + x = & -3 & \\
\underline{-3 \qquad\qquad -3} & & \text{Subtract 3 to isolate } x. \\
x = & -6 &
\end{array}
$$

The solution is -6. To check, always return to the original equation. That catches any possible errors in simplifying. Replacing x with -6 gives

$$5 + 8(-6) - 2 \overset{?}{=} 2(-6) - 3 + 5(-6)$$
$$5 - 48 - 2 \overset{?}{=} -12 - 3 - 30$$
$$-45 = -45 \qquad \text{(True)}$$

 Check Yourself 8

Solve and check.

(a) $3 + 6x + 4 = 8x - 3 - 3x$ **(b)** $5x + 21 + 3x = 20 + 7x - 2$

We may have to apply some of the properties discussed in Section 1.1 in solving equations. Example 9 illustrates the use of the distributive property to clear an equation of parentheses.

Example 9   **Using the Distributive Property and Solving Equations**

< Objective 4 >

Solve

$$2(3x + 4) = 5x - 6$$

Applying the distributive property on the left, we have

$$6x + 8 = 5x - 6$$

NOTE

$2(3x + 4)$

$= 2(3x) + 2(4)$

$= 6x + 8$

We can then proceed as before:

$$6x + 8 = 5x - 6$$
$$\underline{-5x \qquad\quad -5x} \qquad \text{Subtract } 5x.$$
$$x + 8 = \qquad - 6$$
$$\underline{\quad -8 \qquad\qquad - 8} \qquad \text{Subtract } 8.$$
$$x \qquad = \qquad -14$$

The solution is -14. We leave it to you to check this result.

Remember: Always return to the original equation to check.

Check Yourself 9

Solve and check each of the following equations.

(a) $4(5x - 2) = 19x + 4$ (b) $3(5x + 1) = 2(7x - 3) - 4$

Given an expression such as

$-2(x - 5)$

the distributive property can be used to create the equivalent expression

$-2x + 10$

The distribution of a negative number is shown in Example 10.

Example 10   **Distributing a Negative Number**

Solve each of the following equations.

(a) $-2(x - 5) = -3x + 2$

$$-2x + 10 = -3x + 2 \qquad \text{Distribute the } -2.$$
$$\underline{+3x \qquad\qquad +3x} \qquad\quad \text{Add } 3x.$$
$$x + 10 = \qquad\quad 2$$
$$\underline{\quad -10 = \qquad - 10} \qquad\quad \text{Subtract } 10.$$
$$x \qquad = \qquad -8$$

(b) $-3(3x + 5) = -5(2x - 2)$

$$-9x - 15 = -5(2x - 2) \qquad \text{Distribute the } -3.$$
$$-9x - 15 = -10x + 10 \qquad \text{Distribute the } -5.$$
$$\underline{+10x \qquad\qquad +10x} \qquad\quad \text{Add } 10x.$$
$$x - 15 = \qquad\quad 10$$
$$\underline{\quad + 15 \qquad\qquad + 15} \qquad\quad \text{Add } 15.$$
$$x \qquad = \qquad 25$$

Check Yourself 10

Solve each of the following.

(a) $-2(x - 3) = -x + 5$ (b) $-4(2x - 1) = -3(3x + 2)$

When parentheses are preceded only by a negative, or by the minus sign, we say that we have a silent -1. Example 11 illustrates this case.

Example 11   Distributing the Silent -1

Solve

$$-(2x + 3) = -3x + 7$$
$$-1(2x + 3) = -3x + 7$$
$$(-1)(2x) + (-1)(3) = -3x + 7 \qquad \text{Distribute the } -1.$$

$$
\begin{aligned}
-2x - 3 &= -3x + 7 \\
+3x \qquad\quad & +3x \qquad\qquad \text{Add } 3x. \\
\hline
x - 3 &= 7 \\
+ 3 \quad & + 3 \qquad\qquad \text{Add } 3. \\
\hline
x \quad &= 10
\end{aligned}
$$

Check Yourself 11

Solve $-(3x + 2) = -2x - 6$.

Of course, there are many applications that require us to use the addition property to solve an equation. Consider the consumer application in the next example.

Example 12   A Consumer Application

NOTE

Applications should always be answered with a full sentence.

An appliance store is having a sale on washers and dryers. They are charging $999 for a washer and dryer combination. If the washer sells for $649, how much is a customer paying for the dryer as part of the combination?

Let d be the cost of the dryer and solve the equation $d + 649 = 999$ to answer the question.

$$
\begin{aligned}
d + 649 &= 999 \\
- 649 \quad & -649 \qquad\qquad \text{Subtract 649 from both sides.} \\
\hline
d \qquad &= 350
\end{aligned}
$$

The dryer adds $350 to the price.

Check Yourself 12

Of 18,540 votes cast in the school board election, 11,320 went to Carla. How many votes did her opponent Marco receive? Who won the election?

Let m be the number of votes Marco received and solve the equation $11,320 + m = 18,540$ in order to answer the questions.

Check Yourself ANSWERS

1. **(a)** 4 is not a solution; **(b)** 6 is a solution
2. **(a)** An equation that is not linear; **(b)** linear equation; **(c)** expression;
(d) linear equation; **(e)** an equation that is not linear   **3.** 9   **4.** 7
5. 3   **6.** −3   **7. (a)** 7; **(b)** −6   **8. (a)** −10; **(b)** −3   **9. (a)** 12;
(b) −13   **10. (a)** 1; **(b)** −10   **11.** 4
12. Marco received 7,220 votes; Carla won the election.

Reading Your Text

The following fill-in-the-blank exercises are designed to ensure that you understand some of the key vocabulary used in this section.

SECTION 2.1

(a) You should do your homework while you are still _____.

(b) An _____ is a mathematical statement that two expressions are equal.

(c) A solution for an equation is a value for the variable that makes the equation a _____ statement.

(d) Equivalent equations have exactly the same _____.

Name _____

Section _____ Date _____

Answers

1. _____ 2. _____

3. _____ 4. _____

5. _____ 6. _____

7. _____ 8. _____

9. _____ 10. _____

11. _____ 12. _____

13. _____ 14. _____

15. _____ 16. _____

17. _____ 18. _____

19. _____ 20. _____

21. _____ 22. _____

23. _____

24. _____

25. _____

26. _____

Basic Skills | Advanced Skills | Vocational-Technical Applications | Calculator/Computer | Above and Beyond | Getting Ready

< Objective 1 >

Is the number shown in parentheses a solution for the given equation?

1. $x + 7 = 12$ (5)

2. $x + 2 = 11$ (8)

3. $x - 15 = 6$ (−21)

4. $x - 11 = 5$ (16)

5. $5 - x = 2$ (4)

6. $10 - x = 7$ (3)

7. $8 - x = 5$ (−3)

8. $5 - x = 6$ (−3)

9. $3x + 4 = 13$ (8)

10. $5x + 6 = 31$ (5)

11. $4x - 5 = 7$ (2)

12. $4x - 3 = 9$ (3)

13. $7 - 3x = 10$ (−1)

14. $4 - 5x = 9$ (−2)

15. $4x - 5 = 2x + 3$ (4)

16. $5x + 4 = 2x + 10$ (4)

17. $x + 3 + 2x = 5 + x + 8$ (5)

18. $5x - 3 + 2x = 3 + x - 12$ (−2)

19. $\dfrac{2}{3}x = 9$ (15)

20. $\dfrac{3}{5}x = 24$ (40)

21. $\dfrac{3}{5}x + 5 = 11$ (10)

22. $\dfrac{2}{3}x + 8 = -12$ (−6)

< Objective 2 >

Label each of the following as an expression, a linear equation, or an equation that is not linear.

23. $2x + 1 = 9$

24. $7x + 14$

25. $2x - 8$

26. $5x - 3 = 12$

27. $2x^2 - 8 = 0$

28. $x + 5 = 13$

29. $2x - 8 = 3$

30. $\dfrac{2}{x} - 4 = 3x$

< Objectives 3–4 >

Solve and check the following equations.

31. $x + 9 = 11$

32. $x - 4 = 6$

33. $x - 5 = -9$

34. $x + 11 = 15$

35. $x - 8 = -10$

36. $x + 5 = 2$

37. $x + 4 = -3$

38. $x - 6 = -5$

39. $17 = x + 11$

40. $x + 7 = 0$

41. $4x = 3x + 4$

42. $7x = 6x - 8$

43. $9x = 8x - 12$

44. $9x = 8x + 5$

45. $6x + 3 = 5x$

46. $12x - 6 = 11x$

47. $7x - 5 = 6x$

48. $9x - 7 = 8x$

49. $2x + 3 = x + 5$ > Videos

50. $5x - 6 = 4x + 2$

51. **CRAFTS** Jeremiah had found 50 bones for a Halloween costume. In order to complete his 62-bone costume, how many more does he need?
    Let b be the number of bones he needs and use the equation $b + 50 = 62$ to solve the problem.

52. **BUSINESS AND FINANCE** Four hundred tickets were sold to the opening of an art exhibit. General admission tickets cost $5.50, whereas students were required to pay only $4.50 for tickets. If total ticket sales were $1,950, how many of each type of ticket were sold?
    Let x be the number of general admission tickets sold and $400 - x$ be the number of student tickets sold. Use the equation $5.5x + 4.5(400 - x) = 1,950$ to solve the problem.

Answers

27. _____

28. _____

29. _____

30. _____

31. _____	32. _____
33. _____	34. _____
35. _____	36. _____
37. _____	38. _____
39. _____	40. _____

41. _____

42. _____

43. _____

44. _____

45. _____

46. _____

47. _____

48. _____

49. _____

50. _____

51. _____

52. _____

Answers

53. _____

54. _____

55. _____

56. _____

57. _____

58. _____

59. _____

60. _____

61. _____

62. _____

63. _____

64. _____

65. _____

66. _____

53. BUSINESS AND FINANCE A shop pays $2.25 for each copy of a magazine and sells the magazines for $3.25 each. If the fixed costs associated with the sale of these magazines are $50 per month, how many must the shop sell in order to realize $175 in profit from the magazines?

Let m be the number of magazines they must sell and use the equation $3.25m - 2.25m - 50 = 175$ to solve the problem.

54. NUMBER PROBLEM The sum of a number and 15 is 22. Find the number.

Let x be the number and solve the equation $x + 15 = 22$ to find the number.

55. Which of the following is equivalent to the equation $5x - 7 = 4x - 12$?

(a) $9x = 19$ **(b)** $x - 7 = -12$
(c) $x = -18$ **(d)** $4x - 5 = 8$

56. Which of the following is equivalent to the equation $12x - 6 = 8x + 14$?

(a) $4x - 6 = 14$ **(b)** $x = 20$
(c) $20x = 20$ **(d)** $4x = 8$

57. Which of the following is equivalent to the equation $7x + 5 = 12x - 10$?

(a) $5x = -15$ **(b)** $7x - 5 = 12x$
(c) $-5 = 5x$ **(d)** $7x + 15 = 12x$

58. Which of the following is equivalent to the equation $8x + 5 = 9x - 4$?

(a) $17x = -9$ **(b)** $x = -9$
(c) $8x + 9 = 9x$ **(d)** $9 = 17x$

*Determine whether each of the following statements is **true** or **false**.*

59. Every linear equation with one variable has no more than one solution.

60. Isolating the variable on the right side of an equation results in a negative solution.

| Basic Skills | **Advanced Skills** | Vocational-Technical Applications | Calculator/Computer | Above and Beyond | Getting Ready |

Solve and check the following equations.

61. $4x - \dfrac{3}{5} = 3x + \dfrac{1}{10}$

62. $5\left(x - \dfrac{3}{4}\right) = 4x + \dfrac{3}{8}$

63. $\dfrac{7}{8}(x - 2) = \dfrac{3}{4} - \dfrac{1}{8}x$ > Videos

64. $\dfrac{5}{6}(3x - 2) = \dfrac{3}{2}(x + 1)$

65. $3x - 0.54 = 2(x - 0.15)$

66. $7x + 0.125 = 6x - 0.289$

67. $6x + 3(x - 0.2789) = 4(2x + 0.3912)$

68. $9x - 2(3x - 0.124) = 2x + 0.965$

69. $5x - 7 + 6x - 9 - x = 2x - 8 + 7x$

70. $5x + 8 + 3x - x + 5 = 6x - 3$

71. $5x - (0.345 - x) = 5x + 0.8713$ **72.** $-3(0.234 - x) = 2(x + 0.974)$

73. $3(7x + 2) = 5(4x + 1) + 17$ **74.** $5(5x + 3) = 3(8x - 2) + 4$

> Videos

75. $\dfrac{5}{4}x - 1 = \dfrac{1}{4}x + 7$ **76.** $\dfrac{7x}{5} + 3 = \dfrac{2x}{5} - 8$

77. $\dfrac{9x}{2} - \dfrac{3}{4} = \dfrac{7x}{2} + \dfrac{5}{4}$ **78.** $\dfrac{11}{3}x + \dfrac{1}{6} = \dfrac{8}{3}x + \dfrac{19}{6}$

Basic Skills | Advanced Skills | Vocational-Technical Applications | Calculator/Computer | **Above and Beyond** | Getting Ready
▲

An algebraic equation is a complete sentence. It has a subject and a predicate. For example, the equation $x + 2 = 5$ can be written in English as "two more than a number is five," or "a number added to two is five."

Write an English version of each of the following equations. Be sure that you write complete sentences and that your sentences express the same idea as the equations. Exchange sentences with another student and see whether each other's sentences result in the same equation.

79. $2x - 5 = x + 1$ **80.** $2(x + 2) = 14$

81. $n + 5 = \dfrac{n}{2} - 6$ **82.** $7 - 3a = 5 + a$

83. Complete the following sentence in your own words. "The difference between $3(x - 1) + 4 - 2x$ and $3(x - 1) + 4 = 2x$ is. . . ."

84. "Surprising Results!" Work with other students to try this experiment. Each person should do the following six steps mentally, not telling anyone else what their calculations are:
   (a) Think of a number. **(b)** Add 7.
   (c) Multiply by 3. **(d)** Add 3 more than the original number.
   (e) Divide by 4. **(f)** Subtract the original number.

Answers

67. _____
68. _____
69. _____
70. _____
71. _____
72. _____
73. _____
74. _____
75. _____
76. _____
77. _____
78. _____
79. _____
80. _____
81. _____
82. _____
83. _____
84. _____

Answers

(a) _____

(b) _____

(c) _____

(d) _____

(e) _____

(f) _____

(g) _____

(h) _____

What number do you end up with? Compare your answer with everyone else's. Does everyone have the same answer? Make sure that everyone followed the directions accurately. How do you explain the results? Algebra makes the explanation clear. Work together to do the problem again, using a variable for the number. Make up another series of computations that yields "surprising results."

| Basic Skills | Advanced Skills | Vocational-Technical Applications | Calculator/Computer | Above and Beyond | **Getting Ready** ▲ |

Multiply.

(a) $\left(\dfrac{1}{3}\right)(3)$

(b) $(-6)\left(-\dfrac{1}{6}\right)$

(c) $(7)\left(\dfrac{1}{7}\right)$

(d) $\left(-\dfrac{1}{4}\right)(-4)$

(e) $\left(\dfrac{3}{5}\right)\left(\dfrac{5}{3}\right)$

(f) $\left(\dfrac{7}{8}\right)\left(\dfrac{8}{7}\right)$

(g) $\left(-\dfrac{4}{7}\right)\left(-\dfrac{7}{4}\right)$

(h) $\left(-\dfrac{6}{11}\right)\left(-\dfrac{11}{6}\right)$

Answers

1. Yes   **3.** No   **5.** No   **7.** No   **9.** No   **11.** No   **13.** Yes
15. Yes   **17.** Yes   **19.** No   **21.** Yes   **23.** Linear equation
25. Expression   **27.** An equation that is not linear   **29.** Linear equation
31. 2   **33.** −4   **35.** −2   **37.** −7   **39.** 6   **41.** 4
43. −12   **45.** −3   **47.** 5   **49.** 2   **51.** 12   **53.** 225
55. (b)   **57.** (d)   **59.** True   **61.** $\dfrac{7}{10}$ **63.** $\dfrac{5}{2}$ **65.** 0.24
67. 2.4015   **69.** 8   **71.** 1.2163   **73.** 16   **75.** 8   **77.** 2
79. Above and Beyond   **81.** Above and Beyond   **83.** Above and Beyond
(a) 1   **(b)** 1   **(c)** 1   **(d)** 1   **(e)** 1   **(f)** 1   **(g)** 1   **(h)** 1

2.2

Solving Equations by the Multiplication Property

< 2.2 Objectives >

1 > Use the multiplication property to solve equations

2 > Solve an application involving the multiplication property

Consider a different type of equation. For instance, what if we want to solve the equation

$$6x = 18 \quad ?$$

The addition property of Section 2.1 does not help. We need a second property for solving such equations.

> **Property**
>
> | **The Multiplication Property of Equality** | If $a = b$ then $ac = bc$ with $c \neq 0$ |
> |---|---|
> | | In words, multiplying both sides of an equation by the same nonzero number produces an equivalent equation. |

RECALL

As long as you do the *same* thing to *both* sides of the equation, the "balance" is maintained.

Again, we return to the image of the balance scale. We start with the assumption that a and b have the same weight.

The multiplication property tells us that the scale will be in balance as long as we have the same number of "a weights" as we have of "b weights."

NOTE

The scale represents the equation $5a = 5b$.

We work through some examples, using this second rule.

Example 1   Solving Equations Using the Multiplication Property

Beginning Algebra   The Streeter/Hutchison Series in Mathematics   © The McGraw-Hill Companies. All Rights Reserved.

< Objective 1 >

NOTE

Multiplying both sides by $\frac{1}{6}$ is equivalent to dividing both sides by 6.

NOTE

$\frac{1}{6}(6x) = \left(\frac{1}{6}\cdot 6\right)x$

$= 1\cdot x$ or x

We then have x alone on the left, which is what we want.

Solve

$6x = 18$

Here the variable x is multiplied by 6. So we apply the multiplication property and multiply both sides by $\frac{1}{6}$. Keep in mind that we want an equation of the form

$x = \square$

$\frac{1}{6}(6x) = \frac{1}{6}(18)$

We can now simplify.

$1\cdot x = 3$ or $x = 3$

The solution is 3. To check, replace x with 3:

$6(3) \stackrel{?}{=} 18$

$18 = 18$ (True)

 Check Yourself 1

Solve and check.

$8x = 32$

In Example 1 we solved the equation by multiplying both sides by the reciprocal of the coefficient of the variable.

Example 2 illustrates a slightly different approach to solving an equation by using the multiplication property.

 Example 2   Solving Equations Using the Multiplication Property

NOTE

Because division is defined in terms of multiplication, we can also divide both sides of an equation by the same nonzero number.

Solve

$5x = -35$

The variable x is multiplied by 5. We *divide* both sides by 5 to "undo" that multiplication:

$\frac{5x}{5} = \frac{-35}{5}$ This is the same as multiplying by $\frac{1}{5}$.

$x = -7$ Note that the right side reduces to -7. Be careful with the rules for signs.

We leave it to you to check the solution.

Check Yourself 2

Solve and check.

$7x = -42$

▶ **Example 3**   **Equations with Negative Coefficients**

Solve

$-9x = 54$

In this case, x is multiplied by -9, so we divide both sides by -9 to isolate x on the left:

$$\frac{-9x}{-9} = \frac{54}{-9}$$

$$x = -6$$

The solution is -6. To check:

$(-9)(-6) \overset{?}{=} 54$

$54 = 54$ (True)

Check Yourself 3

Solve and check.

$-10x = -60$

Example 4 illustrates the use of the multiplication property when fractions appear in an equation.

▶ **Example 4**   **Solving Equations That Contain Fractions**

NOTE

$\dfrac{x}{3} = \dfrac{1}{3}x$

(a) Solve

$$\frac{x}{3} = 6$$

Here x is *divided* by 3. We use multiplication to isolate x.

$$3\left(\frac{x}{3}\right) = 3 \cdot 6$$

$$x = 18$$

⎧ This leaves x alone on the
⎨ left because
⎩ $3\left(\dfrac{x}{3}\right) = \dfrac{3}{1} \cdot \dfrac{x}{3} = \dfrac{x}{1} = x$

To check:

$$\left(\frac{18}{3}\right) \overset{?}{=} 6$$

$$6 = 6 \quad \text{(True)}$$

NOTE

$$\frac{x}{5} = \frac{1}{5}x$$

(b) Solve

$$\frac{x}{5} = -9$$

$$5\left(\frac{x}{5}\right) = 5(-9)$$ Because x is divided by 5, multiply both sides by 5.

$$x = -45$$

The solution is -45. To check, we replace x with -45:

$$\left(\frac{-45}{5}\right) \overset{?}{=} -9$$

$$-9 = -9 \quad \text{(True)}$$

The solution is verified.

Check Yourself 4

Solve and check.

(a) $\dfrac{x}{7} = 3$ **(b)** $\dfrac{x}{4} = -8$

When the variable is multiplied by a fraction that has a numerator other than 1, there are two approaches to finding the solution.

Example 5   **Solving Equations Using Reciprocals**

Solve

$$\frac{3}{5}x = 9$$

One approach is to multiply by 5 as the first step.

$$5\left(\frac{3}{5}x\right) = 5 \cdot 9$$

$$3x = 45$$

Now we divide by 3.

$$\frac{3x}{3} = \frac{45}{3}$$

$$x = 15$$

To check:

$$\frac{3}{5}(15) \overset{?}{=} 9$$

$$9 = 9 \quad \text{(True)}$$

RECALL

$\frac{5}{3}$ is the *reciprocal* of $\frac{3}{5}$, and the product of a number and its reciprocal is just 1! So

$$\left(\frac{5}{3}\right)\left(\frac{3}{5}\right) = 1$$

A second approach combines the multiplication and division steps and is generally more efficient. We multiply by $\frac{5}{3}$.

$$\frac{5}{3}\left(\frac{3}{5}x\right) = \frac{5}{3} \cdot 9$$

$$x = \frac{5}{\cancel{3}} \cdot \frac{\cancel{9}^{3}}{1} = 15$$

So $x = 15$, as before.

Check Yourself 5

Solve and check.

$$\frac{2}{3}x = 18$$

You may have to simplify an equation before applying the methods of this section. Example 6 illustrates this procedure.

 Example 6   **Simplifying an Equation**

RECALL

$3x + 5x = (3 + 5)x$
$\qquad\quad = 8x$

Solve and check:

$$3x + 5x = 40$$

Using the distributive property, we can combine the like terms on the left to write

$$8x = 40$$

We can now proceed as before.

$$\frac{8x}{8} = \frac{40}{8} \qquad \text{Divide by 8.}$$

$$x = 5$$

The solution is 5. To check, we return to the original equation. Substituting 5 for x yields

$$3(5) + 5(5) \stackrel{?}{=} 40$$
$$15 + 25 \stackrel{?}{=} 40$$
$$40 = 40 \qquad \text{(True)}$$

Check Yourself 6

Solve and check.

$$7x + 4x = -66$$

As with the addition property, there are many applications that require us to use the multiplication property.

| **Example 7** | An Application Involving the Multiplication Property |

< Objective 2 >

On her first day on the job in a photography lab, Samantha processed all of the film given to her. The following day, her boss gave her four times as much film to process. Over the two days, she processed 60 rolls of film. How many rolls did she process on the first day?

Let x be the number of rolls Samantha processed on her first day and solve the equation $x + 4x = 60$ to answer the question.

$$x + 4x = 60$$
$$5x = 60 \quad \text{Combine like terms first.}$$
$$\frac{1}{5}(5x) = \frac{1}{5}(60) \quad \text{Multiply by } \frac{1}{5}, \text{ to isolate the variable.}$$
$$x = 12$$

Samantha processed 12 rolls of film on her first day.

 Check Yourself 7

On a recent trip to Japan, Marilyn exchanged $1,200 and received 139,812 yen. What exchange rate did she receive?
Let x be the exchange rate and solve the equation $1,200x = 139,812$ to answer the question (to the nearest hundredth).

 Check Yourself ANSWERS

1. 4   **2.** −6   **3.** 6   **4. (a)** 21; **(b)** −32   **5.** 27   **6.** −6
7. She received 116.51 yen for each dollar.

 Reading Your Text

The following fill-in-the-blank exercises are designed to ensure that you understand some of the key vocabulary used in this section.

SECTION 2.2

(a) Multiplying both sides of an equation by the same nonzero number yields an _____ equation.

(b) Division is defined in terms of _____.

(c) Multiplying an equation by 5 is the same as _____ by $\frac{1}{5}$.

(d) The product of a nonzero number and its _____ is 1.

Basic Skills

2.2 exercises

< Objective 1 >

Solve for x and check your result.

1. $5x = 20$

2. $6x = 30$

3. $8x = 48$

4. $6x = -42$

5. $77 = 11x$

6. $66 = 6x$

7. $4x = -16$

8. $-3x = 27$

9. $-9x = 72$ > Videos

10. $10x = -100$

11. $6x = -54$

12. $-7x = 49$

13. $-5x = -15$

14. $52 = -4x$

15. $-42 = 6x$

16. $-7x = -35$

17. $-6x = -54$

18. $-7x = -42$

19. $\dfrac{x}{2} = 4$

20. $\dfrac{x}{3} = 2$

21. $\dfrac{x}{5} = 3$

22. $\dfrac{x}{8} = 5$

23. $5 = \dfrac{x}{8}$

24. $6 = \dfrac{x}{3}$

25. $\dfrac{x}{5} = -4$

26. $\dfrac{x}{7} = -5$

27. $-\dfrac{x}{3} = 8$ > Videos

28. $-\dfrac{x}{6} = -2$

MathZone

Boost your grade at mathzone.com!
> Practice
  Problems
> NetTutor
> Self-Tests
> e-Professors
> Videos

Name _____

Section _____ Date _____

Answers

1. _____ 2. _____

3. _____ 4. _____

5. _____ 6. _____

7. _____ 8. _____

9. _____ 10. _____

11. _____ 12. _____

13. _____ 14. _____

15. _____ 16. _____

17. _____ 18. _____

19. _____ 20. _____

21. _____ 22. _____

23. _____ 24. _____

25. _____ 26. _____

27. _____ 28. _____

Answers

29. _____

30. _____

31. _____

32. _____

33. _____

34. _____

35. _____

36. _____

37. _____

38. _____

39. _____

40. _____

41. _____

42. _____

43. _____

44. _____

29. $\dfrac{2}{3}x = 0.9$

30. $\dfrac{3}{7}x = 15$

31. $\dfrac{3}{4}x = -15$

32. $\dfrac{3}{5}x = 10 - \dfrac{6}{5}$

33. $-\dfrac{6}{5}x = -18$

34. $5x + 4x = 36$

35. $16x - 9x = -16.1$ > Videos

36. $4x - 2x + 7x = 36$

37. **BUSINESS AND FINANCE** Returning from Mexico City, Sung-A exchanged her remaining 450 pesos for $41.70. What exchange rate did she receive?
   Use the equation $450x = 41.70$ to solve this problem (round to the nearest thousandth).
   chapter 2 > Make the Connection

38. **BUSINESS AND FINANCE** Upon arrival in Portugal, Nicolas exchanged $500 and received 417.35 euros (€). What exchange rate did he receive?
   Use the equation $500x = 417.35$ to solve this problem (round to the nearest hundredth).
   chapter 2 > Make the Connection

39. **SCIENCE AND TECHNOLOGY** On Tuesday, there were twice as many patients in the clinic as on Monday. Over the 2-day period, 48 patients were treated. How many patients were treated on Monday?
   Let p be the number of patients who came in on Monday and use the equation $p + 2p = 48$ to answer the question. > Videos

40. **NUMBER PROBLEM** Two-thirds of a number is 46. Find the number.
   Use the equation $\dfrac{2}{3}x = 46$ to solve the problem.

| Basic Skills | **Advanced Skills** | Vocational-Technical Applications | Calculator/Computer | Above and Beyond | Getting Ready |

Certain equations involving decimals can be solved by the methods of this section. For instance, to solve $2.3x = 6.9$, we use the multiplication property to divide both sides of the equation by 2.3. This isolates x on the left, as desired. Use this idea to solve each of the following equations for x.

41. $3.2x = 12.8$

42. $5.1x = -15.3$

43. $-4.5x = 13.5$

44. $-8.2x = -32.8$

45. $1.3x + 2.8x = 12.3$

46. $2.7x + 5.4x = -16.2$

47. $9.3x - 6.2x = 12.4$

48. $12.5x - 7.2x = -21.2$

Basic Skills | Advanced Skills | **Vocational-Technical Applications** | Calculator/Computer | Above and Beyond | Getting Ready
▲

49. INFORMATION TECHNOLOGY A 50-GB-capacity hard drive contains 30 GB of used space. What percent of the hard drive is full?

50. INFORMATION TECHNOLOGY A compression program reduces the size of files and folders by 36%. If a folder contains 17.5 MB, how large will it be after it is compressed?

51. AUTOMOTIVE TECHNOLOGY It is estimated that 8% of rebuilt alternators do not last through the 90-day warranty period. If a parts store had 6 bad alternators returned during the year, how many did they sell?

52. AGRICULTURAL TECHNOLOGY A farmer sold 2,200 bushels of barley on the futures market. Due to a poor harvest, he was able to make only 94% of his bid. How many bushels did he actually harvest?

Basic Skills | Advanced Skills | Vocational-Technical Applications | **Calculator/Computer** | Above and Beyond | Getting Ready
▲

Use your calculator to solve each equation. Round your answers to the nearest hundredth.

53. $230x = 157$

54. $31x = -15$

55. $-29x = 432$

56. $-141x = -3,467$

57. $23.12x = 94.6$

58. $46.1x = -1$

Basic Skills | Advanced Skills | Vocational-Technical Applications | Calculator/Computer | **Above and Beyond** | Getting Ready
▲

59. Describe the difference between the multiplication property and the addition property for solving equations. Give examples of when to use one property or the other.

60. Describe when you should add a quantity to or subtract a quantity from both sides of an equation as opposed to when you should multiply or divide both sides by the same quantity.

Answers

45. _____

46. _____

47. _____

48. _____

49. _____

50. _____

51. _____

52. _____

53. _____

54. _____

55. _____

56. _____

57. _____

58. _____

59. _____

60. _____

Answers

61. _____

62. _____

(a) _____

(b) _____

(c) _____

(d) _____

(e) _____

(f) _____

(g) _____

(h) _____

Motors, Windings, and More! sells every motor, regardless of type, for $2.50. This vendor also has a deal in which customers can choose whether to receive a mark-down or free shipping. Shipping costs are $1.00 per item. If you do not choose the free shipping option, you can deduct 17.5% from your total order (but not the cost of shipping).

61. If you buy six motors, calculate the total cost for each of the two options. Which option is cheaper?

62. Is one option *always* cheaper than the other? Justify your result.

| Basic Skills | Advanced Skills | Vocational-Technical Applications | Calculator/Computer | Above and Beyond | **Getting Ready** |

Use the distributive property to remove the parentheses in the following expressions.

(a) $2(x - 3)$ **(b)** $3(a + 4)$
(c) $5(2b + 1)$ **(d)** $3(3p - 4)$
(e) $7(3x - 4)$ **(f)** $-4(5x + 4)$
(g) $-3(4x - 3)$ **(h)** $-5(3y - 2)$

Answers

1. 4 **3.** 6 **5.** 7 **7.** -4 **9.** -8 **11.** -9 **13.** 3
15. -7 **17.** 9 **19.** 8 **21.** 15 **23.** 40 **25.** -20
27. -24 **29.** 1.35 **31.** -20 **33.** 15 **35.** -2.3
37. 0.093 dollar for each peso **39.** 16 **41.** 4 **43.** -3 **45.** 3
47. 4 **49.** 60% **51.** 75 **53.** 0.68 **55.** -14.90 **57.** 4.09
59. Above and Beyond **61.** Above and Beyond **(a)** $2x - 6$
(b) $3a + 12$ **(c)** $10b + 5$ **(d)** $9p - 12$ **(e)** $21x - 28$
(f) $-20x - 16$ **(g)** $-12x + 9$ **(h)** $-15y + 10$

2.3

Combining the Rules to Solve Equations

< 2.3 Objectives >

1 > Combine the addition and multiplication properties to solve an equation

2 > Solve equations containing parentheses

3 > Solve equations containing fractions

4 > Recognize identities and contradictions

In each example thus far, either the addition property or the multiplication property was used to solve an equation. Often, finding a solution will require the use of both properties.

Example 1	Solving Equations

< Objective 1 >

> CAUTION

Use the addition property before applying the multiplication property. That is, do not divide by 4 until after you have added 5!

(a) Solve

$$4x - 5 = 7$$

Here x is *multiplied* by 4. The result, $4x$, then has 5 subtracted from it (or -5 added to it) on the left side of the equation. These two operations mean that both properties must be applied to solve the equation.

Because there is only one variable term, we start by adding 5 to both sides:

The first step is to *isolate* the variable term, $4x$, on one side of the equation.

$$
\begin{array}{rl}
4x - 5 =& 7 \\
+\ 5 & +5 \\
\hline
4x\ \ \ \ =& 12
\end{array}
$$

The first step is to *isolate* the variable term, $4x$, on one side of the equation.

We now divide both sides by 4:

$$\frac{4x}{4} = \frac{12}{4} \qquad \text{Next, } isolate \text{ the variable } x.$$

$$x = 3$$

The solution is 3. To check, replace x with 3 in the original equation. Be careful to follow the rules for the order of operations.

$$4(3) - 5 \stackrel{?}{=} 7$$
$$12 - 5 \stackrel{?}{=} 7$$
$$7 = 7 \qquad \text{(True)}$$

213

(b) Solve

$$3x + 8 = -4$$
$$\underline{-8 \qquad -8}$$ Subtract 8 from both sides.
$$3x = -12$$

NOTES

Isolate the variable term, $3x$.

Isolate the variable.

Now divide both sides by 3 to isolate x.

$$\frac{3x}{3} = \frac{-12}{3}$$
$$x = -4$$

The solution is -4. We leave it to you to check this result.

Check Yourself 1

Solve and check.

(a) $6x + 9 = -15$ **(b)** $5x - 8 = 7$

The variable may appear in any position in an equation. Just apply the rules carefully as you try to write an equivalent equation, and you will find the solution.

▶   **Example 2**   **Solving Equations**

NOTE

$\dfrac{-2}{-2} = 1$, so we divide by -2 to isolate x.

Solve

$$3 - 2x = 9$$
$$\underline{-3 -3}$$ First, subtract 3 from both sides.
$$-2x = 6$$

Now divide both sides by -2. This leaves x alone on the left.

$$\frac{-2x}{-2} = \frac{6}{-2}$$
$$x = -3$$

The solution is -3. We leave it to you to check this result.

Check Yourself 2

Solve and check.

$10 - 3x = 1$

You may also have to combine multiplication with addition or subtraction to solve an equation. Consider Example 3.

Example 3   **Solving Equations**

(a) Solve

$$\frac{x}{5} - 3 = 4$$

To get the variable term $\frac{x}{5}$ alone, we first add 3 to both sides.

NOTE

A *variable term* is a term that has a variable as a factor.

$$\begin{array}{r} \frac{x}{5} - 3 = 4 \\ \underline{+3 +3} \\ \frac{x}{5} = 7 \end{array}$$

To undo the division, multiply both sides of the equation by 5.

$$5\left(\frac{x}{5}\right) = 5 \cdot 7$$
$$x = 35$$

The solution is 35. Return to the original equation to check the result.

$$\frac{(35)}{5} - 3 \overset{?}{=} 4$$
$$7 - 3 \overset{?}{=} 4$$
$$4 = 4 \qquad \text{(True)}$$

(b) Solve

$$\frac{2}{3}x + 5 = 13$$
$$\underline{-5 -5} \qquad \text{First, subtract 5 from both sides.}$$
$$\frac{2}{3}x = 8$$

Now multiply both sides by $\frac{3}{2}$, the reciprocal of $\frac{2}{3}$.

$$\frac{3}{2}\left(\frac{2}{3}x\right) = \left(\frac{3}{2}\right)8$$

or

$$x = 12$$

The solution is 12. We leave it to you to check this result.

Check Yourself 3

Solve and check.

(a) $\dfrac{x}{6} + 5 = 3$ (b) $\dfrac{3}{4}x - 8 = 10$

In Section 2.1, you learned how to solve certain equations when the variable appeared on both sides. Example 4 shows you how to extend that work when using the multiplication and addition properties of equality.

| Example 4 | Solving an Equation |

Solve

$$6x - 4 = 3x - 2$$

We begin by bringing all the variable terms to one side. To do this, we subtract $3x$ from both sides of the equation. This removes the variable term from the right side.

$$\begin{array}{rrr} 6x - 4 = & 3x - 2 \\ -3x & -3x \\ \hline 3x - 4 = & -2 \end{array}$$

We now isolate the variable term by adding 4 to both sides.

$$\begin{array}{rcr} 3x - 4 = & -2 \\ +4 & +4 \\ \hline 3x = & 2 \end{array}$$

Finally, divide by 3.

$$\frac{3x}{3} = \frac{2}{3}$$

$$x = \frac{2}{3}$$

Check:

$$6\left(\frac{2}{3}\right) - 4 \stackrel{?}{=} 3\left(\frac{2}{3}\right) - 2$$

$$4 - 4 \stackrel{?}{=} 2 - 2$$

$$0 = 0 \quad \text{(True)}$$

The basic idea is to use the two properties to form an equivalent equation with the x isolated. Here we subtracted $3x$ and then added 4. You can do these steps in either order. Try it for yourself the other way. In either case, the multiplication property is then used as the *last step* in finding the solution.

Check Yourself 4

Solve and check.

$$7x - 5 = 3x + 5$$

Next, we look at two approaches to solving equations in which the coefficient on the right side is greater than the coefficient on the left side.

| Example 5 | Solving an Equation (Two Methods) |

Solve $4x - 8 = 7x + 7$.

Method 1

$$
\begin{array}{rcl}
4x - 8 = & 7x + 7 \\
-7x & -7x \\
\hline
-3x - 8 = & +7 \\
+8 & +8 \\
\hline
-3x = & 15
\end{array}
$$

Bring the variable terms to the same (left) side.

Isolate the variable term.

$$
\frac{-3x}{-3} = \frac{15}{-3}
$$

$$
x = -5
$$

Isolate the variable.

We let you check this result.

To avoid a negative coefficient (in this example, -3), some students prefer a different approach.

This time we work toward having the number on the *left* and the x term on the *right,* or $\square = x$.

Method 2

NOTE

It is usually easier to isolate the variable term on the side that results in a positive coefficient.

$$
\begin{array}{rcl}
4x - 8 = & 7x + 7 \\
-4x & -4x \\
\hline
-8 = & 3x + 7 \\
-7 & -7 \\
\hline
-15 = & 3x
\end{array}
$$

Bring the variable terms to the same (right) side.

Isolate the variable term.

$$
\frac{-15}{3} = \frac{3x}{3}
$$

$$
-5 = x
$$

Isolate the variable.

Because $-5 = x$ and $x = -5$ are equivalent equations, it really makes no difference; the solution is still -5! You can use whichever approach you prefer.

 Check Yourself 5

Solve $5x + 3 = 9x - 21$ by finding equivalent equations of the form $x = \square$ and $\square = x$ to compare the two methods of finding the solution.

It may also be necessary to remove grouping symbols to solve an equation.

Example 6   **Solving Equations That Contain Parentheses**

< Objective 2 >

NOTE

$5(x - 3)$
$= 5(x + (-3))$
$= 5x + (-15)$

Solve.

$$5(x - 3) - 2x = x + 7 \quad \text{Apply the distributive property.}$$
$$5x - 15 - 2x = x + 7 \quad \text{Combine like terms.}$$
$$3x - 15 = x + 7$$

We now have an equation that we can solve by the usual methods. First, bring the variable terms to one side, then isolate the variable term, and finally, isolate the variable.

$$
\begin{array}{rcl}
3x - 15 &=& x + 7 \\
-x && -x \\
\hline
2x - 15 &=& 7 \\
+ 15 && + 15 \\
\hline
\dfrac{2x}{2} &=& \dfrac{22}{2} \\
x &=& 11
\end{array}
$$

Subtract x to bring the variable terms to the same side.

Add 15 to isolate the variable term.

Divide by 2 to isolate the variable.

The solution is 11. To check, substitute 11 for x in the original equation. Again note the use of our rules for the order of operations.

$$5((11) - 3) - 2(11) \stackrel{?}{=} (11) + 7 \quad \text{Simplify terms in parentheses.}$$
$$5 \cdot 8 - 2 \cdot 11 \stackrel{?}{=} 11 + 7 \quad \text{Multiply.}$$
$$40 - 22 \stackrel{?}{=} 11 + 7 \quad \text{Add and subtract.}$$
$$18 = 18 \quad \text{A true statement}$$

 Check Yourself 6

Solve and check.

$7(x + 5) - 3x = x - 7$

We now look at equations that contain fractions with different denominators. To solve an equation involving fractions, the first step is to multiply both sides of the equation by the **least common multiple (LCM)** of all denominators in the equation. Recall that the **LCM** of a set of numbers is the *smallest* number into which all the numbers divide evenly.

 Example 7   **Solving an Equation That Contains Fractions**

< Objective 3 >

Solve

$$\frac{x}{2} - \frac{2}{3} = \frac{5}{6}$$

First, multiply each side by 6, the LCM of 2, 3, and 6.

$$6\left(\frac{x}{2} - \frac{2}{3}\right) = 6\left(\frac{5}{6}\right)$$ Apply the distributive property.

$$6\left(\frac{x}{2}\right) - 6\left(\frac{2}{3}\right) = 6\left(\frac{5}{6}\right)$$ Simplify.

$$3x - 4 = 5$$

Next, isolate the variable term on the left side.

$$3x = 9$$
$$x = 3$$

The solution can be checked by returning to the original equation.

 Check Yourself 7

Solve and check.

$$\frac{x}{4} - \frac{4}{5} = \frac{19}{20}$$

| **Example 8** | **Solving an Equation That Contains Fractions** |

Solve

$$\frac{2x - 1}{5} + 1 = \frac{x}{2}$$

First multiply each side by 10, the LCM of 5 and 2.

$$10\left(\frac{2x - 1}{5} + 1\right) = 10\left(\frac{x}{2}\right)$$ Apply the distributive property on the left and simplify.

$$10\left(\frac{2x - 1}{5}\right) + 10(1) = 10\left(\frac{x}{2}\right)$$

$$2(2x - 1) + 10 = 5x$$

$$4x - 2 + 10 = 5x$$

$$4x + 8 = 5x$$ Next, isolate x on the right side.

$$8 = x$$ The solution to the original equation is 8.

 Check Yourself 8

Solve and check.

$$\frac{3x + 1}{4} - 2 = \frac{x + 1}{3}$$

An equation that is true for any value of x is called an **identity.**

 Example 9   Solving an Equation

< Objective 4 >

NOTE

We could ask the question "For what values of x does $-6 = -6$?"

Solve the equation $2(x - 3) = 2x - 6$.

$$2(x - 3) = 2x - 6$$
$$2x - 6 = 2x - 6$$
$$\underline{-2x \qquad -2x}$$
$$-6 = -6$$

The statement $-6 = -6$ is true for any value of x. The original equation is an identity. This means that all real numbers are solutions.

 Check Yourself 9

Solve the equation $3(x - 4) - 2x = x - 12$.

There are also equations for which there are no solutions. We call such equations contradictions.

 Example 10   Solving an Equation

NOTE

We could ask the question "For what values of x does $-15 = 1$?"

Solve the equation $3(2x - 5) - 4x = 2x + 1$.

$$3(2x - 5) - 4x = 2x + 1$$
$$6x - 15 - 4x = 2x + 1$$
$$2x - 15 = 2x + 1$$
$$\underline{-2x \qquad -2x}$$
$$-15 = 1$$

These two numbers are never equal. The original equation has no solutions.

 Check Yourself 10

Solve the equation $2(x - 5) + x = 3x - 3$.

A series of steps that solve a problem are called an **algorithm.** The following algorithm can be used to solve a linear equation.

Solving Linear Equations

If no variable remains after step 3, determine whether the equation is an identity or whether it has no solutions.

Step 1	Use the distributive property to remove any grouping symbols.
Step 2	Combine like terms on each side of the equation.
Step 3	Add or subtract variable terms to bring the variable term to one side of the equation.
Step 4	Add or subtract numbers to isolate the variable term.
Step 5	Multiply by the reciprocal of the coefficient to isolate the variable.

Check Yourself ANSWERS

1. (a) -4; **(b)** 3 **2.** 3 **3. (a)** -12; **(b)** 24 **4.** $\dfrac{5}{2}$ **5.** 6 **6.** -14

7. 7 **8.** 5 **9.** The equation is an identity, so x can be any real number.
10. There are no solutions.

Reading Your Text

The following fill-in-the-blank exercises are designed to ensure that you understand some of the key vocabulary used in this section.

SECTION 2.3

(a) The first goal for solving an equation is to _____ the variable term on one side of the equation.

(b) Apply the _____ property before applying the multiplication property.

(c) Always return to the _____ equation to check your result.

(d) It is usually easiest to clear the _____ by multiplying both sides by the LCM when solving an equation with unlike fractions.

2.3 exercises

Name _____

Section _____ Date _____

Answers

1. _____	2. _____
3. _____	4. _____
5. _____	6. _____
7. _____	8. _____
9. _____	10. _____
11. _____	12. _____
13. _____	14. _____
15. _____	16. _____
17. _____	18. _____
19. _____	20. _____
21. _____	22. _____
23. _____	24. _____
25. _____	26. _____
27. _____	28. _____

< Objectives 1–3 >

Solve for x and check your result.

1. $3x + 2 = 14$

2. $3x - 1 = 17$

3. $3x - 2 = 7$

4. $7x + 9 = 37$

5. $4x + 7 = 35$

6. $7x - 8 = 13$

7. $2x + 9 = 5$

8. $6x + 25 = -5$

9. $4 - 7x = 18$

10. $8 - 5x = -7$

11. $5 - 3x = 11$

12. $5 - 4x = 25$

 > Videos

13. $\dfrac{x}{2} + 1 = 5$

14. $\dfrac{x}{5} - 3 = 2$

15. $\dfrac{x}{5} - 3 = 4$

16. $\dfrac{x}{5} + 3 = 8$

17. $\dfrac{2}{3}x + 5 = 17$

18. $\dfrac{3}{4}x - 5 = 4$

19. $\dfrac{3}{4}x - 2 = 16$

20. $\dfrac{5}{7}x + 4 = 14$

21. $5x = 2x + 9$

22. $7x = 18 - 2x$

23. $3x = 10 - 2x$

24. $11x = 7x + 20$

25. $9x + 2 = 3x + 38$

26. $8x - 3 = 4x + 17$

27. $4x - 8 = x - 14$

28. $6x - 5 = 3x - 29$

29. $5x + 7 = 2x - 3$

30. $9x + 7 = 5x - 3$

31. $7x - 3 = 9x + 5$

32. $5x - 2 = 8x - 11$

33. $5x + 4 = 7x - 8$

34. $2x + 23 = 6x - 5$

35. $2x - 3 + 5x = 7 + 4x + 2$

36. $8x - 7 - 2x = 2 + 4x - 5$

37. $6x + 7 - 4x = 8 + 7x - 26$

 > Videos

38. $7x - 2 - 3x = 5 + 8x + 13$

39. $9x - 2 + 7x + 13 = 10x - 13$

40. $5x + 3 + 6x - 11 = 8x + 25$

41. $2(x + 3) = 8$

42. $-3(x - 1) = 4(x + 2) + 2$

 > Videos

43. $7(2x - 1) - 5x = x + 25$

44. $9(3x + 2) - 10x = 12x - 7$

< Objective 4 >

45. $5(x + 1) - 4x = x - 5$

 > Videos

46. $-4(2x - 3) = -8x + 5$

47. $6x - 4x + 1 = 12 + 2x - 11$

48. $-2x + 5x - 9 = 3(x - 4) - 5$

49. $-4(x + 2) - 11 = 2(-2x - 3) - 13$

50. $4(-x - 2) + 5 = -2(2x + 7)$

In exercises 51–54, find the length of each side of the figure for the given perimeter.

51.

x $2x - 2$ $x + 2$

$P = 24$ in.

52.

$3x - 4$ x

$P = 32$ cm

53.

$3x - 1$ $3x$ $2x - 1$ $x + 2$

$P = 90$ in.

54.

$4x + 5$ $3x - 2$

$P = 34$ cm

Answers

29. _____ 30. _____

31. _____ 32. _____

33. _____ 34. _____

35. _____ 36. _____

37. _____ 38. _____

39. _____ 40. _____

41. _____ 42. _____

43. _____ 44. _____

45. _____

46. _____

47. _____

48. _____

49. _____

50. _____

51. _____

52. _____

53. _____

54. _____

Answers

55. _____

56. _____

57. _____

58. _____

59. _____

60. _____

61. _____

62. _____

63. _____

64. _____

65. _____

66. _____

67. _____

68. _____

69. _____

70. _____

71. _____

72. _____

| Basic Skills | **Advanced Skills** | Vocational-Technical Applications | Calculator/Computer | Above and Beyond | Getting Ready |

Solve each equation and check your solution.

55. $3x + 2(4x - 3) = 6x - 9$

56. $7x + 3(2x + 5) = 10x + 17$

57. $\dfrac{8}{3}x - 3 = \dfrac{2}{3}x + 15$

 > Videos

58. $\dfrac{12x}{5} + 7 = 31 - \dfrac{3x}{5}$

59. $\dfrac{2x}{5} - \dfrac{x}{3} = \dfrac{7}{15}$

60. $\dfrac{2}{7}x - \dfrac{3}{5}x = \dfrac{6}{35}$

61. $5.3x - 7 = 2.3x + 5$

62. $9.8x + 2 = 3.8x + 20$

63. $\dfrac{5x - 3}{4} - 2 = \dfrac{x}{3}$

64. $\dfrac{6x - 1}{5} - \dfrac{2x}{3} = 3$

65. $3 - (x - 2) = 2x + 1$

66. $4x + 2(3 - 2x) = 4 - 3(2x + 5)$

67. $2(1 - 3x) + 2(5x + 4) = 3 - (4x - 1)$

68. $11x + 5(3 - 2x) = -2(-3x + 2)$

69. $\dfrac{3x + 1}{5} + \dfrac{2x - 2}{3} = x$

70. $\dfrac{1 - x}{4} - \dfrac{2x + 3}{2} = \dfrac{3}{4}$

71. $3x + \dfrac{2x - 3}{3} = \dfrac{3(4x + 1)}{3}$

72. $\dfrac{-3(x - 1)}{2} = \dfrac{2x - 3}{3}$

Answers

73. AGRICULTURAL TECHNOLOGY The estimated yield Y of a field of corn (in bushels per acre) can be found by multiplying the rainfall r, in inches, during the growing season by 16 and then subtracting 15. This relationship can be modeled by the formula

$$Y = 16r - 15$$

If a farmer wants a yield of 159 bushels per acre, then we can write the equation shown to determine the amount of rainfall required.

$$159 = 16r - 15$$

How much rainfall is necessary to achieve a yield of 159 bushels of corn per acre?

74. CONSTRUCTION TECHNOLOGY The number of studs s required to build a wall (with studs spaced 16 inches on center) is equal to one more than $\frac{3}{4}$ times the length w of the wall, in feet. We model this with the formula

$$s = \frac{3}{4}w + 1$$

If a contractor uses 22 studs to build a wall, how long is the wall?

75. ALLIED HEALTH The internal diameter D (in mm) of an endotracheal tube for a child is calculated using the formula

$$D = \frac{t + 16}{4}$$

in which t is the child's age (in years).
How old is a child who requires an endotracheal tube with an internal diameter of 7 mm?

76. MECHANICAL ENGINEERING The number of BTUs required to heat a house is $2\frac{3}{4}$ times the volume of the air in the house (in cubic feet). What is the maximum air volume that can be heated with a 90,000-BTU furnace?

Answers
73.
74.
75.
76.
77.
78.
79.
80.

Use your calculator to solve each equation. Round your answers to the nearest hundredth.

77. $230x - 52 = 191$

78. $321 - 45x = 1{,}021x + 658$

79. $360 - 29(2x + 1) = 2{,}464$

80. $81(x + 26) = 35(86 - 4x)$

Answers

81.

82.

83.

84.

85.

86.

87.

88.

(a)

(b)

(c)

(d)

(e)

(f)

(g)

(h)

81. $23.12x - 34.2 = 34.06$

82. $46.1x + 5.78 = x - 12$

83. $3.2(0.5x - 5.1) = -6.4(9.7x + 15.8)$

84. $x - 11.304(2 - 1.8x) = 2.4x + 3.7$

| Basic Skills | Advanced Skills | Vocational-Technical Applications | Calculator/Computer | **Above and Beyond** ▲ | Getting Ready |

85. Create an equation of the form $ax + b = c$ that has 2 as a solution.

86. Create an equation of the form $ax + b = c$ that has 7 as a solution.

87. The equation $3x = 3x + 5$ has no solution, whereas the equation $7x + 8 = 8$ has zero as a solution. Explain the difference between a solution of zero and no solution.

88. Construct an equation for which every real number is a solution.

| Basic Skills | Advanced Skills | Vocational-Technical Applications | Calculator/Computer | Above and Beyond | **Getting Ready** ▲ |

Divide.

(a) $\dfrac{3b}{3}$

(b) $\dfrac{5x}{5}$

(c) $\dfrac{4xy}{4x}$

(d) $\dfrac{6a^2b}{6a^2}$

(e) $\dfrac{7mn^2}{7n^2}$

(f) $\dfrac{\pi ab}{\pi a}$

(g) $\dfrac{srt}{sr}$

(h) $\dfrac{x^2yz}{x^2z}$

Answers

1. 4   **3.** 3   **5.** 7   **7.** -2 **9.** -2 **11.** -2 **13.** 8

15. 35   **17.** 18   **19.** 24   **21.** 3   **23.** 2   **25.** 6   **27.** -2

29. $-\dfrac{10}{3}$ **31.** -4 **33.** 6   **35.** 4   **37.** 5   **39.** -4 **41.** 1

43. 4   **45.** No solution   **47.** Identity   **49.** Identity

51. 6 in., 8 in., 10 in.   **53.** 12 in., 19 in., 29 in., 30 in.   **55.** $-\dfrac{3}{5}$

57. 9   **59.** 7   **61.** 4   **63.** 3   **65.** $\dfrac{4}{3}$ **67.** $-\dfrac{3}{4}$ **69.** $\dfrac{7}{4}$

71. -6 **73.** $10\dfrac{7}{8}$ in.   **75.** 12 yr old   **77.** 1.06   **79.** -36.78

81. 2.95   **83.** -1.33 **85.** Above and Beyond   **87.** Above and Beyond

(a) b **(b)** x **(c)** y **(d)** b **(e)** m **(f)** b **(g)** t **(h)** y

2.4

Formulas and Problem Solving

< 2.4 Objectives >

1 > Solve a literal equation for one of its variables

2 > Solve an application involving a literal equation

3 > Translate a word phrase to an expression or an equation

4 > Use an equation to solve an application

Formulas are extremely useful tools in any field in which mathematics is applied. Formulas are simply equations that express a relationship between more than one letter or variable. You are no doubt familiar with all kinds of formulas, such as

$$A = \frac{1}{2} bh$$ The area of a triangle

$$I = Prt$$ Interest

$$V = \pi r^2 h$$ The volume of a cylinder

A formula is also called a **literal equation** because it involves several letters or variables. For instance, our first formula or literal equation, $A = \frac{1}{2} bh$, involves the three variables A (for area), b (for base), and h (for height).

Unfortunately, formulas are not always given in the form needed to solve a particular problem. In such cases, we use algebra to change the formula to a more useful equivalent equation solved for a particular variable. The steps used in the process are very similar to those you used in solving linear equations. Consider an example.

| Example 1 | Solving a Literal Equation Involving a Triangle |

< Objective 1 >

Suppose that we know the area A and the base b of a triangle and want to find its height h.

We are given

$$A = \frac{1}{2} bh$$

NOTE

$$2\left(\frac{1}{2} bh\right) = \left(2 \cdot \frac{1}{2}\right)(bh)$$
$$= 1(bh)$$
$$= bh$$

We need to find an equivalent equation with h, the unknown, by itself on one side. We can think of $\frac{1}{2} b$ as the **coefficient** of h. We can remove the two *factors* of that coefficient, $\frac{1}{2}$ and b, separately.

$$2A = 2\left(\frac{1}{2} bh\right)$$ Multiply both sides by 2 to clear the equation of fractions.

227

or

$$2A = bh$$

$$\frac{2A}{b} = \frac{bh}{b} \qquad \text{Divide by } b \text{ to isolate } h.$$

$$\frac{2A}{b} = h$$

or

$$h = \frac{2A}{b} \qquad \text{Reverse the sides to write } h \text{ on the left.}$$

NOTE

Here ☐ means an expression containing all the numbers or variables *other* than *h*.

We now have the height *h* in terms of the area *A* and the base *b*. This is called **solving the equation for *h*** and means that we are rewriting the formula as an equivalent equation of the form

$$h = \square.$$

Check Yourself 1

Solve $V = \frac{1}{3}Bh$ for *h*.

You have already learned the methods needed to solve most literal equations or formulas for some specified variable. As Example 1 illustrates, the rules of Sections 2.1 and 2.2 are applied in exactly the same way as they were applied to equations with one variable.

You may have to apply both the addition and the multiplication properties when solving a formula for a specified variable. Example 2 illustrates this process.

| Example 2 | Solving a Literal Equation |

(a) Solve $y = mx + b$ for *x*.

Remember that we want to end up with *x* alone on one side of the equation. Start by subtracting *b* from both sides to "undo" the addition on the right.

$$\begin{array}{rcl} y & = & mx + b \\ \underline{-b} & & \underline{-b} \\ y - b & = & mx \end{array}$$

If we now divide both sides by *m*, then *x* will be alone on the right-hand side.

$$\frac{y - b}{m} = \frac{mx}{m}$$

$$\frac{y - b}{m} = x$$

or

$$x = \frac{y - b}{m}$$

The Streeter/Hutchison Series in Mathematics   Beginning Algebra

(b) Solve $3x + 2y = 12$ for y.

Begin by isolating the y term.

$$\begin{array}{rl} 3x + 2y = & 12 \\ \underline{-3x \qquad\quad -3x} & \\ 2y = & -3x + 12 \end{array}$$

Then, isolate y by dividing by its coefficient.

$$\frac{2y}{2} = \frac{-3x + 12}{2}$$

$$y = \frac{-3x + 12}{2}$$

Often, in a situation like this, we use the distributive property to separate the terms on the right-hand side of the equation.

$$y = \frac{-3x + 12}{2}$$

$$= \frac{-3x}{2} + \frac{12}{2}$$

$$= \frac{-3}{2}x + 6 = -\frac{3}{2}x + 6$$

NOTE

v and v_0 represent distinct quantities.

Check Yourself 2

(a) Solve $v = v_0 + gt$ for t.   **(b)** Solve $4x - 3y = 8$ for x.

Here is a summary of the steps illustrated by our examples.

Step by Step

Solving a Formula or Literal Equation		
Step 1	If necessary, multiply both sides of the equation by the same term to clear it of fractions.
Step 2	Add or subtract the same term on each side of the equation so that all terms involving the variable that you are solving for are on one side of the equation and all other terms are on the other side.
Step 3	Divide both sides of the equation by the coefficient of the variable that you are solving for.

Look at one more example using these steps.

Example 3   **Solving a Literal Equation Involving Money**

NOTE

This is a formula for the *amount* of money in an account after interest has been earned.

Solve $A = P + Prt$ for r.

$$\begin{array}{rl} A \quad = & P + Prt \\ \underline{-P \quad -P} & \\ A - P = & Prt \end{array}$$

Subtracting P from both sides leaves the term involving r alone on the right.

$$\frac{A - P}{Pt} = \frac{Prt}{Pt}$$ Dividing both sides by Pt isolates r on the right.

$$\frac{A - P}{Pt} = r$$

or

$$r = \frac{A - P}{Pt}$$

Check Yourself 3

Solve $2x + 3y = 6$ for y.

Now look at an application of solving a literal equation.

Example 4   **Solving a Literal Equation Involving Money**

< Objective 2 >

Suppose that the amount in an account, 3 years after a principal of $5,000 was invested, is $6,050. What was the interest rate?

From Example 3,

$$A = P + Prt$$

in which A is the amount in the account, P is the principal, r is the interest rate, and t is the time that the money has been invested. By the result of Example 3 we have

$$r = \frac{A - P}{Pt}$$

and we can substitute the known values into this equation.

NOTE

Do you see the advantage of having our equation solved for the desired variable?

$$r = \frac{(6,050) - (5,000)}{(5,000)(3)}$$

$$= \frac{1,050}{15,000} = 0.07 = 7\%$$ The interest rate was 7%.

Check Yourself 4

Suppose that the amount in an account, 4 years after a principal of $3,000 was invested, is $3,480. What was the interest rate?

The main reason for learning how to set up and solve algebraic equations is so that we can use them to solve word problems and applications. In fact, algebraic equations were *invented* to make solving word problems much easier. The first word problems that we know about are over 4,000 years old. They were literally "written in stone," on Babylonian tablets, about 500 years before the first algebraic equation made its appearance.

Before algebra, people solved word problems primarily by "guess-and-check," which is a method of finding unknown numbers by using trial and error in a logical way. Example 5 shows how to solve a word problem using this method. We refer to this method as substitution.

Example 5	Solving a Word Problem by Substitution

The sum of two consecutive integers is 37. Find the two integers.

If the two integers were 20 and 21, their sum would be 41, which is more than 37, so the integers must be smaller. If the integers were 15 and 16, the sum would be 31. More trials yield that the sum of 18 and 19 is 37.

Check Yourself 5

The sum of two consecutive integers is 91. Find the two integers.

Most word problems are not so easily solved by the guess-and-check method. For more complicated word problems, we use a five-step procedure. This step-by-step approach will, with practice, allow you to organize your work. Organization is the key to solving word problems. Here are the five steps.

Step by Step

To Solve Word Problems

Step 1   Read the problem carefully. Then reread it to decide what you are asked to find.

Step 2   Choose a letter to represent one of the unknowns in the problem. Then represent all other unknowns of the problem with expressions that use the same letter.

Step 3   Translate the problem to the language of algebra to form an equation.

Step 4   Solve the equation and answer the question of the original problem using a complete sentence and appropriate units.

Step 5   Check your solution by returning to the original problem.

The third step is usually the hardest part. We must translate words to the language of algebra. Before we look at a complete example, the following table may help you review that translation step.

Translating Words to Algebra

Words	Algebra
The sum of x and y	$x + y$
3 plus a	$3 + a$ or $a + 3$
5 more than m	$m + 5$
b increased by 7	$b + 7$
The difference between x and y	$x - y$
4 less than a	$a - 4$
s decreased by 8	$s - 8$
The product of x and y	$x \cdot y$ or xy
5 times a	$5 \cdot a$ or $5a$
Twice m	$2m$
The quotient of x and y	$\dfrac{x}{y}$
a divided by 6	$\dfrac{a}{6}$
One-half of b	$\dfrac{b}{2}$ or $\dfrac{1}{2}b$

Here are some typical examples of translating phrases to algebra to help you review.

▶ **Example 6**	**Translating Statements**

< Objective 3 >

Translate each statement to an algebraic expression.

(a) The sum of a and twice b $a + 2b$

                Sum   Twice b

(b) 5 times m increased by 1 $5m + 1$

              5 times m Increased by 1

(c) 5 less than 3 times x $3x - 5$

           3 times x 5 less than

(d) The product of x and y, divided by 3 $\dfrac{xy}{3}$

                              The product of x and y

                              Divided by 3

 Check Yourself 6

Translate to algebra.

(a) 2 more than twice x **(b)** 4 less than 5 times n

(c) The product of twice a and b **(d)** The sum of s and t, divided by 5

Now we work through a complete example. Although this problem could be solved by substitution, it is presented here to help you practice the five-step approach.

| Example 7 | Solving an Application |

< Objective 4 >

NOTE

The word *is* usually translates into an equal sign, $=$.

NOTE

Always return to the *original problem* to check your result and *not* to the equation of step 3. This prevents many errors!

The sum of a number and 5 is 17. What is the number?

Step 1   *Read carefully.*   You must find the unknown number.

Step 2   *Choose letters or variables.*   Let x represent the unknown number. There are no other unknowns.

Step 3   *Translate.*

The sum of
$$x + 5 = 17$$
is

Step 4   *Solve.*

$$
\begin{array}{r}
x + 5 = 17 \\
-5 \quad -5 \qquad \text{Subtract 5.} \\
\hline
x = 12
\end{array}
$$

So the number is 12.

Step 5   *Check.*   Is the sum of 12 and 5 equal to 17? Yes ($12 + 5 = 17$).

 Check Yourself 7

The sum of a number and 8 is 35. What is the number?

Property

Consecutive Integers

Consecutive integers are integers that follow one another, such as 10, 11, and 12. To represent them in algebra:

If x is an integer, then $x + 1$ is the next consecutive integer, $x + 2$ is the next, and so on.

We need this idea in Example 8.

| Example 8 | Solving an Application |

The sum of two consecutive integers is 41. What are the two integers?

Step 1   We want to find the two consecutive integers.

Step 2   Let x be the first integer. Then $x + 1$ must be the next.

RECALL

Read the problem carefully. What do you need to find? Assign letters to the unknown or unknowns. Write an equation.

Step 3

The first integer   The second integer

$$x + (x + 1) = 41$$

The sum   Is

Step 4

$$x + x + 1 = 41$$
$$2x + 1 = 41$$
$$2x = 40$$
$$x = 20$$

The first integer (x) is 20, and the next integer ($x + 1$) is 21.

Step 5   The sum of the two integers 20 and 21 is 41.

Check Yourself 8

The sum of three consecutive integers is 51. What are the three integers?

Sometimes algebra is used to reconstruct missing information. Example 9 does just that with some election information.

Example 9	Solving an Application

There were 55 more yes votes than no votes on an election measure. If 735 votes were cast in all, how many yes votes were there? How many no votes?

Step 1   We want to find the number of yes votes and the number of no votes.

Step 2   Let x be the number of no votes. Then

$$x + 55$$

55 more than x

is the number of yes votes.

NOTES

Solve the equation.

Check.

NOTES

What do you need to find?

Assign letters to the unknowns.

NOTES

Write an equation.

Solve the equation.

Check.

Step 3

$$x + x + 55 = 735$$

No votes   Yes votes

Step 4

$$x + x + 55 = 735$$
$$2x + 55 = 735$$
$$2x = 680$$
$$x = 340$$
$$\text{No votes } (x) = 340$$
$$\text{Yes votes } (x + 55) = 395$$

Step 5   340 no votes plus 395 yes votes equals 735 total votes. The solution checks.

Check Yourself 9

Francine earns $120 per month more than Rob. If they earn a total of $2,680 per month, what are their monthly salaries?

Similar methods allow you to solve a variety of word problems. Example 10 includes three unknown quantities but uses the same basic solution steps.

| Example 10 | Solving an Application |

NOTE

There are other choices for x, but choosing the smallest quantity usually gives the easiest equation to write and solve.

Juan worked twice as many hours as Jerry. Marcia worked 3 more hours than Jerry. If they worked a total of 31 hours, find out how many hours each worked.

Step 1   We want to find the hours each worked, so there are three unknowns.

Step 2   Let x be the hours that Jerry worked.

Twice Jerry's hours

Then $2x$ is Juan's hours worked

3 more hours than Jerry worked

and $x + 3$ is Marcia's hours.

Step 3

Jerry   Juan   Marcia

$$x \; + \; 2x \; + \; (x + 3) = 31$$

Sum of their hours

Step 4

$$x + 2x + x + 3 = 31$$
$$4x + 3 = 31$$
$$4x = 28$$
$$x = 7$$

Jerry's hours $(x) = 7$

Juan's hours $(2x) = 14$

Marcia's hours $(x + 3) = 10$

Step 5   The sum of their hours $(7 + 14 + 10)$ is 31, and the solution is verified.

 Check Yourself 10

> Paul jogged half as many miles (mi) as Lucy and 7 less than Isaac. If the three ran a total of 23 mi, how far did each person run?

 Check Yourself ANSWERS

1. $h = \dfrac{3V}{B}$ **2.** (a) $t = \dfrac{v - v_0}{g}$; (b) $x = \dfrac{3}{4}y + 2$

3. $y = \dfrac{6 - 2x}{3}$ or $y = -\dfrac{2}{3}x + 2$ **4.** 4%   **5.** 45 and 46

6. (a) $2x + 2$; (b) $5n - 4$; (c) $2ab$; (d) $\dfrac{s + t}{5}$

7. The equation is $x + 8 = 35$. The number is 27.

8. The equation is $x + x + 1 + x + 2 = 51$. The integers are 16, 17, and 18.

9. The equation is $x + x + 120 = 2,680$. Rob's salary is \$1,280 and Francine's is \$1,400.   **10.** Paul: 4 mi; Lucy: 8 mi; Isaac: 11 mi

---

Reading Your Text ◀

The following fill-in-the-blank exercises are designed to ensure that you understand some of the key vocabulary used in this section.

SECTION 2.4

(a) _____ are also called literal equations.

(b) Isolating one variable in a formula is also called _____ the equation for that variable.

(c) The first step to solving an application is to _____ the problem carefully.

(d) In an application, the word *is* usually translates to an _____ sign in algebra.

< Objective 1 >

Solve each literal equation for the indicated variable.

1. $P = 4s$ (for s)   Perimeter of a square

2. $V = Bh$ (for B)   Volume of a prism

3. $E = IR$ (for R)   Voltage in an electric circuit

4. $I = Prt$ (for r)   Simple interest

5. $V = LWH$ (for H)   Volume of a rectangular solid

6. $V = \pi r^2 h$ (for h)   Volume of a cylinder

7. $A + B + C = 180$ (for B)   Measure of angles in a triangle

8. $P = I^2 R$ (for R)   Power in an electric circuit

9. $ax + b = 0$ (for x)   Linear equation in one variable

10. $y = mx + b$ (for m)   Slope-intercept form for a line

11. $s = \dfrac{1}{2}gt^2$ (for g)   Distance

12. $K = \dfrac{1}{2}mv^2$ (for m)   Energy

13. $x + 5y = 15$ (for y)   Linear equation in two variables

14. $2x + 3y = 6$ (for x)   Linear equation in two variables

15. $P = 2L + 2W$ (for L)   Perimeter of a rectangle

16. $ax + by = c$ (for y)   Linear equation in two variables

17. $V = \dfrac{KT}{P}$ (for T)   Volume of a gas

18. $V = \dfrac{1}{3}\pi r^2 h$ (for h)   Volume of a cone

19. $x = \dfrac{a + b}{2}$ (for b)   Mean of two numbers

> Videos

Name _____

Section _____ Date _____

Answers

1. _____   2. _____

3. _____   4. _____

5. _____   6. _____

7. _____

8. _____   9. _____

10. _____   11. _____

12. _____

13. _____

14. _____

15. _____

16. _____

17. _____

18. _____

19. _____

Answers

20. _____

21. _____

22. _____

23. _____

24. _____

25. _____

26. _____

27. _____

28. _____

29. _____

30. _____

20. $D = \dfrac{C - s}{n}$ (for s)   Depreciation

21. $F = \dfrac{9}{5}C + 32$ (for C)   Celsius/Fahrenheit

22. $A = P + Prt$ (for t)   Amount at simple interest

23. $S = 2\pi r^2 + 2\pi rh$ (for h)   Total surface area of a cylinder

24. $A = \dfrac{1}{2}h(B + b)$ (for b)   Area of a trapezoid

< Objective 2 >

25. **GEOMETRY** A rectangular solid has a base with length 8 cm and width 5 cm. If the volume of the solid is 120 cm³, find the height of the solid. (See exercise 5.)

26. **GEOMETRY** A cylinder has a radius of 4 in. If the volume of the cylinder is 48π in.³, what is the height of the cylinder? (See exercise 6.)

27. **BUSINESS AND FINANCE** A principal of $3,000 was invested in a savings account for 3 years. If the interest earned for the period was $450, what was the interest rate? (See exercise 4.)

28. **GEOMETRY** If the perimeter of a rectangle is 60 ft and the width is 12 ft, find its length. (See exercise 15.)

29. **SCIENCE AND MEDICINE** The high temperature in New York for a particular day was reported at 77°F. How would the same temperature have been given in degrees Celsius? (See exercise 21.)

30. **CRAFTS** Rose's garden is in the shape of a trapezoid. If the height of the trapezoid is 16 m, one base is 20 m, and the area is 224 m², find the length of the other base. (See exercise 24.)

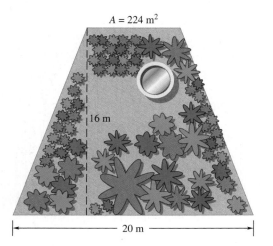

$A = 224$ m²

16 m

20 m

< Objective 3 >

Translate each statement to an algebraic equation. Let x represent the number in each case.

31. 3 more than a number is 7.

32. 5 less than a number is 12.

33. 7 less than 3 times a number is twice that same number. > Videos

34. 4 more than 5 times a number is 6 times that same number.

35. 2 times the sum of a number and 5 is 18 more than that same number.

36. 3 times the sum of a number and 7 is 4 times that same number.

37. 3 more than twice a number is 7.

38. 5 less than 3 times a number is 25.

39. 7 less than 4 times a number is 41.

40. 10 more than twice a number is 44.

41. 5 more than two-thirds of a number is 21.

42. 3 less than three-fourths of a number is 24.

43. 3 times a number is 12 more than that number.

44. 5 times a number is 8 less than that number.

< Objective 4 >

Solve the following word problems. Be sure to label the unknowns and to show the equation you use for the solution.

45. NUMBER PROBLEM The sum of a number and 7 is 33. What is the number?

46. NUMBER PROBLEM The sum of a number and 15 is 22. What is the number?

47. NUMBER PROBLEM The sum of a number and -15 is 7. What is the number?

48. NUMBER PROBLEM The sum of a number and -8 is 17. What is the number?

49. SOCIAL SCIENCE In an election, the winning candidate has 1,840 votes. If the total number of votes cast was 3,260, how many votes did the losing candidate receive?

50. BUSINESS AND FINANCE Mike and Stefanie work at the same company and make a total of $2,760 per month. If Stefanie makes $1,400 per month, how much does Mike earn every month?

Answers

31. _____

32. _____

33. _____

34. _____

35. _____

36. _____

37. _____

38. _____

39. _____

40. _____

41. _____

42. _____

43. _____

44. _____

45. _____

46. _____

47. _____

48. _____

49. _____

50. _____

Answers

51. _____

52. _____

53. _____

54. _____

55. _____

56. _____

57. _____

58. _____

59. _____

60. _____

61. _____

62. _____

63. _____

64. _____

65. _____

66. _____

67. _____

68. _____

51. **NUMBER PROBLEM** The sum of twice a number and 5 is 35. What is the number?

52. **NUMBER PROBLEM** 3 times a number, increased by 8, is 50. Find the number.

53. **NUMBER PROBLEM** 5 times a number, minus 12, is 78. Find the number.

54. **NUMBER PROBLEM** 4 times a number, decreased by 20, is 44. What is the number?

55. **NUMBER PROBLEM** The sum of two consecutive integers is 47. Find the two integers.

56. **NUMBER PROBLEM** The sum of two consecutive integers is 145. Find the two integers.

57. **NUMBER PROBLEM** The sum of three consecutive integers is 63. What are the three integers?

58. **NUMBER PROBLEM** If the sum of three consecutive integers is 93, find the three integers.

> Videos

59. **NUMBER PROBLEM** The sum of two consecutive even integers is 66. What are the two integers? (*Hint:* Consecutive even integers such as 10, 12, and 14 can be represented by x, $x + 2$, $x + 4$, and so on.)

60. **NUMBER PROBLEM** If the sum of two consecutive even integers is 114, find the two integers.

61. **NUMBER PROBLEM** If the sum of two consecutive odd integers is 52, what are the two integers? (*Hint:* Consecutive odd integers such as 21, 23, and 25 can be represented by x, $x + 2$, $x + 4$, and so on.)

62. **NUMBER PROBLEM** The sum of two consecutive odd integers is 88. Find the two integers.

63. **NUMBER PROBLEM** The sum of three consecutive odd integers is 63. What are the three integers?

64. **NUMBER PROBLEM** The sum of three consecutive even integers is 126. What are the three integers?

65. **NUMBER PROBLEM** The sum of four consecutive integers is 86. What are the four integers?

66. **NUMBER PROBLEM** The sum of four consecutive integers is 62. What are the four integers?

67. **NUMBER PROBLEM** 4 times an integer is 9 more than 3 times the next consecutive integer. What are the two integers?

68. **NUMBER PROBLEM** 4 times an integer is 30 less than 5 times the next consecutive even integer. Find the two integers.

69. SOCIAL SCIENCE In an election, the winning candidate had 160 more votes than the loser. If the total number of votes cast was 3,260, how many votes did each candidate receive?

70. BUSINESS AND FINANCE Jody earns $140 more per month than Frank. If their monthly salaries total $2,760, what amount does each earn?

71. BUSINESS AND FINANCE A washer-dryer combination costs $650. If the washer costs $70 more than the dryer, what does each appliance cost?

72. CRAFTS Yuri has a board that is 98 in. long. He wishes to cut the board into two pieces so that one piece will be 10 in. longer than the other. What should the length of each piece be?

73. SOCIAL SCIENCE Yan Ling is 1 year less than twice as old as his sister. If the sum of their ages is 14 years, how old is Yan Ling?

74. SOCIAL SCIENCE Diane is twice as old as her brother Dan. If the sum of their ages is 27 years, how old are Diane and her brother?

75. SOCIAL SCIENCE Maritza is 3 years less than 4 times as old as her daughter. If the sum of their ages is 37, how old is Maritza?

76. SOCIAL SCIENCE Mrs. Jackson is 2 years more than 3 times as old as her son. If the difference between their ages is 22 years, how old is Mrs. Jackson?

Answers

69. _____

70. _____

71. _____

72. _____

73. _____

74. _____

75. _____

76. _____

Answers

77. _____

78. _____

79. _____

80. _____

81. _____

82. _____

83. _____

84. _____

85. _____

77. **BUSINESS AND FINANCE** On her vacation in Europe, Jovita's expenses for food and lodging were $60 less than twice as much as her airfare. If she spent $2,400 in all, what was her airfare?

> chapter 2 > Make the Connection

78. **BUSINESS AND FINANCE** Rachel earns $6,000 less than twice as much as Tom. If their two incomes total $48,000, how much does each earn?

79. **STATISTICS** There are 99 students registered in three sections of algebra. There are twice as many students in the 10 A.M. section as the 8 A.M. section and 7 more students at 12 P.M. than at 8 A.M. How many students are in each section?

80. **BUSINESS AND FINANCE** The Randolphs used 12 more gal of fuel oil in October than in September and twice as much oil in November as in September. If they used 132 gal for the 3 months, how much was used during each month?

Basic Skills | Advanced Skills | **Vocational-Technical Applications** | Calculator/Computer | Above and Beyond | Getting Ready

▲

---

81. **MECHANICAL ENGINEERING** A motor's horsepower (hp) is approximated by the equation

$$hp = \frac{6.2832 \cdot T \cdot (rpm)}{33,000}$$

in which T is the torque of the motor and (rpm) is its revolutions per minute.
Find the rpm required to produce 240 hp in a motor that produces 380 foot-pounds of torque (nearest hundredth).

82. **MECHANICAL ENGINEERING** In a planetary gear, the size and number of teeth must satisfy the equation

$$Cx = By(F - 1)$$

Calculate the number of teeth y needed if $C = 9$ in., $x = 14$ teeth, $B = 2$ in., and $F = 8$.

83. **ELECTRICAL ENGINEERING** Power dissipation, in watts, is given by the quotient of the square of the voltage and the resistance.

   (a) Express the given relationship with a formula.
   (b) Determine the power dissipation when 13.2 volts pass through a 220-Ω resistor (nearest thousandth).

84. **INFORMATION TECHNOLOGY** The total distance around a circular ring network in a metropolitan area is 100 mi. What is the diameter of the ring network (three decimal places)?

85. **ALLIED HEALTH** A patient enters treatment with an abdominal tumor weighing 32 g. Each day, chemotherapy reduces the size of the tumor by 2.33 g. Therefore, a formula to describe the mass m of the tumor after t days of treatment is

$$m = 32 - 2.33t$$

   (a) How much does the tumor weigh after one week of treatment?
   (b) When will the tumor weigh less than 10 g?
   (c) How many days of chemotherapy are required to eliminate the tumor?

86. **ALLIED HEALTH** Yohimbine is used to reverse the effects of xylazine in deer. The recommended dose is 0.125 mg per kilogram of a deer's weight.

   (a) Write a formula that expresses the required dosage level d for a deer of weight w.

   (b) How much yohimbine should be administered to a 15-kg fawn?

   (c) What size deer requires a 5.0-mg dosage?

ELECTRONICS TECHNOLOGY *Temperature sensors output voltage at a certain temperature. The output voltage varies with respect to temperature. For a particular sensor, the output voltage V for a given Celsius temperature C is given by*

$V = 0.28C + 2.2$

87. Determine the output voltage at 0°C.

88. Determine the output voltage at 22°C.

89. Determine the temperature if the sensor outputs 14.8 V.

90. At what temperature is there no voltage output (two decimal places)?

Basic Skills | Advanced Skills | Vocational-Technical Applications | Calculator/Computer | **Above and Beyond** | Getting Ready
▲

91. "I make $2.50 an hour more in my new job." If $x =$ the amount I used to make per hour and $y =$ the amount I now make, which equation(s) below say the same thing as the statement above? Explain your choice(s) by translating the equation into English and comparing with the original statement.

   (a) $x + y = 2.50$

   (b) $x - y = 2.50$

   (c) $x + 2.50 = y$

   (d) $2.50 + y = x$

   (e) $y - x = 2.50$

   (f) $2.50 - x = y$

92. "The river rose 4 feet above flood stage last night." If $a =$ the river's height at flood stage and $b =$ the river's height now (the morning after), which equation(s) below say the same thing as the statement? Explain your choice(s) by translating the equations into English and comparing with the original statement.

   (a) $a + b = 4$

   (b) $b - 4 = a$

   (c) $a - 4 = b$

   (d) $a + 4 = b$

   (e) $b + 4 = b$

   (f) $b - a = 4$

93. Maxine lives in Pittsburgh, Pennsylvania, and pays 8.33 cents per kilowatt hour (kWh) for electricity. During the 6 months of cold winter weather, her household uses about 1,500 kWh of electric power per month. During the two hottest summer months, the usage is also high because the family uses electricity to run an air conditioner. During these summer months, the usage is 1,200 kWh per month; the rest of the year, usage averages 900 kWh per month.

   (a) Write an expression for the total yearly electric bill.

   (b) Maxine is considering spending $2,000 for more insulation for her home so that it is less expensive to heat and to cool. The insulation company

Answers

86. _____

87. _____

88. _____

89. _____

90. _____

91. _____

92. _____

93. _____

The Streeter/Hutchison Series in Mathematics   Beginning Algebra

Answers

claims that "with proper installation the insulation will reduce your heating and cooling bills by 25 percent." If Maxine invests the money in insulation, how long will it take her to get her money back by saving on her electric bill? Write to her about what information she needs to answer this question. Give her your opinion about how long it will take to save $2,000 on heating and cooling bills, and explain your reasoning. What is your advice to Maxine?

| Basic Skills | Advanced Skills | Vocational-Technical Applications | Calculator/Computer | Above and Beyond | **Getting Ready** |

Perform the indicated operations.

(a) $4(8 - 6) + 7$

(b) $3(8 - 5) + 10 \div 2$

(c) $4(6 - 8) \div 4 + 5$

(d) $8(7 - 3 \cdot 4) + 12(5 - 3)$

(e) $-2(16 \div 4 \cdot 2) - 5(3 - 5)$

(f) $8 \cdot (4 \cdot 5 - 2)(10 \div 2 \cdot 5)$

(g) $-7(13 - 4 \cdot 2) \div (25 - 10 \cdot 2)$

(h) $2(-25 + 5 \cdot 3) \cdot (4 - 3 \cdot 2)$

Answers

1. $s = \dfrac{P}{4}$ **3.** $R = \dfrac{E}{I}$ **5.** $H = \dfrac{V}{LW}$ **7.** $B = 180 - A - C$

9. $x = -\dfrac{b}{a}$ **11.** $g = \dfrac{2s}{t^2}$ **13.** $y = \dfrac{15 - x}{5}$ or $y = -\dfrac{1}{5}x + 3$

15. $L = \dfrac{P - 2W}{2}$ or $L = \dfrac{P}{2} - W$ **17.** $T = \dfrac{PV}{K}$ **19.** $b = 2x - a$

21. $C = \dfrac{5}{9}(F - 32)$ or $C = \dfrac{5(F - 32)}{9}$

23. $h = \dfrac{S - 2\pi r^2}{2\pi r}$ or $h = \dfrac{S}{2\pi r} - r$ **25.** 3 cm   **27.** 5%   **29.** 25°C

31. $x + 3 = 7$ **33.** $3x - 7 = 2x$ **35.** $2(x + 5) = x + 18$

37. $2x + 3 = 7$ **39.** $4x - 7 = 41$ **41.** $\dfrac{2}{3}x + 5 = 21$

43. $3x = x + 12$ **45.** $x + 7 = 33; 26$ **47.** $x - 15 = 7; 22$

49. $x + 1,840 = 3,260; 1,420$ **51.** $2x + 5 = 35; 15$ **53.** $5x - 12 = 78; 18$

55. $x + x + 1 = 47; 23, 24$ **57.** $x + x + 1 + x + 2 = 63; 20, 21, 22$

59. $x + x + 2 = 66; 32, 34$ **61.** $x + x + 2 = 52; 25, 27$

63. $x + x + 2 + x + 4 = 63; 19, 21, 23$

65. $x + x + 1 + x + 2 + x + 3 = 86; 20, 21, 22, 23$

67. $4x = 3(x + 1) + 9; 12, 13$ **69.** $x + x + 160 = 3,260; 1,550, 1,710$

71. $x + x + 70 = 650$; Washer, $360; dryer, $290

73. $x + 2x - 1 = 14$; 9 years old   **75.** $x + 4x - 3 = 37$; 29 years old

77. $x + 2x - 60 = 2,400; \$820$

79. $x + 2x + x + 7 = 99$; 8 A.M.: 23, 10 A.M.: 46, 12 P.M.: 30

81. 3,317.12 rpm   **83.** (a) $D = \dfrac{V^2}{R}$; (b) 0.792

85. (a) 15.69 g; (b) 10 days; (c) 14 days   **87.** 2.2 V   **89.** 45°C

91. Above and Beyond   **93.** Above and Beyond   (a) 15   (b) 14

(c) 3   (d) −16   (e) −6   (f) 3,600   (g) −7   (h) 40

2.5

Applications of Linear Equations

< 2.5 Objectives >

1 > Set up and solve an application

2 > Solve geometry problems

3 > Solve mixture problems

4 > Solve motion problems

5 > Identify the elements of a percent problem

6 > Solve applications involving percents

We now have all the tools needed to solve problems that can be modeled by linear equations. Before moving to real-world applications, we look at a number problem to review the five-step process for solving word problems outlined in the previous section.

| ▶ | **Example 1** | Solving an Application—The Five-Step Process |

< Objective 1 >

One number is 5 more than a second number. The sum of the smaller number multiplied by 3 and the larger number times 4 is 104. Find the two numbers.

Step 1 What are you asked to find? You must find the two numbers.

Step 2 Represent the unknowns. Let x be the smaller number. Then

$$x + 5$$

is the larger number.

NOTES

In step 2, "5 more than" x translates to $x + 5$.

The parentheses are *essential* in writing the correct equation.

Step 3 Write an equation.

$$3x + 4(x + 5) = 104$$

3 times the smaller   Plus   4 times the larger

Step 4 Solve the equation.

$$3x + 4(x + 5) = 104$$
$$3x + 4x + 20 = 104$$
$$7x + 20 = 104$$
$$7x = 84$$
$$x = 12$$

The smaller number (x) is 12, and the larger number ($x + 5$) is 17.

Step 5 Check the solution: 12 is the smaller number, and 17 is the larger number.

$$3 \cdot (12) + 4 \cdot ((12) + 5) = 104 \quad \text{(True)}$$

245

Check Yourself 1

One number is 4 more than another. If 6 times the smaller minus 4 times the larger is 4, what are the two numbers?

The solutions for many problems from geometry will also yield equations involving parentheses. Consider Example 2.

| Example 2 | Solving a Geometry Application |

< Objective 2 >

The length of a rectangle is 1 cm less than 3 times the width. If the perimeter is 54 cm, find the dimensions of the rectangle.

Step 1   You want to find the dimensions (the width and length).

Step 2   Let x be the width.

Then $3x - 1$ is the length.

3 times   1 less than
the width

NOTE

When working with geometric figures, you should always draw a sketch of the problem, including the labels assigned in step 2.

Length $3x - 1$

Width
x

Step 3   To write an equation, we use this formula for the perimeter of a rectangle.

$$P = 2W + 2L$$

So

$$2x + 2(3x - 1) = 54$$

Twice the   Twice the   Perimeter
width   length

Step 4   Solve the equation.

$$2x + 2(3x - 1) = 54$$
$$2x + 6x - 2 = 54$$
$$8x = 56$$
$$x = 7$$

RECALL

Be sure to return to the original statement of the problem when checking your result.

The width x is 7 cm, and the length, $3x - 1$, is 20 cm. We leave step 5, the check, to you.

Check Yourself 2

The length of a rectangle is 5 in. more than twice the width. If the perimeter of the rectangle is 76 in., what are the dimensions of the rectangle?

Often, we need parentheses to set up a *mixture problem*. Mixture problems involve combining things that have a different value, rate, or strength, as shown in Example 3.

| **Example** 3 | **Solving a Mixture Problem**

< Objective 3 >

NOTE

We subtract *x*, the number of general admission tickets, from 400, the total number of tickets, to find the number of student tickets.

Four hundred tickets were sold for a school play. General admission tickets were $4, and student tickets were $3. If the total ticket sales were $1,350, how many of each type of ticket were sold?

Step 1 You want to find the number of each type of ticket sold.

Step 2 Let x be the number of general admission tickets.

Then $\underline{400 - x}$ student tickets were sold.

$\uparrow$

400 tickets were
sold in all.

Step 3 The revenue from each kind of ticket is found by multiplying the price of the ticket by the number sold.

General admission tickets: $4x$ $4 for each of the *x* tickets

Student tickets: $3(400 - x)$ $3 for each of the $400 - x$ tickets

So to form an equation, we have

$$4x + \underline{3(400 - x)} = 1,350$$

$\uparrow$ $\uparrow$ $\uparrow$

Revenue   Revenue   Total
from   from   revenue
general   student
admission   tickets
tickets

Step 4 Solve the equation.

$$4x + 3(400 - x) = 1,350$$
$$4x + 1,200 - 3x = 1,350$$
$$x + 1,200 = 1,350$$
$$x = 150$$

So 150 general admission and $400 - 150$ or 250 student tickets were sold. We leave the check to you.

Check Yourself 3

Beth bought 40¢ stamps and 3¢ stamps at the post office. If she purchased 92 stamps at a total cost of $22, how many of each kind did she buy?

The next group of applications that we look at are *motion problems*. They involve a distance traveled, a rate or speed, and time. To solve motion problems, we need a relationship among these three quantities.

> **CAUTION**

Make your units consistent. If a rate is given in *miles per hour,* then the time must be given in *hours* and the distance in *miles.*

Suppose you travel at a rate of 50 mi/h on a highway for 6 h. How far (what distance) will you have gone? To find the distance, you multiply:

$$(50 \text{ mi/h})(6 \text{ h}) = 300 \text{ mi}$$

↑ Speed or rate   ↑ Time   ↑ Distance

Property

Relationship for Motion Problems	In general, if r is a rate, t is the time, and d is the distance traveled, then $$d = r \cdot t$$

This is the key relationship, and it will be used in all motion problems. We apply this relationship in Example 4.

Example 4   Solving a Motion Problem

< Objective 4 >

On Friday morning Ricardo drove from his house to the beach in 4 h. In coming back on Sunday afternoon, heavy traffic slowed his speed by 10 mi/h, and the trip took 5 h. What was his average speed (rate) in each direction?

Step 1   We want the speed or rate in each direction.

Step 2   Let x be Ricardo's speed to the beach. Then $x - 10$ is his return speed.
   It is always a good idea to sketch the given information in a motion problem. Here we would have

Going   ——— x mi/h for 4 h ———→

Returning   ←——— $(x - 10)$ mi/h for 5 h ———

NOTE

Distance (going) = distance (returning)
or
Time · rate (going) = time · rate (returning)

Step 3   Because we know that the distance is the same each way, we can write an equation, using the fact that the product of the rate and the time each way must be the same.
   So

$$\underbrace{4x}_{\substack{\text{Time} \cdot \text{rate} \\ \text{(going)}}} = \underbrace{5(x - 10)}_{\substack{\text{Time} \cdot \text{rate} \\ \text{(returning)}}}$$

An alternate method is to use a chart, which can help summarize the given information. We begin by filling in the information given in the problem.

	Distance	Rate	Time
Going		x	4
Returning		$x - 10$	5

Now we fill in the missing information. Here we use the fact that $d = rt$ to complete the chart.

	Distance	Rate	Time
Going	$4x$	x	4
Returning	$5(x - 10)$	$x - 10$	5

From here we set the two distances equal to each other and solve as before.

Step 4   Solve.

$$4x = 5(x - 10)$$
$$4x = 5x - 50$$
$$-x = -50$$
$$x = 50 \text{ mi/h}$$

So Ricardo's rate going to the beach was 50 mi/h, and his rate returning was 40 mi/h.

Step 5   To check, you should verify that the product of the time and the rate is the same in each direction.

NOTE

x was his rate going,
$x - 10$ was his rate returning.

Check Yourself 4

A plane made a flight (with the wind) between two towns in 2 h. Returning against the wind, the plane's speed was 60 mi/h slower, and the flight took 3 h. What was the plane's speed in each direction?

Example 5 illustrates another way of using the distance relationship.

Example 5 | **Solving a Motion Problem**

Katy leaves Las Vegas for Los Angeles at 10 A.M., driving at 50 mi/h. At 11 A.M. Jensen leaves Los Angeles for Las Vegas, driving at 55 mi/h along the same route. If the cities are 260 mi apart, at what time will Katy and Jensen meet?

Step 1   Find the time that Katy travels until they meet.

Step 2   Let x be Katy's time.

Then $x - 1$ is Jensen's time.

Jensen left 1 h later.

Again, you should draw a sketch of the given information.

Step 3 To write an equation, we again need the relationship $d = rt$. From this equation, we can write

Katy's distance $= 50x$

Jensen's distance $= 55(x - 1)$

As before, we can use a chart to solve.

	Distance	Rate	Time
Katy	$50x$	50	x
Jensen	$55(x - 1)$	55	$x - 1$

From the original problem, the sum of the distances is 260 mi, so

$$50x + 55(x - 1) = 260$$

Step 4

$$50x + 55(x - 1) = 260$$
$$50x + 55x - 55 = 260$$
$$105x - 55 = 260$$
$$105x = 315$$
$$x = 3 \text{ h}$$

Finally, because Katy left at 10 A.M., the two will meet at 1 P.M. We leave the check of this result to you.

Check Yourself 5

At noon a jogger leaves one point, running at 8 mi/h. One hour later a bicyclist leaves the same point, traveling at 20 mi/h in the opposite direction. At what time will they be 36 mi apart?

The final type of problem we look at in this section involves percents. Percents come up in more applications than nearly any other type of problem, so it is important that you become comfortable modeling and solving percent problems.

Every complete percent statement has three parts that need to be identified. We call these parts the *base,* the *rate,* and the *amount.* Here are definitions for each of these terms.

Definition

Base, Amount, and Rate

The **base** is the whole in a problem. It is the standard used for comparison.

The **amount** is the part of the whole being compared to the base.

The **rate** is the ratio of the amount to the base. It is usually written as a percent.

The following examples provide some practice in determining the parts of a percent problem.

Example 6 — Identifying the Parts of a Percent Problem

< Objective 5 >

NOTES

The base is usually the quantity we begin with.

We will solve this type of problem for the unknown amount.

In each case, identify the base, the amount, and the rate.

(a) 50% of 480 is 240.

The base in this problem is 480.

The amount is 240. This is being compared to the base.

The rate is 50%. It is the percent.

(b) Delia borrows $10,000 for 1 year at 11.49% interest. How much interest will she pay?

The base is the beginning amount, $10,000.

In this case, the amount is the interest she will pay. The amount is *unknown*.

The rate is given by the percent, 11.49%.

Check Yourself 6

Identify the base, the amount, and the rate in each case.

(a) 150 is 25% of what number?
(b) Steffen earned $120 in interest from a CD account that paid 8% interest when he invested $1,500 for one year.

NOTE

To solve problems involving percents, we write the rate as a decimal or fraction.

As we said, every percent problem consists of these three parts: base, amount, and rate. In nearly every such problem, one of these parts is unknown. Solving a percent problem is a matter of identifying and finding the missing part. To do this, we use the *percent relationship*.

Property

The Percent Relationship

In a percent statement, the amount is equal to the product of the rate and the base. We can write this as a formula with B equal to the base, A the amount, and R the rate.

$$A = R \cdot B$$

Now we are ready to solve percent problems. We begin with some straightforward ones and work our way to more involved applications. In all cases, your first step should be to identify the parts of the percent relationship.

Example 7 — Solving Percent Problems

< Objective 6 >

(a) 84 is 5% of what number?

5% is the rate and 84 is the amount. The base is unknown.

We substitute these values into the percent-relationship equation and solve.

$$\underbrace{(84)}_{A = \text{Amount}} = \underbrace{(0.05)}_{R = \text{Rate}} \cdot \underbrace{B}_{\substack{B = \text{Unknown} \\ \text{Base}}}$$ Write the rate as a decimal.

$$\frac{84}{0.05} = B$$ Divide by 0.05 to isolate the variable.

$$1,680 = B$$

Answer the question using a sentence: 84 is 5% of 1,680.

(b) Delia borrows \$10,000 for 1 year at 11.49% interest. How much interest will she pay?

From Example 6(b), we know that the missing element is the amount.

$$A = R \cdot B$$
$$= (0.1149) \cdot (10,000)$$
$$= 1,149$$

Delia's interest payment comes to \$1,149 after one year.

Check Yourself 7

Solve each problem.

(a) 32 is what percent of 128?

(b) If you invest \$5,000 for one year at $8\frac{1}{2}$%, how much interest will you earn?

We conclude this section with some more involved percent applications.

Example 8   Solving Percent Applications

(a) A state adds a 7.25% sales tax to the price of most goods. If a 30-GB iPod is listed for \$299, how much will it cost after the sales tax has been added?

This problem is similar to the application in Example 7(b), in that we are missing the amount. There is the further complication that we need to add the sales tax to the original price.

If we use the price, including tax, as the unknown amount, then the rate is

$$R = 107.25\% = 1.0725$$

The base is the list price, $B = \$299$.

As before, we use the percent relationship to solve the problem.

$$A = R \cdot B$$
$$= (1.0725) \cdot (299)$$
$$= 320.6775$$

Because our answer refers to money, we round to two decimal places. The iPod sells for \$320.68, after the sales tax has been included.

(b) A store sells a certain Kicker amplifier model for a car stereo system for \$249.95. If the store pays \$199.95 for the amplifier, what is its markup percentage for the item (to the nearest whole percent)?

The base is given by the wholesale price, $B = \$199.95$.

RECALL

To round to the nearest whole percent (two decimal places), we need to divide to a third decimal place.

In this case, though, the amount is not the selling price, but rather, the difference between the selling price and the wholesale price.

$$A = 249.95 - 199.95 = 50$$

Therefore, in this problem we are missing the rate.
Once we have the amount, we can use the percent relationship, as before.

$$A = R \cdot B$$
$$(50) = R \cdot (199.95) \qquad \text{Isolate the variable.}$$
$$\frac{50}{199.95} = R$$
$$0.250 = R$$

The store marked up the amplifier by 25%.

Check Yourself 8

(a) In order to make room for the new fall line of merchandise, a proprietor offers to discount all existing stock by 15%. How much would you pay for a Fendi handbag that the store usually sells for $229?

(b) A grocery store adds a 30% markup to the wholesale price of an item to determine the selling price. If the store sells a half-gallon container of orange juice for $2.99, what is the wholesale price of the orange juice?

Check Yourself ANSWERS

1. The numbers are 10 and 14.   **2.** The width is 11 in. and the length is 27 in.
3. Beth bought fifty-two 40¢ stamps and forty 3¢ stamps.   **4.** The plane flew at a rate of 180 mi/h with the wind and 120 mi/h against the wind.
5. At 2 P.M. the jogger and the bicyclist will be 36 mi apart.
6. **(a)** $B =$ unknown, $A = 150$, $R = 25\%$; **(b)** $B = \$1,500$, $A = \$120$, $R = 8\%$
7. **(a)** 25%; **(b)** $425   **8.** **(a)** $194.65; **(b)** $2.30

Reading Your Text

The following fill-in-the-blank exercises are designed to ensure that you understand some of the key vocabulary used in this section.

SECTION 2.5

(a) Always try to _____ a sketch of the figures when solving geometry applications.

(b) _____ problems involve combining things that have a different value, rate, or strength.

(c) In a percent problem, the rate is the ratio of the _____ to the base.

(d) To solve a percent problem, begin by _____ the parts of the percent relationship.

MathZone

Boost your grade at mathzone.com!

> Practice   > Self-Tests
  Problems
             > e-Professors
> NetTutor   > Videos

Name _____

Section _____ Date _____

Answers

1. _____

2. _____

3. _____

4. _____

5. _____

6. _____

7. _____

8. _____

9. _____

10. _____

< Objective 1 >

Solve the following word problems. Be sure to show the equation you use for the solution.

1. **NUMBER PROBLEM** One number is 8 more than another. If the sum of the smaller number and twice the larger number is 46, find the two numbers.

2. **NUMBER PROBLEM** One number is 3 less than another. If 4 times the smaller number minus 3 times the larger number is 4, find the two numbers.

3. **NUMBER PROBLEM** One number is 7 less than another. If 4 times the smaller number plus 2 times the larger number is 62, find the two numbers.

4. **NUMBER PROBLEM** One number is 10 more than another. If the sum of twice the smaller number and 3 times the larger number is 55, find the two numbers. > Videos

5. **NUMBER PROBLEM** Find two consecutive integers such that the sum of twice the first integer and 3 times the second integer is 28. (*Hint:* If x represents the first integer, $x + 1$ represents the next consecutive integer.)

6. **NUMBER PROBLEM** Find two consecutive odd integers such that 3 times the first integer is 5 more than twice the second. (*Hint:* If x represents the first integer, $x + 2$ represents the next consecutive odd integer.)

< Objective 2 >

7. **GEOMETRY** The length of a rectangle is 1 in. more than twice its width. If the perimeter of the rectangle is 74 in., find the dimensions of the rectangle.

8. **GEOMETRY** The length of a rectangle is 5 cm less than 3 times its width. If the perimeter of the rectangle is 46 cm, find the dimensions of the rectangle. > Videos

9. **GEOMETRY** The length of a rectangular garden is 4 m more than 3 times its width. The perimeter of the garden is 56 m. What are the dimensions of the garden?

10. **GEOMETRY** The length of a rectangular playing field is 5 ft less than twice its width. If the perimeter of the playing field is 230 ft, find the length and width of the field.

11. **Geometry** The base of an isosceles triangle is 3 cm less than the length of the equal sides. If the perimeter of the triangle is 36 cm, find the length of each of the sides.

12. **Geometry** The length of one of the equal legs of an isosceles triangle is 3 in. less than twice the length of the base. If the perimeter is 29 in., find the length of each of the sides.

< Objective 3 >

13. **Business and Finance** Tickets for a play cost $8 for the main floor and $6 in the balcony. If the total receipts from 500 tickets were $3,600, how many of each type of ticket were sold?

14. **Business and Finance** Tickets for a basketball tournament were $6 for students and $9 for nonstudents. Total sales were $10,500, and 250 more student tickets were sold than nonstudent tickets. How many of each type of ticket were sold?

> Videos

15. **Business and Finance** Maria bought 56 stamps at the post office in 37¢ and 20¢ denominations. If she paid $18 for the stamps, how many of each denomination did she buy?

16. **Business and Finance** A bank teller had a total of 125 $10 bills and $20 bills to start the day. If the value of the bills was $1,650, how many of each denomination did he have?

17. **Business and Finance** Tickets for a train excursion were $120 for a sleeping room, $80 for a berth, and $50 for a coach seat. The total ticket sales were $8,600. If there were 20 more berth tickets sold than sleeping room tickets and 3 times as many coach tickets as sleeping room tickets, how many of each type of ticket were sold?

18. **Business and Finance** Admission for a college baseball game is $6 for box seats, $5 for the grandstand, and $3 for the bleachers. The total receipts for one evening were $9,000. There were 100 more grandstand tickets sold than box seat tickets. Twice as many bleacher tickets were sold as box seat tickets. How many tickets of each type were sold?

< Objective 4 >

19. **Science and Medicine** Patrick drove 3 h to attend a meeting. On the return trip, his speed was 10 mi/h less and the trip took 4 h. What was his speed each way?

Answers

11. _____

12. _____

13. _____

14. _____

15. _____

16. _____

17. _____

18. _____

19. _____

Answers

20. _____

21. _____

22. _____

23. _____

24. _____

25. _____

26. _____

27. _____

28. _____

20. **SCIENCE AND MEDICINE** A bicyclist rode into the country for 5 h. In returning, her speed was 5 mi/h faster and the trip took 4 h. What was her speed each way?

21. **SCIENCE AND MEDICINE** A car leaves a city and goes north at a rate of 50 mi/h at 2 P.M. One hour later a second car leaves, traveling south at a rate of 40 mi/h. At what time will the two cars be 320 mi apart?

> Videos

22. **SCIENCE AND MEDICINE** A bus leaves a station at 1 P.M., traveling west at an average rate of 44 mi/h. One hour later a second bus leaves the same station, traveling east at a rate of 48 mi/h. At what time will the two buses be 274 mi apart?

23. **SCIENCE AND MEDICINE** At 8:00 A.M., Catherine leaves on a trip at 45 mi/h. One hour later, Max decides to join her and leaves along the same route, traveling at 54 mi/h. When will Max catch up with Catherine?

24. **SCIENCE AND MEDICINE** Martina leaves home at 9 A.M., bicycling at a rate of 24 mi/h. Two hours later, John leaves, driving at the rate of 48 mi/h. At what time will John catch up with Martina?

25. **SCIENCE AND MEDICINE** Mika leaves Boston for Baltimore at 10:00 A.M., traveling at 45 mi/h. One hour later, Hiroko leaves Baltimore for Boston on the same route, traveling at 50 mi/h. If the two cities are 425 mi apart, when will Mika and Hiroko meet?

26. **SCIENCE AND MEDICINE** A train leaves town A for town B, traveling at 35 mi/h. At the same time, a second train leaves town B for town A at 45 mi/h. If the two towns are 320 mi apart, how long will it take for the two trains to meet?

27. **BUSINESS AND FINANCE** There are a total of 500 Douglas fir and hemlock trees in a section of forest bought by Hoodoo Logging Co. The company paid an average of $250 for each Douglas fir and $300 for each hemlock. If the company paid $132,000 for the trees, how many of each kind did the company buy?

28. **BUSINESS AND FINANCE** There are 850 Douglas fir and ponderosa pine trees in a section of forest bought by Sawz Logging Co. The company paid an average of $300 for each Douglas fir and $225 for each ponderosa pine. If the company paid $217,500 for the trees, how many of each kind did the company buy?

The Streeter/Hutchison Series in Mathematics   Beginning Algebra

< Objective 5 >

Identify the indicated quantity in each statement.

29. The rate in the statement "23% of 400 is 92."

30. The base in the statement "40% of 600 is 240."

31. The amount in the statement "200 is 40% of 500."

32. The rate in the statement "480 is 60% of 800."

33. The base in the statement "16% of 350 is 56."

34. The amount in the statement "150 is 75% of 200."

Identify the rate, the base, and the amount in the following applications. Do not solve the applications at this point.

35. BUSINESS AND FINANCE Jan has a 5% commission rate on all her sales. If she sells $40,000 worth of merchandise in 1 month, what commission will she earn?

> Videos

36. BUSINESS AND FINANCE 22% of Shirley's monthly salary is deducted for withholding. If those deductions total $209, what is her salary?

37. SCIENCE AND MEDICINE In a chemistry class of 30 students, 5 received a grade of A. What percent of the students received A's?

38. BUSINESS AND FINANCE A can of mixed nuts contains 80% peanuts. If the can holds 16 oz, how many ounces of peanuts does it contain?

39. STATISTICS A college had 9,000 students at the start of a school year. If there is an enrollment increase of 6% by the beginning of the next year, how many additional students will there be?

40. BUSINESS AND FINANCE Paul invested $5,000 in a time deposit. What interest will he earn for 1 year if the interest rate is 6.5%?

Answers

29. _____

30. _____

31. _____

32. _____

33. _____

34. _____

35. _____

36. _____

37. _____

38. _____

39. _____

40. _____

Answers

41. _____

42. _____

43. _____

44. _____

45. _____

46. _____

47. _____

< Objective 6 >

Solve each of the following applications.

41. BUSINESS AND FINANCE What interest will you pay on a $3,400 loan for 1 year if the interest rate is 12%?

42. SCIENCE AND MEDICINE A chemist has 300 milliliters (mL) of solution that is 18% acid. How many milliliters of acid are in the solution?

300 mL

43. BUSINESS AND FINANCE Roberto has 26% of his pay withheld for deductions. If he earns $550 per week, what amount is withheld?

44. BUSINESS AND FINANCE A real estate agent's commission rate is 6%. What will the amount of the commission be on the sale for a $185,000 home?

45. BUSINESS AND FINANCE If a salesman is paid a $140 commission on the sale of a $2,800 sailboat, what is his commission rate?

46. BUSINESS AND FINANCE Ms. Jordan has been given a loan of $2,500 for 1 year. If the interest charged is $275, what is the interest rate on the loan?

47. BUSINESS AND FINANCE Joan was charged $18 interest for 1 month on a $1,200 credit card balance. What was the monthly interest rate?

48. **SCIENCE AND MEDICINE** There is 117 grams (g) of acid in 900 g of a solution of acid and water. What percent of the solution is acid?

49. **STATISTICS** On a test, Alice had 80% of the problems right. If she had 20 problems correct, how many questions were on the test?

50. **BUSINESS AND FINANCE** A state sales tax rate is 3.5%. If the tax on a purchase is $7, what was the amount of the purchase?

51. **BUSINESS AND FINANCE** If a house sells for $125,000 and the commission rate is $6\frac{1}{2}$%, how much will the salesperson make for the sale?

52. **STATISTICS** Marla needs 70% on a final test to receive a C for a course. If the exam has 120 questions, how many questions must she answer correctly?

> Videos

53. **SOCIAL SCIENCE** A study has shown that 102 of the 1,200 people in the workforce of a small town are unemployed. What is the town's unemployment rate?

54. **STATISTICS** A survey of 400 people found that 66 were left-handed. What percent of those surveyed were left-handed?

55. **STATISTICS** Of 60 people who start a training program, 45 complete the course. What is the dropout rate?

56. **BUSINESS AND FINANCE** In a shipment of 250 parts, 40 are found to be defective. What percent of the parts are faulty?

57. **STATISTICS** In a recent survey, 65% of those responding were in favor of a freeway improvement project. If 780 people were in favor of the project, how many people responded to the survey?

58. **STATISTICS** A college finds that 42% of the students taking a foreign language are enrolled in Spanish. If 1,512 students are taking Spanish, how many foreign language students are there?

Answers

48. _____

49. _____

50. _____

51. _____

52. _____

53. _____

54. _____

55. _____

56. _____

57. _____

58. _____

Answers

59. _____

60. _____

61. _____

62. _____

63. _____

64. _____

65. _____

66. _____

67. _____

59. BUSINESS AND FINANCE An appliance dealer marks up refrigerators 22% (based on cost). If the cost of one model was $600, what will its selling price be?

60. STATISTICS A school had 900 students at the start of a school year. If there is an enrollment increase of 7% by the beginning of the next year, what is the new enrollment?

61. BUSINESS AND FINANCE A home lot purchased for $125,000 increased in value by 25% over 3 years. What was the lot's value at the end of the period?

62. BUSINESS AND FINANCE New cars depreciate an average of 28% in their first year of use. What would an $18,000 car be worth after 1 year?

63. STATISTICS A school's enrollment was up from 950 students in 1 year to 1,064 students in the next. What was the rate of increase?

64. BUSINESS AND FINANCE Under a new contract, the salary for a position increases from $31,000 to $33,635. What rate of increase does this represent?

65. BUSINESS AND FINANCE The price of a new van has increased $4,830, which amounts to a 14% increase. What was the price of the van before the increase?

66. BUSINESS AND FINANCE A television set is marked down $75, for a sale. If this is a 12.5% decrease from the original price, what was the selling price before the sale?

67. STATISTICS A company had 66 fewer employees in July 2005 than in July 2004. If this represents a 5.5% decrease, how many employees did the company have in July 2004?

68. **BUSINESS AND FINANCE** Carlotta received a monthly raise of $162.50. If this represented a 6.5% increase, what was her monthly salary before the raise?

69. **BUSINESS AND FINANCE** A pair of shorts, advertised for $48.75, is being sold at 25% off the original price. What was the original price?

70. **BUSINESS AND FINANCE** If the total bill at a restaurant, including a 15% tip, is $65.32, what was the cost of the meal alone?

The chart below gives U.S.-Mexico trade data from 2000 to 2005. Use this information for exercises 71–74.

U.S. Trade with Mexico, 2000 to 2005 (in millions of dollars)			
Year	Exports	Imports	Trade Balance
2000	$111,349	$135,926	−$24,577
2001	101,297	131,338	−30,041
2002	97,470	134,616	−37,146
2003	97,412	138,060	−40,648
2004	110,835	155,902	−45,067
2005	120,049	170,198	−50,149

Source: U.S. Census Bureau, Foreign Trade Division.

71. What was the percent increase (to the nearest whole percent) of exports from 2000 to 2005?

72. What was the percent increase (to the nearest whole percent) of imports from 2000 to 2005?

73. By what percent (to the nearest whole percent) did imports exceed exports in 2000? 2005?

74. By what percent (to the nearest whole percent) did the trade imbalance increase between 2000 and 2005?

Answers

68. _____

69. _____

70. _____

71. _____

72. _____

73. _____

74. _____

chapter 2 > Make the Connection

Answers

75. _____

76. _____

77. _____

78. _____

79. _____

80. _____

75. **STATISTICS** In 1990, there were an estimated 145.0 million passenger cars registered in the United States. The total number of vehicles registered in the United States for 1990 was estimated at 194.5 million. What percent of the vehicles registered were passenger cars (to the nearest tenth)?

76. **STATISTICS** Gasoline accounts for 85% of the motor fuel consumed in the United States every day. If 8,882 thousand barrels (bbl) of motor fuel are consumed each day, how much gasoline is consumed each day in the United States (to the nearest gallon)?

77. **STATISTICS** In 1999, transportation accounted for 63% of U.S. petroleum consumption. Assuming that same rate applies now, and 10.85 million bbl of petroleum are used each day for transportation in the United States, what is the total daily petroleum consumption by all sources in the United States (to the nearest hundredth)?

78. **STATISTICS** Each year, 540 million metric tons (t) of carbon dioxide are added to the atmosphere by the United States. Burning gasoline and other transportation fuels is responsible for 35% of the carbon dioxide emissions in the United States. How much carbon dioxide is emitted each year by the burning of transportation fuels in the United States?

| Basic Skills | Advanced Skills | Vocational-Technical Applications | Calculator/Computer | **Above and Beyond** | Getting Ready |

79. There is a universally accepted "order of operations" used to simplify expressions. Explain how the order of operations is used in solving equations. Be sure to use complete sentences.

80. A common mistake when solving equations is the following:

The equation: $2(x - 2) = x + 3$
First step in solving: $2x - 2 = x + 3$

Write a clear explanation of what error has been made. What could be done to avoid this error?

81. Another very common mistake is in the equation below:

The equation: $6x - (x + 3) = 5 + 2x$

First step in solving: $6x - x + 3 = 5 + 2x$

Write a clear explanation of what error has been made and what could be done to avoid the mistake.

82. Write an algebraic equation for the English statement "Subtract 5 from the sum of x and 7 times 3 and the result is 20." Compare your equation with those of other students. Did you all write the same equation? Are all the equations correct even though they don't look alike? Do all the equations have the same solution? What is wrong? The English statement is *ambiguous*. Write another English statement that leads correctly to more than one algebraic equation. Exchange with another student and see whether the other student thinks the statement is ambiguous. Notice that the algebra is *not* ambiguous!

Basic Skills | Advanced Skills | Vocational-Technical Applications | Calculator/Computer | Above and Beyond | **Getting Ready**

Locate each of the following numbers on the number line.

(a) 4

(b) −5

(c) −3

(d) 2

(e) $-\dfrac{7}{2}$

(f) $\dfrac{2}{3}$

(g) 2.5

(h) −1.1

Answers

Answers

81. _____

82. _____

(a) _____

(b) _____

(c) _____

(d) _____

(e) _____

(f) _____

(g) _____

(h) _____

Answers

1. 10, 18   **3.** 8, 15   **5.** 5, 6   **7.** 12 in., 25 in.   **9.** 6 m, 22 m
11. 13-cm legs, 10-cm base   **13.** 200 $6 tickets, 300 $8 tickets
15. 40 37¢ stamps, 16 20¢ stamps   **17.** 60 coach, 40 berth, 20 sleeping room
19. 40 mi/h, 30 mi/h   **21.** 6 P.M.   **23.** 2 P.M.   **25.** 3 P.M.
27. 360 Douglas firs, 140 hemlocks   **29.** 23%   **31.** 200   **33.** 350
35. $R = 5\%, B = \$40{,}000, A = $ unknown   **37.** $R = $ unknown, $B = 30, A = 5$
39. $R = 6\%, B = 9{,}000, A = $ unknown   **41.** $408   **43.** $143   **45.** 5%
47. 1.5%   **49.** 25 questions   **51.** $8,125   **53.** 8.5%   **55.** 25%
57. 1,200 people   **59.** $732   **61.** $156,250   **63.** 12%
65. $34,500   **67.** 1,200 employees   **69.** $65   **71.** 8%
73. 22%; 42%   **75.** ≈74.6%   **77.** ≈17.22 million bbl
79. Above and Beyond   **81.** Above and Beyond
(a)–(h)

2.6

Inequalities—An Introduction

< 2.6 Objectives >

1 > Use inequality notation

2 > Graph the solution set of an inequality

3 > Solve an inequality and graph the solution set

4 > Solve an application using inequalities

As we pointed out earlier, an equation is a statement that two expressions are equal. In algebra, an **inequality** is a statement that one expression is less than or greater than another. Four new symbols are used in writing inequalities. The use of two of them is illustrated in Example 1.

 Example 1 Reading the Inequality Symbol

< Objective 1 >

$5 < 8$ is an inequality read "5 is less than 8."

$9 > 6$ is an inequality read "9 is greater than 6."

NOTE

To help you remember, the "arrowhead" always points toward the smaller quantity.

Check Yourself 1

Fill in the blanks using the symbols $<$ and $>$.

(a) 12 _____ 8   (b) 20 _____ 25

Like an equation, an inequality can be represented by a balance scale. Note that, in each case, the inequality arrow points to the side that is "lighter."

$2x < 4x - 3$

NOTE

The $2x$ side is less than the $4x - 3$ side, so it is "lighter."

$5x - 6 > 9$

Just as was the case with equations, inequalities that involve variables may be either true or false depending on the value that we give to the variable. For instance, consider the inequality

$x < 6$

$$\text{If } x = \begin{cases} 3 & 3 < 6 \text{ is true} \\ 5 & 5 < 6 \text{ is true} \\ -10 & -10 < 6 \text{ is true} \\ 8 & 8 < 6 \text{ is false} \end{cases}$$

Therefore, 3, 5, and -10 are some *solutions* for the inequality $x < 6$; they make the inequality a true statement. You should see that 8 is *not* a solution. We call the set of all solutions the **solution set** for the inequality. Of course, there are many possible solutions.

Because there are so many solutions (an infinite number, in fact), we certainly do not want to try to list them all! A convenient way to show the solution set of an inequality is with a number line.

 Example 2 | **Solving Inequalities**

< Objective 2 >

NOTE

The colored arrow indicates the direction of the *solution set.*

To graph the solution set for the inequality $x < 6$, we want to include all real numbers that are "less than" 6. This means all numbers *to the left* of 6 on the number line. We start at 6 and draw an arrow extending left, as shown:

Note: The **open circle** at 6 means that we do not include 6 in the solution set (6 is not less than itself). The colored arrow shows all the numbers in the solution set, with the arrowhead indicating that the solution set continues indefinitely to the left.

 Check Yourself 2

Graph the solution set of $x < -2$.

Two other symbols are used in writing inequalities. They are used with inequalities such as

$x \geq 5$ and $x \leq 2$

Here $x \geq 5$ is really a combination of the two statements $x > 5$ and $x = 5$. It is read "x is greater than or equal to 5." The solution set includes 5 in this case.

The inequality $x \leq 2$ combines the statements $x < 2$ and $x = 2$. It is read "x is less than or equal to 2."

 Example 3   Graphing Inequalities

The solution set for $x \geq 5$ is graphed as follows.

 Check Yourself 3

Graph the solution sets.

(a) $x \leq -4$ (b) $x \geq 3$

You have learned how to graph the solution sets of some simple inequalities, such as $x < 8$ or $x \geq 10$. Now we look at more complicated inequalities, such as

$$2x - 3 < x + 4$$

This is called a **linear inequality in one variable.** Only one variable is involved in the inequality, and it appears only to the first power. Fortunately, the methods used to solve this type of inequality are very similar to those we used earlier in this chapter to solve linear equations in one variable. Here is our first property for inequalities.

Property

The Addition Property of Inequality

If $a < b$ then $a + c < b + c$

In words, adding the same quantity to both sides of an inequality gives an **equivalent inequality.**

Again, we can use the idea of a balance scale to see the significance of this property. If we add the same weight to both sides of an unbalanced scale, it stays unbalanced.

Example 4   **Solving Inequalities**

< Objective 3 >

NOTE

The inequality is solved when an equivalent inequality has the form

$x < \square$ or $x > \square$

Solve and graph the solution set for $x - 8 < 7$.

To solve $x - 8 < 7$, add 8 to both sides of the inequality by the addition property.

$$
\begin{array}{rcl}
x - 8 & < & 7 \\
+\,8 & & +8 \\
\hline
x & < & 15 \qquad \text{(The inequality is solved.)}
\end{array}
$$

The graph of the solution set is   ←———|———————◇———→
                                         0   15

Check Yourself 4

Solve and graph the solution set.

$x - 9 > -3$

As with equations, the addition property allows us to subtract the same quantity from both sides of an inequality.

Example 5   **Solving Inequalities**

NOTE

We subtracted $3x$ and then added 2 to both sides. If these steps are done in the reverse order, the resulting inequality is the same.

Solve and graph the solution set for $4x - 2 \geq 3x + 5$.

First, we subtract $3x$ from both sides of the inequality.

$$
\begin{array}{rcl}
4x - 2 & \geq & 3x + 5 \\
-3x & & -3x \\
\hline
x - 2 & \geq & 5 \\
+\,2 & & +2 \qquad \text{Now we add 2 to both sides.} \\
\hline
x & \geq & 7
\end{array}
$$

The graph of the solution set is   ←———|————————◆———→
                                         0   7

Check Yourself 5

Solve and graph the solution set.

$7x - 8 \leq 6x + 2$

We also need a rule for multiplying on both sides of an inequality. Here we have to be a bit careful. There is a difference between the multiplication property for inequalities and that for equations. Look at the following:

$2 < 7$ (A true inequality)

Multiply both sides by 3.

$$
\begin{array}{l}
2 < 7 \\
3 \cdot 2 < 3 \cdot 7 \\
\quad 6 < 21 \qquad \text{(A true inequality)}
\end{array}
$$

Now we multiply both sides of the original inequality by -3.

$$2 < 7$$
$$(-3)(2) < (-3)(7)$$
$$-6 < -21 \qquad \text{(Not a true inequality)}$$

But, if we try this,

$$2 < 7$$

Change the direction
of the inequality:
$< \text{ becomes } >$.

$$(-3)(2) > (-3)(7)$$

(This is now a true
inequality.)

$$-6 > -21$$

This suggests that multiplying both sides of an inequality by a negative number changes the direction of the inequality.

We can state the following general property.

Property

The Multiplication Property of Inequality

If $a < b$ then $ac < bc$ if $c > 0$

and $ac > bc$ if $c < 0$

In words, multiplying both sides of an inequality by the same *positive* number gives an equivalent inequality.

When both sides of an inequality are multiplied by the same *negative* number, it is necessary to *reverse the direction* of the inequality to give an equivalent inequality.

> **NOTE**
>
> Because division is defined in terms of multiplication, this rule applies to division, as well.

▶ **Example 6**   **Solving and Graphing Inequalities**

(a) Solve and graph the solution set for $5x < 30$.

Multiplying both sides of the inequality by $\dfrac{1}{5}$ gives

$$\frac{1}{5}(5x) < \frac{1}{5}(30)$$

Simplifying, we have

$$x < 6$$

The graph of the solution set is

(b) Solve and graph the solution set for $-4x \geq 28$.

In this case we want to multiply both sides of the inequality by $-\dfrac{1}{4}$ to leave x alone on the left.

$$\left(-\frac{1}{4}\right)(-4x) \leq \left(-\frac{1}{4}\right)(28)$$

Reverse the direction of the inequality because you are multiplying by a negative number!

or $x \leq -7$

The graph of the solution set is

Check Yourself 6

Solve and graph the solution sets.

(a) $7x > 35$ **(b)** $-8x \leq 48$

Example 7 illustrates the use of the multiplication property when fractions are involved in an inequality.

Example 7   **Solving and Graphing Inequalities**

(a) Solve and graph the solution set for

$$\frac{x}{4} > 3$$

Here we multiply both sides of the inequality by 4. This isolates x on the left.

$$4\left(\frac{x}{4}\right) > 4(3)$$

$$x > 12$$

The graph of the solution set is

(b) Solve and graph the solution set for

$$-\frac{x}{6} \geq -3$$

In this case, we multiply both sides of the inequality by -6:

$$(-6)\left(-\frac{x}{6}\right) \leq (-6)(-3)$$

$$x \leq 18$$

The graph of the solution set is

NOTE

We reverse the direction of the inequality because we are multiplying by a negative number.

Check Yourself 7

Solve and graph the solution sets.

(a) $\frac{x}{5} \leq 4$ **(b)** $-\frac{x}{3} < -7$

Example 8   **Solving and Graphing Inequalities**

(a) Solve and graph the solution set for $5x - 3 < 2x$.

$$
\begin{array}{rl}
5x - 3 < & 2x \\
-2x \quad\quad & -2x \\
\hline
3x - 3 < & 0 \\
+ 3 \quad & +3 \\
\hline
3x \quad < & 3
\end{array}
$$

Bring the variable terms to the same (left) side.

Isolate the variable term.

Next *divide* both sides by 3.

$$\frac{3x}{3} < \frac{3}{3}$$

$$x < 1$$

The graph of the solution set is

(b) Solve and graph the solution set for $2 - 5x < 7$.

$$\begin{array}{rcc}
2 - 5x < & 7 & \\
-2 & -2 & \text{Add } -2. \\
\hline
-5x < & 5 &
\end{array}$$

$$\frac{-5x}{-5} > \frac{5}{-5} \qquad \text{Divide by } -5. \text{ Be sure to reverse the direction of the inequality.}$$

or $\qquad x > -1$

The graph is

Check Yourself 8

Solve and graph the solution sets.

(a) $4x + 9 \geq x$ $\qquad\qquad\qquad$ **(b)** $5 - 6x < 41$

As with equations, we collect all variable terms on one side and all constant terms on the other.

Example 9   Solving and Graphing Inequalities

Solve and graph the solution set for $5x - 5 \geq 3x + 4$.

$$\begin{array}{rcc}
5x - 5 \geq & 3x + 4 & \\
-3x & -3x & \text{Bring the variable terms to the same (left) side.} \\
\hline
2x - 5 \geq & 4 & \\
+5 & +5 & \text{Isolate the variable term.} \\
\hline
2x \quad \geq & 9 &
\end{array}$$

$$\frac{2x}{2} \geq \frac{9}{2} \qquad \text{Isolate the variable.}$$

$$x \geq \frac{9}{2}$$

The graph of the solution set is $\qquad$

Check Yourself 9

Solve and graph the solution set.

$$8x + 3 < 4x - 13$$

Be especially careful when negative coefficients occur in the process of solving.

Example 10 | **Solving and Graphing Inequalities**

Solve and graph the solution set for $2x + 4 < 5x - 2$.

$$
\begin{array}{rcl}
2x + 4 & < & 5x - 2 \\
-5x & & -5x \\
\hline
-3x + 4 & < & -2 \\
-4 & & -4 \\
\hline
-3x & < & -6 \\
\dfrac{-3x}{-3} & > & \dfrac{-6}{-3} \\
x & > & 2
\end{array}
$$

Bring the variable terms to the same (left) side.

Isolate the variable term.

Isolate the variable. Be sure to reverse the direction of the inequality when you divide by a negative number.

The graph of the solution set is

Check Yourself 10

Solve and graph the solution set.

$$5x + 12 \geq 10x - 8$$

Solving inequalities may also require the distributive property.

Example 11 | **Solving and Graphing Inequalities**

Solve and graph the solution set for

$$5(x - 2) \geq -8$$

Applying the distributive property on the left yields

$$5x - 10 \geq -8$$

Solving as before yields

$$
\begin{array}{rcl}
5x - 10 & \geq & -8 \\
+10 & & +10 \\
\hline
5x & \geq & 2
\end{array}
$$

Add 10.

$$\text{or} \qquad x \geq \frac{2}{5}$$

Divide by 5.

The graph of the solution set is

$$
\begin{array}{c}
\longleftarrow \quad \overset{|}{} \quad \overset{\diamond}{\phantom{\frac{2}{5}}} \longrightarrow \\
0 \qquad \frac{2}{5}
\end{array}
$$

Check Yourself 11

Solve and graph the solution set.

$4(x + 3) < 9$

Some applications are solved by using an inequality instead of an equation. Example 12 illustrates such an application.

Example 12 | Solving an Inequality Application

< Objective 4 >

NOTE

The *mean* of a data set is its arithmetic average.

Mohammed needs a mean score of 92 or higher on four tests to get an A. So far his scores are 94, 89, and 88. What score on the fourth test will get him an A?

NOTES

What do you need to find?

Assign a letter to the unknown.

Write an inequality.

Solve the inequality.

Step 1 We are looking for the score that will, when combined with the other scores, give Mohammed an A.

Step 2 Let x represent a fourth-test score that will get him an A.

Step 3 The inequality will have the mean on the left side, which must be greater than or equal to the 92 on the right.

$$\frac{94 + 89 + 88 + x}{4} \geq 92$$

Step 4 First, multiply both sides by 4:

$94 + 89 + 88 + x \geq 368$

Then add the test scores:

$183 + 88 + x \geq 368$

$271 + x \geq 368$

Subtracting 271 from both sides,

$x \geq 97$ Mohammed needs to score 97 or higher to earn an A.

Step 5 To check the solution, we find the mean of the four test scores, 94, 89, 88, and 97.

$$\frac{94 + 89 + 88 + (97)}{4} = \frac{368}{4} = 92$$

Check Yourself 12

Felicia needs a mean score of at least 75 on five tests to get a passing grade in her health class. On her first four tests she has scores of 68, 79, 71, and 70. What score on the fifth test will give her a passing grade?

The following outline (or algorithm) summarizes our work in this section.

Step by Step

Solving Linear Inequalities

Step 1 Perform operations, as needed, to write an equivalent inequality without any grouping symbols, and combine any like terms appearing on either side of the inequality.

Step 2 Apply the addition property to write an equivalent inequality with the variable term on one side of the inequality and the number on the other.

Step 3 Apply the multiplication property to write an equivalent inequality with the variable isolated on one side of the inequality. Be sure to reverse the direction of the inequality if you multiply or divide by a negative number. The set of solutions derived in step 3 can then be graphed on a number line.

Check Yourself ANSWERS

1. (a) $>$; (b) $<$ **2.**

3. (a)   ; (b)

4. $x > 6$ **5.** $x \leq 10$

6. (a) $x > 5$; (b) $x \geq -6$

7. (a) $x \leq 20$; (b) $x > 21$

8. (a) $x \geq -3$; (b) $x > -6$

9. $x < -4$ **10.** $x \leq 4$

11. $x < -\dfrac{3}{4}$ **12.** 87 or greater

Reading Your Text

The following fill-in-the-blank exercises are designed to ensure that you understand some of the key vocabulary used in this section.

SECTION 2.6

(a) A statement that one expression is less than another is an _____.

(b) In an inequality, the "arrowhead" always points to the _____ quantity.

(c) A filled-in or closed circle on a number line indicates that the number is part of the _____ set.

(d) When multiplying both sides of an inequality by a _____ number, remember to switch the direction of the inequality symbol.

2.6 exercises

< Objective 1 >

Complete the statements, using the symbol < or >.

1. 9 _____ 6

2. 9 _____ 8

3. 7 _____ −2

4. 0 _____ −5

5. 0 _____ 4

6. −12 _____ −7

7. −2 _____ −5 > Videos

8. −4 _____ −11

Write each inequality in words.

9. $x < 3$

10. $x \leq -5$

11. $x \geq -4$

12. $x < -2$

13. $-5 \leq x$

14. $2 < x$

< Objective 2 >

Graph the solution set of each of the following inequalities.

15. $x > 2$

16. $x < -3$

17. $x < 10$

18. $x > 4$

19. $x > 1$

20. $x < -2$

21. $x < 8$

22. $x > 5$

Answers

1. _____   2. _____

3. _____   4. _____

5. _____   6. _____

7. _____   8. _____

9. _____

10. _____

11. _____

12. _____

13. _____

14. _____

15. _____

16. _____

17. _____

18. _____

19. _____

20. _____

21. _____

22. _____

Answers

23. _____

24. _____

25. _____

26. _____

27. _____

28. _____

29. _____   30. _____

31. _____   32. _____

33. _____   34. _____

35. _____   36. _____

37. _____

38. _____

39. _____

40. _____

41. _____

42. _____

43. _____

44. _____

45. _____

46. _____

23. $x > -7$

24. $x < -4$

25. $x \geq 11$

26. $x \geq 0$

> Videos

27. $x < 0$

28. $x \leq -3$

< Objective 3 >

Solve and graph the solution set of each of the following inequalities.

29. $x + 9 < 22$

30. $x + 5 \leq 4$

31. $x + 8 \geq 10$

32. $x - 14 > -17$

33. $5x < 4x + 7$

34. $3x \geq 2x - 4$

35. $6x - 8 \leq 5x$

36. $3x + 2 > 2x$

37. $6x + 5 \geq 5x + 19$

38. $5x + 2 \leq 4x - 6$

39. $7x + 5 < 6x - 4$

40. $8x - 7 > 7x + 3$

41. $4x \leq 12$

42. $5x > 20$

43. $5x > -35$

44. $8x \leq -24$

45. $-6x \geq 18$

46. $-9x < 45$

The Streeter/Hutchison Series in Mathematics   Beginning Algebra

47. $-12x < -72$

48. $-12x \geq -48$

49. $\dfrac{x}{4} > 5$

50. $\dfrac{x}{3} \leq -3$

51. $-\dfrac{x}{2} \geq -3$ > Videos

52. $-\dfrac{x}{4} < 5$

53. $\dfrac{2x}{3} < 6$

54. $\dfrac{3x}{4} \geq -9$

55. $6x > 3x + 12$

56. $4x \leq x - 9$

57. $5x - 2 > 3x$

58. $7x + 3 \geq 2x$

59. $3 - 2x > 5$

60. $-7 - 5x \leq 18$

61. $2x \geq 5x + 18$

62. $3x < 7x - 28$

63. $5x - 3 \leq 3x + 15$

64. $8x + 7 > 5x + 34$

65. $11x + 8 > 4x - 6$

66. $10x - 5 \leq 8x - 25$

67. $7x - 5 < 3x + 2$

68. $5x - 2 \geq 2x - 7$

69. $5x + 7 > 8x - 17$

70. $4x - 3 \leq 9x + 27$

Answers

47.

48.

49.

50.

51.

52.

53.

54.

55.

56.

57.

58.

59.

60.

61.

62.

63.

64.

65.

66.

67.

68.

69.

70.

Answers

71. _____

72. _____

73. _____

74. _____

75. _____

76. _____

77. _____

78. _____

79. _____

80. _____

81. _____

82. _____

71. $3x - 2 \leq 5x + 3$

<--------------------------->

72. $2x + 3 > 8x - 2$

<--------------------------->

73. $4(x + 7) \leq 2x + 31$

<--------------------------->

74. $7(x - 3) > 5x - 14$

<--------------------------->

75. $2(x - 7) > 5x - 12$

<--------------------------->

76. $3(x + 4) \leq 7x + 7$

<--------------------------->

< Objective 4 >

77. SOCIAL SCIENCE There are fewer than 1,000 wild giant pandas left in the bamboo forests of China. Write an inequality expressing this relationship.

78. SCIENCE AND MEDICINE Let C represent the amount of Canadian forest and M represent the amount of Mexican forest. Write an inequality showing the relationship of the forests of Mexico and Canada if Canada contains at least 9 times as much forest as Mexico.

79. STATISTICS To pass a course with a grade of B or better, Liza must have an average of 80 or more. Her grades on three tests are 72, 81, and 79. Write an inequality representing the score that Liza must get on the fourth test to obtain a B average or better for the course.

80. STATISTICS Sam must have an average of 70 or more in his summer course to obtain a grade of C. His first three test grades were 75, 63, and 68. Write an inequality representing the score that Sam must get on the last test to get a C grade.
> Videos

81. BUSINESS AND FINANCE Juanita is a salesperson for a manufacturing company. She may choose to receive $500 or 5% commission on her sales as payment for her work. How much does she need to sell to make the 5% offer a better deal?

82. BUSINESS AND FINANCE The cost for a long-distance telephone call is $0.36 for the first minute and $0.21 for each additional minute or portion thereof. Write an inequality representing the number of minutes a person could talk without exceeding $3.

83. GEOMETRY The perimeter of a rectangle is to be no greater than 250 cm and the length must be 105 cm. Find the maximum width of the rectangle.

```
|←——————— 105 cm ———————→|
┌─────────────────────────────┐   ↑
│                             │  x cm
└─────────────────────────────┘   ↓
```

84. STATISTICS Sarah bowled 136 and 189 in her first two games. What must she bowl in her third game to have an average of at least 170?

Basic Skills | **Advanced Skills** | Vocational-Technical Applications | Calculator/Computer | Above and Beyond | Getting Ready
▲

Translate the following statements into inequalities. Let x represent the number in each case.

85. 6 more than a number is greater than 5.

86. 3 less than a number is less than or equal to 5.

87. 4 less than twice a number is less than or equal to 7.

88. 10 more than a number is greater than negative 2.

89. 4 times a number, decreased by 15, is greater than that number.

90. 2 times a number, increased by 28, is less than or equal to 6 times that number.

Match each inequality on the right with a statement on the left.

91. x is nonnegative   **(a)** $x \geq 0$

92. x is negative   **(b)** $x \geq 5$

93. x is no more than 5   **(c)** $x \leq 5$

94. x is positive   **(d)** $x > 0$

95. x is at least 5   **(e)** $x < 5$

96. x is less than 5   **(f)** $x < 0$

Basic Skills | Advanced Skills | Vocational-Technical Applications | Calculator/Computer | **Above and Beyond** | Getting Ready
▲

97. You are the office manager for a small company. You need to acquire a new copier for the office. You find a suitable one that leases for $250 a month from the copy machine company. It costs 2.5¢ per copy to run the machine. You purchase paper for $3.50 a ream (500 sheets). If your copying budget is no more than $950 per month, is this machine a good choice? Write a brief recommendation to the purchasing department. Use equations and inequalities to explain your recommendation.

Answers

83. _____

84. _____

85. _____

86. _____

87. _____

88. _____

89. _____

90. _____

91. _____

92. _____

93. _____

94. _____

95. _____

96. _____

97. _____

Answers

98.

_____

98. Your aunt calls to ask for your help in making a decision about buying a new refrigerator. She says that she found two that seem to fit her needs, and both are supposed to last at least 14 years, according to *Consumer Reports*. The initial cost for one refrigerator is $712, but it uses only 88 kilowatt-hours (kWh) per month. The other refrigerator costs $519 and uses an estimated 100 kWh per month. You do not know the price of electricity per kilowatt-hour where your aunt lives, so you will have to decide what prices in cents per kilowatt-hour will make the first refrigerator cheaper to run during its 14 years of expected usefulness. Write your aunt a letter explaining what you did to calculate this cost, and tell her to make her decision based on how the kilowatt-hour rate she has to pay in her area compares with your estimation.

Answers

1. $9 > 6$ **3.** $7 > -2$ **5.** $0 < 4$ **7.** $-2 > -5$ **9.** x is less than 3
11. x is greater than or equal to -4 **13.** -5 is less than or equal to x

15.

17.

19.

21.

23.

25.

27.

29. $x < 13$

31. $x \geq 2$

33. $x < 7$

35. $x \leq 8$

37. $x \geq 14$

39. $x < -9$

41. $x \leq 3$

43. $x > -7$

45. $x \leq -3$

47. $x > 6$

49. $x > 20$

51. $x \leq 6$

53. $x < 9$

55. $x > 4$

57. $x > 1$

59. $x < -1$

61. $x \leq -6$

63. $x \leq 9$

65. $x > -2$

67. $x < \dfrac{7}{4}$

69. $x < 8$

71. $x \geq -\dfrac{5}{2}$

73. $x \leq \dfrac{3}{2}$

75. $x < -\dfrac{2}{3}$

77. $P < 1,000$ **79.** $x \geq 88$

81. More than $10,000 **83.** 20 cm **85.** $x + 6 > 5$ **87.** $2x - 4 \leq 7$
89. $4x - 15 > x$ **91.** (a) **93.** (c) **95.** (b)
97. Above and Beyond

Definition/Procedure	Example	Reference
Solving Equations by the Addition Property		Section 2.1
Equation		
A mathematical statement that two expressions are equal	$2x - 3 = 5$ is an equation.	*p. 187*
Solution		
A value for a variable that makes an equation a true statement	4 is a solution for the above equation because $2(4) - 3 = 5$.	*p. 188*
Equivalent Equations		
Equations that have exactly the same solutions	$2x - 3 = 5$ and $x = 4$ are equivalent equations.	*p. 190*
The Addition Property of Equality		
If $a = b$, then $a + c = b + c$.	If $2x - 3 = 7$, then $2x - 3 + 3 = 7 + 3$.	*p. 190*
Solving Equations by the Multiplication Property		Section 2.2
The Multiplication Property of Equality		
If $a = b$, then $a \cdot c = b \cdot c$.	If $\dfrac{1}{2}x = 7$, then $2\left(\dfrac{1}{2}x\right) = 2(7)$.	*p. 203*

Combining the Rules to Solve Equations · Section 2.3

Solving Linear Equations

The steps of solving a linear equation are as follows:

1. Use the distributive property to remove any grouping symbols. Then simplify by combining like terms.
2. Add or subtract the same term on each side of the equation until the variable term is on one side and a number is on the other.
3. Multiply or divide both sides of the equation by the same nonzero number so that the variable is alone on one side of the equation.
4. Check the solution in the original equation.

Solve:

$$
\begin{aligned}
3(x - 2) + 4x &= 3x + 14 \\
3x - 6 + 4x &= 3x + 14 \\
7x - 6 &= 3x + 14 \\
-3x &\ -3x \\
\hline
4x - 6 &= 14 \\
+6 &\ + 6 \\
\hline
4x &= 20 \\
\frac{4x}{4} &= \frac{20}{4} \\
x &= 5
\end{aligned}
$$

p. 221

Continued

Definition/Procedure	Example	Reference

Formulas and Problem Solving

Section 2.4

Literal Equation

An equation that involves more than one letter or variable	$a = \dfrac{2b + c}{3}$	*p. 227*

Solving Literal Equations

1. Multiply both sides of the equation by the same term to clear it of fractions.
2. Add or subtract the same term on both sides of the equation so that all terms containing the variable you are solving for are on one side.
3. Divide both sides by the coefficient of the variable that you are solving for.

Solve for b:

$$a = \frac{2b + c}{3}$$

$$3a = \left(\frac{2b + c}{3}\right)3$$

$$3a = 2b + c$$

$$3a - c = 2b$$

$$\frac{3a - c}{2} = b$$

p. 229

Applications of Linear Equations

Section 2.5

The base is the whole in a percent statement.	14 is 25% of 56. 56 is the base.	*p. 250*
The amount is the part being compared to the base.	14 is the amount.	*p. 250*
The rate is the ratio of the amount to the base.	25% is the rate.	*p. 250*
The **percent relationship** is given by $A = R \cdot B$ Amount $=$ Rate $\cdot$ Base	$\underbrace{14}_{A\,=\,\text{Amount}} = \underbrace{0.25}_{\substack{R\,=\,\text{Rate} \\ \text{in decimal} \\ \text{form}}} \cdot \underbrace{56}_{B\,=\,\text{Base}}$	*p. 251*

Inequalities—An Introduction

Section 2.6

Inequality

A statement that one quantity is less than (or greater than) another. Four symbols are used: $a < b$ $a > b$ $a \leq b$ $a \geq b$ ↑ ↑ ↑ ↑ a is less than b a is greater than b a is less than or equal to b a is greater than or equal to b	$-4 < -1$ $x^2 + 1 \geq x + 1$	*p. 264*

Definition/Procedure	Example	Reference

Graphing Inequalities

To graph $x < a$, we use an open circle and an arrow pointing left.

The heavy arrow indicates all numbers less than (or to the left of) a.

The open circle means a is not included in the solution set.

Graph $x < 3$.

p. 265

To graph $x \geq b$, we use a closed circle and an arrow pointing right.

The closed circle means that in this case b is included in the solution set.

Graph $x \geq -1$.

p. 266

Solving Inequalities

An inequality is "solved" when it is in the form $x < \square$ or $x > \square$.

Proceed as in solving equations by using the following properties.

1. If $a < b$, then $a + c < b + c$.

Adding (or subtracting) the same quantity to each side of an inequality gives an equivalent inequality.

2. If $a < b$, then $ac < bc$ when $c > 0$ and $ac > bc$ when $c < 0$.

Multiplying both sides of an inequality by the same *positive number* gives an equivalent inequality. When both sides of an inequality are multiplied by the same *negative number, you must reverse the direction* of the inequality to give an equivalent inequality.

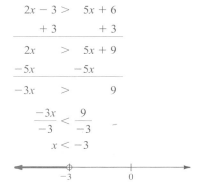

p. 267

This summary exercise set is provided to give you practice with each of the objectives of this chapter. Each exercise is keyed to the appropriate chapter section. When you are finished, you can check your answers to the odd-numbered exercises in the back of the text. If you have difficulty with any of these questions, go back and reread the examples from that section. The answers to the even-numbered exercises appear in the *Instructor's Solutions Manual.* Your instructor will give you guidelines on how best to use these exercises in your instructional setting.

2.1 *Tell whether the number shown in parentheses is a solution for the given equation.*

1. $7x + 2 = 16$ (2)

2. $5x - 8 = 3x + 2$ (4)

3. $7x - 2 = 2x + 8$ (2)

4. $4x + 3 = 2x - 11$ (−7)

5. $x + 5 + 3x = 2 + x + 23$ (6)

6. $\dfrac{2}{3}x - 2 = 10$ (21)

2.1–2.3 *Solve the following equations and check your results.*

7. $x + 5 = 7$

8. $x - 9 = 3$

9. $7 + 6x = 5x$

10. $3x - 9 = 2x$

11. $5x - 3 = 4x + 2$

12. $9x + 2 = 8x - 7$

13. $7x - 5 = 6x - 4$

14. $3 + 4x - 1 = x - 7 + 2x$

15. $4(2x + 3) = 7x + 5$

16. $5(5x - 3) = 6(4x + 1)$

17. $6x = 42$

18. $7x = -28$

19. $-6x = 24$

20. $-9x = -63$

21. $\dfrac{x}{8} = 4$

22. $-\dfrac{x}{3} = -5$

23. $\dfrac{2}{3}x = 18$

24. $\dfrac{3}{4}x = 24$

25. $5x - 3 = 12$

26. $4x + 3 = -13$

27. $7x + 8 = 3x$

28. $3 - 5x = -17$

29. $3x - 7 = x$

30. $2 - 4x = 5$

31. $\dfrac{x}{3} - 5 = 1$

32. $\dfrac{3}{4}x - 2 = 7$

33. $6x - 5 = 3x + 13$

34. $3x + 7 = x - 9$

35. $7x + 4 = 2x + 6$

36. $9x - 8 = 7x - 3$

37. $2x + 7 = 4x - 5$

38. $3x - 15 = 7x - 10$

39. $\dfrac{10}{3}x - 5 = \dfrac{4}{3}x + 7$

40. $\dfrac{11}{4}x - 15 = 5 - \dfrac{5}{4}x$ **41.** $3.7x + 8 = 1.7x + 16$ **42.** $5.4x - 3 = 8.4x + 9$

43. $3x - 2 + 5x = 7 + 2x + 21$ **44.** $8x + 3 - 2x + 5 = 3 - 4x$ **45.** $5(3x - 1) - 6x = 3x - 2$

2.4 *Solve for the indicated variable.*

46. $V = LWH$ (for L)   **47.** $P = 2L + 2W$ (for L)

48. $ax + by = c$ (for y)   **49.** $A = \dfrac{1}{2}bh$ (for h)

50. $A = P + Prt$ (for t)   **51.** $m = \dfrac{n - p}{q}$ (for n)

2.4–2.5 *Solve the following word problems. Be sure to label the unknowns and to show the equation you used.*

52. NUMBER PROBLEM The sum of 3 times a number and 7 is 25. What is the number?

53. NUMBER PROBLEM 5 times a number, decreased by 8, is 32. Find the number.

54. NUMBER PROBLEM If the sum of two consecutive integers is 85, find the two integers.

55. NUMBER PROBLEM The sum of three consecutive odd integers is 57. What are the three integers?

56. BUSINESS AND FINANCE Rafael earns $35 more per week than Andrew. If their weekly salaries total $715, what amount does each earn?

57. NUMBER PROBLEM Larry is 2 years older than Susan, and Nathan is twice as old as Susan. If the sum of their ages is 30 years, find each of their ages.

58. BUSINESS AND FINANCE Joan works on a 4% commission basis. She sold $45,000 in merchandise during 1 month. What was the amount of her commission?

59. BUSINESS AND FINANCE David buys a dishwasher that is marked down $77 from its original price of $350. What is the discount rate?

60. SCIENCE AND MEDICINE A chemist prepares a 400-milliliter (400-mL) acid-water solution. If the solution contains 30 mL of acid, what percent of the solution is acid?

61. BUSINESS AND FINANCE The price of a new compact car has increased $819 over the previous year. If this amounts to a 4.5% increase, what was the price of the car before the increase?

62. BUSINESS AND FINANCE A store advertises, "Buy the red-tagged items at 25% off their listed price." If you buy a coat marked $136, what will you pay for the coat during the sale?

63. BUSINESS AND FINANCE Tom has 6% of his salary deducted for a retirement plan. If that deduction is $168, what is his monthly salary?

64. STATISTICS A college finds that 35% of its science students take biology. If there are 252 biology students, how many science students are there altogether?

65. BUSINESS AND FINANCE A company finds that its advertising costs increased from $72,000 to $76,680 in 1 year. What was the rate of increase?

66. BUSINESS AND FINANCE A savings bank offers 3.25% on 1-year time deposits. If you place $900 in an account, how much will you have at the end of the year?

67. BUSINESS AND FINANCE Maria's company offers her a 4% pay raise. This will amount to a $126 per month increase in her salary. What is her monthly salary before and after the raise?

68. STATISTICS A computer has 8 gigabytes (GB) of storage space. Arlene is going to add 16 GB of storage space. By what percent will the available storage space be increased?

69. STATISTICS A virus scanning program is checking every file for viruses. It has completed 30% of the files in 150 s. How long should it take to check all the files?

70. BUSINESS AND FINANCE If the total bill at a restaurant for 10 people is $572.89, including an 18% tip, what was the cost of the food?

71. BUSINESS AND FINANCE A pair of running shoes is advertised at 30% off the original price for $80.15. What was the original price?

2.6 Solve and graph the solution sets for the following inequalities.

72. $x - 4 \leq 7$

73. $x + 3 > -2$

74. $5x > 4x - 3$

75. $4x \geq -12$

76. $-12x < 36$

77. $-\dfrac{x}{5} \geq 3$

78. $2x \leq 8x - 3$

79. $2x + 3 \geq 9$

80. $4 - 3x > 8$

81. $5x - 2 \leq 4x + 5$

82. $7x + 13 \geq 3x + 19$

83. $4x - 2 < 7x + 16$

This purpose of this chapter test is to help you check your progress so that you can find sections and concepts that you need to review before the next exam. Allow yourself about an hour to take this test. At the end of that hour, check your answers against those given in the back of this text. If you missed any, note the section reference that accompanies the answer. Go back to that section and reread the examples until you have mastered that particular concept.

Answers

Tell whether the number shown in parentheses is a solution for the given equation.

1. $7x - 3 = 25$ (5)

2. $8x - 3 = 5x + 9$ (4)

Solve the following equations and check your results.

3. $x - 7 = 4$

4. $7x - 12 = 6x$

5. $9x - 2 = 8x + 5$

Solve the following equations and check your results.

6. $7x = 49$

7. $\dfrac{1}{4}x = -3$

8. $\dfrac{4}{5}x = 20$

Solve the following equations and check your results.

9. $7x - 5 = 16$

10. $10 - 3x = -2$

11. $7x - 3 = 4x - 5$

12. $\dfrac{3x}{2} - 5 = 4x + \dfrac{5}{8}$

Solve for the indicated variable.

13. $C = 2\pi r$ (for r)

14. $V = \dfrac{1}{3}Bh$ (for h)

15. $3x + 2y = 6$ (for y)

Solve and graph the solution sets for the following inequalities.

16. $x - 5 \le 9$

17. $5 - 3x > 17$

1. _____
2. _____
3. _____
4. _____
5. _____
6. _____
7. _____
8. _____
9. _____
10. _____
11. _____
12. _____
13. _____
14. _____
15. _____
16. _____
17. _____

Answers

18. _____

19. _____

20. _____

21. _____

22. _____

23. _____

24. _____

25. _____

18. $5x + 13 \geq 2x + 17$

19. $2x - 3 < 7x + 2$

Solve the following word problems. Be sure to show the equation you used for the solution.

20. NUMBER PROBLEM 5 times a number, decreased by 7, is 28. What is the number?

21. NUMBER PROBLEM The sum of three consecutive integers is 66. Find the three integers.

22. NUMBER PROBLEM Jan is twice as old as Juwan, and Rick is 5 years older than Jan. If the sum of their ages is 35 years, find each of their ages.

23. GEOMETRY The perimeter of a rectangle is 62 in. If the length of the rectangle is 1 in. more than twice its width, what are the dimensions of the rectangle?

24. BUSINESS AND FINANCE Mrs. Moore made a $450 commission on the sale of a $9,000 pickup truck. What was her commission rate?

25. BUSINESS AND FINANCE Cynthia makes a 5% commission on all her sales. She earned $1,750 in commissions during 1 month. What were her gross sales for the month?

Activity 2 :: Monetary Conversions

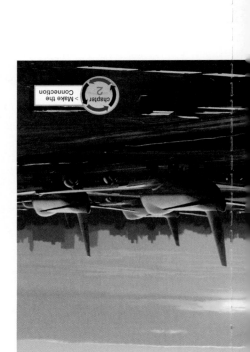

chapter 2 > Make the Connection

Each activity in this text is designed to either enhance your understanding of the topics of the preceding chapter, provide you with a mathematical extension of those topics, or both. The activities can be undertaken by one student, but they are better suited for a small group project. Occasionally it is only through discussion that different facets of the activity become apparent. For material related to this activity, visit the text website at www.mhhe.com/streeter.

In the opener to this chapter, we discussed international travel and using exchange rates to acquire local currency. In this activity, we use these exchange rates to explore the idea of variables. You should recall that a **variable** is a symbol used to represent an unknown quantity or a quantity that varies.

Currency exchange rates are published on a daily basis by many sources such as *Yahoo!Finance* and the *Wall Street Journal*. For instance, on May 20, 2006, the exchange rate for trading US$ for CAN$ was 1.1191. This means that US$1 is equivalent to CAN$1.1191. That is, if you exchanged $100 of U.S. money, you would have received $111.91 in Canadian dollars. We compute this as follows:

$$\text{CAN\$} = \text{Exchange rate} \times \text{US\$}$$

Activity

I.

1. Choose a country that you would like to visit. Use a search engine (as we learned to do in Activity 1) to find the exchange rate between US$ and the currency of your chosen country. Bookmark this page (as described in Activity 1).

2. If you are visiting for only a short time, you may not need too much money. Determine how much of the local currency you will receive in exchange for US$250.

3. If you stay for an extended period, you will need more money. How much would you receive in exchange for US$900?

In part I, we treated the amount (US$) as a *variable*. This quantity varied depending upon our needs. If we visit Canada and let $x =$ the amount exchanged in US$ and $y =$ the amount received in CAN$, then, using the exchange rate previously given, we have the equation

$$y = 1.1191x$$

You may ask, "Isn't the amount of Canadian money received (y) a variable, too?" The answer to this question is yes; in fact, all three quantities are variables. According to *Yahoo!Finance*, the exchange rate for US-CAN currency was 1.372 on December 14, 2001. The exchange rate varies on a daily basis. If we let $r =$ the exchange rate, then we can write our equation as

$$y = rx$$

II.

1. Consider the country you chose to visit in part I. Find the exchange rate for another date and repeat steps 1.2 and 1.3 for this other exchange rate.

2. Choose another nation that you would like to visit. Repeat the steps in part I for this country.

Data Set

Currency	US$	Yen (¥)	Euro (€)	CAN$	U.K. (£)	Aus$
1 US$	1	111.705	0.7833	1.1191	0.5327	1.3181
1 Yen (¥)	0.008952	1	0.007012	0.010018	0.004769	0.0118
1 Euro (€)	1.2766	142.6026	1	1.4286	0.6801	1.6827
1 CAN$	0.8936	99.8213	0.7	1	0.476	1.1779
1 U.K. (£)	1.8772	209.6924	1.4705	2.1007	1	2.4744
1 Aus$	0.7586	84.745	0.5943	0.849	0.4041	1

Source: Yahoo!Finance: 5/20/06.

I.1   We chose to visit Canada and will use the 5/20/06 exchange rate of 1.1191 from the sample data set.

I.2   Exchange rate $\times$ US$ = CAN$

$(1.1191) \cdot (US\$250) = CAN\$279.775$

We would receive $279.78 in Canadian dollars for $250 in U.S. money (round Canadian money to two decimal places).

I.3 $(1.1191) \cdot (US\$900) = CAN\$1,007.19$

II.1   Had we visited Canada on 12/14/01, we would have received an exchange rate of 1.372.

$(1.372) \cdot (US\$250) = CAN\$343$

$(1.372) \cdot (US\$900) = CAN\$1,234.80$

II.2   We chose to visit Japan. The 5/20/06 exchange rate was 111.705 Yen (¥) for each US$.

$(111.705) \cdot (US\$250) = ¥27,926.25$

$(111.705) \cdot (US\$900) = ¥100,534.5$

We would receive 27,926 yen for US$250, and 100,535 yen for US$900.

The following exercises are presented to help you review concepts from earlier chapters. This is meant as review material and not as a comprehensive exam. The answers are presented in the back of the text. Beside each answer is a section reference for the concept. If you have difficulty with any of these exercises, be certain to at least read through the summary related to that section.

Perform the indicated operations.

1. $8 + (-4)$

2. $-7 + (-5)$

3. $6 - (-2)$

4. $-4 - (-7)$

5. $(-6)(3)$

6. $(-11)(-4)$

7. $20 \div (-4)$

8. $(-50) \div (-5)$

9. $0 \div (-26)$

10. $15 \div 0$

Evaluate the expressions if $x = 5$, $y = 2$, $z = -3$, and $w = -4$.

11. $2xy$

12. $2x + 7z$

13. $3z^2$

14. $4(x + 3w)$

15. $\dfrac{2w}{y}$

16. $\dfrac{2x - w}{2y - z}$

Simplify each of the following expressions.

17. $14x^2y - 11x^2y$

18. $2x^3(3x - 5y)$

19. $\dfrac{x^2y - 2xy^2 + 3xy}{xy}$

20. $10x^2 + 5x + 2x^2 - 2x$

Solve the following equations and check your results.

21. $9x - 5 = 8x$

22. $-\dfrac{3}{4}x = 18$

23. $6x - 8 = 2x - 3$

24. $2x + 3 = 7x + 5$

25. $\dfrac{4}{3}x - 6 = 4 - \dfrac{2}{3}x$

Name _____

Section _____ Date _____

Answers

1. _____ 2. _____

3. _____ 4. _____

5. _____ 6. _____

7. _____ 8. _____

9. _____ 10. _____

11. _____ 12. _____

13. _____ 14. _____

15. _____

16. _____

17. _____

18. _____

19. _____

20. _____

21. _____

22. _____

23. _____

24. _____

25. _____

Answers

26. _____

27. _____

28. _____

29. _____

30. _____

31. _____

32. _____

33. _____

34. _____

35. _____

36. _____

37. _____

38. _____

39. _____

40. _____

Solve the following equations for the indicated variable.

26. $I = Prt$ (for r)   **27.** $A = \dfrac{1}{2}bh$ (for h)   **28.** $ax + by = c$ (for y)

Solve and graph the solution sets for the following inequalities.

29. $3x - 5 < 4$

30. $7 - 2x \geq 10$

31. $7x - 2 > 4x + 10$

32. $2x + 5 \leq 8x - 3$

Solve the following word problems. Be sure to show the equation used for the solution.

33. **NUMBER PROBLEM** If 4 times a number decreased by 7 is 45, find that number.

34. **NUMBER PROBLEM** The sum of two consecutive integers is 85. What are those two integers?

35. **NUMBER PROBLEM** If 3 times an integer is 12 more than the next consecutive odd integer, what is that integer?

36. **BUSINESS AND FINANCE** Michelle earns $120 more per week than Dmitri. If their weekly salaries total $720, how much does Michelle earn?

37. **GEOMETRY** The length of a rectangle is 2 cm more than 3 times its width. If the perimeter of the rectangle is 44 cm, what are the dimensions of the rectangle?

38. **GEOMETRY** One side of a triangle is 5 in. longer than the shortest side. The third side is twice the length of the shortest side. If the triangle perimeter is 37 in., find the length of each leg.

39. **BUSINESS AND FINANCE** Jesse paid $1,562.50 in state income tax last year. If his salary was $62,500, what was the rate of tax?

40. **BUSINESS AND FINANCE** A car is marked down from $31,500 to $29,137.50. What was the discount rate?

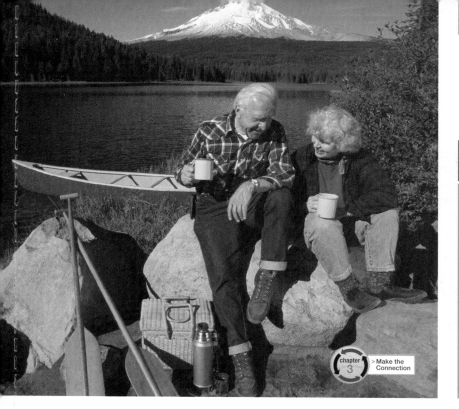

CHAPTER

3

Polynomials

INTRODUCTION

Polynomial equations are used in many disciplines and industries to find solutions and make predictions. For example, aerospace engineers use complex formulas to plan and guide space shuttle flights, and telecommunications engineers use them to improve digital signal processing. Equations expressing relationships among variables play a significant role in building construction, estimating electrical power generation needs and consumption, astronomy, medicine and pharmacological measurements, determining manufacturing costs, and projecting retail revenue.

The field of personal investments and savings presents an opportunity to estimate the future value of savings accounts, Individual Retirement Accounts, and other investment products. In the process we can even explore the power of compound interest.

CHAPTER 3 OUTLINE

3 pretest

Name _____

Section _____ Date _____

This pretest provides a preview of the types of exercises you will encounter in each section of this chapter. The answers for these exercises can be found in the back of the text. If you are working on your own, or ahead of the class, this pretest can help you identify the sections in which you should focus more of your time.

Simplify each of the following expressions. Write your answers with positive exponents only.

Answers

1. _____   2. _____

3. _____   4. _____

5. _____

6. _____

7. _____

8. _____

9. _____

10. _____

11. _____

12. _____

13. _____

14. _____

15. _____

16. _____

17. _____

18. _____

19. _____

20. _____

3.1–3.2   **1.** $x^5 x^7$ **2.** $(2x^3 y^2)(4x^2 y^5)$

3. $\dfrac{9x^5 y^2}{3x^2 y}$ **4.** $(2x^3 y^4)^2 (x^{-4} y^{-6})^0$

5. $\dfrac{(x^{-2})^{-4}}{x^{-8}}$ **6.** $2x^{-5} y^3$

Classify each polynomial as a monomial, binomial, or trinomial.

7. $6x^2 - 7x$ **8.** $-4x^3 + 5x - 9$

3.3   Add.

9. $4x^2 - 7x + 5$ and $-2x^2 + 5x - 7$

Subtract.

10. $-2x^2 + 3x - 1$ from $7x^2 - 8x + 5$

3.4   Multiply.

11. $3xy(4x^2 y^2 - 2xy + 7xy^3)$ **12.** $(3x + 2)(2x - 5)$

13. $(x + 2y)(x - 2y)$ **14.** $(4m + 5)^2$

15. $(3x - 2y)(x^2 - 4xy + 3y^2)$ **16.** $x(3x - 5y)^2$

3.5   Divide.

17. $\dfrac{28x^2 y^3 - 35x^4 y^5}{7x^2 y^2}$ **18.** $\dfrac{x^2 - x - 6}{x - 3}$

19. $\dfrac{x^3 - 2x - 3x^2 + 6}{x^2 - 2}$ **20.** $\dfrac{3x^2 + 7x - 25}{x + 4}$

3.1

Exponents and Polynomials

< 3.1 Objectives >

1 > Use the properties of exponents to simplify expressions

2 > Identify types of polynomials

3 > Find the degree of a polynomial

4 > Write a polynomial in descending-exponent form

5 > Evaluate a polynomial

Overcoming Math Anxiety

Hint #4   Preparing for a Test

Preparing for a test begins on the first day of class. Everything you do in class and at home is part of that preparation. In fact, if you attend class every day, take good notes, and keep up with the homework, then you will already be prepared and not need to "cram" for your exam.

Instead of cramming, here are a few things to focus on in the days before a scheduled test.

1. Study for your exam, but finish studying 24 hours before the test. Make certain to get some rest before taking a test.

2. Study for an exam by going over homework and class notes. Write down all of the problem types, formulas, and definitions that you think might give you trouble on the test.

3. The last item before you finish studying is to take the notes you made in step 2 and transfer the most important ideas to a 3 × 5 (index) card. You should complete this step a full 24 hours before your exam.

4. One hour before your exam, review the information on the 3 × 5 card you made in step 3. You will be surprised at how much you remember about each concept.

5. The biggest obstacle for many students is believing that they can be successful on the test. You can overcome this obstacle easily enough. If you have been completing the homework and keeping up with the classwork, then you should perform quite well on the test. Truly anxious students are often surprised to score well on an exam. These students attribute a good test score to blind luck when it is not luck at all. This is the first sign that they "get it." Enjoy the success!

Recall that exponent notation indicates repeated multiplication; the exponent tells us how many times the base is to be used as a factor.

Exponent
$\downarrow$
$$3^5 = \underbrace{3 \cdot 3 \cdot 3 \cdot 3 \cdot 3}_{5 \text{ factors}} = 243$$
$\uparrow$
Base

Now, we will look at the properties of exponents.

The first property is used when multiplying two values with the same base.

Property

Product Property of Exponents

For any real number a and positive integers m and n,

$$a^m \cdot a^n = a^{m+n}$$

In words, the product of two terms with the same base is the base taken to the power that is the sum of the exponents.

For example, $2^5 \cdot 2^7 = 2^{5+7} = 2^{12}$

The second property is used when dividing two values with the same base.

Property

Quotient Property of Exponents

For any real number a and positive integers m and n, with $m > n$,

$$\frac{a^m}{a^n} = a^{m-n}$$

For example, $\dfrac{2^{12}}{2^7} = 2^{12-7} = 2^5$

NOTE

This means that the base, x^2, is used as a factor 4 times.

Consider the following:

$$(x^2)^4 = x^2 \cdot x^2 \cdot x^2 \cdot x^2 = x^8$$

This leads us to our third property for exponents.

Property

Power to a Power Property of Exponents

For any real number a and positive integers m and n,

$$(a^m)^n = a^{m \cdot n}$$

For example, $(2^3)^2 = 2^{3 \cdot 2} = 2^6$

The use of this new property is illustrated in Example 1.

| Example 1 | **Using the Power to a Power Property of Exponents**

< Objective 1 >

  > C A U T I O N

Be careful! Be sure to distinguish between the correct use of the product property and the power to a power property.

$(x^4)^5 = x^{4\cdot 5} = x^{20}$

but

$x^4 \cdot x^5 = x^{4+5} = x^9$

Simplify each expression.

(a) $(x^4)^5 = x^{4\cdot 5} = x^{20}$

(b) $(2^3)^4 = 2^{3\cdot 4} = 2^{12}$ Multiply the exponents.

  Check Yourself 1

Simplify each expression.

(a) $(m^5)^6$ **(b)** $(m^5)(m^6)$ **(c)** $(3^2)^4$ **(d)** $(3^2)(3^4)$

NOTES

Here the base is $3x$.

We have applied the commutative and associative properties.

Suppose we have a product raised to a power, such as $(3x)^4$. We know that

$$(3x)^4 = (3x)(3x)(3x)(3x)$$
$$= (3 \cdot 3 \cdot 3 \cdot 3)(x \cdot x \cdot x \cdot x)$$
$$= 3^4 \cdot x^4 = 81x^4$$

Note that the power, here 4, has been applied to each factor, 3 and x. In general, we have:

Property

Product to a Power Property of Exponents

For any real numbers a and b and integer m,

$(ab)^m = a^m b^m$

For example, $(3x)^3 = 3^3 \cdot x^3 = 27x^3$

The use of this property is shown in Example 2.

 | Example 2 | **Using the Product to a Power Property of Exponents**

NOTE

$(2x)^5$ and $2x^5$ are entirely different expressions. For $(2x)^5$, the base is $2x$, so we raise each factor to the fifth power. For $2x^5$, the base is x, and so the exponent applies only to x.

Simplify each expression.

(a) $(2x)^5 = 2^5 \cdot x^5 = 32x^5$

(b) $(3ab)^4 = 3^4 \cdot a^4 \cdot b^4 = 81a^4b^4$

(c) $5(-2r)^3 = 5 \cdot (-2)^3 \cdot (r)^3 = 5 \cdot (-8) \cdot r^3 = -40r^3$

  Check Yourself 2

Simplify each expression.

(a) $(3y)^4$ **(b)** $(2mn)^6$ **(c)** $3(4x)^2$ **(d)** $6(-2x)^3$

We may have to use more than one property when simplifying an expression involving exponents. Consider Example 3.

Example 3 | **Using the Properties of Exponents**

NOTE

To help you understand each step of the simplification, we refer to the property being applied. Make a list of the properties now to help you as you work through the remainder of this section and Section 3.2.

Simplify each expression.

(a) $(r^4 s^3)^3 = (r^4)^3 \cdot (s^3)^3$ Product to a power property

$= r^{12} s^9$ Power to a power property

(b) $(3x^2)^2 \cdot (2x^3)^3$

$= 3^2 (x^2)^2 \cdot 2^3 \cdot (x^3)^3$ Product to a power property

$= 9x^4 \cdot 8x^9$ Power to a power property

$= 72x^{13}$ Multiply the coefficients and apply the product property.

(c) $\dfrac{(a^3)^5}{a^4} = \dfrac{a^{15}}{a^4}$ Power to a power property

$= a^{11}$ Quotient property

Check Yourself 3

Simplify each expression.

(a) $(m^5 n^2)^3$ **(b)** $(2p)^4 (-4p^2)^2$ **(c)** $\dfrac{(s^4)^3}{s^5}$

We have one final exponent property to develop. Suppose we have a quotient raised to a power. Consider the following:

$$\left(\frac{x}{3}\right)^3 = \frac{x}{3} \cdot \frac{x}{3} \cdot \frac{x}{3} = \frac{x \cdot x \cdot x}{3 \cdot 3 \cdot 3} = \frac{x^3}{3^3}$$

Note that the power, here 3, has been applied to the numerator x and to the denominator 3. This gives us our fifth property of exponents.

Property

Quotient to a Power Property of Exponents

For any real numbers a and b, when b is not equal to 0, and integer m,

$$\left(\frac{a}{b}\right)^m = \frac{a^m}{b^m}$$

For example,

$$\left(\frac{2}{5}\right)^3 = \frac{2^3}{5^3} = \frac{8}{125}$$

Example 4 illustrates the use of this property. Again note that the other properties may also be applied when simplifying an expression.

The Streeter/Hutchison Series in Mathematics   Beginning Algebra

> **Example 4**   **Using the Quotient to a Power Property of Exponents**

Simplify each expression.

(a) $\left(\dfrac{3}{4}\right)^3 = \dfrac{3^3}{4^3} = \dfrac{27}{64}$ Quotient to a power property

(b) $\left(\dfrac{x^3}{y^2}\right)^4 = \dfrac{(x^3)^4}{(y^2)^4}$ Quotient to a power property

$= \dfrac{x^{12}}{y^8}$ Power to a power property

(c) $\left(\dfrac{r^2 s^3}{t^4}\right)^2 = \dfrac{(r^2 s^3)^2}{(t^4)^2}$ Quotient to a power property

$= \dfrac{(r^2)^2 (s^3)^2}{(t^4)^2}$ Product to a power property

$= \dfrac{r^4 s^6}{t^8}$ Power to a power property

Check Yourself 4

Simplify each expression.

(a) $\left(\dfrac{2}{3}\right)^4$ **(b)** $\left(\dfrac{m^3}{n^4}\right)^5$ **(c)** $\left(\dfrac{a^2 b^3}{c^5}\right)^2$

The following table summarizes the five properties of exponents that were discussed in this section:

Property	General Form	Example
Product	$a^m a^n = a^{m+n}$	$x^2 \cdot x^3 = x^5$
Quotient	$\dfrac{a^m}{a^n} = a^{m-n} \quad (m > n)$	$\dfrac{5^7}{5^3} = 5^4$
Power to a power	$(a^m)^n = a^{mn}$	$(z^5)^4 = z^{20}$
Product to a power	$(ab)^m = a^m b^m$	$(4x)^3 = 4^3 x^3 = 64x^3$
Quotient to a power	$\left(\dfrac{a}{b}\right)^m = \dfrac{a^m}{b^m}$	$\left(\dfrac{2}{3}\right)^3 = \dfrac{2^3}{3^3} = \dfrac{8}{27}$

Our work in this chapter deals with the most common kind of algebraic expression, a *polynomial*. To define a polynomial, we recall our earlier definition of the word *term*.

> **Definition**
>
> **Term**   A **term** is a number or the product of a number and one or more variables.

This definition indicates that constants, such as the number 3, and single variables, such as x, are included.

For example, x^5, $3x$, $-4xy^2$, and 8 are terms. A **polynomial** consists of one or more terms in which the only allowable exponents are the whole numbers, 0, 1, 2, 3, . . . and so on. These terms are connected by addition or subtraction signs.

Definition	
Numerical Coefficient	In each term of a polynomial, the number is called the **numerical coefficient,** or more simply the **coefficient,** of that term.

Example 5   Identifying Polynomials

< Objective 2 >

NOTE

In a polynomial, terms are separated by $+$ and $-$ signs.

State whether each expression is a polynomial. List the terms and the coefficient of each term.

(a) $x + 3$ is a polynomial. The terms are x and 3. The coefficients are 1 and 3.

(b) $3x^2 - 2x + 5$, or $3x^2 + (-2x) + 5$, is also a polynomial. Its terms are $3x^2$, $-2x$, and 5. The coefficients are 3, -2, and 5.

(c) $5x^3 + 2 - \dfrac{3}{x}$ is *not* a polynomial because of the division by x in the third term.

Check Yourself 5

Which of the following are polynomials?

(a) $5x^2$ **(b)** $3y^3 - 2y + \dfrac{5}{y}$ **(c)** $4x^2 - \dfrac{2}{3}x + 3$

Certain polynomials are given special names because of the number of terms that they have.

Definition		
Monomial, Binomial, and Trinomial	A polynomial with one term is called a **monomial.**	The prefix *mono-* means 1.
	A polynomial with two terms is called a **binomial.**	The prefix *bi-* means 2.
	A polynomial with three terms is called a **trinomial.**	The prefix *tri-* means 3.
		There are no special names for polynomials with four or more terms.

Example 6   Identifying Types of Polynomials

(a) $3x^2y$ is a monomial. It has one term.

(b) $2x^3 + 5x$ is a binomial. It has two terms, $2x^3$ and $5x$.

(c) $5x^2 - 4x + 3$, or $5x^2 + (-4x) + 3$, is a trinomial. Its three terms are $5x^2$, $-4x$, and 3.

Check Yourself 6

Classify each of these as a monomial, binomial, or trinomial.

(a) $5x^4 - 2x^3$ **(b)** $4x^7$ **(c)** $2x^2 + 5x - 3$

RECALL

In a polynomial the allowable exponents are the whole numbers 0, 1, 2, 3, and so on. The degree will be a whole number.

We also classify polynomials by their *degree*. The **degree** of a polynomial that has only one variable is the highest power appearing in any one term.

Example 7   Classifying Polynomials by Their Degree

< Objective 3 >

The highest power

(a) $5x^3 - 3x^2 + 4x$ has degree 3.

The highest power

(b) $4x - 5x^4 + 3x^3 + 2$ has degree 4.

NOTE

We will see in the next section that $x^0 = 1$.

(c) $8x$ has degree 1.   (Because $8x = 8x^1$)

(d) 7 has degree 0.   The degree of any nonzero constant expression is zero.

Note: Polynomials can have more than one variable, such as $4x^2y^3 + 5xy^2$. The degree is then the highest sum of the powers in any single term (here $2 + 3$, or 5). In general, we will be working with polynomials in a single variable, such as x.

Check Yourself 7

Find the degree of each polynomial.

(a) $6x^5 - 3x^3 - 2$ **(b)** $5x$ **(c)** $3x^3 + 2x^6 - 1$ **(d)** 9

Working with polynomials is much easier if you get used to writing them in **descending-exponent form** (sometimes called *descending-power form*). This simply means that the term with the highest exponent is written first, then the term with the next highest exponent, and so on.

Example 8   Writing Polynomials in Descending-Exponent Form

< Objective 4 >

The exponents get smaller from left to right.

(a) $5x^7 - 3x^4 + 2x^2$ is in descending-exponent form.

(b) $4x^4 + 5x^6 - 3x^5$ is *not* in descending-exponent form. The polynomial should be written as

$$5x^6 - 3x^5 + 4x^4$$

Notice that the degree of the polynomial is the power of the *first*, or *leading*, term once the polynomial is arranged in descending-exponent form.

Check Yourself 8

Write the following polynomials in descending-exponent form.

(a) $5x^4 - 4x^5 + 7$ (b) $4x^3 + 9x^4 + 6x^8$

A polynomial can represent any number. Its value depends on the value given to the variable.

 Example 9   Evaluating Polynomials

< Objective 5 >

RECALL

Note how the rules for the order of operations are applied. See Section 0.4 for a review.

Given the polynomial

$3x^3 - 2x^2 - 4x + 1$

(a) Find the value of the polynomial when $x = 2$.

To evaluate the polynomial, substitute 2 for x.

$3(2)^3 - 2(2)^2 - 4(2) + 1$
$= 3(8) - 2(4) - 4(2) + 1$
$= 24 - 8 - 8 + 1$
$= 9$

(b) Find the value of the polynomial when $x = -2$.

Now we substitute -2 for x.

$3(-2)^3 - 2(-2)^2 - 4(-2) + 1$
$= 3(-8) - 2(4) - 4(-2) + 1$
$= -24 - 8 + 8 + 1$
$= -23$

> **CAUTION**

Be particularly careful when dealing with powers of negative numbers!

Check Yourself 9

Find the value of the polynomial

$4x^3 - 3x^2 + 2x - 1$

when

(a) $x = 3$ (b) $x = -3$

Polynomials are used in almost every professional field. Many applications are related to predictions and forecasts. In allied health, polynomials can be used to calculate the concentration of a medication in the bloodstream after a given amount of time, as the next example demonstrates.

 Example 10   An Application of Polynomial Evaluation

The concentration of digoxin, a medication prescribed for congestive heart failure, in a patient's bloodstream t hours after injection is given by the polynomial

$-0.0015t^2 + 0.0845t + 0.7170$

where concentration is measured in nanograms per milliliter (ng/mL). Determine the concentration of digoxin in a patient's bloodstream 19 hours after injection.

We are asked to evaluate the polynomial

$$-0.0015t^2 + 0.0845t + 0.7170$$

for the variable value $t = 19$. We substitute 19 for t in the polynomial.

$$-0.0015(19)^2 + 0.0845(19) + 0.7170$$
$$= -0.0015(361) + 1.6055 + 0.7170$$
$$= -0.5415 + 1.6055 + 0.7170$$
$$= 1.781$$

The concentration is 1.781 nanograms per milliliter.

Check Yourself 10

The concentration of a sedative, in micrograms per milliliter (mcg/mL), in a patient's bloodstream t hours after injection is given by the polynomial $-1.35t^2 + 10.81t + 7.38$. Determine the concentration of the sedative in a patient's bloodstream 3.5 hours after injection. Round to the nearest tenth.

Check Yourself ANSWERS

1. **(a)** m^{30}; **(b)** m^{11}; **(c)** 3^8; **(d)** 3^6 **2.** **(a)** $81y^4$; **(b)** $64m^6n^6$; **(c)** $48x^2$;

(d) $-48x^3$ **3.** **(a)** $m^{15}n^6$; **(b)** $256p^8$; **(c)** s^7 **4.** **(a)** $\dfrac{16}{81}$; **(b)** $\dfrac{m^{15}}{n^{20}}$; **(c)** $\dfrac{a^4b^6}{c^{10}}$

5. **(a)** polynomial; **(b)** not a polynomial; **(c)** polynomial   **6.** **(a)** binomial;
(b) monomial; **(c)** trinomial   **7.** **(a)** 5; **(b)** 1; **(c)** 6; **(d)** 0
8. **(a)** $-4x^5 + 5x^4 + 7$; **(b)** $6x^8 + 9x^4 + 4x^3$ **9.** **(a)** 86; **(b)** -142
10. 28.7 mcg/mL

Reading Your Text

The following fill-in-the-blank exercises are designed to ensure that you understand some of the key vocabulary used in this section.

SECTION 3.1

(a) Exponent notation indicates repeated _____.

(b) Given the expression $(3x)^2$, the power is applied to each _____.

(c) A _____ is a number or product of a number and one or more variables.

(d) In each term of a polynomial, the number is called the numerical _____.

3.1 exercises

Name _____

Section _____ Date _____

Answers

1. _____ 2. _____

3. _____ 4. _____

5. _____ 6. _____

7. _____ 8. _____

9. _____ 10. _____

11. _____ 12. _____

13. _____ 14. _____

15. _____ 16. _____

17. _____ 18. _____

19. _____ 20. _____

21. _____ 22. _____

23. _____ 24. _____

25. _____ 26. _____

27. _____ 28. _____

Basic Skills | Advanced Skills | Vocational-Technical Applications | Calculator/Computer | Above and Beyond | Getting Ready

< Objective 1 >

Use the power to a power property of exponents to simplify each of the following expressions.

1. $(x^2)^3$

2. $(a^5)^3$

3. $(m^4)^4$

4. $(p^7)^2$

5. $(2^4)^2$

6. $(3^3)^2$

7. $(5^3)^5$

8. $(7^2)^4$

Use the five properties of exponents to simplify each of the following expressions.

9. $(3x)^3$

10. $(4m)^2$

11. $(2xy)^4$

12. $(5pq)^3$

13. $\left(\dfrac{3}{4}\right)^2$

14. $\left(\dfrac{2}{3}\right)^3$

15. $\left(\dfrac{x}{5}\right)^3$

16. $\left(\dfrac{a}{2}\right)^5$

17. $(2x^2)^4$

18. $(3y^2)^5$

19. $(a^8b^6)^2$

20. $(p^3q^4)^2$

21. $(4x^2y)^3$

22. $(4m^4n^4)^2$ > Videos

23. $(3m^2)^4(-2m^3)^2$

24. $(-2y^4)^3(4y^3)^2$

25. $\dfrac{(x^4)^3}{x^2}$

26. $\dfrac{(m^5)^3}{m^6}$

27. $\dfrac{(s^3)^2(s^2)^3}{(s^5)^2}$

28. $\dfrac{(y^5)^3(y^3)^2}{(y^4)^4}$ > Videos

29. $\left(\dfrac{m^3}{n^2}\right)^3$

30. $\left(\dfrac{a^4}{b^3}\right)^4$

31. $\left(\dfrac{a^3 b^2}{c^4}\right)^2$

32. $\left(\dfrac{x^5 y^2}{z^4}\right)^3$

Complete each of the following statements with **never, sometimes,** *or* **always.**

33. A polynomial is _____ a trinomial.

34. A trinomial is _____ a polynomial.

35. The product of two monomials is _____ a monomial.

36. A term is _____ a binomial.

< Objective 2 >

Which of the following expressions are polynomials?

37. $7x^3$

38. $5x^3 - \dfrac{3}{x}$

39. 7

40. $4x^3 + x$

41. $\dfrac{3 + x}{x^2}$

42. $5a^2 - 2a + 7$

For each of the following polynomials, list the terms and the coefficients.

43. $2x^2 - 3x$

44. $5x^3 + x$

45. $4x^3 - 3x + 2$ > Videos

46. $7x^2$

Classify each of the following as a **monomial, binomial,** *or* **trinomial** *where possible.*

47. $7x^3 - 3x^2$

48. $4x^7$

49. $7y^2 + 4y + 5$

50. $2x^2 + \dfrac{1}{3}xy + y^2$

51. $2x^4 - 3x^2 + 5x - 2$

52. $x^4 + \dfrac{5}{x} + 7$

Answers

29. _____ 30. _____

31. _____ 32. _____

33. _____ 34. _____

35. _____

36. _____

37. _____

38. _____

39. _____

40. _____

41. _____

42. _____

43. _____

44. _____

45. _____

46. _____

47. _____

48. _____

49. _____

50. _____

51. _____

52. _____

Answers

53. _____

54. _____

55. _____

56. _____

57. _____

58. _____

59. _____

60. _____

61. _____

62. _____

63. _____

64. _____

65. _____

66. _____

67. _____ 68. _____

69. _____ 70. _____

71. _____ 72. _____

73. _____ 74. _____

75. _____ 76. _____

77. _____ 78. _____

53. $6y^8$

54. $4x^4 - 2x^2 + \dfrac{3}{4}x - 7$

55. $x^5 - \dfrac{3}{x^2}$

56. $4x^2 - 9$

< Objectives 3–4 >

Arrange in descending-exponent form if necessary, and give the degree of each polynomial.

57. $4x^5 - 3x^2$

58. $5x^2 + 3x^3 + 4$

59. $7x^7 - 5x^9 + 4x^3$

60. $2 + x$

61. $4x$

62. $x^{17} - 3x^4$

63. $5x^2 - 3x^5 + x^6 - 7$ > Videos

64. 5

< Objective 5 >

Evaluate each of the following polynomials for the given values of the variable.

65. $6x + 1$, $x = 1$ and $x = -1$

66. $5x - 5$, $x = 2$ and $x = -2$

67. $x^3 - 2x$, $x = 2$ and $x = -2$

68. $3x^2 + 7$, $x = 3$ and $x = -3$

69. $3x^2 + 4x - 2$, $x = 4$ and $x = -4$

70. $2x^2 - 5x + 1$, $x = 2$ and $x = -2$

71. $-x^2 - 2x + 3$, $x = 1$ and $x = -3$

72. $-x^2 - 5x - 6$, $x = -3$ and $x = -2$

< Objectives 1–5 >

*Determine whether each of the following statements is **always true, sometimes true,** or **never true.***

73. A monomial is a polynomial.

74. A binomial is a trinomial.

75. The degree of a trinomial is 3.

76. A trinomial has three terms.

77. A polynomial has four or more terms.

78. A binomial must have two coefficients.

Basic Skills | Advanced Skills | Vocational-Technical Applications | Calculator/Computer | **Above and Beyond** | Getting Ready

▲

Solve the following problems.

79. Write x^{12} as a power of x^2.

80. Write y^{15} as a power of y^3.

81. Write a^{16} as a power of a^2.

82. Write m^{20} as a power of m^5.

83. Write each of the following as a power of 8. (Remember that $8 = 2^3$.)
$2^{12}, 2^{18}, (2^5)^3, (2^7)^6$

84. Write each of the following as a power of 9.
$3^8, 3^{14}, (3^5)^8, (3^4)^7$

85. What expression raised to the third power is $-8x^6y^9z^{15}$?

86. What expression raised to the fourth power is $81x^{12}y^8z^{16}$?

The formula $(1 + R)^y = G$ gives us useful information about the growth of a population. Here R is the rate of growth expressed as a decimal, y is the time in years, and G is the growth factor. If a country has a 2% growth rate for 35 years, then its population will double:

$(1.02)^{35} \approx 2$

87. SOCIAL SCIENCE

    (a) With a 2% growth rate, how many doublings will occur in 105 years? How much larger will the country's population be to the nearest whole number?

    (b) The less-developed countries of the world had an average growth rate of 2% in 1986. If their total population was 3.8 billion, what will their population be in 105 years if this rate remains unchanged?

88. SOCIAL SCIENCE The United States has a growth rate of 0.7%. What will be its growth factor after 35 years (to the nearest percent)?

89. Write an explanation of why $(x^3)(x^4)$ is *not* x^{12}.

90. Your algebra study partners are confused. "Why isn't $x^2 \cdot x^3 = 2x^5$?" they ask you. Write an explanation that will convince them.

Answers

79. _____

80. _____

81. _____

82. _____

83. _____

84. _____

85. _____

86. _____

87. _____

88. _____

89. _____

90. _____

Answers

91. _____

92. _____

93. _____

94. _____

95. _____

96. _____

97. _____

98. _____

99. _____

100. _____

101. _____

102. _____

103. _____

104. _____

105. _____

106. _____

(a) _____ (b) _____

(c) _____ (d) _____

(e) _____ (f) _____

(g) _____ (h) _____

Capital italic letters such as P and Q are often used to name polynomials. For example, we might write $P(x) = 3x^3 - 5x^2 + 2$ in which $P(x)$ is read "P of x." The notation permits a convenient shorthand. We write $P(2)$, read "P of 2," to indicate the value of the polynomial when $x = 2$. Here

$$P(2) = 3(2)^3 - 5(2)^2 + 2$$
$$= 3 \cdot 8 - 5 \cdot 4 + 2$$
$$= 6$$

Use the preceding information in the following problems.

If $P(x) = x^3 - 2x^2 + 5$ and $Q(x) = 2x^2 + 3$, find:

91. $P(1)$ **92.** $P(-1)$

93. $Q(2)$ **94.** $Q(-2)$

95. $P(3)$ **96.** $Q(-3)$

97. $P(0)$ **98.** $Q(0)$

99. $P(2) + Q(-1)$ **100.** $P(-2) + Q(3)$

101. $P(3) - Q(-3) \div Q(0)$ **102.** $Q(-2) \div Q(2) \cdot P(0)$

103. $\left| Q(4) \right| - \left| P(4) \right|$ **104.** $\dfrac{P(-1) + Q(0)}{P(0)}$

105. BUSINESS AND FINANCE The cost, in dollars, of typing a term paper is given as 3 times the number of pages plus 20. Use y as the number of pages to be typed and write a polynomial to describe this cost. Find the cost of typing a 50-page paper.

106. BUSINESS AND FINANCE The cost, in dollars, of making suits is described as 20 times the number of suits plus 150. Use s as the number of suits and write a polynomial to describe this cost. Find the cost of making seven suits.

Basic Skills | Advanced Skills | Vocational-Technical Applications | Calculator/Computer | Above and Beyond | **Getting Ready** ▲

Simplify each of the following fractions.

(a) $\dfrac{2^3}{2^5}$ (b) $\dfrac{3^7}{3^{10}}$ (c) $\dfrac{4^3}{4^9}$ (d) $\dfrac{5^4}{5^8}$

(e) $\dfrac{2^3}{2^3}$ (f) $\dfrac{3^5}{3^5}$ (g) $\dfrac{4^7}{4^7}$ (h) $\dfrac{5^{10}}{5^{10}}$

Answers

1. x^6 **3.** m^{16} **5.** 2^8 **7.** 5^{15} **9.** $27x^3$ **11.** $16x^4y^4$

13. $\dfrac{9}{16}$ **15.** $\dfrac{x^3}{125}$ **17.** $16x^8$ **19.** $a^{16}b^{12}$ **21.** $64x^6y^3$

23. $324m^{14}$ **25.** x^{10} **27.** s^2 **29.** $\dfrac{m^9}{n^6}$ **31.** $\dfrac{a^6b^4}{c^8}$

33. sometimes   **35.** always   **37.** Polynomial   **39.** Polynomial
41. Not a polynomial   **43.** $2x^2, -3x; 2, -3$ **45.** $4x^3, -3x, 2; 4, -3, 2$
47. Binomial   **49.** Trinomial   **51.** Not classified   **53.** Monomial
55. Not a polynomial   **57.** $4x^5 - 3x^2; 5$ **59.** $-5x^9 + 7x^7 + 4x^3; 9$
61. $4x; 1$ **63.** $x^6 - 3x^5 + 5x^2 - 7; 6$ **65.** $7, -5$ **67.** $4, -4$
69. $62, 30$ **71.** $0, 0$ **73.** Always   **75.** Sometimes
77. Sometimes   **79.** $(x^2)^6$ **81.** $(a^2)^8$ **83.** $8^4, 8^6, 8^5, 8^{14}$
85. $-2x^2y^3z^5$ **87. (a)** Three doublings, 8 times as large; **(b)** 30.4 billion
89. Above and Beyond   **91.** 4   **93.** 11   **95.** 14   **97.** 5

99. 10   **101.** 7   **103.** -2 **105.** $3y + 20$, \$170   **(a)** $\dfrac{1}{2^2}$

(b) $\dfrac{1}{3^3}$ **(c)** $\dfrac{1}{4^6}$ **(d)** $\dfrac{1}{5^4}$ **(e)** 1   **(f)** 1   **(g)** 1   **(h)** 1

3.2

Negative Exponents and Scientific Notation

< 3.2 Objectives >

1 > Evaluate expressions involving a zero or negative exponent

2 > Simplify expressions involving a zero or negative exponent

3 > Write a decimal number in scientific notation

4 > Solve an application using scientific notation

In Section 3.1, we discussed exponents. We now want to extend our exponent notation to include 0 and negative integers as exponents.

First, what do we do with x^0? It will help to look at a problem that gives us x^0 as a result. What if the numerator and denominator of a fraction have the same base raised to the same power and we extend our division rule? For example,

$$\frac{a^5}{a^5} = a^{5-5} = a^0$$

But from our experience with fractions we know that

$$\frac{a^5}{a^5} = 1$$

By comparing these equations, it seems reasonable to make the following definition:

RECALL

By the quotient property,
$$\frac{a^m}{a^n} = a^{m-n}$$
when $m > n$. Here m and n are *both* 5, so $m = n$.

Definition

Zero Power

For any nonzero number a,

$a^0 = 1$

In words, any expression, except 0, raised to the 0 power is 1.

Example 1 illustrates the use of this definition.

Example 1 | Raising Expressions to the Zero Power

< Objective 1 >

> CAUTION

In part **(d)** the 0 exponent applies only to the x and *not* to the factor 6, because the base is x.

Evaluate each expression. Assume all variables are nonzero.

(a) $5^0 = 1$

(b) $27^0 = 1$

(c) $(x^2 y)^0 = 1$ if $x \neq 0$ and $y \neq 0$

(d) $6x^0 = 6 \cdot 1 = 6$ if $x \neq 0$

310

Check Yourself 1

Evaluate each expression. Assume all variables are nonzero.

(a) 7^0 (b) $(-8)^0$ (c) $(xy^3)^0$ (d) $3x^0$

Before we introduce the next property, we will look at some examples that use the properties of Section 3.1.

Example 2   **Evaluating Expressions**

Evaluate each expression.

(a) $\dfrac{5^6}{5^2}$ From our earlier work, we get $5^{6-2} = 5^4 = 625$.

(b) $\dfrac{5^2}{5^6}$

$$\frac{5^2}{5^6} = \frac{5 \cdot 5}{5 \cdot 5 \cdot 5 \cdot 5 \cdot 5 \cdot 5} = \frac{1}{5^4} = \frac{1}{625}$$

(c) $\dfrac{10^3}{10^9} = \dfrac{10 \cdot 10 \cdot 10}{10 \cdot 10 \cdot 10 \cdot 10 \cdot 10 \cdot 10 \cdot 10 \cdot 10 \cdot 10} = \dfrac{1}{10^6}$ or $\dfrac{1}{1,000,000}$

Check Yourself 2

Evaluate each expression.

(a) $\dfrac{5^9}{5^6}$ (b) $\dfrac{5^6}{5^9}$ (c) $\dfrac{10^6}{10^{10}}$

NOTES

Divide the numerator and denominator by the two common factors of x.

John Wallis (1616–1703), an English mathematician, was the first to fully discuss the meaning of 0 and negative exponents.

The quotient property of exponents allows us to define a negative exponent. Suppose that the exponent in the denominator is *greater than* the exponent in the numerator. Consider the expression $\dfrac{x^2}{x^5}$.

Our previous work with fractions tells us that

$$\frac{x^2}{x^5} = \frac{x \cdot x}{x \cdot x \cdot x \cdot x \cdot x} = \frac{1}{x^3}$$

However, if we extend the quotient property to let n be greater than m, we have

$$\frac{x^2}{x^5} = x^{2-5} = x^{-3}$$

Now, by comparing these equations, it seems reasonable to define x^{-3} as $\dfrac{1}{x^3}$.

In general, we have this result:

Definition

Negative Powers

For any number a, $a \neq 0$, and any positive integer n,

$$a^{-n} = \frac{1}{a^n}$$

This definition tells us that if we have a base a raised to a negative integer power, such as a^{-5}, we may rewrite this as 1 over the base a raised to a positive integer power: $\frac{1}{a^5}$. We work with this in Example 3.

Example 3   **Rewriting Expressions That Contain Negative Exponents**

< Objective 2 >

Rewrite each expression using only positive exponents. Simplify when possible.

Negative exponent in numerator

(a) $x^{-4} = \dfrac{1}{x^4}$

Positive exponent in denominator

(b) $m^{-7} = \dfrac{1}{m^7}$

(c) $3^{-2} = \dfrac{1}{3^2}$ or $\dfrac{1}{9}$

(d) $\dfrac{1}{10^{-3}} = \dfrac{1}{\left(\frac{1}{10}\right)^3} = \dfrac{1 \cdot \left(\frac{10}{1}\right)^3}{\left(\frac{1}{10}\right)^3 \cdot \left(\frac{10}{1}\right)^3} = \left(\dfrac{10}{1}\right)^3 = 1{,}000$

A negative power in the denominator is equivalent to a positive power in the numerator.

So, $\dfrac{1}{x^{-3}} = x^3$

(e) $2x^{-3} = 2 \cdot \dfrac{1}{x^3} = \dfrac{2}{x^3}$

The -3 exponent applies only to x, because x is the base.

> CAUTION

$2x^{-3}$ is not the same as $(2x)^{-3}$.

(f) $\left(\dfrac{2}{5}\right)^{-1} = \dfrac{1}{\frac{2}{5}} = \dfrac{5}{2}$

(g) $-4x^{-5} = -4 \cdot \dfrac{1}{x^5} = -\dfrac{4}{x^5}$

Check Yourself 3

Write each expression using only positive exponents.

(a) a^{-10} **(b)** 4^{-3} **(c)** $3x^{-2}$ **(d)** $\left(\dfrac{3}{2}\right)^{-2}$

We will now allow negative integers as exponents in our product property for exponents. Consider Example 4.

 Example 4   **Simplifying Expressions Containing Exponents**

> **RECALL**
>
> $a^m \cdot a^n = a^{m+n}$ for *any* integers m and n. So add the exponents.

Simplify (write an equivalent expression that uses only positive exponents).

(a) $x^5 x^{-2} = x^{5+(-2)} = x^3$

Note: An alternative approach would be

$$x^5 x^{-2} = x^5 \cdot \frac{1}{x^2} = \frac{x^5}{x^2} = x^3$$

(b) $a^7 a^{-5} = a^{7+(-5)} = a^2$

(c) $y^5 y^{-9} = y^{5+(-9)} = y^{-4} = \frac{1}{y^4}$

Check Yourself 4

Simplify (write an equivalent expression that uses only positive exponents).

(a) $x^7 x^{-2}$ **(b)** $b^3 b^{-8}$

Example 5 shows that all the properties of exponents introduced in the last section can be extended to expressions with negative exponents.

 Example 5   **Simplifying Expressions Containing Exponents**

Simplify each expression.

(a) $\dfrac{m^{-3}}{m^4} = m^{-3-4}$ Quotient property

$= m^{-7} = \dfrac{1}{m^7}$

> **NOTE**
>
> This could also be done by using the power to a power property first, so
>
> $(2x^4)^{-3} = 2^{-3} \cdot (x^4)^{-3} = 2^{-3} x^{-12}$
>
> $= \dfrac{1}{2^3 x^{12}}$
>
> $= \dfrac{1}{8x^{12}}$

(b) $\dfrac{a^{-2} b^6}{a^5 b^{-4}} = a^{-2-5} b^{6-(-4)}$ Apply the quotient property to each variable.

$= a^{-7} b^{10} = \dfrac{b^{10}}{a^7}$

(c) $(2x^4)^{-3} = \dfrac{1}{(2x^4)^3}$ Definition of the negative exponent

$= \dfrac{1}{2^3 (x^4)^3}$ Product to a power property

$= \dfrac{1}{8x^{12}}$ Power to a power property

(d) $\dfrac{(y^{-2})^4}{(y^3)^{-2}} = \dfrac{y^{-8}}{y^{-6}}$ Power to a power property

$\qquad\qquad\quad = y^{-8-(-6)}$ Quotient property

$\qquad\qquad\quad = y^{-2} = \dfrac{1}{y^2}$

Check Yourself 5

Simplify each expression.

(a) $\dfrac{x^5}{x^{-3}}$ **(b)** $\dfrac{m^3 n^{-5}}{m^{-2} n^3}$ **(c)** $(3a^3)^{-4}$ **(d)** $\dfrac{(r^3)^{-2}}{(r^{-4})^2}$

NOTE

This must equal
2,300,000,000.

NOTE

Consider the following table:

$2.3 = 2.3 \times 10^0$
$23 = 2.3 \times 10^1$
$230 = 2.3 \times 10^2$
$2300 = 2.3 \times 10^3$
$23{,}000 = 2.3 \times 10^4$
$230{,}000 = 2.3 \times 10^5$

> Calculator

Let us now take a look at an important use of exponents, scientific notation.

We begin the discussion with a calculator exercise. On most calculators, if you multiply 2.3 times 1,000, the display will read

2300

Multiply by 1,000 a second time. Now you will see

2300000.

On most calculators, multiplying by 1,000 a third time will result in the display

2.3 09   or   2.3 E09

And multiplying by 1,000 again yields

2.3 12   or   2.3 E12

Can you see what is happening? This is the way calculators display very large numbers. The number on the left is always between 1 and 10, and the number on the right indicates the number of places the decimal point must be moved to the right to put the answer in standard (or decimal) form.

This notation is used frequently in science. It is not uncommon in scientific applications of algebra to find yourself working with very large or very small numbers. Even in the time of Archimedes (287–212 B.C.E.), the study of such numbers was not unusual. Archimedes estimated that the universe was 23,000,000,000,000,000 m in diameter, which is the approximate distance light travels in $2\frac{1}{2}$ years. By comparison, Polaris (the North Star) is actually 680 light-years from Earth. Example 7 will discuss the idea of light-years.

In scientific notation, Archimedes' estimate for the diameter of the universe would be

2.3×10^{16} m

If a number is divided by 1,000 again and again, we get a negative exponent on the calculator. In scientific notation, we use positive exponents to write very large numbers, such as the distance of stars. We use negative exponents to write very small numbers, such as the width of an atom.

Definition

Scientific Notation	Any number written in the form

$$a \times 10^n$$

in which $1 \le a < 10$ and n is an integer, is written in **scientific notation**.

 Example 6   Using Scientific Notation

< Objective 3 >

Write each of the following numbers in scientific notation.

(a) $120{,}000. = 1.2 \times 10^5$
    5 places   The power is 5.

NOTE

The exponent on 10 shows the *number of places* we must move the decimal point. A positive exponent tells us to move right, and a negative exponent indicates a move to the left.

(b) $88{,}000{,}000. = 8.8 \times 10^7$
    7 places   The power is 7.

(c) $520{,}000{,}000. = 5.2 \times 10^8$
    8 places

(d) $4{,}000{,}000{,}000. = 4 \times 10^9$
    9 places

(e) $0.0005 = 5 \times 10^{-4}$ If the decimal point is to be moved to the
    4 places   left, the exponent will be negative.

NOTE

To convert back to standard or decimal form, the process is simply reversed.

(f) $0.0000000081 = 8.1 \times 10^{-9}$
    9 places

 Check Yourself 6

Write in scientific notation.

(a) 212,000,000,000,000,000   **(b)** 0.00079
(c) 5,600,000   **(d)** 0.0000007

Example 7   An Application of Scientific Notation

< Objective 4 >

(a) Light travels at a speed of 3.0×10^8 meters per second (m/s). There are approximately 3.15×10^7 s in a year. How far does light travel in a year?

We multiply the distance traveled in 1 s by the number of seconds in a year. This yields

NOTE

$9.45 \times 10^{15} \approx 10 \times 10^{15} = 10^{16}$

$$(3.0 \times 10^8)(3.15 \times 10^7) = (3.0 \cdot 3.15)(10^8 \cdot 10^7)$$
Multiply the coefficients, and add the exponents.
$$= 9.45 \times 10^{15}$$

For our purposes we round the distance light travels in 1 year to 10^{16} m. This unit is called a **light-year,** and it is used to measure astronomical distances.

(b) The distance from Earth to the star Spica (in Virgo) is 2.2×10^{18} m. How many light-years is Spica from Earth?

$$\frac{2.2 \times 10^{18}}{10^{16}} = 2.2 \times 10^{18-16}$$

$$= 2.2 \times 10^2 = 220 \text{ light-years}$$

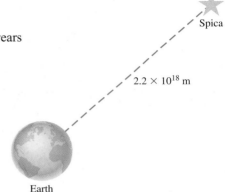

Spica

2.2×10^{18} m

Earth

NOTE

We divide the distance (in meters) by the number of meters in 1 light-year.

 Check Yourself 7

The farthest object that can be seen with the unaided eye is the Andromeda galaxy. This galaxy is 2.3×10^{22} m from Earth. What is this distance in light-years?

 Check Yourself ANSWERS

1. (a) 1; **(b)** 1; **(c)** 1; **(d)** 3   **2. (a)** 125; **(b)** $\dfrac{1}{125}$; **(c)** $\dfrac{1}{10,000}$ **3. (a)** $\dfrac{1}{a^{10}}$;

(b) $\dfrac{1}{4^3}$ or $\dfrac{1}{64}$; **(c)** $\dfrac{3}{x^2}$; **(d)** $\dfrac{4}{9}$ **4. (a)** x^5; **(b)** $\dfrac{1}{b^5}$ **5. (a)** x^8; **(b)** $\dfrac{m^5}{n^8}$; **(c)** $\dfrac{1}{81a^{12}}$;

(d) r^2 **6. (a)** 2.12×10^{17}; **(b)** 7.9×10^{-4}; **(c)** 5.6×10^6; **(d)** 7×10^{-7}

7. 2,300,000 light-years

Reading Your Text

The following fill-in-the-blank exercises are designed to ensure that you understand some of the key vocabulary used in this section.

SECTION 3.2

(a) Any number, except _____, when raised to the 0 power is 1.

(b) John Wallis, an English mathematician, was the first to fully discuss the meaning of zero and _____ exponents.

(c) Spica is actually 220 _____-years from Earth.

(d) _____ travels at a speed of 3.0×10^8 meters per second.

< Objective 1 >

Evaluate (assume the variables are nonzero).

1. 4^0

2. $(-7)^0$

3. $(-29)^0$

4. 75^0

5. $(x^3 y^2)^0$

6. $7m^0$

7. $11x^0$ > Videos

8. $(2a^3 b^7)^0$

9. $(-3p^6 q^8)^0$

10. $-7x^0$

< Objective 2 >

Write each of the following expressions using positive exponents; simplify when possible.

11. b^{-8}

12. p^{-12}

13. 3^{-4}

14. 2^{-5}

15. $\left(\dfrac{1}{5}\right)^{-2}$

16. $\left(\dfrac{1}{4}\right)^{-3}$

17. $\dfrac{1}{10^{-4}}$

18. $\dfrac{1}{10^{-5}}$

19. $5x^{-1}$

20. $3a^{-2}$

21. $(5x)^{-1}$

22. $(3a)^{-2}$

23. $-2x^{-5}$

24. $3x^{-4}$

25. $(-2x)^{-5}$ > Videos

26. $(3x)^{-4}$

Name _____

Section _____ Date _____

Answers

1. _____   2. _____

3. _____   4. _____

5. _____   6. _____

7. _____   8. _____

9. _____   10. _____

11. _____   12. _____

13. _____   14. _____

15. _____   16. _____

17. _____   18. _____

19. _____   20. _____

21. _____   22. _____

23. _____   24. _____

25. _____   26. _____

Answers

27. _____ 28. _____

29. _____ 30. _____

31. _____ 32. _____

33. _____

34. _____

35. _____

36. _____

37. _____

38. _____

39. _____

40. _____

41. _____

42. _____

43. _____

44. _____

45. _____

46. _____

47. _____

48. _____

49. _____

50. _____

Use the product and quotient properties to simplify each of the following expressions. Write your answers with only positive exponents.

27. $a^5 a^3$

28. $m^5 m^7$

29. $x^8 x^{-2}$

30. $a^{12} a^{-8}$

31. $x^0 x^5$

32. $r^{-3} r^0$

33. $\dfrac{a^8}{a^5}$

34. $\dfrac{m^9}{m^4}$

35. $\dfrac{x^7}{x^9}$ > Videos

36. $\dfrac{a^3}{a^{10}}$

Determine whether each of the following statements is **true** *or* **false.**

37. Zero raised to any power is one.

38. One raised to any power is one.

39. When multiplying two terms with the same base, add the exponents to find the power of that base in the product.

40. When multiplying two terms with the same base, multiply the exponents to find the power of that base in the product.

Simplify each of the following expressions. Write your answers with positive exponents only.

41. $\dfrac{x^{-4} yz}{x^{-5} yz}$

42. $\dfrac{p^{-6} q^{-3}}{p^{-3} q^{-6}}$

43. $\dfrac{m^5 n^{-3}}{m^{-4} n^5}$

44. $\dfrac{p^{-3} q^{-2}}{p^4 q^{-3}}$ > Videos

45. $(-2a^{-3})^4$

46. $(3x^2)^{-3}$

47. $(x^{-2} y^3)^{-2}$

48. $(-a^5 b^{-3})^{-3}$

49. $\dfrac{(r^{-2})^3}{r^{-4}}$

50. $\dfrac{(y^3)^{-4}}{y^{-6}}$

Basic Skills | **Advanced Skills** | Vocational-Technical Applications | Calculator/Computer | Above and Beyond | Getting Ready
▲

Simplify each of the following expressions. Write your answers with positive exponents only.

51. $\dfrac{m^{-2}n^3}{m^2n^4}$

52. $\dfrac{c^2d^{-3}}{c^{-4}d^{-5}}$

53. $\dfrac{r^3s^{-3}}{s^4t^{-2}}$

54. $\dfrac{x^3yz^{-2}}{x^{-2}y^{-3}z^4}$

55. $\dfrac{a^{-5}(b^2)^{-3}c^{-1}}{a(b^{-4})^3c^{-1}}$

56. $\dfrac{x^4y^{-3}z}{(xy^2)^{-2}z^{-1}}$

57. $\dfrac{(p^0q^{-2})^{-3}}{p(q^0)^2(p^{-1}q)^0}$

58. $\dfrac{x^{-1}(x^2y^{-2})^{-3}z^{-2}}{xy^{-3}z^0}$

59. $3(2x^{-2})^{-3}$

60. $2b^{-1}(2b^{-3})^{-2}$

61. $ab^{-2}(a^{-3}b^0)^{-2}$

62. $m^{-1}(m^2n^{-3})^2$

63. $2a^6(3a^{-4})^2$

64. $4x^2y^{-1}(2x^{-2}y^3)^{-2}$

65. $\left[c(c^{-2}d^0)^{-2}\right]^3$

66. $\left[x^2y(x^4y^{-3})^{-1}\right]^{-2}$

67. $\dfrac{w(w^2)^{-3}}{(w^2)^{-2}}$

68. $\dfrac{(2n^2)^{-3}}{(2n^{-2})^{-4}}$

69. $\dfrac{a^{-5}(a^2)^{-3}}{a(a^{-4})^3}$

70. $\dfrac{y^2(y^2)^{-2}}{(y^3)^{-2}(y^0)^2}$

< Objective 3 >

In exercises 71–74, express each number in scientific notation.

71. SCIENCE AND MEDICINE The distance from Earth to the Sun: 93,000,000 mi.

> Videos

Answers

51. _____

52. _____

53. _____

54. _____

55. _____

56. _____

57. _____

58. _____

59. _____

60. _____

61. _____

62. _____

63. _____

64. _____

65. _____

66. _____

67. _____

68. _____

69. _____

70. _____

71. _____

Answers

72. _____

73. _____

74. _____

75. _____

76. _____

77. _____

78. _____

79. _____

80. _____

81. _____

82. _____

83. _____

84. _____

85. _____

86. _____

87. _____

88. _____

89. _____

90. _____

91. _____

92. _____

72. **SCIENCE AND MEDICINE** The diameter of a grain of sand: 0.000021 m.

73. **SCIENCE AND MEDICINE** The diameter of the Sun: 130,000,000,000 cm.

74. **SCIENCE AND MEDICINE** The number of molecules in 22.4 L of a gas: 602,000,000,000,000,000,000,000 (Avogadro's number).

75. **SCIENCE AND MEDICINE** The mass of the Sun is approximately 1.99×10^{30} kg. If this were written in standard or decimal form, how many 0's would follow the second 9s digit?

76. **SCIENCE AND MEDICINE** Archimedes estimated the universe to be 2.3×10^{19} millimeters (mm) in diameter. If this number were written in standard or decimal form, how many 0's would follow the digit 3?

In exercises 77 to 80, write each expression in standard notation.

77. 8×10^{-3}

78. 7.5×10^{-6}

79. 2.8×10^{-5}

80. 5.21×10^{-4}

In exercises 81 to 84, write each of the following in scientific notation.

81. 0.0005

82. 0.000003

83. 0.00037

84. 0.000051

In exercises 85 to 88, compute the expressions using scientific notation, and write your answer in that form.

85. $(4 \times 10^{-3})(2 \times 10^{-5})$

86. $(1.5 \times 10^{-6})(4 \times 10^{2})$

87. $\dfrac{9 \times 10^{3}}{3 \times 10^{-2}}$

88. $\dfrac{7.5 \times 10^{-4}}{1.5 \times 10^{2}}$

In exercises 89 to 94, perform the indicated calculations. Write your result in scientific notation.

89. $(2 \times 10^{5})(4 \times 10^{4})$

90. $(2.5 \times 10^{7})(3 \times 10^{5})$

91. $\dfrac{6 \times 10^{9}}{3 \times 10^{7}}$

92. $\dfrac{4.5 \times 10^{12}}{1.5 \times 10^{7}}$

93. $\dfrac{(3.3 \times 10^{15})(6 \times 10^{15})}{(1.1 \times 10^{8})(3 \times 10^{6})}$

94. $\dfrac{(6 \times 10^{12})(3.2 \times 10^{8})}{(1.6 \times 10^{7})(3 \times 10^{2})}$

> Videos

< Objectives 1–2 >

In 1975 the population of Earth was approximately 4 billion and doubling every 35 years. The formula for the population P in year y for this doubling rate is

P (in billions) $= 4 \times 2^{(y-1975)/35}$

95. SOCIAL SCIENCE What was the approximate population of Earth in 1960?

96. SOCIAL SCIENCE What will Earth's population be in 2025?

The U.S. population in 1990 was approximately 250 million, and the average growth rate for the past 30 years gives a doubling time of 66 years. The formula just given for the United States then becomes

P (in millions) $= 250 \times 2^{(y-1990)/66}$

97. SOCIAL SCIENCE What was the approximate population of the United States in 1960?

98. SOCIAL SCIENCE What will the population of the United States be in 2025 if this growth rate continues?

< Objective 4 >

99. SCIENCE AND MEDICINE Megrez, the nearest of the Big Dipper stars, is 6.6×10^{17} m from Earth. Approximately how long does it take light, traveling at $10^{16} \dfrac{\text{m}}{\text{year}}$, to travel from Megrez to Earth?

100. SCIENCE AND MEDICINE Alkaid, the most distant star in the Big Dipper, is 2.1×10^{18} m from Earth. Approximately how long does it take light to travel from Alkaid to Earth?

101. SOCIAL SCIENCE The number of liters of water on Earth is 15,500 followed by 19 zeros. Write this number in scientific notation. Then use the number of liters of water on Earth to find out how much water is available for each person on Earth. The population of Earth is 6 billion.

102. SOCIAL SCIENCE If there are 6×10^{9} people on Earth and there is enough freshwater to provide each person with 8.79×10^{5} L, how much freshwater is on Earth?

103. SOCIAL SCIENCE The United States uses an average of 2.6×10^{6} L of water per person each year. The United States has 3.2×10^{8} people. How many liters of water does the United States use each year?

Answers

93. _____

94. _____

95. _____

96. _____

97. _____

98. _____

99. _____

100. _____

101. _____

102. _____

103. _____

Answers

(a) _____

(b) _____

(c) _____

(d) _____

(e) _____

(f) _____

(g) _____

(h) _____

Combine like terms where possible.

(a) $8m + 7m$ **(b)** $9x - 5x$

(c) $9m^2 - 8m$ **(d)** $8x^2 - 7x^2$

(e) $5c^3 + 15c^3$ **(f)** $9s^3 + 8s^3$

(g) $8c^2 - 6c + 2c^2$ **(h)** $8r^3 - 7r^2 + 5r^3$

Answers

1. 1   **3.** 1   **5.** 1   **7.** 11   **9.** 1   **11.** $\dfrac{1}{b^8}$ **13.** $\dfrac{1}{81}$

15. 25   **17.** 10,000   **19.** $\dfrac{5}{x}$ **21.** $\dfrac{1}{5x}$ **23.** $-\dfrac{2}{x^5}$

25. $-\dfrac{1}{32x^5}$ **27.** a^8 **29.** x^6 **31.** x^5 **33.** a^3 **35.** $\dfrac{1}{x^2}$

37. False   **39.** True   **41.** x **43.** $\dfrac{m^9}{n^8}$ **45.** $\dfrac{16}{a^{12}}$ **47.** $\dfrac{x^4}{y^6}$

49. $\dfrac{1}{r^2}$ **51.** $\dfrac{1}{m^4 n}$ **53.** $\dfrac{r^3 t^2}{s^7}$ **55.** $\dfrac{b^6}{a^6}$ **57.** $\dfrac{q^6}{p}$ **59.** $\dfrac{3x^6}{8}$

61. $\dfrac{a^7}{b^2}$ **63.** $\dfrac{18}{a^2}$ **65.** c^{15} **67.** $\dfrac{1}{w}$ **69.** 1

71. 9.3×10^7 mi   **73.** 1.3×10^{11} cm   **75.** 28   **77.** 0.008

79. 0.000028   **81.** 5×10^{-4} **83.** 3.7×10^{-4} **85.** 8×10^{-8}

87. 3×10^5 **89.** 8×10^9 **91.** 2×10^2 **93.** 6×10^{16}

95. 2.97 billion   **97.** 182 million   **99.** 66 years

101. 1.55×10^{23} L; 2.58×10^{13} L   **103.** 8.32×10^{14} L

(a) $15m$ **(b)** $4x$ **(c)** $9m^2 - 8m$ **(d)** x^2 **(e)** $20c^3$

(f) $17s^3$ **(g)** $10c^2 - 6c$ **(h)** $13r^3 - 7r^2$

3.3

Adding and Subtracting Polynomials

< 3.3 Objectives >

1 > Add two polynomials

2 > Distribute a negative over a binomial

3 > Subtract two polynomials

Addition is always a matter of combining like quantities (two apples plus three apples, four books plus five books, and so on). If you keep that basic idea in mind, adding polynomials will be easy. It is just a matter of combining like terms.

Suppose that you want to add

$$5x^2 + 3x + 4 \quad \text{and} \quad 4x^2 + 5x - 6$$

Parentheses are sometimes used when adding, so for the sum of these polynomials, we can write

$$(5x^2 + 3x + 4) + (4x^2 + 5x - 6)$$

Now what about the parentheses? You can use the following rule.

NOTE

The plus sign between the parentheses indicates the addition.

Property

Removing Signs of Grouping Case 1	When finding the sum of two polynomials, if a plus sign (+) or nothing at all appears in front of parentheses, simply remove the parentheses. No other changes are necessary.

NOTES

Just remove the parentheses. No other changes are necessary.

Note the use of the associative and commutative properties in reordering and regrouping.

Here we use the distributive property. For example,
$5x^2 + 4x^2 = (5 + 4)x^2 = 9x^2$

Now let's return to the addition.

$$(5x^2 + 3x + 4) + (4x^2 + 5x - 6)$$
$$= 5x^2 + 3x + 4 + 4x^2 + 5x - 6$$

Like terms   Like terms   Like terms

Collect like terms. (*Remember:* Like terms have the same variables raised to the same power).

$$= (5x^2 + 4x^2) + (3x + 5x) + (4 - 6)$$

Combine like terms for the result:

$$= 9x^2 + 8x - 2$$

As should be clear, much of this work can be done mentally. You can then write the sum directly by locating like terms and combining. Example 1 illustrates this property.

| Example 1 | Combining Like Terms |

< Objective 1 >

Add $3x - 5$ and $2x + 3$.

Write the sum.

$$(3x - 5) + (2x + 3) = 3x - 5 + 2x + 3 = 5x - 2$$

Like terms   Like terms

  Check Yourself 1

Add $6x^2 + 2x$ and $4x^2 - 7x$.

The same technique is used to find the sum of two trinomials.

| Example 2 | Adding Polynomials Using the Horizontal Method |

Add $4a^2 - 7a + 5$ and $3a^2 + 3a - 4$.

Write the sum.

$$(4a^2 - 7a + 5) + (3a^2 + 3a - 4)$$
$$= 4a^2 - 7a + 5 + 3a^2 + 3a - 4 = 7a^2 - 4a + 1$$

Like terms

Like terms

Like terms

Check Yourself 2

Add $5y^2 - 3y + 7$ and $3y^2 - 5y - 7$.

| Example 3 | Adding Polynomials Using the Horizontal Method |

Add $2x^2 + 7x$ and $4x - 6$.

Write the sum.

$$(2x^2 + 7x) + (4x - 6)$$
$$= 2x^2 + 7x + 4x - 6$$

These are the only like terms;
$2x^2$ and -6 cannot be combined.

$$= 2x^2 + 11x - 6$$

Check Yourself 3

Add $5m^2 + 8$ and $8m^2 - 3m$.

Writing polynomials in descending-exponent form usually makes the work easier.

 Example 4 | **Adding Polynomials Using the Horizontal Method**

Add $3x - 2x^2 + 7$ and $5 + 4x^2 - 3x$.

Write the polynomials in descending-exponent form and then add.

$(-2x^2 + 3x + 7) + (4x^2 - 3x + 5)$
$= 2x^2 + 12$

Check Yourself 4

Add $8 - 5x^2 + 4x$ and $7x - 8 + 8x^2$.

Subtracting polynomials requires another rule for removing signs of grouping.

Property

Removing Signs of Grouping Case 2

When finding the difference of two polynomials, if a minus sign $(-)$ appears in front of a set of parentheses, the parentheses can be removed by changing the sign of each term inside the parentheses.

The use of this rule is illustrated in Example 5.

 Example 5 | **Removing Parentheses**

< Objective 2 >

In each of the following, remove the parentheses.

NOTE

This uses the distributive property, because
$-(2x + 3y) = (-1)(2x + 3y)$
$= -2x - 3y$

(a) $-(2x + 3y) = -2x - 3y$ Change each sign to remove the parentheses.

(b) $m - (5n - 3p) = m - 5n + 3p$

              Sign changes.

(c) $2x - (-3y + z) = 2x + 3y - z$

              Sign changes.

Check Yourself 5

In each expression, remove the parentheses.

(a) $-(3m + 5n)$ (b) $-(5w - 7z)$
(c) $3r - (2s - 5t)$ (d) $5a - (-3b - 2c)$

Subtracting polynomials is now a matter of using the previous rule to remove the parentheses and then combining the like terms. Consider Example 6.

 Example 6   **Subtracting Polynomials Using the Horizontal Method**

< Objective 3 >

NOTE

The expression following "from" is written first in the problem.

(a) Subtract $5x - 3$ from $8x + 2$.

Write

$(8x + 2) - (5x - 3)$

$= 8x + \underbrace{2 - 5x + 3}_{\text{Sign changes}}$ Recall that subtracting $5x$ is the same as adding $-5x$.

$= 3x + 5$

(b) Subtract $4x^2 - 8x + 3$ from $8x^2 + 5x - 3$.

Write

$(8x^2 + 5x - 3) - (4x^2 - 8x + 3)$

$= 8x^2 + 5x - 3 \underbrace{- 4x^2 + 8x - 3}_{\text{Sign changes}}$

$= 4x^2 + 13x - 6$

 Check Yourself 6

(a) Subtract $7x + 3$ from $10x - 7$.
(b) Subtract $5x^2 - 3x + 2$ from $8x^2 - 3x - 6$.

Again, writing all polynomials in descending-exponent form makes locating and combining like terms much easier. Look at Example 7.

 Example 7   **Subtracting Polynomials Using the Horizontal Method**

(a) Subtract $4x^2 - 3x^3 + 5x$ from $8x^3 - 7x + 2x^2$.

Write

$(8x^3 + 2x^2 - 7x) - (-3x^3 + 4x^2 + 5x)$

$= 8x^3 + 2x^2 - 7x \underbrace{+ 3x^3 - 4x^2 - 5x}_{\text{Sign changes}}$

$= 11x^3 - 2x^2 - 12x$

(b) Subtract $8x - 5$ from $-5x + 3x^2$.

Write

$(3x^2 - 5x) - (8x - 5)$

$= 3x^2 \underbrace{- 5x - 8x}_{} + 5$

Only the like terms can be combined.

$= 3x^2 - 13x + 5$

Check Yourself 7

(a) Subtract $7x - 3x^2 + 5$ from $5 - 3x + 4x^2$.
(b) Subtract $3a - 2$ from $5a + 4a^2$.

If you think back to addition and subtraction in arithmetic, you should remember that the work was arranged vertically. That is, the numbers being added or subtracted were placed under one another so that each column represented the same place value. This meant that in adding or subtracting columns you were always dealing with "like quantities."

It is also possible to use a vertical method for adding or subtracting polynomials. First rewrite the polynomials in descending-exponent form, and then arrange them one under another, so that each column contains like terms. Then add or subtract in each column.

| Example 8 | Adding Using the Vertical Method |

Add $2x^2 - 5x$, $3x^2 + 2$, and $6x - 3$.

Like terms are placed in columns.

$$
\begin{array}{r}
2x^2 - 5x \\
3x^2 \quad\;\; + 2 \\
6x - 3 \\
\hline
5x^2 + \;\; x - 1
\end{array}
$$

Check Yourself 8

Add $3x^2 + 5$, $x^2 - 4x$, and $6x + 7$.

Example 9 illustrates subtraction by the vertical method.

| Example 9 | Subtracting Using the Vertical Method |

(a) Subtract $5x - 3$ from $8x - 7$.

Write

$$
\begin{array}{r}
8x - 7 \\
(-)\; (5x - 3) \\
\hline
3x - 4
\end{array}
$$

To subtract, change each sign of $5x - 3$ to get $-5x + 3$ and then add.

$$
\begin{array}{r}
8x - 7 \\
-5x + 3 \\
\hline
3x - 4
\end{array}
$$

The Streeter/Hutchison Series in Mathematics   Beginning Algebra

(b) Subtract $5x^2 - 3x + 4$ from $8x^2 + 5x - 3$.

Write

$$\begin{array}{r} 8x^2 + 5x - 3 \\ (-)\ (5x^2 - 3x + 4) \\ \hline 3x^2 + 8x - 7 \end{array}$$

To subtract, change each sign of $5x^2 - 3x + 4$ to get $-5x^2 + 3x - 4$ and then add.

$$\begin{array}{r} 8x^2 + 5x - 3 \\ -5x^2 + 3x - 4 \\ \hline 3x^2 + 8x - 7 \end{array}$$

Subtracting using the vertical method takes some practice. Take time to study the method carefully. You will use it in long division in Section 3.5.

Check Yourself 9

Subtract, using the vertical method.

(a) $4x^2 - 3x$ from $8x^2 + 2x$ **(b)** $8x^2 + 4x - 3$ from $9x^2 - 5x + 7$

Check Yourself ANSWERS

1. $10x^2 - 5x$ **2.** $8y^2 - 8y$ **3.** $13m^2 - 3m + 8$ **4.** $3x^2 + 11x$
5. (a) $-3m - 5n$; **(b)** $-5w + 7z$; **(c)** $3r - 2s + 5t$; **(d)** $5a + 3b + 2c$
6. (a) $3x - 10$; **(b)** $3x^2 - 8$ **7. (a)** $7x^2 - 10x$; **(b)** $4a^2 + 2a + 2$
8. $4x^2 + 2x + 12$ **9. (a)** $4x^2 + 5x$; **(b)** $x^2 - 9x + 10$

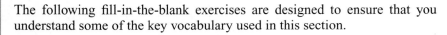

Reading Your Text

The following fill-in-the-blank exercises are designed to ensure that you understand some of the key vocabulary used in this section.

SECTION 3.3

(a) When finding the _____ of two polynomials, if a plus sign appears in front of the parentheses, simply remove the parentheses.

(b) When subtracting, the expression following the word "from" is written _____ in the problem.

(c) In arithmetic, we add vertically so that each column has the same _____ value.

(d) When adding two polynomials, only like _____ are combined.

Basic Skills | Advanced Skills | Vocational-Technical Applications | Calculator/Computer | Above and Beyond | Getting Ready

▲

3.3 exercises

< Objective 1 >

Add.

1. $6a - 5$ and $3a + 9$

2. $9x + 3$ and $3x - 4$

3. $8b^2 - 11b$ and $5b^2 - 7b$

4. $2m^2 + 3m$ and $6m^2 - 8m$

5. $3x^2 - 2x$ and $-5x^2 + 2x$

6. $3p^2 + 5p$ and $-7p^2 - 5p$

7. $2x^2 + 5x - 3$ and $3x^2 - 7x + 4$ > Videos

8. $4d^2 - 8d + 7$ and $5d^2 - 6d - 9$

9. $2b^2 + 8$ and $5b + 8$

10. $4x - 3$ and $3x^2 - 9x$

11. $8y^3 - 5y^2$ and $5y^2 - 2y$

12. $9x^4 - 2x^2$ and $2x^2 + 3$

13. $2a^2 - 4a^3$ and $3a^3 + 2a^2$

14. $9m^3 - 2m$ and $-6m - 4m^3$

15. $4x^2 - 2 + 7x$ and $5 - 8x - 6x^2$

16. $5b^3 - 8b + 2b^2$ and $3b^2 - 7b^3 + 5b$

< Objective 2 >

Remove the parentheses in each of the following expressions and simplify when possible.

17. $-(2a + 3b)$

18. $-(7x - 4y)$

19. $5a - (2b - 3c)$

20. $7x - (4y + 3z)$

21. $9r - (3r + 5s)$

22. $10m - (3m - 2n)$

23. $5p - (-3p + 2q)$ > Videos

24. $8d - (-7c - 2d)$

MathZone

Name _____

Section _____ Date _____

Answers

1. _____	2. _____
3. _____	4. _____
5. _____	6. _____
7. _____	
8. _____	
9. _____	
10. _____	
11. _____	12. _____
13. _____	14. _____
15. _____	
16. _____	
17. _____	18. _____
19. _____	
20. _____	
21. _____	22. _____
23. _____	24. _____

Answers

25. _____ 26. _____

27. _____ 28. _____

29. _____ 30. _____

31. _____

32. _____

33. _____

34. _____

35. _____ 36. _____

37. _____ 38. _____

39. _____

40. _____

41. _____

42. _____

43. _____

44. _____

45. _____

46. _____

47. _____

48. _____

49. _____

50. _____

< Objective 3 >

Subtract.

25. $x + 4$ from $2x - 3$

26. $x - 2$ from $3x + 5$

27. $3m^2 - 2m$ from $4m^2 - 5m$

28. $9a^2 - 5a$ from $11a^2 - 10a$

29. $6y^2 + 5y$ from $4y^2 + 5y$

30. $9n^2 - 4n$ from $7n^2 - 4n$

31. $x^2 - 4x - 3$ from $3x^2 - 5x - 2$

32. $3x^2 - 2x + 4$ from $5x^2 - 8x - 3$

33. $3a + 7$ from $8a^2 - 9a$

34. $3x^3 + x^2$ from $4x^3 - 5x$

35. $4b^2 - 3b$ from $5b - 2b^2$

36. $7y - 3y^2$ from $3y^2 - 2y$

37. $x^2 - 5 - 8x$ from
$3x^2 - 8x + 7$

38. $4x - 2x^2 + 4x^3$ from
$4x^3 + x - 3x^2$

Perform the indicated operations.

39. Subtract $3b + 2$ from the sum of $4b - 2$ and $5b + 3$.

40. Subtract $5m - 7$ from the sum of $2m - 8$ and $9m - 2$.

41. Subtract $3x^2 + 2x - 1$ from the sum of $x^2 + 5x - 2$ and $2x^2 + 7x - 8$.

42. Subtract $4x^2 - 5x - 3$ from the sum of $x^2 - 3x - 7$ and $2x^2 - 2x + 9$.

43. Subtract $2x^2 - 3x$ from the sum of $4x^2 - 5$ and $2x - 7$.

44. Subtract $5a^2 - 3a$ from the sum of $3a - 3$ and $5a^2 + 5$.

45. Subtract the sum of $3y^2 - 3y$ and $5y^2 + 3y$ from $2y^2 - 8y$.

46. Subtract the sum of $7r^3 - 4r^2$ and $-3r^3 + 4r^2$ from $2r^3 + 3r^2$.

< Objective 1 >

Add using the vertical method.

47. $2w^2 + 7$, $3w - 5$, and $4w^2 - 5w$

48. $3x^2 - 4x - 2$, $6x - 3$, and $2x^2 + 8$

49. $3x^2 + 3x - 4$, $4x^2 - 3x - 3$, and $2x^2 - x + 7$

50. $5x^2 + 2x - 4$, $x^2 - 2x - 3$, and $2x^2 - 4x - 3$

Basic Skills | Advanced Skills | **Vocational-Technical Applications** | Calculator/Computer | Above and Beyond | Getting Ready

▲

51. **ALLIED HEALTH** A patient's arterial oxygen content (CaO_2), as a percentage measurement, is calculated using the formula $CaO_2 = 1.34(Hb)(SaO_2) + 0.003PaO_2$, which is based on a patient's hemoglobin content (Hb), as a percentage measurement, arterial oxygen saturation (SaO_2), a percent expressed as a decimal, and arterial oxygen tension (PaO_2), in millimeters of mercury (mm Hg). Similarly, a patient's end-capillary oxygen content (CcO_2), as a percentage measurement, is calculated using the formula $CcO_2 = 1.34(Hb)(SaO_2) + 0.003P_AO_2$, which is based on the alveolar oxygen tension (P_AO_2), in mm Hg, instead of the arterial oxygen tension. Write a simplified formula for the difference between the end-capillary and arterial oxygen contents.

52. **ALLIED HEALTH** A diabetic patient's morning (m) and evening (n) blood glucose levels depend on the number of days (t) since the patient's treatment began and can be approximated by the formulas $m = 0.472t^3 - 5.298t^2 + 11.802t + 93.143$ and $n = -1.083t^3 + 11.464t^2 - 29.524t + 117.429$. Write a formula for the difference (d) in morning and evening blood glucose levels based on the number of days since treatment began.

53. **MANUFACTURING TECHNOLOGY** The shear polynomial for a polymer is

$0.4x^2 - 144x + 318$

After vulcanization of the polymer, the shear factor is increased by

$0.2x^2 - 14x + 144$

Find the shear polynomial for the polymer after vulcanization (add the polynomials).

54. **MANUFACTURING TECHNOLOGY** The moment of inertia of a square object is given by

$$I(s) = \frac{s^4}{12}$$

The moment of inertia for a circular object is approximately given by

$$I(s) = \frac{3.14s^4}{48}$$

Find the moment of inertia of a square with a circular inlay (add the polynomials).

Subtract using the vertical method.

55. $5x^2 - 3x$ from $8x^2 - 9$ 56. $7x^2 + 6x$ from $9x^2 - 3$

Basic Skills | **Advanced Skills** | Vocational-Technical Applications | Calculator/Computer | Above and Beyond | Getting Ready

▲

Perform the indicated operations.

57. $[(9x^2 - 3x + 5) - (3x^2 + 2x - 1)] - (x^2 - 2x - 3)$ ▷ Videos

58. $[(5x^2 + 2x - 3) - (-2x^2 + x - 2)] - (2x^2 + 3x - 5)$

Answers

51. _____

52. _____

53. _____

54. _____

55. _____

56. _____

57. _____

58. _____

Answers

59. _____

60. _____

61. _____

62. _____

63. _____

64. _____

(a) _____

(b) _____

(c) _____

(d) _____

(e) _____

(f) _____

(g) _____

(h) _____

| Basic Skills | Advanced Skills | Vocational-Technical Applications | Calculator/Computer | **Above and Beyond** | Getting Ready |

▲

Find values for a, b, c, and d so that the following equations are true.

59. $3ax^4 - 5x^3 + x^2 - cx + 2 = 9x^4 - bx^3 + x^2 - 2d$

60. $(4ax^3 - 3bx^2 - 10) - 3(x^3 + 4x^2 - cx - d) = x^2 - 6x + 8$

61. **GEOMETRY** A rectangle has sides of $8x + 9$ and $6x - 7$. Find the polynomial that represents its perimeter.

62. **GEOMETRY** A triangle has sides $3x + 7$, $4x - 9$, and $5x + 6$. Find the polynomial that represents its perimeter.

63. **BUSINESS AND FINANCE** The cost of producing x units of an item is $C = 150 + 25x$. The revenue for selling x units is $R = 90x - x^2$. The profit is given by the revenue minus the cost. Find the polynomial that represents profit.

64. **BUSINESS AND FINANCE** The revenue for selling y units is $R = 3y^2 - 2y + 5$ and the cost of producing y units is $C = y^2 + y - 3$. Find the polynomial that represents profit.

| Basic Skills | Advanced Skills | Vocational-Technical Applications | Calculator/Computer | Above and Beyond | **Getting Ready** |

▲

Multiply.

(a) $2^5 \cdot 2^7$ (b) $3^8 \cdot 3^{12}$ (c) $2 \cdot 4^3 \cdot 4^4$

(d) $3 \cdot 5^5 \cdot 5^2$ (e) $4 \cdot 2^5 \cdot 3 \cdot 2$ (f) $6 \cdot 3^2 \cdot 5 \cdot 3^3$

(g) $(-2 \cdot 4^2)(8 \cdot 4^7)$ (h) $(-10 \cdot 5)(-3 \cdot 5^5)$

Answers

1. $9a + 4$ **3.** $13b^2 - 18b$ **5.** $-2x^2$ **7.** $5x^2 - 2x + 1$
9. $2b^2 + 5b + 16$ **11.** $8y^3 - 2y$ **13.** $-a^3 + 4a^2$
15. $-2x^2 - x + 3$ **17.** $-2a - 3b$ **19.** $5a - 2b + 3c$
21. $6r - 5s$ **23.** $8p - 2q$ **25.** $x - 7$ **27.** $m^2 - 3m$ **29.** $-2y^2$
31. $2x^2 - x + 1$ **33.** $8a^2 - 12a - 7$ **35.** $-6b^2 + 8b$ **37.** $2x^2 + 12$
39. $6b - 1$ **41.** $10x - 9$ **43.** $2x^2 + 5x - 12$ **45.** $-6y^2 - 8y$
47. $6w^2 - 2w + 2$ **49.** $9x^2 - x$ **51.** $CcO_2 - CaO_2 = 0.003(P_AO_2 - PaO_2)$
53. $0.6x^2 + 158x + 462$ **55.** $3x^2 + 3x - 9$ **57.** $5x^2 - 3x + 9$
59. $a = 3, b = 5, c = 0, d = -1$ **61.** $28x + 4$ **63.** $-x^2 + 65x - 150$
(a) 2^{12} (b) 3^{20} (c) $2(4)^7$ (d) $3(5)^7$ (e) $12(2)^6$ (f) $30(3)^5$
(g) $-16(4)^9$ (h) $30(5)^6$

3.4

Multiplying Polynomials

< 3.4 Objectives >

1 > Find the product of a monomial and a polynomial

2 > Find the product of two binomials

3 > Find the product of two polynomials

4 > Square a binomial

You have already had some experience in multiplying polynomials. In Section 1.7 we stated the product property of exponents and used that property to find the product of two monomial terms.

Step by Step

| To Find the Product of Monomials | Step 1 | Multiply the coefficients. |
| | Step 2 | Use the product property of exponents to combine the variables. |

 Example 1 **Multiplying Monomials**

< Objective 1 >

Multiply $3x^2y$ and $2x^3y^5$.

Write

$$(3x^2y)(2x^3y^5)$$
$$= \underbrace{(3 \cdot 2)}\underbrace{(x^2 \cdot x^3)}\underbrace{(y \cdot y^5)}$$

Multiply the coefficients.   Add the exponents.

$$= 6x^5y^6$$

RECALL

We used the commutative and associative properties to rewrite the problem.

You might want to review Section 1.1 before going on.

 Check Yourself 1

Multiply.

(a) $(5a^2b)(3a^2b^4)$ **(b)** $(-3xy)(4x^3y^5)$

Our next task is to find the product of a monomial and a polynomial. Here we use the distributive property, which we introduced in Section 1.1. That property leads us to the following rule for multiplication.

333

Property

| To Multiply a Polynomial by a Monomial | Use the distributive property to multiply each term of the polynomial by the monomial. |

 Example 2   Multiplying a Monomial and a Binomial

NOTES

Distributive property:

$a(b + c) = ab + ac$

With practice you will do this step mentally.

(a) Multiply $2x + 3$ by x.

Write

$x(2x + 3)$

$= x \cdot 2x + x \cdot 3$

$= 2x^2 + 3x$

Multiply x by 2x and then by 3 (the terms of the polynomial). That is, "distribute" the multiplication over the sum.

(b) Multiply $2a^3 + 4a$ by $3a^2$.

Write

$3a^2(2a^3 + 4a)$

$= 3a^2 \cdot 2a^3 + 3a^2 \cdot 4a = 6a^5 + 12a^3$

 Check Yourself 2

Multiply.

(a) $2y(y^2 + 3y)$ **(b)** $3w^2(2w^3 + 5w)$

The patterns of Example 2 extend to *any* number of terms.

 Example 3   Multiplying a Monomial and a Polynomial

NOTE

We show all the steps of the process. With practice, you will be able to write the product directly and should try to do so.

Multiply the following.

(a) $3x(4x^3 + 5x^2 + 2)$

$= 3x \cdot 4x^3 + 3x \cdot 5x^2 + 3x \cdot 2 = 12x^4 + 15x^3 + 6x$

(b) $5y^2(2y^3 - 4)$

$= 5y^2 \cdot 2y^3 - 5y^2 \cdot 4 = 10y^5 - 20y^2$

(c) $-5c(4c^2 - 8c)$

$= (-5c)(4c^2) - (-5c)(8c) = -20c^3 + 40c^2$

(d) $3c^2d^2(7cd^2 - 5c^2d^3)$

$= 3c^2d^2 \cdot 7cd^2 - 3c^2d^2 \cdot 5c^2d^3 = 21c^3d^4 - 15c^4d^5$

Check Yourself 3

Multiply.

(a) $3(5a^2 + 2a + 7)$ **(b)** $4x^2(8x^3 - 6)$
(c) $-5m(8m^2 - 5m)$ **(d)** $9a^2b(3a^3b - 6a^2b^4)$

 Example 4   **Multiplying Binomials**

< Objective 2 >

> **NOTE**
>
> This ensures that each term, x and 2, of the first binomial is multiplied by each term, x and 3, of the second binomial.

(a) Multiply $x + 2$ by $x + 3$.

We can think of $x + 2$ as a single quantity and apply the distributive property.

$(x + 2)(x + 3)$ Multiply $x + 2$ by x and then by 3.
$= (x + 2)x + (x + 2)3$
$= x \cdot x + 2 \cdot x + x \cdot 3 + 2 \cdot 3$
$= x^2 + 2x + 3x + 6$
$= x^2 + 5x + 6$

(b) Multiply $a - 3$ by $a - 4$. (Think of $a - 3$ as a single quantity and distribute.)

$(a - 3)(a - 4)$
$= (a - 3)a - (a - 3)(4)$
$= a \cdot a - 3 \cdot a - [(a \cdot 4) - (3 \cdot 4)]$
$= a^2 - 3a - (4a - 12)$ The parentheses are needed here
$= a^2 - 3a - 4a + 12$ because a *minus sign* precedes the binomial.
$= a^2 - 7a + 12$

Check Yourself 4

Multiply.

(a) $(x + 2)(x + 5)$ **(b)** $(y + 5)(y - 6)$

Fortunately, there is a pattern to this kind of multiplication that allows you to write the product of two binomials without going through all these steps. We call it the **FOIL method** of multiplying. The reason for this name will be clear as we look at the process in more detail.

To multiply $(x + 2)(x + 3)$:

> **NOTES**
>
> Remember this by F!
>
> Remember this by O!
>
> Remember this by I!
>
> Remember this by L!

1. $(x + 2)(x + 3)$
$ x \cdot x$ Find the product of the *first* terms of the factors.

2. $(x + 2)(x + 3)$
$ x \cdot 3$ Find the product of the *outer* terms.

3. $(x + 2)(x + 3)$
$ 2 \cdot x$ Find the product of the *inner* terms.

4. $(x + 2)(x + 3)$
$ 2 \cdot 3$ Find the product of the *last* terms.

NOTE

Of course these are the same four terms found in Example 4(a).

Combining the four steps, we have

$$(x + 2)(x + 3)$$
$$= x^2 + 3x + 2x + 6$$
$$= x^2 + 5x + 6$$

   With practice, you can use the FOIL method to write products quickly and easily. Consider Example 5, which illustrates this approach.

 Example 5 | **Using the FOIL Method**

NOTE

It is called FOIL to give you an easy way of remembering the steps: *First, Outer, Inner,* and *Last.*

Find each product using the FOIL method.

(a) $(x + 4)(x + 5)$

$$= x^2 + 5x + 4x + 20$$

$$= x^2 + 9x + 20$$

NOTE

When possible, you should combine the outer and inner products mentally and write just the final product.

(b) $(x - 7)(x + 3)$

Combine the outer and inner products as $-4x$.

$$= x^2 - 4x - 21$$

 Check Yourself 5

Multiply.

(a) $(x + 6)(x + 7)$ **(b)** $(x + 3)(x - 5)$ **(c)** $(x - 2)(x - 8)$

   Using the FOIL method, you can also find the product of binomials with coefficients other than 1 or with more than one variable.

Example 6	Using the FOIL Method

Find each product using the FOIL method.

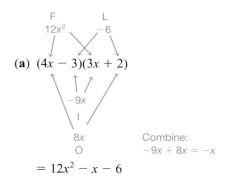

(a) $(4x - 3)(3x + 2)$

Combine:
$-9x + 8x = -x$

$= 12x^2 - x - 6$

(b) $(3x - 5y)(2x - 7y)$

Combine:
$-10xy - 21xy = -31xy$

$= 6x^2 - 31xy + 35y^2$

The following rule summarizes our work in multiplying binomials.

Step by Step

To Multiply Two Binomials

Step 1 Find the first term of the product of the binomials by multiplying the first terms of the binomials (F).

Step 2 Find the middle term of the product as the sum of the outer and inner products (O + I).

Step 3 Find the last term of the product by multiplying the last terms of the binomials (L).

Check Yourself 6

Multiply.

(a) $(5x + 2)(3x - 7)$ **(b)** $(4a - 3b)(5a - 4b)$
(c) $(3m + 5n)(2m + 3n)$

Sometimes, especially with larger polynomials, it is easier to use the vertical method to find their product. This is the same method you originally learned when multiplying two large integers.

Example 7   Multiplying Using the Vertical Method

Use the vertical method to find the product $(3x + 2)(4x - 1)$.
    First, we rewrite the multiplication in vertical form.

$$3x + 2$$
$$4x + (-1)$$

Multiplying the quantity $3x + 2$ by -1 yields

$$\begin{array}{r} 3x + 2 \\ 4x + (-1) \\ \hline -3x + (-2) \end{array}$$

Note that we maintained the columns of the original binomial when we found the product. We will continue with those columns as we multiply by the $4x$ term.

$$\begin{array}{r} 3x + 2 \\ 4x + (-1) \\ \hline -\ 3x + (-2) \\ 12x^2 + 8x \quad\quad \\ \hline 12x^2 + 5x + (-2) \end{array}$$

We could write the product as $(3x + 2)(4x - 1) = 12x^2 + 5x - 2$.

   Check Yourself 7

Use the vertical method to find the product $(5x - 3)(2x + 1)$.

We use the vertical method again in Example 8. This time, we will multiply a binomial and a trinomial. Note that the FOIL method can never work for anything but the product of two binomials.

  Example 8   Using the Vertical Method to Multiply Polynomials

< Objective 3 >

Multiply $x^2 - 5x + 8$ by $x + 3$.

Step 1 $$\begin{array}{r} x^2 - \ 5x + \ 8 \\ x + \ 3 \\ \hline 3x^2 - 15x + 24 \end{array}$$ Multiply each term of $x^2 - 5x + 8$ by 3.

Step 2 $$\begin{array}{r} x^2 - \ 5x + \ 8 \\ x + \ 3 \\ \hline 3x^2 - 15x + 24 \\ x^3 - 5x^2 + \ 8x \quad\quad\quad \\ \hline \end{array}$$ Now multiply each term by x.

Note that this line is shifted over so that like terms are in the same columns.

NOTE

Using the vertical method ensures that each term of one factor multiplies each term of the other. That's why it works!

Step 3

$$
\begin{array}{r}
x^2 - 5x + 8 \\
x + 3 \\
\hline
3x^2 - 15x + 24 \\
x^3 - 5x^2 + 8x \\
\hline
x^3 - 2x^2 - 7x + 24
\end{array}
$$

Now combine like terms to write the product.

Check Yourself 8

Multiply $2x^2 - 5x + 3$ by $3x + 4$.

Certain products occur frequently enough in algebra that it is worth learning special formulas for dealing with them. First, look at the **square of a binomial,** which is the product of two equal binomial factors.

$$(x + y)^2 = (x + y)(x + y)$$
$$= x^2 + 2xy + y^2$$

$$(x - y)^2 = (x - y)(x - y)$$
$$= x^2 - 2xy + y^2$$

The patterns above lead us to the following rule.

Step by Step

To Square a Binomial

Step 1 Find the first term of the square by squaring the first term of the binomial.

Step 2 Find the middle term of the square as twice the product of the two terms of the binomial.

Step 3 Find the last term of the square by squaring the last term of the binomial.

Example 9   **Squaring a Binomial**

< Objective 4 >

> CAUTION

A very common mistake in squaring binomials is to forget the middle term.

(a) $(x + 3)^2 = x^2 + 2 \cdot x \cdot 3 + 3^2$

Square of first term   Twice the product of the two terms   Square of the last term

$$= x^2 + 6x + 9$$

(b) $(3a + 4b)^2 = (3a)^2 + 2(3a)(4b) + (4b)^2$
$$= 9a^2 + 24ab + 16b^2$$

(c) $(y - 5)^2 = y^2 + 2 \cdot y \cdot (-5) + (-5)^2$
$$= y^2 - 10y + 25$$

(d) $(5c - 3d)^2 = (5c)^2 + 2(5c)(-3d) + (-3d)^2$
$$= 25c^2 - 30cd + 9d^2$$

Again we have shown all the steps. With practice you can write just the square.

Check Yourself 9

Simplify.

(a) $(2x + 1)^2$ **(b)** $(4x - 3y)^2$

| Example 10 | Squaring a Binomial |

NOTE

You should see that
$(2 + 3)^2 \neq 2^2 + 3^2$ because
$5^2 \neq 4 + 9.$

Find $(y + 4)^2$.

$(y + 4)^2$ is *not* equal to $y^2 + 4^2$ or $y^2 + 16$

The correct square is

$(y + 4)^2 = y^2 + 8y + 16$

The middle term is twice the product of y and 4.

Check Yourself 10

Simplify.

(a) $(x + 5)^2$ **(b)** $(3a + 2)^2$
(c) $(y - 7)^2$ **(d)** $(5x - 2y)^2$

A second special product will be very important in Chapter 4, which deals with factoring. Suppose the form of a product is

$(x + y)(x - y)$

The two terms differ
only in sign.

Let's see what happens when we multiply these two terms.

$(x + y)(x - y)$
$= x^2 - xy + xy - y^2$

$= 0$

$= x^2 - y^2$

Because the middle term becomes 0, we have the following rule.

Property

Special Product

The product of two binomials that differ only in the sign between the terms is the square of the first term minus the square of the second term.

Following are some examples of this rule.

| Example 11 | Finding a Special Product |

< Objective 2 >

Multiply each pair of binomials.

(a) $(x + 5)(x - 5) = x^2 - 5^2$

Square of the first term   Square of the second term

$$= x^2 - 25$$

RECALL

$(2y)^2 = (2y)(2y)$
$ = 4y^2$

(b) $(x + 2y)(x - 2y) = x^2 - (2y)^2$

Square of the first term   Square of the second term

$$= x^2 - 4y^2$$

(c) $(3m + n)(3m - n) = 9m^2 - n^2$

(d) $(4a - 3b)(4a + 3b) = 16a^2 - 9b^2$

Check Yourself 11

Find the products.

(a) $(a - 6)(a + 6)$ **(b)** $(x - 3y)(x + 3y)$
(c) $(5n + 2p)(5n - 2p)$ **(d)** $(7b - 3c)(7b + 3c)$

When finding the product of three or more factors, it is useful to first look for the pattern in which two binomials differ only in their sign. Finding this product first will make it easier to find the product of all the factors.

| Example 12 | Multiplying Polynomials |

(a) $x(x - 3)(x + 3)$ These binomials differ only in the sign.
$ = x(x^2 - 9)$
$ = x^3 - 9x$

(b) $(x + 1)(x - 5)(x + 5)$ These binomials differ only in the sign.
$ = (x + 1)(x^2 - 25)$ With two binomials, use the FOIL method.
$ = x^3 + x^2 - 25x - 25$

(c) $(2x - 1)(x + 3)(2x + 1)$ These two binomials differ only in the sign of the second term. We can use the commutative property to rearrange the terms.
$ = (x + 3)(2x - 1)(2x + 1)$
$ = (x + 3)(4x^2 - 1)$
$ = 4x^3 + 12x^2 - x - 3$

Check Yourself 12

Multiply.

(a) $3x(x - 5)(x + 5)$

(b) $(x - 4)(2x + 3)(2x - 3)$

(c) $(x - 7)(3x - 1)(x + 7)$

Check Yourself ANSWERS

1. (a) $15a^4b^5$; **(b)** $-12x^4y^6$ **2. (a)** $2y^3 + 6y^2$; **(b)** $6w^5 + 15w^3$
3. (a) $15a^2 + 6a + 21$; **(b)** $32x^5 - 24x^2$; **(c)** $-40m^3 + 25m^2$;
(d) $27a^5b^2 - 54a^4b^5$ **4. (a)** $x^2 + 7x + 10$; **(b)** $y^2 - y - 30$
5. (a) $x^2 + 13x + 42$; **(b)** $x^2 - 2x - 15$; **(c)** $x^2 - 10x + 16$
6. (a) $15x^2 - 29x - 14$; **(b)** $20a^2 - 31ab + 12b^2$; **(c)** $6m^2 + 19mn + 15n^2$
7. $10x^2 - x - 3$ **8.** $6x^3 - 7x^2 - 11x + 12$ **9. (a)** $4x^2 + 4x + 1$;
(b) $16x^2 - 24xy + 9y^2$ **10. (a)** $x^2 + 10x + 25$; **(b)** $9a^2 + 12a + 4$;
(c) $y^2 - 14y + 49$; **(d)** $25x^2 - 20xy + 4y^2$ **11. (a)** $a^2 - 36$; **(b)** $x^2 - 9y^2$;
(c) $25n^2 - 4p^2$; **(d)** $49b^2 - 9c^2$ **12. (a)** $3x^3 - 75x$;
(b) $4x^3 - 16x^2 - 9x + 36$; **(c)** $3x^3 - x^2 - 147x + 49$

Reading Your Text

The following fill-in-the-blank exercises are designed to ensure that you
understand some of the key vocabulary used in this section.

SECTION 3.4

(a) When multiplying monomials, we use the product property of exponents
to combine the _____.

(b) The F in FOIL stands for the product of the _____ terms.

(c) The O in FOIL stands for the product of the _____ terms.

(d) The square of a binomial ALWAYS has exactly _____ terms.

< Objective 1 >

Multiply.

1. $(5x^2)(3x^3)$

2. $(7a^5)(4a^6)$

3. $(-2b^2)(14b^8)$

4. $(14y^4)(-4y^6)$

5. $(-10p^6)(-4p^7)$

6. $(-6m^8)(9m^7)$

7. $(4m^5)(-3m)$

8. $(-5r^7)(-3r)$

9. $(4x^3y^2)(8x^2y)$

10. $(-3r^4s^2)(-7r^2s^5)$

11. $(-3m^5n^2)(2m^4n)$

12. $(7a^3b^5)(-6a^4b)$

13. $5(2x + 6)$

14. $4(7b - 5)$

15. $3a(4a + 5)$

16. $5x(2x - 7)$

17. $3s^2(4s^2 - 7s)$

18. $9a^2(3a^3 + 5a)$

19. $2x(4x^2 - 2x + 1)$

20. $5m(4m^3 - 3m^2 + 2)$

21. $3xy(2x^2y + xy^2 + 5xy)$

22. $5ab^2(ab - 3a + 5b)$

23. $6m^2n(3m^2n - 2mn + mn^2)$

24. $8pq^2(2pq - 3p + 5q)$

> Videos

Name _____

Section _____ Date _____

Answers

1. _____   2. _____

3. _____   4. _____

5. _____   6. _____

7. _____   8. _____

9. _____   10. _____

11. _____   12. _____

13. _____   14. _____

15. _____   16. _____

17. _____

18. _____

19. _____

20. _____

21. _____

22. _____

23. _____

24. _____

Answers

25. _____

26. _____

27. _____

28. _____

29. _____

30. _____

31. _____

32. _____

33. _____

34. _____

35. _____

36. _____

37. _____

38. _____

39. _____

40. _____

41. _____

42. _____

43. _____

44. _____

45. _____

46. _____

47. _____

48. _____

49. _____   50. _____

51. _____   52. _____

53. _____

54. _____

< Objective 2 >

Multiply.

25. $(x + 3)(x + 2)$

26. $(a - 3)(a - 7)$

27. $(m - 5)(m - 9)$

28. $(b + 7)(b + 5)$

29. $(p - 8)(p + 7)$

30. $(x - 10)(x + 9)$

31. $(w + 10)(w + 20)$

32. $(s - 12)(s - 8)$

33. $(3x - 5)(x - 8)$

34. $(w + 5)(4w - 7)$

35. $(2x - 3)(3x + 4)$

36. $(5a + 1)(3a + 7)$

37. $(3a - b)(4a - 9b)$ Videos

38. $(7s - 3t)(3s + 8t)$

39. $(3p - 4q)(7p + 5q)$

40. $(5x - 4y)(2x - y)$

41. $(2x + 5y)(3x + 4y)$

42. $(4x - 5y)(4x + 3y)$

43. $(x + 5)(x + 5)$

44. $(y + 8)(y + 8)$

45. $(y - 9)(y - 9)$

46. $(2a + 3)(2a + 3)$

47. $(6m + n)(6m + n)$

48. $(7b - c)(7b - c)$

49. $(a - 5)(a + 5)$

50. $(x - 7)(x + 7)$

51. $(x - 2y)(x + 2y)$

52. $(7x + y)(7x - y)$

53. $(5s + 3t)(5s - 3t)$

54. $(9c - 4d)(9c + 4d)$

Basic Skills | **Advanced Skills** | Vocational-Technical Applications | Calculator/Computer | Above and Beyond | Getting Ready
▲

Label the following as **true** *or* **false**.

55. $(x + y)^2 = x^2 + y^2$

56. $(x - y)^2 = x^2 - y^2$

57. $(x + y)^2 = x^2 + 2xy + y^2$

58. $(x - y)^2 = x^2 - 2xy + y^2$

59. **GEOMETRY** The length of a rectangle is given by $3x + 5$ cm and the width is given by $2x - 7$ cm. Express the area of the rectangle in terms of x.

60. **GEOMETRY** The base of a triangle measures $3y + 7$ in. and the height is $2y - 3$ in. Express the area of the triangle in terms of y.

Multiply each of the following.

61. $(x + 5)^2$

62. $(y + 9)^2$

63. $(2a - 1)^2$

64. $(3x - 2)^2$

65. $(6m + 1)^2$

66. $(7b - 2)^2$

67. $(3x - y)^2$

68. $(5m + n)^2$

69. $(2r + 5s)^2$

70. $(3a - 4b)^2$

71. $\left(x + \dfrac{1}{2}\right)^2$ Videos

72. $\left(w - \dfrac{1}{4}\right)^2$

73. $(x - 6)(x + 6)$

74. $(y + 8)(y - 8)$

75. $(m + 12)(m - 12)$ Videos

76. $(w - 10)(w + 10)$

77. $\left(x - \dfrac{1}{2}\right)\left(x + \dfrac{1}{2}\right)$

78. $\left(x + \dfrac{2}{3}\right)\left(x - \dfrac{2}{3}\right)$

79. $(p - 0.4)(p + 0.4)$

80. $(m - 0.6)(m + 0.6)$

Answers

55. _____   56. _____

57. _____   58. _____

59. _____

60. _____

61. _____

62. _____

63. _____

64. _____

65. _____

66. _____

67. _____

68. _____

69. _____

70. _____

71. _____

72. _____

73. _____   74. _____

75. _____   76. _____

77. _____   78. _____

79. _____   80. _____

Answers

81.

82.

83.

84.

85.

86.

87.

88.

89.

90.

91.

92.

93.

94.

81. $(a - 3b)(a + 3b)$

82. $(p + 4q)(p - 4q)$

83. $(4r - s)(4r + s)$

84. $(7x - y)(7x + y)$

| Basic Skills | Advanced Skills | Vocational-Technical Applications | Calculator/Computer | **Above and Beyond** | Getting Ready |

Note that $(28)(32) = (30 - 2)(30 + 2) = 900 - 4 = 896.$ *Use this pattern to find each of the following products.*

85. $(49)(51)$

86. $(27)(33)$

87. $(34)(26)$

88. $(98)(102)$

89. $(55)(65)$

90. $(64)(56)$ > Videos

91. **AGRICULTURE** Suppose an orchard is planted with trees in straight rows. If there are $5x - 4$ rows with $5x - 4$ trees in each row, how many trees are there in the orchard?

92. **GEOMETRY** A square has sides of length $3x - 2$ cm. Express the area of the square as a polynomial.

3x − 2 cm

3x − 2 cm

93. Complete the following statement: $(a + b)^2$ is not equal to $a^2 + b^2$ because. . . . But, wait! Isn't $(a + b)^2$ *sometimes* equal to $a^2 + b^2$? What do you think?

94. Is $(a + b)^3$ ever equal to $a^3 + b^3$? Explain.

95. GEOMETRY In the following figures, identify the length, width, and area of the square:

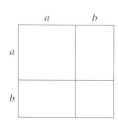

Length = _____

Width = _____

Area = _____

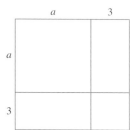

Length = _____

Width = _____

Area = _____

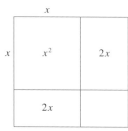

Length = _____

Width = _____

Area = _____

96. GEOMETRY The square shown here is x units on a side. The area is _____.

Draw a picture of what happens when the sides are doubled. The area is _____.

Continue the picture to show what happens when the sides are tripled. The area is _____.

If the sides are quadrupled, the area is _____.

In general, if the sides are multiplied by n, the area is _____.

If each side is increased by 3, the area is increased by _____.

If each side is decreased by 2, the area is decreased by _____.

In general, if each side is increased by n, the area is increased by _____, and if each side is decreased by n, the area is decreased by _____.

Answers

97. _____

98. _____

(a) _____

(b) _____

(c) _____

(d) _____

(e) _____

(f) _____

(g) _____

(h) _____

97. Geometry Find the volume of a rectangular solid whose length measures $(2x + 4)$, width measures $(x + 2)$, and height measures $(x - 3)$.

98. Geometry Neil and Suzanne are building a pool. Their backyard measures $(2x + 3)$ feet by $(2x + 12)$ feet, and the pool will measure $(x + 4)$ feet by $(x + 10)$ feet. If the remainder of the yard will be cement, how many square feet of the backyard will be covered by cement?

Basic Skills | Advanced Skills | Vocational-Technical Applications | Calculator/Computer | Above and Beyond | **Getting Ready**

Divide.

(a) $\dfrac{2x^2}{2x}$ **(b)** $\dfrac{3a^3}{3a}$

(c) $\dfrac{6p^3}{2p^2}$ **(d)** $\dfrac{10m^4}{5m^2}$

(e) $\dfrac{20a^3}{5a^3}$ **(f)** $\dfrac{6x^2y}{3xy}$

(g) $\dfrac{12r^3s^2}{4rs}$ **(h)** $\dfrac{49c^4d^6}{7cd^3}$

Answers

1. $15x^5$ **3.** $-28b^{10}$ **5.** $40p^{13}$ **7.** $-12m^6$ **9.** $32x^5y^3$
11. $-6m^9n^3$ **13.** $10x + 30$ **15.** $12a^2 + 15a$ **17.** $12s^4 - 21s^3$
19. $8x^3 - 4x^2 + 2x$ **21.** $6x^3y^2 + 3x^2y^3 + 15x^2y^2$

23. $18m^4n^2 - 12m^3n^2 + 6m^3n^3$ **25.** $x^2 + 5x + 6$ **27.** $m^2 - 14m + 45$

29. $p^2 - p - 56$ **31.** $w^2 + 30w + 200$ **33.** $3x^2 - 29x + 40$

35. $6x^2 - x - 12$ **37.** $12a^2 - 31ab + 9b^2$ **39.** $21p^2 - 13pq - 20q^2$

41. $6x^2 + 23xy + 20y^2$ **43.** $x^2 + 10x + 25$ **45.** $y^2 - 18y + 81$

47. $36m^2 + 12mn + n^2$ **49.** $a^2 - 25$ **51.** $x^2 - 4y^2$ **53.** $25s^2 - 9t^2$

55. False   **57.** True   **59.** $(6x^2 - 11x - 35)\,\text{cm}^2$ **61.** $x^2 + 10x + 25$

63. $4a^2 - 4a + 1$ **65.** $36m^2 + 12m + 1$ **67.** $9x^2 - 6xy + y^2$

69. $4r^2 + 20rs + 25s^2$ **71.** $x^2 + x + \dfrac{1}{4}$ **73.** $x^2 - 36$ **75.** $m^2 - 144$

77. $x^2 - \dfrac{1}{4}$ **79.** $p^2 - 0.16$ **81.** $a^2 - 9b^2$ **83.** $16r^2 - s^2$

85. 2,499   **87.** 884   **89.** 3,575   **91.** $25x^2 - 40x + 16$

93. Above and Beyond   **95.** Above and Beyond

97. $2x^3 + 2x^2 - 16x - 24$ **(a)** x **(b)** a^2 **(c)** $3p$ **(d)** $2m^2$

(e) 4   **(f)** $2x$ **(g)** $3r^2s$ **(h)** $7c^3d^3$

3.5

Dividing Polynomials

< 3.5 Objectives >

1 > Find the quotient when a polynomial is divided by a monomial

2 > Find the quotient when a polynomial is divided by a binomial

In Section 1.7, we introduced the quotient property of exponents, which was used to divide one monomial by another monomial.

Step by Step

To Divide a Monomial by a Monomial	Step 1	Divide the coefficients.
	Step 2	Use the quotient property of exponents to combine the variables.

Example 1 | **Dividing by a Monomial**

< Objective 1 >

RECALL

The quotient property says: If x is not zero, then

$$\frac{x^m}{x^n} = x^{m-n}$$

Divide: $\frac{8}{2} = 4$

(a) $\dfrac{8x^4}{2x^2} = 4x^{4-2}$

Subtract the exponents.

$$= 4x^2$$

(b) $\dfrac{45a^5b^3}{9a^2b} = 5a^3b^2$

Check Yourself 1

Divide.

(a) $\dfrac{16a^5}{8a^3}$

(b) $\dfrac{28m^4n^3}{7m^3n}$

NOTE

This step depends on the distributive property and the definition of division.

Now look at how this can be extended to divide any polynomial by a monomial. For example, to divide $12a^3 + 8a^2$ by $4a$, proceed as follows:

$$\frac{12a^3 + 8a^2}{4a} = \frac{12a^3}{4a} + \frac{8a^2}{4a}$$

Divide each term in the numerator by the denominator, $4a$.

Now do each division.

$$= 3a^2 + 2a$$

This work leads us to the following rule.

Step by Step

| **To Divide a Polynomial by a Monomial** | Step 1 | Divide each term of the polynomial by the monomial. |
| | Step 2 | Simplify the results. |

Example 2   Dividing by a Monomial

Divide each term by 2.

(a) $\dfrac{4a^2 + 8}{2} = \dfrac{4a^2}{2} + \dfrac{8}{2}$

$\qquad\qquad = 2a^2 + 4$

Divide each term by 6y.

(b) $\dfrac{24y^3 + (-18y^2)}{6y} = \dfrac{24y^3}{6y} + \dfrac{-18y^2}{6y}$

$\qquad\qquad\qquad = 4y^2 - 3y$

Remember the rules for signs in division.

(c) $\dfrac{15x^2 + 10x}{-5x} = \dfrac{15x^2}{-5x} + \dfrac{10x}{-5x}$

$\qquad\qquad\quad = -3x - 2$

NOTE

With practice you can just write the quotient.

(d) $\dfrac{14x^4 + 28x^3 - 21x^2}{7x^2} = \dfrac{14x^4}{7x^2} + \dfrac{28x^3}{7x^2} - \dfrac{21x^2}{7x^2}$

$\qquad\qquad\qquad\qquad = 2x^2 + 4x - 3$

(e) $\dfrac{9a^3b^4 - 6a^2b^3 + 12ab^4}{3ab} = \dfrac{9a^3b^4}{3ab} - \dfrac{6a^2b^3}{3ab} + \dfrac{12ab^4}{3ab}$

$\qquad\qquad\qquad\qquad = 3a^2b^3 - 2ab^2 + 4b^3$

Check Yourself 2

Divide.

(a) $\dfrac{20y^3 - 15y^2}{5y}$

(b) $\dfrac{8a^3 - 12a^2 + 4a}{-4a}$

(c) $\dfrac{16m^4n^3 - 12m^3n^2 + 8mn}{4mn}$

We are now ready to look at dividing one polynomial by another polynomial (with more than one term). The process is very much like long division in arithmetic, as Example 3 illustrates.

    Example 3   **Dividing by a Binomial**

< Objective 2 >

Compare the steps in these two divisions.

Divide $x^2 + 7x + 10$ by $x + 2$.   *Divide 2,176 by 32.*

> **NOTE**
>
> The first term in the dividend, x^2, is divided by the first term in the divisor, x.

Step 1 $x + 2 \overline{)x^2 + 7x + 10}$ Divide x^2 by x to get x.

$$\begin{array}{r} x \\ x+2{\overline{\smash{\big)}\,x^2 + 7x + 10}} \end{array}$$

$$\begin{array}{r} 6 \\ 32{\overline{\smash{\big)}\,2176}} \end{array}$$

Step 2

$$\begin{array}{r} x \\ x+2{\overline{\smash{\big)}\,x^2 + 7x + 10}} \\ \underline{x^2 + 2x} \end{array}$$

$$\begin{array}{r} 6 \\ 32{\overline{\smash{\big)}\,2176}} \\ \underline{192} \end{array}$$

Multiply the divisor, $x + 2$, by x.

> **RECALL**
>
> To subtract $x^2 + 2x$, mentally change the signs to $-x^2 - 2x$ and add. Take your time and be careful here. It's where most errors are made.

Step 3

$$\begin{array}{r} x \\ x+2{\overline{\smash{\big)}\,x^2 + 7x + 10}} \\ \underline{x^2 + 2x} \\ 5x + 10 \end{array}$$

$$\begin{array}{r} 6 \\ 32{\overline{\smash{\big)}\,2176}} \\ \underline{192} \\ 256 \end{array}$$

Subtract and bring down 10.

Step 4

$$\begin{array}{r} x + 5 \\ x+2{\overline{\smash{\big)}\,x^2 + 7x + 10}} \\ \underline{x^2 + 2x} \\ 5x + 10 \end{array}$$

$$\begin{array}{r} 68 \\ 32{\overline{\smash{\big)}\,2176}} \\ \underline{192} \\ 256 \end{array}$$

Divide $5x$ by x to get 5.

> **NOTE**
>
> We repeat the process until the degree of the remainder is less than that of the divisor or until there is no remainder.

Step 5

$$\begin{array}{r} x + 5 \\ x+2{\overline{\smash{\big)}\,x^2 + 7x + 10}} \\ \underline{x^2 + 2x} \\ 5x + 10 \\ \underline{5x + 10} \\ 0 \end{array}$$

$$\begin{array}{r} 68 \\ 32{\overline{\smash{\big)}\,2176}} \\ \underline{192} \\ 256 \\ \underline{256} \\ 0 \end{array}$$

Multiply $x + 2$ by 5 and then subtract.

The quotient is $x + 5$.

   Check Yourself 3

Divide $x^2 + 9x + 20$ by $x + 4$.

In Example 3, we showed all the steps separately to help you see the process. In practice, the work can be shortened.

Example 4   **Dividing by a Binomial**

Divide $x^2 + x - 12$ by $x - 3$.

NOTE

You might want to write out a problem like $408 \div 17$ to compare the steps.

$$
\begin{array}{r}
x + 4 \\
x - 3 \overline{)x^2 + x - 12} \\
x^2 - 3x \\
\hline
4x - 12 \\
4x - 12 \\
\hline
0
\end{array}
$$

Step 1 Divide x^2 by x to get x, the first term of the quotient.
Step 2 Multiply $x - 3$ by x.
Step 3 Subtract and bring down -12. Remember to mentally change the signs to $-x^2 + 3x$ and add.
Step 4 Divide $4x$ by x to get 4, the second term of the quotient.
Step 5 Multiply $x - 3$ by 4 and subtract.

The quotient is $x + 4$.

✓ **Check Yourself 4**

Divide.

$(x^2 + 2x - 24) \div (x - 4)$

You may have a remainder in algebraic long division just as in arithmetic. Consider Example 5.

 Example 5   **Dividing by a Binomial**

Divide $4x^2 - 8x + 11$ by $2x - 3$.

$$
\begin{array}{r}
2x - 1 \quad \text{Quotient}\\
2x - 3 \overline{)4x^2 - 8x + 11} \\
4x^2 - 6x \\
\hline
-2x + 11 \\
-2x + 3 \\
\hline
8
\end{array}
$$

Divisor

↑ Remainder

This result can be written as

$$
\frac{4x^2 - 8x + 11}{2x - 3}
$$

$$
= \underbrace{2x - 1}_{\text{Quotient}} + \frac{8}{2x - 3} \quad \begin{matrix} \leftarrow \text{Remainder} \\ \leftarrow \text{Divisor} \end{matrix}
$$

Check Yourself 5

Divide.

$(6x^2 - 7x + 15) \div (3x - 5)$

The division process shown in our previous examples can be extended to dividends of a higher degree. The steps involved in the division process are exactly the same, as Example 6 illustrates.

Example 6   **Dividing by a Binomial**

Divide $6x^3 + x^2 - 4x - 5$ by $3x - 1$.

$$
\begin{array}{r}
2x^2 + x - 1 \\
3x - 1\overline{)6x^3 + x^2 - 4x - 5} \\
\underline{6x^3 - 2x^2} \\
3x^2 - 4x \\
\underline{3x^2 - x} \\
-3x - 5 \\
\underline{-3x + 1} \\
-6
\end{array}
$$

The result can be written as

$$
\frac{6x^3 + x^2 - 4x - 5}{3x - 1} = 2x^2 + x - 1 + \frac{-6}{3x - 1}
$$

Check Yourself 6

Divide $4x^3 - 2x^2 + 2x + 15$ by $2x + 3$.

Suppose that the dividend is "missing" a term in some power of the variable. You can use 0 as the coefficient for the missing term. Consider Example 7.

Example 7   **Dividing by a Binomial**

Divide $x^3 - 2x^2 + 5$ by $x + 3$.

$$
\begin{array}{r}
x^2 - 5x + 15 \\
x + 3\overline{)x^3 - 2x^2 + 0x + 5} \\
\underline{x^3 + 3x^2} \\
-5x^2 + 0x \\
\underline{-5x^2 - 15x} \\
15x + 5 \\
\underline{15x + 45} \\
-40
\end{array}
$$

Write $0x$ for the "missing" term in x.

This result can be written as

$$\frac{x^3 - 2x^2 + 5}{x + 3} = x^2 - 5x + 15 + \frac{-40}{x + 3}$$

 Check Yourself 7

Divide.

$$(4x^3 + x + 10) \div (2x - 1)$$

You should always arrange the terms of the divisor and dividend in descending-exponent form before starting the long-division process, as illustrated in Example 8.

| Example 8 | Dividing by a Binomial |

Divide $5x^2 - x + x^3 - 5$ by $-1 + x^2$.

Write the divisor as $x^2 - 1$ and the dividend as $x^3 + 5x^2 - x - 5$.

$$
\begin{array}{r}
x + 5 \\
x^2 - 1 \overline{) x^3 + 5x^2 - x - 5} \\
\underline{x^3 - x } \\
5x^2 - 5 \\
\underline{5x^2 - 5} \\
0
\end{array}
$$

Write $x^3 - x$, the product of x and $x^2 - 1$, so that like terms fall in the same columns.

The quotient is $x + 5$.

 Check Yourself 8

Divide:

$$(5x^2 + 10 + 2x^3 + 4x) \div (2 + x^2)$$

 Check Yourself ANSWERS

1. **(a)** $2a^2$; **(b)** $4mn^2$ 2. **(a)** $4y^2 - 3y$; **(b)** $-2a^2 + 3a - 1$;

(c) $4m^3n^2 - 3m^2n + 2$ 3. $x + 5$ 4. $x + 6$ 5. $2x + 1 + \dfrac{20}{3x - 5}$

6. $2x^2 - 4x + 7 + \dfrac{-6}{2x + 3}$ 7. $2x^2 + x + 1 + \dfrac{11}{2x - 1}$ 8. $2x + 5$

Reading Your Text

The following fill-in-the-blank exercises are designed to ensure that you understand some of the key vocabulary used in this section.

SECTION 3.5

(a) When dividing two monomials, we use the quotient property of exponents to combine the _____.

(b) Our final step when dividing a polynomial by a monomial is to _____ the result.

(c) A polynomial with exactly _____ terms is called a trinomial.

(d) When the dividend is missing a term in some power of the variable, we use _____ as a coefficient for that missing term.

Basic Skills | Advanced Skills | Vocational-Technical Applications | Calculator/Computer | Above and Beyond | Getting Ready

3.5 **exercises**

< Objective 1 >

Divide.

1. $\dfrac{18x^6}{9x^2}$

2. $\dfrac{20a^7}{5a^5}$

3. $\dfrac{35m^3n^2}{7mn^2}$

4. $\dfrac{42x^5y^2}{6x^3y}$

5. $\dfrac{3a + 6}{3}$

6. $\dfrac{4x - 8}{4}$

7. $\dfrac{9b^2 - 12}{3}$

8. $\dfrac{10m^2 + 5m}{5}$

9. $\dfrac{16a^3 - 24a^2}{4a}$

10. $\dfrac{9x^3 + 12x^2}{3x}$

11. $\dfrac{12m^2 + 6m}{-3m}$

12. $\dfrac{20b^3 - 25b^2}{-5b}$

13. $\dfrac{18a^4 + 12a^3 - 6a^2}{6a}$

14. $\dfrac{21x^5 - 28x^4 + 14x^3}{7x}$

15. $\dfrac{20x^4y^2 - 15x^2y^3 + 10x^3y}{5x^2y}$

16. $\dfrac{16m^3n^3 + 24m^2n^2 - 40mn^3}{8mn^2}$

17. $\dfrac{27a^5b^5 + 9a^4b^4 - 3a^2b^3}{3a^2b^3}$

18. $\dfrac{7x^5y^5 - 21x^4y^4 + 14x^3y^3}{7x^3y^3}$

19. $\dfrac{3a^6b^4c^2 - 2a^4b^2c + 6a^3b^2c}{a^3b^2c}$

20. $\dfrac{2x^4y^4z^4 + 3x^3y^3z^3 - xy^2z^3}{xy^2z^3}$

< Objective 2 >

Perform the indicated divisions.

21. $\dfrac{x^2 + 5x + 6}{x + 2}$

22. $\dfrac{x^2 + 8x + 15}{x + 3}$

23. $\dfrac{x^2 - x - 20}{x + 4}$

24. $\dfrac{x^2 - 2x - 35}{x + 5}$

MathZone

Boost your grade at mathzone.com!

> Practice Problems
> NetTutor
> Self-Tests
> e-Professors
> Videos

Name _____

Section _____ Date _____

Answers

1. _____   2. _____

3. _____   4. _____

5. _____   6. _____

7. _____   8. _____

9. _____   10. _____

11. _____   12. _____

13. _____

14. _____

15. _____

16. _____

17. _____

18. _____

19. _____

20. _____

21. _____

22. _____

23. _____

24. _____

Answers

25. _____

26. _____

27. _____

28. _____

29. _____

30. _____

31. _____

32. _____

33. _____

34. _____

35. _____

36. _____

37. _____

38. _____

39. _____

40. _____

41. _____

42. _____

43. _____

44. _____

25. $\dfrac{2x^2 - 3x - 5}{x - 3}$

26. $\dfrac{3x^2 + 17x - 12}{x + 6}$

27. $\dfrac{6x^2 - x - 10}{3x - 5}$

28. $\dfrac{4x^2 + 6x - 25}{2x + 7}$

29. $\dfrac{x^3 + x^2 - 4x - 4}{x + 2}$

30. $\dfrac{x^3 - 2x^2 + 4x - 21}{x - 3}$

31. $\dfrac{4x^3 + 7x^2 + 10x + 5}{4x - 1}$

32. $\dfrac{2x^3 - 3x^2 + 4x + 4}{2x + 1}$

33. $\dfrac{x^3 - x^2 + 5}{x - 2}$

34. $\dfrac{x^3 + 4x - 3}{x + 3}$

35. $\dfrac{25x^3 + x}{5x - 2}$

36. $\dfrac{8x^3 - 6x^2 + 2x}{4x + 1}$

37. $\dfrac{2x^2 - 8 - 3x + x^3}{x - 2}$

38. $\dfrac{x^2 - 18x + 2x^3 + 32}{x + 4}$

39. $\dfrac{x^4 - 1}{x - 1}$ > Videos

40. $\dfrac{x^4 + x^2 - 16}{x + 2}$

Basic Skills | **Advanced Skills** | Vocational-Technical Applications | Calculator/Computer | Above and Beyond | Getting Ready

41. $\dfrac{x^3 - 3x^2 - x + 3}{x^2 - 1}$

42. $\dfrac{x^3 + 2x^2 + 3x + 6}{x^2 + 3}$

43. $\dfrac{x^4 + 2x^2 - 2}{x^2 + 3}$ > Videos

44. $\dfrac{x^4 + x^2 - 5}{x^2 - 2}$

45. $\dfrac{y^3 + 1}{y + 1}$

46. $\dfrac{y^3 - 8}{y - 2}$

47. $\dfrac{x^4 - 1}{x^2 - 1}$

48. $\dfrac{x^6 - 1}{x^3 - 1}$

Answers

45. _____

46. _____

47. _____

48. _____

49. _____

50. _____

51. _____

52. _____

53. _____

Basic Skills | Advanced Skills | Vocational-Technical Applications | Calculator/Computer | **Above and Beyond** | Getting Ready
▲

49. Find the value of c so that $\dfrac{y^2 - y + c}{y + 1} = y - 2$.

50. Find the value of c so that $\dfrac{x^3 + x^2 + x + c}{x^2 + 1} = x + 1$.

51. Write a summary of your work with polynomials. Explain how a polynomial is recognized and explain the rules for the arithmetic of polynomials—how to add, subtract, multiply, and divide. What parts of this chapter do you feel you understand very well, and what parts do you still have questions about or feel unsure of? Exchange papers with another student and compare your questions.

52. A funny (and useful) thing about division of polynomials: To find out about it, do this division. Compare your answer with another student's.

$(x - 2)\overline{)2x^2 + 3x - 5}$ Is there a remainder?

Now, evaluate the polynomial $2x^2 + 3x - 5$ when $x = 2$. Is this value the same as the remainder?

Try $(x + 3)\overline{)5x^2 - 2x + 1}$ Is there a remainder?

Evaluate the polynomial $5x^2 - 2x + 1$ when $x = -3$. Is this value the same as the remainder?
    What happens when there is no remainder?

Try $(x - 6)\overline{)3x^3 + 14x^2 - 23x + 6}$ Is the remainder zero?

Evaluate the polynomial $3x^3 + 14x^2 - 23x + 6$ when $x = 6$. Is this value zero? Write a description of the patterns you see. When does the pattern hold? Make up several more examples and test your conjecture.

53. (a) Divide $\dfrac{x^2 - 1}{x - 1}$.   **(b)** Divide $\dfrac{x^3 - 1}{x - 1}$.   **(c)** Divide $\dfrac{x^4 - 1}{x - 1}$.

(d) Based on your results on parts **(a)**, **(b)**, and **(c)**, predict $\dfrac{x^{50} - 1}{x - 1}$.

54. (a) Divide $\dfrac{x^2 + x + 1}{x - 1}$.

(b) Divide $\dfrac{x^3 + x^2 + x + 1}{x - 1}$.

(c) Divide $\dfrac{x^4 + x^3 + x^2 + x + 1}{x - 1}$.

(d) Based on your results to **(a)**, **(b)**, and **(c)**, predict

$$\dfrac{x^{10} + x^9 + x^8 + \cdots + x + 1}{x - 1}.$$

54. _____

Answers

1. $2x^4$ **3.** $5m^2$ **5.** $a + 2$ **7.** $3b^2 - 4$ **9.** $4a^2 - 6a$
11. $-4m - 2$ **13.** $3a^3 + 2a^2 - a$ **15.** $4x^2y - 3y^2 + 2x$
17. $9a^3b^2 + 3a^2b - 1$ **19.** $3a^3b^2c - 2a + 6$ **21.** $x + 3$

23. $x - 5$ **25.** $2x + 3 + \dfrac{4}{x - 3}$ **27.** $2x + 3 + \dfrac{5}{3x - 5}$

29. $x^2 - x - 2$ **31.** $x^2 + 2x + 3 + \dfrac{8}{4x - 1}$

33. $x^2 + x + 2 + \dfrac{9}{x - 2}$ **35.** $5x^2 + 2x + 1 + \dfrac{2}{5x - 2}$

37. $x^2 + 4x + 5 + \dfrac{2}{x - 2}$ **39.** $x^3 + x^2 + x + 1$ **41.** $x - 3$

43. $x^2 - 1 + \dfrac{1}{x^2 + 3}$ **45.** $y^2 - y + 1$ **47.** $x^2 + 1$ **49.** $c = -2$

51. Above and Beyond **53. (a)** $x + 1$; **(b)** $x^2 + x + 1$;
(c) $x^3 + x^2 + x + 1$; **(d)** $x^{49} + x^{48} + \cdots + x + 1$

Definition/Procedure	Example	Reference
Exponents and Polynomials		Section 3.1

Properties of Exponents

$a^m \cdot a^n = a^{m+n}$	Product property	$3^3 \cdot 3^4 = 3^7$	*p.* 296
$\dfrac{a^m}{a^n} = a^{m-n}$	Quotient property	$\dfrac{x^6}{x^2} = x^4$	*p.* 296
$(a^m)^n = a^{mn}$	Power to a power property	$(x^3)^5 = x^{15}$	*p.* 296
$(ab)^m = a^m b^m$	Product to a power property	$(3x)^2 = 9x^2$	*p.* 297
$\left(\dfrac{a}{b}\right)^m = \dfrac{a^m}{b^m}$	Quotient to a power property	$\left(\dfrac{2}{3}\right)^3 = \dfrac{8}{27}$	*p.* 298

Term

A number or the product of a number and variables.

Polynomial

An algebraic expression made up of terms in which the exponents are whole numbers. These terms are connected by plus or minus signs. Each sign ($+$ or $-$) is attached to the term following that sign.

$4x^3 - 3x^2 + 5x$ is a polynomial.
The terms of $4x^3 - 3x^2 + 5x$ are $4x^3$, $-3x^2$, and $5x$.

p. 299

p. 300

Coefficient

In each term of a polynomial, the number is called the *numerical coefficient* or, more simply, the *coefficient,* of that term.

The coefficients of $4x^3 - 3x^2$ are 4 and -3.

p. 300

Types of Polynomials

A polynomial can be classified according to the number of terms it has.

A *mono*mial has one term.
A *bi*nomial has two terms.
A *tri*nomial has three terms.

$2x^3$ is a monomial.
$3x^2 - 7x$ is a binomial.
$5x^5 - 5x^3 + 2$ is a trinomial.

p. 300

Degree

The highest power of the variable appearing in any one term.

The degree of $4x^5 - 5x^3 + 3x$ is 5.

p. 301

Descending-Exponent Form

The form of a polynomial when it is written with the highest-degree term first, the next highest-degree term second, and so on.

$4x^5 - 5x^3 + 3x$ is written in descending-exponent form.

p. 301

Continued

Definition/Procedure	Example	Reference

Negative Exponents and Scientific Notation

Section 3.2

The Zero Power

Any nonzero expression raised to the 0 power equals 1.

$3^0 = 1$

$(5x)^0 = 1$

p. 310

Negative Powers

An expression raised to a negative power equals its reciprocal taken to the absolute value of its power.

$\left(\dfrac{x}{3}\right)^{-4} = \left(\dfrac{3}{x}\right)^4 = \dfrac{3^4}{x^4}$

p. 312

Scientific Notation

Any number written in the form $a \times 10^n$ in which $1 \le a < 10$ and n is an integer, is written in scientific notation.

6.2×10^{23}

p. 315

Adding and Subtracting Polynomials

Section 3.3

Removing Signs of Grouping

1. If a plus sign ($+$) or no sign at all appears in front of parentheses, just remove the parentheses. No other changes are necessary.

$3x + (2x - 3)$
$= 3x + 2x - 3$
$= 5x - 3$

p. 323

2. If a minus sign ($-$) appears in front of parentheses, the parentheses can be removed by changing the sign of each term inside the parentheses.

$2x - (x - 4)$
$= 2x - x + 4$
$= x + 4$

p. 325

Adding Polynomials

Remove the signs of grouping. Then collect and combine any like terms.

$(2x + 3) + (3x - 5)$
$= 2x + 3 + 3x - 5 = 5x - 2$

p. 323

Subtracting Polynomials

Remove the signs of grouping by changing the sign of each term in the polynomial being subtracted. Then combine any like terms.

$(3x^2 + 2x) - (2x^2 + 3x - 1)$
$= 3x^2 + 2x - 2x^2 - 3x + 1$

↑   ↑   ↗
Sign changes

$= 3x^2 - 2x^2 + 2x - 3x + 1$
$= x^2 - x + 1$

p. 326

Multiplying Polynomials

Section 3.4

To Multiply a Polynomial by a Monomial

Multiply each term of the polynomial by the monomial and simplify the results.

$3x(2x + 3)$
$= 3x \cdot 2x + 3x \cdot 3$
$= 6x^2 + 9x$

p. 334

Definition/Procedure	Example	Reference

To Multiply a Binomial by a Binomial

Use the FOIL method:

$$\quad\quad\quad\quad F\quad\quad O\quad\quad I\quad\quad L$$
$$(a + b)(c + d) = a \cdot c + a \cdot d + b \cdot c + b \cdot d$$

$(2x - 3)(3x + 5)$
$= 6x^2 + 10x - 9x - 15$
$\quad\;\; F \quad\;\; O \quad\;\; I \quad\;\; L$
$= 6x^2 + x - 15$

p. 337

To Multiply a Polynomial by a Polynomial

Arrange the polynomials vertically. Multiply each term of the upper polynomial by each term of the lower polynomial and add the results.

$$\begin{array}{r} x^2 - 3x + 5 \\ 2x - 3 \\ \hline -3x^2 + 9x - 15 \\ 2x^3 - 6x^2 + 10x \quad\quad\;\; \\ \hline 2x^3 - 9x^2 + 19x - 15 \end{array}$$

p. 338

The Square of a Binomial

$(a + b)^2 = a^2 + 2ab + b^2$

$(2x - 5)^2$
$= 4x^2 + 2 \cdot 2x \cdot (-5) + 25$
$= 4x^2 - 20x + 25$

p. 339

The Product of Binomials That Differ Only in Sign

Subtract the square of the second term from the square of the first term.
$(a + b)(a - b) = a^2 - b^2$

$(2x - 5y)(2x + 5y)$
$= (2x)^2 - (5y)^2$
$= 4x^2 - 25y^2$

p. 340

Dividing Polynomials

Section 3.5

To Divide a Polynomial by a Monomial

1. Divide each term of the polynomial by the monomial.
2. Simplify the result.

$\dfrac{27x^2y^2 + 9x^3y^4}{3xy^2}$

$= \dfrac{27x^2y^2}{3xy^2} + \dfrac{9x^3y^4}{3xy^2}$

$= 9x + 3x^2y^2$

p. 350

This summary exercise set is provided to give you practice with each of the objectives of this chapter. Each exercise is keyed to the appropriate chapter section. When you are finished, you can check your answer to the odd-numbered exercises against those presented in the back of the text. If you have difficulty with any of these questions, go back and reread the examples from that section. Your instructor will give you guidelines on how best to use these exercises in your instructional setting.

3.1 *Simplify each of the following expressions.*

1. $\dfrac{x^{10}}{x^3}$

2. $\dfrac{a^5}{a^4}$

3. $\dfrac{x^2 \cdot x^3}{x^4}$

4. $\dfrac{m^2 \cdot m^3 \cdot m^4}{m^5}$

5. $\dfrac{18p^7}{9p^5}$

6. $\dfrac{24x^{17}}{8x^{13}}$

7. $\dfrac{30m^7n^5}{6m^2n^3}$

8. $\dfrac{108x^9y^4}{9xy^4}$

9. $\dfrac{48p^5q^3}{6p^3q}$

10. $\dfrac{52a^5b^3c^5}{13a^4c}$

11. $(2ab)^2$

12. $(p^2q^3)^3$

13. $(2x^2y^2)^3(3x^3y)^2$

14. $\left(\dfrac{p^2q^3}{t^4}\right)^2$

15. $\dfrac{(x^5)^2}{(x^3)^3}$

16. $(4w^2t)^2\,(3wt^2)^3$

17. $(y^3)^2(3y^2)^3$

18. $\left(\dfrac{4x^4}{3y}\right)^2$

Find the value of each of the following polynomials for the given value of the variable.

19. $5x + 1; x = -1$

20. $2x^2 + 7x - 5; x = 2$

21. $-x^2 + 3x - 1; x = 6$

22. $4x^2 + 5x + 7; x = -4$

Classify each polynomial as a monomial, binomial, or trinomial, where possible.

23. $5x^3 - 2x^2$

24. $7x^5$

25. $4x^5 - 8x^3 + 5$

26. $x^3 + 2x^2 - 5x + 3$

27. $9a^3 - 18a^2$

Arrange in descending-exponent form, if necessary, and give the degree of each polynomial.

28. $5x^5 + 3x^2$

29. $9x$

30. $6x^2 + 4x^4 + 6$

31. $5 + x$

32. -8

33. $9x^4 - 3x + 7x^6$

3.2 *Evaluate each of the following expressions.*

34. 4^0

35. $(3a)^0$

36. $6x^0$

37. $(3a^4b)^0$

Write using positive exponents. Simplify when possible.

38. x^{-5}

39. 3^{-3}

40. 10^{-4}

41. $4x^{-4}$

42. $\dfrac{x^6}{x^8}$

43. m^7m^{-9}

44. $\dfrac{a^{-4}}{a^{-9}}$

45. $\dfrac{x^2y^{-3}}{x^{-3}y^2}$

46. $(3m^{-3})^2$

47. $\dfrac{(a^4)^{-3}}{(a^{-2})^{-3}}$

In exercises 48 to 50, express each number in scientific notation.

48. The average distance from Earth to the Sun is 150,000,000,000 m.

49. A bat emits a sound with a frequency of 51,000 cycles per second.

50. The diameter of a grain of salt is 0.000062 m.

In exercises 51 to 54, compute the expression using scientific notation and express your answers in that form.

51. $(2.3 \times 10^{-3})(1.4 \times 10^{12})$

52. $(4.8 \times 10^{-10})(6.5 \times 10^{34})$

53. $\dfrac{(8 \times 10^{23})}{(4 \times 10^6)}$

54. $\dfrac{(5.4 \times 10^{-12})}{(4.5 \times 10^{16})}$

3.3 *Add.*

55. $9a^2 - 5a$ and $12a^2 + 3a$

56. $5x^2 + 3x - 5$ and $4x^2 - 6x - 2$

57. $5y^3 - 3y^2$ and $4y + 3y^2$

Subtract.

58. $4x^2 - 3x$ from $8x^2 + 5x$

59. $2x^2 - 5x - 7$ from $7x^2 - 2x + 3$

60. $5x^2 + 3$ from $9x^2 - 4x$

Perform the indicated operations.

61. Subtract $5x - 3$ from the sum of $9x + 2$ and $-3x - 7$.

62. Subtract $5a^2 - 3a$ from the sum of $5a^2 + 2$ and $7a - 7$.

63. Subtract the sum of $16w^2 - 3w$ and $8w + 2$ from $7w^2 - 5w + 2$.

Add using the vertical method.

64. $x^2 + 5x - 3$ and $2x^2 + 4x - 3$

65. $9b^2 - 7$ and $8b + 5$

66. $x^2 + 7$, $3x - 2$, and $4x^2 - 8x$

Subtract using the vertical method.

67. $5x^2 - 3x + 2$ from $7x^2 - 5x - 7$

68. $8m - 7$ from $9m^2 - 7$

3.4 *Multiply.*

69. $(5a^3)(a^2)$

70. $(2x^2)(3x^5)$

71. $(-9p^3)(-6p^2)$

72. $(3a^2b^3)(-7a^3b^4)$

73. $5(3x - 8)$

74. $4a(3a + 7)$

75. $(-5rs)(2r^2s - 5rs)$

76. $7mn(3m^2n - 2mn^2 + 5mn)$

77. $(x + 5)(x + 4)$

78. $(w - 9)(w - 10)$

79. $(a - 7b)(a + 7b)$

80. $(p - 3q)^2$

81. $(a + 4b)(a + 3b)$

82. $(b - 8)(2b + 3)$

83. $(3x - 5y)(2x - 3y)$

84. $(5r + 7s)(3r - 9s)$

85. $(y + 2)(y^2 - 2y + 3)$

86. $(b + 3)(b^2 - 5b - 7)$

87. $(x - 2)(x^2 + 2x + 4)$

88. $(m^2 - 3)(m^2 + 7)$

89. $2x(x + 5)(x - 6)$

90. $a(2a - 5b)(2a - 7b)$

Find the following products.

91. $(x + 7)^2$

92. $(a - 8)^2$

93. $(2w - 5)^2$

94. $(3p + 4)^2$

95. $(a + 7b)^2$

96. $(8x - 3y)^2$

97. $(x - 5)(x + 5)$

98. $(y + 9)(y - 9)$

99. $(2m + 3)(2m - 3)$

100. $(3r - 7)(3r + 7)$

101. $(5r - 2s)(5r + 2s)$

102. $(7a + 3b)(7a - 3b)$

103. $2x(x - 5)^2$

104. $3c(c + 5d)(c - 5d)$

3.5 *Divide.*

105. $\dfrac{9a^5}{3a^2}$

106. $\dfrac{24m^4n^2}{6m^2n}$

107. $\dfrac{15a - 10}{5}$

108. $\dfrac{32a^3 + 24a}{8a}$

109. $\dfrac{9r^2s^3 - 18r^3s^2}{-3rs^2}$

110. $\dfrac{35x^3y^2 - 21x^2y^3 + 14x^3y}{7x^2y}$

Perform the indicated long division.

111. $\dfrac{x^2 - 2x - 15}{x + 3}$

112. $\dfrac{2x^2 + 9x - 35}{2x - 5}$

113. $\dfrac{x^2 - 8x + 17}{x - 5}$

114. $\dfrac{6x^2 - x - 10}{3x + 4}$

115. $\dfrac{6x^3 + 14x^2 - 2x - 6}{6x + 2}$

116. $\dfrac{4x^3 + x + 3}{2x - 1}$

117. $\dfrac{3x^2 + x^3 + 5 + 4x}{x + 2}$

118. $\dfrac{2x^4 - 2x^2 - 10}{x^2 - 3}$

Name _____

Section _____ Date _____

Answers

1. _____

2. _____

3. _____

4. _____

5. _____

6. _____

7. _____

8. _____

9. _____

10. _____

11. _____

12. _____

13. _____

14. _____

15. _____

16. _____

17. _____

18. _____

19. _____

The purpose of this self-test is to help you check your progress and to review for the next in-class exam. Allow yourself about an hour to take this test. At the end of that hour check your answers against those given in the back of the text. Section references accompany the answers. If you missed any questions, go back to those sections and reread the examples until you master the concepts.

Simplify each of the following expressions.

1. $a^5 \cdot a^9$

2. $3x^2y^3 \cdot 5xy^4$

3. $\dfrac{4x^5}{2x^2}$

4. $\dfrac{20a^3b^5}{5a^2b^2}$

5. $(3x^2y)^3$

6. $\left(\dfrac{2w^2}{3t^3}\right)^2$

7. $(2x^3y^2)^4(x^2y^3)^3$

8. Find the value of the polynomial $-3x^2 - 5x + 8$ if $x = -2$.

Classify each of the following polynomials as a **monomial, binomial,** *or* **trinomial.**

9. $6x^2 + 7x$

10. $5x^2 + 8x - 8$

Arrange in descending-exponent form; then give the coefficients and degree of the polynomial.

11. $-3x^2 + 8x^4 - 7$

Evaluate (assume the variables are nonzero).

12. 8^0

13. $6x^0$

Rewrite using positive exponents. Simplify when possible.

14. y^{-5}

15. $3b^{-7}$

16. y^4y^{-8}

17. $\dfrac{p^{-5}}{p^5}$

Add.

18. $3x^2 - 7x + 2$ and $7x^2 - 5x - 9$

19. $7a^2 - 3a$ and $7a^3 + 4a^2$

Subtract.

20. $5x^2 - 2x + 5$ from $8x^2 + 9x - 7$ **21.** $2b^2 + 5$ from $3b^2 - 7b$

22. $5a^2 + a$ from the sum of $3a^2 - 5a$ and $9a^2 - 4a$

Add using the vertical method.

23. $x^2 + 3, 5x - 7,$ and $3x^2 - 2$

Subtract using the vertical method.

24. $3x^2 - 5$ from $5x^2 - 7x$

Multiply.

25. $5ab(3a^2b - 2ab + 4ab^2)$ **26.** $(x - 2)(3x + 7)$

27. $(2x + y)(x^2 + 3xy - 2y^2)$ **28.** $(4x + 3y)(2x - 5y)$

29. $x(3x - y)(4x + 5y)$ **30.** $(3m + 2n)^2$

31. $(a - 7b)(a + 7b)$

Divide.

32. $\dfrac{14x^3y - 21xy^2}{7xy}$ **33.** $\dfrac{20c^3d - 30cd + 45c^2d^2}{5cd}$

34. $(x^2 - 2x - 24) \div (x + 4)$ **35.** $(2x^2 + x + 4) \div (2x - 3)$

36. $(6x^3 - 7x^2 + 3x + 9) \div (3x + 1)$ **37.** $(x^3 - 5x^2 + 9x - 9) \div (x - 1)$

Compute and answer using scientific notation.

38. $(2.1 \times 10^7)(8 \times 10^{12})$ **39.** $(6 \times 10^{-23})(5.2 \times 10^{12})$

40. $\dfrac{7.28 \times 10^3}{1.4 \times 10^{-16}}$

Answers

20. _____

21. _____

22. _____

23. _____

24. _____

25. _____

26. _____

27. _____

28. _____

29. _____

30. _____

31. _____

32. _____

33. _____

34. _____

35. _____

36. _____

37. _____

38. _____

39. _____

40. _____

Activity 3 ::
The Power of Compound Interest

Suppose that when you were born, a wealthy uncle put $500 in the bank for you. He never deposited money again, but the bank paid 5% interest on the money every year on your birthday. How much money was in the bank after 1 year? After 2 years? After 1 year (as you know), the amount is $500 + 500(0.05), which can be written as $500(1 + 0.05) because of the distributive property. $1 + 0.05 = 1.05$, so after 1 year the amount in the bank was 500(1.05). After 2 years, this amount was again multiplied by 1.05. How much is in the bank today? Complete the following chart.

Birthday	Computation	Amount
0 (Day of birth)		$500
1	$500(1.05)	
2	$500(1.05)(1.05)	
3	$500(1.05)(1.05)(1.05)	
4	$500(1.05)^4	
5	$500(1.05)^5	
6		
7		
8		

Write a formula for the amount in the bank on your nth birthday. About how many years does it take for the money to double? How many years for it to double again? Can you see any connection between this and the rules for exponents? Explain why you think there may or may not be a connection.

(a) If the account earned 6% each year, how much more would it accumulate at the end of year 8? Year 21?

(b) Imagine that you start an Individual Retirement Account (IRA) at age 20, contributing $2,500 each year for 5 years (total $12,500) to an account that produces a return of 8% every year. You stop contributing and let the account grow. Using the information from the previous example, calculate the value of the account at age 65.

(c) Imagine that you don't start the IRA until you are 30. In an attempt to catch up, you invest $2,500 into the same account, 8% annual return, each year for 10 years. You then stop contributing and let the account grow. What will its value be at age 65?

(d) What have you discovered as a result of these computations?

The following questions are presented to help you review concepts from earlier chapters. This is meant as a review and not as a comprehensive exam. The answers are presented in the back of the text. Section references accompany the answers. If you have difficulty with any of these questions, be certain to at least read through the summary related to those sections.

Perform the indicated operations.

1. $8 - (-9)$ **2.** $-26 + 32$ **3.** $(-25)(-6)$ **4.** $(-48) \div (-12)$

Evaluate the expressions if $x = -2$, $y = 5$, and $z = -2$.

5. $-5(-3y - 2z)$

6. $\dfrac{3x - 4y}{2z + 5y}$

Use the properties of exponents to simplify each of the following expressions.

7. $(3x^2)^2 (x^3)^4$ **8.** $\left(\dfrac{x^5}{y^3}\right)^2$ **9.** $(2x^3y)^3$

10. $7y^0$ **11.** $(3x^4y^5)^0$

Simplify each expression using positive exponents only.

12. x^{-4} **13.** $3x^{-2}$ **14.** x^5x^{-9} **15.** $\dfrac{x^{-3}}{y^3}$

Simplify each of the following expressions.

16. $21x^5y - 17x^5y$ **17.** $(3x^2 + 4x - 5) - (2x^2 - 3x - 5)$

18. $3x + 2y - x - 4y$ **19.** $(x + 3)(x - 5)$

20. $(x + y)^2$ **21.** $(3x - 4y)^2$

22. $\dfrac{x^2 + 2x - 8}{x - 2}$ **23.** $x(x + y)(x - y)$

Name _____

Section _____ **Date** _____

Answers

1. _____   2. _____

3. _____   4. _____

5. _____

6. _____

7. _____

8. _____

9. _____

10. _____

11. _____

12. _____

13. _____

14. _____

15. _____

16. _____

17. _____

18. _____

19. _____

20. _____

21. _____

22. _____

23. _____

Answers

24. _____

25. _____

26. _____

27. _____

28. _____

29. _____

30. _____

31. _____

32. _____

33. _____

34. _____

Solve the following equations.

24. $7x - 4 = 3x - 12$

25. $3x + 2 = 4x + 4$

26. $\dfrac{3}{4}x - 2 = 5 + \dfrac{2}{3}x$

27. $6(x - 1) - 3(1 - x) = 0$

28. Solve the equation $A = \dfrac{1}{2}(b + B)$ for B.

Solve the following inequalities.

29. $-5x - 7 \le 3x + 9$

30. $-3(x + 5) > -2x + 7$

Solve the following problems.

31. **BUSINESS AND FINANCE** Sam made $10 more than twice what Larry earned in one month. If together they earned $760, how much did each earn that month?

32. **NUMBER PROBLEM** The sum of two consecutive odd integers is 76. Find the two integers.

33. **BUSINESS AND FINANCE** Two-fifths of a woman's income each month goes to taxes. If she pays $848 in taxes each month, what is her monthly income?

34. **BUSINESS AND FINANCE** The retail selling price of a sofa is $806.25. What is the cost to the dealer if she sells at 25% markup on the cost?

Beginning Algebra   The Streeter/Hutchison Series in Mathematics

CHAPTER

4

INTRODUCTION

Developing secret codes is big business because of the widespread use of computers and the Internet. Corporations all over the world sell encryption systems that are supposed to keep data secure and safe.

In 1977, three professors from the Massachusetts Institute of Technology developed an encryption system they called RSA, a name derived from the first letters of their last names. Their security code was based on a number that has *129 digits*. They called the code RSA-129. For the code to be broken, the 129-digit number must be factored into two prime numbers.

A data security company says that people who are using their system are safe because as yet no truly efficient algorithm for finding prime factors of massive numbers has been found, although one may someday exist. This company, hoping to test its encrypting system, now sponsors contests challenging people to factor more very large numbers into two prime numbers. RSA-576 up to RSA-2048 are being worked on now.

The U.S. government does not allow any codes to be used for which it does not have the key. The software firms claim that this prohibition is costing them about $60 billion in lost sales because many companies will not buy an encryption system knowing they can be monitored by the U.S. government.

Factoring

CHAPTER 4 OUTLINE

373

This pretest provides a preview of the types of exercises you will encounter in each section of this chapter. The answers for these exercises can be found in the back of the text. If you are working on your own, or ahead of the class, this pretest can help you identify the sections in which you should focus more of your time.

Answers

1. _____

2. _____

3. _____

4. _____

5. _____

6. _____

7. _____

8. _____

9. _____

10. _____

11. _____

12. _____

13. _____

14. _____

15. _____

16. _____

17. _____

18. _____

19. _____

20. _____

Factor each of the following polynomials.

4.1

1. $15c + 35$

2. $8q^4 - 20q^3$

3. $6x^2 - 12x + 24$

4. $7c^3d^2 - 21cd + 14cd^3$

Factor each of the following trinomials.

4.2

5. $b^2 + 2b - 15$

6. $x^2 + 10x + 24$

7. $x^2 - 14x + 45$

8. $a^2 + 7ab + 12b^2$

Factor each of the following trinomials completely.

4.3

9. $3y^2 + 5y - 12$

10. $5w^2 + 23w + 12$

11. $6x^2 + 5xy - 21y^2$

12. $2x^3 - 7x^2 - 15x$

Factor each of the following polynomials completely.

4.4–4.5

13. $b^2 - 49$

14. $36p^2 - q^2$

15. $9x^2 - 12xy + 4y^2$

16. $27xy^2 - 48x^3$

Solve each of the following equations.

4.6

17. $x^2 - 11x + 28 = 0$

18. $x^2 - 5x = 14$

19. $5x^2 + 7x - 6 = 0$

20. $9p^2 - 18p = 0$

4.1

An Introduction to Factoring

< 4.1 Objectives >

1 > Factor out the greatest common factor (GCF)

2 > Factor out a binomial GCF

3 > Factor a polynomial by grouping terms

> ▶ **Overcoming Math Anxiety**

Hint #5
Working Together
How many of your classmates do you know? Whether you are by nature gregarious or shy, you have much to gain by getting to know your classmates.

1. It is important to have someone to call when you miss class or are unclear on an assignment.

2. Working with another person is almost always beneficial to both people. If you don't understand something, it helps to have someone to ask about it. If you do understand something, nothing cements that understanding more than explaining the idea to another person.

3. Sometimes we need to commiserate. If an assignment is particularly frustrating, it is reassuring to find that it is also frustrating for other students.

4. Have you ever thought you had the right answer, but it doesn't match the answer in the text? Frequently the answers are equivalent, but that's not always easy to see. A different perspective can help you see that. Occasionally there is an error in a textbook (here we are talking about *other* textbooks). In such cases it is wonderfully reassuring to find that someone else has the same answer you do.

In Chapter 3 you were given factors and asked to find a product. We are now going to reverse the process. You will be given a polynomial and asked to find its factors. This is called **factoring.**

We start with an example from arithmetic. To *multiply* $5 \cdot 7$, you write

$$5 \cdot 7 = 35$$

To *factor* 35, you would write

$$35 = 5 \cdot 7$$

Factoring is the *reverse* of multiplication.

Now let's look at factoring in algebra. You have used the distributive property as

$$a(b + c) = ab + ac$$

> **NOTE**
>
> 3 and $x + 5$ are the factors of $3x + 15$.

For instance,

$$3(x + 5) = 3x + 15$$

To use the distributive property in factoring, we reverse that property as

$$ab + ac = a(b + c)$$

The property lets us factor out the common factor a from the terms of $ab + ac$. To use this in factoring, the first step is to see whether each term of the polynomial has a common monomial factor. In our earlier example,

$$3x + 15 = 3 \cdot x + 3 \cdot 5$$

Common factor

So, by the distributive property,

$$3x + 15 = 3(x + 5)$$ The original terms are each divided by the greatest common factor to determine the terms in parentheses.

To check this, multiply $3(x + 5)$.

Multiplying

$$\xrightarrow{}$$
$$3(x + 5) = 3x + 15$$
$$\xleftarrow{}$$

Factoring

> **NOTES**
>
> Again, factoring is the reverse of multiplication.
>
> Factoring out the GCF is the *first* method to try in any of the factoring problems we will discuss.

The first step in factoring is to identify the *greatest common factor* (GCF) of a set of terms. This factor is the product of the largest common numerical coefficient and the largest common factor of each variable.

Definition	
Greatest Common Factor	The **greatest common factor (GCF)** of a polynomial is the factor that is the product of the largest common numerical coefficient factor of the polynomial and each variable with the smallest exponent in any term.

 Example 1   **Finding the GCF**

< Objective 1 >

Find the GCF for each set of terms.

(a) 9 and 12   The largest number that is a factor of both is 3.

(b) 10, 25, 150   The GCF is 5.

(c) x^4 and x^7

$$x^4 = \boxed{x} \cdot \boxed{x} \cdot \boxed{x} \cdot \boxed{x}$$
$$x^7 = \boxed{x} \cdot \boxed{x} \cdot \boxed{x} \cdot \boxed{x} \cdot x \cdot x \cdot x$$

The largest power that divides both terms is x^4.

(d) $12a^3$ and $18a^2$

$$12a^3 = 2 \cdot \boxed{2} \cdot \boxed{3} \cdot \boxed{a} \cdot \boxed{a} \cdot a$$
$$18a^2 = \boxed{2} \cdot \boxed{3} \cdot 3 \cdot \boxed{a} \cdot \boxed{a}$$

The GCF is $6a^2$.

Check Yourself 1

Find the GCF for each set of terms.

(a) 14, 24   **(b)** 9, 27, 81
(c) a^9, a^5 **(d)** $10x^5$, $35x^4$

Step by Step

To Factor a Monomial from a Polynomial	**Step 1** Find the GCF for all the terms.
	Step 2 Use the GCF to factor each term and then apply the distributive property.
	Step 3 Mentally check your factoring by multiplication. Checking your answer is always important and perhaps is never easier than after you have factored.

 Example 2   **Finding the GCF of a Binomial**

(a) Factor $8x^2 + 12x$.

The largest common numerical factor of 8 and 12 is 4, and x is the common variable factor with the largest power. So $4x$ is the GCF. Write

$$8x^2 + 12x = 4x \cdot 2x + 4x \cdot 3$$

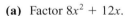
GCF

Now, by the distributive property, we have

$$8x^2 + 12x = 4x(2x + 3)$$

It is always a good idea to check your answer by multiplying to make sure that you get the original polynomial. Try it here. Multiply $4x$ by $2x + 3$.

(b) Factor $6a^4 - 18a^2$.

The GCF in this case is $6a^2$. Write

$$6a^4 + (-18a^2) = 6a^2 \cdot a^2 + 6a^2 \cdot (-3)$$

GCF

Again, using the distributive property yields

$$6a^4 - 18a^2 = 6a^2(a^2 - 3)$$

You should check this by multiplying.

NOTE

It is also true that

$6a^4 + (-18a^2)$
$= 3a(2a^3 + (-6a))$.

However, this is *not completely factored.* Do you see why? You want to find the common monomial factor with the *largest possible* coefficient and the *largest exponent*, in this case $6a^2$.

Check Yourself 2

Factor each of the following polynomials.

(a) $5x + 20$ **(b)** $6x^2 - 24x$ **(c)** $10a^3 - 15a^2$

The process is exactly the same for polynomials with more than two terms. Consider Example 3.

  Example 3   Finding the GCF of a Polynomial

(a) Factor $5x^2 - 10x + 15$.

$$5x^2 - 10x + 15 = 5 \cdot x^2 - 5 \cdot 2x + 5 \cdot 3$$

GCF

$$= 5(x^2 - 2x + 3)$$

(b) Factor $6ab + 9ab^2 - 15a^2$.

$$6ab + 9ab^2 - 15a^2 = 3a \cdot 2b + 3a \cdot 3b^2 - 3a \cdot 5a$$

GCF

$$= 3a(2b + 3b^2 - 5a)$$

(c) Factor $4a^4 + 12a^3 - 20a^2$.

$$4a^4 + 12a^3 - 20a^2 = 4a^2 \cdot a^2 + 4a^2 \cdot 3a - 4a^2 \cdot 5$$

GCF

$$= 4a^2(a^2 + 3a - 5)$$

(d) Factor $6a^2b + 9ab^2 + 3ab$.

Mentally note that 3, a, and b are factors of each term, so

$$6a^2b + 9ab^2 + 3ab = 3ab(2a + 3b + 1)$$

> **NOTES**
>
> The GCF is 5.
>
> The GCF is 3a.
>
> The GCF is 4a^2.
>
> In each of these examples, you should check the result by multiplying the factors.

 Check Yourself 3

Factor each of the following polynomials.

(a) $8b^2 + 16b - 32$ (b) $4xy - 8x^2y + 12x^3$
(c) $7x^4 - 14x^3 + 21x^2$ (d) $5x^2y^2 - 10xy^2 + 15x^2y$

We can have two or more terms that have a binomial factor in common, as is the case in Example 4.

Example 4   Finding a Common Factor

< Objective 2 >

(a) Factor $3x(x + y) + 2(x + y)$.

We see that *the binomial $x + y$ is a common factor* and can be removed.

$$3x(x + y) + 2(x + y)$$
$$= (x + y) \cdot 3x + (x + y) \cdot 2$$
$$= (x + y)(3x + 2)$$

(b) Factor $3x^2(x - y) + 6x(x - y) + 9(x - y)$.

> **NOTE**
>
> Because of the commutative property, the factors can be written in either order.

We note that here the GCF is $3(x - y)$. Factoring as before, we have

$3(x - y)(x^2 + 2x + 3)$.

Check Yourself 4

Completely factor each of the polynomials.

(a) $7a(a - 2b) + 3(a - 2b)$ **(b)** $4x^2(x + y) - 8x(x + y) - 16(x + y)$

Some polynomials can be factored by grouping the terms and finding common factors within each group. Such a process is called *factoring by grouping* and will be explored now.

In Example 4, we looked at the expression

$3x(x + y) + 2(x + y)$

and found that we could factor out the common binomial, $(x + y)$, giving us

$(x + y)(3x + 2)$

That technique will be used in Example 5.

 Example 5 | **Factoring by Grouping Terms**

< Objective 3 >

Suppose we want to factor the polynomial

$ax - ay + bx - by$

As you can see, the polynomial has no common factors. However, look at what happens if we separate the polynomial into *two groups of two terms*.

NOTE

Note that our example has *four* terms. That is a clue for trying the factoring by grouping method.

$ax - ay + bx - by$

$= \underbrace{ax - ay}_{(1)} + \underbrace{bx - by}_{(2)}$

Now *each* group has a common factor, and we can write the polynomial as

$a(x - y) + b(x - y)$

In this form, we can see that $x - y$ is the GCF. Factoring out $x - y$, we get

$a(x - y) + b(x - y) = (x - y)(a + b)$

Check Yourself 5

Use the factoring by grouping method.

$x^2 - 2xy + 3x - 6y$

Be particularly careful of your treatment of algebraic signs when applying the factoring by grouping method. Consider Example 6.

Example 6　　**Factoring by Grouping Terms**

Factor $2x^3 - 3x^2 - 6x + 9$.

We group the polynomial as follows.

$$\underbrace{2x^3 - 3x^2}_{(1)} - \underbrace{6x + 9}_{(2)}$$

Factor out the common factor of -3 from the second two terms.

$$= x^2(2x - 3) - 3(2x - 3)$$
$$= (2x - 3)(x^2 - 3)$$

NOTE

$9 = (-3)(-3)$

✓ Check Yourself 6

Factor by grouping.

$3y^3 + 2y^2 - 6y - 4$

It may also be necessary to change the order of the terms as they are grouped. Look at Example 7.

Example 7　　**Factoring by Grouping Terms**

Factor $x^2 - 6yz + 2xy - 3xz$.

Grouping the terms as before, we have

$$\underbrace{x^2 - 6yz}_{(1)} + \underbrace{2xy - 3xz}_{(2)}$$

Do you see that we have accomplished nothing because there are no common factors in the first group?

We can, however, rearrange the terms to write the original polynomial as

$$\underbrace{x^2 + 2xy}_{(1)} - \underbrace{3xz - 6yz}_{(2)}$$

$$= x(x + 2y) - 3z(x + 2y)$$

We can now factor out the common factor of $x + 2y$ in group (1) and group (2).

$$= (x + 2y)(x - 3z)$$

Note: It is often true that the grouping can be done in more than one way. The factored form will be the same.

✓ Check Yourself 7

We can write the polynomial of Example 7 as

$x^2 - 3xz + 2xy - 6yz$

Factor and verify that the factored form is the same in either case.

The Streeter/Hutchison Series in Mathematics　　Beginning Algebra

Check Yourself ANSWERS

1. (a) 2; **(b)** 9; **(c)** a^5; **(d)** $5x^4$ **2. (a)** $5(x + 4)$; **(b)** $6x(x - 4)$;
(c) $5a^2(2a - 3)$ **3. (a)** $8(b^2 + 2b - 4)$; **(b)** $4x(y - 2xy + 3x^2)$;
(c) $7x^2(x^2 - 2x + 3)$; **(d)** $5xy(xy - 2y + 3x)$ **4. (a)** $(a - 2b)(7a + 3)$;
(b) $4(x + y)(x^2 - 2x - 4)$ **5.** $(x - 2y)(x + 3)$ **6.** $(3y + 2)(y^2 - 2)$
7. $(x - 3z)(x + 2y)$

Reading Your Text

The following fill-in-the-blank exercises are designed to ensure that you understand some of the key vocabulary used in this section.

SECTION 4.1

(a) The GCF of a polynomial is the factor that is the product of the _____ numerical coefficient factor of the polynomial and each variable with the smallest exponent in any term.

(b) To factor a monomial from a polynomial, first find the GCF for all of the _____.

(c) Some polynomials can be factored by grouping the terms and finding common _____ within each group.

(d) It is often true that grouping can be done in more than one way. The factored form will be the _____.

4.1 exercises

MathZone

Boost your grade at mathzone.com!
> Practice Problems
> NetTutor
> Self-Tests
> e-Professors
> Videos

Name _____

Section _____ Date _____

Answers

1. _____   2. _____
3. _____   4. _____
5. _____   6. _____
7. _____   8. _____
9. _____   10. _____
11. _____   12. _____
13. _____   14. _____
15. _____   16. _____
17. _____   18. _____
19. _____   20. _____
21. _____   22. _____
23. _____
24. _____
25. _____
26. _____
27. _____
28. _____

< Objective 1 >

Find the greatest common factor for each of the following sets of terms.

1. $10, 12$

2. $15, 35$

3. $16, 32, 88$

4. $55, 33, 132$

5. x^2, x^5

6. y^7, y^9

7. a^3, a^6, a^9

8. b^4, b^6, b^8

9. $5x^4, 10x^5$

10. $8y^9, 24y^3$

11. $8a^4, 6a^6, 10a^{10}$

12. $9b^3, 6b^5, 12b^4$

13. $9x^2y, 12xy^2, 15x^2y^2$

14. $12a^3b^2, 18a^2b^3, 6a^4b^4$

15. $15ab^3, 10a^2bc, 25b^2c^3$

16. $9x^2, 3xy^3, 6y^3$

17. $15a^2bc^2, 9ab^2c^2, 6a^2b^2c^2$

 > Videos

18. $18x^3y^2z^3, 27x^4y^2z^3, 81xy^2z$

19. $(x + y)^2, (x + y)^3$

20. $12(a + b)^4, 4(a + b)^3$

Factor each of the following polynomials.

21. $8a + 4$

22. $5x - 15$

23. $24m - 32n$

24. $7p - 21q$

25. $12m + 8$

26. $24n - 32$

27. $10s^2 + 5s$

28. $12y^2 - 6y$

29. $12x^2 + 12x$

30. $14b^2 + 14b$

31. $15a^3 - 25a^2$

32. $36b^4 + 24b^2$

33. $6pq + 18p^2q$

34. $8ab - 24ab^2$

35. $6x^2 - 18x + 30$

36. $7a^2 + 21a - 42$

37. $3a^3 + 6a^2 - 12a$

38. $5x^3 - 15x^2 + 25x$

39. $6m + 9mn - 12mn^2$

40. $4s + 6st - 14st^2$

41. $10r^3s^2 + 25r^2s^2 - 15r^2s^3$

42. $28x^2y^3 - 35x^2y^2 + 42x^3y$

 > Videos

43. $9a^5 - 15a^4 + 21a^3 - 27a$

44. $8p^6 - 40p^4 + 24p^3 - 16p^2$

| Basic Skills | **Advanced Skills** | Vocational-Technical Applications | Calculator/Computer | Above and Beyond | Getting Ready |

Complete each of the following statements with **never, sometimes,** *or* **always.**

45. The GCF for two numbers is _____ a prime number.

46. The GCF of a polynomial _____ includes variables.

47. Multiplying the result of factoring will _____ result in the original polynomial.

48. Factoring a negative number from a negative term will _____ result in a negative term.

< Objective 2 >
Factor out the binomial in each expression.

49. $a(a + 2) - 3(a + 2)$

50. $b(b - 5) + 2(b - 5)$

51. $x(x - 2) + 3(x - 2)$ > Videos

52. $y(y + 5) - 3(y + 5)$

Answers

29. _____

30. _____

31. _____

32. _____

33. _____

34. _____

35. _____

36. _____

37. _____

38. _____

39. _____

40. _____

41. _____

42. _____

43. _____

44. _____

45. _____

46. _____

47. _____

48. _____

49. _____

50. _____

51. _____

52. _____

Answers

53. _____

54. _____

55. _____

56. _____

57. _____

58. _____

59. _____

60. _____

61. _____

62. _____

63. _____

64. _____

65. _____

66. _____

67. _____

68. _____

69. _____

70. _____

71. _____

72. _____

73. _____

< Objective 3 >

Factor each polynomial by grouping the first two terms and the last two terms.

53. $x^3 - 4x^2 + 3x - 12$

54. $x^3 - 6x^2 + 2x - 12$

55. $a^3 - 3a^2 + 5a - 15$

56. $6x^3 - 2x^2 + 9x - 3$

57. $10x^3 + 5x^2 - 2x - 1$

58. $x^5 + x^3 - 2x^2 - 2$

59. $x^4 - 2x^3 + 3x - 6$

60. $x^3 - 4x^2 + 2x - 8$

Factor each polynomial completely by factoring out any common factors and then factor by grouping. Do not combine like terms.

61. $3x - 6 + xy - 2y$

62. $2x - 10 + xy - 5y$

63. $ab - ac + b^2 - bc$

64. $ax + 2a + bx + 2b$

65. $3x^2 - 2xy + 3x - 2y$

66. $xy - 5y^2 - x + 5y$

67. $5s^2 + 15st - 2st - 6t^2$

68. $3a^3 + 3ab^2 + 2a^2b + 2b^3$

69. $3x^3 + 6x^2y - x^2y - 2xy^2$

70. $2p^4 + 3p^3q - 2p^3q - 3p^2q^2$

Basic Skills | Advanced Skills | **Vocational-Technical Applications** | Calculator/Computer | Above and Beyond | Getting Ready

▲

71. ALLIED HEALTH A patient's protein secretion amount, in milligrams per day, is recorded over several days. Based on these observations, lab technicians determine that the polynomial $-t^3 - 6t^2 + 11t + 66$ provides a good approximation of the patient's protein secretion amounts t days after testing begins. Factor this polynomial.

72. ALLIED HEALTH The concentration, in micrograms per milliliter $\frac{\mu g}{mL}$ of the antibiotic chloramphenicol is given by $8t^2 - 2t^3$, in which t is the number of hours after the drug is taken. Factor this polynomial.

73. MANUFACTURING TECHNOLOGY Polymer pellets need to be as perfectly round as possible. In order to avoid flat spots from forming during the hardening process, the pellets are kept off a surface by blasts of air. The height of a pellet above the surface t seconds after a blast is given by $v_0t - 4.9t^2$. Factor this expression.

74. INFORMATION TECHNOLOGY The total time to transmit a packet is given by the expression $d + 2p$, in which d is the quotient of the distance and the propagation velocity and p is the quotient of the size of the packet and the information transfer rate. How long will it take to transmit a 1,500-byte packet 10 meters on an Ethernet if the information transfer rate is 100 MB per second and the propagation velocity is 2×10^8 m/s?

| Basic Skills | Advanced Skills | Vocational-Technical Applications | Calculator/Computer | **Above and Beyond** | Getting Ready |

75. The GCF of $2x - 6$ is 2. The GCF of $5x + 10$ is 5. Find the GCF of the product $(2x - 6)(5x + 10)$.

76. The GCF of $3z + 12$ is 3. The GCF of $4z + 8$ is 4. Find the GCF of the product $(3z + 12)(4z + 8)$.

77. The GCF of $2x^3 - 4x$ is $2x$. The GCF of $3x + 6$ is 3. Find the GCF of the product $(2x^3 - 4x)(3x + 6)$.

78. State, in a sentence, the rule that exercises 75 to 77 illustrated.

Find the GCF of each product.

79. $(2a + 8)(3a - 6)$ **80.** $(5b - 10)(2b + 4)$

81. $(2x^2 + 5x)(7x - 14)$ **82.** $(6y^2 - 3y)(y + 7)$

83. GEOMETRY The area of a rectangle with width t is given by $33t - t^2$. Factor the expression and determine the length of the rectangle in terms of t.

84. GEOMETRY The area of a rectangle of length x is given by $3x^2 + 5x$. Find the width of the rectangle.

85. NUMBER PROBLEM For centuries, mathematicians have found factoring numbers into prime factors a fascinating subject. A prime number is a number that cannot be written as a product of any numbers but 1 and itself. The list of primes begins with 2 because 1 is not considered a prime number and then goes on: 3, 5, 7, 11, . . . What are the first 10 primes? What are the primes less than 100? If you list the numbers from 1 to 100 and then cross out all numbers that are multiples of 2, 3, 5, and 7, what is left? Are all the numbers not crossed out prime? Write a paragraph to explain why this might be so. You might want to investigate the Sieve of Eratosthenes, a system from 230 B.C.E. for finding prime numbers.

Answers

74. _____

75. _____

76. _____

77. _____

78. _____

79. _____

80. _____

81. _____

82. _____

83. _____

84. _____

85. _____

Answers

86. _____

87. _____

(a) _____

(b) _____

(c) _____

(d) _____

(e) _____

(f) _____

86. NUMBER PROBLEM If we could make a list of all the prime numbers, what number would be at the end of the list? Because there are an infinite number of prime numbers, there is no "largest prime number." But is there some formula that will give us all the primes? Here are some formulas proposed over the centuries:

$$n^2 + n + 17 \qquad 2n^2 + 29 \qquad n^2 - n + 11$$

In all these expressions, $n = +1, 2, 3, 4, \ldots$, that is, a positive integer beginning with 1. Investigate these expressions with a partner. Do the expressions give prime numbers when they are evaluated for these values of n? Do the expressions give *every* prime in the range of resulting numbers? Can you put in *any* positive number for n?

87. NUMBER PROBLEM How are primes used in coding messages and for security? Work together to decode the messages. The messages are coded using this code: After the numbers are factored into prime factors, the power of 2 gives the number of the letter in the alphabet. This code would be easy for a code breaker to figure out. Can you make up code that would be more difficult to break?

chapter 4 > Make the Connection

(a) 1310720, 229376, 1572864, 1760, 460, 2097152, 336

(b) 786432, 286, 4608, 278528, 1344, 98304, 1835008, 352, 4718592, 5242880

(c) Code a message using this rule. Exchange your message with a partner to decode it.

Basic Skills | Advanced Skills | Vocational-Technical Applications | Calculator/Computer | Above and Beyond | **Getting Ready** ▲

Multiply.

(a) $(a - 1)(a + 4)$ 　　　　(b) $(x - 1)(x + 3)$

(c) $(x - 3)(x - 3)$ 　　　　(d) $(y - 11)(y + 3)$

(e) $(x + 5)(x + 7)$ 　　　　(f) $(y + 1)(y - 13)$

Answers

1. 2 　　**3.** 8 　　**5.** x^2 　　**7.** a^3 　　**9.** $5x^4$ 　　**11.** $2a^4$ 　　**13.** $3xy$
15. $5b$ 　　**17.** $3abc^2$ 　　**19.** $(x + y)^2$ 　　**21.** $4(2a + 1)$ 　　**23.** $8(3m - 4n)$
25. $4(3m + 2)$ 　　**27.** $5s(2s + 1)$ 　　**29.** $12x(x + 1)$ 　　**31.** $5a^2(3a - 5)$
33. $6pq(1 + 3p)$ 　　**35.** $6(x^2 - 3x + 5)$ 　　**37.** $3a(a^2 + 2a - 4)$
39. $3m(2 + 3n - 4n^2)$ 　　**41.** $5r^2s^2(2r + 5 - 3s)$
43. $3a(3a^4 - 5a^3 + 7a^2 - 9)$ 　　**45.** sometimes 　　**47.** always
49. $(a - 3)(a + 2)$ 　　**51.** $(x + 3)(x - 2)$ 　　**53.** $(x - 4)(x^2 + 3)$
55. $(a - 3)(a^2 + 5)$ 　　**57.** $(2x + 1)(5x^2 - 1)$ 　　**59.** $(x - 2)(x^3 + 3)$
61. $(x - 2)(3 + y)$ 　　**63.** $(b - c)(a + b)$ 　　**65.** $(x + 1)(3x - 2y)$
67. $(s + 3t)(5s - 2t)$ 　　**69.** $x(x + 2y)(3x - y)$ 　　**71.** $(t + 6)(-t^2 + 11)$
73. $t(v_0 - 4.9t)$ 　　**75.** 10 　　**77.** $6x$ 　　**79.** 6 　　**81.** $7x$
83. $t(33 - t); 33 - t$ 　　**85.** Above and Beyond 　　**87.** Above and Beyond
(a) $a^2 + 3a - 4$ 　　(b) $x^2 + 2x - 3$ 　　(c) $x^2 - 6x + 9$ 　　(d) $y^2 - 8y - 33$
(e) $x^2 + 12x + 35$ 　　(f) $y^2 - 12y - 13$

4.2 Factoring Trinomials of the Form $x^2 + bx + c$

< 4.2 Objectives >

1 > Factor a trinomial of the form $x^2 + bx + c$

2 > Factor a trinomial containing a common factor

NOTE

The process used to factor here is frequently called the *trial-and-error method.* You'll see the reason for the name as you work through this section.

> CAUTION

Not every trinomial can be written as the product of two binomials.

You learned how to find the product of any two binomials by using the FOIL method in Section 3.4. Because factoring is the reverse of multiplication, we now want to use that pattern to find the factors of certain trinomials.

Recall that when we multiply the binomials $x + 2$ and $x + 3$, our result is

$$(x + 2)(x + 3) = x^2 + 5x + 6$$

The product of the first terms ($x \cdot x$).

The sum of the products of the outer and inner terms ($3x$ and $2x$).

The product of the last terms ($2 \cdot 3$).

Suppose now that you are given $x^2 + 5x + 6$ and want to find its factors. First, you know that the factors of a trinomial may be two binomials. So write

$$x^2 + 5x + 6 = (\quad)(\quad)$$

Because the first term of the trinomial is x^2, the first terms of the binomial factors must be x and x. We now have

$$x^2 + 5x + 6 = (x \quad)(x \quad)$$

The product of the last terms must be 6. Because 6 is positive, the factors must have *like* signs. Here are the possibilities:

$$6 = 1 \cdot 6$$
$$= 2 \cdot 3$$
$$= (-1)(-6)$$
$$= (-2)(-3)$$

This means that the possible factors of the trinomial are

$$(x + 1)(x + 6)$$
$$(x + 2)(x + 3)$$
$$(x - 1)(x - 6)$$
$$(x - 2)(x - 3)$$

How do we tell which is the correct pair? From the FOIL pattern we know that the sum of the outer and inner products must equal the middle term of the trinomial, in this case $5x$. This is the crucial step!

Possible Factorizations	Middle Terms
$(x + 1)(x + 6)$	$7x$
$(x + 2)(x + 3)$	$5x$
$(x - 1)(x - 6)$	$-7x$
$(x - 2)(x - 3)$	$-5x$

The correct middle term!

So we know that the correct factorization is

$$x^2 + 5x + 6 = (x + 2)(x + 3)$$

Are there any clues so far that will make this process quicker? Yes, there is an important one that you may have spotted. We started with a trinomial that had a positive middle term and a positive last term. The negative pairs of factors for 6 led to negative middle terms. So you don't need to bother with the negative factors if the middle term and the last term of the trinomial are both positive.

Example 1	Factoring a Trinomial

< Objective 1 >

(a) Factor $x^2 + 9x + 8$.

Because the middle term and the last term of the trinomial are both positive, consider only the positive factors of 8, that is, $8 = 1 \cdot 8$ or $8 = 2 \cdot 4$.

NOTE

If you are wondering why we didn't list $(x + 8)(x + 1)$ as a possibility, remember that multiplication is commutative. The order doesn't matter!

Possible Factorizations	Middle Terms
$(x + 1)(x + 8)$	$9x$
$(x + 2)(x + 4)$	$6x$

Because the first pair gives the correct middle term,

$$x^2 + 9x + 8 = (x + 1)(x + 8)$$

(b) Factor $x^2 + 12x + 20$.

NOTE

The factor-pairs of 20 are

$20 = 1 \cdot 20$

$\quad = 2 \cdot 10$

$\quad = 4 \cdot 5$

Possible Factorizations	Middle Terms
$(x + 1)(x + 20)$	$21x$
$(x + 2)(x + 10)$	$12x$
$(x + 4)(x + 5)$	$9x$

So

$$x^2 + 12x + 20 = (x + 2)(x + 10)$$

Check Yourself 1

Factor.

(a) $x^2 + 6x + 5$ **(b)** $x^2 + 10x + 16$

What if the middle term of the trinomial is negative but the first and last terms are still positive? Consider

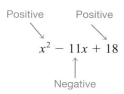

Because we want a negative middle term ($-11x$) and a positive last term, we use *two negative factors* for 18. Recall that the product of two negative numbers is positive.

 Example 2   **Factoring a Trinomial**

(a) Factor $x^2 - 11x + 18$.

NOTE

The negative factors of 18 are

$18 = (-1)(-18)$

$ = (-2)(-9)$

$ = (-3)(-6)$

Possible Factorizations	Middle Terms
$(x - 1)(x - 18)$	$-19x$
$(x - 2)(x - 9)$	$-11x$
$(x - 3)(x - 6)$	$-9x$

So

$x^2 - 11x + 18 = (x - 2)(x - 9)$

(b) Factor $x^2 - 13x + 12$.

NOTE

The negative factors of 12 are

$12 = (-1)(-12)$

$ = (-2)(-6)$

$ = (-3)(-4)$

Possible Factorizations	Middle Terms
$(x - 1)(x - 12)$	$-13x$
$(x - 2)(x - 6)$	$-8x$
$(x - 3)(x - 4)$	$-7x$

So

$x^2 - 13x + 12 = (x - 1)(x - 12)$

A few more clues: We have listed all the possible factors in the above examples. It really isn't necessary. Just work until you find the right pair. Also, with practice much of this work can be done mentally.

Check Yourself 2

Factor.

(a) $x^2 - 10x + 9$ **(b)** $x^2 - 10x + 21$

Now let's look at the process of factoring a trinomial whose last term is negative. For instance, to factor $x^2 + 2x - 15$, we can start as before:

$$x^2 + 2x - 15 = (x \quad ?)(x \quad ?)$$

Note that the product of the last terms must be negative (-15 here). So we must choose factors that have different signs.

What are our choices for the factors of -15?

$$-15 = (1)(-15)$$
$$= (-1)(15)$$
$$= (3)(-5)$$
$$= (-3)(5)$$

This means that the possible factors and the resulting middle terms are

Possible Factorizations	Middle Terms
$(x + 1)(x - 15)$	$-14x$
$(x - 1)(x + 15)$	$14x$
$(x + 3)(x - 5)$	$-2x$
$(x - 3)(x + 5)$	$2x$

NOTE

Another clue: Some students prefer to look at the list of numerical factors rather than looking at the actual algebraic factors. Here you want the pair whose sum is 2, the coefficient of the middle term of the trinomial. That pair is −3 and 5, which leads us to the correct factors.

So $x^2 + 2x - 15 = (x - 3)(x + 5)$.

Let's work through some examples in which the constant term is negative.

Example 3 Factoring a Trinomial

(a) Factor $x^2 - 5x - 6$.

First, list the factors of -6. Of course, one factor will be positive, and one will be negative.

$$-6 = (1)(-6)$$
$$= (-1)(6)$$
$$= (2)(-3)$$
$$= (-2)(3)$$

NOTE

You may be able to pick the factors directly from this list. You want the pair whose sum is −5 (the coefficient of the middle term).

For the trinomial, then, we have

Possible Factorizations	Middle Terms
$(x + 1)(x - 6)$	$-5x$
$(x - 1)(x + 6)$	$5x$
$(x + 2)(x - 3)$	$-x$
$(x - 2)(x + 3)$	x

So $x^2 - 5x - 6 = (x + 1)(x - 6)$.

(b) Factor $x^2 + 8xy - 9y^2$.

The process is similar if two variables are involved in the trinomial. Start with

$$x^2 + 8xy - 9y^2 = (x \quad ?)(x \quad ?).$$

The product of the last terms must be $-9y^2$.

$$-9y^2 = (-y)(9y)$$
$$= (y)(-9y)$$
$$= (3y)(-3y)$$

Possible Factorizations	Middle Terms
$(x - y)(x + 9y)$	$8xy$
$(x + y)(x - 9y)$	$-8xy$
$(x + 3y)(x - 3y)$	0

So $x^2 + 8xy - 9y^2 = (x - y)(x + 9y)$.

Check Yourself 3

Factor.

(a) $x^2 + 7x - 30$ **(b)** $x^2 - 3xy - 10y^2$

As we pointed out in Section 4.1, any time that there is a common factor, that factor should be factored out *before* we try any other factoring technique. Consider Example 4.

Example 4	Factoring a Trinomial

< Objective 2 >

(a) Factor $3x^2 - 21x + 18$.

$3x^2 - 21x + 18 = 3(x^2 - 7x + 6)$ Factor out the common factor of 3.

We now factor the remaining trinomial. For $x^2 - 7x + 6$:

Possible Factorizations	Middle Terms
$(x - 1)(x - 6)$	$-7x$
$(x - 2)(x - 3)$	$-5x$

The correct middle term

> CAUTION

A common mistake is to forget to write the 3 that was factored out as the first step.

So $3x^2 - 21x + 18 = 3(x - 1)(x - 6)$.

(b) Factor $2x^3 + 16x^2 - 40x$.

$2x^3 + 16x^2 - 40x = 2x(x^2 + 8x - 20)$ Factor out the common factor of 2x.

To factor the remaining trinomial, which is $x^2 + 8x - 20$, we have

NOTE

Once we have found the desired middle term, there is no need to continue.

Possible Factorizations	Middle Terms
$(x + 2)(x - 10)$	$-8x$
$(x - 2)(x + 10)$	$8x$

The correct middle term

So $2x^3 + 16x^2 - 40x = 2x(x - 2)(x + 10)$.

Check Yourself 4

Factor.

(a) $3x^2 - 3x - 36$ **(b)** $4x^3 + 24x^2 + 32x$

One further comment: Have you wondered whether all trinomials are factorable? Look at the trinomial

$x^2 + 2x + 6$

The only possible factors are $(x + 1)(x + 6)$ and $(x + 2)(x + 3)$. Neither pair is correct (you should check the middle terms), and so this trinomial does not have factors with integer coefficients. Of course, there are many other trinomials that cannot be factored. Can you find one?

Check Yourself ANSWERS

1. **(a)** $(x + 1)(x + 5)$; **(b)** $(x + 2)(x + 8)$ **2.** **(a)** $(x - 9)(x - 1)$;
(b) $(x - 3)(x - 7)$ **3.** **(a)** $(x + 10)(x - 3)$; **(b)** $(x + 2y)(x - 5y)$
4. **(a)** $3(x - 4)(x + 3)$; **(b)** $4x(x + 2)(x + 4)$

Reading Your Text

The following fill-in-the-blank exercises are designed to ensure that you understand some of the key vocabulary used in this section.

SECTION 4.2

(a) Factoring is the reverse of _____.

(b) From the FOIL pattern, we know that the sum of the inner and outer products must equal the _____ term of the trinomial.

(c) The product of two negative factors is always _____.

(d) Some trinomials do not have _____ with integer coefficients.

MathZone

Boost your grade at mathzone.com!

> Practice Problems
> NetTutor
> Self-Tests
> e-Professors
> Videos

Name _____

Section _____ Date _____

Answers

1. _____   2. _____
3. _____   4. _____
5. _____   6. _____
7. _____   8. _____
9. _____   10. _____
11. _____
12. _____
13. _____
14. _____
15. _____
16. _____
17. _____
18. _____
19. _____
20. _____
21. _____
22. _____
23. _____
24. _____

< Objective 1 >

Complete each of the following statements.

1. $x^2 - 8x + 15 = (x - 3)(\quad)$

2. $y^2 - 3y - 18 = (y - 6)(\quad)$

3. $m^2 + 8m + 12 = (m + 2)(\quad)$

4. $x^2 - 10x + 24 = (x - 6)(\quad)$

5. $p^2 - 8p - 20 = (p + 2)(\quad)$

6. $a^2 + 9a - 36 = (a + 12)(\quad)$

7. $x^2 - 16x + 64 = (x - 8)(\quad)$

8. $w^2 - 12w - 45 = (w + 3)(\quad)$

9. $x^2 - 7xy + 10y^2 = (x - 2y)(\quad)$

 > Videos

10. $a^2 + 18ab + 81b^2 = (a + 9b)(\quad)$

Factor each of the following trinomials completely.

11. $x^2 + 8x + 15$

12. $x^2 - 11x + 24$

13. $x^2 - 11x + 28$

14. $y^2 - y - 20$

15. $s^2 + 13s + 30$

16. $b^2 + 14b + 33$

17. $a^2 - 2a - 48$

18. $x^2 - 17x + 60$

19. $x^2 - 8x + 7$

20. $x^2 + 7x - 18$

21. $m^2 + 3m - 28$

 > Videos

22. $a^2 + 10a + 25$

23. $x^2 - 6x - 40$

24. $x^2 - 11x + 10$

25. $x^2 - 14x + 49$

26. $s^2 - 4s - 32$

27. $p^2 - 10p - 24$

28. $x^2 - 11x - 60$

29. $x^2 + 5x - 66$

30. $a^2 + 2a - 80$

31. $c^2 + 19c + 60$

32. $t^2 - 4t - 60$

33. $x^2 + 7xy + 10y^2$

34. $x^2 - 8xy + 12y^2$

35. $a^2 - ab - 42b^2$ > Videos

36. $m^2 - 8mn + 16n^2$

37. $x^2 - 13xy + 40y^2$

38. $r^2 - 9rs - 36s^2$

39. $x^2 - 2xy - 8y^2$

40. $u^2 + 6uv - 55v^2$

41. $25m^2 + 10mn + n^2$

42. $64m^2 - 16mn + n^2$

Basic Skills | **Advanced Skills** | Vocational-Technical Applications | Calculator/Computer | Above and Beyond | Getting Ready
▲

Determine whether each of the following statements is **true** *or* **false.**

43. Factoring is the reverse of division.

44. From the FOIL pattern, we know that the sum of the inner and outer products must equal the middle term of the trinomial.

45. The sum of two negative factors is always negative.

46. Every trinomial has integer coefficients.

< Objective 2 >

Factor each of the following trinomials completely.

47. $3a^2 - 3a - 126$

48. $2c^2 + 2c - 60$

49. $r^3 + 7r^2 - 18r$

50. $m^3 + 5m^2 - 14m$

Answers

25. _____

26. _____

27. _____

28. _____

29. _____

30. _____

31. _____

32. _____

33. _____

34. _____

35. _____

36. _____

37. _____

38. _____

39. _____

40. _____

41. _____

42. _____

43. _____ 44. _____

45. _____ 46. _____

47. _____

48. _____

49. _____

50. _____

Answers

51. _____

52. _____

53. _____

54. _____

55. _____

56. _____

57. _____

58. _____

59. _____

60. _____

61. _____

62. _____

63. _____

64. _____

65. _____

66. _____

(a) _____

(b) _____

(c) _____

(d) _____

(e) _____

(f) _____

(g) _____

(h) _____

51. $2x^3 - 20x^2 - 48x$

52. $3p^3 + 48p^2 - 108p$

53. $x^2y - 9xy^2 - 36y^3$ > Videos

54. $4s^4 - 20s^3t - 96s^2t^2$

55. $m^3 - 29m^2n + 120mn^2$

56. $2a^3 - 52a^2b + 96ab^2$

Basic Skills | Advanced Skills | **Vocational-Technical Applications** | Calculator/Computer | Above and Beyond | Getting Ready

▲

57. MANUFACTURING TECHNOLOGY The shape of a beam loaded with a single concentrated load is described by the expression $\dfrac{x^2 - 64}{200}$. Factor the numerator, $(x^2 - 64)$.

58. ALLIED HEALTH The concentration, in micrograms per milliliter (mcg/mL), of Vancocin, an antibiotic used to treat peritonitis, is given by the negative of the polynomial $t^2 - 8t - 20$, where t is the number of hours since the drug was administered via intravenous injection. Write this given polynomial in factored form.

Basic Skills | Advanced Skills | Vocational-Technical Applications | Calculator/Computer | **Above and Beyond** | Getting Ready

▲

Find all positive values for k for which each of the following can be factored.

59. $x^2 - kx + 16$ > Videos

60. $x^2 - kx + 17$

61. $x^2 - kx - 5$

62. $x^2 - kx - 7$

63. $x^2 + 3x + k$

64. $x^2 + 5x + k$

65. $x^2 + 2x - k$

66. $x^2 + x - k$

Basic Skills | Advanced Skills | Vocational-Technical Applications | Calculator/Computer | Above and Beyond | **Getting Ready**

▲

Multiply.

(a) $(2x - 1)(2x + 3)$

(b) $(3a - 1)(a + 4)$

(c) $(x - 4)(2x - 3)$

(d) $(2w - 11)(w + 2)$

(e) $(y + 5)(2y + 9)$

(f) $(2x + 1)(x - 12)$

(g) $(p + 9)(2p + 5)$

(h) $(3a - 5)(2a + 4)$

Answers

1. $x - 5$ **3.** $m + 6$ **5.** $p - 10$ **7.** $x - 8$ **9.** $x - 5y$

11. $(x + 3)(x + 5)$ **13.** $(x - 4)(x - 7)$ **15.** $(s + 3)(s + 10)$

17. $(a - 8)(a + 6)$ **19.** $(x - 1)(x - 7)$ **21.** $(m + 7)(m - 4)$

23. $(x + 4)(x - 10)$ **25.** $(x - 7)(x - 7)$ **27.** $(p - 12)(p + 2)$

29. $(x + 11)(x - 6)$ **31.** $(c + 4)(c + 15)$ **33.** $(x + 2y)(x + 5y)$

35. $(a + 6b)(a - 7b)$ **37.** $(x - 5y)(x - 8y)$ **39.** $(x + 2y)(x - 4y)$

41. $(5m + n)(5m + n)$ **43.** False   **45.** True   **47.** $3(a + 6)(a - 7)$

49. $r(r - 2)(r + 9)$ **51.** $2x(x - 12)(x + 2)$ **53.** $y(x + 3y)(x - 12y)$

55. $m(m - 5n)(m - 24n)$ **57.** $(x + 8)(x - 8)$ **59.** 8, 10, or 17

61. 4   **63.** 2   **65.** 3, 8, 15, 24, . . .   **(a)** $4x^2 + 4x - 3$

(b) $3a^2 + 11a - 4$ **(c)** $2x^2 - 11x + 12$ **(d)** $2w^2 - 7w - 22$

(e) $2y^2 + 19y + 45$ **(f)** $2x^2 - 23x - 12$ **(g)** $2p^2 + 23p + 45$

(h) $6a^2 + 2a - 20$

4.3

Factoring Trinomials of the Form $ax^2 + bx + c$

< 4.3 Objectives >

1 > Factor a trinomial of the form $ax^2 + bx + c$

2 > Completely factor a trinomial

3 > Use the *ac* test to determine factorability

4 > Use the results of the *ac* test to factor a trinomial

Factoring trinomials is more time-consuming when the coefficient of the first term is not 1. Look at the following multiplication.

$(5x + 2)(2x + 3) = 10x^2 + 19x + 6$

Factors   Factors
of $10x^2$ of 6

Do you see the additional problem? We must consider all possible factors of the first coefficient (10 in our example) as well as those of the third term (6 in our example).

There is no easy way out! You need to form all possible combinations of factors and then check the middle term until the proper pair is found. If this seems a bit like guesswork, it is. In fact, some call this process factoring by *trial and error.*

We can simplify the work a bit by reviewing the sign patterns found in Section 4.2.

Property

Sign Patterns for Factoring Trinomials

1. If all terms of a trinomial are positive, the signs between the terms in the binomial factors are both plus signs.
2. If the third term of the trinomial is positive and the middle term is negative, the signs between the terms in the binomial factors are both minus signs.
3. If the third term of the trinomial is negative, the signs between the terms in the binomial factors are opposite (one is + and one is −).

| Example 1 | Factoring a Trinomial |

< Objective 1 >

Factor $3x^2 + 14x + 15$.
    First, list the possible factors of 3, the coefficient of the first term.

$3 = 1 \cdot 3$

Now list the factors of 15, the last term.

$15 = 1 \cdot 15$
$ = 3 \cdot 5$

Because the signs of the trinomial are all positive, we know any factors will have the form

The product of the numbers in
the last blanks must be 15.

$(\_x + \_)(\_x + \_)$

The product of the
numbers in the first
blanks must be 3.

So the following are the possible factors and the corresponding middle terms:

Possible Factorizations	Middle Terms
$(x + 1)(3x + 15)$	$18x$
$(x + 15)(3x + 1)$	$46x$
$(3x + 3)(x + 5)$	$18x$
$(3x + 5)(x + 3)$	$14x$

The correct middle term

So

$3x^2 + 14x + 15 = (3x + 5)(x + 3)$

 Check Yourself 1

Factor.

(a) $5x^2 + 14x + 8$ (b) $3x^2 + 20x + 12$

Example 2   **Factoring a Trinomial**

Factor $4x^2 - 11x + 6$.

Because only the middle term is negative, we know the factors have the form

$(\_x - \_)(\_x - \_)$

Both signs are negative.

Now look at the factors of the first coefficient and the last term.

$$4 = 1 \cdot 4 \qquad 6 = 1 \cdot 6$$
$$ = 2 \cdot 2 \qquad = 2 \cdot 3$$

This gives us the possible factors:

Possible Factorizations	Middle Terms
$(x - 1)(4x - 6)$	$-10x$
$(x - 6)(4x - 1)$	$-25x$
$(x - 2)(4x - 3)$	$-11x$

The correct middle term

RECALL

Again, at least mentally, check your work by multiplying the factors.

Note that, in this example, we *stopped* as soon as the correct pair of factors was found. So

$$4x^2 - 11x + 6 = (x - 2)(4x - 3)$$

 Check Yourself 2

Factor.

(a) $2x^2 - 9x + 9$ (b) $6x^2 - 17x + 10$

Next, we will factor a trinomial whose last term is negative.

▶ **Example 3**   **Factoring a Trinomial**

Factor $5x^2 + 6x - 8$.

Because the last term is negative, the factors have the form

$$(\_x + \_)(\_x - \_)$$

Consider the factors of the first coefficient and the last term.

$$5 = 1 \cdot 5 \qquad 8 = 1 \cdot 8$$
$$= 2 \cdot 4$$

The possible factors are then

Possible Factorizations	Middle Terms
$(x + 1)(5x - 8)$	$-3x$
$(x + 8)(5x - 1)$	$39x$
$(5x + 1)(x - 8)$	$-39x$
$(5x + 8)(x - 1)$	$3x$
$(x + 2)(5x - 4)$	$6x$

Again we stop as soon as the correct pair of factors is found.

$$5x^2 + 6x - 8 = (x + 2)(5x - 4)$$

 Check Yourself 3

Factor $4x^2 + 5x - 6$.

The same process is used to factor a trinomial with more than one variable.

▶ **Example 4**   **Factoring a Trinomial**

Factor $6x^2 + 7xy - 10y^2$.

The form of the factors must be

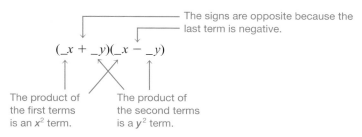

The signs are opposite because the last term is negative.

$$(\_x + \_y)(\_x - \_y)$$

The product of the first terms is an x^2 term.

The product of the second terms is a y^2 term.

Again look at the factors of the first and last coefficients.

$$6 = 1 \cdot 6 \qquad 10 = 1 \cdot 10$$
$$ = 2 \cdot 3 \qquad = 2 \cdot 5$$

NOTE

Be certain that you have a pattern that matches up every possible pair of coefficients.

Possible Factorizations	Middle Terms
$(x + y)(6x - 10y)$	$-4xy$
$(x + 10y)(6x - y)$	$59xy$
$(6x + y)(x - 10y)$	$-59xy$
$(6x + 10y)(x - y)$	$4xy$
$(x + 2y)(6x - 5y)$	$7xy$

Once more, we stop as soon as the correct factors are found.

$$6x^2 + 7xy - 10y^2 = (x + 2y)(6x - 5y)$$

 Check Yourself 4

Factor $15x^2 - 4xy - 4y^2$.

Example 5 illustrates a special kind of trinomial called a *perfect square trinomial*.

Example 5   Factoring a Trinomial

Factor $9x^2 + 12xy + 4y^2$.

Because all terms are positive, the form of the factors must be

$$(\_x + \_y)(\_x + \_y)$$

Consider the factors of the first and last coefficients.

$$9 = 9 \cdot 1 \qquad 4 = 4 \cdot 1$$
$$ = 3 \cdot 3 \qquad = 2 \cdot 2$$

Possible Factorizations	Middle Terms
$(x + y)(9x + 4y)$	$13xy$
$(x + 4y)(9x + y)$	$37xy$
$(3x + 2y)(3x + 2y)$	$12xy$

NOTE

Perfect square trinomials can be factored by using previous methods. Recognizing the special pattern simply saves time.

So

$$9x^2 + 12xy + 4y^2 = (3x + 2y)(3x + 2y)$$
$$= (3x + 2y)^2$$

Square   2(3x)(2y)   Square
of 3x   of 2y

This trinomial is the result of squaring a binomial, thus the special name of perfect square trinomial.

Check Yourself 5

Factor.

(a) $4x^2 + 28x + 49$ (b) $16x^2 - 40xy + 25y^2$

Before looking at Example 6, review one important point from Section 4.2. Recall that when you factor trinomials, you should not forget to look for a common factor as the first step. If there is a common factor, factor it out and then factor the remaining trinomial as before.

 Example 6   **Factoring a Trinomial**

< Objective 2 >

Factor $18x^2 - 18x + 4$.

First look for a common factor in all three terms. Here that factor is 2, so write

$$18x^2 - 18x + 4 = 2(9x^2 - 9x + 2)$$

By our earlier methods, we can factor the remaining trinomial as

$$9x^2 - 9x + 2 = (3x - 1)(3x - 2)$$

NOTE

If you don't see why this is true, use your pencil to work it out before you move on!

So

$$18x^2 - 18x + 4 = 2(3x - 1)(3x - 2)$$

Don't forget the 2 that was factored out!

Check Yourself 6

Factor $16x^2 + 44x - 12$.

Now look at an example in which the common factor includes a variable.

 Example 7   **Factoring a Trinomial**

Factor

$$6x^3 + 10x^2 - 4x$$

The common factor is 2x.

So

$$6x^3 + 10x^2 - 4x = 2x(3x^2 + 5x - 2)$$

Because

$$3x^2 + 5x - 2 = (3x - 1)(x + 2)$$

we have

$$6x^3 + 10x^2 - 4x = 2x(3x - 1)(x + 2)$$

RECALL

Be certain to include the monomial factor.

 Check Yourself 7

Factor $6x^3 - 27x^2 + 30x$.

You have now had a chance to work with a variety of factoring techniques. Your success in factoring polynomials depends on your ability to recognize when to use which technique. Here are some guidelines to help you apply the factoring methods you have studied in this chapter.

Step by Step

Factoring Polynomials

Step 1   Look for a greatest common factor other than 1. If such a factor exists, factor out the GCF.

Step 2   If the polynomial that remains is a *trinomial,* try to factor the trinomial by the trial-and-error methods of Sections 4.2 and 4.3.

Example 8 illustrates the use of this strategy.

Example 8   **Factoring a Trinomial**

NOTE

$m^2 + 4$ cannot be factored any further.

(a) Factor $5m^2n + 20n$.

First, we see that the GCF is $5n$. Factoring it out gives

$$5m^2n + 20n = 5n(m^2 + 4)$$

(b) Factor $3x^3 - 24x^2 + 48x$.

First, we see that the GCF is $3x$. Factoring out $3x$ yields

$$3x^3 - 24x^2 + 48x = 3x(x^2 - 8x + 16)$$
$$= 3x(x - 4)(x - 4) \quad \text{or} \quad 3x(x - 4)^2$$

(c) Factor $8r^2s + 20rs^2 - 12s^3$.

First, the GCF is $4s$, and we can write the original polynomial as

$$8r^2s + 20rs^2 - 12s^3 = 4s(2r^2 + 5rs - 3s^2)$$

Because the remaining polynomial is a trinomial, we can use the trial-and-error method to complete the factoring.

$$8r^2s + 20rs^2 - 12s^3 = 4s(2r - s)(r + 3s)$$

Check Yourself 8

Factor the following polynomials.

(a) $8a^3 + 32a^2b + 32ab^2$ **(b)** $7x^3 + 7x^2y - 42xy^2$
(c) $5m^4 + 15m^3 + 5m^2$

To this point we have used the trial-and-error method to factor trinomials. We have also learned that not all trinomials can be factored. In the remainder of this section we will look at the same kinds of trinomials, but in a slightly different context. We first determine whether a trinomial is factorable. We then use the results of that analysis to factor the trinomial.

Some students prefer the trial-and-error method for factoring because it is generally faster and more intuitive. Other students prefer the method used in the remainder of this section (called the *ac* method) because it yields the answer in a systematic way. We let you determine which method you prefer.

We begin by looking at some factored trinomials.

| Example 9 | Matching Trinomials and Their Factors |

Determine which of the following are true statements.

(a) $x^2 - 2x - 8 = (x - 4)(x + 2)$

This is a true statement. Using the FOIL method, we see that

$$(x - 4)(x + 2) = x^2 + 2x - 4x - 8$$
$$= x^2 - 2x - 8$$

(b) $x^2 + 6x + 5 = (x + 2)(x + 3)$

This is not a true statement.

$$(x + 2)(x + 3) = x^2 + 3x + 2x + 6 = x^2 + 5x + 6$$

(c) $x^2 + 5x - 14 = (x - 2)(x + 7)$

This is true: $(x - 2)(x + 7) = x^2 + 7x - 2x - 14 = x^2 + 5x - 14$

(d) $x^2 - 8x - 15 = (x - 5)(x - 3)$

This is false: $(x - 5)(x - 3) = x^2 - 3x - 5x + 15 = x^2 - 8x + 15$

Check Yourself 9

Determine which of the following are true statements.

(a) $2x^2 - 2x - 3 = (2x - 3)(x + 1)$
(b) $3x^2 + 11x - 4 = (3x - 1)(x + 4)$
(c) $2x^2 - 7x + 3 = (x - 3)(2x - 1)$

The first step in learning to factor a trinomial is to identify its coefficients. So that we are consistent, we first write the trinomial in standard form, $ax^2 + bx + c$, and then label the three coefficients as a, b, and c.

Example 10 | **Identifying the Coefficients of $ax^2 + bx + c$**

First, when necessary, rewrite the trinomial in $ax^2 + bx + c$ form. Then give the values for a, b, and c, in which a is the coefficient of the x^2 term, b is the coefficient of the x term, and c is the constant.

(a) $x^2 - 3x - 18$

$a = 1 \qquad b = -3 \qquad c = -18$

NOTE

The negative sign is attached to the coefficient.

(b) $x^2 - 24x + 23$

$a = 1 \qquad b = -24 \qquad c = 23$

(c) $x^2 + 8 - 11x$

First rewrite the trinomial in descending-exponent order:

$x^2 - 11x + 8$

$a = 1 \qquad b = -11 \qquad c = 8$

Check Yourself 10

First, when necessary, rewrite the trinomials in $ax^2 + bx + c$ form. Then label a, b, and c, in which a is the coefficient of the x^2 term, b is the coefficient of the x term, and c is the constant.

(a) $x^2 + 5x - 14$ **(b)** $x^2 - 18x + 17$ **(c)** $x - 6 + 2x^2$

Not all trinomials can be factored. To discover whether a trinomial is factorable, we try the **ac test.**

Definition

The ac Test

A trinomial of the form $ax^2 + bx + c$ is factorable if (and only if) there are two integers, m and n, such that

$$ac = mn \qquad \text{and} \qquad b = m + n$$

In Example 11 we will look for m and n to determine whether each trinomial is factorable.

 Example 11 | **Using the ac Test**

< Objective 3 >

Use the ac test to determine which of the following trinomials can be factored. Find the values of m and n for each trinomial that can be factored.

(a) $x^2 - 3x - 18$

First, we find the values of a, b, and c, so that we can find ac.

$a = 1 \qquad b = -3 \qquad c = -18$

$ac = 1(-18) = -18 \qquad \text{and} \qquad b = -3$

Then, we look for two numbers, m and n, such that their product is ac and their sum is b. In this case, that means

$$mn = -18 \qquad \text{and} \qquad m + n = -3$$

We now look at all pairs of integers with a product of -18. We then look at the sum of each pair of integers, looking for a sum of -3.

mn	$m + n$
$1(-18) = -18$	$1 + (-18) = -17$
$2(-9) = -18$	$2 + (-9) = -7$
$3(-6) = -18$	$3 + (-6) = -3$
$6(-3) = -18$	
$9(-2) = -18$	
$18(-1) = -18$	

We need to look no further than 3 and -6.

3 and -6 are the two integers with a product of ac and a sum of b. We can say that

$$m = 3 \qquad \text{and} \qquad n = -6$$

Because we found values for m and n, we know that $x^2 - 3x - 18$ is factorable.

(b) $x^2 - 24x + 23$

We find that

$$a = 1 \qquad b = -24 \qquad c = 23$$

$$ac = 1(23) = 23 \qquad \text{and} \qquad b = -24$$

So

$$mn = 23 \qquad \text{and} \qquad m + n = -24$$

We now calculate integer pairs, looking for two numbers with a product of 23 and a sum of -24.

mn	$m + n$
$1(23) = 23$	$1 + 23 = 24$
$-1(-23) = 23$	$-1 + (-23) = -24$

$$m = -1 \qquad \text{and} \qquad n = -23$$

So, $x^2 - 24x + 23$ is factorable.

(c) $x^2 - 11x + 8$

We find that $a = 1$, $b = -11$, and $c = 8$. Therefore, $ac = 8$ and $b = -11$. Thus $mn = 8$ and $m + n = -11$. We calculate integer pairs:

mn	$m + n$
$1(8) = 8$	$1 + 8 = 9$
$2(4) = 8$	$2 + 4 = 6$
$-1(-8) = 8$	$-1 + (-8) = -9$
$-2(-4) = 8$	$-2 + (-4) = -6$

There are no other pairs of integers with a product of 8, and none of these pairs has a sum of -11. The trinomial $x^2 - 11x + 8$ is not factorable.

(d) $2x^2 + 7x - 15$

We find that $a = 2$, $b = 7$, and $c = -15$. Therefore, $ac = 2(-15) = -30$ and $b = 7$. Thus $mn = -30$ and $m + n = 7$. We calculate integer pairs:

mn	$m + n$
$1(-30) = -30$	$1 + (-30) = -29$
$2(-15) = -30$	$2 + (-15) = -13$
$3(-10) = -30$	$3 + (-10) = -7$
$5(-6) = -30$	$5 + (-6) = -1$
$6(-5) = -30$	$6 + (-5) = 1$
$10(-3) = -30$	$10 + (-3) = 7$

There is no need to go any further. We see that 10 and -3 have a product of -30 and a sum of 7, so

$m = 10$ and $n = -3$

Therefore, $2x^2 + 7x - 15$ is factorable.

Check Yourself 11

Use the *ac* test to determine which of the following trinomials can be factored. Find the values of *m* and *n* for each trinomial that can be factored.

(a) $x^2 - 7x + 12$ **(b)** $x^2 + 5x - 14$
(c) $3x^2 - 6x + 7$ **(d)** $2x^2 + x - 6$

It is not always necessary to evaluate all the products and sums to determine whether a trinomial is factorable. You may have noticed patterns and shortcuts that make it easier to find m and n. By all means, use them to help you find m and n. This is essential in mathematical thinking. You are taught a mathematical process that will always work for solving a problem. Such a process is called an **algorithm.** It is very easy to teach a computer to use an algorithm. It is very difficult (some would say impossible) for a computer to have insight. Shortcuts that you discover are *insights.* They may be the most important part of your mathematical education.

So far we have used the results of the *ac* test to determine whether a trinomial is factorable. The results can also be used to help factor the trinomial.

Example 12   **Using the Results of the *ac* Test to Factor**

< Objective 4 >

Rewrite the middle term as the sum of two terms and then factor by grouping.

(a) $x^2 - 3x - 18$

We find that $a = 1$, $b = -3$, and $c = -18$, so $ac = -18$ and $b = -3$. We are looking for two numbers, m and n, where $mn = -18$ and $m + n = -3$. In Example 11, part **(a)**, we looked at every pair of integers whose product (mn) was -18, to find a pair that had a sum ($m + n$) of -3. We found the two integers to be 3 and -6, because $3(-6) = -18$ and $3 + (-6) = -3$, so $m = 3$ and $n = -6$. We now use that result to rewrite the middle term as the sum of $3x$ and $-6x$.

$$x^2 + 3x - 6x - 18$$

We then factor by grouping:

$$= (x^2 + 3x) - (6x + 18)$$
$$x^2 + 3x - 6x - 18 = x(x + 3) - 6(x + 3)$$
$$= (x + 3)(x - 6)$$

(b) $x^2 - 24x + 23$

We use the results from Example 11, part **(b)**, in which we found $m = -1$ and $n = -23$, to rewrite the middle term of the equation.

$$x^2 - 24x + 23 = x^2 - x - 23x + 23$$

Then we factor by grouping:

$$x^2 - x - 23x + 23 = (x^2 - x) - (23x - 23)$$
$$= x(x - 1) - 23(x - 1)$$
$$= (x - 1)(x - 23)$$

(c) $2x^2 + 7x - 15$

From Example 11, part **(d)**, we know that this trinomial is factorable, and $m = 10$ and $n = -3$. We use that result to rewrite the middle term of the trinomial.

$$2x^2 + 7x - 15 = 2x^2 + 10x - 3x - 15$$
$$= (2x^2 + 10x) - (3x + 15)$$
$$= 2x(x + 5) - 3(x + 5)$$
$$= (x + 5)(2x - 3)$$

Note that we did not factor the trinomial in Example 11, part **(c)**, $x^2 - 11x + 8$. Recall that, by the *ac* method, we determined that this trinomial is not factorable.

Check Yourself 12

Use the results of Check Yourself 11 to rewrite the middle term as the sum of two terms and then factor by grouping.

(a) $x^2 - 7x + 12$ **(b)** $x^2 + 5x - 14$ **(c)** $2x^2 + x - 6$

Next, we look at some examples that require us to first find m and n and then factor the trinomial.

Example 13   **Rewriting Middle Terms to Factor**

Rewrite the middle term as the sum of two terms and then factor by grouping.

(a) $2x^2 - 13x - 7$

We find $a = 2$, $b = -13$, and $c = -7$, so $mn = ac = -14$ and $m + n = b = -13$. Therefore,

mn	$m + n$
$1(-14) = -14$	$1 + (-14) = -13$

So, $m = 1$ and $n = -14$. We rewrite the middle term of the trinomial as follows:

$$2x^2 - 13x - 7 = 2x^2 + x - 14x - 7$$
$$= (2x^2 + x) - (14x + 7)$$
$$= x(2x + 1) - 7(2x + 1)$$
$$= (2x + 1)(x - 7)$$

(b) $6x^2 - 5x - 6$

We find that $a = 6$, $b = -5$, and $c = -6$, so $mn = ac = -36$ and $m + n = b = -5$.

mn	$m + n$
$1(-36) = -36$	$1 + (-36) = -35$
$2(-18) = -36$	$2 + (-18) = -16$
$3(-12) = -36$	$3 + (-12) = -9$
$4(-9) \ = -36$	$4 + (-9) \ = -5$

So, $m = 4$ and $n = -9$. We rewrite the middle term of the trinomial:

$$6x^2 - 5x - 6 = 6x^2 + 4x - 9x - 6$$
$$= (6x^2 + 4x) - (9x + 6)$$
$$= 2x(3x + 2) - 3(3x + 2)$$
$$= (3x + 2)(2x - 3)$$

Check Yourself 13

Rewrite the middle term as the sum of two terms and then factor by grouping.

(a) $2x^2 - 7x - 15$ **(b)** $6x^2 - 5x - 4$

Be certain to check trinomials and binomial factors for any common monomial factor. (There is no common factor in the binomial unless it is also a common factor in the original trinomial.) Example 14 shows the factoring out of monomial factors.

▶ Example 14   **Factoring Out Common Factors**

Completely factor the trinomial.

$$3x^2 + 12x - 15$$

We could first factor out the common factor of 3:

$$3x^2 + 12x - 15 = 3(x^2 + 4x - 5)$$

Finding m and n for the trinomial $x^2 + 4x - 5$ yields $mn = -5$ and $m + n = 4$.

mn	$m + n$
$1(-5) = -5$	$1 + (-5) = -4$
$5(-1) = -5$	$-1 + (5) = \ \ 4$

So, $m = 5$ and $n = -1$. This gives us

$$3x^2 + 12x - 15 = 3(x^2 + 4x - 5)$$
$$= 3(x^2 + 5x - x - 5)$$
$$= 3[(x^2 + 5x) - (x + 5)]$$
$$= 3[x(x + 5) - (x + 5)]$$
$$= 3[(x + 5)(x - 1)]$$
$$= 3(x + 5)(x - 1)$$

 Check Yourself 14

Completely factor the trinomial.

$6x^3 + 3x^2 - 18x$

Not all possible product pairs need to be tried to find m and n. A look at the sign pattern of the trinomial will eliminate many of the possibilities. Assuming the leading coefficient is positive, there are four possible sign patterns.

Pattern	Example	Conclusion
1. b and c are both positive.	$2x^2 + 13x + 15$	m and n must both be positive.
2. b is negative and c is positive.	$x^2 - 7x + 12$	m and n must both be negative.
3. b is positive and c is negative.	$x^2 + 3x - 10$	m and n are of opposite signs. (The value with the larger absolute value is positive.)
4. b and c are both negative.	$x^2 - 3x - 10$	m and n are of opposite signs. (The value with the larger absolute value is negative.)

 Check Yourself ANSWERS

1. **(a)** $(5x + 4)(x + 2)$; **(b)** $(3x + 2)(x + 6)$ **2.** **(a)** $(2x - 3)(x - 3)$;
(b) $(6x - 5)(x - 2)$ **3.** $(4x - 3)(x + 2)$ **4.** $(3x - 2y)(5x + 2y)$
5. **(a)** $(2x + 7)^2$; **(b)** $(4x - 5y)^2$ **6.** $4(4x - 1)(x + 3)$
7. $3x(2x - 5)(x - 2)$ **8.** **(a)** $8a(a + 2b)(a + 2b)$; **(b)** $7x(x + 3y)(x - 2y)$;
(c) $5m^2(m^2 + 3m + 1)$ **9.** **(a)** False; **(b)** true; **(c)** true
10. **(a)** $a = 1, b = 5, c = -14$; **(b)** $a = 1, b = -18, c = 17$;
(c) $a = 2, b = 1, c = -6$ **11.** **(a)** Factorable, $m = -3, n = -4$;
(b) factorable, $m = 7, n = -2$; **(c)** not factorable;
(d) factorable, $m = 4, n = -3$ **12.** **(a)** $x^2 - 3x - 4x + 12 = (x - 3)(x - 4)$;
(b) $x^2 + 7x - 2x - 14 = (x + 7)(x - 2)$;
(c) $2x^2 + 4x - 3x - 6 = (x + 2)(2x - 3)$
13. **(a)** $2x^2 - 10x + 3x - 15 = (x - 5)(2x + 3)$;
(b) $6x^2 - 8x + 3x - 4 = (3x - 4)(2x + 1)$ **14.** $3x(2x - 3)(x + 2)$

Reading Your Text

The following fill-in-the-blank exercises are designed to ensure that you understand some of the key vocabulary used in this section.

SECTION 4.3

(a) If all the terms of a trinomial are positive, the signs between the terms in the binomial factors are both _____ signs.

(b) If the third term of a trinomial is negative, the signs between the terms in the binomial factors are _____.

(c) The first step in factoring a polynomial is to factor out the _____.

(d) A mathematical process that will always work for solving a problem is called an _____.

Name _____

Section _____ Date _____

Answers

1. _____
2. _____
3. _____
4. _____
5. _____
6. _____
7. _____
8. _____
9. _____
10. _____
11. _____
12. _____
13. _____
14. _____
15. _____
16. _____
17. _____

< Objective 1 >

Complete each of the following statements.

1. $4x^2 - 4x - 3 = (2x + 1)(\quad)$

2. $3w^2 + 11w - 4 = (w + 4)(\quad)$

3. $6a^2 + 13a + 6 = (2a + 3)(\quad)$

4. $25y^2 - 10y + 1 = (5y - 1)(\quad)$

5. $15x^2 - 16x + 4 = (3x - 2)(\quad)$

6. $6m^2 + 5m - 4 = (3m + 4)(\quad)$

7. $16a^2 + 8ab + b^2 = (4a + b)(\quad)$

8. $6x^2 + 5xy - 4y^2 = (3x + 4y)(\quad)$ > Videos

9. $4m^2 + 5mn - 6n^2 = (m + 2n)(\quad)$

10. $10p^2 - pq - 3q^2 = (5p - 3q)(\quad)$

State whether each of the following is **true** *or* **false.**

11. $x^2 + 2x - 3 = (x + 3)(x - 1)$

12. $y^2 - 3y - 18 = (y - 6)(y + 3)$

13. $x^2 - 10x - 24 = (x - 6)(x + 4)$

14. $a^2 + 9a - 36 = (a - 12)(a + 4)$

15. $x^2 - 16x + 64 = (x - 8)(x - 8)$

16. $w^2 - 12w - 45 = (w - 9)(w - 5)$

17. $25y^2 - 10y + 1 = (5y - 1)(5y + 1)$ > Videos

18. $6x^2 + 5xy - 4y^2 = (6x - 2y)(x + 2y)$

19. $10p^2 - pq - 3q^2 = (5p - 3q)(2p + q)$

20. $6a^2 + 13a + 6 = (2a + 3)(3a + 2)$

For each of the following trinomials, label a, b, and c.

21. $x^2 + 4x - 9$

22. $x^2 + 5x + 11$

23. $x^2 - 3x + 8$

24. $x^2 + 7x - 15$

25. $3x^2 + 5x - 8$

26. $2x^2 + 7x - 9$

27. $4x^2 + 11 + 8x$

28. $5x^2 - 9 + 7x$

29. $5x - 3x^2 - 10$

30. $9x - 7x^2 - 18$

< Objective 3 >

Use the ac test to determine which of the following trinomials can be factored. Find the values of m and n for each trinomial that can be factored.

31. $x^2 + x - 6$

32. $x^2 + 2x - 15$

33. $x^2 + x + 2$

34. $x^2 - 3x + 7$

35. $x^2 - 5x + 6$

36. $x^2 - x + 2$

37. $2x^2 + 5x - 3$

38. $3x^2 - 14x - 5$

39. $6x^2 - 19x + 10$ > Videos

40. $4x^2 + 5x + 6$

Basic Skills | **Advanced Skills** | Vocational-Technical Applications | Calculator/Computer | Above and Beyond | Getting Ready

Complete each of the following statements with **never, sometimes,** *or* **always.**

41. A trinomial with integer coefficients is _____ factorable.

42. If a trinomial with all positive terms is factored, the signs between the terms in the binomial factors will _____ be positive.

Answers

18. _____

19. _____

20. _____

21. _____

22. _____

23. _____

24. _____

25. _____

26. _____

27. _____

28. _____

29. _____

30. _____

31. _____

32. _____

33. _____

34. _____

35. _____

36. _____

37. _____

38. _____

39. _____

40. _____

41. _____

42. _____

Answers

43. _____

44. _____

45. _____

46. _____

47. _____

48. _____

49. _____

50. _____

51. _____

52. _____

53. _____

54. _____

55. _____

56. _____

57. _____

58. _____

59. _____

60. _____

61. _____

62. _____

63. _____

64. _____

65. _____

66. _____

67. _____

68. _____

43. The product of two binomials _____ results in a trinomial.

44. If the GCF for the terms in a polynomial is not 1, it should _____ be factored out first.

< Objectives 2–4 >

By any method, factor each polynomial completely.

45. $x^2 + 8x + 15$

46. $x^2 - 11x + 24$

47. $s^2 + 13s + 30$

48. $b^2 + 14b + 33$

49. $x^2 - 6x - 40$

50. $x^2 - 11x + 10$

51. $p^2 - 10p - 24$

52. $x^2 - 11x - 60$

53. $x^2 + 5x - 66$

54. $a^2 + 2a - 80$

55. $c^2 + 19c + 60$

56. $t^2 - 4t - 60$

57. $n^2 + 5n - 50$

58. $x^2 - 16x + 63$

59. $x^2 + 7xy + 10y^2$

60. $x^2 - 8xy + 12y^2$

61. $a^2 - ab - 42b^2$

62. $m^2 - 8mn + 16n^2$

63. $x^2 - 13xy + 40y^2$

64. $r^2 - 9rs - 36s^2$

65. $6x^2 + 19x + 10$

66. $6x^2 - 7x - 3$

67. $15x^2 + x - 6$

68. $12w^2 + 19w + 4$

69. $6m^2 + 25m - 25$

70. $8x^2 - 6x - 9$

71. $9x^2 - 12x + 4$

72. $20x^2 - 23x + 6$

73. $12x^2 - 8x - 15$

74. $16a^2 + 40a + 25$

75. $3y^2 + 7y - 6$

76. $12x^2 + 11x - 15$

77. $8x^2 - 27x - 20$ > Videos

78. $24v^2 + 5v - 36$

79. $2x^2 + 3xy + y^2$

80. $3x^2 - 5xy + 2y^2$

81. $5a^2 - 8ab - 4b^2$

82. $5x^2 + 7xy - 6y^2$

83. $9x^2 + 4xy - 5y^2$

84. $16x^2 + 32xy + 15y^2$

85. $6m^2 - 17mn + 12n^2$

86. $15x^2 - xy - 6y^2$

87. $36a^2 - 3ab - 5b^2$

88. $3q^2 - 17qr - 6r^2$

89. $x^2 + 4xy + 4y^2$

90. $25b^2 - 80bc + 64c^2$

91. $20x^2 - 20x - 15$

92. $24x^2 - 18x - 6$

93. $8m^2 + 12m + 4$

94. $14x^2 - 20x + 6$

95. $15r^2 - 21rs + 6s^2$

96. $10x^2 + 5xy - 30y^2$

Answers

69. _____

70. _____

71. _____

72. _____

73. _____

74. _____

75. _____

76. _____

77. _____

78. _____

79. _____

80. _____

81. _____

82. _____

83. _____

84. _____

85. _____

86. _____

87. _____

88. _____

89. _____

90. _____

91. _____

92. _____

93. _____

94. _____

95. _____

96. _____

Answers

97. _____

98. _____

99. _____

100. _____

101. _____

102. _____

103. _____

104. _____

105. _____

106. _____

107. _____

108. _____

109. _____

110. _____

111. _____

112. _____

97. $2x^3 - 2x^2 - 4x$

98. $2y^3 + y^2 - 3y$

99. $2y^4 + 5y^3 + 3y^2$ > Videos

100. $4z^3 - 18z^2 - 10z$

101. $36a^3 - 66a^2 + 18a$

Basic Skills I Advanced Skills I **Vocational-Technical Applications** I Calculator/Computer I Above and Beyond I Getting Ready
▲

---

102. ALLIED HEALTH The number of people who are sick t days after the outbreak of a flu epidemic is given by the polynomial

$$50 + 25t - 3t^2$$

Write this polynomial in factored form.

103. MANUFACTURING TECHNOLOGY The bending moment in an overhanging beam is described by the expression

$$218(x^2 - 20x + 36)$$

Factor the $x^2 - 20x + 36$ portion of the expression.

104. MANUFACTURING TECHNOLOGY The flow rate through a hydraulic hose can be found from the equation

$$2Q^2 + Q - 21 = 0$$

Factor the left side of this equation.

Basic Skills I Advanced Skills I Vocational-Technical Applications I Calculator/Computer I **Above and Beyond** I Getting Ready
▲

Find a positive value for k for which each of the following can be factored.

105. $x^2 + kx + 8$

106. $x^2 + kx + 9$

107. $x^2 - kx + 16$

108. $x^2 - kx + 17$

Factor each of the following polynomials completely.

109. $10(x + y)^2 - 11(x + y) - 6$ > Videos

110. $8(a - b)^2 + 14(a - b) - 15$

111. $5(x - 1)^2 - 15(x - 1) - 350$

112. $3(x + 1)^2 - 6(x + 1) - 45$

113. $15 + 29x - 48x^2$

114. $12 + 4a - 21a^2$

115. $-6x^2 + 19x - 15$

116. $-3s^2 - 10s + 8$

Basic Skills | Advanced Skills | Vocational-Technical Applications | Calculator/Computer | Above and Beyond | **Getting Ready** ▲

Multiply.

(a) $(x - 1)(x + 1)$

(b) $(a + 7)(a - 7)$

(c) $(x - y)(x + y)$

(d) $(2x - 5)(2x + 5)$

(e) $(3a - b)(3a + b)$

(f) $(5a - 4b)(5a + 4b)$

Answers

1. $2x - 3$ **3.** $3a + 2$ **5.** $5x - 2$ **7.** $4a + b$ **9.** $4m - 3n$
11. True **13.** False **15.** True **17.** False **19.** True
21. $a = 1, b = 4, c = -9$ **23.** $a = 1, b = -3, c = 8$
25. $a = 3, b = 5, c = -8$ **27.** $a = 4, b = 8, c = 11$
29. $a = -3, b = 5, c = -10$ **31.** Factorable; $3, -2$ **33.** Not factorable
35. Factorable; $-3, -2$ **37.** Factorable; $6, -1$ **39.** Factorable; $-15, -4$
41. sometimes **43.** sometimes **45.** $(x + 3)(x + 5)$
47. $(s + 10)(s + 3)$ **49.** $(x - 10)(x + 4)$ **51.** $(p - 12)(p + 2)$
53. $(x + 11)(x - 6)$ **55.** $(c + 4)(c + 15)$ **57.** $(n + 10)(n - 5)$
59. $(x + 2y)(x + 5y)$ **61.** $(a - 7b)(a + 6b)$ **63.** $(x - 5y)(x - 8y)$
65. $(3x + 2)(2x + 5)$ **67.** $(5x - 3)(3x + 2)$ **69.** $(6m - 5)(m + 5)$
71. $(3x - 2)(3x - 2) = (3x - 2)^2$ **73.** $(6x + 5)(2x - 3)$
75. $(3y - 2)(y + 3)$ **77.** $(8x + 5)(x - 4)$ **79.** $(2x + y)(x + y)$
81. $(5a + 2b)(a - 2b)$ **83.** $(9x - 5y)(x + y)$ **85.** $(3m - 4n)(2m - 3n)$
87. $(12a - 5b)(3a + b)$ **89.** $(x + 2y)^2$ **91.** $5(2x - 3)(2x + 1)$
93. $4(2m + 1)(m + 1)$ **95.** $3(5r - 2s)(r - s)$ **97.** $2x(x - 2)(x + 1)$
99. $y^2(2y + 3)(y + 1)$ **101.** $6a(3a - 1)(2a - 3)$ **103.** $(x - 18)(x - 2)$
105. 6 or 9 **107.** 8 or 10 or 17 **109.** $(5x + 5y + 2)(2x + 2y - 3)$
111. $5(x - 11)(x + 6)$ **113.** $(1 + 3x)(15 - 16x)$ **115.** $(3x - 5)(-2x + 3)$
(a) $x^2 - 1$ **(b)** $a^2 - 49$ **(c)** $x^2 - y^2$ **(d)** $4x^2 - 25$ **(e)** $9a^2 - b^2$
(f) $25a^2 - 16b^2$

Answers

113. _____

114. _____

115. _____

116. _____

(a) _____

(b) _____

(c) _____

(d) _____

(e) _____

(f) _____

Difference of Squares and Perfect Square Trinomials

< 4.4 Objectives >

1 > Factor a binomial that is the difference of two squares

2 > Factor a perfect square trinomial

In Section 3.4, we introduced some special products. Recall the following formula for the product of a sum and difference of two terms:

$$(a + b)(a - b) = a^2 - b^2$$

This also means that a binomial of the form $a^2 - b^2$, called a **difference of two squares,** has as its factors $a + b$ and $a - b$.

To use this idea for factoring, we can write

$$a^2 - b^2 = (a + b)(a - b)$$

A **perfect square** term has a coefficient that is a square (1, 4, 9, 16, 25, 36, etc.), and any variables have exponents that are multiples of 2 (x^2, y^4, z^6, etc.).

Example 1 | Identifying Perfect Square Terms

< Objective 1 >

For each of the following, decide whether it is a perfect square term. If it is, rewrite the expression as an expression squared.

(a) $36x$

(b) $24x^6$

(c) $9x^4$

(d) $64x^6$

(e) $16x^9$

Only parts **(c)** and **(d)** are perfect square terms.

$9x^4 = (3x^2)^2$

$64x^6 = (8x^3)^2$

Check Yourself 1

For each of the following, decide whether it is a perfect square term. If it is, rewrite the expression as an expression squared.

(a) $36x^{12}$ **(b)** $4x^6$ **(c)** $9x^7$ **(d)** $25x^8$ **(e)** $16x^{25}$

In Example 2, we will factor the difference between two perfect square terms.

Example 2	Factoring the Difference of Two Squares

NOTE

You could also write
$(x - 4)(x + 4)$. The order
doesn't matter because
multiplication is commutative.

Factor $x^2 - 16$.

Think $x^2 - 4^2$.

Because $x^2 - 16$ is a difference of squares, we have

$$x^2 - 16 = (x + 4)(x - 4)$$

Check Yourself 2

Factor $m^2 - 49$.

Any time an expression is a difference of two squares, it can be factored.

Example 3	Factoring the Difference of Two Squares

Factor $4a^2 - 9$.

Think $(2a)^2 - 3^2$.

So

$$4a^2 - 9 = (2a)^2 - (3)^2$$
$$= (2a + 3)(2a - 3)$$

Check Yourself 3

Factor $9b^2 - 25$.

The process for factoring a difference of squares does not change when more than one variable is involved.

Example 4	Factoring the Difference of Two Squares

NOTE

Think $(5a)^2 - (4b^2)^2$.

Factor $25a^2 - 16b^4$.

$$25a^2 - 16b^4 = (5a + 4b^2)(5a - 4b^2)$$

Check Yourself 4

Factor $49c^4 - 9d^2$.

We will now consider an example that combines common-term factoring with difference-of-squares factoring. Note that the common factor is always factored out as the *first step*.

Example 5 Removing the GCF First

Factor $32x^2y - 18y^3$.

Note that $2y$ is a common factor, so

$$32x^2y - 18y^3 = 2y(16x^2 - 9y^2)$$

Difference of squares

$$= 2y(4x + 3y)(4x - 3y)$$

NOTE

Step 1
Factor out the GCF.
Step 2
Factor the remaining binomial.

Check Yourself 5

Factor $50a^3 - 8ab^2$.

> **CAUTION**

Note that this is different from the sum of two squares (such as $x^2 + y^2$), which never has real factors.

Recall the following multiplication pattern.

$$(a + b)^2 = a^2 + 2ab + b^2$$

For example,

$$(x + 2)^2 = x^2 + 4x + 4$$
$$(x + 5)^2 = x^2 + 10x + 25$$
$$(2x + 1)^2 = 4x^2 + 4x + 1$$

Recognizing this pattern can simplify the process of factoring perfect square trinomials.

Example 6 Factoring a Perfect Square Trinomial

< Objective 2 >

Factor the trinomial $4x^2 + 12xy + 9y^2$.

Note that this is a perfect square trinomial in which

$$a = 2x \quad \text{and} \quad b = 3y.$$

In factored form, we have

$$4x^2 + 12xy + 9y^2 = (2x + 3y)^2$$

Check Yourself 6

Factor the trinomial $16u^2 + 24uv + 9v^2$.

Recognizing the same pattern can simplify the process of factoring perfect square trinomials in which the second term is negative.

| | **Example** 7 | **Factoring a Perfect Square Trinomial** |

Factor the trinomial $25x^2 - 10xy + y^2$.

This is also a perfect square trinomial, in which

$a = 5x$ and $b = -y$.

In factored form, we have

$$25x^2 - 10xy + y^2 = (5x + (-y))^2 = (5x - y)^2$$

Check Yourself 7

Factor the trinomial $4u^2 - 12uv + 9v^2$.

Check Yourself ANSWERS

1. **(a)** $(6x^6)^2$; **(b)** $(2x^3)^2$; **(d)** $(5x^4)^2$ **2.** $(m + 7)(m - 7)$
3. $(3b + 5)(3b - 5)$ **4.** $(7c^2 + 3d)(7c^2 - 3d)$
5. $2a(5a + 2b)(5a - 2b)$ **6.** $(4u + 3v)^2$ **7.** $(2u - 3v)^2$

Reading Your Text

The following fill-in-the-blank exercises are designed to ensure that you understand some of the key vocabulary used in this section.

SECTION 4.4

(a) A perfect square term has a coefficient that is a square and any variables have exponents that are _____ of 2.

(b) Any time an expression is the difference of two squares, it can be _____.

(c) When factoring, the _____ factor is always factored out as the first step.

(d) Although the difference of two squares can be factored, the _____ of two squares cannot.

4.4 exercises

Name _____

Section _____ Date _____

Answers

1. _____ 2. _____

3. _____ 4. _____

5. _____ 6. _____

7. _____ 8. _____

9. _____ 10. _____

11. _____

12. _____

13. _____

14. _____

15. _____

16. _____

17. _____

18. _____

19. _____

20. _____

21. _____

22. _____

23. _____

24. _____

Basic Skills | Advanced Skills | Vocational-Technical Applications | Calculator/Computer | Above and Beyond | Getting Ready

< Objective 1 >

For each of the following binomials, is the binomial a difference of squares?

1. $3x^2 + 2y^2$

2. $5x^2 - 7y^2$

3. $16a^2 - 25b^2$

4. $9n^2 - 16m^2$

5. $16r^2 + 4$

6. $p^2 - 45$

7. $16a^2 - 12b^3$

8. $9a^2b^2 - 16c^2d^2$

9. $a^2b^2 - 25$ > Videos

10. $4a^3 - b^3$

Factor the following binomials.

11. $m^2 - n^2$

12. $r^2 - 9$

13. $x^2 - 49$

14. $c^2 - d^2$

15. $49 - y^2$

16. $81 - b^2$

17. $9b^2 - 16$

18. $36 - x^2$

19. $16w^2 - 49$

20. $4x^2 - 25$

21. $4s^2 - 9r^2$

22. $64y^2 - x^2$

23. $9w^2 - 49z^2$ > Videos

24. $25x^2 - 81y^2$

25. $16a^2 - 49b^2$

26. $64m^2 - 9n^2$

27. $x^4 - 36$

28. $y^6 - 49$

29. $x^2y^2 - 16$

30. $m^2n^2 - 64$

31. $25 - a^2b^2$

32. $49 - w^2z^2$

33. $81a^2 - 100b^6$

34. $64x^4 - 25y^4$

35. $18x^3 - 2xy^2$ > Videos

36. $50a^2b - 2b^3$

37. $12m^3n - 75mn^3$

38. $63p^4 - 7p^2q^2$

Determine whether each of the following statements is **true** *or* **false.**

39. A perfect square term has a coefficient that is a square and any variables have exponents that are factors of 2.

40. Any time an expression is the difference of two squares, it can be factored.

41. Although the difference of two squares can be factored, the sum of two squares cannot.

42. When factoring, the middle factor is always factored out as the first step.

< Objective 2 >

Determine whether each of the following trinomials is a perfect square. If it is, factor the trinomial.

43. $x^2 - 14x + 49$

44. $x^2 + 9x + 16$

45. $x^2 - 18x - 81$

46. $x^2 + 10x + 25$

47. $x^2 - 18x + 81$

48. $x^2 - 24x + 48$

Answers

25. _____
26. _____
27. _____
28. _____
29. _____
30. _____
31. _____
32. _____
33. _____
34. _____
35. _____
36. _____
37. _____
38. _____
39. _____
40. _____
41. _____
42. _____
43. _____
44. _____
45. _____
46. _____
47. _____
48. _____

Answers

49. _____

50. _____

51. _____

52. _____

53. _____

54. _____

55. _____

56. _____

57. _____

58. _____

59. _____

60. _____

61. _____

62. _____

Factor the following trinomials.

49. $x^2 + 4x + 4$

50. $x^2 + 6x + 9$

51. $x^2 - 10x + 25$

52. $x^2 - 8x + 16$

| Basic Skills | **Advanced Skills** | Vocational-Technical Applications | Calculator/Computer | Above and Beyond | Getting Ready |

53. $4x^2 + 12xy + 9y^2$

54. $16x^2 + 40xy + 25y^2$

55. $9x^2 - 24xy + 16y^2$ Videos

56. $9w^2 - 30wv + 25v^2$

57. $y^3 - 10y^2 + 25y$

58. $12b^3 - 12b^2 + 3b$

| Basic Skills | Advanced Skills | **Vocational-Technical Applications** | Calculator/Computer | Above and Beyond | Getting Ready |

59. **MANUFACTURING TECHNOLOGY** The difference d in the calculated maximum deflection between two similar cantilevered beams is given by the formula

$$d = \left(\frac{w}{8EI}\right)\left(l_1^2 - l_2^2\right)\left(l_2^2 + l_2^2\right)$$

Rewrite the formula in its completely factored form.

60. **MANUFACTURING TECHNOLOGY** The work done W by a steam turbine is given by the formula

$$W = \frac{1}{2}m\left(v_1^2 - v_2^2\right)$$

Factor the right-hand side of this equation.

61. **ALLIED HEALTH** A toxic chemical is introduced into a protozoan culture. The number of deaths per hour is given by the polynomial $338 - 2t^2$, in which t is the number of hours after the chemical is introduced. Factor this expression.

62. **ALLIED HEALTH** Radiation therapy is one technique used to control cancer. After treatment, the total number of cancerous cells, in thousands, can be estimated by $144 - 4t^2$, in which t is the number of days of treatment. Factor this expression.

Factor each expression.

63. $x^2(x + y) - y^2(x + y)$ > Videos

64. $a^2(b - c) - 16b^2(b - c)$

65. $2m^2(m - 2n) - 18n^2(m - 2n)$

66. $3a^3(2a + b) - 27ab^2(2a + b)$

67. Find a value for k so that $kx^2 - 25$ will have the factors $2x + 5$ and $2x - 5$.

68. Find a value for k so that $9m^2 - kn^2$ will have the factors $3m + 7n$ and $3m - 7n$.

69. Find a value for k so that $2x^3 - kxy^2$ will have the factors $2x$, $x - 3y$, and $x + 3y$.

70. Find a value for k so that $20a^3b - kab^3$ will have the factors $5ab$, $2a - 3b$, and $2a + 3b$.

71. Complete the following statement in complete sentences: "To factor a number, you. . . ."

72. Complete this statement: "To factor an algebraic expression into prime factors means. . . ."

Factor.

(a) $2x(3x + 2) - 5(3x + 2)$ **(b)** $3y(y - 4) + 5(y - 4)$
(c) $3x(x + 2y) + y(x + 2y)$ **(d)** $5x(2x - y) - 3(2x - y)$
(e) $4x(2x - 5y) - 3y(2x - 5y)$

Answers

1. No   **3.** Yes   **5.** No   **7.** No   **9.** Yes   **11.** $(m + n)(m - n)$
13. $(x + 7)(x - 7)$ **15.** $(7 + y)(7 - y)$ **17.** $(3b + 4)(3b - 4)$
19. $(4w + 7)(4w - 7)$ **21.** $(2s + 3r)(2s - 3r)$ **23.** $(3w + 7z)(3w - 7z)$

Answers

63. _____
64. _____
65. _____
66. _____
67. _____
68. _____
69. _____
70. _____
71. _____
72. _____
(a) _____
(b) _____
(c) _____
(d) _____
(e) _____

25. $(4a + 7b)(4a - 7b)$ **27.** $(x^2 + 6)(x^2 - 6)$ **29.** $(xy + 4)(xy - 4)$

31. $(5 + ab)(5 - ab)$ **33.** $(9a + 10b^3)(9a - 10b^3)$

35. $2x(3x + y)(3x - y)$ **37.** $3mn(2m + 5n)(2m - 5n)$ **39.** False

41. True   **43.** Yes; $(x - 7)^2$ **45.** No   **47.** Yes; $(x - 9)^2$

49. $(x + 2)^2$ **51.** $(x - 5)^2$ **53.** $(2x + 3y)^2$ **55.** $(3x - 4y)^2$

57. $y(y - 5)^2$ **59.** $d = \left(\dfrac{w}{8EI}\right)(l_1 + l_2)(l_1 - l_2)(l_1^2 + l_2^2)$

61. $2(13 - t)(13 + t)$ **63.** $(x + y)^2(x - y)$

65. $2(m - 2n)(m + 3n)(m - 3n)$ **67.** 4   **69.** 18

71. Above and Beyond   **(a)** $(3x + 2)(2x - 5)$ **(b)** $(y - 4)(3y + 5)$

(c) $(x + 2y)(3x + y)$ **(d)** $(2x - y)(5x - 3)$ **(e)** $(2x - 5y)(4x - 3y)$

4.5

Strategies in Factoring

< 4.5 Objectives >

1 > Recognize factoring patterns

2 > Apply appropriate factoring strategies

In Sections 4.1 to 4.4 you have seen a variety of techniques for factoring polynomials. This section reviews those techniques and presents some guidelines for choosing an appropriate strategy or combination of strategies.

1. Always look for a greatest common factor. If you find a GCF (other than 1), factor out the GCF as your first step.

   To factor $5x^2y - 10xy + 25xy^2$, the GCF is $5xy$, so

$$5x^2y - 10xy + 25xy^2 = 5xy(x - 2 + 5y)$$

2. Now look at the number of terms in the polynomial you are trying to factor.

   (a) If the polynomial is a *binomial,* consider the formula for the difference of two squares. Recall that a sum of squares does not factor over the real numbers.

   (i) To factor $x^2 - 49y^2$, recognize the difference of squares, so

$$x^2 - 49y^2 = (x + 7y)(x - 7y)$$

   (ii) The binomial

$$x^2 + 64$$

   cannot be further factored.

   (b) If the polynomial is a *trinomial,* try to factor the trinomial as a product of two binomials, using trial and error.

   To factor $2x^2 - x - 6$, a consideration of possible factors of the first and last terms of the trinomial will lead to

$$2x^2 - x - 6 = (2x + 3)(x - 2)$$

> **NOTE**
>
> You may prefer to use the *ac* method shown in Section 4.3.

   (c) If the polynomial has *more than three terms,* try factoring by grouping.

   To factor $2x^2 - 3xy + 10x - 15y$, group the first two terms, and then the last two, and factor out common factors.

$$2x^2 - 3xy + 10x - 15y = x(2x - 3y) + 5(2x - 3y)$$

   Now factor out the common factor $(2x - 3y)$.

$$2x^2 - 3xy + 10x - 15y = (2x - 3y)(x + 5)$$

3. You should always factor the given polynomial completely. So after you apply one of the techniques given in part 2, another one may be necessary.

   (a) To factor

$$6x^3 + 22x^2 - 40x$$

   first factor out the common factor of $2x$. So

$$6x^3 + 22x^2 - 40x = 2x(3x^2 + 11x - 20)$$

Now continue to factor the trinomial as before and

$$6x^3 + 22x^2 - 40x = 2x(3x - 4)(x + 5)$$

(b) To factor

$$x^3 - x^2y - 4x + 4y$$

first we proceed by grouping:

$$x^3 - x^2y - 4x + 4y = x^2(x - y) - 4(x - y)$$
$$= (x - y)(x^2 - 4)$$

Now because $x^2 - 4$ is a difference of two squares, we continue to factor and obtain

$$x^3 - x^2y - 4x + 4y = (x - y)(x + 2)(x - 2)$$

Example 1   **Recognizing Factoring Patterns**

< Objective 1 >

For each of the following expressions, state the appropriate first step for factoring the polynomial.

(a) $9x^2 - 18x - 72$

Find the GCF.

(b) $x^2 - 3x + 2xy - 6y$

Group the terms.

(c) $x^4 - 81y^4$

Factor the difference of squares.

(d) $3x^2 + 7x + 2$

Use the *ac* method (or trial and error).

Check Yourself 1

For each of the following expressions, state the appropriate first step for factoring the polynomial.

(a) $5x^2 + 2x - 3$ **(b)** $a^4b^4 - 16$
(c) $3x^2 + 3x - 60$ **(d)** $2a^2 - 5a + 4ab - 10b$

Example 2   **Factoring Polynomials**

< Objective 2 >

For each of the following expressions, completely factor the polynomial.

(a) $9x^2 - 18x - 72$

The GCF is 9.

$$9x^2 - 18x - 72 = 9(x^2 - 2x - 8)$$
$$= 9(x - 4)(x + 2)$$

(b) $x^2 - 3x + 2xy - 6y$

Grouping the terms, we have

$$x^2 - 3x + 2xy - 6y = (x^2 - 3x) + (2xy - 6y)$$
$$= x(x - 3) + 2y(x - 3)$$
$$= (x - 3)(x + 2y)$$

(c) $x^4 - 81y^4$

Factoring the difference of squares, we find

$$x^4 - 81y^4 = (x^2 + 9y^2)(x^2 - 9y^2)$$
$$= (x^2 + 9y^2)(x - 3y)(x + 3y)$$

(d) $3x^2 + 7x + 2$

Using the *ac* method, we find $m = 1$ and $n = 6$.

$$3x^2 + 7x + 2 = 3x^2 + x + 6x + 2$$
$$= (3x^2 + x) + (6x + 2)$$
$$= x(3x + 1) + 2(3x + 1)$$
$$= (3x + 1)(x + 2)$$

Check Yourself 2

For each of the following expressions, completely factor the polynomial.

(a) $5x^2 + 2x - 3$ **(b)** $a^4b^4 - 16$

(c) $3x^2 + 3x - 60$ **(d)** $2a^2 - 5a + 4ab - 10b$

There are other patterns that sometimes occur when factoring. Several of these relate to the factoring of expressions that contain terms that are perfect cubes. The most common are the sum or difference of cubes, shown here.

Factoring the sum of two perfect cubes

$$x^3 + y^3 = (x + y)(x^2 - xy + y^2)$$

Factoring the difference of two perfect cubes

$$x^3 - y^3 = (x - y)(x^2 + xy + y^2)$$

Example 3   **Factoring Expressions Involving Perfect Cube Terms**

Factor each of the following expressions.

(a) $8x^3 + 27y^3$.

$$8x^3 + 27y^3 = (2x)^3 + (3y)^3 \qquad \text{Substitute these values into the given patterns}$$
$$= [(2x) + (3y)][(2x)^2 - (2x)(3y) + (3y)^2] \qquad \text{Simplify}$$
$$= (2x + 3y)(4x^2 - 6xy + 9y^2)$$

(b) $a^3b^3 - 64c^3$

$$a^3b^3 - 64c^3 = (ab)^3 - (4c)^3$$

$$= [(ab) - (4c)][(ab)^2 + (ab)(4c) + (4c)^2]$$

$$= (ab - 4c)(a^2b^2 + 4abc + 16c^2)$$

Check Yourself 3

Factor each of the following expressions

(a) $a^3 + 64b^3c^3$ **(b)** $27x^3 - 8y^3$

Check Yourself ANSWERS

1. (a) ac method (or trial and error); **(b)** factor the difference of squares; **(c)** find the GCF; **(d)** group the terms

2. (a) $(5x - 3)(x + 1)$; **(b)** $(a^2b^2 + 4)(ab - 2)(ab + 2)$; **(c)** $3(x + 5)(x - 4)$; **(d)** $(2a - 5)(a + 2b)$ **3. (a)** $(a + 4bc)(a^2 - 4abc + 16b^2c^2)$; **(b)** $(3x - 2y)(9x^2 - 6xy + 4y^2)$

Reading Your Text

The following fill-in-the-blank exercises are designed to ensure that you understand some of the key vocabulary used in this section.

SECTION 4.5

(a) The first step in factoring requires that we find the _____ of all the terms.

(b) If a polynomial is a _____ , we can consider the formula for the difference of two squares.

(c) You should always factor a given polynomial _____.

(d) If a polynomial has more than three terms, we can try factoring by _____.

Basic Skills | Advanced Skills | Vocational-Technical Applications | Calculator/Computer | Above and Beyond | Getting Ready

< Objectives 1–2 >

Factor each polynomial completely. To begin, state which method should be applied as the first step, given the guidelines of this section. Then continue the exercise and factor each polynomial completely.

1. $x^2 - 3x$

2. $4y^2 - 9$

3. $x^2 - 5x - 24$

4. $8x^3 + 10x$

5. $x(x - y) + 2(x - y)$

6. $5a^2 - 10a + 25$

7. $2x^2y - 6xy + 8y^2$ > Videos

8. $2p - 6q + pq - 3q^2$

9. $y^2 - 13y + 40$

10. $m^3 + 27m^2n$

11. $3b^2 + 17b - 28$ > Videos

12. $3x^2 + 6x - 5xy - 10y$ > Videos

13. $3x^2 - 14xy - 24y^2$

14. $16c^2 - 49d^2$

15. $2a^2 + 11a + 12$

16. $m^3n^3 - mn$

17. $125r^3 + r^2$

18. $(x - y)^2 - 16$

MathZone

Boost your grade at mathzone.com!

> Practice Problems
> NetTutor
> Self-Tests
> e-Professors
> Videos

Name _____

Section _____ Date _____

Answers

1. _____
2. _____
3. _____
4. _____
5. _____
6. _____
7. _____
8. _____
9. _____
10. _____
11. _____
12. _____
13. _____
14. _____
15. _____
16. _____
17. _____
18. _____

Answers

19. _____

20. _____

21. _____

22. _____

23. _____

24. _____

25. _____

26. _____

27. _____

28. _____

29. _____

30. _____

31. _____

32. _____

33. _____

34. _____

35. _____

36. _____

(a) _____ (b) _____

(c) _____ (d) _____

(e) _____ (f) _____

19. $3x^2 - 30x + 63$

20. $3a^2 - 108$

21. $40a^2 + 5$

22. $4p^2 - 8p - 60$

Basic Skills | **Advanced Skills** | Vocational-Technical Applications | Calculator/Computer | Above and Beyond | Getting Ready

23. $2w^2 - 14w - 36$

24. $xy^3 - 9xy$

25. $3a^2b - 48b^3$ > Videos

26. $12b^3 - 86b^2 + 14b$

27. $x^4 - 3x^2 - 10$

28. $m^4 - 9n^4$

29. $8p^3 - q^3r^3$

30. $27x^3 + 125y^3$

Basic Skills | Advanced Skills | Vocational-Technical Applications | Calculator/Computer | **Above and Beyond** | Getting Ready

Factor completely.

31. $(x - 5)^2 - 169$ > Videos

32. $(x - 7)^2 - 81$

33. $x^2 + 4xy + 4y^2 - 16$

34. $9x^2 + 12xy + 4y^2 - 25$

35. $6(x - 2)^2 + 7(x - 2) - 5$

36. $12(x + 1)^2 - 17(x + 1) + 6$

Basic Skills | Advanced Skills | Vocational-Technical Applications | Calculator/Computer | Above and Beyond | **Getting Ready**

Solve.

(a) $x - 5 = 0$ **(b)** $2x - 1 = 0$ **(c)** $3x + 2 = 0$
(d) $x + 4 = 0$ **(e)** $7 - x = 0$ **(f)** $9 - 4x = 0$

Answers

1. GCF, $x(x - 3)$ **3.** Trial and error, $(x - 8)(x + 3)$
5. GCF, $(x + 2)(x - y)$ **7.** GCF, $2y(x^2 - 3x + 4y)$
9. Trial and error, $(y - 5)(y - 8)$ **11.** Trial and error, $(b + 7)(3b - 4)$
13. Trial and error, $(3x + 4y)(x - 6y)$ **15.** Trial and error, $(2a + 3)(a + 4)$
17. GCF, $r^2(125r + 1)$ **19.** GCF, then trial and error, $3(x - 3)(x - 7)$
21. GCF, $5(8a^2 + 1)$ **23.** GCF, then trial and error, $2(w - 9)(w + 2)$
25. GCF, then difference of squares, $3b(a + 4b)(a - 4b)$
27. Trial and error, $(x^2 - 5)(x^2 + 2)$ **29.** $(2p - qr)(4p^2 - 2pqr + q^2r^2)$
31. $(x + 8)(x - 18)$ **33.** $(x + 2y + 4)(x + 2y - 4)$

35. $(2x - 5)(3x - 1)$ **(a)** $x = 5$ **(b)** $x = \dfrac{1}{2}$ **(c)** $x = -\dfrac{2}{3}$

(d) $x = -4$ **(e)** $x = 7$ **(f)** $x = \dfrac{9}{4}$

4.6

Solving Quadratic Equations by Factoring

< 4.6 Objectives >

1 > Solve quadratic equations by factoring

2 > Solve applications of quadratic equations

The factoring techniques you have learned provide us with tools for solving equations that can be written in the form

$$ax^2 + bx + c = 0 \qquad a \neq 0$$

> This is a quadratic equation in one variable, here x. You can recognize such a quadratic equation by the fact that the highest power of the variable x is the second power.

in which a, b, and c are constants.

An equation written in the form $ax^2 + bx + c = 0$ is called a **quadratic equation in standard form.** Using factoring to solve quadratic equations requires the **zero-product principle,** which says that if the product of two factors is 0, then one or both of the factors must be equal to 0. In symbols:

Definition

Zero-Product Principle	If $a \cdot b = 0$, then $a = 0$ or $b = 0$ or $a = b = 0$.

We can now apply this principle to solve quadratic equations.

Example 1 — Solving Equations by Factoring

< Objective 1 >

Solve.

$$x^2 - 3x - 18 = 0$$

Factoring on the left, we have

$$(x - 6)(x + 3) = 0$$

By the zero-product principle, we know that one or both of the factors must be zero. We can then write

$$x - 6 = 0 \qquad \text{or} \qquad x + 3 = 0$$

Solving each equation gives

$$x = 6 \qquad \text{or} \qquad x = -3$$

NOTE

To use the zero-product principle, 0 must be on one side of the equation.

The two solutions are 6 and -3.

Quadratic equations can be checked in the same way as linear equations were checked: by substitution. For instance, if $x = 6$, we have

$$6^2 - 3 \cdot 6 - 18 \overset{?}{=} 0$$

$$36 - 18 - 18 \overset{?}{=} 0$$

$$0 = 0$$

which is a true statement. We leave it to you to check the solution -3.

Check Yourself 1

Solve $x^2 - 9x + 20 = 0$.

Other factoring techniques are also used in solving quadratic equations. Example 2 illustrates this.

 Example 2   **Solving Equations by Factoring**

(a) Solve $x^2 - 5x = 0$.

Again, factor the left side of the equation and apply the zero-product principle.

$$x(x - 5) = 0$$

Now

$$x = 0 \qquad \text{or} \qquad x - 5 = 0$$
$$x = 5$$

> **CAUTION**

A *common mistake* is to forget the statement $x = 0$ when you are solving equations of this type. Be sure to include the *two statements* obtained.

The two solutions are 0 and 5.

(b) Solve $x^2 - 9 = 0$.

Factoring yields

$$(x + 3)(x - 3) = 0$$
$$x + 3 = 0 \qquad \text{or} \qquad x - 3 = 0$$
$$x = -3 \qquad\qquad x = 3$$

NOTE

The symbol $\pm$ is read "plus or minus."

The solutions may be written as $x = \pm 3$.

Check Yourself 2

Solve by factoring.

(a) $x^2 + 8x = 0$ **(b)** $x^2 - 16 = 0$

Example 3 illustrates a crucial point. Our solution technique depends on the zero-product principle, which means that the product of factors *must be equal to* 0. The importance of this is shown now.

Example 3 | Solving Equations by Factoring

> **CAUTION**

Consider the equation

$x(2x - 1) = 3$

Students are sometimes tempted to write

$x = 3$ or $2x - 1 = 3$

This is *not correct*. Instead, subtract 3 from both sides of the equation *as the first step* to write

$x(2x - 1) - 3 = 0$

Then proceed to write the equation in standard form. Only then can you factor and proceed as before.

Solve $2x^2 - x = 3$.

The first step in the solution is to write the equation in standard form (that is, write it so that one side of the equation is 0). So start by adding -3 to both sides of the equation.

Then,

$2x^2 - x - 3 = 0$ Make sure all terms are on one side of the equation. The other side will be 0.

You can now factor and solve by using the zero-product principle.

$$(2x - 3)(x + 1) = 0$$
$$2x - 3 = 0 \quad\text{or}\quad x + 1 = 0$$
$$2x = 3 \qquad\qquad x = -1$$
$$x = \frac{3}{2}$$

The solutions are $\frac{3}{2}$ and -1.

Check Yourself 3

Solve $3x^2 = 5x + 2$.

In all the previous examples, the quadratic equations had two distinct real-number solutions. That may not always be the case, as we shall see.

Example 4 | Solving Equations by Factoring

Solve $x^2 - 6x + 9 = 0$.

Factoring, we have

$$(x - 3)(x - 3) = 0$$

and

$$x - 3 = 0 \quad\text{or}\quad x - 3 = 0$$
$$x = 3 \qquad\qquad x = 3$$

The solution is 3.

A quadratic (or second-degree) equation always has *two* solutions. When an equation such as this one has two solutions that are the same number, we call 3 the **repeated** (or **double**) **solution** of the equation.

Although a quadratic equation always has two solutions, they may not always be real numbers. You will learn more about this in a later course.

Check Yourself 4

Solve $x^2 + 6x + 9 = 0$.

Always examine the quadratic member of an equation for common factors. It will make your work much easier, as Example 5 illustrates.

 Example 5 | **Solving Equations by Factoring**

Solve $3x^2 - 3x - 60 = 0$.

First, note the common factor 3 in the quadratic member of the equation. Factoring out the 3, we have

$$3(x^2 - x - 20) = 0$$

NOTE

Notice the advantage of dividing both sides of the equation by 3. The coefficients in the quadratic expression become smaller and are easier to factor.

Now, because the common factor has no variables, we can divide both sides of the equation by 3.

$$\frac{3(x^2 - x - 20)}{3} = \frac{0}{3}$$

or

$$x^2 - x - 20 = 0$$

We can now factor and solve as before.

$$(x - 5)(x + 4) = 0$$

$$x - 5 = 0 \quad \text{or} \quad x + 4 = 0$$
$$x = 5 \qquad\qquad x = -4$$

 Check Yourself 5

Solve $2x^2 - 10x - 48 = 0$.

Many applications can be solved with quadratic equations.

 Example 6 | **Solving an Application**

< Objective 2 >

The Microhard Corporation has found that the equation

$$P = x^2 - 7x - 94$$

describes the profit P, in thousands of dollars, for every x hundred computers sold. How many computers were sold if the profit was \$50,000?

If the profit was \$50,000, then $P = 50$. We now set up and solve the equation.

NOTE

P is expressed in thousands so the value 50 is substituted for P, not 50,000.

$$50 = x^2 - 7x - 94$$
$$0 = x^2 - 7x - 144$$
$$0 = (x + 9)(x - 16)$$
$$x = -9 \quad \text{or} \quad x = 16$$

They cannot sell a negative number of computers, so $x = 16$. They sold 1,600 computers.

Check Yourself 6

The Gerbil Babyfood Corporation has found that the equation

$P = x^2 - 6x - 7$

describes the profit P, in hundreds of dollars, for every x thousand jars sold. How many jars were sold if the profit was $20,000?

Check Yourself ANSWERS

1. $4, 5$ **2. (a)** $0, -8$; **(b)** $4, -4$ **3.** $-\dfrac{1}{3}, 2$ **4.** -3 **5.** $-3, 8$
6. $9,000$ jars

Reading Your Text

The following fill-in-the-blank exercises are designed to ensure that you understand some of the key vocabulary used in this section.

SECTION 4.6

(a) An equation written in the form $ax^2 + bx + c = 0$ is called a _____ equation in standard form.

(b) Using factoring to solve quadratic equations requires the _____ principle.

(c) To use the zero-product principle, it is important that the product of factors be equal to _____.

(d) When an equation has two solutions that are the same number, we call it a _____ solution.

4.6 exercises

Name _____

Section _____ Date _____

Answers

1. _____	2. _____
3. _____	4. _____
5. _____	6. _____
7. _____	8. _____
9. _____	10. _____
11. _____	12. _____
13. _____	14. _____
15. _____	16. _____
17. _____	18. _____
19. _____	20. _____
21. _____	22. _____
23. _____	24. _____
25. _____	26. _____
27. _____	28. _____
29. _____	30. _____
31. _____	32. _____

< Objective 1 >

Solve each of the following quadratic equations.

1. $(x - 3)(x - 4) = 0$

2. $(x - 7)(x + 1) = 0$

3. $(3x + 1)(x - 6) = 0$

4. $(5x - 4)(x - 6) = 0$

5. $x^2 - 2x - 3 = 0$

6. $x^2 + 5x + 4 = 0$

7. $x^2 - 7x + 6 = 0$

8. $x^2 + 3x - 10 = 0$

9. $x^2 + 8x + 15 = 0$

10. $x^2 - 3x - 18 = 0$

11. $x^2 + 4x - 21 = 0$

12. $x^2 - 12x + 32 = 0$

13. $x^2 - 4x = 12$ > Videos

14. $x^2 + 8x = -15$

15. $x^2 + 5x = 14$

16. $x^2 = 11x - 24$

17. $2x^2 + 5x - 3 = 0$

18. $3x^2 + 7x + 2 = 0$

19. $4x^2 - 24x + 35 = 0$

20. $6x^2 + 11x - 10 = 0$

21. $4x^2 + 11x = -6$

22. $5x^2 + 2x = 3$

23. $5x^2 + 13x = 6$

24. $4x^2 = 13x + 12$ > Videos

25. $x^2 - 2x = 0$

26. $x^2 + 5x = 0$

27. $x^2 = -8x$

28. $x^2 = 7x$

29. $5x^2 - 15x = 0$ > Videos

30. $4x^2 + 20x = 0$

31. $x^2 - 25 = 0$

32. $x^2 = 49$

33. $x^2 = 81$

34. $x^2 = 64$

35. $2x^2 - 18 = 0$

36. $3x^2 - 75 = 0$

37. $3x^2 + 24x + 45 = 0$

38. $4x^2 - 4x = 24$

Basic Skills | **Advanced Skills** | Vocational-Technical Applications | Calculator/Computer | Above and Beyond | Getting Ready

---

39. $2x(3x + 14) = 10$ > Videos

40. $3x(5x + 9) = 6$

41. $(x + 3)(x - 2) = 14$

42. $(x - 5)(x + 2) = 18$

< Objective 2 >

Solve the following problems.

43. **NUMBER PROBLEM** The product of two consecutive integers is 132. Find the two integers.

44. **NUMBER PROBLEM** The product of two consecutive positive even integers is 120. Find the two integers. > Videos

45. **NUMBER PROBLEM** The sum of an integer and its square is 72. What is the integer?

46. **NUMBER PROBLEM** The square of an integer is 56 more than the integer. Find the integer.

47. **GEOMETRY** If the sides of a square are increased by 3 in., the area is increased by 39 in.². What were the dimensions of the original square?

48. **GEOMETRY** If the sides of a square are decreased by 2 cm, the area is decreased by 36 cm². What were the dimensions of the original square?

49. **BUSINESS AND FINANCE** The profit on a small appliance is given by $P = x^2 - 3x - 60$, in which x is the number of appliances sold per day. How many appliances were sold on a day when there was a $20 loss?

50. **BUSINESS AND FINANCE** The relationship between the number of calculators x that a company can sell per month and the price of each calculator p is given by $x = 1{,}700 - 100p$. Find the price at which a calculator should be sold to produce a monthly revenue of $7,000. (*Hint:* Revenue = xp.)

Answers

33. _____

34. _____

35. _____

36. _____

37. _____

38. _____

39. _____

40. _____

41. _____

42. _____

43. _____

44. _____

45. _____

46. _____

47. _____

48. _____

49. _____

50. _____

Answers

51. _____

52. _____

53. _____

54. _____

55. _____

Basic Skills | Advanced Skills | **Vocational-Technical Applications** | Calculator/Computer | Above and Beyond | Getting Ready
▲

51. **ALLIED HEALTH** The concentration, C, in micrograms per milliliter (mcg/mL), of Tobrex, an antibiotic prescribed for burn patients, is given by the equation $C = 12 + t - t^2$, where t is the number of hours since the drug was administered via intravenous injection. Find the value of t when the concentration is $C = 0$.

52. **ALLIED HEALTH** The number of people who are sick t days after the outbreak of a flu epidemic is given by the equation $P = 50 + 25t - 3t^2$. Write the polynomial in factored form. Find the value of t when the number of people is $P = 0$.

53. **MANUFACTURING TECHNOLOGY** The maximum stress for a given allowable strain (deformation) for a certain material is given by the polynomial

$$S = 85.8x - 0.6x^2 - 1,537.2$$

in which x is the allowable strain in micrometers. Find the allowable strain in micrometers when the stress is $S = 0$.
   Hint: Rearrange the polynomial and factor out a common factor of -0.6 first.

Basic Skills | Advanced Skills | Vocational-Technical Applications | Calculator/Computer | **Above and Beyond** | Getting Ready
▲

54. Write a short comparison that explains the difference between $ax^2 + bx + c$ and $ax^2 + bx + c = 0$.

55. When solving quadratic equations, some people try to solve an equation in the manner shown below, but this does not work! Write a paragraph to explain what is wrong with this approach.

$$2x^2 + 7x + 3 = 52$$
$$(2x + 1)(x + 3) = 52$$
$$2x + 1 = 52 \quad \text{or} \quad x + 3 = 52$$
$$x = \frac{51}{2} \quad \text{or} \quad x = 49$$

Answers

1. $3, 4$ 3. $-\dfrac{1}{3}, 6$ 5. $-1, 3$ 7. $1, 6$ 9. $-3, -5$ 11. $-7, 3$

13. $-2, 6$ 15. $-7, 2$ 17. $-3, \dfrac{1}{2}$ 19. $\dfrac{5}{2}, \dfrac{7}{2}$ 21. $-\dfrac{3}{4}, -2$

23. $-3, \dfrac{2}{5}$ 25. $0, 2$ 27. $0, -8$ 29. $0, 3$ 31. $-5, 5$

33. $-9, 9$ 35. $-3, 3$ 37. $-5, -3$ 39. $-5, \dfrac{1}{3}$ 41. $4, -5$

43. $11, 12$ or $-12, -11$ 45. -9 or 8 47. 5 in. by 5 in.   49. 8
51. $t = 4$ hours   53. $x = 21$ or $x = 122$ micrometers
55. Above and Beyond

Definition/Procedure	Example	Reference

An Introduction to Factoring

Common Monomial Factor

Section 4.1

A single term that is a factor of every term of the polynomial.

The greatest common factor (GCF) of a polynomial is the factor that is a product of (a) the largest common numerical factor and (b) each variable with the smallest exponent in any term.

$4x^2$ is the greatest common monomial factor of $8x^4 - 12x^3 + 16x^2$.

p. 376

Factoring a Monomial from a Polynomial

1. Determine the GCF for all terms.
2. Use the GCF to factor each term and then apply the distributive property in the form

$$ab + ac = a(b + c)$$

The greatest common factor

3. Mentally check by multiplication.

$8x^4 - 12x^3 + 16x^2$
$= 4x^2(2x^2 - 3x + 4)$

p. 377

Factoring Trinomials

Sections 4.2 and 4.3

Factoring by Grouping

When there are four terms of a polynomial, factor the first pair and factor the last pair. If these two pairs have a common binomial factor, factor that out. The result will be the product of two binomials.

$4x^2 - 6x + 10x - 15$
$= 2x(2x - 3) + 5(2x - 3)$
$= (2x - 3)(2x + 5)$

p. 379

Trial and Error

To factor a trinomial, find the appropriate sign pattern and then find integer values that yield the appropriate coefficients for the trinomial.

$x^2 - 5x - 24$
$= (x - \ )(x + \ )$
$= (x - 8)(x + 3)$

p. 387

Using the ac Method to Factor

To factor a trinomial, first use the *ac* test to determine factorability. If the trinomial is factorable, the *ac* test will yield two terms (which have as their sum the middle term) that allow the factoring to be completed by using the grouping method.

$x^2 + 3x - 28$
$ac = -28; b = 3$
$mn = -28; m + n = 3$
$m = 7, n = -4$
$x^2 + 7x - 4x - 28$
$= x(x + 7) - 4(x + 7)$
$= (x - 4)(x + 7)$

p. 405

Continued

Definition/Procedure	Example	Reference
Difference of Squares and Perfect Square Trinomials		Section 4.4
Factoring a Difference of Squares		
Use the following form: $a^2 - b^2 = (a + b)(a - b)$	To factor: $16x^2 - 25y^2$: Think: $(4x)^2 - (5y)^2$ so $16x^2 - 25y^2$ $= (4x + 5y)(4x - 5y)$	p. 418
Factoring a Perfect Square Trinomial		
Use the following form: $a^2 + 2ab + b^2 = (a + b)^2$	$4x^2 + 12xy + 9y^2$ $= (2x)^2 + 2(2x)(3y) + (3y)^2$ $= (2x + 3y)^2$	p. 420
Strategies in Factoring		Section 4.5

When factoring a polynomial,
 	
1. Look for the GCF.
2. Consider the number of terms.
  a. If it is a binomial, look for a difference of squares.
  b. If it is trinomial, use the *ac* method or trial and error.
  c. If there are four or more terms, try grouping terms.
3. Be certain that the polynomial is completely factored.
	Given $12x^3 - 86x^2 + 14x$,
	factor out $2x$.
	$2x(6x^2 - 43x + 7)$
	$= 2x(6x - 1)(x - 7)$
		p. 427

Solving Quadratic Equations by Factoring		Section 4.6
1. Add or subtract the necessary terms on both sides of the equation so that the equation is in standard form (set equal to 0). 2. Factor the quadratic expression. 3. Set each factor equal to 0. 4. Solve the resulting equations to find the solutions. 5. Check each solution by substituting in the original equation.	To solve $x^2 + 7x = 30$ $x^2 + 7x - 30 = 0$ $(x + 10)(x - 3) = 0$ $x + 10 = 0$ or $x - 3 = 0$ $x = -10$ and $x = 3$ are solutions.	p. 433

This summary exercise set is provided to give you practice with each of the objectives of this chapter. Each exercise is keyed to the appropriate chapter section. When you are finished, you can check your answers to the odd-numbered exercises against those presented in the back of the text. If you have difficulty with any of these questions, go back and reread the examples from that section. Your instructor will give you guidelines on how best to use these exercises in your instructional setting.

4.1 *Factor each of the following polynomials.*

1. $18a + 24$

2. $9m^2 - 21m$

3. $24s^2t - 16s^2$

4. $18a^2b + 36ab^2$

5. $35s^3 - 28s^2$

6. $3x^3 - 6x^2 + 15x$

7. $18m^2n^2 - 27m^2n + 18m^2n^3$

8. $121x^8y^3 + 77x^6y^3$

9. $8a^2b + 24ab - 16ab^2$

10. $3x^2y - 6xy^3 + 9x^3y - 12xy^2$

11. $x(2x - y) + y(2x - y)$

12. $5(w - 3z) - w(w - 3z)$

4.2 *Factor each of the following trinomials completely.*

13. $x^2 + 9x + 20$

14. $x^2 - 10x + 24$

15. $a^2 - a - 12$

16. $w^2 - 13w + 40$

17. $x^2 + 12x + 36$

18. $r^2 - 9r - 36$

19. $b^2 - 4bc - 21c^2$

20. $m^2n + 4mn - 32n$

21. $m^3 + 2m^2 - 35m$

22. $2x^2 - 2x - 40$

23. $3y^3 - 48y^2 + 189y$

24. $3b^3 - 15b^2 - 42b$

4.3 *Factor each of the following trinomials completely.*

25. $3x^2 + 8x + 5$

26. $5w^2 + 13w - 6$

27. $2b^2 - 9b + 9$

28. $8x^2 + 2x - 3$

29. $10x^2 - 11x + 3$

30. $4a^2 + 7a - 15$

31. $9y^2 - 3yz - 20z^2$

32. $8x^2 + 14xy - 15y^2$

33. $8x^3 - 36x^2 - 20x$

34. $9x^2 - 15x - 6$

35. $6x^3 - 3x^2 - 9x$

36. $5w^2 - 25wz + 30z^2$

4.4 *Factor each of the following completely.*

37. $p^2 - 49$

38. $25a^2 - 16$

39. $m^2 - 9n^2$

40. $16r^2 - 49s^2$

41. $25 - z^2$

42. $a^4 - 16b^2$

43. $25a^2 - 36b^2$

44. $x^6 - 4y^2$

45. $3w^3 - 12wz^2$

46. $9a^4 - 49b^2$

47. $2m^2 - 72n^4$

48. $3w^3z - 12wz^3$

49. $x^2 + 8x + 16$

50. $x^2 - 18x + 81$

51. $4x^2 + 12x + 9$

52. $9x^2 - 12x + 4$

53. $16x^3 + 40x^2 + 25x$

54. $4x^3 - 4x^2 + x$

4.5 *Factor the following polynomials completely.*

55. $x^2 - 4x + 5x - 20$

56. $x^2 + 7x - 2x - 14$

57. $6x^2 + 4x - 15x - 10$

58. $12x^2 - 9x - 28x + 21$

59. $6x^3 + 9x^2 - 4x^2 - 6x$

60. $3x^4 + 6x^3 + 5x^3 + 10x^2$

4.6 *Solve each of the following quadratic equations.*

61. $(x - 1)(2x + 3) = 0$

62. $x^2 - 5x + 6 = 0$

63. $x^2 - 10x = 0$

64. $x^2 = 144$

65. $x^2 - 2x = 15$

66. $3x^2 - 5x - 2 = 0$

67. $4x^2 - 13x + 10 = 0$

68. $2x^2 - 3x = 5$

69. $3x^2 - 9x = 0$

70. $x^2 - 25 = 0$

71. $2x^2 - 32 = 0$

72. $2x^2 - x - 3 = 0$

The purpose of this self-test is to help you check your progress and to review for the next in-class exam. Allow yourself about an hour to take this test. At the end of that hour check your answers against those given in the back of the text. Section references accompany the answers. If you missed any questions, go back to those sections and reread the examples until you master the concepts.

Name _____

Section _____ Date _____

Answers

Factor each of the following polynomials.

1. $12b + 18$

2. $9p^3 - 12p^2$

3. $5x^2 - 10x + 20$

4. $6a^2b - 18ab + 12ab^2$

Factor each of the following polynomials completely.

5. $a^2 - 10a + 25$

6. $64m^2 - n^2$

7. $49x^2 - 16y^2$

8. $32a^2b - 50b^3$

Factor each of the following polynomials completely.

9. $a^2 - 5a - 14$

10. $b^2 + 8b + 15$

11. $x^2 - 11x + 28$

12. $y^2 + 12yz + 20z^2$

13. $x^2 + 2x - 5x - 10$

14. $6x^2 + 2x - 9x - 3$

Factor each of the following polynomials completely.

15. $2x^2 + 15x - 8$

16. $3w^2 + 10w + 7$

17. $8x^2 - 2xy - 3y^2$

18. $6x^3 + 3x^2 - 30x$

1. _____

2. _____

3. _____

4. _____

5. _____

6. _____

7. _____

8. _____

9. _____

10. _____

11. _____

12. _____

13. _____

14. _____

15. _____

16. _____

17. _____

18. _____

Answers

19. _____

20. _____

21. _____

22. _____

23. _____

24. _____

25. _____

Solve each of the following equations for x.

19. $x^2 - 8x + 15 = 0$

20. $x^2 - 3x = 4$

21. $3x^2 + x - 2 = 0$

22. $4x^2 - 12x = 0$

23. $x(x - 4) = 0$

24. $(x - 3)(x - 2) = 30$

25. $x^2 - 14x = -49$

Activity 4 ::
ISBNs and the Check Digit

Each activity in this text is designed to either enhance your understanding of the topics of the preceding chapter, or to provide you with a mathematical extension of those topics, or both. The activities can be undertaken by one student, but they are better suited for a small group project. Occasionally it is only through discussion that different facets of the activity become apparent. For material related to this activity, visit the text website at www.mhhe.com/streeter.

Have you ever noticed the long number, usually accompanied by a bar code, that can be found on the back of a book? This is called the International Standard Book Number (ISBN). Each book has a unique ISBN, which is 10 digits in length. A common form for an ISBN is X-XX-XXXXXX-X.

Each ISBN has four blocks of digits.

- The first block of digits on the left represents the language of the book (0 is used to represent English). This block is usually one digit in length.

- The second block of digits represents the publisher. This block is usually two or three digits in length.

- The third block of digits represents the number assigned to the book by the publishing company. This is usually five or six digits in length.

- The fourth block consists of the check digit.

For the purposes of this activity, we will consider the ISBN 0-07-229654-2, which is the ISBN for *Elementary and Intermediate Algebra* by Hutchison, Bergman, and Hoelzle.

The digit of most interest is the final digit, the **check digit.** When an ISBN is entered into a computer, the computer looks at the check digit to make certain that the numbers were properly entered. We will look at the algorithm used to generate the check digit.

Activity

I. To determine the check digit for the ISBN of a text, follow these steps.

  1. Multiply each of the nine assigned digits by a weighted value. The weighted values are 1 for the first digit from the left, 2 for the second digit, 3 for third digit, etc. In our case, we have

$$1 \cdot 0 + 2 \cdot 0 + 3 \cdot 7 + 4 \cdot 2 + 5 \cdot 2 + 6 \cdot 9 + 7 \cdot 6 + 8 \cdot 5 + 9 \cdot 4 = 211$$

  2. In mathematics, we occasionally are interested in only the remainder after we do division. We refer to the remainder as the **modular** (or mod) of the divisor. The ISBN uses (mod 11) to determine the check digit.

$211 \div 11 = 19$ with a remainder of 2. We say $211 = 2 \pmod{11}$

That is how we get the 2 in the ISBN 0-07-229654-2.

II. If we are given an ISBN and we want to check its validity, we can follow a similar algorithm.

  1. Multiply each of the 10 digits by a weighted value. The weighted values are still 1 for the first digit from the left, 2 for the second digit, etc., but we also multiply 10 times the tenth (check) digit.

In our case, we now have

$$1 \cdot 0 + 2 \cdot 0 + 3 \cdot 7 + 4 \cdot 2 + 5 \cdot 2 + 6 \cdot 9 + 7 \cdot 6 + 8 \cdot 5 + 9 \cdot 4 + 10 \cdot 2$$
$$= 231$$

2. For any valid ISBN, the result will have a remainder of zero when divided by 11. In other words, we will have 0 (mod 11). Note that $231 \div 11 = 21$.

Determine whether each of the following is a valid ISBN.

0-07-038023-6
0-15-249584-2
0-553-34948-1
0-07-000317-3

For those numbers above that are valid, go online and find the books to which they refer.

The following exercises are presented to help you review concepts from earlier chapters. This is meant as a review and not as a comprehensive exam. The answers are presented in the back of the text. Section references accompany the answers. If you have difficulty with any of these exercises, be certain to at least read through the summary related to those sections.

Name _____

Section _____ Date _____

Answers

Perform the indicated operations.

1. $7 - (-10)$

2. $(-34) \div (17)$

Perform each of the indicated operations.

3. $(7x^2 + 5x - 4) + (2x^2 - 6x - 1)$

4. $(3a^2 - 2a) - (7a^2 + 5)$

5. Subtract $4b^2 - 3b$ from the sum of $6b^2 + 5b$ and $4b^2 - 3$.

6. $3rs(5r^2s - 4rs + 6rs^2)$

7. $(2a - b)(3a^2 - ab + b^2)$

8. $\dfrac{7xy^3 - 21x^2y^2 + 14x^3y}{-7xy}$

9. $\dfrac{3a^2 - 10a - 8}{a - 4}$

10. $\dfrac{2x^3 - 8x + 5}{2x + 4}$

Solve the following equation for x.

11. $2 - 4(3x + 1) = 8 - 7x$

Solve the following inequality.

12. $4(x - 7) \le -(x - 5)$

Solve the following equation for the indicated variable.

13. $S = \dfrac{n}{2}(a + t)$ for t

Answers
1. _____
2. _____
3. _____
4. _____
5. _____
6. _____
7. _____
8. _____
9. _____
10. _____
11. _____
12. _____
13. _____

Answers

14. _____

15. _____

16. _____

17. _____

18. _____

19. _____

20. _____

21. _____

22. _____

23. _____

24. _____

25. _____

26. _____

27. _____

28. _____

29. _____

30. _____

Simplify the following expressions.

14. $x^6 x^{11}$

15. $(3x^2y^3)(2x^3y^4)$

16. $(3x^2y^3)^2(-4x^3y^2)^0$

17. $\dfrac{16x^2y^5}{4xy^3}$

18. $(3x^2)^3(2x)^2$

Factor each of the following polynomials completely.

19. $36w^5 - 48w^4$

20. $5x^2y - 15xy + 10xy^2$

21. $25x^2 + 30xy + 9y^2$

22. $4p^3 - 144pq^2$

23. $a^2 + 4a + 3$

24. $2w^3 - 4w^2 - 24w$

25. $3x^2 + 11xy + 6y^2$

Solve each of the following equations.

26. $a^2 - 7a + 12 = 0$

27. $3w^2 - 48 = 0$

28. $15x^2 + 5x = 10$

Solve the following problems.

29. **NUMBER PROBLEM** Twice the square of a positive integer is 12 more than 10 times that integer. What is the integer?

30. **GEOMETRY** The length of a rectangle is 1 in. more than 4 times its width. If the area of the rectangle is 105 in.^{2}, find the dimensions of the rectangle.

Beginning Algebra   The Streeter/Hutchison Series in Mathematics

CHAPTER

5

INTRODUCTION

The House of Representatives is made up of officials elected from congressional districts within each state. The number of representatives a state sends to the House depends on the state's population. The total number of representatives has grown from 106 in 1790 to 435, the maximum number established in 1930. These 435 representatives are apportioned to the 50 states on the basis of population. This apportionment is revised after every decennial (10-year) census.

If a particular state has a population, A, and its number of representatives is equal to a, then $\dfrac{A}{a}$ represents the ratio of people in the state to their total number of representatives in the U.S. House.

A comparison of these ratios for states in 2006 finds Pennsylvania with 652,959 people per representative and Arizona with 717,979—the national average was 687,080 people per representative for 2006. The difference is a result of ratios that do not divide out evenly. Should the numbers be rounded up or down? If they are all rounded down, the total is too small, if rounded up, the total number of representatives would be more than the 435 seats in the House. Because all the states cannot be treated equally, the question of what is fair and how to decide who gets the additional representatives has been debated in Congress since its inception.

Rational Expressions

CHAPTER 5 OUTLINE

451

5 pretest

Name _____

Section _____ Date _____

This pretest provides a preview of the types of exercises you will encounter in each section of this chapter. The answers for these exercises can be found in the back of the text. If you are working on your own, or ahead of the class, this pretest can help you identify the sections in which you should focus more of your time.

Answers

1. _____

2. _____

3. _____

4. _____

5. _____

6. _____

7. _____

8. _____

9. _____

10. _____

11. _____

12. _____

13. _____

14. _____

15. _____

16. _____

Write each fraction in simplest form.

5.1 **1.** $\dfrac{-15a^4b^7}{25a^6b}$ **2.** $\dfrac{x^2 - 16}{2x - 8}$

3. $\dfrac{3x^2 - 2x - 1}{6x^2 + 2x}$

Add or subtract as indicated.

5.3 **4.** $\dfrac{7a}{12} + \dfrac{19a}{12}$ **5.** $\dfrac{5x}{x + 1} + \dfrac{5}{x + 1}$

6. $\dfrac{x^2}{x - 6} - \dfrac{36}{x - 6}$

5.4 **7.** $\dfrac{5}{2w} - \dfrac{3}{w^2}$ **8.** $\dfrac{4}{b - 3} - \dfrac{1}{b}$

9. $\dfrac{2}{3x - 3} - \dfrac{5}{2x - 2}$ **10.** $\dfrac{4x}{x^2 - 8x + 15} + \dfrac{6}{x - 3}$

Multiply or divide as indicated.

5.2 **11.** $\dfrac{-4a^2}{6ab^3} \cdot \dfrac{3ab^2}{-4ab}$ **12.** $\dfrac{x^2 + 5x + 4}{2x^2 + 2x} \cdot \dfrac{x^2 - x - 12}{x^2 - 16}$

13. $\dfrac{8b^4}{5bc} \div \dfrac{12b^2c^2}{15bc^3}$ **14.** $\dfrac{x^2y + 2xy^2}{x^2 - 4y^2} \div \dfrac{4x^2y}{x^2 - xy - 2y^2}$

Simplify each complex fraction.

5.5 **15.** $\dfrac{\frac{x^3}{16}}{\frac{x^2}{24}}$ **16.** $\dfrac{2 - \frac{x}{y}}{4 - \frac{x^2}{y^2}}$

What values for x, if any, must be excluded in the following algebraic fractions?

5.6 **17.** $\dfrac{5}{x-3}$

18. $\dfrac{4}{x^2-3x-10}$

Solve the following equations for x.

5.6 **19.** $\dfrac{x}{4}-\dfrac{x}{5}=2$

20. $\dfrac{x}{x-2}+1=\dfrac{x+4}{x-2}$

21. $\dfrac{7}{x}-\dfrac{1}{x-3}=\dfrac{9}{x^2-3x}$

22. $\dfrac{x-3}{8}=\dfrac{x-2}{10}$

Solve the following applications.

5.7 **23.** **NUMBER PROBLEM** One number is 4 times another. If the sum of their reciprocals is $\dfrac{1}{4}$, find the two numbers.

24. **SCIENCE AND MEDICINE** Mark drove 240 mi to visit Sandra. Returning by a shorter route, he found that the trip was only 200 mi, but traffic slowed his speed by 8 mi/h. If the two trips took exactly the same time, what was his rate each way?

25. **NUMBER PROBLEM** A 55-ft cable is to be cut into two pieces whose lengths have the ratio 3 to 8. Find the lengths of the two pieces.

Answers

17. _____

18. _____

19. _____

20. _____

21. _____

22. _____

23. _____

24. _____

25. _____

5.1

Simplifying Rational Expressions

< 5.1 Objectives >

1 > Find the GCF for two monomials and simplify
a rational expression

2 > Find the GCF for two polynomials and simplify
a rational expression

Much of our work with rational expressions (also called algebraic fractions) will be
similar to your work in arithmetic. For instance, in algebra, as in arithmetic, many fractions name the same number. Remember from Chapter 0 that

$$\frac{1}{4} = \frac{1 \cdot 2}{4 \cdot 2} = \frac{2}{8}$$

and

$$\frac{1}{4} = \frac{1 \cdot 3}{4 \cdot 3} = \frac{3}{12}$$

So $\frac{1}{4}, \frac{2}{8}$, and $\frac{3}{12}$ all name the same number; they are called **equivalent fractions.**
These examples illustrate what is called the **Fundamental Principle of Fractions.** In
algebra it becomes

Property

Fundamental Principle of Rational Expressions	For polynomials P, Q, and R, $$\frac{P}{Q} = \frac{PR}{QR} \qquad \text{when } Q \neq 0 \text{ and } R \neq 0$$

NOTE

A rational expression is
sometimes called an
algebraic fraction, or simply
a fraction.

This principle allows us to multiply or divide the numerator and denominator of a fraction by the same nonzero polynomial. The result will be an expression that is equivalent to the original one.

Our objective in this section is to simplify rational expressions by using the fundamental principle. In algebra, as in arithmetic, to write a fraction in simplest form,
you divide the numerator and denominator of the fraction by their greatest common
factor (GCF). The numerator and denominator of the resulting fraction will have no
common factors other than 1, and the fraction is then in **simplest form.** The following
rule summarizes this procedure.

Step by Step

| To Write Rational Expressions in Simplest Form | Step 1 | Factor the numerator and denominator. |
| | Step 2 | Divide the numerator and denominator by the GCF. The resulting fraction will be in lowest terms. |

NOTE

Step 2 uses the Fundamental Principle of Fractions. The GCF is R in the Fundamental Principle of Rational Expressions rule.

In Example 1, we simplify both numeric and algebraic fractions using the steps provided above.

Example 1   Writing Fractions in Simplest Form

< Objective 1 >

RECALL

This is the same as dividing both the numerator and denominator of $\frac{18}{30}$ by 6.

(a) Write $\dfrac{18}{30}$ in simplest form.

$$\frac{18}{30} = \frac{2 \cdot 3 \cdot 3}{2 \cdot 3 \cdot 5} = \frac{\cancel{2} \cdot \cancel{3} \cdot 3}{\cancel{2} \cdot \cancel{3} \cdot 5} = \frac{3}{5}$$

Divide by the GCF. The slash lines indicate that we have divided the numerator and denominator by 2 and by 3.

(b) Write $\dfrac{4x^3}{6x}$ in simplest form.

$$\frac{4x^3}{6x} = \frac{\cancel{2} \cdot 2 \cdot \cancel{x} \cdot x \cdot x}{\cancel{2} \cdot 3 \cdot \cancel{x}} = \frac{2x^2}{3}$$

(c) Write $\dfrac{15x^3y^2}{20xy^4}$ in simplest form.

$$\frac{15x^3y^2}{20xy^4} = \frac{3 \cdot \cancel{5} \cdot \cancel{x} \cdot x \cdot x \cdot \cancel{y} \cdot \cancel{y}}{2 \cdot 2 \cdot \cancel{5} \cdot \cancel{x} \cdot \cancel{y} \cdot \cancel{y} \cdot y \cdot y} = \frac{3x^2}{4y^2}$$

We can also simplify directly by finding the GCF. In this case, we have

$$\frac{15x^3y^2}{20xy^4} = \frac{(5xy^2)(3x^2)}{(5xy^2)(4y^2)} = \frac{3x^2}{4y^2}$$

NOTE

With practice you will be able to simplify these terms without writing out the factorizations.

(d) Write $\dfrac{3a^2b}{9a^3b^2}$ in simplest form.

$$\frac{3a^2b}{9a^3b^2} = \frac{(3a^2b)}{(3a^2b)(3ab)} = \frac{1}{3ab}$$

(e) Write $\dfrac{10a^5b^4}{2a^2b^3}$ in simplest form.

$$\frac{10a^5b^4}{2a^2b^3} = \frac{(2a^2b^3)(5a^3b)}{(2a^2b^3)} = \frac{(5a^3b)}{1} = 5a^3b$$

NOTE

Most of the methods of this chapter build on our factoring work of the last chapter.

Check Yourself 1

Write each fraction in simplest form.

(a) $\dfrac{30}{66}$ (b) $\dfrac{5x^4}{15x}$ (c) $\dfrac{12xy^4}{18x^3y^2}$

(d) $\dfrac{5m^2n}{10m^3n^3}$ (e) $\dfrac{12a^4b^6}{2a^3b^4}$

In simplifying arithmetic fractions, common factors are generally easy to recognize. With rational expressions, the factoring techniques you studied in Chapter 4 are used as the *first step* in determining those factors.

▶ | **Example 2** | **Writing Fractions in Simplest Form**

< Objective 2 >

Write each fraction in simplest form.

(a) $\dfrac{2x - 4}{x^2 - 4} = \dfrac{2(x - 2)}{(x + 2)(x - 2)}$ Factor the numerator and denominator.

$= \dfrac{2\cancel{(x - 2)}}{(x + 2)\cancel{(x - 2)}}$ Divide by the GCF $x - 2$. The slash lines indicate that we have divided by that common factor.

$= \dfrac{2}{x + 2}$

NOTE

$3x^2 - 3$
$= 3(x^2 - 1)$
$= 3(x - 1)(x + 1)$

(b) $\dfrac{3x^2 - 3}{x^2 - 2x - 3} = \dfrac{3(x - 1)\cancel{(x + 1)}}{(x - 3)\cancel{(x + 1)}}$

$= \dfrac{3(x - 1)}{x - 3}$

 > CAUTION

Pick any value, other than 0, for x and substitute. You will quickly see that

$\dfrac{x + 2}{x + 1} \neq \dfrac{2}{1}$

For example, if $x = 4$,

$\dfrac{4 + 2}{4 + 1} = \dfrac{6}{5}$

(c) $\dfrac{2x^2 + x - 6}{2x^2 - x - 3} = \dfrac{(x + 2)\cancel{(2x - 3)}}{(x + 1)\cancel{(2x - 3)}}$

$= \dfrac{x + 2}{x + 1}$

Be careful! The expression $\dfrac{x + 2}{x + 1}$ is already in simplest form. Students are often tempted to divide as follows:

$\dfrac{\cancel{x} + 2}{\cancel{x} + 1}$ is *not equal to* $\dfrac{2}{1}$

The *x*'s are *terms* in the numerator and denominator. They *cannot* be divided out. Only *factors* can be divided. The fraction

$$\frac{x + 2}{x + 1}$$

is in its simplest form.

Check Yourself 2

Write each fraction in simplest form.

(a) $\dfrac{5x - 15}{x^2 - 9}$ (b) $\dfrac{a^2 - 5a + 6}{3a^2 - 6a}$

(c) $\dfrac{3x^2 + 14x - 5}{3x^2 + 2x - 1}$ (d) $\dfrac{5p - 15}{p^2 - 4}$

Remember the rules for signs in division. The quotient of a positive number and a negative number is always negative. Thus there are three equivalent ways to write such a quotient. For instance,

$$\frac{-2}{3} = \frac{2}{-3} = -\frac{2}{3}$$

The quotient of two positive numbers or two negative numbers is always positive. For example,

$$\frac{-2}{-3} = \frac{2}{3}$$

Example 3	Writing Fractions in Simplest Form

Write each fraction in simplest form.

(a) $\dfrac{6x^2}{-3xy} = \dfrac{2 \cdot \cancel{3} \cdot \cancel{x} \cdot x}{(-1) \cdot \cancel{3} \cdot \cancel{x} \cdot y} = \dfrac{2x}{-y} = -\dfrac{2x}{y}$

(b) $\dfrac{-5a^2b}{-10b^2} = \dfrac{(\cancel{-1}) \cdot \cancel{5} \cdot a \cdot a \cdot \cancel{b}}{(\cancel{-1}) \cdot 2 \cdot \cancel{5} \cdot \cancel{b} \cdot b} = \dfrac{a^2}{2b}$

Check Yourself 3

Write each fraction in simplest form.

(a) $\dfrac{8x^3y}{-4xy^2}$ (b) $\dfrac{-16a^4b^2}{-12a^2b^5}$

It is sometimes necessary to factor out a monomial before simplifying the fraction.

Example 4 | **Writing Fractions in Simplest Form**

Write each fraction in simplest form.

(a) $\dfrac{6x^2 + 2x}{2x^2 + 12x} = \dfrac{2x(3x + 1)}{2x(x + 6)} = \dfrac{3x + 1}{x + 6}$

(b) $\dfrac{x^2 - 4}{x^2 + 6x + 8} = \dfrac{(x + 2)(x - 2)}{(x + 2)(x + 4)} = \dfrac{x - 2}{x + 4}$

(c) $\dfrac{x + 3}{x^2 + 7x + 12} = \dfrac{x + 3}{(x + 3)(x + 4)} = \dfrac{1}{x + 4}$

Check Yourself 4

Simplify each fraction.

(a) $\dfrac{3x^3 - 6x^2}{9x^4 - 3x^2}$ (b) $\dfrac{x^2 - 9}{x^2 - 12x + 27}$

Simplifying certain rational expressions will be easier with the following result. First, verify for yourself that

$5 - 8 = -(8 - 5)$

In general, it is true that

$a - b = -(b - a)$

or, by dividing both sides of the equation by $b - a$,

$\dfrac{a - b}{b - a} = \dfrac{-(b - a)}{b - a}$

So dividing by $b - a$ on the right, we have

$\dfrac{a - b}{b - a} = -1$

Example 5 | **Writing Rational Expressions in Simplest Form**

Write each fraction in simplest form.

(a) $\dfrac{2x - 4}{4 - x^2} = \dfrac{2(x - 2)}{(2 + x)(2 - x)}$ This is equal to -1.

$= \dfrac{2(-1)}{2 + x}$

$= \dfrac{-2}{2 + x}$

(b) $\dfrac{9 - x^2}{x^2 + 2x - 15} = \dfrac{(3 + x)(3 - x)}{(x + 5)(x - 3)}$ This is equal to -1.

$= \dfrac{(3 + x)(-1)}{x + 5}$

$= \dfrac{-x - 3}{x + 5}$

 Check Yourself 5

Write each fraction in simplest form.

(a) $\dfrac{3x - 9}{9 - x^2}$ **(b)** $\dfrac{x^2 - 6x - 27}{81 - x^2}$

 Check Yourself ANSWERS

1. (a) $\dfrac{5}{11}$; **(b)** $\dfrac{x^3}{3}$; **(c)** $\dfrac{2y^2}{3x^2}$; **(d)** $\dfrac{1}{2mn^2}$; **(e)** $6ab^2$ **2. (a)** $\dfrac{5}{x + 3}$; **(b)** $\dfrac{a - 3}{3a}$;

(c) $\dfrac{x + 5}{x + 1}$; **(d)** $\dfrac{5(p - 3)}{(p + 2)(p - 2)}$ **3. (a)** $-\dfrac{2x^2}{y}$; **(b)** $\dfrac{4a^2}{3b^3}$

4. (a) $\dfrac{x - 2}{3x^2 - 1}$; **(b)** $\dfrac{x + 3}{x - 9}$ **5. (a)** $-\dfrac{3}{x + 3}$; **(b)** $\dfrac{-x - 3}{x + 9}$

Reading Your Text

The following fill-in-the-blank exercises are designed to ensure that you understand some of the key vocabulary used in this section.

SECTION 5.1

(a) Fractions that name the same number are called _____ fractions.

(b) The fundamental principle allows us to multiply the numerator and denominator of a fraction by the same _____ polynomial.

(c) When the numerator and denominator of a fraction have no common factors other than 1, it is said to be in _____ form.

(d) The quotient of a positive number and a negative number is always _____.

5.1 exercises

Name _____

Section _____ Date _____

Answers

1. _____   2. _____

3. _____   4. _____

5. _____   6. _____

7. _____   8. _____

9. _____   10. _____

11. _____   12. _____

13. _____   14. _____

15. _____   16. _____

17. _____

18. _____

19. _____

20. _____

21. _____

22. _____

< Objective 1 >

Write each fraction in simplest form.

1. $\dfrac{16}{24}$

2. $\dfrac{56}{64}$

3. $\dfrac{80}{180}$

4. $\dfrac{18}{30}$

5. $\dfrac{4x^5}{6x^2}$

6. $\dfrac{10x^2}{15x^4}$

7. $\dfrac{9x^3}{27x^6}$

8. $\dfrac{25w^6}{20w^2}$

9. $\dfrac{10a^2b^5}{25ab^2}$

10. $\dfrac{18x^4y^3}{24x^2y^3}$

11. $\dfrac{42x^3y}{14xy^3}$

12. $\dfrac{18pq}{45p^2q^2}$

13. $\dfrac{2xyw^2}{6x^2y^3w^3}$ > Videos

14. $\dfrac{3c^2d^2}{6bc^3d^3}$

15. $\dfrac{10x^5y^5}{2x^3y^4}$

16. $\dfrac{3bc^6d^3}{bc^3d}$

17. $\dfrac{-4m^3n}{6mn^2}$ > Videos

18. $\dfrac{-15x^3y^3}{-20xy^4}$

19. $\dfrac{-8ab^3}{-16a^3b}$

20. $\dfrac{14x^2y}{-21xy^4}$

21. $\dfrac{8r^2s^3t}{-16rs^4t^3}$

22. $\dfrac{-10a^3b^2c^3}{15ab^4c}$

Complete each of the following statements with **never**, **sometimes**, or **always**.

23. The quotient of two negative values is _____ negative.

24. The expression $\dfrac{x + 2}{x + 1}$ is _____ equal to zero.

25. The expression $\dfrac{a - b}{b - a}$ is _____ equal to -1 when $a \neq b$.

26. The quotient of a positive value and a negative value is _____ negative.

< Objective 2 >

Write each expression in simplest form.

27. $\dfrac{3x + 18}{5x + 30}$

28. $\dfrac{4x - 28}{5x - 35}$

29. $\dfrac{6a - 24}{a^2 - 16}$

30. $\dfrac{5x - 5}{x^2 - 4}$

31. $\dfrac{x^2 + 3x + 2}{5x + 10}$ > Videos

32. $\dfrac{4w^2 - 20w}{w^2 - 2w - 15}$

33. $\dfrac{2m^2 + 3m - 5}{2m^2 + 11m + 15}$

34. $\dfrac{6x^2 - x - 2}{3x^2 - 5x + 2}$

35. $\dfrac{p^2 + 2pq - 15q^2}{p^2 - 25q^2}$

36. $\dfrac{4r^2 - 25s^2}{2r^2 + 3rs - 20s^2}$

37. $\dfrac{y - 7}{7 - y}$ > Videos

38. $\dfrac{5 - y}{y - 5}$

39. $\dfrac{2x - 10}{25 - x^2}$

40. $\dfrac{3a - 12}{16 - a^2}$

41. $\dfrac{25 - a^2}{a^2 + a - 30}$

42. $\dfrac{2x^2 - 7x + 3}{9 - x^2}$

43. $\dfrac{x^2 + xy - 6y^2}{4y^2 - x^2}$

44. $\dfrac{16z^2 - w^2}{2w^2 - 5wz - 12z^2}$

45. $\dfrac{x^2 + 4x + 4}{x + 2}$

46. $\dfrac{4x^2 + 12x + 9}{2x + 3}$

Answers

23. _____ 24. _____

25. _____ 26. _____

27. _____ 28. _____

29. _____

30. _____

31. _____

32. _____

33. _____

34. _____

35. _____

36. _____

37. _____

38. _____

39. _____

40. _____

41. _____

42. _____

43. _____

44. _____

45. _____

46. _____

Answers

47. _____

48. _____

49. _____

50. _____

51. _____

52. _____

53. _____

54. _____

55. _____

Basic Skills | **Advanced Skills** | Vocational-Technical Applications | Calculator/Computer | Above and Beyond | Getting Ready

47. $\dfrac{xy - 2y + 4x - 8}{2y + 6 - xy - 3x}$ > Videos   **48.** $\dfrac{ab - 3a + 5b - 15}{15 + 3a^2 - 5b - a^2b}$

49. GEOMETRY The area of the rectangle is represented by $6x^2 + 19x + 10$. What is the length?

$3x + 2$

50. GEOMETRY The volume of the box is represented by $(x^2 + 5x + 6)(x + 5)$. Find the polynomial that represents the area of the bottom of the box.

$x + 2$

Basic Skills | Advanced Skills | Vocational-Technical Applications | Calculator/Computer | **Above and Beyond** | Getting Ready

51. To work with rational expressions correctly, it is important to understand the difference between a *factor* and a *term* of an expression. In your own words, write definitions for both, explaining the difference between the two.

52. Give some examples of terms and factors in rational expressions and explain how both are affected when a fraction is simplified.

53. Show how the following rational expression can be simplified:

$$\frac{x^2 - 9}{4x + 12}$$

Note that your simplified fraction is equivalent to the given fraction. Are there other rational expressions equivalent to this one? Write another rational expression that you think is equivalent to this one. Exchange papers with another student. Do you agree that the other student's fraction is equivalent to yours? Why or why not?

54. Explain the reasoning involved in each step when simplifying the fraction $\dfrac{42}{56}$.

55. Describe why $\dfrac{3}{5}$ and $\dfrac{27}{45}$ are *equivalent fractions*.

Basic Skills | Advanced Skills | Vocational-Technical Applications | Calculator/Computer | Above and Beyond | **Getting Ready**
▲

Perform the indicated operations.

(a) $\dfrac{2}{3} \cdot \dfrac{4}{5}$

(b) $\dfrac{5}{6} \cdot \dfrac{4}{11}$

(c) $\dfrac{4}{7} \div \dfrac{8}{5}$

(d) $\dfrac{1}{6} \div \dfrac{7}{9}$

(e) $\dfrac{5}{8} \cdot \dfrac{16}{15}$

(f) $\dfrac{15}{21} \div \dfrac{10}{7}$

(g) $\dfrac{15}{8} \cdot \dfrac{24}{25}$

(h) $\dfrac{28}{16} \div \dfrac{21}{20}$

Answers

1. $\dfrac{2}{3}$ **3.** $\dfrac{4}{9}$ **5.** $\dfrac{2x^3}{3}$ **7.** $\dfrac{1}{3x^3}$ **9.** $\dfrac{2ab^3}{5}$ **11.** $\dfrac{3x^2}{y^2}$

13. $\dfrac{1}{3xy^2w}$ **15.** $5x^2y$ **17.** $-\dfrac{2m^2}{3n}$ **19.** $\dfrac{b^2}{2a^2}$ **21.** $-\dfrac{r}{2st^2}$

23. never   **25.** always   **27.** $\dfrac{3}{5}$ **29.** $\dfrac{6}{a+4}$ **31.** $\dfrac{x+1}{5}$

33. $\dfrac{m-1}{m+3}$ **35.** $\dfrac{p-3q}{p-5q}$ **37.** -1 **39.** $-\dfrac{2}{x+5}$

41. $\dfrac{-a-5}{a+6} = -\dfrac{a+5}{a+6}$ **43.** $\dfrac{-x-3y}{2y+x} = -\dfrac{x+3y}{2y+x}$ **45.** $x+2$

47. $-\dfrac{(y+4)}{y+3}$ **49.** $2x+5$ **51.** Above and Beyond

53. Above and Beyond   **55.** Above and Beyond

(a) $\dfrac{8}{15}$ **(b)** $\dfrac{10}{33}$ **(c)** $\dfrac{5}{14}$ **(d)** $\dfrac{3}{14}$ **(e)** $\dfrac{2}{3}$ **(f)** $\dfrac{1}{2}$

(g) $\dfrac{9}{5}$ **(h)** $\dfrac{5}{3}$

Answers

(a) _____

(b) _____

(c) _____

(d) _____

(e) _____

(f) _____

(g) _____

(h) _____

5.2

Multiplying and Dividing Rational Expressions

< 5.2 Objectives >

1 > Write the product of two rational expressions in simplest form

2 > Write the quotient of two rational expressions in simplest form

In arithmetic, you found the product of two fractions by multiplying the numerators and the denominators. For example,

$$\frac{2}{5} \cdot \frac{3}{7} = \frac{2 \cdot 3}{5 \cdot 7} = \frac{6}{35}$$

In symbols, we have

Property

Multiplying Rational Expressions	$\dfrac{P}{Q} \cdot \dfrac{R}{S} = \dfrac{PR}{QS}$ when $Q \neq 0$ and $S \neq 0$ P, Q, R, and S again represent polynomials.

NOTE

Divide by the common factors of 3 and 4. The alternative is to multiply *first:*

$$\frac{3}{8} \cdot \frac{4}{9} = \frac{12}{72}$$

and then use the GCF to reduce to lowest terms

$$\frac{12}{72} = \frac{1}{6}$$

It is easier to divide the numerator and denominator by any common factors *before* multiplying. Consider the following.

$$\frac{3}{8} \cdot \frac{4}{9} = \frac{\overset{1}{\cancel{3}} \cdot \overset{1}{\cancel{4}}}{\underset{2}{\cancel{8}} \cdot \underset{3}{\cancel{9}}} = \frac{1}{6}$$

In algebra, we multiply fractions in exactly the same way.

Step by Step

To Multiply Rational Expressions	Step 1	Factor the numerators and denominators.
	Step 2	Write the product of the factors of the numerators over the product of the factors of the denominators.
	Step 3	Divide the numerator and denominator by any common factors.

Example 1 illustrates this process.

Example 1   **Multiplying Rational Expressions**

< Objective 1 >

Multiply the following fractions.

NOTES

Divide by the common factors of 5, x^2, and y.

Divide by the common factors of 3, x, and $x - 3$.

RECALL

$$\frac{2 - x}{x - 2} = \frac{-(x - 2)}{x - 2} = -1$$

NOTE

Divide by the common factors of $x - 4$, x, and 3.

(a) $\dfrac{2x^3}{5y^2} \cdot \dfrac{10y}{3x^2} = \dfrac{2x^3 \cdot 10y}{5y^2 \cdot 3x^2} = \dfrac{20x^3y}{15x^2y^2} = \dfrac{4x}{3y}$

(b) $\dfrac{x}{x^2 - 3x} \cdot \dfrac{6x - 18}{9x} = \dfrac{x}{x(x - 3)} \cdot \dfrac{6(x - 3)}{9x}$

      Factor

$$= \frac{\overset{1}{\cancel{x}} \cdot \overset{2}{\cancel{6}}\overset{1}{(\cancel{x - 3})}}{\underset{1}{\cancel{x}}\underset{1}{(\cancel{x - 3})} \cdot \underset{3}{\cancel{9}}x} = \frac{2}{3x}$$

(c) $\dfrac{4}{x^2 - 2x} \cdot \dfrac{10 - 5x}{8} = \dfrac{4}{x(x - 2)} \cdot \dfrac{5(2 - x)}{8}$

$$= \frac{\overset{1}{\cancel{4}} \cdot 5\overset{-1}{\cancel{(2 - x)}}}{x\underset{1}{\cancel{(x - 2)}} \cdot \underset{2}{\cancel{8}}} = \frac{-5}{2x}$$

(d) $\dfrac{x^2 - 2x - 8}{3x^2} \cdot \dfrac{6x}{3x - 12} = \dfrac{(x - 4)(x + 2) \cdot 6x}{3x^2 \cdot 3(x - 4)}$

$$= \frac{\cancel{(x - 4)}(x + 2) \cdot \overset{2}{\cancel{6}}x}{\underset{x}{3x^2} \cdot 3\underset{1}{\cancel{(x - 4)}}}$$

$$= \frac{2(x + 2)}{3x}$$

(e) $\dfrac{x^2 - y^2}{5x - 5y} \cdot \dfrac{10xy}{x^2 + 2xy + y^2} = \dfrac{(x - y)(x + y) \cdot 10xy}{5(x - y) \cdot (x + y)(x + y)}$

$$= \frac{\overset{1}{\cancel{(x - y)}}\overset{1}{\cancel{(x + y)}} \cdot \overset{2}{\cancel{10}}xy}{\underset{1}{\cancel{5}}\underset{1}{\cancel{(x - y)}} \cdot \underset{1}{\cancel{(x + y)}}(x + y)}$$

$$= \frac{2xy}{x + y}$$

Check Yourself 1

Multiply the following fractions.

(a) $\dfrac{3x^2}{5y^2} \cdot \dfrac{10y^5}{15x^3}$ **(b)** $\dfrac{5x + 15}{x} \cdot \dfrac{2x^2}{x^2 + 3x}$ **(c)** $\dfrac{x}{2x - 6} \cdot \dfrac{3x - x^2}{2}$

(d) $\dfrac{3x - 15}{6x^2} \cdot \dfrac{2x}{x^2 - 25}$ **(e)** $\dfrac{x^2 - 5x - 14}{4x^2} \cdot \dfrac{8x}{x^2 - 49}$

RECALL

$\dfrac{6}{5}$ is the reciprocal of $\dfrac{5}{6}$.

You can also use your experience from arithmetic in dividing fractions. Recall that, to divide fractions, we *invert the divisor* (the *second* fraction) and multiply. For example,

$$\frac{2}{3} \div \frac{5}{6} = \frac{2}{3} \cdot \frac{6}{5} = \frac{2 \cdot 6}{3 \cdot 5} = \frac{12}{15} = \frac{4}{5}$$

In symbols, we have

Property

Dividing Rational Expressions

$$\frac{P}{Q} \div \frac{R}{S} = \frac{P}{Q} \cdot \frac{S}{R} = \frac{PS}{QR}$$

Once more P, Q, R, and S are polynomials.

when $Q \neq 0$, $R \neq 0$, and $S \neq 0$.

We divide rational expressions in exactly the same way.

Step by Step

To Divide Rational Expressions

Step 1   Invert the divisor and change the operation to multiplication.
Step 2   Proceed, using the steps for multiplying rational expressions.

Example 2 illustrates this approach.

Example 2   **Dividing Rational Expressions**

< Objective 2 >

Divide the following fractions.

(a) $\dfrac{6}{x^2} \div \dfrac{9}{x^3} = \dfrac{6}{x^2} \cdot \dfrac{x^3}{9}$ Invert the divisor and multiply.

$$= \frac{\overset{2}{\cancel{6}}\,\overset{x}{\cancel{x^3}}}{\underset{3}{\cancel{9}}\,\underset{1}{\cancel{x^2}}}$$

No simplification can be done until the divisor is inverted. Then divide by the common factors of 3 and x^2.

$$= \frac{2x}{3}$$

(b) $\dfrac{3x^2y}{8xy^3} \div \dfrac{9x^3}{4y^4} = \dfrac{3x^2y}{8xy^3} \cdot \dfrac{4y^4}{9x^3}$

$$= \frac{y^2}{6x^2}$$

NOTE

Factor all numerators and denominators *before* dividing out any common factors.

(c) $\dfrac{2x + 4y}{9x - 18y} \div \dfrac{4x + 8y}{3x - 6y} = \dfrac{2x + 4y}{9x - 18y} \cdot \dfrac{3x - 6y}{4x + 8y}$

$$= \frac{\overset{1}{\cancel{2}}\overset{1}{\cancel{(x + 2y)}} \cdot \overset{1}{\cancel{3}}\overset{1}{\cancel{(x - 2y)}}}{\underset{3}{\cancel{9}}\underset{1}{\cancel{(x - 2y)}} \cdot \underset{2}{\cancel{4}}\underset{1}{\cancel{(x + 2y)}}}$$

$$= \frac{1}{6}$$

(d) $\dfrac{x^2 - x - 6}{2x - 6} \div \dfrac{x^2 - 4}{4x^2} = \dfrac{x^2 - x - 6}{2x - 6} \cdot \dfrac{4x^2}{x^2 - 4}$

$$= \dfrac{\overset{1}{\cancel{(x-3)}}\overset{1}{\cancel{(x+2)}} \cdot \overset{2}{\cancel{4}}x^2}{\underset{1}{\cancel{2}}\underset{1}{\cancel{(x-3)}} \cdot \underset{1}{\cancel{(x+2)}}(x - 2)}$$

$$= \dfrac{2x^2}{x - 2}$$

Check Yourself 2

Divide the following fractions.

(a) $\dfrac{4}{x^5} \div \dfrac{12}{x^3}$

(b) $\dfrac{5xy^2}{7x^3y} \div \dfrac{10y^2}{14x^3}$

(c) $\dfrac{3x - 9y}{2x + 10y} \div \dfrac{x^2 - 3xy}{4x + 20y}$

(d) $\dfrac{x^2 - 9}{4x} \div \dfrac{x^2 - 2x - 15}{2x - 10}$

Check Yourself ANSWERS

1. (a) $\dfrac{2y^3}{5x}$; **(b)** 10; **(c)** $\dfrac{-x^2}{4}$; **(d)** $\dfrac{1}{x(x + 5)}$; **(e)** $\dfrac{2(x + 2)}{x(x + 7)}$

2. (a) $\dfrac{1}{3x^2}$; **(b)** $\dfrac{x}{y}$; **(c)** $\dfrac{6}{x}$; **(d)** $\dfrac{x - 3}{2x}$

Reading Your Text

The following fill-in-the-blank exercises are designed to ensure that you understand some of the key vocabulary used in this section.

SECTION 5.2

(a) In arithmetic, we find the product of two fractions by _____ the numerators and the denominators.

(b) The first step when multiplying rational expressions is to _____ the numerators and the denominators.

(c) When dividing two rational expressions, _____ the divisor and multiply.

(d) When dividing rational expressions, the divisor cannot equal _____.

Name _____

Section _____ Date _____

Answers

1. _____ 2. _____

3. _____ 4. _____

5. _____ 6. _____

7. _____ 8. _____

9. _____ 10. _____

11. _____ 12. _____

13. _____ 14. _____

15. _____ 16. _____

17. _____ 18. _____

19. _____ 20. _____

21. _____ 22. _____

23. _____ 24. _____

25. _____ 26. _____

< Objective 1 >

Multiply the following fractions.

1. $\dfrac{3}{7} \cdot \dfrac{14}{27}$

2. $\dfrac{9}{20} \cdot \dfrac{5}{36}$

3. $\dfrac{x}{2} \cdot \dfrac{y}{6}$

4. $\dfrac{w}{2} \cdot \dfrac{5}{14}$

5. $\dfrac{3a}{2} \cdot \dfrac{4}{a^2}$

6. $\dfrac{5x^3}{3x} \cdot \dfrac{9}{20x}$

7. $\dfrac{3x^3y}{10xy^3} \cdot \dfrac{5xy^2}{9xy^2}$

8. $\dfrac{8xy^5}{5x^3y^2} \cdot \dfrac{15y^2}{16xy^3}$

9. $\dfrac{3a^2b^3}{2ab} \cdot \dfrac{8a^3b}{6ab^3}$

10. $\dfrac{4x^4y^3}{8xy^3} \cdot \dfrac{12xy}{6x^3}$

11. $\dfrac{x^2y^3}{5a^3b} \cdot \dfrac{10ab^3}{3x^3}$

12. $\dfrac{9a^4b^{10}}{xy^3} \cdot \dfrac{2xy}{6b^7}$

13. $\dfrac{-4ab^2}{15a^3} \cdot \dfrac{25ab}{-16b^3}$

14. $\dfrac{-7xy^2}{12x^2y} \cdot \dfrac{24x^3y^5}{-21x^2y^7}$

15. $\dfrac{-3m^3n}{10mn^3} \cdot \dfrac{5mn^2}{-9mn^3}$ > Videos

16. $\dfrac{3x}{2x-6} \cdot \dfrac{x^2-3x}{6}$

17. $\dfrac{x^2+5x}{3x^2} \cdot \dfrac{10x}{5x+25}$

18. $\dfrac{x^2-3x-10}{5x} \cdot \dfrac{15x^2}{3x-15}$

19. $\dfrac{m^2-4m-21}{3m^2} \cdot \dfrac{m^2+7m}{m^2-49}$ > Videos

20. $\dfrac{2x^2-x-3}{3x^2+7x+4} \cdot \dfrac{3x^2-11x-20}{4x^2-9}$

21. $\dfrac{4r^2-1}{2r^2-9r-5} \cdot \dfrac{3r^2-13r-10}{9r^2-4}$

22. $\dfrac{a^2+ab}{2a^2-ab-3b^2} \cdot \dfrac{4a^2-9b^2}{5a^2-4ab}$

23. $\dfrac{x^2-4y^2}{x^2-xy-6y^2} \cdot \dfrac{7x^2-21xy}{5x-10y}$

24. $\dfrac{a^2-9b^2}{a^2+ab-6b^2} \cdot \dfrac{6a^2-12ab}{7a-21b}$

25. $\dfrac{2x-6}{x^2+2x} \cdot \dfrac{3x}{3-x}$

26. $\dfrac{3x-15}{x^2+3x} \cdot \dfrac{4x}{5-x}$

< Objective 2 >

Divide the following fractions.

27. $\dfrac{5}{8} \div \dfrac{15}{16}$

28. $\dfrac{4}{9} \div \dfrac{12}{18}$

29. $\dfrac{5}{x^2} \div \dfrac{10}{x}$

30. $\dfrac{w^2}{3} \div \dfrac{w}{9}$

31. $\dfrac{4x^2y^2}{9x^3} \div \dfrac{8y^2}{27xy}$

32. $\dfrac{8x^3y}{27xy^3} \div \dfrac{16x^3y}{45y}$

33. $\dfrac{3x + 6}{8} \div \dfrac{5x + 10}{6}$

34. $\dfrac{x^2 - 2x}{4x} \div \dfrac{6x - 12}{8}$

35. $\dfrac{4a - 12}{5a + 15} \div \dfrac{8a^2}{a^2 + 3a}$

36. $\dfrac{6p - 18}{9p} \div \dfrac{3p - 9}{p^2 + 2p}$

Determine whether each of the following statements is **true** *or* **false.**

37. The product of three negative values is negative.

38. Order of operations states that we multiply and divide before applying powers.

39. Division by zero results in a quotient of zero.

40. A fraction can always be simplified if the expression in the numerator contains the denominator.

| Basic Skills | **Advanced Skills** | Vocational-Technical Applications | Calculator/Computer | Above and Beyond | Getting Ready |

Divide each of the following.

41. $\dfrac{x^2 + 2x - 8}{9x^2} \div \dfrac{x^2 - 16}{3x - 12}$ > Videos

42. $\dfrac{16x}{4x^2 - 16} \div \dfrac{4x - 24}{x^2 - 4x - 12}$

43. $\dfrac{x^2 - 9}{2x^2 - 6x} \div \dfrac{2x^2 + 5x - 3}{4x^2 - 1}$

44. $\dfrac{2m^2 - 5m - 7}{4m^2 - 9} \div \dfrac{5m^2 + 5m}{2m^2 + 3m}$

45. $\dfrac{a^2 - 9b^2}{4a^2 + 12ab} \div \dfrac{a^2 - ab - 6b^2}{12ab}$

46. $\dfrac{r^2 + 2rs - 15s^2}{r^3 + 5r^2s} \div \dfrac{r^2 - 9s^2}{5r^3}$

Answers

27. _____

28. _____

29. _____

30. _____

31. _____

32. _____

33. _____

34. _____

35. _____

36. _____

37. _____

38. _____

39. _____

40. _____

41. _____

42. _____

43. _____

44. _____

45. _____

46. _____

Answers

47. _____

48. _____

49. _____

50. _____

51. _____

52. _____

53. _____

54. _____

55. _____

56. _____

57. _____

47. $\dfrac{x^2 - 16y^2}{3x^2 - 12xy} \div (x^2 + 4xy)$

48. $\dfrac{p^2 - 4pq - 21q^2}{4p - 28q} \div (2p^2 + 6pq)$

49. $\dfrac{x - 7}{2x + 6} \div \dfrac{21 - 3x}{x^2 + 3x}$

50. $\dfrac{x - 4}{x^2 + 2x} \div \dfrac{16 - 4x}{3x + 6}$

Basic Skills | Advanced Skills | Vocational-Technical Applications | Calculator/Computer | **Above and Beyond** | Getting Ready

▲

Perform the indicated operations.

51. $\dfrac{x^2 + 5x}{3x - 6} \cdot \dfrac{x^2 - 4}{3x^2 + 15x} \cdot \dfrac{6x}{x^2 + 6x + 8}$

52. $\dfrac{m^2 - n^2}{m^2 - mn} \cdot \dfrac{6m}{2m^2 + mn - n^2} \cdot \dfrac{8m - 4n}{12m^2 + 12mn}$

53. $\dfrac{x^2 - 2x - 8}{2x - 8} \cdot \dfrac{x^2 + 5x}{x^2 + 5x + 6} \div \dfrac{x^2 + 2x - 15}{x^2 - 9}$

> Videos

54. $\dfrac{14x - 7}{x^2 + 3x - 4} \cdot \dfrac{x^2 + 6x + 8}{2x^2 + 5x - 3} \div \dfrac{x^2 + 2x}{x^2 + 2x - 3}$

55. **SCIENCE AND MEDICINE** Herbicides constitute $\dfrac{2}{3}$ of all pesticides used in the United States. Insecticides are $\dfrac{1}{4}$ of all pesticides used in the United States. The ratio of herbicides to insecticides used in the United States can be written $\dfrac{2}{3} \div \dfrac{1}{4}$. Write this ratio in simplest form.

56. **SCIENCE AND MEDICINE** Fungicides account for $\dfrac{1}{10}$ of the pesticides used in the United States. Insecticides account for $\dfrac{1}{4}$ of all the pesticides used in the United States. The ratio of fungicides to insecticides used in the United States can be written $\dfrac{1}{10} \div \dfrac{1}{4}$. Write this ratio in simplest form.

57. **SCIENCE AND MEDICINE** The ratio of insecticides to herbicides applied to wheat, soybeans, corn, and cotton can be expressed as $\dfrac{7}{10} \div \dfrac{4}{5}$. Simplify this ratio.

58. GEOMETRY Find the area of the rectangle shown.

$$\frac{2x-4}{x-1}$$

$$\frac{3x-2}{x-2}$$

Perform the indicated operations.

(a) $\dfrac{3}{10} + \dfrac{4}{10}$ **(b)** $\dfrac{5}{8} - \dfrac{4}{8}$ **(c)** $\dfrac{5}{12} - \dfrac{1}{12}$ **(d)** $\dfrac{7}{16} + \dfrac{3}{16}$

(e) $\dfrac{7}{20} + \dfrac{9}{20}$ **(f)** $\dfrac{13}{8} - \dfrac{5}{8}$ **(g)** $\dfrac{11}{6} - \dfrac{2}{6}$ **(h)** $\dfrac{5}{9} + \dfrac{7}{9}$

Answers

1. $\dfrac{2}{9}$ **3.** $\dfrac{xy}{12}$ **5.** $\dfrac{6}{a}$ **7.** $\dfrac{x^2}{6y^2}$ **9.** $2a^3$ **11.** $\dfrac{2y^3b^2}{3xa^2}$ **13.** $\dfrac{5}{12a}$

15. $\dfrac{m^2}{6n^3}$ **17.** $\dfrac{2}{3}$ **19.** $\dfrac{m+3}{3m}$ **21.** $\dfrac{2r-1}{3r-2}$ **23.** $\dfrac{7x}{5}$ **25.** $\dfrac{-6}{x+2}$

27. $\dfrac{2}{3}$ **29.** $\dfrac{1}{2x}$ **31.** $\dfrac{3y}{2}$ **33.** $\dfrac{9}{20}$ **35.** $\dfrac{a-3}{10a}$ **37.** True

39. False **41.** $\dfrac{x-2}{3x^2}$ **43.** $\dfrac{2x+1}{2x}$ **45.** $\dfrac{3b}{a+2b}$ **47.** $\dfrac{1}{3x^2}$

49. $\dfrac{-x}{6}$ **51.** $\dfrac{2x}{3(x+4)}$ **53.** $\dfrac{x}{2}$ **55.** $\dfrac{8}{3}$ **57.** $\dfrac{7}{8}$ **(a)** $\dfrac{7}{10}$

(b) $\dfrac{1}{8}$ **(c)** $\dfrac{1}{3}$ **(d)** $\dfrac{5}{8}$ **(e)** $\dfrac{4}{5}$ **(f)** 1 **(g)** $\dfrac{3}{2}$ **(h)** $\dfrac{4}{3}$

Answers

58. _____

(a) _____

(b) _____

(c) _____

(d) _____

(e) _____

(f) _____

(g) _____

(h) _____

5.3

Adding and Subtracting Like Rational Expressions

< 5.3 Objectives >

1 > Write the sum or difference of two rational expressions whose numerator and denominator are monomials

2 > Write the sum or difference of two rational expressions whose numerator and denominator are polynomials

You probably remember from arithmetic that **like fractions** are fractions that have the same denominator. The same is true in algebra.

$\dfrac{2}{5}, \dfrac{12}{5},$ and $\dfrac{4}{5}$ are like fractions.

$\dfrac{x}{3(x+y)}, \dfrac{y}{3(x+y)},$ and $\dfrac{z-5}{3(x+y)}$ are like fractions.

NOTE

The fractions have different denominators.

$\dfrac{3x}{2}, \dfrac{x}{4},$ and $\dfrac{3x}{8}$ are unlike fractions.

$\dfrac{3}{x}, \dfrac{2}{x^2},$ and $\dfrac{x+1}{x^3}$ are unlike fractions.

In arithmetic, the sum or difference of like fractions is found by adding or subtracting the numerators and writing the result over the common denominator. For example,

$$\frac{3}{11} + \frac{5}{11} = \frac{3+5}{11} = \frac{8}{11}$$

In symbols, we have

Property

To Add or Subtract Like Rational Expressions	$\dfrac{P}{R} + \dfrac{Q}{R} = \dfrac{P+Q}{R} \qquad R \neq 0$
	$\dfrac{P}{R} - \dfrac{Q}{R} = \dfrac{P-Q}{R} \qquad R \neq 0$

Adding or subtracting like rational expressions is just as straightforward. You can use the following steps.

Step by Step

To Add or Subtract Like Rational Expressions	Step 1	Add or subtract the numerators.
	Step 2	Write the sum or difference over the common denominator.
	Step 3	Write the resulting fraction in simplest form.

Example 1   **Adding and Subtracting Rational Expressions**

< Objective 1 >

Add or subtract as indicated. Express your results in simplest form.

Add the numerators.

(a) $\dfrac{2x}{15} + \dfrac{x}{15} = \dfrac{2x + x}{15}$

$= \dfrac{3x}{15} = \dfrac{x}{5}$

Simplify.

Subtract the numerators.

(b) $\dfrac{5y}{6} - \dfrac{y}{6} = \dfrac{5y - y}{6}$

$= \dfrac{4y}{6} = \dfrac{2y}{3}$

Simplify.

(c) $\dfrac{3}{x} + \dfrac{5}{x} = \dfrac{3 + 5}{x} = \dfrac{8}{x}$

(d) $\dfrac{9b}{a^2} - \dfrac{7b}{a^2} = \dfrac{9b - 7b}{a^2} = \dfrac{2b}{a^2}$

(e) $\dfrac{7}{2ab} - \dfrac{5}{2ab} = \dfrac{7 - 5}{2ab}$

$= \dfrac{2}{2ab}$

$= \dfrac{1}{ab}$

Check Yourself 1

Add or subtract as indicated.

(a) $\dfrac{3a}{10} + \dfrac{2a}{10}$ (b) $\dfrac{7b}{8} - \dfrac{3b}{8}$ (c) $\dfrac{4}{x} + \dfrac{3}{x}$ (d) $\dfrac{5}{3xy} - \dfrac{2}{3xy}$

If polynomials are involved in the numerators or denominators, the process is exactly the same.

Example 2   **Adding and Subtracting Rational Expressions**

< Objective 2 >

Add or subtract as indicated. Express your results in simplest form.

(a) $\dfrac{5}{x + 3} + \dfrac{2}{x + 3} = \dfrac{5 + 2}{x + 3} = \dfrac{7}{x + 3}$

(b) $\dfrac{4x}{x-4} - \dfrac{16}{x-4} = \dfrac{4x-16}{x-4}$

Factor and simplify.

$$= \dfrac{4\cancel{(x-4)}}{\cancel{x-4}} = 4$$

(c) $\dfrac{a-b}{3} + \dfrac{2a+b}{3} = \dfrac{(a-b)+(2a+b)}{3}$

$$= \dfrac{a-b+2a+b}{3}$$

$$= \dfrac{\cancel{3}a}{\cancel{3}} = a$$

Notice what happens if parentheses are not used for the second numerator:

$(3x+y)-x-3y$
$= 3x+y-x-3y$
$= 2x-2y$

We get a different (and wrong) result!

Be sure to enclose the second numerator in parentheses!

(d) $\dfrac{3x+y}{2x} - \dfrac{x-3y}{2x} = \dfrac{(3x+y)-(x-3y)}{2x}$

Change both signs.

$$= \dfrac{3x+y-x+3y}{2x}$$

$$= \dfrac{2x+4y}{2x}$$

$$= \dfrac{\cancel{2}(x+2y)}{\cancel{2}x}$$ Factor and divide by the common factor of 2.

$$= \dfrac{x+2y}{x}$$

(e) $\dfrac{3x-5}{x^2+x-2} - \dfrac{2x-4}{x^2+x-2} = \dfrac{(3x-5)-(2x-4)}{x^2+x-2}$ Put the second numerator in parentheses.

Change both signs.

$$= \dfrac{3x-5-2x+4}{x^2+x-2}$$

$$= \dfrac{x-1}{x^2+x-2}$$

$$= \dfrac{\cancel{(x-1)}}{(x+2)\cancel{(x-1)}}$$ Factor and divide by the common factor of $x-1$.

$$= \dfrac{1}{x+2}$$

(f) $\dfrac{2x + 7y}{x + 3y} - \dfrac{x + 4y}{x + 3y} = \dfrac{(2x + 7y) - (x + 4y)}{x + 3y}$

— Change both signs.

$$= \dfrac{2x + 7y - x - 4y}{x + 3y}$$

$$= \dfrac{x + 3y}{x + 3y} = 1$$

 Check Yourself 2

Add or subtract as indicated.

(a) $\dfrac{4}{x - 5} - \dfrac{2}{x - 5}$ (b) $\dfrac{3x}{x + 3} + \dfrac{9}{x + 3}$

(c) $\dfrac{5x - y}{3y} - \dfrac{2x - 4y}{3y}$ (d) $\dfrac{5x + 8}{x^2 - 2x - 15} - \dfrac{4x + 5}{x^2 - 2x - 15}$

 Check Yourself ANSWERS

1. (a) $\dfrac{a}{2}$; (b) $\dfrac{b}{2}$; (c) $\dfrac{7}{x}$; (d) $\dfrac{1}{xy}$ **2.** (a) $\dfrac{2}{x - 5}$; (b) 3; (c) $\dfrac{x + y}{y}$; (d) $\dfrac{1}{x - 5}$

Reading Your Text

The following fill-in-the-blank exercises are designed to ensure that you understand some of the key vocabulary used in this section.

SECTION 5.3

(a) Fractions with the same denominator are called _____ fractions.

(b) When adding rational expressions, the final step is to write the result in _____ form.

(c) When subtracting fractions, the second numerator is enclosed in _____ before subtracting.

(d) Rational expressions can be simplified if the numerator and denominator have a common _____.

5.3 exercises

Name _____

Section _____ Date _____

Answers

1. _____	2. _____
3. _____	4. _____
5. _____	6. _____
7. _____	8. _____
9. _____	10. _____
11. _____	12. _____
13. _____	14. _____
15. _____	16. _____
17. _____	18. _____
19. _____	20. _____
21. _____	22. _____
23. _____	24. _____
25. _____	26. _____

< Objectives 1–2 >

Add or subtract as indicated. Express your results in simplest form.

1. $\dfrac{7}{18} + \dfrac{5}{18}$

2. $\dfrac{5}{18} - \dfrac{2}{18}$

3. $\dfrac{13}{16} - \dfrac{9}{16}$

4. $\dfrac{5}{12} + \dfrac{11}{12}$

5. $\dfrac{x}{8} + \dfrac{3x}{8}$

6. $\dfrac{5y}{16} + \dfrac{7y}{16}$

7. $\dfrac{7a}{10} - \dfrac{3a}{10}$

8. $\dfrac{5x}{12} - \dfrac{x}{12}$

9. $\dfrac{5}{x} + \dfrac{3}{x}$

10. $\dfrac{9}{y} - \dfrac{3}{y}$

11. $\dfrac{8}{w} - \dfrac{2}{w}$

12. $\dfrac{7}{z} + \dfrac{9}{z}$

13. $\dfrac{2}{xy} + \dfrac{3}{xy}$

14. $\dfrac{8}{ab} + \dfrac{4}{ab}$

15. $\dfrac{2}{3cd} + \dfrac{4}{3cd}$

16. $\dfrac{5}{4cd} + \dfrac{11}{4cd}$

17. $\dfrac{7}{x-5} + \dfrac{9}{x-5}$

18. $\dfrac{11}{x+7} - \dfrac{4}{x+7}$

19. $\dfrac{2x}{x-2} - \dfrac{4}{x-2}$ > Videos

20. $\dfrac{7w}{w+3} + \dfrac{21}{w+3}$

21. $\dfrac{8p}{p+4} + \dfrac{32}{p+4}$

22. $\dfrac{5a}{a-3} - \dfrac{15}{a-3}$

23. $\dfrac{x^2}{x+4} + \dfrac{3x-4}{x+4}$ > Videos

24. $\dfrac{x^2}{x-3} - \dfrac{9}{x-3}$

25. $\dfrac{m^2}{m-5} - \dfrac{25}{m-5}$

26. $\dfrac{s^2}{s+3} + \dfrac{2s-3}{s+3}$

*Complete each of the following statements with **never, sometimes,** or **always.***

27. The sum of two negative values is _____ negative.

28. The sum of a negative value and a positive value is _____ negative.

29. The difference of two negative values is _____ negative.

30. The difference of two positive values is _____ negative.

| Basic Skills | **Advanced Skills** | Vocational-Technical Applications | Calculator/Computer | Above and Beyond | Getting Ready |

Add or subtract as indicated.

31. $\dfrac{4m + 7}{6m} - \dfrac{2m + 5}{6m}$ > Videos

32. $\dfrac{6x - y}{4y} - \dfrac{2x + 3y}{4y}$

33. $\dfrac{4w - 7}{w - 5} - \dfrac{2w + 3}{w - 5}$

34. $\dfrac{3b - 8}{b - 6} + \dfrac{b - 16}{b - 6}$

35. $\dfrac{x - 7}{x^2 - x - 6} + \dfrac{2x - 2}{x^2 - x - 6}$

36. $\dfrac{5a - 12}{a^2 - 8a + 15} - \dfrac{3a - 2}{a^2 - 8a + 15}$

37. $\dfrac{y^2}{2y + 8} + \dfrac{3y - 4}{2y + 8}$ > Videos

38. $\dfrac{x^2}{4x - 12} - \dfrac{9}{4x - 12}$

39. $\dfrac{7w}{w + 3} + \dfrac{21}{w + 3}$

40. $\dfrac{2x}{x - 3} - \dfrac{6}{x - 3}$

41. $\dfrac{x^2}{x^2 + x - 6} - \dfrac{6}{(x + 3)(x - 2)} + \dfrac{x}{(x^2 + x - 6)}$ > Videos

42. $\dfrac{-12}{x^2 + x - 12} + \dfrac{x^2}{(x + 4)(x - 3)} + \dfrac{x}{x^2 + x - 12}$

Answers

27. _____

28. _____

29. _____

30. _____

31. _____

32. _____

33. _____

34. _____

35. _____

36. _____

37. _____

38. _____

39. _____

40. _____

41. _____

42. _____

The Streeter/Hutchison Series in Mathematics   Beginning Algebra

Answers

43. _____

44. _____

(a) _____

(b) _____

(c) _____

(d) _____

(e) _____

(f) _____

(g) _____

(h) _____

43. GEOMETRY Find the perimeter of the given figure.

$$\frac{2x}{x+3}$$

$$\frac{6}{x+3}$$

44. GEOMETRY Find the perimeter of the given figure.

$$\frac{x}{2x-5}$$

$$\frac{8}{2x-5}$$

| Basic Skills | Advanced Skills | Vocational-Technical Applications | Calculator/Computer | Above and Beyond | **Getting Ready** |

(a) $\dfrac{3}{4} + \dfrac{1}{2}$ (b) $\dfrac{5}{6} - \dfrac{2}{3}$

(c) $\dfrac{7}{10} - \dfrac{3}{5}$ (d) $\dfrac{5}{8} + \dfrac{3}{4}$

(e) $\dfrac{5}{6} + \dfrac{3}{8}$ (f) $\dfrac{7}{8} - \dfrac{3}{5}$

(g) $\dfrac{9}{10} - \dfrac{2}{15}$ (h) $\dfrac{5}{12} + \dfrac{7}{18}$

Answers

1. $\dfrac{2}{3}$ **3.** $\dfrac{1}{4}$ **5.** $\dfrac{x}{2}$ **7.** $\dfrac{2a}{5}$ **9.** $\dfrac{8}{x}$ **11.** $\dfrac{6}{w}$ **13.** $\dfrac{5}{xy}$

15. $\dfrac{2}{cd}$ **17.** $\dfrac{16}{x-5}$ **19.** 2   **21.** 8   **23.** $x-1$ **25.** $m+5$

27. always   **29.** sometimes   **31.** $\dfrac{m+1}{3m}$ **33.** 2   **35.** $\dfrac{3}{x+2}$

37. $\dfrac{y-1}{2}$ **39.** 7   **41.** 1   **43.** 4   (a) $\dfrac{5}{4}$ (b) $\dfrac{1}{6}$

(c) $\dfrac{1}{10}$ (d) $\dfrac{11}{8}$ (e) $\dfrac{29}{24}$ (f) $\dfrac{11}{40}$ (g) $\dfrac{23}{30}$ (h) $\dfrac{29}{36}$

5.4

Adding and Subtracting Unlike Rational Expressions

< 5.4 Objectives >

1 > Write the sum of two unlike rational expressions in simplest form

2 > Write the difference of two unlike rational expressions in simplest form

Adding or subtracting **unlike rational expressions** (fractions that do not have the same denominator) requires a bit more work than adding or subtracting the like rational expressions of Section 5.3. When the denominators are not the same, we must use the idea of the *least common denominator* (LCD). Each fraction is "built up" to an equivalent fraction having the LCD as a denominator. You can then add or subtract as before.

| Example 1 | Finding the LCD and Adding Fractions |

< Objective 1 >

Add $\dfrac{5}{9} + \dfrac{1}{6}$.

Step 1 To find the LCD, factor each denominator.

$9 = 3 \cdot 3$ ⟵ 3 appears twice.

$6 = 2 \cdot 3$

To form the LCD, include each factor the greatest number of times it appears in any single denominator. In this example, use one 2, because 2 appears only once in the factorization of 6. Use two 3s, because 3 appears twice in the factorization of 9. Thus the LCD for the fractions is $2 \cdot 3 \cdot 3 = 18$.

Step 2 "Build up" each fraction to an equivalent fraction with the LCD as the denominator. Do this by multiplying the numerator and denominator of the given fractions by the same number.

$$\frac{5}{9} = \frac{5 \cdot 2}{9 \cdot 2} = \frac{10}{18}$$

$$\frac{1}{6} = \frac{1 \cdot 3}{6 \cdot 3} = \frac{3}{18}$$

NOTE

Do you see that this uses the fundamental principle in the following form?

$$\frac{P}{Q} = \frac{PR}{QR}$$

Step 3 Add the fractions.

$$\frac{5}{9} + \frac{1}{6} = \frac{10}{18} + \frac{3}{18} = \frac{13}{18}$$

$\dfrac{13}{18}$ is in simplest form and so we are done!

Check Yourself 1

Add the fractions.

(a) $\dfrac{1}{6} + \dfrac{3}{8}$

(b) $\dfrac{3}{10} + \dfrac{4}{15}$

The process of finding the sum or difference is exactly the same in algebra as it is in arithmetic. We can summarize the steps with the following rule.

Step by Step

To Add or Subtract Unlike Rational Expressions	**Step 1**	Find the least common denominator of all the fractions.
	Step 2	Convert each fraction to an equivalent fraction with the LCD as a denominator.
	Step 3	Add or subtract the like fractions formed in step 2.
	Step 4	Write the sum or difference in simplest form.

Example 2   **Adding and Subtracting Unlike Rational Expressions**

< Objectives 1–2 >

(a) Add $\dfrac{3}{2x} + \dfrac{4}{x^2}$.

Step 1   Factor the denominators.

$2x = 2 \cdot x$
$x^2 = x \cdot x$

The LCD must contain the factors 2 and x. The factor x must appear *twice* because it appears twice as a factor in the second denominator.

The LCD is $2 \cdot x \cdot x$, or $2x^2$.

NOTE

Although the product of the denominators will be a common denominator, it is not necessarily the *least* common denominator (LCD).

Step 2

$\dfrac{3}{2x} = \dfrac{3 \cdot x}{2x \cdot x} = \dfrac{3x}{2x^2}$

$\dfrac{4}{x^2} = \dfrac{4 \cdot 2}{x^2 \cdot 2} = \dfrac{8}{2x^2}$

Step 3

$\dfrac{3}{2x} + \dfrac{4}{x^2} = \dfrac{3x}{2x^2} + \dfrac{8}{2x^2}$

$= \dfrac{3x + 8}{2x^2}$

The sum is in simplest form.

(b) Subtract $\dfrac{4}{3x^2} - \dfrac{3}{2x^3}$.

Step 1   Factor the denominators.

$3x^2 = 3 \cdot x \cdot x$

$2x^3 = 2 \cdot x \cdot x \cdot x$

The LCD must contain the factors 2, 3, and x. The LCD is

$2 \cdot 3 \cdot x \cdot x \cdot x$ or $6x^3$ The factor x must appear
3 times. Do you see why?

Step 2

$\dfrac{4}{3x^2} = \dfrac{4 \cdot 2x}{3x^2 \cdot 2x} = \dfrac{8x}{6x^3}$

$\dfrac{3}{2x^3} = \dfrac{3 \cdot 3}{2x^3 \cdot 3} = \dfrac{9}{6x^3}$

Step 3

$\dfrac{4}{3x^2} - \dfrac{3}{2x^3} = \dfrac{8x}{6x^3} - \dfrac{9}{6x^3}$

$= \dfrac{8x - 9}{6x^3}$

The difference is in simplest form.

Check Yourself 2

Add or subtract as indicated.

(a) $\dfrac{5}{x^2} + \dfrac{3}{x^3}$ **(b)** $\dfrac{3}{5x} - \dfrac{1}{4x^2}$

We can also add fractions with more than one variable in the denominator. Example 3 illustrates this type of sum.

Example 3   **Adding Unlike Rational Expressions**

Add $\dfrac{2}{3x^2y} + \dfrac{3}{4x^3}$.

Step 1   Factor the denominators.

$3x^2y = 3 \cdot x \cdot x \cdot y$

$4x^3 = 2 \cdot 2 \cdot x \cdot x \cdot x$

The LCD is $12x^3y$. Do you see why?

Step 2

$$\frac{2}{3x^2y} = \frac{2 \cdot 4x}{3x^2y \cdot 4x} = \frac{8x}{12x^3y}$$

$$\frac{3}{4x^3} = \frac{3 \cdot 3y}{4x^3 \cdot 3y} = \frac{9y}{12x^3y}$$

Step 3

$$\frac{2}{3x^2y} + \frac{3}{4x^3} = \frac{8x}{12x^3y} + \frac{9y}{12x^3y}$$

$$= \frac{8x + 9y}{12x^3y}$$

NOTE

The y in the numerator and that in the denominator cannot be divided out because y is not a factor of the numerator.

Check Yourself 3

Add.

$$\frac{2}{3x^2y} + \frac{1}{6xy^2}$$

Rational expressions with binomials in the denominator can also be added by taking the approach shown in Example 3. Example 4 illustrates this approach with binomials.

Example 4 | **Adding and Subtracting Unlike Rational Expressions**

(a) Add $\dfrac{5}{x} + \dfrac{2}{x-1}$.

Step 1 The LCD must have factors of x and $x - 1$. The LCD is $x(x - 1)$.

Step 2

$$\frac{5}{x} = \frac{5(x-1)}{x(x-1)}$$

$$\frac{2}{x-1} = \frac{2x}{(x-1)x} = \frac{2x}{x(x-1)}$$

Step 3

$$\frac{5}{x} + \frac{2}{x-1} = \frac{5(x-1)}{x(x-1)} + \frac{2x}{x(x-1)}$$

$$= \frac{5x - 5 + 2x}{x(x-1)}$$

$$= \frac{7x - 5}{x(x-1)}$$

The Streeter/Hutchison Series in Mathematics   Beginning Algebra

(b) Subtract $\dfrac{3}{x-2} - \dfrac{4}{x+2}$.

Step 1 The LCD must have factors of $x-2$ and $x+2$. The LCD is $(x-2)(x+2)$.

Step 2

$$\frac{3}{x-2} = \frac{3(x+2)}{(x-2)(x+2)}$$

$$\frac{4}{x+2} = \frac{4(x-2)}{(x+2)(x-2)}$$

Step 3

$$\frac{3}{x-2} - \frac{4}{x+2} = \frac{3(x+2) - 4(x-2)}{(x+2)(x-2)}$$

Note that the x-term becomes negative and the constant term becomes positive.

$$= \frac{3x+6-4x+8}{(x+2)(x-2)}$$

$$= \frac{-x+14}{(x+2)(x-2)}$$

Check Yourself 4

Add or subtract as indicated.

(a) $\dfrac{3}{x+2} + \dfrac{5}{x}$ **(b)** $\dfrac{4}{x+3} - \dfrac{2}{x-3}$

Example 5 shows how factoring must sometimes be used in forming the LCD.

Example 5 | **Adding and Subtracting Unlike Rational Expressions**

(a) Add $\dfrac{3}{2x-2} + \dfrac{5}{3x-3}$.

Step 1 Factor the denominators.

$$2x - 2 = 2(x-1)$$
$$3x - 3 = 3(x-1)$$

> **CAUTION**

$x-1$ is not used twice in forming the LCD.

The LCD must have factors of 2, 3, and $x-1$. The LCD is $2 \cdot 3(x-1)$, or $6(x-1)$.

Step 2

$$\frac{3}{2x-2} = \frac{3}{2(x-1)} = \frac{3 \cdot 3}{2(x-1) \cdot 3} = \frac{9}{6(x-1)}$$

$$\frac{5}{3x-3} = \frac{5}{3(x-1)} = \frac{5 \cdot 2}{3(x-1) \cdot 2} = \frac{10}{6(x-1)}$$

Step 3

$$\frac{3}{2x-2} + \frac{5}{3x-3} = \frac{9}{6(x-1)} + \frac{10}{6(x-1)}$$

$$= \frac{9+10}{6(x-1)}$$

$$= \frac{19}{6(x-1)}$$

(b) Subtract $\dfrac{3}{2x-4} - \dfrac{6}{x^2-4}$.

Step 1   Factor the denominators.

$$2x - 4 = 2(x - 2)$$
$$x^2 - 4 = (x + 2)(x - 2)$$

The LCD must have factors of 2, $x - 2$, and $x + 2$. The LCD is $2(x - 2)(x + 2)$.

NOTES

Multiply numerator and denominator by $x + 2$.

Multiply numerator and denominator by 2.

Step 2

$$\frac{3}{2x-4} = \frac{3}{2(x-2)} = \frac{3(x+2)}{2(x-2)(x+2)}$$

$$\frac{6}{x^2-4} = \frac{6}{(x+2)(x-2)} = \frac{6 \cdot 2}{2(x+2)(x-2)}$$

$$= \frac{12}{2(x+2)(x-2)}$$

Step 3

$$\frac{3}{2x-4} - \frac{6}{x^2-4} = \frac{3(x+2) - 12}{2(x-2)(x+2)}$$

$$= \frac{3x+6-12}{2(x-2)(x+2)}$$

$$= \frac{3x-6}{2(x-2)(x+2)}$$

NOTE

Remove the parentheses and combine like terms in the numerator.

Step 4   Simplify the difference.

$$\frac{3x-6}{2(x-2)(x+2)} = \frac{\overset{1}{3(x-2)}}{2\underset{1}{(x-2)}(x+2)}$$

$$= \frac{3}{2(x+2)}$$

NOTE

Factor the numerator and divide by the common factor, $x - 2$.

(c) Subtract $\dfrac{5}{x^2-1} - \dfrac{2}{x^2+2x+1}$.

Step 1   Factor the denominators.

$$x^2 - 1 = (x - 1)(x + 1)$$
$$x^2 + 2x + 1 = (x + 1)(x + 1)$$

The LCD is $(x - 1)(x + 1)(x + 1)$.

This factor is needed twice.

Step 2

$$\frac{5}{(x-1)(x+1)} = \frac{5(x+1)}{(x-1)(x+1)(x+1)}$$

$$\frac{2}{(x+1)(x+1)} = \frac{2(x-1)}{(x+1)(x+1)(x-1)}$$

Step 3

$$\frac{5}{x^2-1} - \frac{2}{x^2+2x+1} = \frac{5(x+1)-2(x-1)}{(x-1)(x+1)(x+1)}$$

$$= \frac{5x+5-2x+2}{(x-1)(x+1)(x+1)}$$

$$= \frac{3x+7}{(x-1)(x+1)(x+1)}$$

NOTE

Remove the parentheses and simplify in the numerator.

Check Yourself 5

Add or subtract as indicated.

(a) $\dfrac{5}{2x+2} + \dfrac{1}{5x+5}$ **(b)** $\dfrac{3}{x^2-9} - \dfrac{1}{2x-6}$

(c) $\dfrac{4}{x^2-x-2} - \dfrac{3}{x^2+4x+3}$

Recall from Section 5.1 that

$$a - b = -(b - a)$$

Let's see how this can be used in adding or subtracting rational expressions.

    | Example 6 | Adding Unlike Rational Expressions |

Add $\dfrac{4}{x-5} + \dfrac{2}{5-x}$.

Rather than try a denominator of $(x-5)(5-x)$, we rewrite one of the denominators.

$$\frac{4}{x-5} + \frac{2}{5-x} = \frac{4}{x-5} + \frac{2}{-(x-5)}$$

$$= \frac{4}{x-5} + \frac{-2}{x-5}$$

NOTE

Replace $5 - x$ with $-(x - 5)$. We now use the fact that

$$\frac{a}{-b} = \frac{-a}{b}$$

The LCD is now $x - 5$, and we can combine the rational expressions as

$$\frac{4-2}{x-5} = \frac{2}{x-5}$$

Check Yourself 6

Subtract the fractions.

$$\frac{3}{x-3} - \frac{1}{3-x}$$

Check Yourself ANSWERS

1. (a) $\dfrac{13}{24}$; **(b)** $\dfrac{17}{30}$ **2. (a)** $\dfrac{5x+3}{x^3}$; **(b)** $\dfrac{12x-5}{20x^2}$ **3.** $\dfrac{4y+x}{6x^2y^2}$

4. (a) $\dfrac{8x+10}{x(x+2)}$; **(b)** $\dfrac{2x-18}{(x+3)(x-3)}$ **5. (a)** $\dfrac{27}{10(x+1)}$; **(b)** $\dfrac{-1}{2(x+3)}$;

(c) $\dfrac{x+18}{(x+1)(x-2)(x+3)}$ **6.** $\dfrac{4}{x-3}$

Reading Your Text

The following fill-in-the-blank exercises are designed to ensure that you understand some of the key vocabulary used in this section.

SECTION 5.4

(a) Fractions that do not have the same denominator are called unlike _____ expressions.

(b) When adding unlike fractions, it is necessary to find a _____ denominator.

(c) The final step in subtracting fractions is to write the difference in _____ form.

(d) The expression $a - b$ is the _____ of the expression $b - a$.

5.4 exercises

< Objectives 1–2 >

Add or subtract as indicated. Express your result in simplest form.

1. $\dfrac{3}{7} + \dfrac{5}{6}$

2. $\dfrac{7}{12} - \dfrac{4}{9}$

3. $\dfrac{13}{25} - \dfrac{7}{20}$

4. $\dfrac{3}{5} + \dfrac{7}{9}$

5. $\dfrac{y}{4} + \dfrac{3y}{5}$

6. $\dfrac{5x}{6} - \dfrac{2x}{3}$

7. $\dfrac{7a}{3} - \dfrac{a}{7}$

8. $\dfrac{3m}{4} + \dfrac{m}{9}$

9. $\dfrac{3}{x} - \dfrac{4}{5}$

10. $\dfrac{5}{x} + \dfrac{2}{3}$

11. $\dfrac{5}{a} + \dfrac{a}{5}$

12. $\dfrac{y}{3} - \dfrac{3}{y}$

13. $\dfrac{5}{m} + \dfrac{3}{m^2}$

14. $\dfrac{4}{x^2} - \dfrac{3}{x}$

15. $\dfrac{2}{x^2} - \dfrac{5}{7x}$ > Videos

16. $\dfrac{7}{3w} + \dfrac{5}{w^3}$

17. $\dfrac{7}{9s} + \dfrac{5}{s^2}$

18. $\dfrac{11}{x^2} - \dfrac{5}{7x}$

19. $\dfrac{3}{4b^2} + \dfrac{5}{3b^3}$

20. $\dfrac{4}{5x^3} - \dfrac{3}{2x^2}$

MathZone

Boost your grade at mathzone.com!
> Practice Problems   > Self-Tests
> NetTutor   > e-Professors
                    > Videos

Name _____

Section _____ Date _____

Answers

1. _____   2. _____

3. _____   4. _____

5. _____   6. _____

7. _____   8. _____

9. _____   10. _____

11. _____   12. _____

13. _____   14. _____

15. _____   16. _____

17. _____   18. _____

19. _____   20. _____

Answers

21. _____

22. _____

23. _____

24. _____

25. _____

26. _____

27. _____

28. _____

29. _____

30. _____

31. _____

32. _____

33. _____

34. _____

35. _____

36. _____

37. _____

38. _____

21. $\dfrac{x}{x + 2} + \dfrac{2}{5}$

22. $\dfrac{3}{4} - \dfrac{a}{a - 1}$

23. $\dfrac{y}{y - 4} - \dfrac{3}{4}$

24. $\dfrac{m}{m + 3} + \dfrac{2}{3}$

25. $\dfrac{4}{x} + \dfrac{3}{x + 1}$

26. $\dfrac{2}{x} - \dfrac{1}{x - 2}$

27. $\dfrac{5}{a - 1} - \dfrac{2}{a}$

28. $\dfrac{4}{x + 2} + \dfrac{3}{x}$

29. $\dfrac{4}{2x - 3} + \dfrac{2}{3x}$

30. $\dfrac{7}{2y - 1} - \dfrac{3}{2y}$

31. $\dfrac{2}{x + 1} + \dfrac{3}{x + 3}$

32. $\dfrac{5}{x - 1} + \dfrac{2}{x + 2}$

33. $\dfrac{4}{y - 2} - \dfrac{1}{y + 1}$ › Videos

34. $\dfrac{5}{x + 4} - \dfrac{3}{x - 1}$

Determine whether each of the following statements is **true** *or* **false.**

35. The expression $a - b$ is the opposite of the expression $b - a$.

36. The expression $a - b$ is the opposite of the expression $a + b$.

37. You must find the greatest common factor in order to add unlike fractions.

38. To add two like fractions, add the denominators and place the sum under the common numerator.

Basic Skills | **Advanced Skills** | Vocational-Technical Applications | Calculator/Computer | Above and Beyond | Getting Ready

39. $\dfrac{x}{x+4} - \dfrac{2}{3x+12}$

40. $\dfrac{x}{x-3} + \dfrac{5}{2x-6}$

41. $\dfrac{4}{5x-10} - \dfrac{1}{3x-6}$

42. $\dfrac{2}{3w+3} + \dfrac{5}{2w+2}$

43. $\dfrac{7}{3c+6} - \dfrac{2c}{7c+14}$

44. $\dfrac{5}{3c-12} + \dfrac{4c}{5c-20}$

45. $\dfrac{y-1}{y+1} - \dfrac{y}{3y+3}$

46. $\dfrac{x+2}{x-2} - \dfrac{x}{3x-6}$

47. $\dfrac{3}{x^2-4} + \dfrac{2}{x+2}$

48. $\dfrac{4}{x-2} + \dfrac{3}{x^2-x-2}$

49. $\dfrac{3x}{x^2-3x+2} - \dfrac{1}{x-2}$

50. $\dfrac{a}{a^2-1} - \dfrac{4}{a+1}$

51. $\dfrac{2x}{x^2-5x+6} + \dfrac{4}{x-2}$ > Videos

52. $\dfrac{7a}{a^2+a-12} - \dfrac{4}{a+4}$

53. $\dfrac{2}{3x-3} - \dfrac{1}{4x+4}$

54. $\dfrac{2}{5w+10} - \dfrac{3}{2w-4}$

55. $\dfrac{4}{3a-9} - \dfrac{3}{2a+4}$

56. $\dfrac{2}{3b-6} + \dfrac{3}{4b+8}$

57. $\dfrac{5}{x^2-16} - \dfrac{3}{x^2-x-12}$

Answers

39. _____

40. _____

41. _____

42. _____

43. _____

44. _____

45. _____

46. _____

47. _____

48. _____

49. _____

50. _____

51. _____

52. _____

53. _____

54. _____

55. _____

56. _____

57. _____

Answers

58. _____

59. _____

60. _____

61. _____

62. _____

63. _____

64. _____

65. _____

66. _____

67. _____

68. _____

69. _____

70. _____

58. $\dfrac{3}{x^2 + 4x + 3} - \dfrac{1}{x^2 - 9}$

59. $\dfrac{2}{y^2 + y - 6} + \dfrac{3y}{y^2 - 2y - 15}$

 > Videos

60. $\dfrac{2a}{a^2 - a - 12} - \dfrac{3}{a^2 - 2a - 8}$

61. $\dfrac{6x}{x^2 - 9} - \dfrac{5x}{x^2 + x - 6}$

62. $\dfrac{4y}{y^2 + 6y + 5} + \dfrac{2y}{y^2 - 1}$

63. $\dfrac{3}{a - 7} + \dfrac{2}{7 - a}$

64. $\dfrac{5}{x - 5} - \dfrac{3}{5 - x}$

65. $\dfrac{2x}{2x - 3} - \dfrac{1}{3 - 2x}$

 > Videos

66. $\dfrac{9m}{3m - 1} + \dfrac{3}{1 - 3m}$

Basic Skills | Advanced Skills | Vocational-Technical Applications | Calculator/Computer | **Above and Beyond** | Getting Ready

Add or subtract as indicated.

67. $\dfrac{1}{a - 3} - \dfrac{1}{a + 3} + \dfrac{2a}{a^2 - 9}$

68. $\dfrac{1}{p + 1} + \dfrac{1}{p - 3} - \dfrac{4}{p^2 - 2p - 3}$

69. $\dfrac{2x^2 + 3x}{x^2 - 2x - 63} + \dfrac{7 - x}{x^2 - 2x - 63} - \dfrac{x^2 - 3x + 21}{x^2 - 2x - 63}$

70. $-\dfrac{3 - 2x^2}{x^2 - 9x + 20} - \dfrac{4x^2 + 2x + 1}{x^2 - 9x + 20} + \dfrac{2x^2 + 3x}{x^2 - 9x + 20}$

71. NUMBER PROBLEM Use a rational expression to represent the sum of the reciprocals of two consecutive even integers.

72. NUMBER PROBLEM One number is two less than another. Use a rational expression to represent the sum of the reciprocals of the two numbers.

73. GEOMETRY Refer to the rectangle in the figure. Find an expression that represents its perimeter.

74. GEOMETRY Refer to the triangle in the figure. Find an expression that represents its perimeter.

| Basic Skills | Advanced Skills | Vocational-Technical Applications | Calculator/Computer | Above and Beyond | **Getting Ready** |

Solve each of the following equations.

(a) $x + 7 = 10$ **(b)** $x - 8 = -3$

(c) $4x - 5 = 2$ **(d)** $5x + 8 = 4$

(e) $4(x - 2) - 5 = 7$ **(f)** $4(2x + 1) - 3 = -23$

(g) $3(5x - 2) - 4(4x + 1) = -20$ **(h)** $5(3x + 2) - 3(4x - 3) = -2$

Answers

1. $\dfrac{53}{42}$ **3.** $\dfrac{17}{100}$ **5.** $\dfrac{17y}{20}$ **7.** $\dfrac{46a}{21}$ **9.** $\dfrac{15 - 4x}{5x}$ **11.** $\dfrac{25 + a^2}{5a}$

13. $\dfrac{5m + 3}{m^2}$ **15.** $\dfrac{14 - 5x}{7x^2}$ **17.** $\dfrac{7s + 45}{9s^2}$ **19.** $\dfrac{9b + 20}{12b^3}$

21. $\dfrac{7x + 4}{5(x + 2)}$ **23.** $\dfrac{y + 12}{4(y - 4)}$ **25.** $\dfrac{7x + 4}{x(x + 1)}$ **27.** $\dfrac{3a + 2}{a(a - 1)}$

29. $\dfrac{2(8x - 3)}{3x(2x - 3)}$ **31.** $\dfrac{5x + 9}{(x + 1)(x + 3)}$ **33.** $\dfrac{3(y + 2)}{(y - 2)(y + 1)}$

35. True   **37.** False   **39.** $\dfrac{3x - 2}{3(x + 4)}$ **41.** $\dfrac{7}{15(x - 2)}$

43. $\dfrac{49 - 6c}{21(c + 2)}$ **45.** $\dfrac{2y - 3}{3(y + 1)}$ **47.** $\dfrac{2x - 1}{(x - 2)(x + 2)}$

49. $\dfrac{2x + 1}{(x - 1)(x - 2)}$ **51.** $\dfrac{6}{x - 3}$ **53.** $\dfrac{5x + 11}{12(x - 1)(x + 1)}$

55. $\dfrac{-a + 43}{6(a - 3)(a + 2)}$ **57.** $\dfrac{2x + 3}{(x + 4)(x - 4)(x + 3)}$

59. $\dfrac{3y^2 - 4y - 10}{(y + 3)(y - 2)(y - 5)}$ **61.** $\dfrac{x}{(x - 3)(x - 2)}$ **63.** $\dfrac{1}{a - 7}$

65. $\dfrac{2x + 1}{2x - 3}$ **67.** $\dfrac{2}{a - 3}$ **69.** $\dfrac{x - 2}{x - 9}$ **71.** $\dfrac{2x + 2}{x(x + 2)}$

73. $\dfrac{2(6x^2 + 5x + 21)}{5(3x + 1)}$ **(a)** 3   **(b)** 5   **(c)** $\dfrac{7}{4}$ **(d)** $\dfrac{-4}{5}$

(e) 5   **(f)** -3 **(g)** 10   **(h)** -7

Complex Rational Expressions

< 5.5 Objectives >

1 > Simplify a complex arithmetic fraction

2 > Simplify a complex rational expression

Recall, from arithmetic, the manner in which you were taught to divide fractions. The rule was referred to as invert-and-multiply. We will see why this rule works.

$$\frac{3}{5} \div \frac{2}{3}$$

We can write

$$\frac{3}{5} \div \frac{2}{3} = \frac{\dfrac{3}{5}}{\dfrac{2}{3}} = \frac{\dfrac{3}{5} \cdot \dfrac{3}{2}}{\dfrac{2}{3} \cdot \dfrac{3}{2}}$$ We are multiplying by 1.

Interpret the division as a fraction.

$$= \frac{\dfrac{3}{5} \cdot \dfrac{3}{2}}{1}$$

$$\frac{2}{3} \cdot \frac{3}{2} = 1$$

$$= \frac{3}{5} \cdot \frac{3}{2}$$

We then have

$$\frac{3}{5} \div \frac{2}{3} = \frac{3}{5} \cdot \frac{3}{2} = \frac{9}{10}$$

By comparing these expressions, you should see the rule for dividing fractions. Invert the fraction that follows the division symbol and multiply.

A fraction that has a fraction in its numerator, in its denominator, or in both is called a **complex fraction.** For example, the following are complex fractions.

$$\frac{\dfrac{5}{6}}{\dfrac{3}{4}}, \qquad \frac{\dfrac{4}{x}}{\dfrac{3}{x^2}}, \qquad \text{and} \qquad \frac{\dfrac{a+2}{3}}{\dfrac{a-2}{5}}$$

Remember that we can always multiply the numerator and the denominator of a fraction by the same nonzero term.

RECALL

This is the Fundamental Principle of Fractions.

$$\frac{P}{Q} = \frac{P \cdot R}{Q \cdot R} \qquad \text{in which } Q \neq 0 \text{ and } R \neq 0$$

To simplify a complex fraction, multiply the numerator and denominator by the LCD of all fractions that appear within the complex fraction.

> ▶   **Example 1**   Simplifying Complex Fractions

< Objective 1 >

Simplify $\dfrac{\dfrac{3}{4}}{\dfrac{5}{8}}$.

The LCD of $\dfrac{3}{4}$ and $\dfrac{5}{8}$ is 8. So multiply the numerator and denominator by 8.

$$\frac{\dfrac{3}{4}}{\dfrac{5}{8}} = \frac{\dfrac{3}{4} \cdot 8}{\dfrac{5}{8} \cdot 8} = \frac{3 \cdot 2}{5 \cdot 1} = \frac{6}{5}$$

Check Yourself 1

Simplify.

(a) $\dfrac{\dfrac{4}{7}}{\dfrac{3}{7}}$

(b) $\dfrac{\dfrac{3}{8}}{\dfrac{5}{6}}$

The same method can be used to simplify a complex fraction when variables are involved in the expression. Consider Example 2.

> ▶   **Example 2**   Simplifying Complex Rational Expressions

< Objective 2 >

Simplify $\dfrac{\dfrac{5}{x}}{\dfrac{10}{x^2}}$.

NOTE

Be sure to write the result in simplest form.

The LCD of $\dfrac{5}{x}$ and $\dfrac{10}{x^2}$ is x^2, so multiply the numerator and denominator by x^2.

$$\frac{\dfrac{5}{x}}{\dfrac{10}{x^2}} = \frac{\left(\dfrac{5}{x}\right)x^2}{\left(\dfrac{10}{x^2}\right)x^2} = \frac{5x}{10} = \frac{x}{2}$$

Check Yourself 2

Simplify.

(a) $\dfrac{\dfrac{6}{x^3}}{\dfrac{9}{x^2}}$ (b) $\dfrac{\dfrac{m^4}{15}}{\dfrac{m^3}{20}}$

We may also have a sum or a difference in the numerator or denominator of a complex fraction. The simplification steps are exactly the same. Consider Example 3.

Example 3   **Simplifying Complex Fractions**

Simplify $\dfrac{1 + \dfrac{x}{y}}{1 - \dfrac{x}{y}}$.

The LCD of 1, $\dfrac{x}{y}$, 1, and $\dfrac{x}{y}$ is y, so multiply the numerator and denominator by y.

NOTE

We use the distributive property to multiply *each term* in the numerator and in the denominator by y.

$$\frac{1 + \dfrac{x}{y}}{1 - \dfrac{x}{y}} = \frac{\left(1 + \dfrac{x}{y}\right)y}{\left(1 - \dfrac{x}{y}\right)y} = \frac{1 \cdot y + \dfrac{x}{y} \cdot y}{1 \cdot y - \dfrac{x}{y} \cdot y}$$

$$= \frac{y + x}{y - x}$$

Check Yourself 3

Simplify.

$$\frac{\dfrac{x}{y} - 2}{\dfrac{x}{y} + 2}$$

The following algorithm summarizes our work to this point with simplifying complex fractions.

Step by Step

To Simplify Complex Rational Expressions

Step 1   Multiply the numerator and denominator of the complex rational expression by the LCD of all the fractions that appear within the complex rational expression.

Step 2   Write the resulting fraction in simplest form.

A second method for simplifying complex fractions uses the fact that

$$\frac{\dfrac{P}{Q}}{\dfrac{R}{S}} = \frac{P}{Q} \div \frac{R}{S} = \frac{P}{Q} \cdot \frac{S}{R}$$

To use this method, we must write the numerator and denominator of the complex fraction as single fractions. We can then divide the numerator by the denominator as before.

In Example 4, we use this method to simplify the complex rational expression we saw in Example 3.

| ▶ | **Example 4** | **Simplifying Complex Fractions** |

Simplify $\dfrac{1 + \dfrac{x}{y}}{1 - \dfrac{x}{y}}$.

To use this method, we rewrite both the numerator and the denominator as single fractions.

$$\frac{1 + \dfrac{x}{y}}{1 - \dfrac{x}{y}} = \frac{\dfrac{y}{y} + \dfrac{x}{y}}{\dfrac{y}{y} - \dfrac{x}{y}} = \frac{\dfrac{y + x}{y}}{\dfrac{y - x}{y}}$$

Now we invert and multiply.

$$\frac{\dfrac{y + x}{y}}{\dfrac{y - x}{y}} = \frac{y + x}{y} \cdot \frac{y}{y - x} = \frac{y + x}{y - x}$$

Not surprisingly, we have the same result as we found in Example 3.

Check Yourself 4

Simplify using the second method $\dfrac{\dfrac{x}{y} - 2}{\dfrac{x}{y} + 2}$.

Check Yourself ANSWERS

1. (a) $\dfrac{4}{3}$; (b) $\dfrac{9}{20}$ **2.** (a) $\dfrac{2}{3x}$; (b) $\dfrac{4m}{3}$ **3.** $\dfrac{x - 2y}{x + 2y}$ **4.** $\dfrac{x - 2y}{x + 2y}$

Reading Your Text

The following fill-in-the-blank exercises are designed to ensure that you understand some of the key vocabulary used in this section.

SECTION 5.5

(a) The rule for dividing fractions is referred to as _____ and multiply.

(b) We can always multiply the numerator and denominator of a fraction by the same _____ term.

(c) A fraction that has a _____ in its numerator, in its denominator, or in both is called a complex fraction.

(d) To simplify a complex fraction, multiply the numerator and denominator by the _____ of all fractions that appear within the complex fraction.

5.5 exercises

Name _____

Section _____ Date _____

Answers

< Objectives 1–2 >

Simplify each complex fraction.

1. $\dfrac{\dfrac{2}{3}}{\dfrac{6}{8}}$

2. $\dfrac{\dfrac{5}{6}}{\dfrac{10}{15}}$

3. $\dfrac{1 + \dfrac{1}{2}}{2 + \dfrac{1}{4}}$ > Videos

4. $\dfrac{1 + \dfrac{3}{4}}{2 - \dfrac{1}{8}}$

5. $\dfrac{\dfrac{x}{8}}{\dfrac{x^2}{4}}$

6. $\dfrac{\dfrac{m^2}{10}}{\dfrac{m^3}{15}}$

7. $\dfrac{\dfrac{3}{a}}{\dfrac{2}{a^2}}$

8. $\dfrac{\dfrac{6}{x^2}}{\dfrac{9}{x^3}}$

9. $\dfrac{\dfrac{y + 1}{y}}{\dfrac{y - 1}{2y}}$ > Videos

10. $\dfrac{\dfrac{w + 3}{4w}}{\dfrac{w - 3}{2w}}$

11. $\dfrac{2 - \dfrac{1}{x}}{2 + \dfrac{1}{x}}$ > Videos

12. $\dfrac{3 + \dfrac{1}{a}}{3 - \dfrac{1}{a}}$

13. $\dfrac{3 - \dfrac{x}{y}}{\dfrac{6}{y}}$

14. $\dfrac{2 + \dfrac{x}{y}}{\dfrac{4}{y}}$

15. $\dfrac{\dfrac{x^2}{y^2} - 1}{\dfrac{x}{y} + 1}$

16. $\dfrac{\dfrac{a}{b} + 2}{\dfrac{a^2}{b^2} - 4}$

17. $\dfrac{1 + \dfrac{3}{x} - \dfrac{4}{x^2}}{1 + \dfrac{2}{x} - \dfrac{3}{x^2}}$

18. $\dfrac{1 - \dfrac{2}{r} - \dfrac{8}{r^2}}{1 - \dfrac{1}{r} - \dfrac{6}{r^2}}$

19. $\dfrac{\dfrac{2}{x} - \dfrac{1}{xy}}{\dfrac{1}{xy} + \dfrac{2}{y}}$

20. $\dfrac{\dfrac{1}{xy} + \dfrac{2}{x}}{\dfrac{3}{y} - \dfrac{1}{xy}}$

21. $\dfrac{\dfrac{2}{x - 1} + 1}{1 - \dfrac{3}{x - 1}}$

22. $\dfrac{\dfrac{3}{a + 2} - 1}{1 + \dfrac{2}{a + 2}}$

23. $\dfrac{1 - \dfrac{1}{y - 1}}{y - \dfrac{8}{y + 2}}$

24. $\dfrac{1 + \dfrac{1}{x + 2}}{x - \dfrac{18}{x - 3}}$

25. $1 + \dfrac{1}{1 + \dfrac{1}{x}}$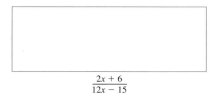

26. $1 + \dfrac{1}{1 - \dfrac{1}{y}}$

27. GEOMETRY The area of the rectangle shown here is $\dfrac{2}{3}$. Find the width.

$\dfrac{2x + 6}{12x - 15}$

28. GEOMETRY The area of the rectangle shown here is $\dfrac{2(3x - 2)}{x - 1}$. Find the width.

$\dfrac{3x - 2}{x - 2}$

Answers

17. _____

18. _____

19. _____

20. _____

21. _____

22. _____

23. _____

24. _____

25. _____

26. _____

27. _____

28. _____

Answers

29. _____

30. _____

29. Complex fractions have some interesting patterns. Work with a partner to evaluate each complex fraction in the sequence below. This is an interesting sequence of fractions because the numerators and denominators are a famous sequence of whole numbers, and the fractions get closer and closer to a number called "the golden mean."

$$1, \quad 1 + \frac{1}{1}, \quad 1 + \frac{1}{1 + \frac{1}{1}}, \quad 1 + \frac{1}{1 + \frac{1}{1 + \frac{1}{1}}}, \quad 1 + \frac{1}{1 + \frac{1}{1 + \frac{1}{1 + \frac{1}{1}}}}, \ldots$$

After you have evaluated these first five, you no doubt will see a pattern in the resulting fractions that allows you to go on indefinitely without having to evaluate more complex fractions. Write each of these fractions as decimals. Write your observations about the sequence of fractions and about the sequence of decimal fractions.

30. The following inequality is used when the U.S. House of Representatives seats are apportioned (see the chapter opener for more information).

$$\frac{\dfrac{E}{e} - \dfrac{A}{a+1}}{\dfrac{A}{a+1}} < \frac{\dfrac{A}{a} - \dfrac{E}{e+1}}{\dfrac{E}{e+1}}$$

Show that this inequality can be simplified to

$$\frac{A}{\sqrt{a(a+1)}} > \frac{E}{\sqrt{e(e+1)}}.$$

Here, A and E represent the populations of two states of the United States, and a and e are the number of representatives each of these two states have in the U.S. House of Representatives.

Mathematicians have shown that there are situations in which the method for apportionment described in the chapter's introduction does not work, and a state may not even get its basic quota of representatives. They give the table below of a hypothetical seven states and their populations as an example.

State	Population	Exact Quota	Number of Reps.
A	325	1.625	2
B	788	3.940	4
C	548	2.740	3
D	562	2.810	3
E	4,263	21.315	21
F	3,219	16.095	15
G	295	1.475	2
Total	10,000	50	50

In this case, the total population of all states is 10,000, and there are 50 representatives in all, so there should be no more than 10,000/50 or 200 people per representative. The quotas are found by dividing the population by 200. Whether a state, A, should get an additional representative before another state, E, should get one is decided in this method by using the simplified inequality below. If the ratio

$$\frac{A}{\sqrt{a(a+1)}} > \frac{E}{\sqrt{e(e+1)}}$$

is true, then A gets an extra representative before E does.

(a) If you go through the process of comparing the inequality above for each pair of states, state F loses a representative to state G. Do you see how this happens? Will state F complain?

(b) Alexander Hamilton, one of the signers of the Constitution, proposed that the extra representative positions be given one at a time to states with the largest remainder until all the "extra" positions were filled. How would this affect the table? Do you agree or disagree?

Basic Skills | Advanced Skills | Vocational-Technical Applications | Calculator/Computer | Above and Beyond | **Getting Ready** ▲

Divide each of the following.

(a) $\dfrac{3}{5} \div 6$ **(b)** $\dfrac{2}{7} \div \dfrac{3}{7}$

(c) $\dfrac{11}{4} \div \dfrac{5}{6}$ **(d)** $\dfrac{5}{8} \div \dfrac{10}{3}$

(e) $\dfrac{15}{4} \div \dfrac{5}{8}$ **(f)** $\dfrac{9}{8} \div \dfrac{3}{4}$

Answers

1. $\dfrac{8}{9}$ **3.** $\dfrac{2}{3}$ **5.** $\dfrac{1}{2x}$ **7.** $\dfrac{3a}{2}$ **9.** $\dfrac{2(y+1)}{y-1}$ **11.** $\dfrac{2x-1}{2x+1}$

13. $\dfrac{3y-x}{6}$ **15.** $\dfrac{x-y}{y}$ **17.** $\dfrac{x+4}{x+3}$ **19.** $\dfrac{2y-1}{1+2x}$ **21.** $\dfrac{x+1}{x-4}$

23. $\dfrac{y+2}{(y-1)(y+4)}$ **25.** $\dfrac{2x+1}{x+1}$ **27.** $\dfrac{4x-5}{x+3}$

29. Above and Beyond   **(a)** $\dfrac{1}{10}$ **(b)** $\dfrac{2}{3}$ **(c)** $\dfrac{33}{10}$ **(d)** $\dfrac{3}{16}$

(e) 6 **(f)** $\dfrac{3}{2}$

5.6

Equations Involving Rational Expressions

< 5.6 Objectives >

1 > Solve a rational equation with integer denominators

2 > Determine the excluded values for the variables of a rational expression

3 > Solve a rational equation

4 > Solve a proportion for an unknown

In Chapter 2, you learned how to solve a variety of equations. We now want to extend that work to the solution of **rational equations,** which are equations that involve rational expressions.

To solve a rational equation, we multiply each term of the equation by the LCD of any fractions in the equation. The resulting equation should be equivalent to the original equation and be cleared of all fractions.

Example 1

Solving Equations with Integer Denominators

< Objective 1 >

NOTE

This equation has three terms: $\dfrac{x}{2}$, $-\dfrac{1}{3}$, and $\dfrac{2x + 3}{6}$.
The sign of the term is not used to find the LCD.

Solve

$$\frac{x}{2} - \frac{1}{3} = \frac{2x + 3}{6}$$

The LCD for $\dfrac{x}{2}$, $\dfrac{1}{3}$, and $\dfrac{2x + 3}{6}$ is 6. Multiply both sides of the equation by 6. Using the distributive property, we multiply *each* term by 6.

$$6 \cdot \frac{x}{2} - 6 \cdot \frac{1}{3} = 6\left(\frac{2x + 3}{6}\right)$$

or

$$3x - 2 = 2x + 3$$

Solving as before, we have

$$3x - 2x = 3 + 2 \qquad \text{or} \qquad x = 5$$

To check, substitute 5 for x in the *original* equation.

$$\frac{(5)}{2} - \frac{1}{3} \stackrel{?}{=} \frac{2(5) + 3}{6}$$

$$\frac{13}{6} \stackrel{\checkmark}{=} \frac{13}{6} \qquad \text{(True)}$$

> C A U T I O N

Be careful! Many students have difficulty because they don't distinguish between adding and subtracting *expressions* (as we did in Sections 5.3 and 5.4) and solving equations. In the **expression**

$$\frac{x+1}{2} + \frac{x}{3}$$

we want to add the two fractions to form a single fraction. In the **equation**

$$\frac{x+1}{2} = \frac{x}{3} + 1$$

we want to solve for x.

Check Yourself 1

Solve and check.

$$\frac{x}{4} - \frac{1}{6} = \frac{4x-5}{12}$$

In Example 1, all the denominators were integers. What happens when we allow variables in the denominator? Recall that, for any fraction, the denominator must not be equal to zero. When a fraction has a variable in the denominator, we must exclude any value for the variable that results in division by zero.

Example 2   Finding Excluded Values for *x*

< Objective 2 >

In the following rational expressions, what values for x must be excluded?

(a) $\dfrac{x}{5}$ Here x can have any value, so none needs be excluded.

(b) $\dfrac{3}{x}$ If $x = 0$, then $\dfrac{3}{x}$ is undefined; 0 is the excluded value.

(c) $\dfrac{5}{x-2}$ If $x = 2$, then $\dfrac{5}{x-2} = \dfrac{5}{(2)-2} = \dfrac{5}{0}$, which is undefined, so 2 is the excluded value.

Check Yourself 2

What values for *x*, if any, must be excluded?

(a) $\dfrac{x}{7}$ **(b)** $\dfrac{5}{x}$ **(c)** $\dfrac{7}{x-5}$

If the denominator of a rational expression contains a product of two or more variable factors, the zero-product principle must be used to determine the excluded values for the variable.

In some cases, you have to factor the denominator to see the restrictions on the values for the variable.

| **Example 3** | Finding Excluded Values for *x* |

What values for x must be excluded in each fraction?

(a) $\dfrac{3}{x^2 - 6x - 16}$

Factoring the denominator, we have

$$\frac{3}{x^2 - 6x - 16} = \frac{3}{(x - 8)(x + 2)}$$

Letting $x - 8 = 0$ or $x + 2 = 0$, we see that 8 and -2 make the denominator 0, so both 8 and -2 must be excluded.

(b) $\dfrac{3}{x^2 + 2x - 48}$

The denominator is zero when

$$x^2 + 2x - 48 = 0$$

Factoring, we find

$$(x - 6)(x + 8) = 0$$

The denominator is zero when

$$x = 6 \qquad \text{or} \qquad x = -8$$

The excluded values are 6 and -8.

Check Yourself 3

What values for *x* must be excluded in the following fractions?

(a) $\dfrac{5}{x^2 - 3x - 10}$ 　　　　　　　　**(b)** $\dfrac{7}{x^2 + 5x - 14}$

The steps for solving an equation involving fractions are summarized in the following rule.

Step by Step

To Solve a Rational Equation		
	Step 1	Remove the fractions in the equation by multiplying each term by the LCD of all the fractions.
	Step 2	Solve the equation resulting from step 1 by the methods of Sections 2.3 and 4.6.
	Step 3	Check your solution in the *original equation*.

We can also solve rational equations with variables in the denominator by using the above algorithm. Example 4 illustrates this approach.

| Example 4 | Solving Rational Equations |

< Objective 3 >

Solve

$$\frac{7}{4x} - \frac{3}{x^2} = \frac{1}{2x^2}$$

NOTE

The factor x appears twice in the LCD.

The LCD of the three terms in the equation is $4x^2$, so we multiply both sides of the equation by $4x^2$.

$$4x^2 \cdot \frac{7}{4x} - 4x^2 \cdot \frac{3}{x^2} = 4x^2 \cdot \frac{1}{2x^2}$$

Simplifying, we have

$$7x - 12 = 2$$
$$7x = 14$$
$$x = 2$$

We leave the check to you. Be sure to return to the original equation.

Check Yourself 4

Solve and check.

$$\frac{5}{2x} - \frac{4}{x^2} = \frac{7}{2x^2}$$

The process of solving rational equations is exactly the same when there are binomials in the denominators.

| Example 5 | Solving Rational Equations |

NOTES

There are three terms.

Each of the terms is multiplied by $x - 3$.

(a) Solve

$$\frac{x}{x-3} - 2 = \frac{1}{x-3}$$

The LCD is $x - 3$, so we multiply each side (every term) by $x - 3$.

$$(x-3) \cdot \left(\frac{x}{x-3}\right) - 2(x-3) = (x-3) \cdot \left(\frac{1}{x-3}\right)$$

Simplifying, we have

$$x - 2(x-3) = 1$$
$$x - 2x + 6 = 1$$
$$-x = -5$$
$$x = 5$$

> CAUTION

Be careful of the signs!

To check, substitute 5 for x in the original equation.

$$\frac{(5)}{(5)-3} - 2 \overset{?}{=} \frac{1}{(5)-3}$$

$$\frac{5}{2} - 2 \overset{?}{=} \frac{1}{2}$$

$$\frac{1}{2} \overset{\checkmark}{=} \frac{1}{2}$$

(b) Solve

$$\frac{3}{x-3} - \frac{7}{x+3} = \frac{2}{x^2-9}$$

In factored form, the three denominators are $x-3$, $x+3$, and $(x+3)(x-3)$. This means that the LCD is $(x+3)(x-3)$, and so we multiply:

$$\overset{1}{\cancel{(x-3)}}(x+3)\left(\frac{3}{\underset{1}{\cancel{x-3}}}\right) - \overset{1}{\cancel{(x+3)}}(x-3)\left(\frac{7}{\underset{1}{\cancel{x+3}}}\right) = \overset{1}{\cancel{(x+3)}}\overset{1}{\cancel{(x-3)}}\left(\frac{2}{\underset{1}{\cancel{x^2-9}}}\right)$$

Simplifying, we have

$$3(x+3) - 7(x-3) = 2$$

$$3x + 9 - 7x + 21 = 2$$

$$-4x + 30 = 2$$

$$-4x = -28$$

$$x = 7$$

Check Yourself 5

Solve and check.

(a) $\dfrac{x}{x-5} - 2 = \dfrac{2}{x-5}$ **(b)** $\dfrac{4}{x-4} - \dfrac{3}{x+1} = \dfrac{5}{x^2-3x-4}$

You should be aware that some rational equations have no solutions. Example 6 shows that possibility.

Example 6   **Solving Rational Equations**

Solve

$$\frac{x}{x-2} - 7 = \frac{2}{x-2}$$

The LCD is $x-2$, and so we multiply each side (every term) by $x-2$.

$$\cancel{(x-2)}\left(\frac{x}{\cancel{x-2}}\right) - 7(x-2) = \cancel{(x-2)}\left(\frac{2}{\cancel{x-2}}\right)$$

Simplifying, we have

$$x - 7x + 14 = 2$$
$$-6x = -12$$
$$x = 2$$

Now, when we try to check our result, we have

$$\frac{(2)}{(2) - 2} - 7 \stackrel{?}{=} \frac{2}{(2) - 2} \qquad \text{or} \qquad \frac{2}{0} - 7 \stackrel{?}{=} \frac{2}{0}$$

These terms are undefined.

What went wrong? Remember that two of the terms in our original equation were $\dfrac{x}{x - 2}$ and $\dfrac{2}{x - 2}$. The variable *x* cannot have the value 2 because 2 is an excluded value (it makes the denominator 0). So our original equation has *no solution*.

Check Yourself 6

Solve, if possible.

$$\frac{x}{x + 3} - 6 = \frac{-3}{x + 3}$$

Equations involving fractions may also lead to quadratic equations, as Example 7 illustrates.

 Example 7   **Solving Rational Equations**

Solve

$$\frac{x}{x - 4} = \frac{15}{x - 3} - \frac{2x}{x^2 - 7x + 12}$$

The LCD is $(x - 4)(x - 3)$. Multiply each side (every term) by $(x - 4)(x - 3)$.

$$\frac{x}{(x - 4)}(x - 4)(x - 3) = \frac{15}{(x - 3)}(x - 4)(x - 3) - \frac{2x}{(x - 4)(x - 3)}(x - 4)(x - 3)$$

Simplifying, we have

$$x(x - 3) = 15(x - 4) - 2x$$

Multiply to remove the parentheses:

$$x^2 - 3x = 15x - 60 - 2x$$

In standard form, the equation is

$$x^2 - 16x + 60 = 0$$

or

$(x - 6)(x - 10) = 0$

Setting the factors to 0, we have

$x - 6 = 0$ or $x - 10 = 0$

$x = 6$ $x = 10$

So $x = 6$ and $x = 10$ are possible solutions. We leave the check of *each* solution to you.

Check Yourself 7

Solve and check.

$$\frac{3x}{x + 2} - \frac{2}{x + 3} = \frac{36}{x^2 + 5x + 6}$$

The following equation is a special kind of equation involving fractions:

$$\frac{135}{t} = \frac{180}{t + 1}$$

An equation of the form $\dfrac{a}{b} = \dfrac{c}{d}$ is said to be in **proportion form,** or, more simply, it is called a **proportion.** This type of equation occurs often enough in algebra that it is worth developing some special methods for its solution. First, we need some definitions.

A **ratio** is a means of comparing two quantities. A ratio can be written as a fraction. For instance, the ratio of 2 to 3 can be written as $\dfrac{2}{3}$. A statement that two ratios are equal is called a *proportion.* A proportion has the form

$$\frac{a}{b} = \frac{c}{d}$$ In this proportion, a and d are called the **extremes** of the proportion, and b and c are called the **means.**

NOTE

bd is the LCD of the denominators.

A useful property of proportions is easily developed. If

$$\frac{a}{b} = \frac{c}{d}$$ We multiply both sides by $b \cdot d$.

$$\left(\frac{a}{b}\right)bd = \left(\frac{c}{d}\right)bd \qquad \text{or} \qquad ad = bc$$

Property

Proportions

If $\dfrac{a}{b} = \dfrac{c}{d}$, then $ad = bc$.

In words:

In any proportion, the product of the extremes (*ad*) is equal to the product of the means (*bc*).

Because a proportion is a special kind of rational equation, this rule gives us an alternative approach to solving equations that are in the proportion form.

Example 8   **Solving a Proportion for an Unknown**

< Objective 4 >

Solve each equation for x.

(a) $\dfrac{x}{5} = \dfrac{12}{15}$

Set the product of the extremes equal to the product of the means.

NOTE

The extremes are x and 15.
The means are 5 and 12.

$15x = 5 \cdot 12$

$15x = 60$

$x = 4$

Our solution is 4. You can check as before, by substituting in the original proportion.

(b) $\dfrac{x + 3}{10} = \dfrac{x}{7}$

Set the product of the extremes equal to the product of the means. Be certain to use parentheses with a numerator with more than one term.

$7(x + 3) = 10x$

$7x + 21 = 10x$

$21 = 3x$

$7 = x$

We leave it to you to check this result.

Check Yourself 8

Solve each equation for *x*.

(a) $\dfrac{x}{8} = \dfrac{3}{4}$ **(b)** $\dfrac{x - 1}{9} = \dfrac{x + 1}{12}$

As the examples of this section illustrated, *whenever* an equation involves rational expressions, the *first step* of the solution is to clear the equation of fractions by multiplication.

The following algorithm summarizes our work in solving equations that involve rational expressions.

Step by Step

To Solve an Equation Involving Fractions

Step 1	Remove the fractions appearing in the equation by multiplying each side (every term) by the LCD of all the fractions.
Step 2	Solve the equation resulting from step 1. If the equation is linear, use the methods of Section 2.3 for the solution. If the equation is quadratic, use the methods of Section 4.6.
Step 3	Check all solutions by substitution in the *original equation*. Be sure to discard any *extraneous* solutions, that is, solutions that result in a zero denominator in the original equation.

Check Yourself ANSWERS

1. $x = 3$ **2.** **(a)** None; **(b)** 0; **(c)** 5   **3.** **(a)** $-2, 5$; **(b)** $-7, 2$ **4.** $x = 3$

5. **(a)** $x = 8$; **(b)** $x = -11$ **6.** No solution   **7.** $x = -5$ or $x = \dfrac{8}{3}$

8. **(a)** $x = 6$; **(b)** $x = 7$

Reading Your Text

The following fill-in-the-blank exercises are designed to ensure that you understand some of the key vocabulary used in this section.

SECTION 5.6

(a) Rational equations are equations that involve rational _____.

(b) To solve a rational equation, we multiply each term by the _____ of any fractions in the equation.

(c) If the denominator of a rational equation contains a product of two or more variable factors, the zero-product principle is used to determine the _____ values for the variable.

(d) The final step in solving a rational equation is to check the solution in the _____ equation.

5.6 exercises

< Objective 2 >

What values for x, if any, must be excluded in each of the following algebraic fractions?

1. $\dfrac{x}{15}$

2. $\dfrac{8}{x}$

3. $\dfrac{17}{x}$

4. $\dfrac{x}{8}$

5. $\dfrac{3}{x-2}$

6. $\dfrac{x-1}{5}$

7. $\dfrac{-5}{x+4}$

8. $\dfrac{4}{x+3}$

9. $\dfrac{x-5}{2}$

10. $\dfrac{x-1}{x-5}$

11. $\dfrac{3x}{(x+1)(x-2)}$

12. $\dfrac{5x}{(x-3)(x+7)}$

13. $\dfrac{x-1}{(2x-1)(x+3)}$

14. $\dfrac{x+3}{(3x+1)(x-2)}$

15. $\dfrac{7}{x^2-9}$

16. $\dfrac{5x}{x^2+x-2}$

17. $\dfrac{x+3}{x^2-7x+12}$

18. $\dfrac{3x-4}{x^2-49}$

19. $\dfrac{2x-1}{3x^2+x-2}$ > Videos

20. $\dfrac{3x+1}{4x^2-11x+6}$

< Objectives 1 and 3 >

Solve and check each of the following equations for x.

21. $\dfrac{x}{2}+3=6$

22. $\dfrac{x}{3}-2=1$

23. $\dfrac{x}{2}-\dfrac{x}{3}=2$

24. $\dfrac{x}{6}-\dfrac{x}{8}=1$

MathZone

Boost your grade at mathzone.com!
> Practice Problems
> NetTutor
> Self-Tests
> e-Professors
> Videos

Name _____

Section _____ Date _____

Answers

1. _____   2. _____

3. _____   4. _____

5. _____   6. _____

7. _____   8. _____

9. _____   10. _____

11. _____   12. _____

13. _____   14. _____

15. _____   16. _____

17. _____   18. _____

19. _____   20. _____

21. _____

22. _____

23. _____

24. _____

Answers

25. _____

26. _____

27. _____

28. _____

29. _____

30. _____

31. _____

32. _____

33. _____

34. _____

35. _____

36. _____

37. _____

38. _____

39. _____

40. _____

41. _____

42. _____

43. _____

44. _____

45. _____

46. _____

47. _____

48. _____

25. $\dfrac{x}{5} - \dfrac{1}{3} = \dfrac{x - 7}{3}$

26. $\dfrac{x}{6} + \dfrac{3}{4} = \dfrac{x - 1}{4}$

27. $\dfrac{x}{4} - \dfrac{1}{5} = \dfrac{4x + 3}{20}$

28. $\dfrac{x}{12} - \dfrac{1}{6} = \dfrac{2x - 7}{12}$

29. $\dfrac{3}{x} + 2 = \dfrac{7}{x}$

30. $\dfrac{4}{x} - 3 = \dfrac{16}{x}$

31. $\dfrac{4}{x} + \dfrac{3}{4} = \dfrac{10}{x}$ > Videos

32. $\dfrac{3}{x} = \dfrac{5}{3} - \dfrac{7}{x}$

33. $\dfrac{5}{2x} - \dfrac{1}{x} = \dfrac{9}{2x^2}$

34. $\dfrac{4}{3x} + \dfrac{1}{x} = \dfrac{14}{3x^2}$

35. $\dfrac{2}{x - 3} + 1 = \dfrac{7}{x - 3}$

36. $\dfrac{x}{x + 1} + 2 = \dfrac{14}{x + 1}$

Complete each of the following statements with **never, sometimes,** *or* **always.**

37. The value of the term $\dfrac{5}{x}$, $x \neq 0$, is _____ equal to 1.

38. The value of the term $\dfrac{0}{x}$, $x \neq 0$, is _____ equal to 1.

39. The value of the term $\dfrac{5}{x}$, $x \neq 0$, is _____ equal to 0.

40. The value of the term $\dfrac{0}{x}$, $x \neq 0$, is _____ equal to 0.

Solve and check each of the following for x.

41. $\dfrac{2}{x + 3} + \dfrac{1}{2} = \dfrac{x + 6}{x + 3}$

42. $\dfrac{6}{x - 5} - \dfrac{2}{3} = \dfrac{x - 9}{x - 5}$

43. $\dfrac{x}{3x + 12} + \dfrac{x - 1}{x + 4} = \dfrac{5}{3}$ > Videos

44. $\dfrac{x}{4x - 12} - \dfrac{x - 4}{x - 3} = \dfrac{1}{8}$

45. $\dfrac{x}{x - 3} - 2 = \dfrac{3}{x - 3}$

46. $\dfrac{x}{x - 5} + 2 = \dfrac{5}{x - 5}$

47. $\dfrac{x - 1}{x + 3} - \dfrac{x - 3}{x} = \dfrac{3}{x^2 + 3x}$

48. $\dfrac{x}{x - 2} - \dfrac{x + 1}{x} = \dfrac{8}{x^2 - 2x}$

49. $\dfrac{1}{x-2} - \dfrac{2}{x+2} = \dfrac{2}{x^2-4}$

50. $\dfrac{1}{x+4} + \dfrac{1}{x-4} = \dfrac{12}{x^2-16}$

51. $\dfrac{5}{x-4} = \dfrac{1}{x+2} - \dfrac{2}{x^2-2x-8}$

52. $\dfrac{11}{x+2} = \dfrac{5}{x^2-x-6} + \dfrac{1}{x-3}$

53. $\dfrac{3}{x-1} - \dfrac{1}{x+9} = \dfrac{18}{x^2+8x-9}$

54. $\dfrac{2}{x+2} = \dfrac{3}{x+6} + \dfrac{9}{x^2+8x+12}$

55. $\dfrac{3}{x+3} + \dfrac{25}{x^2+x-6} = \dfrac{5}{x-2}$

56. $\dfrac{5}{x+6} + \dfrac{2}{x^2+7x+6} = \dfrac{3}{x+1}$

57. $\dfrac{7}{x-5} - \dfrac{3}{x+5} = \dfrac{40}{x^2-25}$

58. $\dfrac{3}{x-3} - \dfrac{18}{x^2-9} = \dfrac{5}{x+3}$

59. $\dfrac{2x}{x-3} + \dfrac{2}{x-5} = \dfrac{3x}{x^2-8x+15}$

> Videos

60. $\dfrac{x}{x-4} = \dfrac{5x}{x^2-x-12} - \dfrac{3}{x+3}$

61. $\dfrac{2x}{x+2} = \dfrac{5}{x^2-x-6} - \dfrac{1}{x-3}$

62. $\dfrac{3x}{x-1} = \dfrac{2}{x-2} - \dfrac{2}{x^2-3x+2}$

63. $\dfrac{7}{x-2} + \dfrac{16}{x+3} = 3$

64. $\dfrac{5}{x-2} + \dfrac{6}{x+2} = 2$

65. $\dfrac{11}{x-3} - 1 = \dfrac{10}{x+3}$

66. $\dfrac{17}{x-4} - 2 = \dfrac{10}{x+2}$

< Objective 4 >

Solve each of the following equations for x.

67. $\dfrac{x}{11} = \dfrac{12}{33}$

68. $\dfrac{4}{x} = \dfrac{16}{20}$

69. $\dfrac{5}{8} = \dfrac{20}{x}$

70. $\dfrac{x}{10} = \dfrac{9}{30}$

71. $\dfrac{x+1}{5} = \dfrac{20}{25}$

72. $\dfrac{2}{5} = \dfrac{x-2}{20}$

Answers

49. _____

50. _____

51. _____

52. _____

53. _____

54. _____

55. _____

56. _____

57. _____

58. _____

59. _____

60. _____

61. _____

62. _____

63. _____

64. _____

65. _____

66. _____

67. _____

68. _____

69. _____

70. _____

71. _____

72. _____

Answers

73. _____

74. _____

75. _____

76. _____

77. _____

78. _____

79. _____

80. _____

81. _____

82. _____

(a) _____

(b) _____

(c) _____

(d) _____

(e) _____

(f) _____

73. $\dfrac{3}{5} = \dfrac{x-1}{20}$

74. $\dfrac{5}{x-3} = \dfrac{15}{21}$

Basic Skills | **Advanced Skills** | Vocational-Technical Applications | Calculator/Computer | Above and Beyond | Getting Ready

▲

---

75. $\dfrac{x}{6} = \dfrac{x+5}{16}$

76. $\dfrac{x-2}{x+2} = \dfrac{12}{20}$ › Videos

77. $\dfrac{x}{x+7} = \dfrac{10}{17}$

78. $\dfrac{x}{10} = \dfrac{x+6}{30}$

79. $\dfrac{2}{x-1} = \dfrac{6}{x+9}$

80. $\dfrac{3}{x-3} = \dfrac{4}{x-5}$

81. $\dfrac{1}{x+3} = \dfrac{7}{x^2-9}$

82. $\dfrac{1}{x+5} = \dfrac{4}{x^2+3x-10}$

Basic Skills | Advanced Skills | Vocational-Technical Applications | Calculator/Computer | Above and Beyond | **Getting Ready**

▲

Write each of the following phrases using symbols. Use the variable x to represent the number in each case.

   (a) One-fourth of a number added to four-fifths of the same number

   (b) 6 times a number, decreased by 12

   (c) The quotient when 5 more than a number is divided by 6

   (d) Three times the length of a side of a rectangle, decreased by 4

   (e) A distance traveled divided by 5

   (f) The speed of a truck that is 5 mi/h slower than a car

Answers

1. None   **3.** 0   **5.** 2   **7.** -4 **9.** None   **11.** $-1, 2$

13. $-3, \dfrac{1}{2}$ **15.** $-3, 3$ **17.** $3, 4$ **19.** $-1, \dfrac{2}{3}$ **21.** 6

23. 12   **25.** 15   **27.** 7   **29.** 2   **31.** 8   **33.** 3

35. 8   **37.** sometimes   **39.** never   **41.** -5 **43.** -23

45. No solution   **47.** 6   **49.** 4   **51.** -4 **53.** -5

55. No solution   **57.** $-\dfrac{5}{2}$ **59.** $-\dfrac{1}{2}, 6$ **61.** $-\dfrac{1}{2}$ **63.** $-\dfrac{1}{3}, 7$

65. $-8, 9$ **67.** 4   **69.** 32   **71.** 3   **73.** 13   **75.** 3   **77.** 10

79. 6   **81.** 10   **(a)** $\dfrac{1}{4}x + \dfrac{4}{5}x$ **(b)** $6x - 12$ **(c)** $\dfrac{x+5}{6}$

(d) $3x - 4$ **(e)** $\dfrac{x}{5}$ **(f)** $x - 5$

5.7

Applications of Rational Expressions

< 5.7 Objectives >

1 > Solve a word problem that leads to a rational equation

2 > Use a proportion to solve a word problem

Many word problems lead to rational equations that can be solved using the methods of Section 5.6. The five steps in solving word problems are, of course, the same as you saw earlier.

(▶) **Example 1**	**Solving a Numerical Application**

< Objective 1 >

If one-third of a number is added to three-fourths of that same number, the sum is 26. Find the number.

Step 1 Read the problem carefully. You want to find the unknown number.

Step 2 Choose a letter to represent the unknown. Let x be the unknown number.

Step 3 Form an equation.

$$\frac{1}{3}x + \frac{3}{4}x = 26$$

One-third   Three-fourths
of number   of number

Step 4 Solve the equation. Multiply each side (every term) of the equation by 12, the LCD.

$$12 \cdot \frac{1}{3}x + 12 \cdot \frac{3}{4}x = 12 \cdot 26$$

Simplifying yields

$$4x + 9x = 312$$
$$13x = 312$$
$$x = 24$$

The number is 24.

NOTE

Be sure to answer the question raised in the problem.

Step 5 Check your solution by returning to the *original problem*. If the number is 24, we have

$$\frac{1}{3}(24) + \frac{3}{4}(24) = 8 + 18 = 26$$

and the solution is verified.

515

Check Yourself 1

The sum of two-fifths of a number and one-half of that number is 18. Find the number.

Number problems that involve reciprocals can be solved by using rational equations. Example 2 illustrates this approach.

Example 2	Solving a Numerical Application

One number is twice another number. If the sum of their reciprocals is $\dfrac{3}{10}$, what are the two numbers?

Step 1   You want to find the two numbers.

Step 2   Let x be one number. Then $2x$ is the other number.

Twice the first

Step 3

RECALL

The reciprocal of a fraction is the fraction obtained by switching the numerator and denominator.

$$\frac{1}{x} + \frac{1}{2x} = \frac{3}{10}$$

The reciprocal of the first number, x

The reciprocal of the second number, $2x$

Step 4   The LCD of the fractions is $10x$, so we multiply by $10x$.

$$10x\left(\frac{1}{x}\right) + 10x\left(\frac{1}{2x}\right) = 10x\left(\frac{3}{10}\right)$$

Simplifying, we have

$$10 + 5 = 3x$$
$$15 = 3x$$
$$5 = x$$

The numbers are 5 and 10.

NOTE

x is one number, and $2x$ is the other.

Step 5

Again check the result by returning to the original problem. If the numbers are 5 and 10, we have

$$\frac{1}{(5)} + \frac{1}{2(5)} = \frac{2+1}{10} = \frac{3}{10}$$

The sum of the reciprocals is $\dfrac{3}{10}$.

Check Yourself 2

One number is 3 times another. If the sum of their reciprocals is $\dfrac{2}{9}$, find the two numbers.

Motion problems often involve rational expressions. Recall that the key equation for solving all motion problems relates the distance traveled, the speed or rate, and the time:

Definition

Motion Problem Relationships

$d = r \cdot t$

Often we use this equation in different forms by solving for r or for t.

$$r = \frac{d}{t} \qquad \text{and} \qquad t = \frac{d}{r}$$

Example 3   Solving an Application Involving $r = \dfrac{d}{t}$

Vince took 2 h longer to drive 225 mi than he did on a trip of 135 mi. If his speed was the same both times, how long did each trip take?

------ 225 miles

135 miles — — —

NOTE

It is often helpful to choose your variable to "suggest" the unknown quantity—here t for time.

Step 1   You want to find the times taken for the 225-mi trip and for the 135-mi trip.

Step 2   Let t be the time for the 135-mi trip (in hours).

$\overbrace{}^{\text{2 h longer}}$

Then $t + 2$ is the time for the 225-mi trip.

It is often helpful to arrange the information in tabular form such as that shown.

RECALL

Rate is distance divided by time. The rate column is formed by using that relationship.

	Distance	Rate	Time
135-mi trip	135	$\dfrac{135}{t}$	t
225-mi trip	225	$\dfrac{225}{t + 2}$	$t + 2$

Beginning Algebra   The Streeter/Hutchison Series in Mathematics   © The McGraw-Hill Companies. All Rights Reserved.

NOTE

The equation is in proportion form. So we could solve by setting the product of the means equal to the product of the extremes.

Step 3   In forming the equation, remember that the speed (or rate) for each trip was the same. That is the *key* idea. We can equate the rates for the two trips that were found in step 2. The two rates are shown in the rightmost column of the table. Thus we can write

$$\frac{135}{t} = \frac{225}{t + 2}$$

Step 4   To solve the equation from step 3, multiply each side by $t(t + 2)$, the LCD of the fractions.

$$\cancel{t}(t + 2)\left(\frac{135}{\cancel{t}}\right) = t\cancel{(t + 2)}\left(\frac{225}{\cancel{t + 2}}\right)$$

Simplifying, we have

$$135(t + 2) = 225t$$
$$135t + 270 = 225t$$
$$270 = 90t$$
$$t = 3 \text{ h}$$

The time for the 135-mi trip was 3 h, and the time for the 225-mi trip was 5 h. We leave it to you to check this result.

Check Yourself 3

Cynthia took 2 h longer to bicycle 75 mi than she did on a trip of 45 mi. If her speed was the same each time, find the time for each trip.

Example 4 uses the $t = \dfrac{d}{r}$ form of the $d = r \cdot t$ relationship to find the speed.

| ▶ | **Example 4** | Solving an Application Involving Distance, Rate, and Time |

A train makes a trip of 300 mi in the same time that a bus can travel 250 mi. If the speed of the train is 10 mi/h faster than the speed of the bus, find the speed of each.

Step 1   You want to find the speeds of the train and of the bus.

Step 2   Let r be the speed (or rate) of the bus (in miles per hour).

Then $\underline{r + 10}$ is the rate of the train.
    10 mi/h faster

Again, form a chart of the information.

RECALL

Time is distance divided by rate. Here the rightmost column is found by using that relationship.

	Distance	Rate	Time
Train	300	$r + 10$	$\dfrac{300}{r + 10}$
Bus	250	r	$\dfrac{250}{r}$

Step 3   To form an equation, remember that the times for the train and bus are the same. We can equate the expressions for time found in step 2. Again, working from the rightmost column, we have

$$\frac{300}{r + 10} = \frac{250}{r}$$

Step 4   We multiply each term by $r(r + 10)$, the LCD of the fractions.

$$\cancel{r}(r + 10)\left(\frac{250}{\cancel{r}}\right) = r\cancel{(r + 10)}\left(\frac{300}{\cancel{r + 10}}\right)$$

Simplifying, we have

$$250(r + 10) = 300r$$
$$250r + 2500 = 300r$$
$$2500 = 50r$$
$$r = 50 \text{ mi/h}$$

The rate of the bus is 50 mi/h, and the rate of the train is 60 mi/h. You can check this result.

RECALL

Remember to find the rates of both vehicles.

Check Yourself 4

A car makes a trip of 280 mi in the same time that a truck travels 245 mi. If the speed of the truck is 5 mi/h slower than that of the car, find the speed of each.

Example 5 involves fractions in decimal form. Mixture problems often use percentages, and those percentages can be written as decimals. Example 5 illustrates this method.

   Example 5 | **Solving an Application Involving Solutions**

A solution of antifreeze is 20% alcohol. How much pure alcohol must be added to 12 quarts (qt) of the solution to make a 40% solution?

Step 1   You want to find the number of quarts of pure alcohol that must be added.

Step 2   Let x be the number of quarts of pure alcohol to be added.

Step 3   To form our equation, note that the amount of alcohol present before mixing *must be the same* as the amount in the combined solution.

A picture will help.

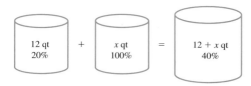

NOTE

Express the percentages as decimals in the equation.

So

$$\underbrace{12(0.20)}_{} + \underbrace{x(1.00)}_{} = \underbrace{(12 + x)(0.40)}_{}$$

The amount of alcohol in the first solution (20% is 0.20)   The amount of pure alcohol ("pure" is 100%, or 1.00)   The amount of alcohol in the mixture

Step 4   Most students prefer to clear the decimals at this stage. Multiplying by 100 moves the decimal point *two places to the right*. We then have

$$12(20) + x(100) = (12 + x)(40)$$
$$240 + 100x = 480 + 40x$$
$$60x = 240$$
$$x = 4 \text{ qt}$$

We should add 4 qt of pure alcohol.

Check Yourself 5

How much pure alcohol must be added to 500 cm³ of a 40% alcohol mixture to make a solution that is 80% alcohol?

There are many types of applications that lead to proportions in their solution. Typically these applications involve a common ratio, such as miles to gallons or miles to hours, and they can be solved with three basic steps.

Step by Step

To Solve an Application Using a Proportion

Step 1   Assign a variable to represent the unknown quantity.

Step 2   Write a proportion, using the known and unknown quantities. Be sure each ratio involves the same units.

Step 3   Solve the proportion written in step 2 for the unknown quantity.

Example 6 illustrates this approach.

Example 6   Solving an Application Using a Proportion

< Objective 2 >

A car uses 3 gal of gas to travel 105 mi. At that mileage rate, how many gallons will be used on a trip of 385 mi?

Step 1   Assign a variable to represent the unknown quantity. Let x be the number of gallons of gas that will be used on the 385-mi trip.

Step 2   Write a proportion. Note that the ratio of miles to gallons must stay the same.

Miles   Miles

$$\frac{105}{3} = \frac{385}{x}$$

Gallons   Gallons

Step 3 Solve the proportion. The product of the extremes is equal to the product of the means.

$$105x = 3 \cdot 385$$
$$105x = 1{,}155$$
$$\frac{105x}{105} = \frac{1{,}155}{105}$$
$$x = 11 \text{ gal}$$

So 11 gal of gas will be used for the 385-mi trip.

NOTE

To verify your solution, return to the original problem and check that the two ratios are equivalent.

Check Yourself 6

A car uses 8 L of gasoline in traveling 100 km. At that rate, how many liters of gas will be used on a trip of 250 km?

Proportions can also be used to solve problems in which a quantity is divided using a specific ratio. Example 7 shows how.

Example 7   **Solving an Application Using a Proportion**

A 60-in.-long piece of wire is to be cut into two pieces whose lengths have the ratio 5 to 7. Find the length of each piece.

Step 1   Let x represent the length of the shorter piece. Then $60 - x$ is the length of the longer piece.

RECALL

A picture of the problem always helps.

```
      Shorter                    Longer
        ↓                          ↓
  |←——— x ———→|←——— 60 − x ———→|
  |←——————————— 60 ———————————→|
```

Step 2   The two pieces have the ratio $\dfrac{5}{7}$, so

$$\frac{x}{60 - x} = \frac{5}{7}$$

NOTE

On the left and right, we have the ratio of the length of the shorter piece to that of the longer piece.

Step 3   Solving as before, we get

$$7x = (60 - x)5$$
$$7x = 300 - 5x$$
$$12x = 300$$
$$x = 25 \quad \text{(Shorter piece)}$$
$$60 - x = 35 \quad \text{(Longer piece)}$$

The pieces have lengths of 25 in. and 35 in.

Check Yourself 7

A 21-ft-long board is to be cut into two pieces so that the ratio of their lengths is 3 to 4. Find the lengths of the two pieces.

Check Yourself ANSWERS

1. The number is 20.   **2.** The numbers are 6 and 18.
3. 75-mi trip: 5 h; 45-mi trip: 3 h   **4.** Car: 40 mi/h; truck: 35 mi/h
5. 1,000 cm^3 **6.** 20 L   **7.** 9 ft; 12 ft

Reading Your Text

The following fill-in-the-blank exercises are designed to ensure that you understand some of the key vocabulary used in this section.

SECTION 5.7

(a) When solving a rational equation, the solution is checked by returning to the _____ problem.

(b) The key equation for solving motion problems relates the distance traveled, the speed, and the _____.

(c) Time is distance divided by _____.

(d) To solve an application using a proportion, first assign a _____ to represent the unknown quantity.

< Objectives 1–2 >

Solve the following word problems.

1. **NUMBER PROBLEM** If two-thirds of a number is added to one-half of that number, the sum is 35. Find the number.

2. **NUMBER PROBLEM** If one-third of a number is subtracted from three-fourths of that number, the difference is 15. What is the number?

3. **NUMBER PROBLEM** If one-fourth of a number is subtracted from two-fifths of the number, the difference is 3. Find the number.

4. **NUMBER PROBLEM** If five-sixths of a number is added to one-fifth of the number, the sum is 31. What is the number?

5. **NUMBER PROBLEM** If one-third of an integer is added to one-half of the next consecutive integer, the sum is 13. What are the two integers?

6. **NUMBER PROBLEM** If one-half of one integer is subtracted from three-fifths of the next consecutive integer, the difference is 3. What are the two integers?

7. **NUMBER PROBLEM** One number is twice another number. If the sum of their reciprocals is $\frac{1}{4}$, find the two numbers.

8. **NUMBER PROBLEM** One number is 3 times another. If the sum of their reciprocals is $\frac{1}{6}$, find the two numbers.

9. **NUMBER PROBLEM** One number is 4 times another. If the sum of their reciprocals is $\frac{5}{12}$, find the two numbers.

10. **NUMBER PROBLEM** One number is 3 times another. If the sum of their reciprocals is $\frac{4}{15}$, what are the two numbers?

11. **NUMBER PROBLEM** One number is 5 times another number. If the sum of their reciprocals is $\frac{6}{35}$, what are the two numbers?

12. **NUMBER PROBLEM** One number is 4 times another. The sum of their reciprocals is $\frac{5}{24}$. What are the two numbers?

13. **NUMBER PROBLEM** If the reciprocal of 5 times a number is subtracted from the reciprocal of that number, the result is $\frac{4}{25}$. What is the number?

14. **NUMBER PROBLEM** If the reciprocal of a number is added to 4 times the reciprocal of that number, the result is $\frac{5}{9}$. Find the number.

MathZone

Boost your grade at mathzone.com!
> Practice Problems
> NetTutor
> Self-Tests
> e-Professors
> Videos

Name _____

Section _____ Date _____

Answers

1. _____
2. _____
3. _____
4. _____
5. _____
6. _____
7. _____
8. _____
9. _____
10. _____
11. _____
12. _____
13. _____
14. _____

Answers

15. _____

16. _____

17. _____

18. _____

19. _____

20. _____

21. _____

22. _____

23. _____

24. _____

15. **SCIENCE AND MEDICINE** Lee can ride his bicycle 50 mi in the same time it takes him to drive 125 mi. If his driving rate is 30 mi/h faster than his rate bicycling, find each rate.

16. **SCIENCE AND MEDICINE** Tina can run 12 mi in the same time it takes her to bicycle 72 mi. If her bicycling rate is 20 mi/h faster than her running rate, find each rate.

17. **SCIENCE AND MEDICINE** An express bus can travel 275 mi in the same time that it takes a local bus to travel 225 mi. If the rate of the express bus is 10 mi/h faster than that of the local bus, find the rate for each bus.

18. **SCIENCE AND MEDICINE** A passenger train can travel 325 mi in the same time a freight train takes to travel 200 mi. If the speed of the passenger train is 25 mi/h faster than the speed of the freight train, find the speed of each.

19. **SCIENCE AND MEDICINE** A light plane took 1 h longer to travel 450 mi on the first portion of a trip than it took to fly 300 mi on the second. If the speed was the same for each portion, what was the flying time for each part of the trip?

20. **SCIENCE AND MEDICINE** A small business jet took 1 h longer to fly 810 mi on the first part of a flight than to fly 540 mi on the second portion. If the jet's rate was the same for each leg of the flight, what was the flying time for each leg?

21. **SCIENCE AND MEDICINE** Charles took 2 h longer to drive 240 mi on the first day of a vacation trip than to drive 144 mi on the second day. If his average driving rate was the same on both days, what was his driving time for each of the days?

22. **SCIENCE AND MEDICINE** Ariana took 2 h longer to drive 360 mi on the first day of a trip than she took to drive 270 mi on the second day. If her speed was the same on both days, what was the driving time each day?

23. **SCIENCE AND MEDICINE** An airplane took 3 h longer to fly 1,200 mi than it took for a flight of 480 mi. If the plane's rate was the same on each trip, what was the time of each flight?

24. **SCIENCE AND MEDICINE** A train travels 80 mi in the same time that a light plane can travel 280 mi. If the speed of the plane is 100 mi/h faster than that of the train, find each of the rates.

25. **SCIENCE AND MEDICINE** Jan and Tariq took a canoeing trip, traveling 6 mi upstream against a 2-mi/h current. They then returned to the same point downstream. If their entire trip took 4 h, how fast can they paddle in still water? [*Hint:* If *r* is their rate (in miles per hour) in still water, their rate upstream is $r - 2$ and their rate downstream is $r + 2$.]

26. **SCIENCE AND MEDICINE** A plane flies 720 mi against a steady 30-mi/h headwind and then returns to the same point with the wind. If the entire trip takes 10 h, what is the plane's speed in still air?

27. **SCIENCE AND MEDICINE** How much pure alcohol must be added to 40 oz of a 25% solution to produce a mixture that is 40% alcohol?

28. **SCIENCE AND MEDICINE** How many centiliters (cL) of pure acid must be added to 200 cL of a 40% acid solution to produce a 50% solution?

29. **SCIENCE AND MEDICINE** A speed of 60 mi/h corresponds to 88 ft/s. If a light plane's speed is 150 mi/h, what is its speed in feet per second?

30. **BUSINESS AND FINANCE** If 342 cups of coffee can be made from 9 lb of coffee, how many cups can be made from 6 lb of coffee?

31. **SOCIAL SCIENCE** A car uses 5 gal of gasoline on a trip of 160 mi. At the same mileage rate, how much gasoline will a 384-mi trip require?

32. **SOCIAL SCIENCE** A car uses 12 L of gasoline in traveling 150 km. At that rate, how many liters of gasoline will be used in a trip of 400 km?

33. **BUSINESS AND FINANCE** Sveta earns $13,500 commission in 20 weeks in her new sales position. At that rate, how much will she earn in 1 year (52 weeks)?

34. **BUSINESS AND FINANCE** Kevin earned $165 interest for 1 year on an investment of $1,500. At the same rate, what amount of interest would be earned by an investment of $2,500?

35. **SOCIAL SCIENCE** A company is selling a natural insect control that mixes ladybug beetles and praying mantises in the ratio of 7 to 4. If there are a total of 110 insects per package, how many of each type of insect is in a package?

Answers

25. _____

26. _____

27. _____

28. _____

29. _____

30. _____

31. _____

32. _____

33. _____

34. _____

35. _____

Answers

36. _____

37. _____

38. _____

36. SOCIAL SCIENCE A woman casts a 4-ft shadow. At the same time, a 72-ft building casts a 48-ft shadow. How tall is the woman?

37. BUSINESS AND FINANCE A brother and sister are to divide an inheritance of $12,000 in the ratio of 2 to 3. What amount will each receive?

38. BUSINESS AND FINANCE In Bucks County, the property tax rate is $25.32 per $1,000 of assessed value. If a house and property have a value of $128,000, find the tax the owner will have to pay.

Assessed Value: $128,000

Tax Rate: $25.32 per $1,000

Answers

1. 30   **3.** 20   **5.** 15, 16   **7.** 6, 12   **9.** 3, 12   **11.** 7, 35
13. 5   **15.** 20 mi/h bicycling, 50 mi/h driving   **17.** Express 55 mi/h, local 45 mi/h   **19.** 3 h, 2 h   **21.** 5 h, 3 h   **23.** 5 h, 2 h
25. 4 mi/h   **27.** 10 oz   **29.** 220 ft/s   **31.** 12 gal   **33.** $35,100
35. 70 ladybugs, 40 praying mantises   **37.** Brother $4,800, sister $7,200

Beginning Algebra   The Streeter/Hutchison Series in Mathematics

Definition/Procedure	Example	Reference

Simplifying Rational Expressions

Section 5.1

Rational Expressions

These have the form

Numerator

Fraction bar $\longrightarrow \dfrac{P}{Q}$

Denominator

in which P and Q are polynomials and Q cannot have the value 0.

$\dfrac{x^2 - 3x}{x - 2}$ is a rational expression. The variable x cannot have the value 2.

p. 454

Writing in Simplest Form

A fraction is in simplest form if its numerator and denominator have no common factors other than 1.
To write in simplest form:

1. Factor the numerator and denominator.
2. Divide the numerator and denominator by all common factors. The resulting fraction will be in simplest form.

$\dfrac{x + 2}{x - 1}$ is in simplest form.

$\dfrac{x^2 - 4}{x^2 - 2x - 8}$

$= \dfrac{(x - 2)(x + 2)}{(x - 4)(x + 2)}$

$= \dfrac{(x - 2)\cancel{(x + 2)}}{(x - 4)\cancel{(x + 2)}}$

$= \dfrac{x - 2}{x - 4}$

p. 455

Multiplying and Dividing Rational Expressions

Section 5.2

Multiplying Rational Expressions

$\dfrac{P}{Q} \cdot \dfrac{R}{S} = \dfrac{PR}{QS}$

in which $Q \neq 0$ and $S \neq 0$.

$\dfrac{2}{3} \cdot \dfrac{4}{5} = \dfrac{2 \cdot 4}{3 \cdot 5} = \dfrac{8}{15}$

p. 464

Multiplying Rational Expressions

Step 1 Factor the numerators and denominators.

Step 2 Write the product of the factors of the numerators over the product of the factors of the denominators.

Step 3 Divide the numerator and denominator by any common factors.

$\dfrac{2x - 4}{x^2 - 4} \cdot \dfrac{x^2 + 2x}{6x + 18}$

$= \dfrac{2(x - 2) \cdot x(x + 2)}{(x - 2)(x + 2) \cdot 6(x + 3)}$

$= \dfrac{2\cancel{(x - 2)} \cdot x\cancel{(x + 2)}}{\cancel{(x - 2)}\cancel{(x + 2)} \cdot 6(x + 3)}$

$= \dfrac{x}{3(x + 3)}$

p. 464

Continued

Definition/Procedure	Example	Reference

Dividing Rational Expressions

$$\frac{P}{Q} \div \frac{R}{S} = \frac{P}{Q} \cdot \frac{S}{R}$$

in which $Q \neq 0$, $R \neq 0$, and $S \neq 0$. In words, invert the divisor (the second fraction) and multiply.

$$\frac{4}{9} \div \frac{8}{12}$$
$$= \frac{4}{9} \cdot \frac{12}{8} = \frac{2}{3}$$
$$\frac{3x}{2x - 6} \div \frac{9x^2}{x^2 - 9}$$
$$= \frac{3x}{2x - 6} \cdot \frac{x^2 - 9}{9x^2}$$
$$= \frac{3x \cdot (x + 3)(x - 3)}{2(x - 3) \cdot 9x^2}$$
$$= \frac{x + 3}{6x}$$

p. 466

Adding and Subtracting Like Rational Expressions

Like Rational Expressions

1. Add or subtract the numerators.
2. Write the sum or difference over the common denominator.
3. Write the resulting fraction in simplest form.

$$\frac{2x}{x^2 + 3x} + \frac{6}{x^2 + 3x}$$
$$= \frac{2x + 6}{x^2 + 3x}$$
$$= \frac{2(x + 3)}{x(x + 3)} = \frac{2}{x}$$

Section 5.3

p. 472

Adding and Subtracting Unlike Rational Expressions

The Least Common Denominator

Finding the LCD:

1. Factor each denominator.
2. Write each factor the greatest number of times it appears in any single denominator.
3. The LCD is the product of the factors found in step 2.

For $\dfrac{2}{x^2 + 2x + 1}$

and $\dfrac{3}{x^2 + x}$

Factor:
$$x^2 + 2x + 1 = (x + 1)(x + 1)$$
$$x^2 + x = x(x + 1)$$
The LCD is $x(x + 1)(x + 1)$.

Section 5.4

p. 479

Definition/Procedure	Example	Reference

Unlike Rational Expressions

To add or subtract unlike rational expressions:

1. Find the LCD.
2. Convert each rational expression to an equivalent rational expression with the LCD as a common denominator.
3. Add or subtract the like rational expressions formed.
4. Write the sum or difference in simplest form.

$$\frac{2}{x^2 + 2x + 1} - \frac{3}{x^2 + x}$$

$$= \frac{2x}{x(x + 1)(x + 1)}$$

$$- \frac{3(x + 1)}{x(x + 1)(x + 1)}$$

$$= \frac{2x - 3x - 3}{x(x + 1)(x + 1)}$$

$$= \frac{-x - 3}{x(x + 1)(x + 1)}$$

p. 480

Complex Rational Expressions

Section 5.5

Simplifying Complex Fractions

$$\frac{\dfrac{a}{b}}{\dfrac{c}{d}} = \frac{a}{b} \div \frac{c}{d} = \frac{a}{b} \cdot \frac{d}{c}$$

$$\frac{\dfrac{3}{8}}{\dfrac{5}{6}} = \frac{3}{8} \div \frac{5}{6}$$

$$= \frac{3}{\overset{}{\underset{4}{8}}} \cdot \frac{\overset{3}{6}}{5}$$

$$= \frac{9}{20}$$

p. 495

Equations Involving Rational Expressions

1. Remove the fractions in the equation by multiplying each term by the LCD of all the fractions.
2. Solve the equation resulting from step 1.
3. Check your solution using the original equation. Discard any extraneous solutions.

$$2 - \frac{4}{x} = \frac{2}{3}$$

LCD: $3x$

$$2(3x) - \frac{4}{x}(3x) = \frac{2}{3}(3x)$$

$$6x - 12 = 2x$$

$$4x = 12$$

$$x = 3$$

Section 5.6

p. 504

This summary exercise set is provided to give you practice with each of the objectives of this chapter. Each exercise is keyed to the appropriate chapter section. When you are finished, you can check your answers to the odd-numbered exercises against those presented in the back of the text. If you have difficulty with any of these questions, go back and reread the examples from that section. Your instructor will give you guidelines on how best to use these exercises in your instructional setting.

5.1 *Write each fraction in simplest form.*

1. $\dfrac{6a^2}{9a^3}$

2. $\dfrac{-12x^4y^3}{18x^2y^2}$

3. $\dfrac{w^2 - 25}{2w - 8}$

4. $\dfrac{3x^2 + 11x - 4}{2x^2 + 11x + 12}$

5. $\dfrac{m^2 - 2m - 3}{9 - m^2}$

6. $\dfrac{3c^2 - 2cd - d^2}{6c^2 + 2cd}$

5.2 *Multiply or divide as indicated.*

7. $\dfrac{6x}{5} \cdot \dfrac{10}{18x^2}$

8. $\dfrac{-2a^2}{ab^3} \cdot \dfrac{3ab^2}{-4ab}$

9. $\dfrac{2x + 6}{x^2 - 9} \cdot \dfrac{x^2 - 3x}{4}$

10. $\dfrac{a^2 + 5a + 4}{2a^2 + 2a} \cdot \dfrac{a^2 - a - 12}{a^2 - 16}$

11. $\dfrac{3p}{5} \div \dfrac{9p^2}{10}$

12. $\dfrac{8m^3}{5mn} \div \dfrac{12m^2n^2}{15mn^3}$

13. $\dfrac{x^2 + 7x + 10}{x^2 + 5x} \div \dfrac{x^2 - 4}{2x^2 - 7x + 6}$

14. $\dfrac{2w^2 + 11w - 21}{w^2 - 49} \div (4w - 6)$

15. $\dfrac{a^2b + 2ab^2}{a^2 - 4b^2} \div \dfrac{4a^2b}{a^2 - ab - 2b^2}$

16. $\dfrac{2x^2 + 6x}{4x} \cdot \dfrac{6x + 12}{x^2 + 2x - 3} \div \dfrac{x^2 - 4}{x^2 - 3x + 2}$

5.3 *Add or subtract as indicated.*

17. $\dfrac{x}{9} + \dfrac{2x}{9}$

18. $\dfrac{7a}{15} - \dfrac{2a}{15}$

19. $\dfrac{8}{x + 2} + \dfrac{3}{x + 2}$

20. $\dfrac{y - 2}{5} - \dfrac{2y + 3}{5}$

21. $\dfrac{7r - 3s}{4r} + \dfrac{r - s}{4r}$

22. $\dfrac{x^2}{x - 4} - \dfrac{16}{x - 4}$

23. $\dfrac{5w - 6}{w - 4} - \dfrac{3w + 2}{w - 4}$

24. $\dfrac{x + 3}{x^2 - 2x - 8} + \dfrac{2x + 3}{x^2 - 2x - 8}$

The Streeter/Hutchison Series in Mathematics   Beginning Algebra

5.4 *Add or subtract as indicated.*

25. $\dfrac{5x}{6} + \dfrac{x}{3}$

26. $\dfrac{3y}{10} - \dfrac{2y}{5}$

27. $\dfrac{5}{2m} - \dfrac{3}{m^2}$

28. $\dfrac{x}{x-3} - \dfrac{2}{3}$

29. $\dfrac{4}{x-3} - \dfrac{1}{x}$

30. $\dfrac{2}{s+5} + \dfrac{3}{s+1}$

31. $\dfrac{5}{w-5} - \dfrac{2}{w-3}$

32. $\dfrac{4x}{2x-1} + \dfrac{2}{1-2x}$

33. $\dfrac{2}{3x-3} - \dfrac{5}{2x-2}$

34. $\dfrac{4y}{y^2-8y+15} + \dfrac{6}{y-3}$

35. $\dfrac{3a}{a^2+5a+4} + \dfrac{2a}{a^2-1}$

36. $\dfrac{3x}{x^2+2x-8} - \dfrac{1}{x-2} + \dfrac{1}{x+4}$

5.5 *Simplify the complex fractions.*

37. $\dfrac{\dfrac{x^2}{12}}{\dfrac{x^3}{8}}$

38. $\dfrac{3 + \dfrac{1}{a}}{3 - \dfrac{1}{a}}$

39. $\dfrac{1 + \dfrac{x}{y}}{1 - \dfrac{x}{y}}$

40. $\dfrac{1 + \dfrac{1}{p}}{p^2 - 1}$

41. $\dfrac{\dfrac{1}{m} - \dfrac{1}{n}}{\dfrac{1}{m} + \dfrac{1}{n}}$

42. $\dfrac{2 - \dfrac{x}{y}}{4 - \dfrac{x^2}{y^2}}$

43. $\dfrac{\dfrac{2}{a+1} + 1}{1 - \dfrac{4}{a+1}}$

44. $\dfrac{\dfrac{a}{b} - 1 - \dfrac{2b}{a}}{\dfrac{1}{b^2} - \dfrac{1}{a^2}}$

5.6 *What values for x, if any, must be excluded in the following rational expressions?*

45. $\dfrac{x}{5}$

46. $\dfrac{3}{x - 4}$

47. $\dfrac{2}{(x + 1)(x - 2)}$

48. $\dfrac{7}{x^2 - 16}$

49. $\dfrac{x - 1}{x^2 + 3x + 2}$

50. $\dfrac{2x + 3}{3x^2 + x - 2}$

Solve the following proportions for x.

51. $\dfrac{x - 3}{8} = \dfrac{x - 2}{10}$

52. $\dfrac{1}{x - 3} = \dfrac{7}{x^2 - x - 6}$

Solve the following equations for x.

53. $\dfrac{x}{4} - \dfrac{x}{5} = 2$

54. $\dfrac{13}{4x} + \dfrac{3}{x^2} = \dfrac{5}{2x}$

55. $\dfrac{x}{x - 2} + 1 = \dfrac{x + 4}{x - 2}$

56. $\dfrac{x}{x - 4} - 3 = \dfrac{4}{x - 4}$

57. $\dfrac{x}{2x - 6} - \dfrac{x - 4}{x - 3} = \dfrac{1}{8}$

58. $\dfrac{7}{x} - \dfrac{1}{x - 3} = \dfrac{9}{x^2 - 3x}$

59. $\dfrac{x}{x - 5} = \dfrac{3x}{x^2 - 7x + 10} + \dfrac{8}{x - 2}$

60. $\dfrac{6}{x + 5} + 1 = \dfrac{3}{x - 5}$

61. $\dfrac{24}{x + 2} - 2 = \dfrac{2}{x - 3}$

5.7 *Solve the following applications.*

62. Number Problem If two-fifths of a number is added to one-half of that number, the sum is 27. Find the number.

63. Number Problem One number is 3 times another. If the sum of their reciprocals is $\dfrac{1}{3}$, what are the two numbers?

64. Number Problem If the reciprocal of 4 times a number is subtracted from the reciprocal of that number, the result is $\dfrac{1}{8}$. What is the number?

65. SCIENCE AND MEDICINE Robert made a trip of 240 mi. Returning by a different route, he found that the distance was only 200 mi, but traffic slowed his speed by 8 mi/h. If the trip took the same amount of time in both directions, what was Robert's rate each way?

66. SCIENCE AND MEDICINE On the first day of a vacation trip, Jovita drove 225 mi. On the second day it took her 1 h longer to drive 270 mi. If her average speed was the same on both days, how long did she drive each day?

67. SCIENCE AND MEDICINE A light plane flies 700 mi against a steady 20-mi/h headwind and then returns, with the wind, to the same point. If the entire trip took 12 h, what was the speed of the plane in still air?

68. SCIENCE AND MEDICINE How much pure alcohol should be added to 300 mL of a 30% solution to obtain a 40% solution?

69. SCIENCE AND MEDICINE A chemist has a 10% acid solution and a 40% solution. How much of the 40% solution should be added to 300 mL of the 10% solution to produce a mixture with a concentration of 20%?

70. BUSINESS AND FINANCE Melina wants to invest a total of $10,800 in two types of savings accounts. If she wants the ratio of the amounts deposited in the two accounts to be 4 to 5, what amount should she invest in each account?

Name _____

Section _____ Date _____

Answers

The purpose of this self-test is to help you check your progress and to review for the next in-class exam. Allow yourself about an hour to take this test. At the end of that hour check your answers against those given in the back of the text. Section references accompany the answers. If you missed any questions, go back to those sections and reread the examples until you master the concepts.

1. _____

2. _____

3. _____

4. _____

5. _____

6. _____

7. _____

8. _____

9. _____

10. _____

11. _____

12. _____

13. _____

14. _____

15. _____

16. _____

Write each fraction in simplest form.

1. $\dfrac{-21x^5y^3}{28xy^5}$

2. $\dfrac{4a - 24}{a^2 - 6a}$

3. $\dfrac{3x^2 + x - 2}{3x^2 - 8x + 4}$

Add or subtract as indicated.

4. $\dfrac{3a}{8} + \dfrac{5a}{8}$

5. $\dfrac{2x}{x + 3} + \dfrac{6}{x + 3}$

6. $\dfrac{7x - 3}{x - 2} - \dfrac{2x + 7}{x - 2}$

7. $\dfrac{x}{3} + \dfrac{4x}{5}$

8. $\dfrac{3}{s} - \dfrac{2}{s^2}$

9. $\dfrac{5}{x - 2} - \dfrac{1}{x + 3}$

10. $\dfrac{6}{w - 2} + \dfrac{9w}{w^2 - 7w + 10}$

Multiply or divide as indicated.

11. $\dfrac{3pq^2}{5pq^3} \cdot \dfrac{20p^2q}{21q}$

12. $\dfrac{x^2 - 3x}{5x^2} \cdot \dfrac{10x}{x^2 - 4x + 3}$

13. $\dfrac{2x^2}{3xy} \div \dfrac{8x^2y}{9xy}$

14. $\dfrac{3m - 9}{m^2 - 2m} \div \dfrac{m^2 - m - 6}{m^2 - 4}$

Simplify the complex fractions.

15. $\dfrac{\dfrac{x^2}{18}}{\dfrac{x^3}{12}}$

16. $\dfrac{2 - \dfrac{m}{n}}{4 - \dfrac{m^2}{n^2}}$

What values for x, if any, must be excluded in the following rational expressions?

17. $\dfrac{8}{x - 4}$

18. $\dfrac{3}{x^2 - 9}$

Solve the following equations for x.

19. $\dfrac{x}{3} - \dfrac{x}{4} = 3$

20. $\dfrac{5}{x} - \dfrac{x - 3}{x + 2} = \dfrac{22}{x^2 + 2x}$

Solve the following proportions.

21. $\dfrac{x - 1}{5} = \dfrac{x + 2}{8}$

22. $\dfrac{2x - 1}{7} = \dfrac{x}{4}$

Solve the following applications.

23. NUMBER PROBLEM One number is 3 times another. If the sum of their reciprocals is $\dfrac{1}{3}$, find the two numbers.

24. SCIENCE AND MEDICINE Mark drove 250 mi to visit Sandra. Returning by a shorter route, he found that the trip was only 225 mi, but traffic slowed his speed by 5 mi/h. If the two trips took exactly the same time, what was his rate each way?

25. CONSTRUCTION A cable that is 55 ft long is to be cut into two pieces whose lengths have the ratio 4 to 7. Find the lengths of the two pieces.

Answers

17. _____

18. _____

19. _____

20. _____

21. _____

22. _____

23. _____

24. _____

25. _____

Activity 5 ::
Determining State Apportionment

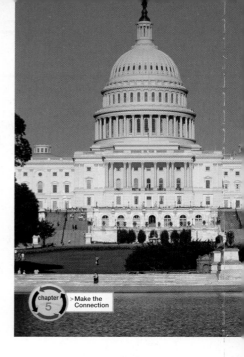

The introduction to this chapter referred to the ratio of the people in a particular state to their total number of representatives in the U.S. House based on the 2000 census. It was noted that the ratio of the total population of the country to the 435 representatives in Congress should equal the state apportionment if it is fair. That is, $\dfrac{A}{a} = \dfrac{P}{r}$, where A is the population of the state, a is the number of representatives for that state, P is the total population of the U.S., and r is the total number of representatives in Congress (435).

Pick 5 states (your own included) and search the Internet to find the following.

1. Determine the year 2000 population of each state.

2. Note the number of representatives for each state and any increase or decrease.

3. Find the number of people per representative for each state.

4. Compare that with the national average of the number of people per representative.

5. Solve the rational equation $\dfrac{A}{a} = \dfrac{P}{r}$ for a. For each state substitute the number values for the variables, A, P, and r. Find a. Based on your findings which states have

   (a) a greater number of representatives than they should (that is, the number has been rounded up), and

   (b) which states have a smaller number of representative than they should (that is, the number has been rounded down)?

You can find out more about apportionment counts and how they are determined from the U.S. Census website.

The following questions are presented to help you review concepts from earlier chapters. This is meant as a review and not as a comprehensive exam. The answers are presented in the back of the text. Section references accompany the answers. If you have difficulty with any of these questions, be certain to at least read through the summary related to those sections.

Name _____

Section _____ Date _____

Answers

Perform the indicated operation.

1. $x^2y - 4xy - x^2y + 2xy$

2. $\dfrac{12a^3b}{9ab}$

3. $(5x^2 - 2x + 1) - (3x^2 + 3x - 5)$

4. $(5a^2 + 6a) - (2a^2 - 1)$

5. $4 + 3(7 - 4)^2$

6. $|3 - 5| - |-4 + 3|$

Multiply.

7. $(x - 2y)(2x + 3y)$

8. $(x + 7)(x + 4)$

Divide.

9. $(2x^2 + 3x - 1) \div (x + 2)$

10. $(x^2 - 5) \div (x - 1)$

Solve each equation and check your results.

11. $4x - 3 = 2x + 5$

12. $2 - 3(2x + 1) = 11$

Factor each polynomial completely.

13. $x^2 - 5x - 14$

14. $3m^2n - 6mn^2 + 9mn$

15. $a^2 - 9b^2$

16. $2x^3 - 28x^2 + 96x$

Solve the following word problems. Show the equation used for each solution.

17. NUMBER PROBLEM 2 more than 4 times a number is 30. Find the number.

1. _____
2. _____
3. _____
4. _____
5. _____
6. _____
7. _____
8. _____
9. _____
10. _____
11. _____
12. _____
13. _____
14. _____
15. _____
16. _____
17. _____

Answers

18. _____

19. _____

20. _____

21. _____

22. _____

23. _____

24. _____

25. _____

26. _____

27. _____

28. _____

29. _____

30. _____

18. NUMBER PROBLEM If the reciprocal of 4 times a number is subtracted from the reciprocal of that number, the result is $\dfrac{3}{16}$. What is the number?

19. SCIENCE AND MEDICINE A speed of 60 mi/h corresponds to 88 ft/s. If a race car is traveling at 180 mi/h, what is its speed in feet per second?

20. GEOMETRY The length of a rectangle is 3 in. less than twice its width. If the area of the rectangle is 35 in.^{2}, find the dimensions of the rectangle.

Write each rational expression in simplest form.

21. $\dfrac{m^2 - 4m}{3m - 12}$

22. $\dfrac{a^2 - 49}{3a^2 + 22a + 7}$

Perform the indicated operations.

23. $\dfrac{4}{3r} + \dfrac{1}{2r^2}$

24. $\dfrac{2}{x - 3} - \dfrac{5}{3x + 9}$

25. $\dfrac{3x^2 + 9x}{x^2 - 9} \cdot \dfrac{2x^2 - 9x + 9}{2x^3 - 3x^2}$

26. $\dfrac{4w^2 - 25}{2w^2 - 5w} \div (6w + 15)$

Simplify the complex rational expressions.

27. $\dfrac{1 - \dfrac{1}{x}}{2 + \dfrac{1}{x}}$

28. $\dfrac{3 - \dfrac{m}{n}}{9 - \dfrac{m^2}{n^2}}$

Solve the following equations for x.

29. $\dfrac{5}{3x} + \dfrac{1}{x^2} = \dfrac{5}{2x}$

30. $\dfrac{10}{x - 3} - 2 = \dfrac{5}{x + 3}$

CHAPTER

6

chapter 6 > Make the Connection

INTRODUCTION

Graphs are used to discern patterns that may be difficult to see when looking at a list of numbers or other kinds of data. The word *graph* comes from Latin and Greek roots and means "to draw a picture." A graph in mathematics is a picture of a relationship between variables. Graphs are used in every field in which numbers are used.

In the field of pediatric medicine there has been controversy about the use of human growth hormone to help children whose growth has been impeded by health problems. The reason for the controversy is that many doctors are giving this expensive drug therapy to children who are simply shorter than average or shorter than their parents want them to be. The determination of which children are healthy but small in stature and which children have health defects that keep them from growing is an issue that has been vigorously argued in medical research.

Measures used to distinguish between the two groups include blood tests and age and height measurements. These measurements are graphed and monitored over several years to gauge the child's rate of growth. If this rate of growth slows to below 4.5 centimeters per year, this indicates that there may be a problem. The graph can also indicate whether the child's size is within a range considered normal at each age of the child's life.

An Introduction to Graphing

CHAPTER 6 OUTLINE

6 pretest

Name _____

Section _____ Date _____

This pretest provides a preview of the types of exercises you will encounter in each section of this chapter. The answers for these exercises can be found in the back of the text. If you are working on your own, or ahead of the class, this pretest can help you find the sections in which you should focus more of your time.

Answers

1. _____

2. _____

3. _____

4. _____

5. _____

6. _____

7. _____

8. _____

Determine which of the ordered pairs are solutions for the given equations.

6.1   **1.** $x - y = 12$; $(15, 3), (9, 6), (18, 6)$

2. $3x + 2y = 6$; $(1, 2), (0, 3), (2, 0)$

3. Complete the ordered pairs so that each is a solution for the given equation.
   $2x + y = 5$; $(1, \ ), (0, \ ), (\ , 11)$

4. Find three solutions for each of the following equations.
   $2x - 3y = 8$ $6x + y = 11$

6.2   **5.** Give the coordinates of the points graphed below.

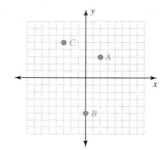

6. Plot the points with the given coordinates: $S(-1, 2)$, $T(3, 0)$.

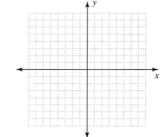

Graph each of the following equations.

6.3   **7.** $x + y = 5$

8. $y = \dfrac{1}{2}x - 1$

9. $y = -2$

Find the slope of the line through the following pairs of points.

6.4 **10.** $(-2, 5)$ and $(5, 12)$ **11.** $(-1, -3)$ and $(7, -11)$

12. Find the constant of variation k if y varies directly with x, and $y = 56$ when $x = 7$.

6.5 **STATISTICS** The total commercial fishery landings at the Cameron, LA, port (in millions of pounds, rounded) for each of the years 1997 to 2001 is shown in the table below. Use this table to answer exercises 13 and 14.

Commercial Fishery Landings; Cameron, LA

Year	1997	1998	1999	2000	2001
Landings	380	257	406	415	324

Source: National Marine Fisheries Service, Fisheries Statistics and Economics Division.

13. Construct a line graph displaying the given data.

14. Predict the commercial fishery landings at the Cameron, LA, port for 2002.

6.4 **15.** **BUSINESS AND FINANCE** Pete's commission varies directly with the number of appliances he sells. If his commission last month was $800 and he sold 20 appliances, what would his salary be if he sold 25 appliances?

Answers

9. _____

10. _____

11. _____

12. _____

13. _____

14. _____

15. _____

6.1

Solutions of Equations in Two Variables

< 6.1 Objectives >

1 > Find solutions for an equation in two variables

2 > Use ordered-pair notation to write solutions for equations in two variables

We discussed finding solutions for equations in Chapter 2. Recall that a solution is a value for the variable that "satisfies" the equation or makes the equation a true statement. For instance, we know that 4 is a solution of the equation

$$2x + 5 = 13$$

We know this is true because, when we replace x with 4, we have

$$2(4) + 5 \stackrel{?}{=} 13$$
$$8 + 5 \stackrel{?}{=} 13$$
$$13 = 13 \qquad \text{A true statement}$$

RECALL

An equation is two expressions connected by an equal sign.

We now want to consider **equations in two variables.** In this chapter we study equations of the form $Ax + By = C$, in which A and B are not both 0. Such equations are called **linear equations,** and are said to be in **standard form.** An example is

$$x + y = 5$$

What will a solution look like? It is not going to be a single number, because there are two variables. Here a solution is a pair of numbers—one value for each of the variables, x and y. Suppose that x has the value 3. In the equation $x + y = 5$, you can substitute 3 for x.

$$(3) + y = 5$$

Solving for y gives

$$y = 2$$

NOTE

The solution of an equation in two variables "pairs" two numbers, one for x and one for y.

So the pair of values $x = 3$ and $y = 2$ satisfies the equation because

$$3 + 2 = 5$$

The pair of numbers that satisfies an equation is called a **solution** for the equation in two variables.

How many such pairs are there? Choose any value for x (or for y). You can always find the other *paired* or *corresponding* value in an equation of this form. We say that there are an *infinite* number of pairs that satisfy the equation. Each of these pairs is a solution. We find some other solutions for the equation $x + y = 5$ in Example 1.

| **Example** 1 | Solving for Corresponding Values |

< Objective 1 >

For the equation $x + y = 5$, find (a) y if $x = 5$ and (b) x if $y = 4$.

(a) If $x = 5$

$(5) + y = 5$ or $y = 0$

(b) If $y = 4$,

$x + (4) = 5$ or $x = 1$

So the pairs $x = 5, y = 0$ and $x = 1, y = 4$ are both solutions.

 Check Yourself 1

For the equation $2x + 3y = 26$,

(a) If $x = 4, y = ?$ (b) If $y = 0, x = ?$

To simplify writing the pairs that satisfy an equation, we use **ordered-pair notation.** The numbers are written in parentheses and are separated by a comma. For example, we know that the values $x = 3$ and $y = 2$ satisfy the equation $x + y = 5$. So we write the pair as

The first number of the pair is *always* the value for x and is called the **x-coordinate.** The second number of the pair is *always* the value for y and is the **y-coordinate.**

Using this ordered-pair notation, we can say that $(3, 2)$, $(5, 0)$, and $(1, 4)$ are all *solutions* for the equation $x + y = 5$. Each pair gives values for x and y that satisfy the equation.

> **CAUTION**

$(3, 2)$ means $x = 3$ and $y = 2$.
$(2, 3)$ means $x = 2$ and $y = 3$.
$(3, 2)$ and $(2, 3)$ are different, which is why we call them *ordered pairs*.

| **Example** 2 | Identifying Solutions of Two-Variable Equations |

< Objective 2 >

Which of the ordered pairs (a) $(2, 5)$, (b) $(5, -1)$, and (c) $(3, 4)$ are solutions for the equation $2x + y = 9$?

(a) To check whether $(2, 5)$ is a solution, let $x = 2$ and $y = 5$ and see whether the equation is satisfied.

$2x + y = 9$ The original equation

$x \quad y$

$2(2) + (5) \stackrel{?}{=} 9$ Substitute 2 for x and 5 for y.

$4 + 5 \stackrel{?}{=} 9$

$9 = 9$ A true statement

$(2, 5)$ is a solution for the equation.

NOTE

$(2, 5)$ is a solution because a
true statement results.

(b) For $(5, -1)$, let $x = 5$ and $y = -1$.

$$2(5) + (-1) \stackrel{?}{=} 9$$
$$10 - 1 \stackrel{?}{=} 9$$
$$9 = 9 \qquad \text{A true statement}$$

So $(5, -1)$ is a solution.

(c) For $(3, 4)$, let $x = 3$ and $y = 4$. Then

$$2(3) + (4) \stackrel{?}{=} 9$$
$$6 + 4 \stackrel{?}{=} 9$$
$$10 \stackrel{?}{=} 9 \qquad \textit{Not} \text{ a true statement}$$

So $(3, 4)$ is *not* a solution for the equation.

Check Yourself 2

Which of the ordered pairs $(3, 4)$, $(4, 3)$, $(1, -2)$, and $(0, -5)$ are solutions for the following equation?

$$3x - y = 5$$

If the equation contains only one variable, then the missing variable can take on any value.

| Example 3 | Identifying Solutions of One-Variable Equations |

NOTE

Think of this equation as

$$1 \cdot x + 0 \cdot y = 2$$

Which of the ordered pairs $(2, 0)$, $(0, 2)$, $(5, 2)$, $(2, 5)$, and $(2, -1)$ are solutions for the equation $x = 2$?

A solution is any ordered pair in which the x-coordinate is 2. That makes $(2, 0)$, $(2, 5)$, and $(2, -1)$ solutions for the given equation.

Check Yourself 3

Which of the ordered pairs $(3, 0)$, $(0, 3)$, $(3, 3)$, $(-1, 3)$, and $(3, -1)$ are solutions for the equation $y = 3$?

Remember that, when an ordered pair is presented, the first number is always the x-coordinate and the second number is always the y-coordinate.

| Example 4 | Completing Ordered-Pair Solutions |

NOTE

The x-coordinate is also called the **abscissa** and the y-coordinate the **ordinate**.

Complete the ordered pairs (a) $(9, \ )$, (b) $(\ , -1)$, (c) $(0, \ )$, and (d) $(\ , 0)$ for the equation $x - 3y = 6$.

(a) The first number, 9, appearing in $(9, \ )$ represents the x-value. To complete the pair $(9, \ )$, substitute 9 for x and then solve for y.

$$(9) - 3y = 6$$
$$-3y = -3$$
$$y = 1$$

$(9, 1)$ is a solution.

(b) To complete the pair (, −1), let y be −1 and solve for x.

$$x - 3(-1) = 6$$
$$x + 3 = 6$$
$$x = 3$$

$(3, -1)$ is a solution.

(c) To complete the pair $(0, \ )$, let x be 0.

$$(0) - 3y = 6$$
$$-3y = 6$$
$$y = -2$$

$(0, -2)$ is a solution.

(d) To complete the pair (, 0), let y be 0.

$$x - 3(0) = 6$$
$$x - 0 = 6$$
$$x = 6$$

$(6, 0)$ is a solution.

Check Yourself 4

Complete the given ordered pairs so that each is a solution for the equation $2x + 5y = 10$.

(10,),   (, 4),   (0,),   and   (, 0)

Example 5 | Finding Some Solutions of a Two-Variable Equation

Find four solutions for the equation

$$2x + y = 8$$

In this case the values used to form the solutions are *up to you.* You can assign any value for x (or for y). We demonstrate with some possible choices.

Solution with $x = 2$:

$$2(2) + y = 8$$
$$4 + y = 8$$
$$y = 4$$

$(2, 4)$ is a solution.

Solution with $y = 6$:

$$2x + (6) = 8$$
$$2x = 2$$
$$x = 1$$

$(1, 6)$ is a solution.

NOTE

Generally, you want to pick values for x (or for y) so that the resulting equation in one variable is easy to solve.

NOTE

The solutions (0, 8) and (4, 0) have special significance when in graphing. They are also easy to find!

Solution with $x = 0$:

$2(0) + y = 8$

$\quad\quad y = 8$

(0, 8) is a solution.

Solution with $y = 0$:

$2x + (0) = 8$

$\quad\; 2x = 8$

$\quad\;\; x = 4$

(4, 0) is a solution.

Check Yourself 5

Find four solutions for $x - 3y = 12$.

Applications involving two-variable equations are fairly common.

▶   **Example 6**   **Applications of Two-Variable Equations**

NOTE

We will look at variable and fixed costs in more detail in Section 7.1.

Suppose that it costs the manufacturer $1.25 for each stapler that is produced. In addition, fixed costs (related to staplers) are $110 per day.

(a) Write an equation relating the total daily costs C to the number x of staplers produced in a day.

Because each stapler costs $1.25 to produce, the cost of producing staplers is $1.25x$. Adding the fixed cost to this gives us an equation for the total daily costs.

$C = 1.25x + 110$

(b) What is the total cost of producing 500 staplers in a day?

We substitute 500 for x in the equation from part **(a)** and calculate the total cost.

$C = 1.25(500) + 110$

$\quad = 625 + 110$

$\quad = 735$

It costs the manufacturer a total of $735 to produce 500 staplers in one day.

RECALL

Divide both sides by 1.25.

$$\frac{1,000}{1.25} = \frac{\cancel{1.25}x}{\cancel{1.25}}$$

$800 = x$

(c) How many staplers can be produced for $1,110?

In this case, we substitute 1,110 for C in the equation from part **(a)** and solve for x.

$(1,110) = 1.25x + 110$ Subtract 110 from both sides.

$\quad\; 1,000 = 1.25x$ Divide both sides by 1.25.

$\quad\quad\; 800 = x$

800 staplers can be produced at a cost of $1,110.

Check Yourself 6

Suppose that the stapler manufacturer earns a profit of $1.80 on each stapler shipped. However, it costs $120 to operate each day.

(a) Write an equation relating the daily profit P to the number x of staplers shipped in a day.
(b) What is the total profit of shipping 500 staplers in a day?
(c) How many staplers need to be shipped to produce a profit of $1,500?

We close this section with an application from the field of medicine.

Example 7	Vocational Application

For a particular patient, the weight w, in grams, of a tumor is related to the number of days d of chemotherapy treatment by the following equation:

$$w = -1.75d + 25$$

(a) What was the original weight of the tumor?

The original weight of the tumor will be the value of w when $d = 0$. Substituting 0 for d in the equation gives

$$w = -1.75(0) + 25$$
$$= 25$$

The original weight of the tumor was 25 grams.

(b) How many days of chemotherapy are required to eliminate the tumor?
The tumor will be eliminated when the weight (w) is 0. So

$$(0) = -1.75d + 25$$
$$-25 = -1.75d$$
$$d \approx 14.3$$

It will take about 14.3 days to eliminate the tumor.

Check Yourself 7

For a particular patient, the weight (w), in grams, of a tumor is related to the number of days (d) of chemotherapy treatment by the following equation:

$$w = -1.6d + 32$$

(a) Find the original weight of the tumor.
(b) Determine the number of days of chemotherapy required to eliminate the tumor.

Check Yourself ANSWERS

1. **(a)** $y = 6$; **(b)** $x = 13$ 2. $(3, 4)$, $(1, -2)$, and $(0, -5)$ are solutions
3. $(0, 3)$, $(3, 3)$, and $(-1, 3)$ are solutions   4. $(10, -2)$, $(-5, 4)$, $(0, 2)$, and $(5, 0)$
5. $(6, -2)$, $(3, -3)$, $(0, -4)$, and $(12, 0)$ are four possibilities
6. **(a)** $P = 1.80x - 120$; **(b)** \$780; **(c)** 900   7. **(a)** 32 g; **(b)** 20 days

Reading Your Text

The following fill-in-the-blank exercises are designed to ensure that you understand some of the key vocabulary used in this section.

SECTION 6.1

(a) A _____ is a value for the variable that makes the equation a true statement.

(b) An equation is two _____ connected by an equal sign.

(c) A _____ of numbers that satisfies an equation is called a solution for the equation in two variables.

(d) To simplify writing the pairs that satisfy an equation, we use _____ notation.

Basic Skills | Advanced Skills | Vocational-Technical Applications | Calculator/Computer | Above and Beyond | Getting Ready

▲

6.1 exercises

< Objectives 1–2 >

Determine which of the ordered pairs are solutions for the given equation.

1. $x + y = 6$ $\quad$ $(4, 2), (-2, 4), (0, 6), (-3, 9)$

2. $x - y = 12$ $\quad$ $(13, 1), (13, -1), (12, 0), (6, 6)$

3. $2x - y = 8$ $\quad$ $(5, 2), (4, 0), (0, 8), (6, 4)$

4. $x + 5y = 20$ $\quad$ $(10, -2), (10, 2), (20, 0), (25, -1)$

5. $3x - 2y = 12$ $\quad$ $(4, 0), \left(\dfrac{2}{3}, -5\right), (0, 6), \left(5, \dfrac{3}{2}\right)$ > Videos

6. $3x + 4y = 12$ $\quad$ $(-4, 0), \left(\dfrac{2}{3}, \dfrac{5}{2}\right), (0, 3), \left(\dfrac{2}{3}, 2\right)$

7. $y = 4x$ $\quad$ $(0, 0), (1, 3), (2, 8), (8, 2)$

8. $y = 2x - 1$ $\quad$ $(0, -2), (0, -1), \left(\dfrac{1}{2}, 0\right), (3, -5)$

9. $x = 3$ $\quad$ $(3, 5), (0, 3), (3, 0), (3, 7)$

10. $y = 5$ $\quad$ $(0, 5), (3, 5), (-2, -5), (5, 5)$

Complete the ordered pairs so that each is a solution for the given equation.

11. $x + y = 12$ $\quad$ $(4,\ ), (\ , 5), (0,\ ), (\ , 0)$

12. $x - y = 7$ $\quad$ $(\ , 4), (15,\ ), (0,\ ), (\ , 0)$

13. $3x - 2y = 12$ $\quad$ $(\ , 0), (\ , -6), (2,\ ), (\ , 3)$ > Videos

14. $2x + 5y = 20$ $\quad$ $(0,\ ), (5,\ ), (\ , 0), (\ , 6)$

15. $y = 3x + 9$ $\quad$ $(\ , 0), \left(\dfrac{2}{3},\ \right), (0,\ ), \left(-\dfrac{2}{3},\ \right)$

Name _____

Section _____ Date _____

Answers

1. _____

2. _____

3. _____

4. _____

5. _____

6. _____

7. _____

8. _____

9. _____

10. _____

11. _____

12. _____

13. _____

14. _____

15. _____

Answers

16. _____

17. _____

18. _____

19. _____

20. _____

21. _____

22. _____

23. _____

24. _____

25. _____

26. _____

27. _____

28. _____

29. _____

30. _____

31. _____

32. _____

16. $3x + 4y = 12$ $(0,\ ), \left(\ , \dfrac{3}{4}\right), (\ , 0), \left(\dfrac{8}{3},\ \right)$

17. $y = 3x - 4$ $(0,\ ), (\ , 5), (\ , 0), \left(\dfrac{5}{3},\ \right)$

18. $y = -2x + 5$ $(0,\ ), (\ , 5), \left(\dfrac{3}{2},\ \right), (\ , 1)$

Find four solutions for each of the following equations. **Note:** *Your answers may vary from those shown in the answer section.*

19. $x - y = 7$ > Videos

20. $x + y = 18$

21. $2x - y = 6$

22. $3x - y = 12$

23. $2x - 5y = 10$

24. $2x + 7y = 14$

25. $y = 2x + 3$

26. $y = 8x - 5$

27. $x = -5$ > Videos

28. $y = 8$

29. **BUSINESS AND FINANCE** When an employee produces x units per hour, the hourly wage in dollars is given by $y = 0.75x + 8$. What are the hourly wages for the following numbers of units: 2, 5, 10, 15, and 20?

30. **SCIENCE AND MEDICINE** Celsius temperature readings can be converted to Fahrenheit readings using the formula $F = \dfrac{9}{5}C + 32$. What is the Fahrenheit temperature that corresponds to each of the following Celsius temperatures: $-10, 0, 15, 100$?

31. **GEOMETRY** The perimeter of a square is given by $P = 4s$. What are the perimeters of the squares whose sides are 5, 10, 12, and 15 cm?

32. **BUSINESS AND FINANCE** When x units are sold, the price of each unit (in dollars) is given by $p = \dfrac{-x}{2} + 75$. Find the unit price when the following quantities are sold: 2, 7, 9, 11.

33. **STATISTICS** The number of programs for the disabled in the United States from 1993 to 1997 is approximated by the equation $y = 162x + 4{,}365$ in which x is the number of years after 1992. Complete the following table.

x	1	2	3	4	6
y					

Answers

33. _____

34. _____

35. _____

36. _____

37. _____

38. _____

34. **BUSINESS AND FINANCE** Your monthly pay as a car salesperson is determined using the equation $S = 200x + 1{,}500$ in which x is the number of cars you can sell each month.

    (a) Complete the following table.

x	12	15	17	18
S				

    (b) You are offered a job at a salary of $56,400 per year. How many cars would you have to sell per month to equal this salary?

Determine whether each of the following statements is **true** *or* **false.**

35. When finding solutions for the equation $1 \cdot x + 0 \cdot y = 5$, you can choose any number for x.

36. When finding solutions for the equation $1 \cdot x + 1 \cdot y = 5$, you can choose any number for x.

Complete each of the following statements with **never, sometimes,** *or* **always.**

37. If (a, b) is a solution to a particular two-variable equation, then (b, a) is _____ a solution to the same equation.

38. There are _____ an infinite number of solutions to a two-variable equation in standard form.

Answers

39. _____

40. _____

41. _____

42. _____

43. _____

44. _____

45. _____

46. _____

47. _____

48. _____

49. _____

50. _____

51. _____

52. _____

53. _____

54. _____

For the following exercises, find two solutions for each equation. **Note:** *Your answers may vary from those shown in the answer section.*

39. $\dfrac{1}{2}x + \dfrac{1}{3}y = 1$

40. $\dfrac{1}{3}x - \dfrac{1}{4}y = 1$

41. $0.3x + 0.5y = 2$

42. $0.6x - 0.2y = 5$

43. $\dfrac{3}{4}x - \dfrac{2}{5}y = 6$ > Videos

44. $\dfrac{4}{5}x + \dfrac{2}{3}y = 8$

45. $0.4x - 0.7y = 3$

46. $0.8x + 0.9y = 2$

An equation in three variables has an ordered triple as a solution. For example, $(1, 2, 2)$ is a solution to the equation $x + 2y - z = 3$. Complete the ordered-triple solutions for each equation.

47. $x + y + z = 0$ $(2, -3, \ )$

48. $2x + y + z = 2$ $(\ , -1, 3)$

49. $x + y + z = 0$ $(1, \ , 5)$

50. $x + y - z = 1$ $(4, \ , 3)$

51. $2x + y + z = 2$ $(-2, \ , 1)$

52. $x + y - z = 1$ $(-2, 1, \ )$

53. **ALLIED HEALTH** The recommended dosage (d), in milligrams (mg) of the antibiotic ampicillin sodium for children weighing less than 40 kilograms is given by the linear equation $d = 7.5w$, where w represents the child's weight in kilograms (kg).

(a) What dose should be given to a child weighing 30 kg?

(b) What size child requires a dose of 150 mg?

 chapter 6 > Make the Connection

54. **ALLIED HEALTH** The recommended dosage (d), in micrograms (mcg) of Neupogen, a medication given to bone marrow transplant patients, is given by the linear equation $d = 8w$, where w represents the patient's weight in kilograms (kg).

(a) What dose should be given to a male patient weighing 92 kg?

(b) What size patient requires a dose of 250 mcg?

55. MANUFACTURING TECHNOLOGY The number of board feet b of lumber in a 2×6 of length L feet is given by the equation

$$b = \frac{8.25}{144}L$$

Determine the number of board feet in 2×6 boards of length 12 ft, 16 ft, and 20 ft. Round to the nearest hundredth.

56. MANUFACTURING TECHNOLOGY The number of studs s (16 inches on center) required to build a wall that is L feet long is given by the formula

$$s = \frac{3}{4}L + 1$$

Determine the number of studs required to build walls of length 12 ft, 20 ft, and 24 ft.

57. MECHANICAL ENGINEERING The force that a coil exerts on an object is related to the distance that the coil is pulled from its natural position. The formula to describe this is $F = kx$. If $k = 72$ pounds per foot for a certain coil, determine the force exerted if $x = 3$ ft or $x = 5$ ft.

58. MECHANICAL ENGINEERING If a machine is to be operated under water, it must be designed to handle the pressure (p) measured in pounds, which depends on the depth (d), measured in feet, of the water. The relationship is approximated by the formula $p = 59d + 13$. Determine the pressure at depths of 10 ft, 20 ft, and 30 ft.

Basic Skills | Advanced Skills | Vocational-Technical Applications | Calculator/Computer | **Above and Beyond** | Getting Ready
▲

59. You now have had practice solving equations with one variable and equations with two variables. Compare equations with one variable to equations with two variables. How are they alike? How are they different?

60. Each of the following sentences describes pairs of numbers that are related. After completing the sentences in parts **(a)** to **(g)**, write two of your own sentences in parts **(h)** and **(i)**.

(a) The *number of hours you work* determines the *amount you are* _____.

(b) The *number of gallons of gasoline* you put in your car determines *the amount you* _____.

(c) The *amount of the* _____ in a restaurant is related to *the amount of the tip.*

(d) The *sale amount of a purchase in a store* determines _____.

(e) The *age of an automobile* is related to _____.

(f) The *amount of electricity you use in a month* determines _____.

(g) The *cost of food for a family of four* and _____.

Think of two more related pairs:

(h) _____.

(i) _____.

Answers

55. _____

56. _____

57. _____

58. _____

59. _____

60. _____

Answers

(a) _____

(b) _____

(c) _____

(d) _____

(e) _____

(f) _____

(g) _____

(h) _____

(i) _____

(j) _____

Plot points with the following coordinates on the number line shown.

(a) -3 **(b)** 7 **(c)** 0 **(d)** -8 **(e)** $\dfrac{3}{2}$

Give the coordinate of each of the following points.

(f) A **(g)** B **(h)** C **(i)** D **(j)** E

Answers

1. $(4, 2), (0, 6), (-3, 9)$ **3.** $(5, 2), (4, 0), (6, 4)$

5. $(4, 0), \left(\dfrac{2}{3}, -5\right), \left(5, \dfrac{3}{2}\right)$ **7.** $(0, 0), (2, 8)$

9. $(3, 5), (3, 0), (3, 7)$ **11.** $(4, 8), (7, 5), (0, 12), (12, 0)$

13. $(4, 0), (0, -6), (2, -3), (6, 3)$ **15.** $(-3, 0), \left(\dfrac{2}{3}, 11\right), (0, 9), \left(-\dfrac{2}{3}, 7\right)$

17. $(0, -4), (3, 5), \left(\dfrac{4}{3}, 0\right), \left(\dfrac{5}{3}, 1\right)$ **19.** $(0, -7), (2, -5), (4, -3), (6, -1)$

21. $(0, -6), (3, 0), (6, 6), (9, 12)$ **23.** $(-5, -4), (0, -2), (5, 0), (10, 2)$

25. $(0, 3), (1, 5), (2, 7), (3, 9)$ **27.** $(-5, 0), (-5, 1), (-5, 2), (-5, 3)$

29. \$9.50, \$11.75, \$15.50, \$19.25, \$23 **31.** 20 cm, 40 cm, 48 cm, 60 cm

33. 4,527, 4,689, 4,851, 5,013, 5,337 **35.** False **37.** sometimes

39. $(2, 0), (0, 3)$ **41.** $(0, 4), \left(\dfrac{20}{3}, 0\right)$ **43.** $(8, 0), (0, -15)$

45. $\left(\dfrac{15}{2}, 0\right), \left(0, -\dfrac{30}{7}\right)$ **47.** $(2, -3, 1)$ **49.** $(1, -6, 5)$ **51.** $(-2, 5, 1)$

53. **(a)** 225 mg; **(b)** 20 kg **55.** 0.69 bd ft, 0.92 bd ft, 1.15 bd ft

57. 216 lb, 360 lb **59.** Above and Beyond

(a)–(e).

(f) -7 **(g)** -4 **(h)** 4 **(i)** 8 **(j)** $\dfrac{19}{2}$

6.2

The Rectangular Coordinate System

< 6.2 Objectives >

1 > Give the coordinates of a set of points in the plane

2 > Graph the points corresponding to a set of ordered pairs

In Section 6.1, we saw that ordered pairs could be used to write the solutions of equations in two variables. The next step is to graph those ordered pairs as points in a plane.

Because there are two numbers (one for x and one for y), we need two number lines. We draw one line horizontally, and the other is drawn vertically; their point of intersection (at their respective zero points) is called the *origin*. The horizontal line is called the **x-axis,** and the vertical line is called the **y-axis.** Together the lines form the **rectangular coordinate system.**

The axes divide the plane into four regions called **quadrants,** which are numbered (usually by Roman numerals) counterclockwise from the upper right.

NOTE

This system is also called the **Cartesian coordinate system,** named in honor of its inventor, René Descartes (1596–1650), a French mathematician and philosopher.

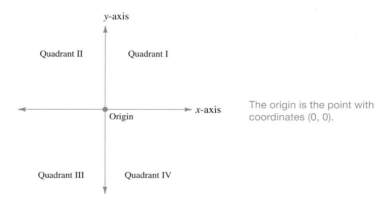

Quadrant II   Quadrant I

Origin x-axis

Quadrant III   Quadrant IV

The origin is the point with coordinates (0, 0).

We now want to establish correspondences between ordered pairs of numbers (x, y) and points in the plane. For any ordered pair

(x, y)

x-coordinate y-coordinate

the following are true:

1. If the x-coordinate is

Positive, the point corresponding to that pair is located x units to the *right* of the y-axis.

Negative, the point is x units to the *left* of the y-axis.

Zero, the point is on the y-axis.

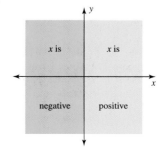

555

2. If the *y*-coordinate is

Positive, the point is *y* units *above* the *x*-axis.

Negative, the point is *y* units *below* the *x*-axis.

Zero, the point is on the *x*-axis.

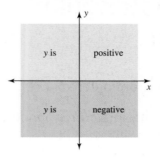

Putting this together we see the following relationship:

In Quadrant I, *x* is positive and *y* is positive.

In Quadrant II, *x* is negative and *y* is positive.

In Quadrant III, *x* is negative and *y* is negative.

In Quadrant IV, *x* is positive and *y* is negative.

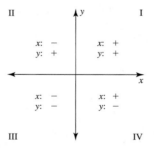

Example 1 illustrates how to use these guidelines to give coordinates to points in the plane.

▶ **Example 1**   **Identifying the Coordinates for a Given Point**

< Objective 1 >

Give the coordinates for the given point.

(a)

RECALL

The *x*-coordinate gives the *horizontal* distance from the *y*-axis. The *y*-coordinate gives the *vertical* distance from the *x*-axis.

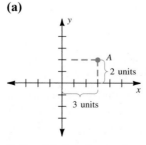

Point *A* is 3 units to the *right* of the *y*-axis and 2 units *above* the *x*-axis. Point *A* has coordinates (3, 2).

(b)

Point *B* is 2 units to the *right* of the *y*-axis and 4 units *below* the *x*-axis. Point *B* has coordinates $(2, -4)$.

(c)

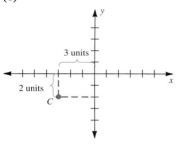

Point *C* is 3 units to the *left* of the *y*-axis and 2 units *below* the *x*-axis. *C* has coordinates $(-3, -2)$.

(d)

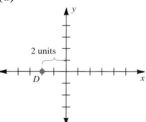

Point *D* is 2 units to the *left* of the *y*-axis and *on* the *x*-axis. Point *D* has coordinates $(-2, 0)$.

 Check Yourself 1

Give the coordinates of points *P, Q, R,* and *S*.

P _____

Q _____

R _____

S _____

Reversing the previous process will allow us to graph (or plot) a point in the plane given the coordinates of the point. You can use the following steps.

Step by Step

To Graph a Point in the Plane	Step 1	Start at the origin.
	Step 2	Move right or left according to the value of the *x*-coordinate.
	Step 3	Move up or down according to the value of the *y*-coordinate.

▶ **Example 2**   **Graphing Points**

< Objective 2 >

NOTE

The graphing of individual points is sometimes called **point plotting**.

(a) Graph the point corresponding to the ordered pair (4, 3).

Move 4 units to the right on the *x*-axis. Then move 3 units up from the point you stopped at on the *x*-axis. This locates the point corresponding to (4, 3).

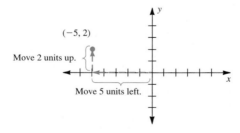

(b) Graph the point corresponding to the ordered pair (−5, 2).

In this case move 5 units *left* (because the *x*-coordinate is negative) and then 2 units *up*.

(c) Graph the point corresponding to (−4, −2).

Here move 4 units *left* and then 2 units *down* (the *y*-coordinate is negative).

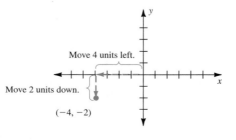

NOTE

Any point on an axis has 0 for one of its coordinates.

(d) Graph the point corresponding to $(0, -3)$.

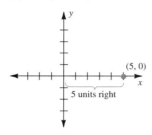

There is *no* horizontal movement because the *x*-coordinate is 0. Move 3 units *down*.

3 units down

$(0, -3)$

(e) Graph the point corresponding to $(5, 0)$.

Move 5 units *right*. The desired point is on the *x*-axis because the *y*-coordinate is 0.

$(5, 0)$

5 units right

Check Yourself 2

Graph the points corresponding to $M(4, 3)$, $N(-2, 4)$, $P(-5, -3)$, and $Q(0, -3)$.

Example 3 gives an application from the field of manufacturing.

▶   **Example 3**   **Vocational Application**

A computer-aided design (CAD) operator has located three corners of a rectangle. The corners are at $(5, 9)$, $(-2, 9)$, and $(5, 2)$. Find the location of the fourth corner.

We plot the three indicated points on graph paper.

The fourth corner must lie directly underneath the point $(-2, 9)$, so the x-coordinate must be -2. The corner must lie on the same horizontal as the point $(5, 2)$, so the y-coordinate must be 2. Therefore the coordinates of the fourth corner are $(-2, 2)$.

Check Yourself 3

A CAD operator has located three corners of a rectangle. The corners are at $(-3, 4)$, $(6, 4)$, and $(-3, -7)$. Find the location of the fourth corner.

Check Yourself ANSWERS

1. $P(4, 5)$, $Q(0, 6)$, $R(-4, -4)$, and $S(2, -5)$

2.

3. $(6, -7)$

Reading Your Text

The following fill-in-the-blank exercises are designed to ensure that you understand some of the key vocabulary used in this section.

SECTION 6.2

(a) The point of intersection of the x-axis and y-axis is called the _____.

(b) The axes divide the plane into four regions called _____.

(c) If the y-coordinate is _____, the point is y units below the x-axis.

(d) Any point on an axis will have _____ for one of its coordinates.

Basic Skills | Advanced Skills | Vocational-Technical Applications | Calculator/Computer | Above and Beyond | Getting Ready

6.2 exercises

< Objective 1 >

Give the coordinates of the points graphed below.

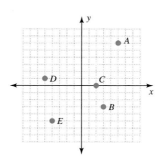

1. *A*

2. *B*

3. *C* > Videos

4. *D*

5. *E* > Videos

Give the coordinates of the points graphed below.

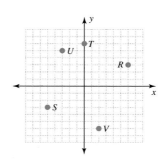

6. *R*

7. *S*

8. *T*

9. *U*

10. *V*

< Objective 2 >

Plot points with the following coordinates on the graph below.

11. $M(5, 3)$

12. $N(0, -3)$

13. $P(-2, 6)$

14. $Q(5, 0)$

15. $R(-4, -6)$

16. $S(-3, -4)$

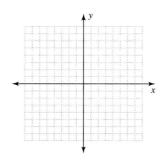

Plot points with the following coordinates on the given graph.

17. $F(-3, -1)$

18. $G(4, 3)$

19. $H(5, -2)$

20. $I(-3, 0)$

MathZone

Boost your grade at mathzone.com!
> Practice Problems
> NetTutor
> Self-Tests
> e-Professors
> Videos

Name _____

Section _____ Date _____

Answers

1. _____   2. _____

3. _____   4. _____

5. _____   6. _____

7. _____   8. _____

9. _____   10. _____

11. _____

12. _____

13. _____

14. _____

15. _____

16. _____

17. _____

18. _____

19. _____

20. _____

Answers

21. _____

22. _____

23. _____

21. $J(-5, 3)$ > Videos

22. $K(0, 6)$

23. **SCIENCE AND MEDICINE** A local plastics company is sponsoring a plastics recycling contest for the local community. The focus of the contest is collecting plastic milk, juice, and water jugs. The company will award $200 plus the current market price of the jugs collected to the group that collects the most jugs in a single month. The number of jugs collected and the amount of money won can be represented as an ordered pair.

(a) In April, group A collected 1,500 lb of jugs to win first place. The prize for the month was $350. On the following graph, x represents the pounds of jugs and y represents the amount of money that the group won. Graph the point that represents the winner for April.

(b) In May, group B collected 2,300 lb of jugs to win first place. The prize for the month was $430. Graph the point that represents the May winner on the same axes you used in part **(a)**.

(c) In June, group C collected 1,200 lb of jugs to win the contest. The prize for the month was $320. Graph the point that represents the June winner on the same axes as used before.

24. STATISTICS The table gives the hours x that Damien studied for five different math exams and the resulting grades y. Plot the data given in the table.

x	4	5	5	2	6
y	83	89	93	75	95

Answers

24. _____

25. _____

26. _____

25. SCIENCE AND MEDICINE The table gives the average temperature y (in degrees Fahrenheit) for each of the first 6 months of the year, x. The months are numbered 1 through 6, with 1 corresponding to January. Plot the data given in the table.

> Videos

x	1	2	3	4	5	6
y	4	14	26	33	42	51

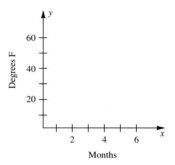

26. BUSINESS AND FINANCE The table gives the total salary of a salesperson, y, for each of the four quarters of the year, x. Plot the data given in the table.

x	1	2	3	4
y	$6,000	$5,000	$8,000	$9,000

Answers

27. _____

28. _____

29. _____

30. _____

27. STATISTICS The table shows the number of runs scored by the Anaheim Angels in each game of the 2002 World Series.

Game	1	2	3	4	5	6	7
Runs	3	11	10	3	4	6	4

Source: Major League Baseball.

Plot the data given in the table.

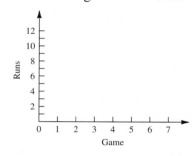

28. STATISTICS The table shows the number of wins and total points for the five teams in the Atlantic Division of the National Hockey League in the early part of a recent season.

Team	Wins	Points
New Jersey Devils	5	12
Philadelphia Flyers	4	10
New York Rangers	4	9
Pittsburgh Penguins	2	6
New York Islanders	2	5

Plot the data given in the table.

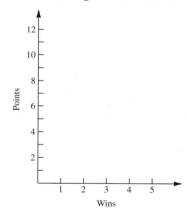

Determine whether each of the following statements is **true** *or* **false.**

29. If the x- and y-coordinates of an ordered pair are both negative, then the plotted point must lie in Quadrant III.

30. If the y-coordinate of an ordered pair is 0, then the plotted point must lie on the y-axis.

Complete each of the following statements with **never, sometimes,** *or* **always.**

31. The ordered pair (a, b) is _____ equal to the ordered pair (b, a).

32. If, in the ordered pair (a, b), a and b have different signs, then the point (a, b) is _____ in the second quadrant.

Answers

31. _____

32. _____

33. _____

34. _____

35. _____

36. _____

| Basic Skills | **Advanced Skills** | Vocational-Technical Applications | Calculator/Computer | Above and Beyond | Getting Ready |

33. Plot points with coordinates $(2, 3)$, $(3, 4)$, and $(4, 5)$ on the given graph. What do you observe? Can you give the coordinates of another point with the same property?

> Videos

34. Plot points with coordinates $(-1, 4)$, $(0, 3)$, and $(1, 2)$ on the given graph. What do you observe? Can you give the coordinates of another point with the same property?

For exercises 35–38, do the following:

   (a) Give the coordinates of the plotted points.

   (b) Describe in words the relationship between the y-coordinate and the x-coordinate.

   (c) Write an equation for the relationship described in part **(b).**

35.

36.

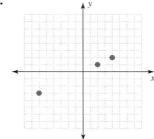

Answers

37.

38.

39.

40.

37.

38.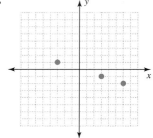

In exercises 39–40, plot the points given in the tables.

39. ALLIED HEALTH Medical lab technicians analyzed several concentrations, in milligrams per deciliter (mg/dL), of glucose solutions to determine the percent transmittance, which measures the percent of light that filters through the solution. The results are summarized in the table.

Glucose concentration (mg/dL)	0	80	160	240	320	400
Percent transmittance (% T)	100	62	40	25	15	10

40. ALLIED HEALTH Daniel's weight, in pounds, has been recorded at various well-baby checkups. The results are summarized in the table.

Age (months)	0	0.5	1	2	7	9
Weight (pounds)	7.8	7.14	9.25	12.5	20.25	21.25

41. **ELECTRONICS** A certain project requires the use of a solenoid, a device that uses an applied electromagnetic force to cause mechanical force. Typically, a conductor such as wire is coiled and current is applied, creating an electromagnet. The magnetic field induced by the energized coil attracts a piece of ferrous material (iron), creating mechanical movement.

Plot the force (in newtons), y, versus applied voltage (in volts), x, of a solenoid using the values given in the table.

x	5	10	15	20
y	0.12	0.24	0.36	0.49

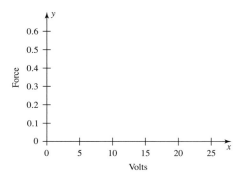

Answers

41. _____

42. _____

43. _____

42. **MANUFACTURING TECHNOLOGY** The temperature and pressure relationship for a coolant is described by the table.

Temperature (°F)	−10	10	30	50	70	90
Pressure (psi)	4.6	14.9	28.3	47.1	71.1	99.2

(a) Graph the points.

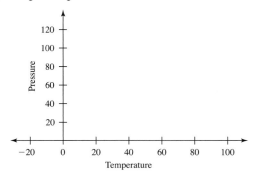

(b) Predict what the pressure will be when the temperature is 60°F.

(c) At what temperature would you expect the coolant to be when the pressure is 37 psi?

43. **AUTOMOTIVE TECHNOLOGY** The table lists the travel time and the distance for several business trips.

Travel time (hours)	6	2	7	9	4
Distance (miles)	320	90	410	465	235

Answers

44. _____

45. _____

Plot these points on a graph.

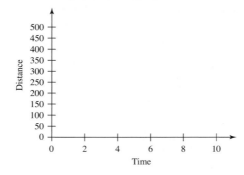

44. **MANUFACTURING TECHNOLOGY** The layout of a jobsite is shown here.

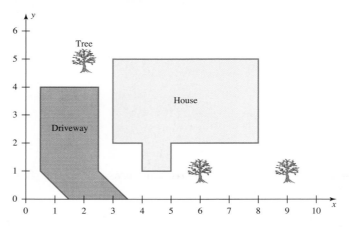

(a) What are the coordinates for each corner of the house?

(b) What is located at (6, 1)?

45. The map shown here uses letters and numbers to label a grid that helps to locate a city. For instance, Salem is located at E4.

(a) Find the coordinates for the following: White Swan, Newport, and Wheeler.

(b) What cities correspond to the following coordinates: A2, F4, and A5?

46. How would you describe a rectangular coordinate system? Explain what information is needed to locate a point in a coordinate system.

47. Some newspapers have a special day that they devote to automobile ads. Use this special section or the Sunday classified ads from your local newspaper to find all the want ads for a particular automobile model. Make a list of the model year and asking price for 10 ads, being sure to get a variety of ages for this model. After collecting the information, make a graph of the age and the asking price for the car.

   Describe your graph, including an explanation of how you decided which variable to put on the vertical axis and which on the horizontal axis. What trends or other information are given by the graph?

Solve each of the following equations.

(a) $2x - 2 = 6$ (b) $2 - 5x = 12$
(c) $7y + 10 = -11$ (d) $-3 + 5x = 1$
(e) $6 - 3x = 8$ (f) $-4y + 6 = 3$

Answers

1. $(5, 6)$ **3.** $(2, 0)$ **5.** $(-4, -5)$ **7.** $(-5, -3)$ **9.** $(-3, 5)$
11–21.

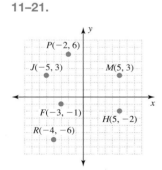

23. (a) $(1,500, 350)$; **(b)** $(2,300, 430)$; **(c)** $(1,200, 320)$
25.   **27.**

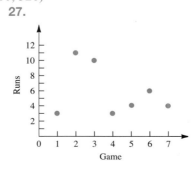

Answers

46.

47.

(a)

(b)

(c)

(d)

(e)

(f)

29. True   **31.** sometimes
33. The points lie on a line; e.g., (1, 2)

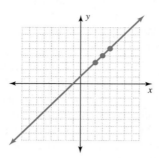

35. (a) $(-2, -4), (1, 2), (3, 6)$; **(b)** The y-value is twice the x-value; **(c)** $y = 2x$
37. (a) $(-2, 6), (-1, 3), (1, -3)$; **(b)** The y-value is -3 times the x-value;
(c) $y = -3x$
39.

41.

43.

45. (a) A7, F2, C2; **(b)** Oysterville, Sweet Home, Mineral
47. Above and Beyond

(a) 4   **(b)** -2 **(c)** -3 **(d)** $\dfrac{4}{5}$ **(e)** $-\dfrac{2}{3}$ **(f)** $\dfrac{3}{4}$

6.3

Graphing Linear Equations

< 6.3 Objectives >

1 > Graph a linear equation by plotting points

2 > Graph a linear equation that results in a vertical or horizontal line

3 > Graph a linear equation by the intercept method

4 > Graph a linear equation by solving the equation for y

We are now ready to combine our work in Sections 6.1 and 6.2. In Section 6.1 you learned to write solutions of equations in two variables as ordered pairs. Then, in Section 6.2, these ordered pairs were graphed in the plane. Putting these ideas together will help us to graph equations. Example 1 illustrates this approach.

| Example 1 | Graphing a Linear Equation |

< Objective 1 >

Graph $x + 2y = 4$.

Step 1 Find some solutions for $x + 2y = 4$. To find solutions, we choose any convenient values for x, say, $x = 0$, $x = 2$, and $x = 4$. Given these values for x, we can substitute and then solve for the corresponding value for y. So

If $x = 0$, then $y = 2$, so $(0, 2)$ is a solution.
If $x = 2$, then $y = 1$, so $(2, 1)$ is a solution.
If $x = 4$, then $y = 0$, so $(4, 0)$ is a solution.

A handy way to show this information is in a table such as this:

x	y
0	2
2	1
4	0

Step 2 We now graph the solutions found in step 1.

$x + 2y = 4$

NOTE

We are going to find *three* solutions for the equation. We'll point out why shortly.

NOTE

The table is just a convenient way to display the information. It is the same as writing (0, 2), (2, 1), and (4, 0).

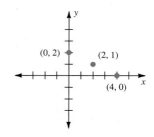

x	y
0	2
2	1
4	0

What pattern do you see? It appears that the three points lie on a straight line, which is in fact the case.

Step 3   Draw a straight line through the three points graphed in step 2.

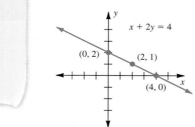

The line shown is the **graph** of the equation $x + 2y = 4$. It represents *all* of the ordered pairs that are solutions (an infinite number) for that equation.

Every ordered pair that is a solution lies on this line. Any point on the line will have coordinates that are a solution for the equation.

Note: Why did we suggest finding *three* solutions in step 1? Two points determine a line, so technically you need only two. The third point that we find is a check to catch any possible errors.

   Check Yourself 1

Graph $2x - y = 6$, using the steps shown in Example 1.

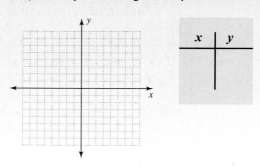

In Section 6.1, we mentioned that an equation that can be written in the form

$$Ax + By = C$$

in which A, B, and C are real numbers and A and B are not both 0 is called a **linear equation in two variables.** The graph of this equation is a *straight line.*

The steps for graphing follow.

Step by Step

To Graph a Linear Equation

Step 1   Find at least three solutions for the equation and put your results in tabular form.

Step 2   Graph the solutions found in step 1.

Step 3   Draw a straight line through the points determined in step 2 to form the graph of the equation.

▶ **Example 2**   **Graphing a Linear Equation**

Graph $y = 3x$.

Step 1   Some solutions are

x	y
0	0
1	3
2	6

NOTE

Let $x = 0$, 1, and 2, and substitute to determine the corresponding y-values. Again the choices for x are simply convenient. Other values for x would serve the same purpose.

Step 2   Graph the points.

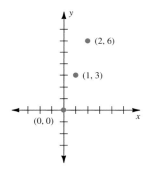

Step 3   Draw a line through the points.

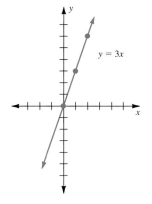

NOTE

Connecting any pair of these points produces the same line.

Check Yourself 2

Graph the equation $y = -2x$ after completing the table of values.

x	y
0	
1	
2	

Let's work through another example of graphing a line from its equation.

▶ **Example 3**   **Graphing a Linear Equation**

Graph $y = 2x + 3$.

Step 1   Some solutions are

x	y
0	3
1	5
2	7

Step 2   Graph the points corresponding to these values.

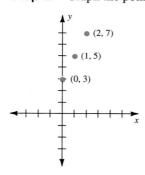

Step 3   Draw a line through the points.

Check Yourself 3

Graph the equation $y = 3x - 2$ after completing the table of values.

x	y
0	
1	
2	

When graphing equations, particularly if fractions are involved, a careful choice of values for x can simplify the process. Consider Example 4.

▶ **Example 4**   **Graphing a Linear Equation**

Graph

$$y = \frac{3}{2}x - 2$$

As before, we want to find solutions for the given equation by picking convenient values for x. Note that in this case, choosing *multiples of* 2 will avoid fractional values for y and make the plotting of those solutions much easier. For instance, here we might choose values of -2, 0, and 2 for x.

Step 1   If $x = -2$:

$$y = \frac{3}{2}(-2) - 2$$
$$= -3 - 2 = -5$$

If $x = 0$:

$$y = \frac{3}{2}(0) - 2$$
$$= 0 - 2 = -2$$

If $x = 2$:

$$y = \frac{3}{2}(2) - 2$$
$$= 3 - 2 = 1$$

In tabular form, the solutions are

x	y
-2	-5
0	-2
2	1

NOTE

Suppose we do *not* choose a multiple of 2, say, $x = 3$. Then

$$y = \frac{3}{2}(3) - 2$$
$$= \frac{9}{2} - 2$$
$$= \frac{5}{2}$$

$\left(3, \dfrac{5}{2}\right)$ is still a valid solution, but we must graph a point with fractions as coordinates.

Step 2   Graph the points determined in the table.

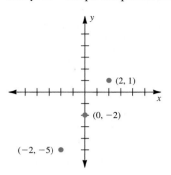

Step 3   Draw a line through the points.

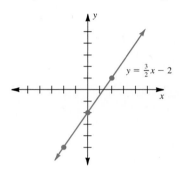

Check Yourself 4

Graph the equation $y = -\dfrac{1}{3}x + 3$ after completing the table of values.

x	y
−3	
0	
3	

Some special cases of linear equations are illustrated in Examples 5 and 6.

Example 5   **Graphing an Equation That Results in a Vertical Line**

< Objective 2 >

Graph $x = 3$.

The equation $x = 3$ is equivalent to $1 \cdot x + 0 \cdot y = 3$, or $x + 0(y) = 3$. Some solutions follow.

If $y = 1$:	If $y = 4$:	If $y = -2$:
$x + 0(1) = 3$	$x + 0(4) = 3$	$x + 0(-2) = 3$
$x = 3$	$x = 3$	$x = 3$

In tabular form,

x	y
3	1
3	4
3	-2

What do you observe? The variable x has the value 3, regardless of the value of y. Look at the following graph.

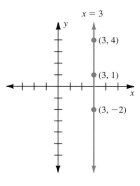

The graph of $x = 3$ is a vertical line crossing the x-axis at $(3, 0)$.

Note that graphing (or plotting) points in this case is not really necessary. Simply recognize that the graph of $x = 3$ *must* be a vertical line (parallel to the y-axis) that intercepts the x-axis at $(3, 0)$.

Check Yourself 5

Graph the equation $x = -2$.

Example 6 is a related example involving a horizontal line.

> **Example 6**   **Graphing an Equation That Results in a Horizontal Line**

Graph $y = 4$.

Because $y = 4$ is equivalent to $0 \cdot x + 1 \cdot y = 4$, or $0(x) + y = 4$, any value for x paired with 4 for y will form a solution. A table of values might be

x	y
-2	4
0	4
2	4

Here is the graph.

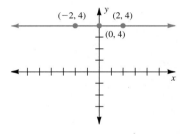

This time the graph is a horizontal line that crosses the y-axis at $(0, 4)$. Again you do not need to graph points. The graph of $y = 4$ *must* be horizontal (parallel to the x-axis) and intercepts the y-axis at $(0, 4)$.

 Check Yourself 6

Graph the equation $y = -3$.

The following box summarizes our work in Examples 5 and 6.

Property

Vertical and Horizontal Lines

1. The graph of $x = a$ is a *vertical line* crossing the x-axis at $(a, 0)$.

2. The graph of $y = b$ is a *horizontal line* crossing the y-axis at $(0, b)$.

To simplify the graphing of certain linear equations, some students prefer the **intercept method** of graphing. This method makes use of the fact that the solutions that are easiest to find are those with an x-coordinate or a y-coordinate of 0. For instance, to graph the equation

$$4x + 3y = 12$$

first, let $x = 0$ and then solve for y.

$$4(0) + 3y = 12$$
$$3y = 12$$
$$y = 4$$

So $(0, 4)$ is one solution. Now we let $y = 0$ and solve for x.

$$4x + 3(0) = 12$$
$$4x = 12$$
$$x = 3$$

A second solution is $(3, 0)$.

The two points corresponding to these solutions can now be used to graph the equation.

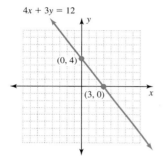

The ordered pair $(3, 0)$ is called the **x-intercept,** and the ordered pair $(0, 4)$ is the **y-intercept** of the graph. Using these points to draw the graph gives the name to this method. Here is a second example of graphing by the intercept method.

> **NOTE**
>
> With practice, this can be done mentally, which is the big advantage of this method.

> **RECALL**
>
> Only two points are needed to graph a line. A third point is used only as a check.

> **NOTE**
>
> The intercepts are the points where the line crosses the x- and y-axes.

| ▶ | Example 7 | Using the Intercept Method to Graph a Line |

< Objective 3 >

Graph $3x - 5y = 15$, using the intercept method.

To find the x-intercept, let $y = 0$.

$$3x - 5(0) = 15$$
$$x = 5$$

The x-intercept is $(5, 0)$.

To find the y-intercept, let $x = 0$.

$$3(0) - 5y = 15$$
$$y = -3$$

The y-intercept is $(0, -3)$.

So (5, 0) and (0, −3) are solutions for the equation, and we can use the corresponding points to graph the equation.

Check Yourself 7

Graph 4*x* + 5*y* = 20, using the intercept method.

NOTE

Finding a third "checkpoint" is always a good idea.

This all looks quite easy, and for many equations it is. What are the drawbacks? For one, you don't have a third checkpoint, and it is possible for errors to occur. You can, of course, still find a third point (other than the two intercepts) to be sure your graph is correct. A second difficulty arises when the *x*- and *y*-intercepts are very close to one another (or are actually the same point—the origin). For instance, if we have the equation

$$3x + 2y = 1$$

the intercepts are $\left(\dfrac{1}{3}, 0\right)$ and $\left(0, \dfrac{1}{2}\right)$. It is difficult to draw a line accurately through these intercepts, so choose other solutions farther away from the origin for your points.

We summarize the steps of graphing by the intercept method for appropriate equations.

Step by Step

Graphing a Line by the Intercept Method	Step 1	To find the *x*-intercept: Let *y* = 0, then solve for *x*.
	Step 2	To find the *y*-intercept: Let *x* = 0, then solve for *y*.
	Step 3	Graph the *x*- and *y*-intercepts.
	Step 4	Draw a straight line through the intercepts.

A third method of graphing linear equations involves **solving the equation for y.** The reason we use this extra step is that it often will make finding solutions for the equation much easier.

| | **Example 8** | Graphing a Linear Equation by Solving for y |

< Objective 4 >

Graph $2x + 3y = 6$.

Rather than finding solutions for the equation in this form, we solve for y.

$2x + 3y = 6$

$$3y = 6 - 2x \qquad \text{Subtract } 2x.$$

$$y = \frac{6 - 2x}{3} \qquad \text{Divide by 3.}$$

or $\quad y = 2 - \frac{2}{3}x$

Now find your solutions by picking convenient values for x.

If $x = -3$:

$$y = 2 - \frac{2}{3}(-3)$$

$$= 2 + 2 = 4$$

So $(-3, 4)$ is a solution.

If $x = 0$:

$$y = 2 - \frac{2}{3}(0)$$

$$= 2$$

So $(0, 2)$ is a solution.

If $x = 3$:

$$y = 2 - \frac{2}{3}(3)$$

$$= 2 - 2 = 0$$

So $(3, 0)$ is a solution.

We can now plot the points that correspond to these solutions and form the graph of the equation as before.

x	y
-3	4
0	2
3	0

Check Yourself 8

Graph the equation $5x + 2y = 10$. Solve for y to determine solutions.

In applications, we often use graphs that do not fit nicely into a standard 8-by-8 grid. In these cases, we should show the portion of the graph that displays the interesting properties of the graph. We may need to scale the x- or y-axis to meet our needs or even set the axes so that it seems they do not intersect at the origin.

When we scale the axes, it is important to include numbers on the axes at convenient grid lines. If we set the axes so that they do not seem to intersect at the origin, we include a mark to indicate this. Both of these situations are illustrated in Example 9.

Example 9	Graphing in Nonstandard Windows

NOTE

In business, the constant, 2,500, is called the **fixed cost.** The coefficient, 45, is referred to as the **variable cost.**

The cost, y, to produce x CD players is given by the equation $y = 45x + 2,500$. Graph the cost equation, with appropriately scaled and set axes.

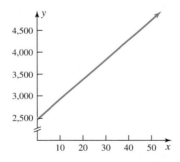

The y-intercept is $(0, 2,500)$. We find more points to plot by creating a table:

x	10	20	30	40	50
y	2,950	3,400	3,850	4,300	4,750

Check Yourself 9

Graph the cost equation given by $y = 60x + 1,200$, with appropriately scaled and set axes.

The following is an example of an application from the field of medicine.

 Example 10   Vocational Application

The arterial oxygen tension (P_aO_2), in millimeters of mercury (mm Hg), of a patient can be estimated based on the patient's age (A), in years. If the patient is lying down, use the equation $P_aO_2 = 103.5 - 0.42A$. Draw the graph of this equation, using appropriately scaled and set axes.

We begin by creating a table. Using a calculator here is very helpful.

A	0	10	20	30	40	50	60	70	80
P_aO_2	103.5	99.3	95.1	90.9	86.7	82.5	78.3	74.1	69.9

Seeing these values allows us to decide upon the vertical axis scaling. We scale from 60 to 110 and include a mark to show a break in the axis. We estimate the locations of these coordinates, and draw the line.

 Check Yourself 10

The arterial oxygen tension (P_aO_2), in millimeters of mercury (mm Hg), of a patient can be estimated based on the patient's age (A), in years. If the patient is seated, use the equation $P_aO_2 = 104.2 - 0.27A$. Draw the graph of this equation, using appropriately scaled and set axes.

Check Yourself ANSWERS

1.

x	y
1	-4
2	-2
3	0

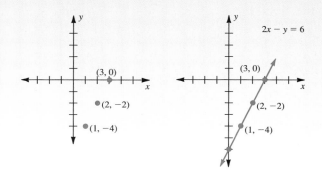

2.

x	y
0	0
1	−2
2	−4

$y = -2x$

3.

x	y
0	−2
1	1
2	4

$y = 3x - 2$

4.

x	y
−3	4
0	3
3	2

$y = -\frac{1}{3}x + 3$

5. $x = -2$

6.

$y = -3$

7. $4x + 5y = 20$

(0, 4)

(5, 0)

8.

x	y
0	5
2	0
4	−5

$y = -\frac{5}{2}x + 5$

9.

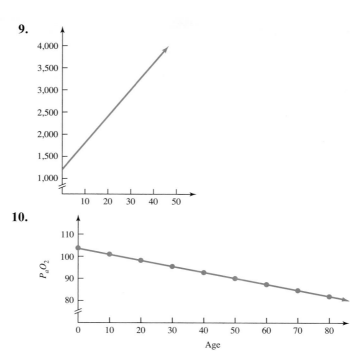

10.

Reading Your Text

The following fill-in-the-blank exercises are designed to ensure that you understand some of the key vocabulary used in this section.

SECTION 6.3

(a) An equation that can be written in the form $Ax + By = C$ is called a _____ equation in two variables.

(b) The graph of $x = a$ is a _____ line crossing the x-axis at $(a, 0)$.

(c) The graph of $y = b$ is a _____ line crossing the y-axis at $(0, b)$.

(d) The _____ method makes use of the fact that the solutions easiest to find are those with an x-coordinate or a y-coordinate of 0.

MathZone

Boost your grade at mathzone.com!

> Practice Problems
> NetTutor
> Self-Tests
> e-Professors
> Videos

Name _____

Section _____ Date _____

Answers

1. _____

2. _____

3. _____

4. _____

5. _____

6. _____

7. _____

8. _____

< Objectives 1–2 >

Graph each of the following equations.

1. $x + y = 6$

2. $x - y = 5$

3. $x - y = -3$ 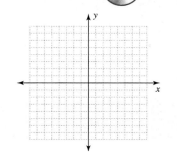 > Videos

4. $x + y = -3$

5. $2x + y = 2$

6. $x - 2y = 6$

7. $3x + y = 0$

8. $3x - y = 6$

9. $x + 4y = 8$

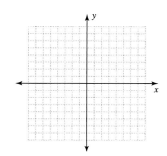

10. $2x - 3y = 6$

11. $y = 5x$

12. $y = -4x$

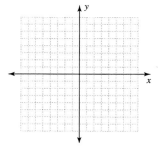

13. $y = 2x - 1$

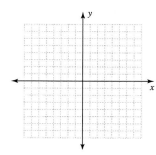

14. $y = 4x + 3$

15. $y = -3x + 1$ ▷ Videos

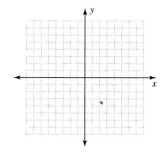

16. $y = -3x - 3$

Answers

9. _____

10. _____

11. _____

12. _____

13. _____

14. _____

15. _____

16. _____

Answers

17. _____

18. _____

19. _____

20. _____

21. _____

22. _____

23. _____

24. _____

17. $y = \dfrac{1}{3}x$

18. $y = -\dfrac{1}{4}x$

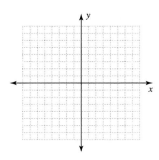

19. $y = \dfrac{2}{3}x - 3$

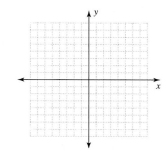

20. $y = \dfrac{3}{4}x + 2$

21. $x = 5$

22. $y = -3$

23. $y = 1$

24. $x = -2$

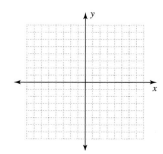

< Objective 3 >

Graph each of the following equations using the intercept method.

25. $x - 2y = 4$

26. $6x + y = 6$

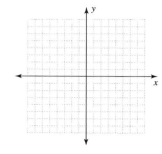

27. $5x + 2y = 10$

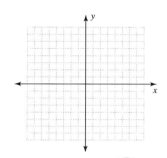

28. $2x + 3y = 6$

29. $3x + 5y = 15$ > Videos

30. $4x + 3y = 12$

< Objective 4 >

Graph each of the following equations by first solving for y.

31. $x + 3y = 6$ > Videos

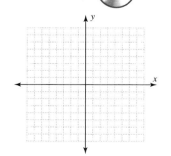

32. $x - 2y = 6$

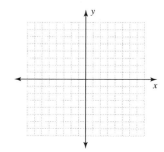

Answers

25. _____

26. _____

27. _____

28. _____

29. _____

30. _____

31. _____

32. _____

Answers

33. _____

34. _____

35. _____

36. _____

37. _____

38. _____

33. $3x + 4y = 12$

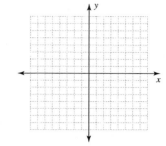

34. $2x - 3y = 12$

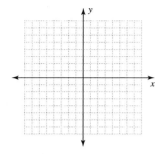

35. $5x - 4y = 20$

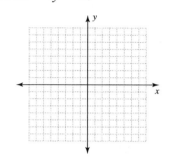

36. $7x + 3y = 21$

37. SCIENCE AND MEDICINE The equation $y = 0.10x + 200$ describes the amount of winnings a group earns for collecting plastic jugs in the recycling contest described in exercise 23 at the end of Section 6.2. Sketch the graph of the line on the given coordinate system.

38. BUSINESS AND FINANCE A car rental agency charges $40 per day for a certain type of car, plus 15¢ per mile driven. The equation $y = 0.15x + 40$ indicates the charge, y, for a day's rental where x miles are driven. Sketch the graph of this equation on the given axes.

39. BUSINESS AND FINANCE A high school class wants to raise some money by re-cycling newspapers. The class decides to rent a truck for a weekend and to collect the newspapers from homes in the neighborhood. The market price for recycled newsprint is currently $11 per ton. The equation $y = 11x - 100$ describes the amount of money the class will make, in which y is the amount of money made in dollars, x is the number of tons of newsprint collected, and 100 is the cost in dollars to rent the truck.

Answers

39. _____

40. _____

(a) Using the given axes, draw a graph that represents the relationship between newsprint collected and money earned.

(b) The truck costs the class $100. How many tons of newspapers must the class collect to break even on this project?

(c) If the class members collect 16 tons of newsprint, how much money will they earn?

(d) Six months later the price of newsprint is $17 a ton, and the cost to rent the truck has risen to $125. Write the equation that describes the amount of money the class might make at that time.

40. BUSINESS AND FINANCE A car rental agency charges $30 per day for a certain type of car, plus 20¢ per mile driven. The equation $y = 0.20x + 30$ indicates the charge, y, for a day's rental where x miles are driven. Sketch the graph of this equation on the given axes.

Answers

41. _____

42. _____

43. _____

44. _____

45. _____

46. _____

41. **BUSINESS AND FINANCE** The cost of producing x items is given by $C = mx + b$, in which b is the fixed cost and m is the variable cost (the cost of producing one more item).

   (a) If the fixed cost is $200 and the variable cost is $15, write the cost equation.

   (b) Graph the cost equation on the given axes.

42. **BUSINESS AND FINANCE** The cost of producing x items is given by $C = mx + b$, in which b is the fixed cost and m is the variable cost (the cost of producing one more item).

   (a) If the fixed cost is $150 and the variable cost is $20, write the cost equation.

   (b) Graph the cost equation on the given axes.

Determine whether each of the following statements is **true** *or* **false.**

43. If the ordered pair (x, y) is a solution to an equation, then the point (x, y) is always on the graph of the equation.

44. The graph of a linear equation must pass through the origin.

Complete each of the following statements with **never, sometimes,** *or* **always.**

45. If the graph of a linear equation $Ax + By = C$ passes through the origin, then C _____ equals zero.

46. The graph of a linear equation _____ has an x-intercept.

Basic Skills | **Advanced Skills** | Vocational-Technical Applications | Calculator/Computer | Above and Beyond | Getting Ready

▲

Write an equation that describes the following relationships between x and y. Then graph each relationship.

47. y is twice x.

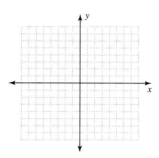

48. y is 2 less than x.

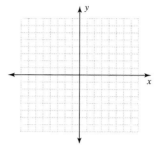

49. y is 3 less than 3 times x.

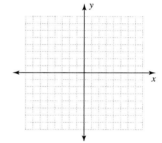

50. y is 4 more than twice x.

51. The difference of x and the product of 4 and y is 12.

52. The difference of twice x and y is 6.

Answers

47. _____

48. _____

49. _____

50. _____

51. _____

52. _____

Answers

53. _____

54. _____

55. _____

56. _____

57. _____

Graph each pair of equations on the same axes. Give the coordinates of the point where the lines intersect.

53. $x + y = 4$
$x - y = 2$

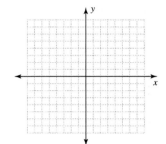

54. $x - y = 3$
$x + y = 5$

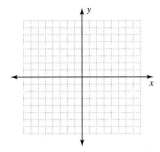

Graph each set of equations on the same coordinate system. Do the lines intersect? What are the y-intercepts?

55. $y = 3x$
$y = 3x + 4$
$y = 3x - 5$

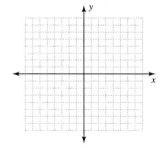

56. $y = -2x$
$y = -2x + 3$
$y = -2x - 5$

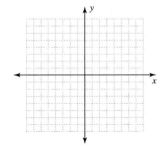

Basic Skills | Advanced Skills | **Vocational-Technical Applications** | Calculator/Computer | Above and Beyond | Getting Ready

57. **ALLIED HEALTH** The weight (w), in kilograms, of a tumor is related to the number of days (d) of chemotherapy treatment by the linear equation $w = -1.75d + 25$. Sketch a graph of this linear equation.

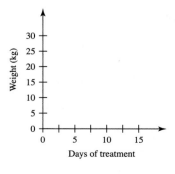

58. **ALLIED HEALTH** The weight (w), in kilograms, of an tumor is related to the number of days (d) of chemotherapy treatment by the linear equation $w = -1.6d + 32$. Sketch a graph of the linear equation.

59. **MECHANICAL ENGINEERING** The force that a coil exerts on an object is related to the distance that the coil is pulled from its natural position. The formula to describe this is $F = kx$. Graph this relationship for a coil with $k = 72$ pounds per foot.

60. **MECHANICAL ENGINEERING** If a machine is to be operated under water, it must be designed to handle the pressure (p), which depends on the depth (d) of the water. The relationship is approximated by the formula $p = 59d + 13$. Graph the relationship between pressure and depth.

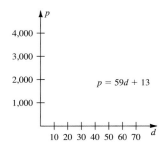

61. **MANUFACTURING TECHNOLOGY** The number of studs, s (16 inches on center), required to build a wall of length L (in feet) is given by the formula

$$s = \frac{3}{4}L + 1$$

Answers

58. _____

59. _____

60. _____

61. _____

Answers

62. _____

63. _____

64. _____

Graph the number of studs as a function of the length of the wall.

$$s = \frac{3}{4}L + 1$$

62. **MANUFACTURING TECHNOLOGY** The number of board feet of lumber, b, in a 2×6 of length L (in feet) is given by the equation

$$b = \frac{8.25}{144}L$$

Graph the number of board feet in terms of the length of the board.

$$b = \frac{8.25}{144}L$$

Basic Skills | Advanced Skills | Vocational-Technical Applications | Calculator/Computer | **Above and Beyond** | Getting Ready

In exercises 63 and 64:

(a) Graph both given equations on the same coordinate system.

(b) Describe any observations you can make concerning the two graphs and their equations. How do the two graphs relate to each other?

63. $y = -3x$

$$y = \frac{1}{3}x$$

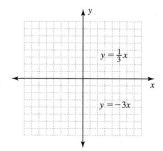

64. $y = \frac{2}{5}x$

$$y = -\frac{5}{2}x$$

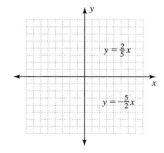

Basic Skills | Advanced Skills | Vocational-Technical Applications | Calculator/Computer | Above and Beyond | **Getting Ready** ▲

Evaluate the following expressions.

(a) $\dfrac{7 - 3}{8 - 4}$

(b) $\dfrac{-9 - 5}{-4 - 3}$

(c) $\dfrac{4 - (-2)}{6 - 2}$

(d) $\dfrac{-4 - (-4)}{8 - 2}$

Answers

(a) _____

(b) _____

(c) _____

(d) _____

Answers

1. $x + y = 6$

3. $x - y = -3$

5. $2x + y = 2$

7. $3x + y = 0$

9. $x + 4y = 8$

11. $y = 5x$

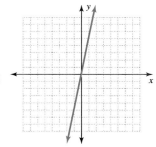

13. $y = 2x - 1$

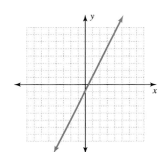

15. $y = -3x + 1$

17. $y = \dfrac{1}{3} x$

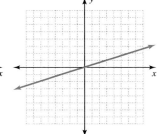

19. $y = \dfrac{2}{3}x - 3$

21. $x = 5$

23. $y = 1$

25. $x - 2y = 4$

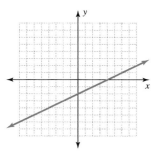

27. $5x + 2y = 10$

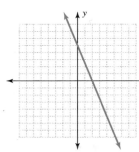

29. $3x + 5y = 15$

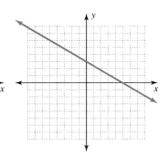

31. $y = 2 - \dfrac{x}{3}$

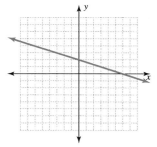

33. $y = 3 - \dfrac{3}{4}x$

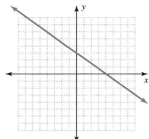

35. $y = -5 + \dfrac{5}{4}x$

37.

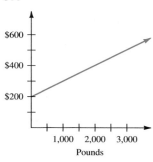

Pounds

39. (a) See graph below;

(b) $\dfrac{100}{11}$ or ≈ 9 tons; **(c)** \$76;

(d) $y = 17x - 125$

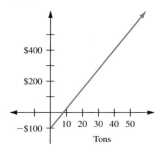

Tons

41. (a) $C = 15x + 200$;

(b)

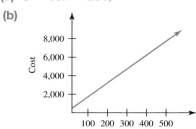

43. True

45. always

47. $y = 2x$

49. $y = 3x - 3$

51. $x - 4y = 12$

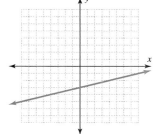

53. $(3, 1)$

55. The lines do not intersect.
The y-intercepts are $(0, 0)$, $(0, 4)$, and $(0, -5)$.

57.

59.

$F = 72x$

61.

63.

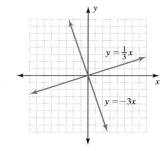

(a) 1   **(b)** 2   **(c)** $\dfrac{3}{2}$ **(d)** 0

6.4

The Slope of a Line

< 6.4 Objectives >

1 > Find the slope of a line through two given points

2 > Find the slopes of vertical and horizontal lines

3 > Find the slope of a line from its graph

4 > Write and graph the equation for a direct-variation relationship

RECALL

An equation such as $y = 2x + 3$ is a *linear equation in two variables*. Its graph is always a straight line.

In Section 6.3 we saw that the graph of an equation such as

$$y = 2x + 3$$

is a straight line. In this section we want to develop an important idea related to the equation of a line and its graph, called the **slope** of a line. The slope of a line is a numerical measure of the "steepness," or inclination, of that line.

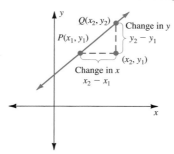

NOTES

x_1 is read "x sub 1," x_2 is read "x sub 2," and so on. The 1 in x_1 and the 2 in x_2 are called **subscripts**.

The difference $x_2 - x_1$ is sometimes called the **run** between points P and Q. The difference $y_2 - y_1$ is called the **rise**. So the slope may be thought of as "rise over run."

To find the slope of a line, we first let $P(x_1, y_1)$ and $Q(x_2, y_2)$ be any two distinct points on that line. The **horizontal change** (or the change in x) between the points is $x_2 - x_1$. The **vertical change** (or the change in y) between the points is $y_2 - y_1$.

We call the ratio of the vertical change, $y_2 - y_1$, to the horizontal change, $x_2 - x_1$, the *slope* of the line as we move along the line from P to Q. That ratio is usually denoted by the letter m, and so we have the following formula:

Definition

The Slope of a Line

If $P(x_1, y_1)$ and $Q(x_2, y_2)$ are any two points on a line, then m, the slope of the line, is given by

$$m = \frac{\text{vertical change}}{\text{horizontal change}} = \frac{y_2 - y_1}{x_2 - x_1} \qquad \text{when } x_2 \neq x_1$$

This definition provides exactly the numerical measure of "steepness" that we want. If a line "rises" as we move from left to right, the slope is positive—the steeper the line, the larger the numerical value of the slope. If the line "falls" from left to right, the slope is negative.

Let's proceed to some examples.

| | **Example 1** | Finding the Slope Given Two Points |

< Objective 1 >

Find the slope of the line containing points with coordinates (1, 2) and (5, 4).

Let $P(x_1, y_1) = (1, 2)$ and $Q(x_2, y_2) = (5, 4)$. By the definition of slope, we have

$$m = \frac{y_2 - y_1}{x_2 - x_1} = \frac{(4) - (2)}{(5) - (1)} = \frac{2}{4} = \frac{1}{2}$$

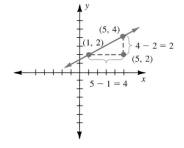

Note: We would have found the same slope if we had reversed P and Q and subtracted in the other order. In that case, $P(x_1, y_1) = (5, 4)$ and $Q(x_2, y_2) = (1, 2)$, so

$$m = \frac{(2) - (4)}{(1) - (5)} = \frac{-2}{-4} = \frac{1}{2}$$

It makes no difference which point is labeled (x_1, y_1) and which is (x_2, y_2). The resulting slope is the same. You must simply stay with your choice once it is made and *not* reverse the order of the subtraction in your calculations.

Check Yourself 1

Find the slope of the line containing points with coordinates (2, 3) and (5, 5).

By now you should be comfortable subtracting negative numbers. We apply that skill to finding a slope.

| | **Example 2** | Finding the Slope |

Find the slope of the line containing points with the coordinates $(-1, -2)$ and $(3, 6)$.

Again, applying the definition, we have

$$m = \frac{(6) - (-2)}{(3) - (-1)} = \frac{6 + 2}{3 + 1} = \frac{8}{4} = 2$$

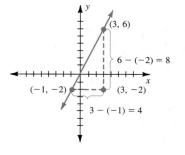

The next figure compares the slopes found in Example 1 and this example. Line l_1, from Example 1, has slope $\frac{1}{2}$. Line l_2, from this example, has slope 2. Do you see the idea of slope measuring steepness? The greater the slope, the more steeply the line is inclined upward.

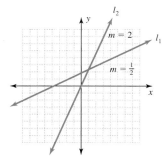

Check Yourself 2

Find the slope of the line containing points with coordinates (−1, 2) and (2, 7). Draw a sketch of this line and the line of Check Yourself 1 on the same set of axes. Compare the lines and the two slopes.

Next, we look at lines with a negative slope.

Example 3   **Finding a Negative Slope**

Find the slope of the line containing points with coordinates $(-2, 3)$ and $(1, -3)$.
By the definition,

$$m = \frac{(-3) - (3)}{(1) - (-2)} = \frac{-6}{3} = -2$$

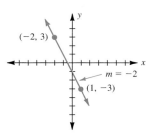

This line has a *negative* slope. The line *falls* as we move from left to right.

Check Yourself 3

Find the slope of the line containing points with coordinates (−1, 3) and (1, −3).

We have seen that lines with positive slope rise from left to right and lines with negative slope fall from left to right. What about lines with a slope of zero? A line with a slope of 0 is especially important in mathematics.

| Example 4 | Finding the Slope of a Horizontal Line |

< Objective 2 >

Find the slope of the line containing points with coordinates $(-5, 2)$ and $(3, 2)$.
   By the definition,

$$m = \frac{(2) - (2)}{(3) - (-5)} = \frac{0}{8} = 0$$

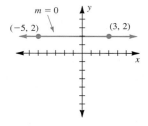

The slope of the line is 0. In fact, that is the case for any horizontal line. Because any two points on the line have the same y-coordinate, the vertical change $y_2 - y_1$ must always be 0, and so the resulting slope is 0.

Check Yourself 4

Find the slope of the line containing points with coordinates $(-2, -4)$ and $(3, -4)$.

Because division by 0 is undefined, it is possible to have a line with an undefined slope.

| Example 5 | Finding the Slope of a Vertical Line |

Find the slope of the line containing points with coordinates $(2, -5)$ and $(2, 5)$.
   By the definition,

$$m = \frac{(5) - (-5)}{(2) - (2)} = \frac{10}{0}$$ Remember that division by 0 is undefined.

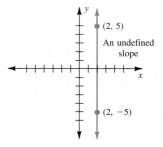

We say that the vertical line has an undefined slope. On a vertical line, any two points have the same x-coordinate. This means that the horizontal change $x_2 - x_1$ must always be 0, and because division by 0 is undefined, the slope of a vertical line will always be undefined.

Check Yourself 5

Find the slope of the line containing points with the coordinates (−3, −5) and (−3, 2).

Given the graph of a line, we can find the slope of that line. Example 6 illustrates this.

Example 6 | **Finding the Slope from the Graph**

< Objective 3 >

Find the slope of the graphed line.

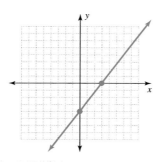

We can find the slope by identifying any two points. It is usually easiest to use the x- and y-intercepts. In this case, those intercepts are (3, 0) and (0, −4).

Using the definition of slope, we find

$$m = \frac{(0) - (-4)}{(3) - (0)} = \frac{4}{3}$$

The slope of the line is $\frac{4}{3}$.

Check Yourself 6

Find the slope of the graphed line.

In Section 6.3, we saw that a line can be drawn from two ordered pairs. Given equations of the form $y = ax$, it is fairly easy to find two ordered pairs. In Example 7, we will use those ordered pairs to find the graph of the equation.

> **Example 7**   Graphing an Equation of the Form $y = ax$

(a) Sketch the graph of the equation $y = -2x$.

From the following table, we know that the ordered pairs $(0, 0)$ and $(1, -2)$ are solutions to the equation.

x	y
0	0
1	-2

The graph is displayed here.

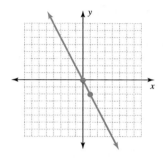

Note that the slope of the line that passes through the points $(0, 0)$ and $(1, -2)$ is

$$m = \frac{(0) - (-2)}{(0) - (1)} = \frac{2}{-1} = -2$$

(b) Sketch the graph of the equation $y = \frac{1}{3}x$.

From the following table, we know that the ordered pairs $(0, 0)$ and $(3, 1)$ are solutions to the equation.

x	y
0	0
3	1

The graph is displayed here.

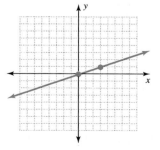

Note that the slope of the line that passes through the points (0, 0) and (3, 1) is

$$m = \frac{(0) - (1)}{(0) - (3)} = \frac{1}{3}$$

Check Yourself 7

Sketch the graph of the equation $y = -\frac{1}{2}x$.

In Example 7, we noted that the slope of the line for the equation $y = -2x$ is -2, and the slope of the line for the equation $y = \frac{1}{3}x$ is $\frac{1}{3}$. This leads us to the following observation.

The slope of a line for an equation of the form $y = mx$ is always m. Because m is the slope, we generally write the form as

$$y = mx$$

Note that (0, 0) is a solution for any equation of this form. As a result, the line for an equation of the form $y = mx$ always passes through the origin.

The following sketch summarizes the results of our previous examples.

Four lines are illustrated in the figure. Note that:

1. The slope of a line that rises from left to right is positive.

2. The slope of a line that falls from left to right is negative.

3. The slope of a horizontal line is 0.

4. A vertical line has an undefined slope.

There are numerous applications involving lines that pass through the origin. Consider the following scenario. Pedro makes $25 an hour as an electrician. If he works for 1 h, he makes $25; if he works for 2 h, he makes $50; and so on. We say his total pay **varies directly** with the number of hours worked. This type of situation occurs so frequently that we use special terminology to describe it.

Definition

Direct Variation

If y is a constant multiple of x, we write

$$y = kx \qquad \text{in which } k \text{ is a positive constant.}$$

We say that y *varies directly* with x, or that y is *directly proportional* to x. The constant k is called the **constant of variation**.

Example 8   Writing an Equation for Direct Variation

< Objective 4 >

Marina earns $9 an hour as a tutor. Write the equation that describes the relationship between the number of hours she works and her pay.

Her pay (P) is equal to the rate of pay (r) times the number of hours worked (h), so

$$P = r \cdot h \qquad \text{or} \qquad P = 9h$$

    Check Yourself 8

Sorina is driving at a constant rate of 50 mi/h. Write the equation that shows the distance she travels (d) in h hours.

RECALL

k is the constant of variation.

If two things vary directly and values are given for x and y, we can find k. This property is illustrated in Example 9.

    Example 9   Finding the Constant of Variation

NOTE

The direct variation equation is $y = kx$. (6, 30) is one ordered pair that satisfies the equation.

If y varies directly with x, and $y = 30$ when $x = 6$, find k.
    Because y varies directly with x, we know from the definition that

$$y = kx$$

We need to find k. We do this by substituting 30 for y and 6 for x.

$$(30) = k(6) \qquad \text{or} \qquad k = 5$$

    Check Yourself 9

If y varies directly with x, and $y = 100$ when $x = 25$, find the constant of variation.

The graph for a linear equation of direct variation always passes through the origin. Example 10 illustrates this.

Example 10 | **Graphing an Equation of Direct Variation**

Let y vary directly with x, with a constant of variation $k = 3.5$. Graph the equation of variation.

The equation of variation is $y = 3.5x$, so the graph has a slope of 3.5. Two points that satisfy the relationship are $(0, 0)$ and $(2, 7)$.

 Check Yourself 10

Let y vary directly with x, with a constant of variation $k = \dfrac{7}{3}$. Graph the equation of variation.

The following is an example from the field of electronics.

Example 11 | **Vocational Application**

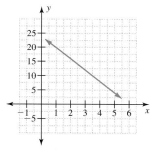

The graph depicts the relationship between the position of a linear potentiometer (variable resistor) and the output voltage of some DC source. Consider the potentiometer to be a slider control, possibly to control volume of a speaker or the speed of a motor.

The linear position of the potentiometer is represented on the x-axis, and the resulting output voltage is represented on the y-axis. At the 2-cm position, the output voltage measured with a voltmeter is 16 VDC. At a position of 3.5 cm, the measured output is 10 VDC.

What is the slope of the resulting line?

We see that we have two ordered pairs: $(2, 16)$ and $(3.5, 10)$. Using our formula for slope, we have

$$m = \frac{16 - 10}{2 - 3.5} = \frac{6}{-1.5} = -4$$

The slope is -4.

 Check Yourself 11

The same potentiometer described in Example 11 is used in another circuit. This time, though, when at position 0 cm, the output voltage is 12 volts. At position 5 cm, the output voltage is 3 volts. Draw a graph using the new data and determine the slope.

Check Yourself ANSWERS

1. $m = \dfrac{2}{3}$ **2.** $m = \dfrac{5}{3}$

3. $m = -3$ **4.** $m = 0$ **5.** m is undefined   **6.** $m = -\dfrac{2}{5}$

7.

[Note: Your second point could have been $(-2, 1)$ or $(2, -1)$.]

8. $d = 50h$ **9.** $k = 4$

10.

11. Slope $= -\dfrac{9}{5}$

Reading Your Text

The following fill-in-the-blank exercises are designed to ensure that you understand some of the key vocabulary used in this section.

SECTION 6.4

(a) The _____ of a line gives us a numerical measure of the steepness of that line.

(b) The difference $x_2 - x_1$ is sometimes called the _____ between points P and Q.

(c) We say that a vertical line has an _____ slope.

(d) When y varies directly with x, we write $y = kx$, and k is called the constant of _____.

6.4 exercises

< Objective 1 >

Find the slope of the line through the following pairs of points.

1. $(5, 7)$ and $(9, 11)$

2. $(4, 9)$ and $(8, 17)$

3. $(-2, 3)$ and $(3, 7)$

4. $(-3, -4)$ and $(3, -2)$

5. $(3, 3)$ and $(5, 0)$ 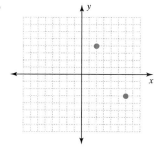 > Videos

6. $(-2, 4)$ and $(3, 1)$

7. $(5, -4)$ and $(5, 2)$

8. $(-5, 4)$ and $(2, 4)$

9. $(-4, -2)$ and $(3, 3)$

10. $(-5, -3)$ and $(-5, 2)$

11. $(-1, 7)$ and $(2, 3)$

12. $(-4, -2)$ and $(6, 4)$

In exercises 13–18, two points are shown. Find the slope of the line through the given points.

13.

14.

15.

16.

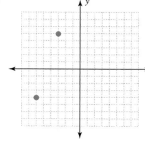

MathZone

Boost your grade at mathzone.com!

> Practice   > Self-Tests
  Problems
            > e-Professors
> NetTutor
            > Videos

Name _____

Section _____ Date _____

Answers

1. _____

2. _____

3. _____

4. _____

5. _____

6. _____

7. _____

8. _____

9. _____

10. _____

11. _____

12. _____

13. _____

14. _____

15. _____

16. _____

Answers

17. _____

18. _____

19. _____

20. _____

21. _____

22. _____

23. _____

24. _____

17.

18.

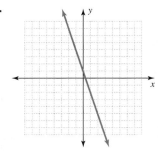
> Videos

< Objectives 2–3 >

In exercises 19–22, find the slope of the lines graphed.

19.

20.

> Videos

21.

22.

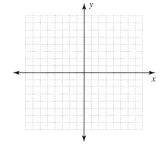

Sketch the graph of the following equations.

23. $y = -4x$

24. $y = 3x$

25. $y = \dfrac{2}{3}x$

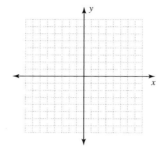

26. $y = -\dfrac{3}{4}x$

25. _____

26. _____

27. $y = \dfrac{5}{4}x$

28. $y = -\dfrac{4}{5}x$

 > Videos

27. _____

28. _____

29. _____

30. _____

31. _____

32. _____

33. _____

34. _____

In exercises 29–32, find the constant of variation k.

29. y varies directly with x; $y = 54$ when $x = 6$.

30. m varies directly with n; $m = 144$ when $n = 8$.

31. y varies directly with x; $y = 2,100$ when $x = 600$.

32. y varies directly with x; $y = 400$ when $x = 1,000$.

In exercises 33–36, y varies directly with x and the value of k is given. Graph the equation of variation.

33. $k = 2$

34. $k = 4$

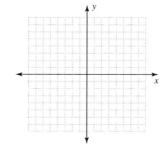

Answers

35. _____

36. _____

37. _____

38. _____

39. _____

40. _____

35. $k = 2.5$

36. $k = 2.2$

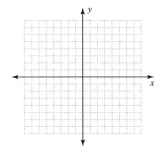

37. BUSINESS AND FINANCE Robin earns $12 per hour. Write an equation that shows how much she makes (S) in h hours.

38. BUSINESS AND FINANCE Kwang earns $11.50 per hour. Write an equation that shows how much he earns (S) in h hours.

39. BUSINESS AND FINANCE At a factory that makes grinding wheels, Kalila makes $0.20 for each wheel completed. Sketch the equation of direct variation.

40. BUSINESS AND FINANCE Palmer makes $1.25 per page for each page that he types. Sketch the equation of direct variation.

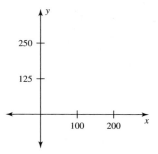

41. BUSINESS AND FINANCE Josephine works part-time in a local video store. Her salary varies directly as the number of hours worked. Last week she earned $43.20 for working 8 hours. This week she earned $118.80. How many hours did she work this week?

42. BUSINESS AND FINANCE The revenue for a sandwich shop is directly proportional to its advertising budget. When the owner spent $2,000 a month on advertising, the revenue was $120,000. If the revenue is now $180,000, how much is the owner spending on advertising?

Determine whether each of the following statements is **true** *or* **false.**

43. The slope of a horizontal line is undefined.

44. A negative slope means that the line falls from left to right.

Complete each of the following statements with **never, sometimes,** *or* **always.**

45. The graph of $y = mx$ _____ passes through the origin.

46. If y varies directly with x, the constant of variation k is _____ negative.

Basic Skills | **Advanced Skills** | Vocational-Technical Applications | Calculator/Computer | Above and Beyond | Getting Ready
▲

47. Consider the equation $y = 2x - 5$.

  (a) Complete the following table:

> Videos

x	y
3	
4	

  (b) Use the ordered pairs found in part **(a)** to calculate the slope of the line.
  (c) What do you observe concerning the slope found in part **(b)** and the given equation?

Answers

41. _____

42. _____

43. _____

44. _____

45. _____

46. _____

47. _____

Answers

48. _____

49. _____

50. _____

51. _____

52. _____

53. _____

54. _____

48. Repeat exercise 47 for $y = \dfrac{3}{2}x + 5$ and

x	y
2	
4	

49. Repeat exercise 47 for $y = -\dfrac{1}{3}x + 2$ and

x	y
3	
6	

50. Repeat exercise 47 for $y = -4x - 6$ and

x	y
−1	
−3	

51. Consider the equation $y = 2x + 3$.

(a) Complete the following table of values.

Point	x	y
A	5	
B	6	
C	7	
D	8	
E	9	

(b) As the x-coordinate changes by 1 (for example, as you move from point A to point B), by how much do the corresponding y-coordinates change?

(c) Is your answer to part (b) the same if you move from B to C? from C to D? from D to E?

(d) Describe the "growth rate" of the line using these observations. Complete the following statement: When the x-value grows by 1 unit, the y-value _____.

52. Repeat exercise 51 using $y = 2x + 5$.

53. Repeat exercise 51 using $y = -4x + 50$.

54. Repeat exercise 51 using $y = -4x + 40$.

In the following exercises, (a) plot the given point; (b) using the given slope, move from the point plotted in (a) to plot a new point; (c) draw the line that passes through the points plotted in (a) and (b).

55. $(3, 1)$, $m = 2$

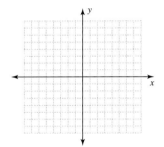

56. $(-1, 4)$, $m = -2$

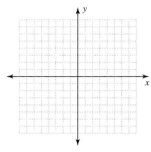

57. $(-2, -1)$, $m = -4$

58. $(-3, 5)$, $m = 2$

Answers

55. _____

56. _____

57. _____

58. _____

59. _____

Basic Skills | Advanced Skills | **Vocational-Technical Applications** | Calculator/Computer | Above and Beyond | Getting Ready

59. ALLIED HEALTH The recommended dosage, d, in milligrams (mg) of the antibiotic ampicillin sodium for children weighing less than 40 kilograms is given by the linear equation $d = 7.5w$, where w represents the child's weight in kilograms (kg). Sketch a graph of the linear equation.

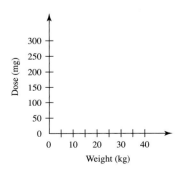

60. ALLIED HEALTH The recommended dosage, d, in micrograms (mcg) of Neupogen, a medication given to bone marrow transplant patients, is given by the linear equation $d = 8w$, where w represents the patient's weight in kilograms (kg). Sketch a graph of the linear equation.

61. MANUFACTURING TECHNOLOGY The following graph shows the bending moment in a wood beam.

Calculate the slope of the moment graph:

(a) Between 0 and 4 feet

(b) Between 4 and 11 feet

(c) Between 11 and 19 feet

(d) Between 19 and 24 feet

Basic Skills | Advanced Skills | Vocational-Technical Applications | Calculator/Computer | **Above and Beyond** | Getting Ready

62. Summarize the results of exercises 51 to 54. In particular, how does the concept of "growth rate" connect to the concept of slope?

63. Complete the statement: "The difference between undefined slope and zero slope is . . ."

64. Complete the statement: "The slope of a line tells you . . ."

Basic Skills | Advanced Skills | Vocational-Technical Applications | Calculator/Computer | Above and Beyond | **Getting Ready** ▲

Write each number as a percent.

(a) 0.06

(b) 0.375

(c) 2.4

(d) $\dfrac{43}{100}$

(e) $\dfrac{2}{5}$

(f) $2\dfrac{2}{3}$

Answers

1. 1   **3.** $\dfrac{4}{5}$ **5.** $-\dfrac{3}{2}$ **7.** Undefined   **9.** $\dfrac{5}{7}$ **11.** $-\dfrac{4}{3}$

13. 2   **15.** -2 **17.** 0   **19.** 2   **21.** $\dfrac{1}{3}$

23. $y = -4x$

25. $y = \dfrac{2}{3}x$

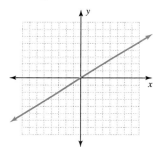

27. $y = \dfrac{5}{4}x$

29. 9

31. 3.5

33.

35.

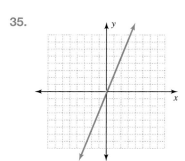

Answers

(a) _____

(b) _____

(c) _____

(d) _____

(e) _____

(f) _____

37. $S = 12h$ **39.**

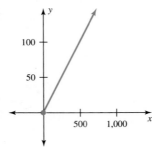

41. 22 h   **43.** False   **45.** always

47. **(a)** $(3, 1)$, $(4, 3)$; **(b)** 2; **(c)** slope equals coefficient of x

49. **(a)** $(3, 1)$, $(6, 0)$; **(b)** $-\dfrac{1}{3}$; **(c)** slope equals coefficient of x

51. **(a)** $(5, 13)$, $(6, 15)$, $(7, 17)$, $(8, 19)$, $(9, 21)$; **(b)** 2; **(c)** yes; **(d)** increases by 2

53. **(a)** $(5, 30)$, $(6, 26)$, $(7, 22)$, $(8, 18)$, $(9, 14)$; **(b)** 4; **(c)** yes; **(d)** decreases by 4

55.

57.

59.

61. **(a)** 6,250 foot-pounds per foot;
(b) 2,857.14 foot-pounds per foot;
(c) $-2,500$ foot-pounds per foot;
(d) $-5,000$ foot-pounds per foot

63. **(a)** Above and Beyond

(a) 6%   **(b)** 37.5%   **(c)** 240%   **(d)** 43%   **(e)** 40%   **(f)** $266\dfrac{2}{3}\%$

6.5

Reading Graphs*

< 6.5 Objectives >

1 > Read and interpret a table

2 > Read and interpret a pie chart

3 > Read different types of bar graphs

4 > Read and interpret a line graph

5 > Create and make a prediction from a line graph

NOTE

Spreadsheets and databases are commonly used software tools for working with tables of data.

NOTE

Antarctica has been omitted from this table.

A **table** is a display of information in parallel rows or columns. Tables can be used anywhere that information is to be summarized.

Following is a table describing land area and world population. Each entry in a table is called a **cell.** This table will be used for Examples 1 and 2.

Continent or Region	Land Area (1,000 mi²)	Population (millions)		
		1900	1950	2000
North America	9,400	106	221	305
South America	6,900	38	111	515
Europe	3,800	400	392	510
Asia (including Russia)	17,400	932	1,591	4,178
Africa	11,700	118	229	889
Oceania (including Australia)	3,300	6	12	32
World totals	52,500	1,600	2,556	6,429

Source: Bureau of the Census: U.S. Dept. of Commerce.

Example 1 | Reading a Table

< Objective 1 >

From the land area and world population table above, find each of the following.

(a) What was the population of Africa in 1950?

* This section is included for those instructors who have it in their curricula. It is optional in the sense that it is not required for the remainder of the text.

Continent or Region	Land Area (1,000 mi²)	Population (millions)		
		1900	1950	2000
North America	9,400	106	221	305
South America	6,900	38	111	515
Europe	3,800	400	392	510
Asia (including Russia)	17,400	932	1,591	4,178
Africa	11,700	118	229	889
Oceania (including Australia)	3,300	6	12	32
World totals	52,500	1,600	2,556	6,429

Source: Bureau of the Census: U.S. Dept. of Commerce.

Looking at the cell in the row labeled Africa and the column labeled 1950, we find the number 229. Because we are told the population is given in millions, we conclude that the population of Africa in 1950 was 229,000,000.

(b) What is the land area of Asia, in square miles?

The cell in the row Asia and column Land Area reads 17,400. The land area is given in 1,000 mi² units, so the actual land area of Asia is 17,400,000 mi².

Check Yourself 1

Use the land area and world population table to answer each question.

(a) What was the population of South America in 1900?
(b) What is the land area of Europe?

We frequently use a table to find information not explicitly given in the table. We will use the land area and world population table again to illustrate this point.

| ▶ | Example 2 | Interpreting a Table |

Use the land area and world population table to answer each of the following questions.

(a) By what percentage did the population of Africa change between 1950 and 2000?

In Example 1, we found that the 1950 population of Africa was 229 million. The 2000 population of Africa was 889 million, or 660 million more people than in 1950. The percentage change is

$$\frac{660,000,000}{229,000,000} \approx 2.882 = 288.2\%$$

So, the population of Africa increased by about 288.2% between 1950 and 2000.

(b) What is the land area of Asia as a percentage of the earth's total land area (excluding Antarctica)?

In Example 1 we concluded that Asia is 17,400,000 mi². We find that there are 52,500,000 mi² of land area (excluding Antarctica) from the row labeled World totals. Therefore, the percentage of the earth's total land area made up by Asia is

$$\frac{17,400,000}{52,500,000} \approx 33.1\%$$

(c) What was the density of the population (people per square mile) in North America in 2000?

North America had a population of 305 million in 2000. With a land area of 9,400,000 mi², the population density of North America is given by

$$\frac{305,000,000}{9,400,000} \approx 32.4 \text{ people per square mile}$$

Check Yourself 2

Use the land area and world population table to answer each of the following questions.

(a) By what percentage did the population of South America change between 1900 and 1950?

(b) What is the land area of Europe as a percentage of the earth's total land area (excluding Antarctica)?

(c) Did the world population increase by a greater percentage between 1900 and 1950 or between 1950 and 2000?

NOTE

Pie charts are sometimes called *circle graphs*.

If we need to present the relationship between two sets of data in only a column or two of a table, it is often easier to use pictures. A **graph** is a diagram that represents the connection between two or more quantities.

The first graph we will look at is called a **pie chart.** We use a circle to represent some total that interests us. Wedges (or sectors) are drawn in the circle to show how much of the whole each part makes up.

| Example 3 | Reading a Pie Chart |

< Objective 2 >

The following pie chart represents the results of a survey that asked students how they get to school most often.

NOTE

The total of the percentages of the wedges is always 100%.

Bus
30%

Car
55%

Walk
15%

(a) What percentage of the students walk to school?

We see that 15% walk to school.

(b) What percentage of the students do not arrive by car?

Because 55% arrive by car, there are 100% − 55%, or 45%, who do not.

Check Yourself 3

This pie chart represents the results of a survey that asked students whether they bought lunch, brought it, or skipped lunch altogether.

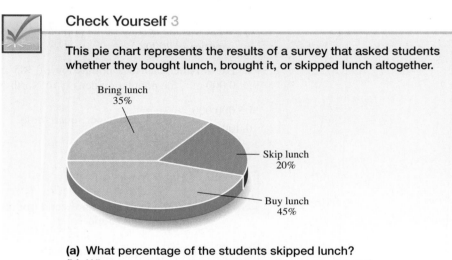

Bring lunch
35%

Skip lunch
20%

Buy lunch
45%

(a) What percentage of the students skipped lunch?
(b) What percentage of the students did not buy lunch?

If we know what the whole pie represents, we can also find out more about what each wedge represents, as illustrated by Example 4.

Example 4   Interpreting a Pie Chart

This pie chart shows how Sarah spent her $12,000 college scholarship.

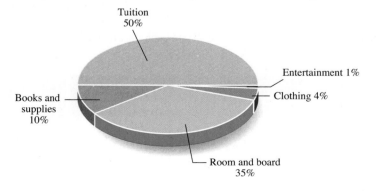

Tuition
50%

Entertainment 1%

Clothing 4%

Books and
supplies
10%

Room and board
35%

(a) How much did she spend on tuition?

She spent 50% of her $12,000 scholarship, or $6,000, on tuition.

(b) How much did she spend on clothing and entertainment?

Together, 5% of the money was spent on clothing and entertainment, and 0.05 ×
12,000 = 600. Therefore, $600 was spent on clothing and entertainment.

Check Yourself 4

**This pie chart shows how Rebecca spends an average 24-hour
school day.**

(a) How many hours does she spend sleeping each day?
**(b) How many hours does she spend altogether studying
and in class?**

A **bar graph** provides yet another way to present information. It shows the relationship between two sets of data.

| Example 5 | Reading a Bar Graph |

< Objective 3 >

The following bar graph represents the number of cars sold in the United States in each
year listed (sales are in thousands).

(a) How many cars were sold in the United States in 1990?

We frequently have to estimate our answer when reading a bar graph. In this case,
there were approximately 9,500,000 cars sold in 1990.

(b) In which year listed did the most car sales occur?

The tallest bar occurs in 1985; therefore, more cars were sold in 1985 than in any
other year represented on the graph.

NOTE

Remember that sales
numbers are in thousands.

Check Yourself 5

The following bar graph represents the response to a recent Gallup poll that asked people to name their favorite spectator sport.

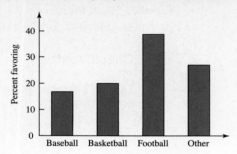

(a) Find the percentage of people for whom football is their favorite spectator sport.

(b) Of the three major sports listed, which was named as a favorite by the fewest people?

When we use bar graphs to display additional information, we often use different colors for different bars. With such graphs, it is necessary to include a legend. A **legend** is a key describing what each color or shade of bar represents.

 Example 6   Reading a Bar Graph

The following bar graph compares the number of cars sold in the world and in the United States in each year listed (sales are in thousands).

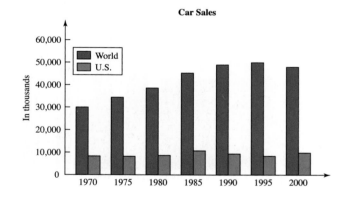

(a) How many cars were sold in the world in 1985? In the United States in 1985?

The legend tells us that the blue bar represents worldwide sales and the pink bar represents U.S. sales.

It would appear that there were about 45,000,000 cars sold in the world in 1985. Approximately 11,000,000 of these were sold in the United States.

(b) What percentage of global sales did the United States account for in 1985?

$$\frac{11}{45} \approx 0.244 = 24.4\%$$

Check Yourself 6

The following bar graph represents the average student age at Berndt Community College.

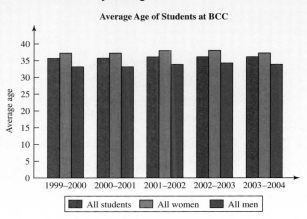

(a) What was the average age of female students in 2003–2004?
(b) Did the average age of female students increase or decrease between 2002–2003 and 2003–2004?
(c) Who tends to be older, male students or female students?

Another useful type of graph is called a **line graph.** In a line graph, one set of data is usually related to time. Line graphs give us a way of visualizing changes to something over time.

Example 7 | **Reading and Interpreting a Line Graph**

< Objective 4 >

The following line graph represents the number of Social Security beneficiaries in each of the years listed.

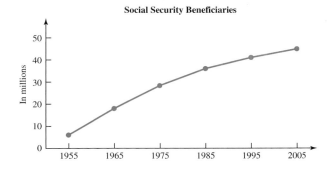

(a) How many Social Security beneficiaries were there in 1975?

We find the point on the line graph that lies above 1975. We then determine that this point lies at about 28 on the vertical axis.

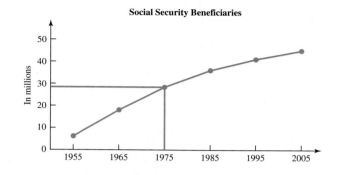

Therefore, we conclude that there were approximately 28 million Social Security beneficiaries in 1975.

(b) How many beneficiaries were there in 1990?

The line connecting 1985 and 1995 passes through 1990 at about 39 million.

(c) Approximate the mean number of annual beneficiaries between 1955 and 2005.

We estimate the number of beneficiaries in each of the years 1955, 1965, 1975, 1985, 1995, and 2005 to be 6 million, 18 million, 28 million, 36 million, 41 million, and 45 million, respectively.

Therefore, we calculate the mean number of beneficiaries as

$$\frac{6 + 18 + 28 + 36 + 41 + 45}{6} = \frac{174}{6} = 29$$

So, the mean number of annual beneficiaries is 29 million.

> **RECALL**
>
> The mean is computed by adding all of the values in a set and dividing by the number of values in the set.

Check Yourself 7

The following graph indicates the high temperatures in Baltimore, Maryland, for a week in September.

(a) What was the high temperature on Friday?
(b) Find the mean high temperature for that week.

We use many of the techniques learned in Section 6.2 to create a line graph. We start with the rectangular coordinate system. Generally, we make the x-axis represent the time quantity and scale the y-axis accordingly.

The Streeter/Hutchison Series in Mathematics Beginning Algebra

For each time given, we plot the point corresponding to whatever it is we are measuring. Finally, we "connect the dots." That is, we draw a line segment from each point to the next point immediately to the right.

| | Example 8 | Constructing a Line Graph |

< Objective 5 >

Construct a line graph to summarize the data shown.

FBI: Larceny-Theft Cases (in hundred thousands)	
Year	**Thefts**
1985	69
1990	79
1995	80
2000	70
2005	64

Source: Bureau of Justice Statistics

We set, scale, and label our axes appropriately. Then, we plot the points given in the table, with x as the year and y as the number of thefts.

Finally, we sketch line segments from point to point (that is, we "connect the dots").

Check Yourself 8

Construct a line graph based on the following table of data describing the cost of a first-class postage stamp, by year.

First-Class Postage Stamps (in cents)	
Year	**Cost**
1960	4
1965	5
1970	6
1975	10
1980	15
1985	22
1990	25
1995	32
2000	33
2005	37

Source: United States Postal Service

An important feature of line graphs is that they allow us to (cautiously) predict results beyond the data given. To do this, we extend the sketched curve as needed.

Example 9   **Making a Prediction from a Line Graph**

Use the line graph in Example 8 to predict the number of larceny-theft cases that will be reported to the FBI in 2010.

We extend the line segment connecting the points at 2000 and 2005 to 2010.

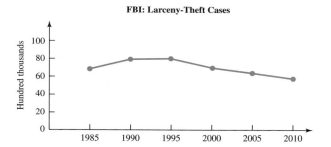

FBI: Larceny-Theft Cases

We predict that there will be 58 hundred thousand, or 5,800,000, larceny-theft cases reported to the FBI in 2010.

Check Yourself 9

Use the postage stamp line graph constructed in Check Yourself 8 to predict the cost of a first-class postage stamp in the year 2010.

You should not use a line graph to predict occurrences too far out of range of the original data (at least not if you are looking for an accurate prediction). To illustrate this, consider the following.

Between the ages of 3 and 10 months, most kittens gain about 1 lb per month. Therefore, if a kitten weighs 2 lb at 3 months, it would be reasonable to predict that the kitten will weigh 9 lb at 10 months. We could even guess that the kitten might weigh around 11 lb after 1 year.

The Streeter/Hutchison Series in Mathematics   Beginning Algebra

But, if we extrapolate further, we would then predict that this same kitten will weigh 59 lb when it is 5 years old. This is, of course, ridiculous. The problem is that our initial data describe only a small segment (within the first year). When we try to move too far beyond the original data, we run into difficulties because the original data no longer apply to our prediction.

The following is an example from the field of manufacturing.

Example 10   **Vocational Application**

The following pie chart shows the breakdown of the costs of producing a part.

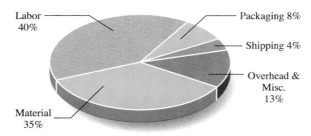

(a) What percentage of the cost of producing a product is the actual "material" for the product?

We see in the chart that 35% of the production cost goes to "material."

(b) For a part that costs $12.40 to produce, how much would you expect the labor cost to be?

Since 40% of production costs go to "labor," we find 40% of $12.40:

$0.40 \times 12.40 = 4.96$

So, we would expect the labor cost to be $4.96.

(c) What is the largest expense in producing a part?

From the chart, we see that the category with the greatest percentage is "labor."

(d) A part uses $4.30 in materials. What would you expect the total cost of producing the part to be?

From the chart, we can write that 35% of the total cost is $4.30. Writing this as an equation, we have

$0.35c = 4.30$

$c \approx 12.29$

We would expect the total cost of producing the part to be $12.29.

 Check Yourself 10

Using the pie chart from Example 10, find:

(a) the shipping cost if the total production cost is $17.89;
(b) the total production cost if the "overhead and miscellaneous" cost is $3.15.

Check Yourself ANSWERS

1. (a) 38,000,000; **(b)** 3,800,000 mi^2 **2. (a)** 192.1%; **(b)** 7.2%;
(c) 1950–2000 (152% vs. 60%)   **3. (a)** 20%; **(b)** 55%   **4. (a)** 6 h;
(b) 14.4 h   **5. (a)** 38%; **(b)** baseball   **6. (a)** 37 years; **(b)** it decreased;
(c) female students   **7. (a)** 88°F; **(b)** 86°F

8.

9. 41¢

10. (a) $0.72;
    (b) $24.23

Reading Your Text

The following fill-in-the-blank exercises are designed to ensure that you understand some of the key vocabulary used in this section.

SECTION 6.5

(a) A _____ is a display of information in parallel rows or columns.

(b) Pie charts are sometimes called _____ graphs.

(c) In a bar graph, a _____ is a key describing what each color or shade of bar represents.

(d) In a _____ graph, one set of data is usually related to time.

< Objective 1 >

Use the following table for exercises 1–8.

World Motor Vehicle Production, 1950–1997

| Year | Production (in thousands) | | | | | |
	United States	Canada	Europe	Japan	Other	World Total
1997	12,119	2,571	17,773	10,975	10,024	53,463
1996	11,799	2,397	17,550	10,346	9,241	51,332
1995	11,985	2,408	17,045	10,196	8,349	49,983
1994	12,263	2,321	16,195	10,554	8,167	49,500
1993	10,898	2,246	15,208	11,228	7,205	46,785
1992	9,729	1,961	17,628	12,499	6,269	48,088
1991	8,811	1,888	17,804	13,245	5,180	46,928
1990	9,783	1,928	18,866	13,487	4,496	48,554
1985	11,653	1,933	16,113	12,271	2,939	44,909
1980	8,010	1,324	15,496	11,043	2,692	38,565
1970	8,284	1,160	13,049	5,289	1,637	29,419
1960	7,905	398	6,837	482	866	16,488
1950	8,006	388	1,991	32	160	10,577

Note: As far as can be determined, production refers to vehicles locally manufactured.
Source: American Automobile Manufacturers Assn.

1. What was the motor vehicle production in Japan in 1950? 1997?

2. What was the motor vehicle production in countries outside the United States in 1950? 1997?

3. What was the percentage increase in motor vehicle production in the United States from 1950 to 1997? > Videos

4. What was the percentage increase in motor vehicle production in countries outside the United States from 1950 to 1997?

5. What percentage of world motor vehicle production occurred in Japan in 1997?

6. What percentage of world motor vehicle production occurred in the United States in 1997?

7. What percentage of world motor vehicle production occurred outside the United States and Japan in 1997?

8. Between 1950 and 1997, did the production of motor vehicles increase by a greater percentage in Canada or Europe?

MathZone

Boost your grade at mathzone.com!
> Practice Problems
> NetTutor
> Self-Tests
> e-Professors
> Videos

Name _____

Section _____ Date _____

Answers

1. _____

2. _____

3. _____

4. _____

5. _____

6. _____

7. _____

8. _____

Answers

9. _____

10. _____

11. _____

12. _____

13. _____

14. _____

15. _____

16. _____

17. _____

18. _____

< Objective 2 >

The following pie chart represents the way a new company ships its goods. Use the information presented to complete exercises 9 to 12.

9. What percentage is shipped by air freight?

10. What percentage is shipped by truck?

11. What percentage is shipped by truck or second-day air freight?

12. If the company shipped a total of 550 items last month, how many were shipped using second-day air freight?

< Objective 3 >

In exercises 13 to 18, use the given bar graph representing the number of bankruptcy filings during a recent 5-year period.

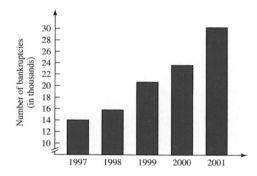

13. How many people filed for bankruptcy in 1998?

14. How many people filed for bankruptcy in 2001?

15. What was the increase in filings from 1999 to 2001?

16. What was the increase in filings from 1997 to 2001?

17. Which year had the greatest increase in filings?

18. In which year did the greatest percentage of increase in filings occur?

< Objective 4 >

Use the line graph showing ticket sales for the last 6 months of the year to answer exercises 19 and 20.

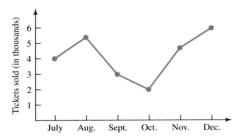

19. What month had the greatest number of ticket sales?

20. Between what two months did the greatest decrease in ticket sales occur?

< Objective 5 >

Use the following table to complete exercises 21 and 22.

Year	Population of United States (in millions)
1950	151
1960	179
1970	203
1980	227
1990	249
2000	281

21. Create a line graph to present the data given in the table.

22. Use the line graph to predict the U.S. population in the year 2010.

Determine whether each of the following statements is **true** *or* **false.**

23. A pie chart is the best tool for visualizing changes to something over time.

24. Caution must be used when making predictions beyond the range of the given data.

Answers

19.
_____

20.
_____

21.
_____

22.
_____

23.
_____

24.
_____

Beginning Algebra   The Streeter/Hutchison Series in Mathematics   © The McGraw-Hill Companies. All Rights Reserved.

Answers

25. _____

26. _____

27. _____

28. _____

Complete each of the following statements with **never, sometimes,** *or* **always.**

25. In a pie chart, the total of the percentages of the wedges is _____ 100%.

26. You should _____ use a line graph to predict occurrences that are far out of the range of the original data.

Basic Skills | Advanced Skills | **Vocational-Technical Applications** | Calculator/Computer | Above and Beyond | Getting Ready

▲

27. SCIENCE AND MEDICINE The following table gives the number of live births, broken down by the race of the mother, in the United States from 1994 to 1999.

Year	Total Births	White	African American	American Indian	Asian or Pacific Islander	Hispanic
1999	3,959,417	3,132,501	605,970	40,170	180,776	764,339
1998	3,941,553	3,118,727	609,902	40,272	172,652	734,661
1997	3,880,894	3,072,640	599,913	38,572	169,769	709,767
1996	3,891,494	3,093,057	594,781	37,880	165,776	701,339
1995	3,899,589	3,098,885	603,139	37,278	160,287	679,768
1994	3,952,767	3,121,004	636,391	37,740	157,632	665,026

Source: National Vital Statistics Reports, Volume 52, Number 10, December 17, 2003.

(a) Which categories showed a steady increase in live births from 1994 to 1999?

(b) What percentage of the live births were to Hispanic mothers in 1999?

(c) What percentage of the live births were to white mothers in 1996?

(d) Determine the percentage increase or decrease in live births from 1994 to 1999 for African American mothers.

28. ELECTRONICS The following graph shows typical home energy uses in the United States.

Home Energy Use
Based on national averages

Lighting cooking and other appliances 33%

Water heating 14%

Refrigerator 9%

Heating and cooling 44%

From "Small Electric Wind Systems: A U.S. Consumer's Guide," May 2001, Revised October 2002. U.S. Department of Energy, Office of Energy Efficiency and Renewable Energy, Wind and Hydropower Technologies Program.

Assume that a household uses 10,600 kWh (kilowatt-hours) of electricity annually. Calculate the energy (in kWh) used annually for each category.

29. **AUTOMOTIVE TECHNOLOGY** This table shows the effects of adjusting the spark advance on the horsepower and the exhaust temperature of an engine.

Spark Advance	Brake Horsepower	Exhaust Temperature
10°	59 hp	1,340°F
20°	76 hp	1,275°F
30°	84 hp	1,250°F
40°	87 hp	1,255°F
50°	82 hp	1,300°F

(a) Which degree of spark advance results in the greatest horsepower?

(b) Which degree of spark advance results in the lowest exhaust temperature?

30. **MECHANICAL ENGINEERING** The following table shows properties of five different metals.

Metal	Density (g/cm³)	Melting point (°C)
Iron	7.87	1,538
Aluminum	2.699	660.4
Copper	8.93	1,084.9
Tin	5.765	231.9
Titanium	4.507	1,668

(a) What is the density of titanium?

(b) What is the difference in melting point between copper and iron?

(c) Which metal has the highest melting point?

31. **BUSINESS AND FINANCE** The following table shows sales figures for cars sold by year.

Year Manufactured	Sales (1,000s of vehicles)
2000	43.4
2001	29.1
2002	36.8
2003	19.7
2004	28.1

Create a line graph to display these data.

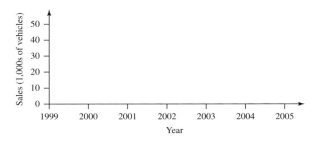

Answers

29. _____

30. _____

31. _____

32.
_____

32. MANUFACTURING TECHNOLOGY Use the table to answer the following questions.

Type of wood	Tensile strength (parallel to grain), psi	Tensile strength (perpendicular to grain), psi	Compressive strength (parallel to grain), psi	Compressive strength (perpendicular to grain), psi
Elm	17,500	660	5,520	690
Maple	15,700	1,100	7,830	1,470
Oak	11,300	940	6,200	810
Cedar	6,600	320	6,020	920
Pine	10,600	310	4,800	440
Spruce	8,600	370	5,610	580

(a) What is the compressive strength of oak parallel to the grain?

(b) What is the difference in tensile strength of pine between use parallel to the grain and use perpendicular to the grain?

(c) What is the difference in tensile strength parallel to the grain between elm and pine?

(d) Make a general statement about the best way to use wood in order to obtain the greatest strength.

Answers

1. 32,000; 10,975,000   **3.** 51.4%   **5.** 20.5%   **7.** 56.8%   **9.** 55%
11. 85%   **13.** 16,000   **15.** 9,000   **17.** 2001   **19.** December
21.

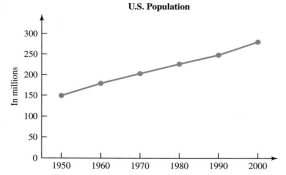

23. False   **25.** always   **27. (a)** Hispanic and Asian or Pacific Islander;
(b) 19.3%; **(c)** 79.5%; **(d)** −4.78%   **29. (a)** 40°; **(b)** 30°
31.

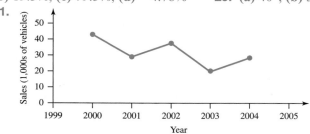

Definition/Procedure	Example	Reference
Solutions of Equations in Two Variables		Section 6.1
Solutions of Equations		
A pair of values that satisfies the equation. Solutions for equations in two variables are written as *ordered pairs*. An ordered pair has the form $$(x, y)$$	If $2x - y = 10$, $(6, 2)$ is a solution for the equation, because substituting 6 for x and 2 for y gives a true statement.	*p. 542*
The Rectangular Coordinate System		Section 6.2
The Rectangular Coordinate System		
A system formed by two perpendicular axes that intersect at a point called the **origin.** The horizontal line is called the *x*-axis. The vertical line is called the *y*-axis.		*p. 555*
Graphing Points from Ordered Pairs		
The coordinates of an ordered pair allow you to associate a point in the plane with the ordered pair. To graph a point in the plane,	To graph the point corresponding to (2, 3):	*p. 558*
1. Start at the origin.		
2. Move right or left according to the value of the *x*-coordinate: to the right if x is positive or to the left if x is negative.		
3. Then move up or down according to the value of the *y*-coordinate: up if y is positive or down if y is negative.		
Graphing Linear Equations		Section 6.3
Linear Equation		
An equation that can be written in the form $Ax + By = C$ in which A and B are not both 0.	$2x - 3y = 4$ is a linear equation.	*p. 572*
Graphing Linear Equations		
1. Find at least three solutions for the equation, and put your results in tabular form.		*p. 573*
2. Graph the solutions found in step 1.		
3. Draw a straight line through the points determined in step 2 to form the graph of the equation.		

Continued

Definition/Procedure	Example	Reference

The Slope of a Line

Slope

The slope of a line gives a numerical measure of the steepness of the line. The slope m of a line containing the distinct points in the plane $P(x_1, y_1)$ and $Q(x_2, y_2)$ is given by

$$m = \frac{\text{vert change}}{\text{horiz change}} = \frac{y_2 - y_1}{x_2 - x_1} \quad \text{when } x_2 \neq x_1$$

Section 6.4

To find the slope of the line through $(-2, -3)$ and $(4, 6)$:

$$m = \frac{(6) - (-3)}{(4) - (-2)}$$

$$= \frac{6 + 3}{4 + 2} = \frac{9}{6} = \frac{3}{2}$$

p. 601

The Graph of $y = mx$

A line passing through the origin with slope m.

To graph $y = -4x$:

1. Plot $(0, 0)$

2. $\dfrac{\text{Rise}}{\text{Run}} = -4 = \dfrac{-4}{1}$

p. 607

Direct Variation

If y is a constant positive multiple of x, we write $y = kx$, where $k > 0$,

and say y varies directly as x.

If $y = 20$ when $x = 5$, and y varies directly as x,

$(20) = k(5)$

$k = 4 \qquad y = 4x$

p. 607

Reading Graphs

Section 6.5

Tables

A **table** is a rectangular display of information or data.

p. 621

Graphs

Graph A diagram that relates two different pieces of information.

pp. 623–631

Pie chart A graph that shows the component parts of a whole.

Bar graph One of the most common types of graph. It relates the amounts of items to each other.

Line graph A graph in which one of the axes is usually related to time. We can use line graphs to make **predictions** about events.

This summary exercise set is provided to give you practice with each of the objectives of this chapter. Each exercise is keyed to the appropriate chapter section. When you are finished, you can check your answers to the odd-numbered exercises against those presented in the back of the text. If you have difficulty with any of these questions, go back and reread the examples from that section. Your instructor will give you guidelines on how best to use these exercises in your instructional setting.

6.1 *Tell whether the number shown in parentheses is a solution for the given equation.*

1. $7x + 2 = 16$ (2)

2. $5x - 8 = 3x + 2$ (4)

3. $7x - 2 = 2x + 8$ (2)

4. $4x + 3 = 2x - 11$ (-7)

5. $x + 5 + 3x = 2 + x + 23$ (6)

6. $\dfrac{2}{3}x - 2 = 10$ (21)

Determine which of the ordered pairs are solutions for the given equations.

7. $x - y = 6$ $(6, 0), (3, 3), (3, -3), (0, -6)$

8. $2x + y = 8$ $(4, 0), (2, 2), (2, 4), (4, 2)$

9. $2x + 3y = 6$ $(3, 0), (6, 2), (-3, 4), (0, 2)$

10. $2x - 5y = 10$ $(5, 0), \left(\dfrac{5}{2}, -1\right), \left(2, \dfrac{2}{5}\right), (0, -2)$

Complete the ordered pairs so that each is a solution for the given equation.

11. $x + y = 8$ $(4,\ ), (\ , 8), (8,\ ), (6,\ )$

12. $x - 2y = 10$ $(0,\ ), (12,\ ), (\ , -2), (8,\ )$

13. $2x + 3y = 6$ $(3,\ ), (6,\ ), (\ , -4), (-3,\ )$

14. $y = 3x + 4$ $(2,\ ), (\ , 7), \left(\dfrac{1}{3},\ \right), \left(\dfrac{4}{3},\ \right)$

Find four solutions for each of the following equations.

15. $x + y = 10$

16. $2x + y = 8$

17. $2x - 3y = 6$

18. $y = -\dfrac{3}{2}x + 2$

6.2 *Give the coordinates of the points in the following graph.*

19. *A*

20. *B*

21. *E*

22. *F*

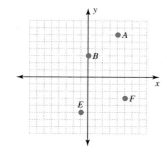

Plot points with the coordinates shown.

23. $P(6, 0)$

24. $Q(5, 4)$

25. $T(-2, 4)$

26. $U(4, -2)$

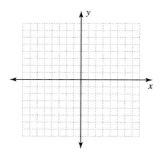

6.3 *Graph each of the following equations.*

27. $x + y = 5$

28. $x - y = 6$

29. $y = 2x$

30. $y = -3x$

31. $y = \dfrac{3}{2}x$

32. $y = 3x + 2$

33. $y = 2x - 3$

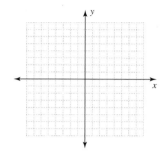

34. $y = -3x + 4$

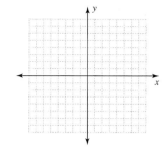

35. $y = \dfrac{2}{3}x + 2$

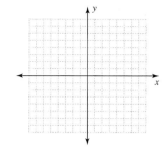

36. $3x - y = 3$

37. $2x + y = 6$

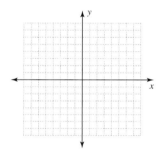

38. $3x + 2y = 12$

39. $3x - 4y = 12$

40. $x = 3$

41. $y = -2$

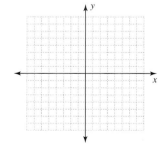

Graph each of the following equations.

42. $5x - 3y = 15$

43. $4x + 3y = 12$

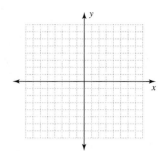

Graph each equation by first solving for y.

44. $2x + y = 6$

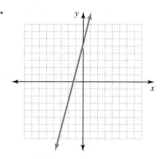

45. $3x + 2y = 6$

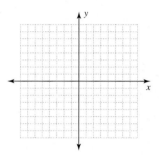

6.4 *Find the slope of the line through each of the following pairs of points.*

46. $(3, 4)$ and $(5, 8)$

47. $(-2, 3)$ and $(1, -6)$

48. $(-2, 5)$ and $(2, 3)$

49. $(-5, -2)$ and $(1, 2)$

50. $(-2, 6)$ and $(5, 6)$

51. $(-3, 2)$ and $(-1, -3)$

52. $(-3, -6)$ and $(5, -2)$

53. $(-6, -2)$ and $(-6, 3)$

In exercises 54–57, find the slope of the line graphed.

54.

55.

56.

57.

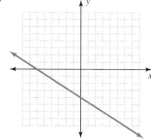

Graph each of the following equations.

58. $y = 6x$

59. $y = -6x$

60. $y = \dfrac{2}{5}x$

61. $y = -\dfrac{3}{4}x$

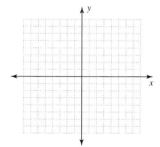

Solve for k, the constant of variation.

62. y varies directly as x; $y = 20$ when $x = 40$

63. y varies directly as x; $y = 5$ when $x = 3$

In exercises 64 and 65, y varies directly with x and the value of k is given. Graph the equation of variation.

64. $k = 4$

65. $k = -3.5$

6.5 *Use the following table describing technology available in public schools to answer exercises 66–69.*

Technology in U.S. Public Schools, 1995–1998				
	Number of Schools			
Technology	**1995**	**1996**	**1997**	**1998**
Schools with modems[1]	30,768	37,889	40,876	61,930
Elementary	16,010	20,250	22,234	35,066
Junior high	5,652	6,929	7,417	10,996
Senior high[2]	8,790	10,277	10,781	14,540
Schools with networks[1]	24,604	29,875	32,299	49,178
Elementary	11,693	14,868	16,441	26,422
Junior high	4,599	5,590	6,035	9,003
Senior high[2]	8,159	9,166	9,565	12,853
Schools with CD-ROMs[1]	34,480	43,499	46,388	64,200
Elementary	18,343	24,353	26,377	37,908
Junior high	6,510	7,952	8,410	11,023
Senior high[2]	9,327	10,756	11,140	13,985
Schools with Internet access[1]	NA	14,211	35,762	60,224
Elementary	NA	7,608	21,026	34,195
Junior high	NA	2,707	5,752	10,888
Senior high[2]	NA	3,736	8,984	13,829

NA = Not applicable. (1) Includes schools for special and adult education, not shown separate with grade spans of K–3, K–5, K–6, K–8, and K–12. (2) Includes schools with grade spans of technical and alternative high schools and schools with grade spans of 7–12, 9–12, and 10.

Source: Quality Education Data, Inc., Denver, CO.

66. What is the increase in the number of schools with modems from 1995 to 1998?

67. How many senior high schools had either modems, networks, or CD-ROMs in 1998?

68. What is the percent increase in public schools with Internet access from 1996 to 1998?

69. What is the percent increase in elementary schools who have either modems, networks, or Internet access from 1996 to 1998?

6.5 *Use the following pie charts describing U.S. car sales by size in* 1993 *and* 2003 *to answer questions* 70–73.

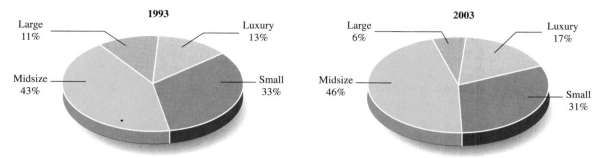

70. What was the percentage of midsize and small cars sold in 2003?

71. What was the percentage of large and luxury cars sold in 2003?

72. Sales of which type of car increased the most between 1993 and 2003?

73. If 21,303,000 U.S. cars were sold in 1993, how many small cars were sold?

In exercises 74 and 75, use the following graph showing enrollment at Berndt Community College.

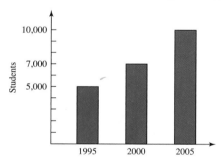

74. How many more students were enrolled in 2005 than in 1995?

75. What was the percent increase from 1995 to 2000?

6.5 *Use the following line graph to complete exercises 76–78.*

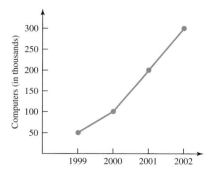

76. How many more personal computers were sold in 2002 than in 1999?

77. What was the percent increase in sales from 1999 to 2002?

78. Predict the sales of personal computers in the year 2003.

79. Use the following table to create a line graph comparing annual gross income (in thousands) of 30-year-olds with their years of formal education.

Years of Education	Gross Income
8	16
10	21
12	23
14	28
16	31

The purpose of this self-test is to help you check your progress and to review for the next in-class exam. Allow yourself about an hour to take this test. At the end of that hour check your answers against those given in the back of the text. Section references accompany the answers. If you missed any questions, go back to those sections and reread the examples until you master the concepts.

Name _____

Section _____ Date _____

Answers

Determine which of the ordered pairs are solutions for the given equations.

1. $x + y = 9$; $(3, 6), (9, 0), (3, 2)$

2. $4x - y = 16$; $(4, 0), (3, -1), (5, 4)$

Complete the ordered pairs so that each is a solution for the given equation.

3. $x + 3y = 12$; $(3,\ ), (\ , 2), (9,\ )$

4. $4x + 3y = 12$; $(3,\ ), (\ , 4), (\ , 3)$

Find four solutions for each of the following equations.

5. $x - y = 7$ **6.** $5x - 6y = 30$

1. _____

2. _____

3. _____

4. _____

5. _____

6. _____

7. _____

8. _____

9. _____

10. _____

11. _____

12. _____

Give the coordinates of the points in the following graph.

7. A

8. B

9. C

Plot points with the coordinates shown.

10. $S(1, -2)$

11. $T(0, 3)$

12. $U(-2, -3)$

Answers

13. _____

14. _____

15. _____

16. _____

17. _____

18. _____

Graph each of the following equations.

13. $x + y = 4$

14. $y = 3x$

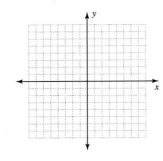

15. $y = \dfrac{3}{4}x - 4$

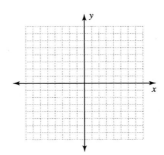

16. $x + 3y = 6$

17. $2x + 5y = 10$

18. $y = -4$

Find the slope of the line through the following pairs of points.

19. $(-3, 5)$ and $(2, 10)$

20. $(-2, 6)$ and $(2, 9)$

21. $(4, 6)$ and $(4, 8)$

22. $(7, 9)$ and $(3, 9)$

23. Find the slope of the line graphed.

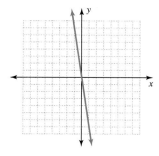

24. Graph the equation $y = 2x$.

25. Solve for the constant of variation if y varies directly with x and $y = 35$ when $x = 7$.

The pie chart below represents the portion of the $40 million tourism industry spent by tourists from each country. Use the chart to complete exercises 26 and 27.

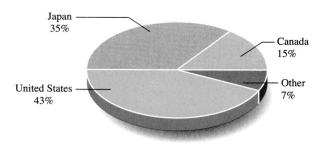

Japan
35%

Canada
15%

Other
7%

United States
43%

26. What percentage of the total tourism dollars is accounted for by Canada?

27. How many dollars does the United States account for?

Answers

19. _____

20. _____

21. _____

22. _____

23. _____

24. _____

25. _____

26. _____

27. _____

Answers

28. _____

29. _____

30. _____

Use the following line graph to answer exercises 28–30.

28. How many fewer family doctors were there in the United States in 1990 than in 1980?

29. What was the total change in the number of family doctors between 1980 and 2000?

30. In what 5-year period was the decrease in family doctors the greatest? What was the decrease?

The Streeter/Hutchison Series in Mathematics   Beginning Algebra

Activity 6 ::
Graphing with a Calculator

The graphing calculator is a tool that can be used to help you solve many different kinds of problems. This activity will walk you through several features of the TI-83 Plus. By the time you complete this activity, you will be able to graph equations, change the viewing window to better accommodate a graph, and look at a table of values that represents some of the solutions for an equation. The first portion of this activity will demonstrate how you can create the graph of an equation. The features described here can be found on most graphing calculators. See your calculator manual to learn how to get your particular calculator model to perform this activity.

Menus and Graphing

1. To graph the equation $y = 2x + 3$ on a graphing calculator, follow these steps.

   (a) Press the $\boxed{Y =}$ key.

   (b) Type $2x + 3$ at the Y_1 prompt. (This represents the first equation. You can type up to 10 separate equations.) Use the $\boxed{X, T, \theta, n}$ key for the variable.

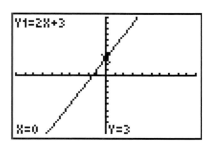

   (c) Press the $\boxed{GRAPH}$ key to see the graph.

   (d) Press the $\boxed{TRACE}$ key to display the equation. Once you have selected the $\boxed{TRACE}$ key, you can use the left and right arrows of the calculator to move the cursor along the line. Experiment with this movement. Look at the coordinates at the bottom of the display screen as you move along the line.

Frequently, we can learn more about an equation if we look at a different section of the graph than the one offered on the display screen. The portion of the graph displayed is called the **window.** The second portion of the activity explains how this window can be changed.

2. Press the ⌐WINDOW⌐ key. The **standard** graphing screen is shown.

    Xmin = left edge of screen
    Xmax = right edge of screen
      Xscl = scale given by each tick mark on x-axis
     Ymin = bottom edge of screen
    Ymax = top edge of screen
      Yscl = scale given by each tick mark on y-axis
      Xres = resolution (do not alter this)

```
WINDOW
 Xmin=-10
 Xmax=10
 Xscl=1
 Ymin=-10
 Ymax=10
 Yscl=1
 Xres=1
```

Note: To turn the scales off, enter a 0 for Xscl or Yscl. Do this when the intervals used are very large.

By changing the values for Xmin, Xmax, Ymin, and Ymax, you can adjust the viewing window. Change the viewing window so that Xmin = 0, Xmax = 40, Ymin = 0, and Ymax = 10. Again, press ⌐GRAPH⌐. Notice that the tick marks along the x-axis are now much closer together. Changing Xscl from 1 to 5 will improve the display. Try it.

Sometimes we can learn something important about a graph by zooming in or zooming out. The third portion of this activity discusses this feature of the TI-83 Plus.

3.

(a) Press the ⌐ZOOM⌐ key. There are 10 options. Use the ▼ key to scroll down.

```
ZOOM MEMORY
1:ZBox
2:Zoom In
3:Zoom Out
4:ZDecimal
5:ZSquare
6:ZStandard
7↓ZTrig
```
```
ZOOM MEMORY
4↑ZDecimal
5:ZSquare
6:ZStandard
7:ZTrig
8:ZInteger
9:ZoomStat
0:ZoomFit
```

(b) Selecting the first option, ZBox, allows the user to enlarge the graph within a specified rectangle.

(i) Graph the equation $y = x^2 + x - 1$ in the standard window. *Note:* To type in the exponent, use the ⌐x^2⌐ key or the ⌐∧⌐ key.

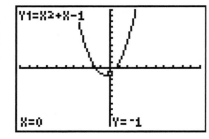

(ii) When ZBox is selected, a blinking "+" cursor will appear in the graph window. Use the arrow keys to move the cursor to where you would like a corner of the screen to be; then press the ENTER key.

(iii) Use the arrow keys to trace out the box containing the desired portion of the graph. Do not press the ENTER key until you have reached the diagonal corner and a full box is on your screen.

After using the down arrow   *After using the right arrow*

After pressing the ENTER *key a second time*

Now the desired portion of a graph can be seen more clearly.

The Zbox feature is especially useful when analyzing the roots (*x*-intercepts) of an equation.

(c) Another feature that allows us to focus is ZoomIn. Press the ZoomIn button in the Zoom menu. Place the cursor in the center of the portion of the graph you are interested in and press the ENTER key. The window will reset with the cursor at the center of a zoomed-in view.

(d) ZoomOut works like ZoomIn, except that it sets the view larger (that is, it zooms out) to enable you to see a larger portion of the graph.

(e) ZStandard sets the window to the standard window. This is a quick and convenient way to reset the viewing window.

(f) ZSquare recalculates the view so that one horizontal unit is the same length as one vertical unit. This is sometimes necessary to get an accurate view of a graph since the width of the calculator screen is greater than its height.

Home Screen

This is where all the basic computations take place. To get to the home screen from any other screen, press 2nd , Mode . This accesses the QUIT feature. To clear the home screen of calculations, press the CLEAR key (once or twice).

Tables

The final feature that we will look at here is Table. Enter the function $y = 2x + 3$ into the Y = menu. Then press 2nd , WINDOW to access the TBLSET menu. Set the table as shown here and press 2nd , GRAPH to access the TABLE feature. You will see the screen shown here.

The following exercises are presented to help you review concepts from earlier chapters. This is meant as a review and not as a comprehensive exam. The answers are presented in the back of the text. Section references accompany the answers. If you have difficulty with any of these exercises, be certain to at least read through the summary related to those sections.

Answers

Perform the indicated operations.

1. $9 + (-6)$

2. $-4 - (-9)$

3. $25 - (-12)$

4. $-32 + (-21)$

5. $(-23)(-3)$

6. $(12)(-10)$

7. $30 \div (-6)$

8. $(-24) \div (-8)$

Evaluate the expressions if $x = -3$, $y = 4$, *and* $z = -5$.

9. $3x^2y$

10. $-3z - 3y$

11. $-3(-2y + 3z)$

12. $\dfrac{3y - 2x}{5y + 6x}$

Solve the following equations and check your results.

13. $5x - 2 = 2x - 6$

14. $3(x - 2) = 2(3x + 1)$

15. $\dfrac{5}{6}x - 3 = 2 + \dfrac{1}{3}x$

16. $4(2 - x) + 9 = 7 + 6x$

17. Solve the equation $F = \dfrac{9}{5}C + 32$ for C.

Solve the following inequalities.

18. $4x - 9 < 7$

19. $-5x + 15 \geq 2x - 6$

1. _____

2. _____

3. _____

4. _____

5. _____

6. _____

7. _____

8. _____

9. _____

10. _____

11. _____

12. _____

13. _____

14. _____

15. _____

16. _____

17. _____

18. _____

19. _____

Answers

Beginning Algebra   The Streeter/Hutchison Series in Mathematics   © The McGraw-Hill Companies. All Rights Reserved.

20. _____

21. _____

22. _____

23. _____

24. _____

25. _____

26. _____

27. _____

28. _____

29. _____

30. _____

31. _____

32. _____

33. _____

34. _____

35. _____

36. _____

37. _____

38. _____

Use the properties of exponents to simplify the following expressions and write the results with positive exponents.

20. $(x^2y^3)^{-2}$

21. $\dfrac{x^3y^2}{x^4y^{-3}}$

22. $(x^6y^{-3})^0$

Perform the indicated operation for each of the following polynomials.

23. Add $2x^2 + 4x - 6$ and $3x^2 - 4x - 4$.

24. Subtract $3a^2 - 2a + 5$ from the sum of $a^2 + 3a - 2$ and $-5a^2 + 2a + 9$.

Evaluate each polynomial for the indicated variable value.

25. $2x^2 - 5x + 7$ for $x = 4$

26. $x^3 + 3x^2 - 7x + 8$ for $x = -2$

Multiply the following polynomials.

27. $(3x - 5y)(2x + 4y)$

28. $3x(x - 3)(2x + 5)$

29. $(2a + 7b)(2a - 7b)$

Completely factor each polynomial.

30. $12p^2n^2 + 20pn^2 - 16pn^3$

31. $y^3 - 3y^2 - 5y + 15$

32. $9a^2b - 49b$

33. $6x^2 - 2x - 4$

34. $6a^2 + 7ab - 3b^2$

Solve each of the following by factoring.

35. $x^2 - 8x - 33 = 0$

36. $35x^2 - 38x = -8$

Simplify each of the following rational expressions.

37. $\dfrac{-35a^4b^5}{21ab^7}$

38. $\dfrac{2w^2 - w - 6}{2w^2 + 9w + 9}$

Add or subtract as indicated. Simplify your answer.

39. $\dfrac{2}{a-5} - \dfrac{1}{a}$

40. $\dfrac{2w}{w^2-9w+20} + \dfrac{8}{w-4}$

Multiply or divide as indicated.

41. $\dfrac{4xy^3}{5xy^2} \cdot \dfrac{15x^3y}{16y^2}$

42. $\dfrac{m^2-3m}{m^2-9} \div \dfrac{4m^2}{m^2-m-12}$

Solve each of the following equations.

43. $\dfrac{w}{w-2} + 1 = \dfrac{w+4}{w-2}$

44. $\dfrac{7}{x} - \dfrac{1}{x-3} = \dfrac{9}{x^2-3x}$

Graph each of the following equations.

45. $3x + 4y = 12$

46. $y = -7$

47. $x = 2y$

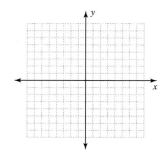

48. Find the slope of the line passing through the points $(-3, 5)$ and $(1, 13)$.

Answers

39. _____

40. _____

41. _____

42. _____

43. _____

44. _____

45. _____

46. _____

47. _____

48. _____

Answers

49. _____

50. _____

51. _____

52. _____

53. _____

49. Graph the equation of variation if $k = 4$.

50. Find the constant of variation if y varies directly with x, and if $y = 450$ when $x = 15$.

Solve the following problems.

51. The length of a rectangle is 3 in. less than twice its width. If the perimeter is 24 in., find the dimensions of the rectangle.

52. The sum of three consecutive odd integers is 129. Find the three integers.

53. The carpet outlet is selling rug remnants at 25% off. If the sale price is $150, what was the original price?

chapter 7 > Make the Connection

INTRODUCTION

In the pharmaceutical-making process, great caution must be exercised to ensure that the medicines and drugs are pure and contain precisely what is indicated on the label. The quality control division of the pharmaceutical company is responsible for guaranteeing such purity.

A lab technician working in quality control must run a series of tests on samples of every ingredient, even simple ingredients such as salt (NaCl). One such test is a measure of how much weight is lost as a sample is dried. The technician must set up a 3-hour procedure that involves cleaning and drying bottles and stoppers and then weighing them while they are empty and again when they contain samples of the substance to be heated and dried.

The pharmaceutical company may have a standard of acceptability for this substance. For instance, the substance may not be acceptable if the loss of weight from drying is greater than 10%. The technician would then use an inequality to calculate acceptable weight loss. In the following inequality, the right side represents the percent of weight loss from drying.

$$10 \geq \frac{W_g - W_f}{W_g - T} \cdot 100$$

We will examine inequalities further in this chapter.

Graphing and Inequalities

CHAPTER 7 OUTLINE

Name _____

Section _____ Date _____

This pretest provides a preview of the types of exercises you will encounter in each section of this chapter. The answers for these exercises can be found in the back of the text. If you are working on your own, or ahead of the class, this pretest can help you find the sections in which you should focus more of your time.

Answers

1. _____

2. _____

3. _____

4. _____

5. _____

6. _____

7. _____

8. _____

9. _____

10. _____

11. _____

12. _____

7.1   **1.** Find the slope and y-intercept of the line given by $y = -2x + 3$.

2. Find the slope and y-intercept from the equation $5x - 4y = 20$.

Write the equation of the line with the given slope and y-intercept. Then graph each line.

3. Slope 5 and y-intercept $(0, -2)$ **4.** Slope -4 and y-intercept $(0, 6)$

7.2   Determine whether the following pairs of lines are parallel, perpendicular, or neither.

5. L_1 through $(3, 7)$ and $(1, 11)$ **6.** L_1 with equation $y = 3x + 9$
     L_2 through $(-5, -1)$ and $(6, -23)$ L_2 with equation $3y + x = 9$

7.3   **7.** Find the equation of the line through $(-5, 8)$ and parallel to the line $4x + y = 8$.

Write the equation of the line L satisfying each of the following sets of geometric conditions.

8. L passes through $(5, -6)$ and is perpendicular to $3x - 5y = 15$.

9. L has y-intercept $(0, -3)$ and is parallel to $-3x + 5y = -15$.

7.4   Graph each of the following inequalities.

10. $2x + y < 5$ **11.** $-3x + 5y \geq 15$

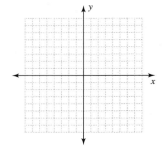

7.5   Evaluate $f(x)$ as indicated.

12. $f(x) = -2x^3 - 6x - 4x^2 + 9$; find $f(-1)$ and $f(1)$.

7.1

The Slope-Intercept Form

< 7.1 Objectives >

1 > Find the slope and y-intercept from the equation of a line

2 > Given the slope and y-intercept, write the equation of a line

3 > Use the slope and y-intercept to graph a line

4 > Solve an application involving slope-intercept form

In Chapter 6, we used two points to find the slope of a line. In this chapter we use the slope and y-intercept to sketch the graph of an equation.

First, we want to consider finding the equation of a line when its slope and y-intercept are known.

Suppose that the y-intercept of a line is $(0, b)$. Then the point at which the line crosses the y-axis has coordinates $(0, b)$, as shown in the sketch at left.

Now, using any other point (x, y) on the line and using our definition of slope, we can write

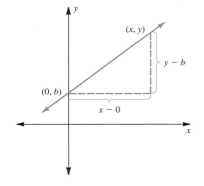

Change in y

$$m = \frac{y - b}{x - 0}$$

Change in x

or

$$m = \frac{y - b}{x}$$

Multiplying both sides of this equation by x, we have

$$mx = y - b$$

Finally, adding b to both sides of this equation gives

$$mx + b = y$$

or

$$y = mx + b$$

We can summarize this discussion as follows:

NOTE

In this form, the equation is *solved for y*. The coefficient of x gives the slope of the line, and the constant term gives the y-intercept.

Definition

The Slope-Intercept Form for a Line	An equation of the line with slope m and y-intercept $(0, b)$ is $y = mx + b$

Example 1   **Finding the Slope and y-Intercept**

< Objective 1 >

Find the slope and y-intercept of the equation

$$y = -\frac{2}{3}x - 5$$

$$\underset{m}{\uparrow} \quad \underset{b}{\uparrow}$$

The slope of the line is $-\frac{2}{3}$; the y-intercept is $(0, -5)$.

>
>
> **Check Yourself 1**
>
> **Find the slope and y-intercept for the graph of each of the following equations.**
>
> **(a)** $y = -3x - 7$ **(b)** $y = \frac{3}{4}x + 5$

As Example 2 illustrates, we may have to solve for y as the first step in determining the slope and the y-intercept for the graph of an equation.

Example 2   **Finding the Slope and y-Intercept**

NOTE

If we write the equation as

$$y = \frac{-3x + 6}{2}$$

it is more difficult to identify the slope and the intercept.

Find the slope and y-intercept of the equation

$$3x + 2y = 6$$

First, we must solve the equation for y.

$$3x + 2y = 6$$
$$2y = -3x + 6 \qquad \text{Subtract } 3x \text{ from both sides.}$$
$$y = -\frac{3}{2}x + 3 \qquad \text{Divide each term by 2.}$$

The equation is now in slope-intercept form. The slope is $-\frac{3}{2}$, and the y-intercept is $(0, 3)$.

> **Check Yourself 2**
>
> **Find the slope and y-intercept for the graph of the equation**
>
> $$2x - 5y = 10$$

As we mentioned earlier, knowing certain properties of a line (namely, its slope and y-intercept) will also allow us to write the equation of the line by using the slope-intercept form. Example 3 illustrates this approach.

| Example 3 | Writing the Equation of a Line |

< Objective 2 >

Write the equation of a line with slope $-\dfrac{3}{4}$ and y-intercept $(0, -3)$.

We know that $m = -\dfrac{3}{4}$ and $b = -3$. In this case,

$$y = -\dfrac{\overset{m}{\downarrow}}{4}x + \overset{b}{(-3)}$$

or

$$y = -\dfrac{3}{4}x - 3$$

which is the desired equation.

Check Yourself 3

Write the equation of a line with the following:

(a) slope -2 and y-intercept $(0, 7)$

(b) slope $\dfrac{2}{3}$ and y-intercept $(0, -3)$

We can also use the slope and y-intercept of a line in drawing its graph. Consider Example 4.

| ▶ Example 4 | Graphing a Line Using the Slope and y-intercept |

< Objective 3 >

Graph the line with slope $\dfrac{2}{3}$ and y-intercept $(0, 2)$.

Because the y-intercept is $(0, 2)$, we begin by plotting the point $(0, 2)$. Because the horizontal change (or run) is 3, we move 3 units to the right *from that y-intercept.* Then because the vertical change (or rise) is 2, we move 2 units up to locate another point on the desired graph. Note that we will have located that second point at $(3, 4)$. The final step is simply to draw a line through that point and the y-intercept.

NOTE

$m = \dfrac{2}{3} = \dfrac{\text{rise}}{\text{run}}$

The line rises from left to right because the slope is positive.

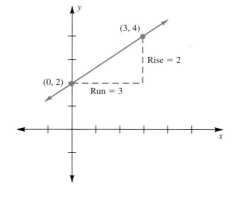

The equation of this line is $y = \dfrac{2}{3}x + 2$.

Check Yourself 4

Graph the equation of a line with slope $\frac{3}{5}$ and y-intercept $(0, -2)$.

When creating a graph such as the one drawn in Example 4, it is a good idea to plot several points. This is quick and easy using the idea of slope. Recall that we plotted $(0, 2)$ and then used the slope of $\frac{2}{3}$ to plot $(3, 4)$: We moved 3 to the right and up 2.

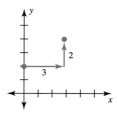

Now we simply continue this process from that point: run 3 and rise 2, repeatedly:

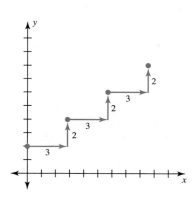

We may also move to the left and down from $(0, 2)$: run -3 and rise -2.

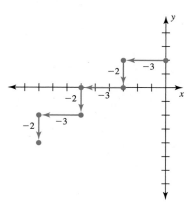

Example 5   Graphing a Line Using the Slope and *y*-intercept

Sketch the graph of $y = -2x + 5$.

We begin by noting that the *y*-intercept is (0, 5) and the slope is -2. To view the slope as $\dfrac{\text{rise}}{\text{run}}$, we may write -2 either as $\dfrac{-2}{1}$ or $\dfrac{2}{-1}$. Now we plot (0, 5) and then use the slope to plot more points. Using $m = \dfrac{-2}{1}$, we run 1 and then rise -2; that is, as we run 1, we drop 2. We repeat this action several times to plot the following:

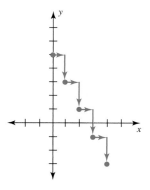

If we use $m = \dfrac{2}{-1}$, we run -1 and then rise 2; that is, as we run 1 *to the left,* we go up 2.

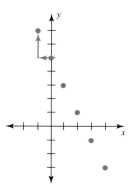

The completed graph is shown here:

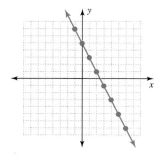

With practice, you will be able to quickly plot many points and sketch an accurate line.

Check Yourself 5

Sketch the graph of $y = -3x + 1$.

Graphing by Using the Slope-Intercept Form

Step 1   Write the original equation of the line in slope-intercept form.
Step 2   Determine the slope m and the y-intercept $(0, b)$.
Step 3   Plot the y-intercept at $(0, b)$.
Step 4   Use m (the change in y over the change in x) to determine other points on the desired line.
Step 5   Draw a line through the points determined in steps 1–4 to complete the graph.

You have now seen two methods for graphing lines: the slope-intercept method (this section) and the intercept method (Section 6.3). When you graph a linear equation, you first should decide which is the appropriate method.

Example 6   **Selecting an Appropriate Graphing Method**

Decide which of the two methods for graphing lines—the intercept method or the slope-intercept method—is more appropriate for graphing equations **(a)**, **(b)**, and **(c)**.

(a) $2x - 5y = 10$

Because both intercepts are easy to find, you should choose the intercept method to graph this equation.

(b) $2x + y = 6$

This equation can be graphed quickly by either method. As it is written, you might choose the intercept method. It can, however, be rewritten as $y = -2x + 6$. In that case the slope-intercept method is more appropriate.

(c) $y = \dfrac{1}{4}x - 4$

Because the equation is in slope-intercept form, that is the more appropriate method to choose.

Check Yourself 6

Which would be more appropriate for graphing each equation, the intercept method or the slope-intercept method?

(a) $x + y = -2$ **(b)** $3x - 2y = 12$ **(c)** $y = -\dfrac{1}{2}x - 6$

It is important to understand what the slope of a linear equation can tell us about the relationship between x and y. For example, if $y = 2x - 5$, we note that the slope is 2. Writing this as $\frac{2}{1}$, we can see the following: If x increases by 1 unit, then y increases by 2 units. Consider the next example.

| Example 7 | Interpreting the Slope |

Interpret the slope for each of the following equations.

(a) $y = -3x + 4$

Write the slope as $\frac{-3}{1}$. If x increases by 1 unit, then y decreases by 3 units.

(b) $y = 0.8x + 6.2$

Writing the slope, 0.8, as $\frac{0.8}{1}$, we can say that as x increases by 1 unit, y increases by 0.8 (eight-tenths of a unit).

(c) $C = 25n + 120$

Write the slope as $\frac{25}{1}$. As n increases by 1, C increases by 25.

 Check Yourself 7

Interpret the slope for each equation.

(a) $y = \frac{2}{3}x - 7$ **(b)** $y = -5x - 8$

Beginning Algebra

The Streeter/Hutchison Series in Mathematics

© The McGraw-Hill Companies. All Rights Reserved.

NOTES

The y-intercept b in the equation $y = mx + b$ is the constant term.

Variable costs are sometimes called *marginal costs*.

The slope-intercept form lends itself easily to applications involving linear equations. Consider the cost of manufacturing a product; there are really two costs involved.

Fixed costs are those costs that are independent of the number of units produced. That is, no matter how many (or how few) items are produced, fixed costs remain the same. Fixed costs include the cost of leasing property, insurance costs, and labor costs for salaried personnel. If we write an equation relating total costs to the number of units produced, then the fixed costs are represented by the constant term.

Variable costs are those costs that change depending on the number of units produced. These include the costs of materials, labor costs for hourly employees, many utility costs, and service costs. In the slope-intercept equation, variable costs correspond to the slope m because the dollar amount increases each time another unit is produced.

| Example 8 | An Application of Slope-Intercept Equations |

< Objective 4 >

NOTE

As x (the number of stereos produced) increases by 1, the cost y increases by 26. This says that the slope is 26.

S-Bar Electronics determines that the cost to produce each stereo is $26. In addition, the cost to keep its factory open each month is $3,500.

(a) Write an equation relating the total monthly cost of producing stereos to the number of stereos produced.

The y-intercept represents the fixed costs, so we have $b = 3,500$. The slope m is given by the variable costs, which are $26 per stereo. Putting this together gives us the equation

$y = 26x + 3,500$

(b) Use the cost equation to determine the total cost of producing 320 stereos in a month.

In the cost equation, we substitute 320 for x and calculate y:

$$y = 26(320) + 3,500$$
$$= 11,820$$

So it costs S-Bar Electronics $11,820 to produce 320 stereos in a month.

Check Yourself 8

A manager at the chic new restaurant Sweet Eats determines that the average dinner costs the restaurant $18 to produce. In addition, it costs the restaurant $620 to stay open each evening.

(a) Write an equation relating the total nightly cost of operation to the number of dinners served.
(b) How much does it cost the restaurant to serve 50 dinners in an evening?

The following is an application from the field of medicine.

Example 9	Vocational Application

The ideal body weight (*IBW*), in pounds, for a man is a linear function of his height (*H*), in inches, given by the formula $IBW = 106 + 6(H - 60)$.

(a) Determine the slope of this equation.

If we simplify the right-hand side of the equation, we get

$$106 + 6(H - 60) = 106 + 6H - 360$$
$$= 6H - 254$$

The slope is 6.

(b) Interpret this slope verbally.

Because the units of *IBW* are pounds and the units of *H* are inches, the units attached to the slope are $\dfrac{\text{pounds}}{\text{inch}}$. So we can say that for every additional inch in a man's height, his ideal body weight increases by 6 pounds.

Check Yourself 9

The ideal body weight (*IBW*), in pounds, for a woman is a linear function of her height (*H*), in inches, given by the formula *IBW* = 105 + 5(*H* − 60).

(a) Determine the slope of this function.
(b) Interpret this slope verbally.

Check Yourself ANSWERS

1. (a) Slope is -3, y-intercept is $(0, -7)$; **(b)** Slope is $\dfrac{3}{4}$, y-intercept is $(0, 5)$

2. The slope is $\dfrac{2}{5}$; the y-intercept is $(0, -2)$

3. (a) $y = -2x + 7$; **(b)** $y = \dfrac{2}{3}x - 3$

4.

5.

6. (a) Either; **(b)** intercept; **(c)** slope-intercept

7. (a) When x increases by 3 units, y increases by 2 units. Alternatively, when x increases by 1 unit, y increases by $\dfrac{2}{3}$ of a unit; **(b)** When x increases by 1 unit, y decreases by 5 units.

8. (a) $y = 18x + 620$; **(b)** $\$1,520$

9. (a) The slope is 5; **(b)** For each additional inch in a woman's height, her ideal body weight increases by 5 pounds.

Reading Your Text

The following fill-in-the-blank exercises are designed to ensure that you understand some of the key vocabulary used in this section.

SECTION 7.1

(a) In the slope-intercept form, the coefficient of x gives the _____ of the line.

(b) _____ costs are those costs that are independent of the number of units produced.

(c) In the slope-intercept equation, _____ costs correspond to the slope m.

(d) Variable costs are sometimes called _____ costs.

Name _____

Section _____ Date _____

Answers

1. _____
2. _____
3. _____
4. _____
5. _____
6. _____
7. _____
8. _____
9. _____
10. _____
11. _____
12. _____
13. _____
14. _____

Basic Skills | Advanced Skills | Vocational-Technical Applications | Calculator/Computer | Above and Beyond | Getting Ready

< Objective 1 >

Find the slope and y-intercept of the line represented by each of the following equations.

1. $y = 3x + 5$

2. $y = -7x + 3$

3. $y = -2x - 5$

4. $y = 5x - 2$

5. $y = \frac{3}{4}x + 1$

6. $y = -4x$

7. $y = \frac{2}{3}x$

8. $y = -\frac{3}{5}x - 2$

9. $4x + 3y = 12$

10. $2x + 5y = 10$

11. $y = 9$

12. $2x - 3y = 6$

13. $3x - 2y = 8$ > Videos

14. $x = 5$

< Objectives 2–3 >

Write the equation of the line with given slope and y-intercept. Then graph each line using the slope and y-intercept.

Answers

15. Slope: 3; *y*-intercept: (0, 5)

16. Slope: −2; *y*-intercept: (0, 4)

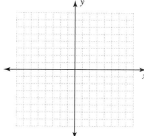

17. Slope: −3; *y*-intercept: (0, 4)

 > Videos

18. Slope: 5; *y*-intercept: (0, −2)

19. Slope: $\frac{1}{2}$; *y*-intercept: (0, −2)

20. Slope: $-\frac{3}{4}$; *y*-intercept: (0, 8)

21. Slope: $-\frac{2}{3}$; *y*-intercept: (0, 0)

22. Slope: $\frac{2}{3}$; *y*-intercept: (0, −2)

 > Videos

15. _____

16. _____

17. _____

18. _____

19. _____

20. _____

21. _____

22. _____

Answers

23. _____

24. _____

25. _____

26. _____

27. _____

28. _____

29. _____

30. _____

23. Slope: $\dfrac{3}{4}$; y-intercept: $(0, 3)$

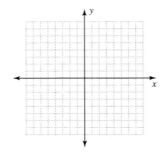

24. Slope: -3; y-intercept: $(0, 0)$

In exercises 25–32, match the graph with one of the following equations.

  (a) $y = 2x$ **(b)** $y = x + 1$ **(c)** $y = -x + 3$ **(d)** $y = 2x + 1$

  (e) $y = -3x - 2$ **(f)** $y = \dfrac{2}{3}x + 1$ **(g)** $y = -\dfrac{3}{4}x + 1$

  (h) $y = -4x$

25.

26.

27.

28.

29.

30.

31.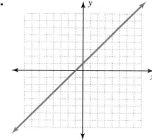

32.

< Objective 4 >

33. **SOCIAL SCIENCE** The equation $y = 0.10x + 200$ describes the award money in a recycling contest. What are the slope and the y-intercept for this equation?

34. **BUSINESS AND FINANCE** The equation $y = 15x - 100$ describes the amount of money a high school class might earn from a paper drive. What are the slope and y-intercept for this equation?

35. **SCIENCE AND MEDICINE** On a certain February day in Philadelphia, the temperature at 6:00 A.M. was 10°F. By 2:00 P.M. the temperature was up to 26°F. What was the hourly rate of temperature change? > Videos

36. **CONSTRUCTION** A roof rises 8.75 ft over a horizontal distance of 15.09 ft. Find the slope of the roof to the nearest hundredth.

37. **SCIENCE AND MEDICINE** An airplane covered 15 mi of its route while decreasing its altitude by 24,000 ft. Find the slope of the line of descent that was followed (1 mi = 5,280 ft). Round to the nearest hundredth.

38. **TECHNOLOGY** Driving down a mountain, Tom finds that he has descended 1,800 ft in elevation by the time he is 3.25 mi horizontally away from the top of the mountain. Find the slope of his descent to the nearest hundredth.

Determine whether each of the following statements is **true** *or* **false.**

39. The slope of a line through the origin can be zero.

40. A line with undefined slope is the same as a line with a slope of zero.

Complete each of the following statements with **never, sometimes,** *or* **always.**

41. Lines _____ have exactly one x-intercept.

42. The y-intercept of a line through the origin is _____ zero.

Answers

31. _____

32. _____

33. _____

34. _____

35. _____

36. _____

37. _____

38. _____

39. _____

40. _____

41. _____

42. _____

Answers

43. _____

44. _____

45. _____

46. _____

47. _____

48. _____

49. _____

50. _____

| Basic Skills | **Advanced Skills** | Vocational-Technical Applications | Calculator/Computer | Above and Beyond | Getting Ready |

In which quadrant(s) are there no solutions for each line?

43. $y = -x + 1$ > Videos

44. $y = 3x + 2$

45. $y = -2x - 5$

46. $y = -5x - 7$

47. $y = 3$

48. $x = -2$

49. On the same graph, sketch the following lines:

$$y = 2x - 1 \quad \text{and} \quad y = 2x + 3$$

What do you observe about these graphs? Will the lines intersect?

50. Repeat exercise 49 using

$$y = -2x + 4 \quad \text{and} \quad y = -2x + 1$$

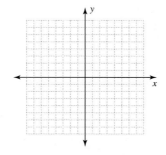

51. On the same graph, sketch the following lines:

$$y = \frac{2}{3}x \quad \text{and} \quad y = -\frac{3}{2}x$$

What do you observe concerning these graphs? Find the product of the slopes of these two lines.

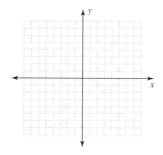

52. Repeat exercise 51 using

$$y = \frac{4}{3}x \quad \text{and} \quad y = -\frac{3}{4}x$$

53. Based on exercises 51 and 52, write the equation of a line that is perpendicular to $y = \frac{3}{5}x$.

54. **ALLIED HEALTH** The arterial oxygen tension (P_aO_2), in millimeters of mercury (mm Hg), of a patient (lying down) can be estimated based on the patient's age (A), in years. Use the equation $P_aO_2 = 103.5 - 0.42A$.

   (a) Determine the slope of the line for this equation.

   (b) Interpret this slope verbally.

55. **ALLIED HEALTH** The arterial oxygen tension (P_aO_2), in millimeters of mercury (mm Hg), of a patient (seated) can be estimated based on the patient's age (A), in years. Use the equation $P_aO_2 = 104.2 - 0.27A$.

   (a) Determine the slope of the line for this equation.

   (b) Interpret this slope verbally.

Answers

51. _____

52. _____

53. _____

54. _____

55. _____

Answers

56. _____

57. _____

58. _____

59. _____

60. _____

61. _____

56. **AGRICULTURAL TECHNOLOGY** The yield Y (in bushels) per acre for a cornfield is estimated from the amount of rainfall R (in inches) using the formula

$$Y = \frac{43,560}{8,000}R$$

(a) What is the slope of the line for this equation?

(b) Interpret the slope.

57. **AGRICULTURAL TECHNOLOGY** During a period of the summer, the growth of corn plants follows a linear pattern that is approximated by this equation:

$$h = 1.77d + 24.92$$

where h is the height (in inches) of the corn, and d is the number of days that have passed.

(a) What is the slope of the line for this equation?

(b) Interpret the slope.

Basic Skills	Advanced Skills	Vocational-Technical Applications	Calculator/Computer	**Above and Beyond**	Getting Ready

58. Complete the following statement: "The difference between undefined slope and zero slope is . . ."

59. Complete the following: "The slope of a line tells you . . ."

60. **STATISTICS** In a study on nutrition conducted in 1984, 18 normal adults aged 23 to 61 years old were measured for body fat, which is given as percentage of weight. The mean (average) body fat percentage for women 40 years old was 28.6% and for women 53 years old was 38.4%. Work with a partner to decide how to show this information by plotting points on a graph. Try to find a linear equation that will tell you percentage of body fat based on a woman's age. What does your equation give for 20 years of age? For 60? Do you think a linear model works well for predicting body fat percentage in women as they age?

chapter 7 > Make the Connection

61. **BUSINESS AND FINANCE** On two occasions last month, Sam Johnson rented a car on a business trip. Both times it was the same model from the same company, and both times it was in San Francisco. Sam now has to fill out an expense account form and needs to know how much he was charged per mile and the base rate. On both occasions he dropped the car at the airport booth and just got the total charge, not the details. All Sam knows is that he was charged $210 for 625 mi on the first occasion and $133.50 for 370 mi on the second trip. Sam has called accounting to ask for help. Plot these two points on a graph and draw the line that goes through them. What question does the slope of the line answer for Sam? How does the y-intercept help? Write a memo to Sam explaining the answers to his question and how a knowledge of algebra and graphing has helped you find the answers.

Basic Skills | Advanced Skills | Vocational-Technical Applications | Calculator/Computer | Above and Beyond | **Getting Ready**
▲

Find the slope of the line connecting the given points.

(a) $(-4, 6)$ and $(3, 20)$ **(b)** $(2, 8)$ and $(-6, -8)$

(c) $(5, -7)$ and $(-5, 3)$ **(d)** $(2, 8)$ and $(2, 5)$

(e) $(6, 9)$ and $(3, 9)$ **(f)** $(4, 6)$ and $(-4, -2)$

Answers

1. Slope 3, y-intercept $(0, 5)$ **3.** Slope -2, y-intercept $(0, -5)$

5. Slope $\dfrac{3}{4}$, y-intercept $(0, 1)$ **7.** Slope $\dfrac{2}{3}$, y-intercept $(0, 0)$

9. Slope $-\dfrac{4}{3}$, y-intercept $(0, 4)$ **11.** Slope 0, y-intercept $(0, 9)$

13. Slope $\dfrac{3}{2}$, y-intercept $(0, -4)$

15. $y = 3x + 5$ **17.** $y = -3x + 4$

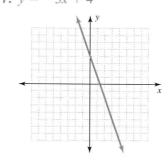

19. $y = \dfrac{1}{2}x - 2$ **21.** $y = -\dfrac{2}{3}x$

 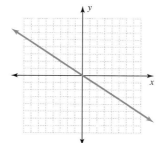

23. $y = \dfrac{3}{4}x + 3$

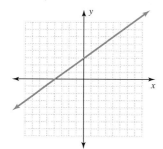

Answers

(a) _____

(b) _____

(c) _____

(d) _____

(e) _____

(f) _____

25. g   **27.** e   **29.** h   **31.** c   **33.** Slope: 0.10; y-intercept: (0, 200)

35. 2°F/h   **37.** −0.30   **39.** True   **41.** sometimes

43. III   **45.** 1   **47.** III and IV

49.

Parallel lines; no

51.

Perpendicular lines; −1

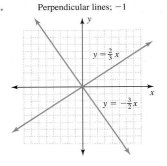

53. $y = -\dfrac{5}{3}x$ **55. (a)** $-0.27; \dfrac{\text{mm Hg}}{\text{yr}}$ **(b)** For every additional year in a patient's age, the arterial oxygen tension decreases by 0.27 mm Hg

57. (a) 1.77; **(b)** For each additional day, the height increases by 1.77 inches

59. Above and Beyond   **61.** Above and Beyond

(a) 2   **(b)** 2   **(c)** −1   **(d)** Undefined   **(e)** 0   **(f)** 1

7.2

Parallel and Perpendicular Lines

< 7.2 Objectives >

1 > Determine whether two lines are parallel

2 > Determine whether two lines are perpendicular

3 > Find the slope of a line parallel or perpendicular to a given line

For most inexperienced drivers, the most difficult driving maneuver to master is parallel parking. What is parallel parking? It is the act of backing into a curbside space so that the car's tires are parallel to the curb.

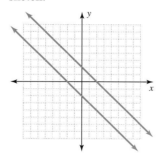

How can you tell that you've done a good job of parallel parking? Most people check to see that both the front tires and the back tires are the same distance (8 in. or so) from the curb. This is checking to be certain that the car is parallel to the curb.

How can we tell that two equations represent parallel lines? Look at the following sketch.

If two lines are parallel, they have the same slope. If their equations are in slope-intercept form, you simply compare the slopes.

| Example 1 | Determining That Two Lines Are Parallel |

< Objective 1 >

Which two equations represent parallel lines?

(a) $y = 2x + 3$
                                               (b) $y = -\dfrac{1}{2}x - 5$

(c) $y = -2x + \dfrac{1}{2}$ **(d)** $y = -2x - 9$

Because **(c)** and **(d)** both have a slope of -2, the lines are parallel.

Check Yourself 1

Which two equations represent parallel lines?

(a) $y = -5x + 5$ **(b)** $y = \dfrac{1}{5}x - 5$

(c) $y = -5x + \dfrac{1}{2}$ **(d)** $y = 5x - 9$

More formally, we can state the following about parallel lines.

Property

Slopes of Parallel Lines

For nonvertical lines L_1 and L_2, if line L_1 has slope m_1 and line L_2 has slope m_2, then

L_1 is parallel to L_2 if, and only if, $m_1 = m_2$

Note: All vertical lines are parallel to each other.

As we discovered in Chapter 6, we can find the slope of a line from any two points on the line.

     Example 2   Determining Whether Two Lines Are Parallel

NOTE

Parallel lines do not intersect unless, of course, L_1 and L_2 are actually the *same line*. In Example 2 a quick sketch shows that the lines are distinct.

Are lines L_1 through $(2, 3)$ and $(4, 6)$ and L_2 through $(-4, 2)$ and $(0, 8)$ parallel, or do they intersect?

$$m_1 = \frac{6 - 3}{4 - 2} = \frac{3}{2}$$

$$m_2 = \frac{8 - 2}{0 - (-4)} = \frac{6}{4} = \frac{3}{2}$$

Because the slopes of the lines are equal, the lines are parallel. They do *not* intersect.

Check Yourself 2

Are lines L_1 through $(-2, -1)$ and $(1, 4)$ and L_2 through $(-3, 4)$ and $(0, 8)$ parallel, or do they intersect?

The Streeter/Hutchison Series in Mathematics   Beginning Algebra

Many important characteristics of lines are evident from a city map.

Note that 4th Street and 5th Street are parallel. Just as these streets never meet, it is true that two distinct parallel lines will never meet.

Recall that the point at which two lines meet is called their **intersection.** This is also true with two streets. We call the common area of the two streets the intersection.

In this case, the two streets meet at right angles. When two lines meet at right angles, we say that they are **perpendicular.**

Property

Slopes of Perpendicular Lines

For nonvertical lines L_1 and L_2, if line L_1 has slope m_1 and line L_2 has slope m_2, then

L_1 is perpendicular to L_2 if, and only if, $m_1 = -\dfrac{1}{m_2}$

or equivalently

$m_1 \cdot m_2 = -1$

Note: Horizontal lines are perpendicular to vertical lines.

⊙   **Example 3**   Determining That Two Lines Are Perpendicular

< Objective 2 >

Which two equations represent perpendicular lines?

(a) $y = 2x + 3$

(b) $y = -\dfrac{1}{2}x - 5$

(c) $y = -2x + \dfrac{1}{2}$

(d) $y = -2x - 9$

Because the product of the slopes for **(a)** and **(b)** is

$$2\left(-\dfrac{1}{2}\right) = -1$$

these two lines are perpendicular. Note that none of the other pairs of slopes have a product of -1.

Check Yourself 3

Which two equations represent perpendicular lines?

(a) $y = -5x + 5$ **(b)** $y = -\dfrac{1}{5}x - 5$

(c) $y = -5x + \dfrac{1}{2}$ **(d)** $y = 5x - 9$

Example 4	Determining Whether Two Lines Are Perpendicular

NOTE

$$\left(\dfrac{4}{3}\right)\left(-\dfrac{3}{4}\right) = -1$$

Are lines L_1 through points $(-2, 3)$ and $(1, 7)$ and L_2 through points $(2, 4)$ and $(6, 1)$ perpendicular?

$$m_1 = \dfrac{7 - 3}{1 - (-2)} = \dfrac{4}{3}$$

$$m_2 = \dfrac{1 - 4}{6 - 2} = -\dfrac{3}{4}$$

Because the slopes are negative reciprocals, the lines are perpendicular.

Check Yourself 4

Are lines L_1 through points (1, 3) and (4, 1) and L_2 through points (-2, 4) and (2, 10) perpendicular?

If we already have a line, we can use its slope to determine the slope of other lines that are parallel or perpendicular to the given line.

| Example 5 | Finding Parallel and Perpendicular Slopes |

< Objective 3 >

RECALL

The slope-intercept equation of a line is given by $y = mx + b$, in which m represents the slope and b gives the y-intercept.

(a) Find the slope of all lines parallel to the line given by $5x + y = 1$.

We begin by writing the equation for the given line in slope-intercept form. We do this by isolating y.

$$5x + y = 1$$
$$y = -5x + 1$$

In this form, we see that the slope of this line is -5. Therefore, all lines parallel to the given line have a slope of -5.

(b) Find the slope of all lines perpendicular to the line given by $5x + y = 1$.

From part **(a)**, we know the slope of the given line is -5. Using our property of perpendicular lines, we have

$$m_1 = -5$$
$$m_2 = -\frac{1}{m_1} = -\frac{1}{(-5)} = \frac{1}{5}$$

All lines perpendicular to the given line have a slope of $\frac{1}{5}$.

    Check Yourself 5

(a) Find the slope of all lines parallel to the line given by $x - 5y = 3$.
(b) Find the slope of all lines perpendicular to the line given by $x - 5y = 3$.

We can also use the slope-intercept form to determine whether the graphs of given equations are parallel, perpendicular, or neither.

| ▶ Example 6 | Verifying That Two Lines Are Parallel |

Show that the graphs of $3x + 2y = 4$ and $6x + 4y = 12$ are parallel lines.
First, we solve each equation for y:

$$3x + 2y = 4$$
$$2y = -3x + 4$$
$$y = -\frac{3}{2}x + 2$$
$$6x + 4y = 12$$
$$4y = -6x + 12$$
$$y = -\frac{3}{2}x + 3$$

NOTE

The slopes are the same, but the y-intercepts are different. Therefore, the lines are distinct.

Because the two lines have the same slope, here $-\frac{3}{2}$, the lines are parallel.

Check Yourself 6

Show that the graphs of the equations

$$-3x + 2y = 4 \quad \text{and} \quad 2x + 3y = 9$$

are perpendicular lines.

Many professions require people to sketch plans of one sort or another. In particular, architects are often required to sketch their designs on a rectangular coordinate system. This makes determining the relationship of adjacent objects such as walls and ceilings a fairly straightforward process, as seen in Example 7.

Example 7   **Vocational Application**

Design plans for a project need to be checked by the architect Nicolas. On the sketch, one line passes through the points (3, 6) and (7, 3). A second line also passes through (7, 3), as well as through the point (13, 11). Should Nicolas approve these plans if the two lines are supposed to be perpendicular?

To determine whether the two lines are perpendicular, we compute the slope of each. The slope of the first line is given by

$$m_1 = \frac{\text{change in } y}{\text{change in } x} = \frac{y_2 - y_1}{x_2 - x_1}$$

$$= \frac{3 - 6}{7 - 3} = \frac{-3}{4}$$

$$= -\frac{3}{4}$$

The slope of the second line is found similarly.

$$m_2 = \frac{11 - 3}{13 - 7} = \frac{8}{6}$$

$$= \frac{4}{3}$$

The product of the two slopes is

$$m_1 \cdot m_2 = \left(-\frac{3}{4}\right)\left(\frac{4}{3}\right) = -1$$

so the lines are perpendicular. Nicolas should approve the plans.

Check Yourself 7

On another portion of Nicolas's plans is a line through the points (4, 9) and (6, 8). Is this line parallel to the first line [through the points (3, 6) and (7, 3)]?

Check Yourself ANSWERS

1. (a) and (c)   **2.** The lines intersect.

3. (b) and (d)   **4.** The lines are perpendicular.

5. (a) $\dfrac{1}{5}$; **(b)** -5

6. $y = \dfrac{3}{2}x + 2$ **7.** The lines are not parallel.

$y = -\dfrac{2}{3}x + 3$

$\left(\dfrac{3}{2}\right)\left(-\dfrac{2}{3}\right) = -1$

Reading Your Text

The following fill-in-the-blank exercises are designed to ensure that you understand some of the key vocabulary used in this section.

SECTION 7.2

(a) If two lines are _____, they have the same slope.

(b) The point at which two lines meet is called their _____.

(c) When two lines meet at right angles, we say that they are _____.

(d) _____ lines are perpendicular to vertical lines.

Name _____

Section _____ Date _____

Answers

1. _____
2. _____
3. _____
4. _____
5. _____
6. _____
7. _____
8. _____

< Objective 1 >

In exercises 1–4, determine which two equations represent parallel lines.

1. (a) $y = -4x + 5$ (b) $y = 4x + 5$

    (c) $y = \dfrac{1}{4}x + 5$ (d) $y = -4x + 9$

2. (a) $y = 3x - 5$ (b) $y = -3x + 5$

    (c) $y = 3x + 2$ (d) $y = -\dfrac{1}{3}x - 5$

3. (a) $y = \dfrac{2}{3}x + 3$ (b) $y = -\dfrac{3}{2}x - 6$

    (c) $y = 4x + 12$ (d) $y = 4x - 3$

4. (a) $y = \dfrac{9}{4}x - 3$ (b) $y = \dfrac{4}{9}x + 7$

    (c) $y = \dfrac{4}{9}x - 7$ (d) $y = -\dfrac{9}{4}x + 7$

< Objective 2 >

In exercises 5–8, determine which two equations represent perpendicular lines.

5. (a) $y = 6x - 3$ (b) $y = \dfrac{1}{6}x + 3$

    (c) $y = -\dfrac{1}{6}x + 3$ (d) $y = \dfrac{1}{6}x - 3$

6. (a) $y = \dfrac{2}{3}x - 8$ (b) $y = -\dfrac{2}{3}x - 6$

    (c) $y = \dfrac{2}{3}x - 6$ (d) $y = \dfrac{3}{2}x - 6$

7. (a) $y = \dfrac{1}{3}x - 9$ (b) $y = 3x - 9$

    (c) $y = \dfrac{1}{3}x + 9$ (d) $y = -\dfrac{1}{3}x + 9$

8. (a) $y = \dfrac{5}{9}x - 6$ (b) $y = 6x - \dfrac{5}{9}$

    (c) $y = -\dfrac{1}{6}x + \dfrac{5}{9}$ (d) $y = \dfrac{1}{6}x - \dfrac{5}{9}$

Are the following pairs of lines **parallel, perpendicular,** *or* **neither?**

9. L_1 through $(-2, -3)$ and $(4, 3)$
   L_2 through $(3, 5)$ and $(5, 7)$

10. L_1 through $(-2, 4)$ and $(1, 8)$
   L_2 through $(-1, -1)$ and $(-5, 2)$

11. L_1 through $(8, 5)$ and $(3, -2)$
   L_2 through $(-2, 4)$ and $(4, -1)$

12. L_1 through $(-2, -3)$ and $(3, -1)$
   L_2 through $(-3, 1)$ and $(7, 5)$

13. L_1 with equation $x - 3y = 6$
   L_2 with equation $3x + y = 3$

 Videos

14. L_1 with equation $x + 2y = 4$
   L_2 with equation $2x + 4y = 5$

< Objective 3 >

15. Find the slope of any line parallel to the line through points $(-2, 3)$ and $(4, 5)$.

16. Find the slope of any line perpendicular to the line through points $(0, 5)$ and $(-3, -4)$.

17. **CONSTRUCTION** Floor plans for a building have the four corners of a room located at the points $(2, 3)$, $(11, 6)$, $(-3, 18)$, and $(8, 21)$. Determine whether the side through the points $(2, 3)$ and $(11, 6)$ is parallel to the side through the points $(-3, 18)$ and $(8, 21)$.

18. **CONSTRUCTION** For the floor plans given in exercise 17, determine whether the side through the points $(2, 3)$ and $(11, 6)$ is perpendicular to the side through the points $(2, 3)$ and $(-3, 18)$.

19. **CONSTRUCTION** Floor plans for a house have the four corners of a room located at the points $(6, 4)$, $(8, -3)$, $(-1, -6)$, and $(-3, 1)$. Determine whether the side through the points $(-1, -6)$ and $(8, -3)$ is perpendicular to the side through the points $(8, -3)$ and $(6, 4)$.

20. **CONSTRUCTION** For the floor plans given in exercise 19, determine whether the side through the points $(-1, -6)$ and $(8, -3)$ is parallel to the side through the points $(-3, 1)$ and $(6, 4)$.

Answers

9. _____

10. _____

11. _____

12. _____

13. _____

14. _____

15. _____

16. _____

17. _____

18. _____

19. _____

20. _____

Answers

21. _____

22. _____

23. _____

24. _____

25. _____

26. _____

27. _____

28. _____

29. _____

30. _____

31. _____

21. **CONSTRUCTION** Determine whether or not the room described in exercise 17 is a rectangle.

22. **CONSTRUCTION** Determine whether or not the room described in exercise 19 is a rectangle.

Determine whether each of the following statements is **true** *or* **false.**

23. Vertical lines are not parallel to each other.

24. If the slope of one line is $\dfrac{c}{d}$, then the slope of a perpendicular line is $-\dfrac{d}{c}$.

Complete each of the following statements with **never, sometimes,** *or* **always.**

25. Two lines that are perpendicular _____ pass through the origin.

26. Parallel lines _____ intersect.

Basic Skills | **Advanced Skills** | Vocational-Technical Applications | Calculator/Computer | Above and Beyond | Getting Ready

27. A line passing through $(-1, 2)$ and $(4, y)$ is parallel to a line with slope 2. What is the value of y?

> Videos

28. A line passing through $(2, 3)$ and $(5, y)$ is perpendicular to a line with slope $\dfrac{3}{4}$. What is the value of y?

In exercises 29–31, use the concept of slope to determine whether the given figure is a parallelogram or a rectangle.

29.

30.

31.

In exercises 32–34, use the concept of slope to determine whether the given figure is a right triangle (i.e., does the triangle contain a right angle?).

> Videos

32.

33.

Answers

32. _____

33. _____

34. _____

35. _____

36. _____

37. _____

38. _____

34.

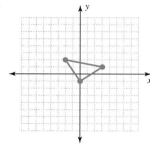

In exercises 35 and 36, use the concept of slope to draw a line perpendicular to the given line segment, passing through the marked point.

35.

36.

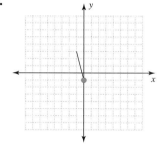

| Basic Skills | Advanced Skills | Vocational-Technical Applications | Calculator/Computer | **Above and Beyond** | Getting Ready |

Work with a partner on the following.

37. On a piece of graph paper, draw a line segment connecting the points (0, 0) and (5, 2). Describe a procedure for drawing a square, where this line segment is one of the sides. Be sure to identify the other corners of the square, and explain why you believe the resulting figure is a square. Is there more than one way to accomplish this? Compare your solution to that of another group.

38. Follow the directions of exercise 37 using the points (−3, 1) and (2, −7).

Answers

(a) _____

(b) _____

(c) _____

(d) _____

(e) _____

(f) _____

Solve the following equations for y.

(a) $3x + 4y = 12$ **(b)** $x - y = -5$ **(c)** $2x + 6 = 4y$

(d) $y + 5 = x$ **(e)** $5x - 2y = 10$ **(f)** $-y + 3x = -5$

Answers

1. a and d   **3.** c and d   **5.** a and c   **7.** b and d   **9.** Parallel

11. Neither   **13.** Perpendicular   **15.** $\dfrac{1}{3}$ **17.** Not parallel

19. Not perpendicular   **21.** Not a rectangle   **23.** False

25. sometimes   **27.** 12   **29.** Parallelogram

31. Rectangle   **33.** Yes   **35.**

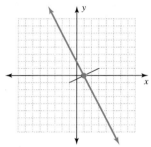

37. Above and Beyond

(a) $y = -\dfrac{3}{4}x + 3$ **(b)** $y = x + 5$ **(c)** $y = \dfrac{1}{2}x + \dfrac{3}{2}$

(d) $y = x - 5$ **(e)** $y = \dfrac{5}{2}x - 5$ **(f)** $y = 3x + 5$

7.3
The Point-Slope Form

< 7.3 Objectives >

1 > Given a point and the slope, find the equation of a line

2 > Given two points, find the equation of a line

3 > Find the equation of a line from given geometric conditions

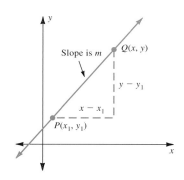

NOTE

We use subscripts (x_1, y_1) to indicate a fixed point on the line.

In mathematics it is often useful to be able to write the equation of a line, given its slope and *any* point on the line. In this section, we will derive a third form for the equation of a line for this purpose.

Suppose that a line has slope m and that it passes through the known point $P(x_1, y_1)$. Let $Q(x, y)$ be any other point on the line. Once again we can use the definition of slope and write

$$m = \frac{y - y_1}{x - x_1}$$

Multiplying both sides by $x - x_1$ gives

$$m(x - x_1) = y - y_1$$

or

$$y - y_1 = m(x - x_1)$$

This equation is called the *point-slope form* for the equation of a line, and all points lying on the line [including (x_1, y_1)] satisfy this equation. We can state the following general result.

Property

Point-Slope Form for the Equation of a Line	The equation of a line with slope m that passes through point (x_1, y_1) is given by $$y - y_1 = m(x - x_1)$$

Example 1   Finding the Equation of a Line

< Objective 1 >

Write the equation for the line that passes through point $(3, -1)$ with a slope of 3.
Letting $(x_1, y_1) = (3, -1)$ and $m = 3$ in point-slope form, we have

$$y - (-1) = 3(x - 3)$$

or

$$y + 1 = 3x - 9$$

We can write the final result in slope-intercept form as

$$y = 3x - 10$$

Check Yourself 1

Write the equation of the line that passes through point $(-2, -4)$ with a slope of $\dfrac{3}{2}$. Write your result in slope-intercept form.

We know that two points determine a line, so it is natural that we should be able to write the equation of a line passing through two given points. Using the point-slope form together with the slope formula allows us to write such an equation.

Example 2   **Finding the Equation of a Line**

< Objective 2 >

Write the equation of the line passing through $(2, 4)$ and $(4, 7)$.

First, we find m, the slope of the line. Here

$$m = \frac{7 - 4}{4 - 2} = \frac{3}{2}$$

Now we apply the point-slope form with $m = \dfrac{3}{2}$ and $(x_1, y_1) = (2, 4)$:

NOTE

We could just as well have chosen to let

$(x_1, y_1) = (4, 7)$

The resulting equation is the same in either case. Take time to verify this for yourself.

$$y - 4 = \frac{3}{2}(x - 2)$$

$$y - 4 = \frac{3}{2}x - 3$$

$$y = \frac{3}{2}x + 1$$

 Check Yourself 2

Write the equation of the line passing through $(-2, 5)$ and $(1, 3)$. Write your result in slope-intercept form.

A line with slope zero is a horizontal line. A line with an undefined slope is vertical. Example 3 illustrates the equations of such lines.

Example 3   **Finding the Equations of Horizontal and Vertical Lines**

< Objective 3 >

(a) Find the equation of a line passing through $(7, -2)$ with a slope of zero.

We could find the equation by letting $m = 0$. Substituting the ordered pair $(7, -2)$ into the slope-intercept form, we can solve for b.

$$y = mx + b$$
$$-2 = 0(7) + b$$
$$-2 = b$$

So,

$$y = 0 \cdot x - 2 \quad \text{or} \quad y = -2$$

It is far easier to remember that any line with a zero slope is a horizontal line and has the form

$$y = b$$

The value for b is the y-coordinate for the given point.

(b) Find the equation of a line with undefined slope passing through $(4, -5)$.

A line with undefined slope is vertical. It always has the form $x = a$, in which a is the x-coordinate for the given point. The equation is

$$x = 4$$

RECALL

The equation of a line with undefined slope passing through the point (x_1, y_1) is given by $x = x_1$.

Check Yourself 3

(a) Find the equation of a line with zero slope that passes through point $(-3, 5)$.

(b) Find the equation of a line passing through $(-3, -6)$ with undefined slope.

Alternate methods for finding the equation of a line through two points exist and have particular significance in other fields of mathematics, such as statistics. Example 4 shows such an alternate approach.

Example 4   **Finding the Equation of a Line**

Write the equation of the line through points $(-2, 3)$ and $(4, 5)$.
First, we find m, as before:

$$m = \frac{5 - 3}{4 - (-2)} = \frac{2}{6} = \frac{1}{3}$$

We now make use of the slope-intercept equation, but in a slightly different form.
Because $y = mx + b$, we can write

$$b = y - mx$$

Now letting $x = -2$, $y = 3$, and $m = \dfrac{1}{3}$, we can calculate b:

$$b = 3 - \left(\frac{1}{3}\right)(-2)$$

$$= 3 + \frac{2}{3} = \frac{11}{3}$$

NOTE

We substitute these values because the line must pass through $(-2, 3)$.

With $m = \dfrac{1}{3}$ and $b = \dfrac{11}{3}$, we can apply the slope-intercept form to write the equation of the desired line. We have

$$y = \frac{1}{3}x + \frac{11}{3}$$

Check Yourself 4

Repeat the Check Yourself 2 exercise, using the technique illustrated in Example 4.

Whichever method we choose, it is an easy matter to check our equation to see if it is correct. In Example 4, we found the equation $y = \dfrac{1}{3}x + \dfrac{11}{3}$. Both points, $(-2, 3)$ and $(4, 5)$, must lie on the line, so each point must satisfy the equation. To check, we can simply substitute an x-value into the equation, and see if we obtain the expected y-value. Using $(-2, 3)$, let $x = -2$:

$$y = \frac{1}{3}(-2) + \frac{11}{3} = -\frac{2}{3} + \frac{11}{3} = \frac{9}{3} = 3 \qquad \text{(the expected } y\text{-value)}$$

Using $(4, 5)$, let $x = 4$:

$$y = \frac{1}{3}(4) + \frac{11}{3} = \frac{4}{3} + \frac{11}{3} = \frac{15}{3} = 5 \qquad \text{(the expected } y\text{-value)}$$

We now can write the equation of a line given appropriate geometric conditions, such as a point on the line and the slope of that line. In some applications the slope may be given not directly but through specified parallel or perpendicular lines.

| | **Example 5** | **Finding the Equation of a Line** |

Find the equation of the line passing through $(-4, -3)$ and parallel to the line determined by $3x + 4y = 12$.

First, we find the slope of the given parallel line, as before:

NOTE

The slope of the given line is $-\dfrac{3}{4}$.

$$3x + 4y = 12$$
$$4y = -3x + 12$$
$$y = -\frac{3}{4}x + 3$$

The desired line is parallel, so its slope is also $-\dfrac{3}{4}$. We can use the point-slope form to write the required equation:

NOTE

The line must pass through $(-4, -3)$, so let $(x_1, y_1) = (-4, -3)$.

$$y - (-3) = -\frac{3}{4}[x - (-4)]$$

This simplifies to

$$y = -\frac{3}{4}x - 6$$

and we have our equation in slope-intercept form.

Check Yourself 5

Find the equation of the line passing through $(2, -1)$ and parallel to the line with equation $3x - y = 2$.

Let us look at another example using geometric information. This time we must apply what we know about perpendicular lines.

| Example 6 | Finding the Equation of a Line |

NOTE

This is true provided that the lines are not horizontal or vertical.

Find the equation of the line passing through (5, 4) and perpendicular to the line with equation $2x - 5y = 10$.

Recall that the slopes of perpendicular lines are negative reciprocals of each other. First we find the slope of the given line:

$2x - 5y = 10$ Solve for y.

$2x - 10 = 5y$

$\dfrac{2}{5}x - 2 = y$ The slope is $\dfrac{2}{5}$.

Since the slope of the given line is $\dfrac{2}{5}$, the slope of the equation we are creating must be $-\dfrac{5}{2}$. Because the desired line must pass through (5, 4), we may use the point-slope form as follows:

NOTE

We usually try to leave the equation in slope-intercept form.

$y - 4 = -\dfrac{5}{2}(x - 5)$

$y - 4 = -\dfrac{5}{2}x + \dfrac{25}{2}$

$y = -\dfrac{5}{2}x + \dfrac{33}{2}$

Check Yourself 6

Find the equation of the line passing through (−2, 3) and perpendicular to the line with equation $4x + y = 7$.

The following chart summarizes the various forms of the equation of a line.

Form	Equation for Line L	Conditions
Standard	$Ax + By = C$	Constants A and B cannot both be zero.
Slope-intercept	$y = mx + b$	Line L has y-intercept $(0, b)$ with slope m.
Point-slope	$y - y_1 = m(x - x_1)$	Line L passes through point (x_1, y_1) with slope m.
Horizontal	$y = a$	Slope is zero.
Vertical	$x = b$	Slope is undefined.

In real-world applications, we rarely begin with the equation of a line. Rather, we usually have points relating two variables based on actual data. If we believe that the two variables are linearly related, we can provide a line through the data points. This line can then be used to determine other points and make predictions.

> **Example 7**   **Vocational Application**

A marketing firm spent $12,400 in advertisements in January for Alexa's Used Car Emporium. February sales at Alexa's were $341,000. In August, the firm spent $8,600 on advertisements, and sales in the following month were $265,000.

(a) Write a linear equation relating the amount spent on advertisements x with sales in the following month y. Write your answer in slope-intercept form.

We identify the two points $(12,400, 341,000)$ and $(8,600, 265,000)$. The slope of the line through these two points is

$$m = \frac{265,000 - 341,000}{8,600 - 12,400}$$

$$= \frac{-76,000}{-3,800}$$

$$= 20$$

NOTE

This indicates that for each additional dollar spent on advertisements, sales increase by $20 in the following month.

We may choose either point to use with the point-slope formula. Here, we will let

$$(x_1, y_1) = (12,400, 341,000)$$

which gives

$$y - 341,000 = 20(x - 12,400)$$

We solve this equation for y and simplify to write it in slope-intercept form.

$$y - 341,000 = 20(x - 12,400)$$
$$y - 341,000 = 20x - 20 \cdot 12,400$$
$$y - 341,000 = 20x - 248,000$$
$$y = 20x - 248,000 + 341,000$$
$$y = 20x + 93,000$$

(b) Use the equation found in part **(a)** to predict the sales amount if $10,000 is spent on advertisements in one month.

We evaluate the slope-intercept equation found in part **(a)** with $x = 10,000$.

$$y = 20(10,000) + 93,000$$
$$= 200,000 + 93,000$$
$$= 293,000$$

So Alexa can assume that if she spends $10,000 on advertisements one month, she will record $293,000 in sales in the following month.

Check Yourself 7

In 2001, the average cost of tuition and fees at public 4-year colleges was $3,351. By 2005, the average cost had risen to $4,630.

(a) Assuming that the relationship between the cost y and the year x is linear, determine an equation relating the cost to the year (write your answer in slope-intercept form). (*Hint:* If you let x be the number of years since 2000, so that $x = 1$ corresponds to 2001, then $x = 5$ corresponds to 2005.)
(b) Use the equation found in part (a) to predict the average cost of tuition and fees at public 4-year colleges in 2012.

Source: The Chronicle of Higher Education

Check Yourself ANSWERS

1. $y = \dfrac{3}{2}x - 1$ **2.** $y = -\dfrac{2}{3}x + \dfrac{11}{3}$ **3.** **(a)** $y = 5$; **(b)** $x = -3$

4. $y = -\dfrac{2}{3}x + \dfrac{11}{3}$ **5.** $y = 3x - 7$ **6.** $y = \dfrac{1}{4}x + \dfrac{7}{2}$

7. **(a)** $y = 319.75x + 3{,}031.25$; **(b)** \$6,868.25

Reading Your Text

The following fill-in-the-blank exercises are designed to ensure that you understand some of the key vocabulary used in this section.

SECTION 7.3

(a) The equation $y - y_1 = m(x - x_1)$ is called the _____ form for the equation of a line.

(b) A line with slope zero is a _____ line.

(c) A line with undefined slope is _____.

(d) The slopes of perpendicular lines (that are not horizontal or vertical) are negative _____ of each other.

7.3 exercises

Name _____

Section _____ Date _____

Answers

1. _____
2. _____
3. _____
4. _____ 5. _____
6. _____
7. _____
8. _____
9. _____ 10. _____
11. _____
12. _____
13. _____
14. _____ 15. _____
16. _____
17. _____ 18. _____
19. _____
20. _____ 21. _____
22. _____

< Objective 1 >

Write the equation of the line passing through each of the given points with the indicated slope. Give your results in slope-intercept form, where possible.

1. $(0, 2)$, $m = 3$

2. $(0, -4)$, $m = -2$

3. $(0, 2)$, $m = \dfrac{3}{2}$

4. $(0, -3)$, $m = -2$

5. $(0, 4)$, $m = 0$

6. $(0, 5)$, $m = -\dfrac{3}{5}$

7. $(0, -5)$, $m = \dfrac{5}{4}$

8. $(0, -4)$, $m = -\dfrac{3}{4}$

9. $(1, 2)$, $m = 3$

10. $(-1, 2)$, $m = 3$

11. $(-2, -3)$, $m = -3$

12. $(1, -4)$, $m = -4$

13. $(5, -3)$, $m = \dfrac{2}{5}$

14. $(4, 3)$, $m = 0$

15. $(2, -3)$, m is undefined

16. $(2, -5)$, $m = \dfrac{1}{4}$

< Objective 2 >

Write the equation of the line passing through each of the given pairs of points. Write your result in slope-intercept form, where possible.

17. $(2, 3)$ and $(5, 6)$

18. $(3, -2)$ and $(6, 4)$

19. $(-2, -3)$ and $(2, 0)$

20. $(-1, 3)$ and $(4, -2)$

21. $(-3, 2)$ and $(4, 2)$

22. $(-5, 3)$ and $(4, 1)$

23. (2, 0) and (0, −3)   **24.** (2, −3) and (2, 4)

25. (0, 4) and (−2, −1)   **26.** (−4, 1) and (3, 1)

< Objective 3 >

Write the equation of the line L satisfying the given geometric conditions.

27. *L* has slope 4 and *y*-intercept (0, −2).

28. *L* has slope $-\dfrac{2}{3}$ and *y*-intercept (0, 4).

29. *L* has *x*-intercept (4, 0) and *y*-intercept (0, 2).

30. *L* has *x*-intercept (−2, 0) and slope $\dfrac{3}{4}$.

31. *L* has *y*-intercept (0, 4) and a slope of zero.

32. *L* has *x*-intercept (−2, 0) and an undefined slope.

33. *L* passes through point (3, 2) with a slope of 5.

34. *L* passes through point (−2, −4) with a slope of $-\dfrac{3}{2}$.

35. **SCIENCE AND MEDICINE** A temperature of 10°C corresponds to a temperature of 50°F. Also 40°C corresponds to 104°F. Find the linear equation relating *F* and *C*.

36. **BUSINESS AND FINANCE** In planning for a new item, a manufacturer assumes that the number of items produced *x* and the cost in dollars *C* of producing these items are related by a linear equation. Projections are that 100 items will cost $10,000 to produce and that 300 items will cost $22,000 to produce. Find the equation that relates *C* and *x*.

37. **TECHNOLOGY** A word processing station was purchased by a company for $10,000. After 4 years it is estimated that the value of the station will be $4,000. If the value in dollars *V* and the time the station has been in use *t* are related by a linear equation, find the equation that relates *V* and *t*.

Answers

23. _____

24. _____

25. _____

26. _____

27. _____

28. _____

29. _____

30. _____

31. _____

32. _____

33. _____

34. _____

35. _____

36. _____

37. _____

Answers

38. _____

39. _____

40. _____

41. _____

42. _____

43. _____

44. _____

45. _____

46. _____

38. **BUSINESS AND FINANCE** Two years after an expansion, a company had sales of $42,000. Four years later the sales were $102,000. Assuming that the sales in dollars S and the time in years t are related by a linear equation, find the equation relating S and t.

39. **BUSINESS AND FINANCE** Two years after purchase, a certain car was valued at $31,000. After 5 years, the car was valued at $20,500. Assume that the car depreciates in a linear manner.

   (a) Find the linear equation that relates the car's value V and the number of years t since it was purchased.
   (b) Give the initial purchase price of the car.
   (c) Interpret the slope.
   (d) Predict the value of the car 10 years after purchase.

40. **BUSINESS AND FINANCE** Three years after purchase, a certain car was valued at $16,200. After 7 years, the car was valued at $13,800. Assume that the car depreciates in a linear manner.

   (a) Find the linear equation that relates the car's value V and the number of years t since it was purchased.
   (b) Give the initial purchase price of the car.
   (c) Interpret the slope.
   (d) Predict the value of the car 10 years after purchase.

Determine whether each of the following statements is **true** *or* **false.**

41. Given two points, there is exactly one line that will pass through them.

42. Given a line, there is exactly one other line that is perpendicular to it.

Complete each of the following statements with **never, sometimes,** *or* **always.**

43. You can _____ write the equation of a line if you know two points that are on the line.

44. The equation of a line can _____ be put into slope-intercept form.

| Basic Skills | **Advanced Skills** | Vocational-Technical Applications | Calculator/Computer | Above and Beyond | Getting Ready |

In exercises 45–58, write the equation of the line L that satisfies the given geometric conditions.

45. L has y-intercept $(0, 3)$ and is parallel to the line with equation $y = 3x - 5$.

46. L has y-intercept $(0, -3)$ and is parallel to the line with equation $y = \frac{2}{3}x + 1$.

47. L has y-intercept $(0, 4)$ and is perpendicular to the line with equation $y = -2x + 1$.

48. L has y-intercept $(0, 2)$ and is parallel to the line with equation $y = -1$.

49. L has y-intercept $(0, 3)$ and is parallel to the line with equation $y = 2$.

50. L has y-intercept $(0, 2)$ and is perpendicular to the line with equation $2x - 3y = 6$.

51. L passes through point $(-3, 2)$ and is parallel to the line with equation $y = 2x - 3$.

52. L passes through point $(-4, 3)$ and is parallel to the line with equation $y = -2x + 1$.

53. L passes through point $(3, 2)$ and is parallel to the line with equation $y = \frac{4}{3}x + 4$.

54. L passes through point $(-2, -1)$ and is perpendicular to the line with equation $y = 3x + 1$.

55. L passes through point $(5, -2)$ and is perpendicular to the line with equation $y = -3x - 2$.

56. L passes through point $(3, 4)$ and is perpendicular to the line with equation $y = -\frac{3}{5}x + 2$.

57. L passes through $(-2, 1)$ and is parallel to the line with equation $x + 2y = 4$.

58. L passes through $(-3, 5)$ and is parallel to the x-axis.

Beginning Algebra   The Streeter/Hutchison Series in Mathematics

Answers

47. _____

48. _____

49. _____

50. _____

51. _____

52. _____

53. _____

54. _____

55. _____

56. _____

57. _____

58. _____

59. _____

60. _____

61. _____

Basic Skills | Advanced Skills | **Vocational-Technical Applications** | Calculator/Computer | Above and Beyond | Getting Ready
▲

59. **AUTOMOTIVE TECHNOLOGY** The cost of a 3-hour car repair is $100. The cost of an 8-hour repair is $250. Write a linear equation to determine the price of a repair based on the number of hours.

60. **CONSTRUCTION** A wall that is 28 feet long requires 22 studs, while a wall that is 44 feet long requires 34 studs. Create the equation that relates the number of studs, s, to the length of the wall, L.

61. **MECHANICAL ENGINEERING** A piece of metal at 20°C is 64 mm long. The same piece of metal expands to 67 mm at 220°C. Find the linear equation that relates the length of the piece to the temperature.

Answers

62. _____

63. _____

64. _____

(a) _____

(b) _____

(c) _____

(d) _____

(e) _____

(f) _____

(g) _____

(h) _____

62. ALLIED HEALTH The absorbance of a solution, or the amount of light that is absorbed by the solution, is linearly related to the solution's concentration. A standard supply of glucose has a concentration of 200 milligrams per milliliter (mg/mL) and an absorbance value of 0.420, as determined by a spectrophotometer. A second supply of glucose has a concentration of 110 mg/mL and an absorbance value of 0.24. Write an equation relating the absorbance (A) to its concentration (C), in mg/mL.

chapter 7 > Make the Connection

Basic Skills | Advanced Skills | Vocational-Technical Applications | Calculator/Computer | **Above and Beyond** | Getting Ready

63. Describe the process for finding the equation of a line if you are given two points on the line.

64. How would you find the equation of a line if you were given the slope and the *x-intercept*?

Basic Skills | Advanced Skills | Vocational-Technical Applications | Calculator/Computer | Above and Beyond | **Getting Ready**

Graph each of the following inequalities on a number line.

(a) $x < 3$

(b) $x \geq -2$

(c) $2x \leq 8$

(d) $3x \geq -9$

(e) $-3x < 12$

(f) $-2x \leq 10$

(g) $\dfrac{2}{3}x \leq 4$

(h) $-\dfrac{3}{4}x \geq 6$

Answers

1. $y = 3x + 2$ **3.** $y = \dfrac{3}{2}x + 2$ **5.** $y = 4$ **7.** $y = \dfrac{5}{4}x - 5$

9. $y = 3x - 1$ **11.** $y = -3x - 9$ **13.** $y = \dfrac{2}{5}x - 5$ **15.** $x = 2$

17. $y = x + 1$ **19.** $y = \dfrac{3}{4}x - \dfrac{3}{2}$ **21.** $y = 2$ **23.** $y = \dfrac{3}{2}x - 3$

25. $y = \dfrac{5}{2}x + 4$ **27.** $y = 4x - 2$ **29.** $y = -\dfrac{1}{2}x + 2$ **31.** $y = 4$

33. $y = 5x - 13$ **35.** $F = \dfrac{9}{5}C + 32$ **37.** $V = -1{,}500t + 10{,}000$

39. (a) $V = -3{,}500t + 38{,}000$; **(b)** \$38,000; **(c)** For each additional year, the value of the car drops by \$3,500; **(d)** \$3,000

41. True   **43.** always   **45.** $y = 3x + 3$ **47.** $y = \dfrac{1}{2}x + 4$

49. $y = 3$ **51.** $y = 2x + 8$ **53.** $y = \dfrac{4}{3}x - 2$ **55.** $y = \dfrac{1}{3}x - \dfrac{11}{3}$

57. $y = -\dfrac{1}{2}x$ **59.** $C = 30h + 10$ **61.** $L = 0.015T + 63.7$

63. Above and Beyond

(a) (number line: open circle at 3; marks at 0, 3)

(b) (number line: filled circle at −2; marks at −2, 0)

(c) (number line: filled circle at 4; marks at 0, 4)

(d) (number line: filled circle at −3; marks at −3, 0)

(e) (number line: open circle at −4; marks at −4, 0)

(f) (number line: filled circle at −5; marks at −5, 0)

(g) (number line: filled circle at 6; marks at 0, 6)

(h) (number line: filled circle at −8; marks at −8, 0)

7.4

Graphing Linear Inequalities

< 7.4 Objective >

1 > Graph a linear inequality in two variables

In Section 2.6 you learned to graph inequalities in one variable on a number line. We now want to extend our work with graphing to include linear inequalities in two variables. We begin with a definition.

> **Definition**
>
> **Linear Inequality in Two Variables**
>
> An inequality that can be written in the form
>
> $Ax + By < C$
>
> in which A and B are not both 0, is called a **linear inequality in two variables.**

> **NOTE**
>
> The inequality symbols $\leq$, $>$, and $\geq$ can also be used.

Some examples of linear inequalities in two variables are

$$x + 3y > 6 \qquad y \leq 3x + 1 \qquad 2x - y \geq 3$$

The *graph* of a linear inequality is always a region (actually a half-plane) of the plane whose boundary is a straight line. Let's look at an example of graphing such an inequality.

> **Example 1** Graphing a Linear Inequality

< Objective 1 >

> **NOTE**
>
> The dotted line indicates that the points on the line $2x + y = 4$ are *not* part of the solution to the inequality $2x + y < 4$.
>
>

Graph $2x + y < 4$.

First, replace the inequality symbol ($<$) with an equal sign. We then have $2x + y = 4$. This equation forms the **boundary line** of the graph of the original inequality. You can graph the line by any of the methods discussed earlier.

The boundary line for our inequality is shown at the left. We see that the boundary line separates the plane into two regions, each of which is called a **half-plane.** We call $2x + y = 4$ the boundary line because:

1. Any point that is on one side of the boundary is a solution for the inequality. For example, test $(-1, -1)$:

$$2(-1) + (-1) \overset{?}{<} 4$$
$$-3 < 4 \qquad \text{A true statement}$$

2. Any point that is on the other side of the boundary line *fails* to be a solution. For example, test $(3, 3)$:

$$2(3) + (3) \overset{?}{<} 4$$
$$9 < 4 \qquad \text{A false statement}$$

Beginning Algebra   The Streeter/Hutchison Series in Mathematics   © The McGraw-Hill Companies. All Rights Reserved.

3. Any point that is precisely on the boundary line *fails* in this case, because the symbol of the inequality is $<$. For example, test $(2, 0)$:

$$2(2) + (0) < 4$$

$$4 < 4 \qquad \text{A false statement}$$

So, once the boundary line is graphed, we need to choose the correct half-plane. Choose any convenient test point not on the boundary line. The origin $(0, 0)$ is a good choice because it results in an easy calculation.

Substitute $x = 0$ and $y = 0$ into the inequality.

$$2 \cdot (0) + (0) \overset{?}{<} 4$$

$$0 + 0 \overset{?}{<} 4$$

$$0 < 4 \qquad \text{A true statement}$$

Because the inequality is *true* for the test point, we shade the half-plane containing that test point (the origin). The origin and all other points *below* the boundary line then represent solutions for our original inequality.

Check Yourself 1

Graph the inequality $x + 3y < 3$.

The process is similar when the boundary line is included in the solution.

 Example 2 | **Graphing a Linear Inequality**

Graph $4x - 3y \geq 12$.

First, graph the boundary line $4x - 3y = 12$.

Note: When equality *is included* ($\leq$ or $\geq$), use a *solid line* for the graph of the boundary line. This means the line is included in the graph of the linear inequality, because every point on the line represents a solution.

The graph of our boundary line (a solid line here) is shown on the figure.

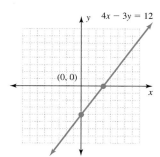

Again, we use $(0, 0)$ as a convenient test point. Substituting 0 for x and for y in the original inequality, we have

$$4 \cdot (0) - 3 \cdot (0) \overset{?}{\geq} 12$$

$$0 \geq 12 \qquad \text{A false statement}$$

Because the inequality is *false* for the test point, we shade the half-plane that does *not* contain that test point, here (0, 0).

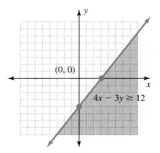

Check Yourself 2

Graph the inequality $3x + 2y \geq 6$.

Example 3	Graphing a Linear Inequality with One Variable

Graph $x \leq 5$.

The boundary line is $x = 5$. Its graph is a solid line because equality is included. Using (0, 0) as a test point, we substitute 0 for x with the result

$0 \leq 5$ A true statement

Because the inequality is *true* for the test point, we shade the half-plane containing the origin.

Check Yourself 3

Graph the inequality $y < 2$.

As we mentioned earlier, we may have to use a point other than the origin as our test point. Example 4 illustrates this approach.

> ▶ **Example 4**   **Graphing a Linear Inequality through the Origin**

Graph $2x + 3y < 0$.

The boundary line is $2x + 3y = 0$. Its graph is shown on the figure.

NOTE

We use a dotted line for our boundary line because equality is not included.

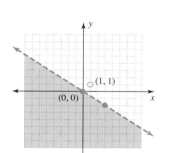

NOTE

We develop the idea of graphing "bounded" regions more fully in Chapter 8.

We cannot use $(0, 0)$ as our test point in this case. Do you see why?

Choose any other point *not* on the line. We chose $(1, 1)$ as a test point. Substituting 1 for x and 1 for y gives

$$2 \cdot (1) + 3 \cdot (1) \overset{?}{<} 0$$
$$2 + 3 \overset{?}{<} 0$$
$$5 < 0 \qquad \text{A false statement}$$

Because the inequality is *false* at our test point, we shade the half-plane *not* containing $(1, 1)$. This is shown in the graph in the margin.

  Check Yourself 4

Graph the inequality $x - 2y < 0$.

The following steps summarize our work in graphing linear inequalities in two variables.

Step by Step

To Graph a Linear Inequality

Step 1   Replace the inequality symbol with an equal sign to form the equation of the boundary line of the graph.

Step 2   Graph the boundary line. Use a dotted line if equality is not included ($<$ or $>$). Use a solid line if equality is included ($\leq$ or $\geq$).

Step 3   Choose any convenient test point *not* on the line.

Step 4   If the inequality is *true* at the checkpoint, shade the half-plane including the test point. If the inequality is *false* at the checkpoint, shade the half-plane not including the test point.

Linear inequalities and their graphs may be used to represent *feasible regions* in applications. These regions include all points that satisfy some set of conditions determined by the application.

Example 5   **Vocational Application**

A hospital food service can serve at most 1,000 meals per day. Patients on a normal diet receive three meals per day and patients on a special diet receive four meals per day.

(a) Write a linear inequality that describes the number of patients that can be served in a day and sketch its graph.

Let x be the number of people served a normal diet. Then $3x$ represents the number of meals served to the people on a normal diet. Let y be the number of people served a special diet. Then $4y$ represents the number of meals served to the people on the special diet. We know that the total number of meals served is at most 1,000. Writing this as an inequality gives

$$3x + 4y \le 1,000$$

or

$$y \le -\frac{3}{4}x + 250 \qquad \text{We solved the inequality for } y.$$

We need to graph this inequality only in the first quadrant. Do you see why?

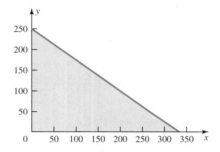

(b) Can the hospital food service serve 100 patients on a normal diet and 100 patients on the special diet in a day?

We can substitute 100 for both x and y in the inequality found in part **(a)** and see if a true statement results.

$$y \le -\frac{3}{4}x + 250$$

$$(100) \stackrel{?}{\le} -\frac{3}{4}(100) + 250$$

$$100 \stackrel{?}{\le} -75 + 250$$

$$100 \le 175 \qquad \text{True!}$$

Since this final inequality is true, we conclude that the hospital food service can serve 100 people on each type of diet in a single day. Graphically, we see that the point (100, 100) is in the solution region.

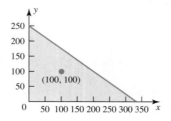

(c) If the hospital serves 200 people on a normal diet, then what is the maximum number of people who can be served the special diet?

Substitute 200 for x and calculate y.

$$y \leq -\frac{3}{4}x + 250$$

$$y \leq -\frac{3}{4}(200) + 250$$

$$y \leq -150 + 250$$

$$y \leq 100$$

We conclude that if the hospital serves 200 meals to patients on a normal diet, they can serve, at most, 100 patients on the special diet.

Check Yourself 5

A manufacturer produces a standard and a deluxe model of 13-in. television sets. The standard model requires 12 h to produce, while the deluxe model requires 18 h to produce. The manufacturer has a total of 360 h of labor available in a week.

(a) Write a linear inequality to represent the number of each type of television set the manufacturer can produce in a week and graph the inequality. (*Hint:* You need to graph the inequality only in the first quadrant.)

(b) If the manufacturer needs to produce 16 standard models one week, what is the largest number of deluxe models that can be produced in that week?

Check Yourself ANSWERS

1.

2.

3.

$y < 2$

4.

$x - 2y < 0$

5. (a) $12x + 18y \leq 360$ or $y \leq -\dfrac{2}{3}x + 20$; **(b)** nine deluxe models

---

Reading Your Text

The following fill-in-the-blank exercises are designed to ensure that you understand some of the key vocabulary used in this section.

SECTION 7.4

(a) The boundary line separates the plane into two regions, each of which is called a _____.

(b) A _____ boundary line indicates that the points on the line are *not* part of the solutions to the inequality.

(c) The _____ is usually a good choice for a convenient test point.

(d) A _____ region includes all points that satisfy some set of conditions determined by an application.

Basic Skills | Advanced Skills | Vocational-Technical Applications | Calculator/Computer | Above and Beyond | Getting Ready

7.4 exercises

< Objective 1 >

In exercises 1–8, we have graphed the boundary line for the linear inequality. Determine the correct half-plane in each case, and complete the graph.

1. $x + y < 5$

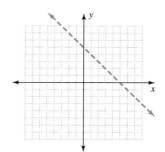

2. $x - y \geq 4$

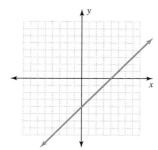

3. $x - 2y \geq 4$

4. $2x + y < 6$

5. $x \leq -3$

6. $y \geq 2x$

7. $y < 2x - 6$

8. $y > 3$

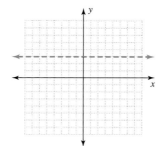

MathZone

Boost your grade at mathzone.com!
> Practice Problems
> NetTutor
> Self-Tests
> e-Professors
> Videos

Name _____

Section _____ Date _____

Answers

1. _____

2. _____

3. _____

4. _____

5. _____

6. _____

7. _____

8. _____

Answers

9. _____

10. _____

11. _____

12. _____

13. _____

14. _____

15. _____

16. _____

Graph each of the following inequalities.

9. $x + y < 3$

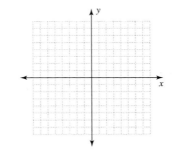

10. $x - y \geq 4$

11. $x - y \leq 5$

12. $x + y > 5$

13. $2x + y < 6$

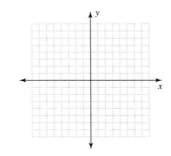

14. $3x + y \geq 6$

15. $x \leq 3$

> Videos

16. $4x + y \geq 4$

17. $x - 5y < 5$

18. $x < 3$

19. $y < -4$

20. $4x + 3y > 12$

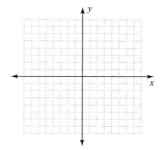

21. $2x - 3y \geq 6$

> Videos

22. $x \geq -2$

23. $3x + 2y \geq 0$

24. $3x + 5y < 15$

Answers

17. _____

18. _____

19. _____

20. _____

21. _____

22. _____

23. _____

24. _____

Answers

25. _____

26. _____

27. _____

28. _____

29. _____

30. _____

31. _____

32. _____

25. $5x + 2y > 10$

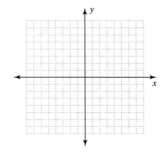

26. $x - 3y \geq 0$

27. $y \leq 2x$

> Videos

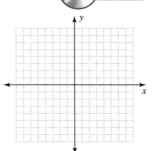

28. $3x - 4y < 12$

29. $y > 2x - 3$

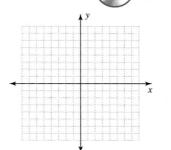

30. $y \geq -2x$

31. $y < -2x - 3$

> Videos

32. $y \leq 3x + 4$

Determine whether each of the following statements is **true** *or* **false.**

33. The boundary line for the solutions to $Ax + By < C$ is always a straight line.

34. The solution region always contains the origin $(0, 0)$.

Complete each of the following statements with **never, sometimes,** *or* **always.**

35. The origin should _____ be used as a test point when finding the region of solutions.

36. The boundary line should _____ be drawn as a dotted line.

Answers

33. _____

34. _____

35. _____

36. _____

37. _____

38. _____

39. _____

40. _____

41. _____

| Basic Skills | **Advanced Skills** | Vocational-Technical Applications | Calculator/Computer | Above and Beyond | Getting Ready |

Graph each of the following inequalities.

37. $2(x + y) - x > 6$

38. $3(x + y) - 2y < 3$

39. $4(x + y) - 3(x + y) \leq 5$

40. $5(2x + y) - 4(2x + y) \geq 4$

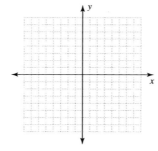

41. **BUSINESS AND FINANCE** Suppose you have two part-time jobs. One is at a video store that pays $9 per hour and the other is at a convenience store that pays $8 per hour. Between the two jobs, you want to earn at least $240 per week. Write an inequality that shows the various number of hours you can work at each job.

chapter 7 > Make the Connection

Answers

42. _____

43. _____

44. _____

45. _____

42. NUMBER PROBLEM You have at least $30 in change in your drawer, consisting of dimes and quarters. Write an inequality that shows the different number of coins in your drawer.

Basic Skills | Advanced Skills | **Vocational-Technical Applications** | Calculator/Computer | Above and Beyond | Getting Ready

▲

43. MANUFACTURING TECHNOLOGY A manufacturer produces 2-slice toasters and 4-slice toasters. The 2-slice type requires 8 hours to produce, and the 4-slice type requires 10 hours to produce. The manufacturer has at most 400 hours of labor available in a week.

chapter 7 > Make the Connection

(a) Write a linear inequality to represent the number of each type of toaster the manufacturer can produce in a week.

(b) Graph the inequality. You need to draw the graph only in the first quadrant.

(c) Is it feasible to produce 20 2-slice toasters and 30 4-slice toasters in a week?

44. MANUFACTURING TECHNOLOGY A certain company produces standard clock radios and deluxe clock radios. It costs the company $15 to produce each standard clock radio and $20 to produce a deluxe model. The company's budget limits production costs to a maximum of $3,000 per day.

chapter 7 > Make the Connection

(a) Write a linear inequality to represent the number of each type of clock radio the company can produce in a day.

(b) Graph the inequality. You need to draw the graph only in the first quadrant.

(c) Is it feasible to produce 80 of each type of clock radio in a day?

Basic Skills | Advanced Skills | Vocational-Technical Applications | Calculator/Computer | **Above and Beyond** | Getting Ready

▲

45. BUSINESS AND FINANCE Linda Williams has just begun a nursery business and seeks your advice. She has limited funds to spend and wants to stock two kinds of fruit-bearing plants. She lives in the northeastern part of Texas and thinks that blueberry bushes and peach trees would sell well there. Linda can buy blueberry bushes from a supplier for $2.50 each and young peach trees for $5.50 each. She wants to know what combination she should buy and to keep her outlay to $500 or less. Write an inequality and draw a graph to depict what combinations of blueberry bushes and peach trees she can buy for the amount of money she has. Explain the graph and her options.

Answers

46. _____

46. **Statistics** After reading an article on the front page of *The New York Times* titled "You Have to Be Good at Algebra to Figure Out the Best Deal for Long Distance," Rafaella De La Cruz decided to apply her skills in algebra to try to decide between two competing long-distance companies. It was difficult at first to get the companies to explain their charge policies. They both kept repeating that they were 25% cheaper than their competition. Finally, Rafaella found someone who explained that the charge depended on when she called, where she called, how long she talked, and how often she called. "Too many variables!" she exclaimed. So she decided to ask one company what they charged as a base amount, just for using the service.

Company A said that they charged $5 for the privilege of using their long-distance service, whether or not she made any phone calls, and that because of this fee they were able to allow her to call anywhere in the United States after 6 P.M. for only $0.15 a minute. Complete this table of charges based on this company's plan:

Total Minutes Long Distance in 1 Month (After 6 P.M.)	Total Charge
0 minutes	
10 minutes	
30 minutes	
60 minutes	
120 minutes	

Use this table to make a graph of the monthly charges from Company A based on the number of minutes of long distance.

Rafaella wanted to compare this offer to Company B, which she was currently using. She looked at her phone bill and saw that one month she had been charged $7.50 for 30 minutes and another month she had been charged $11.25 for 45 minutes of long-distance calling. These calls were made after 6 P.M. to her relatives in Indiana and Arizona. Draw a graph on the same set of axes you made for Company A's figures. Use your graph and what you know about linear inequalities to advise Rafaella about which company is better.

Answers

(a) _____

(b) _____

(c) _____

(d) _____

(e) _____

(f) _____

(g) _____

(h) _____

Evaluate each expression for the given variable value.

(a) $2x + 1$ $(x = 2)$ **(b)** $2x + 1$ $(x = -2)$

(c) $3 - 2x$ $(x = 1)$ **(d)** $3 - 2x$ $(x = -1)$

(e) $x^2 - 2$ $(x = 2)$ **(f)** $x^2 - 2$ $(x = -2)$

(g) $x^2 + 5$ $(x = 1)$ **(h)** $x^2 + 5$ $(x = -1)$

Answers

1. $x + y < 5$

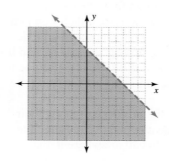

3. $x - 2y \geq 4$

5. $x \leq -3$

7. $y < 2x - 6$

9. $x + y < 3$

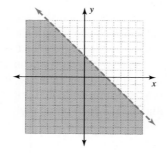

11. $x - y \leq 5$

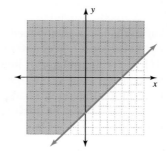

13. $2x + y < 6$

15. $x \leq 3$

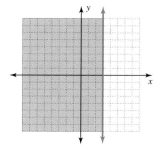

17. $x - 5y < 5$

19. $y < -4$

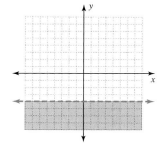

21. $2x - 3y \geq 6$

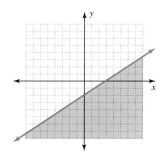

23. $3x + 2y \geq 0$

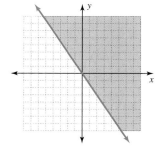

25. $5x + 2y > 10$

27. $y \leq 2x$

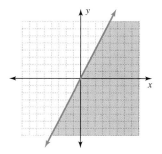

29. $y > 2x - 3$

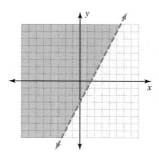

31. $y < -2x - 3$

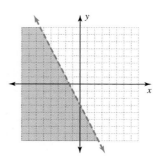

33. True

35. sometimes

37. $x + 2y > 6$

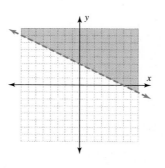

39. $x + y \leq 5$

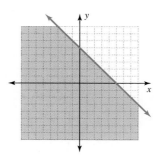

41. $9x + 8y \geq 240$

43. **(a)** $8x + 10y \leq 400$;

(b)

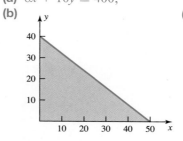

(c) no

45. Above and Beyond

(a) 5   **(b)** −3   **(c)** 1   **(d)** 5   **(e)** 2   **(f)** 2   **(g)** 6   **(h)** 6

7.5

An Introduction to Functions

< 7.5 Objectives >

1 > Evaluate an expression

2 > Evaluate a function

3 > Express the equation of a line as a linear function

4 > Graph a linear function

Variables can be used to represent unknown real numbers. Together with the operations of addition, subtraction, multiplication, division, and exponentiation, numbers and variables form expressions such as

$$3x + 5 \qquad 7x - 4 \qquad x^2 - 3x - 10 \qquad x^4 - 2x^2 + 3x + 4$$

Four different actions can be taken with expressions. We can

1. Substitute values for the variable(s) and **evaluate the expression.**
2. Rewrite an expression as some simpler equivalent expression. This is called **simplifying the expression.**
3. Set two expressions equal to each other and **solve for the stated variable.**
4. Set two expressions equal to each other and **graph the equation.**

Throughout this book, everything we do involves one of these four actions. We now return our focus to the first item, evaluating expressions. As we saw in Section 1.5, expressions can be evaluated for an indicated value of the variable(s). Example 1 illustrates.

| Example 1 | Evaluating Expressions |

< Objective 1 >

Evaluate the expression $x^4 - 2x^2 + 3x + 4$ for the indicated value of x.

(a) $x = 0$

Substituting 0 for x in the expression yields

$$(0)^4 - 2(0)^2 + 3(0) + 4 = 0 - 0 + 0 + 4$$
$$= 4$$

(b) $x = 2$

Substituting 2 for x in the expression yields

$$(2)^4 - 2(2)^2 + 3(2) + 4 = 16 - 8 + 6 + 4$$
$$= 18$$

723

(c) $x = -1$

Substituting -1 for x in the expression yields

$$(-1)^4 - 2(-1)^2 + 3(-1) + 4 = 1 - 2 - 3 + 4$$
$$= 0$$

Check Yourself 1

Evaluate the expression $2x^3 - 3x^2 + 3x + 1$ for the indicated value of x.

(a) $x = 0$ **(b)** $x = 1$ **(c)** $x = -2$

We could design a machine whose function would be to crank out the value of an expression for each given value of x. We could call this machine something simple such as f. Our *function* machine might look like this.

$x \longrightarrow$

Function
f

$2x^3 + 3x^2 - 5x - 1$

NOTE

-1 is called the *input value*, 5 is called the *output value*.

For example, when we put -1 into the machine, the machine substitutes -1 for x in the expression, and 5 comes out the other end because

$$2(-1)^3 + 3(-1)^2 - 5(-1) - 1 = -2 + 3 + 5 - 1 = 5$$

In fact, the idea of the function machine is very useful in mathematics. Your graphing calculator can be used as a function machine. You can enter the expression into the calculator as Y_1 and then evaluate Y_1 for different values of x.

Generally, in mathematics, we do not write $Y_1 = 2x^3 + 3x^2 - 5x - 1$. Instead, we write $f(x) = 2x^3 + 3x^2 - 5x - 1$, which is read as "$f$ of x is equal to. . . ." Instead of calling f a function machine, we say that f is a function of x. The greatest benefit to this notation is that it lets us easily note the input value of x along with the output of the function. Instead of "Evaluate y for $x = 4$" we say "Find $f(4)$." This means that, given the function f, $f(c)$ designates the value of the function when the variable is equal to c.

$>$ **C A U T I O N**

$f(x)$ does not mean f times x.

(▶)   **Example** 2   Evaluating a Function for Different Inputs

$<$ Objective 2 $>$

Given $f(x) = x^3 - 3x^2 + x + 5$, find the following:

(a) $f(0)$

Substituting 0 for x in the expression, we get

$$f(0) = (0)^3 - 3(0)^2 + (0) + 5$$
$$= 5$$

(b) $f(-3)$

Substituting -3 for x in the expression, we get

$$f(-3) = (-3)^3 - 3(-3)^2 + (-3) + 5$$
$$= -27 - 27 - 3 + 5$$
$$= -52$$

(c) $f\left(\dfrac{1}{2}\right)$

Substituting $\dfrac{1}{2}$ for x in the expression, we get

$$f\left(\frac{1}{2}\right) = \left(\frac{1}{2}\right)^3 - 3\left(\frac{1}{2}\right)^2 + \left(\frac{1}{2}\right) + 5$$
$$= \frac{1}{8} - 3\left(\frac{1}{4}\right) + \frac{1}{2} + 5$$
$$= \frac{1}{8} - \frac{3}{4} + \frac{1}{2} + 5$$
$$= \frac{1}{8} - \frac{6}{8} + \frac{4}{8} + \frac{40}{8}$$
$$= \frac{39}{8}$$

Check Yourself 2

Given $f(x) = 2x^3 - x^2 + 3x - 2$, find the following.

(a) $f(0)$ **(b)** $f(3)$ **(c)** $f\left(-\dfrac{1}{2}\right)$

In Example 2 we obtained $f(0) = 5$. We can describe this by saying that when 0 is the input, 5 is the output. In part **(b)**, we found $f(-3) = -52$; that is, when the input is -3, the output is -52. In general, the input value is x and the output value is $f(x)$.

Given a function f, the pair of numbers $(x, f(x))$ is very significant. We always write them in that order. The ordered pair consists of the x-value first and the function value at that x (the $f(x)$) second.

Example 3   **Finding Ordered Pairs**

NOTE

Think of the ordered pairs of a function as (input, output).

Given the function $f(x) = 2x^2 - 3x + 5$, find the ordered pair $(x, f(x))$ associated with each given value for x.

(a) $x = 0$

$f(0) = 5$

so the ordered pair is $(0, 5)$.

(b) $x = -1$

$f(-1) = 2(-1)^2 - 3(-1) + 5 = 10$

The ordered pair is $(-1, 10)$.

(c) $x = \dfrac{1}{4}$

$f\left(\dfrac{1}{4}\right) = 2\left(\dfrac{1}{16}\right) - 3\left(\dfrac{1}{4}\right) + 5 = \dfrac{35}{8}$

The ordered pair is $\left(\dfrac{1}{4}, \dfrac{35}{8}\right)$.

Check Yourself 3

Given $f(x) = 2x^3 - x^2 + 3x - 2$, find the ordered pair associated with each given value of x.

(a) $x = 0$ **(b)** $x = 3$ **(c)** $x = -\dfrac{1}{2}$

Any linear equation $Ax + By = C$ in which $B \neq 0$ can be written using function notation. Because the process for finding a value for y given a particular value for x is the same as finding $f(x)$ given some value for x, we know that both y and $f(x)$ always yield the same number given some value for x.

Consider the equation $y = x + 1$. If $x = 1$, then we determine y by substituting 1 for x in the equation:

$y = (1) + 1 = 2$

We then express this using ordered-pair notation $(x, y) = (1, 2)$ as a solution to the equation $y = x + 1$.

If we write the function $f(x) = x + 1$ and evaluate $f(1)$, we get

$f(1) = (1) + 1 = 2$

So, $(x, f(x)) = (1, 2)$ is an ordered pair associated with the function.

This leads us to the conclusion that $y = f(x)$ represents the relationship between function notation and equations in two variables.

To write an equation in two variables, x and y, as a function, follow these two steps.

1. Solve the equation for y, if possible.

2. Replace y with $f(x)$.

If we are unable to complete step 1, then we cannot write y as a function of x. With linear equations, $Ax + By = C$, this would happen if $B = 0$. That is, the equation for a vertical line $x = a$ does not represent a function.

On the other hand, *all* nonvertical lines may be written as functions.

Example 4   Writing Equations as Functions

< Objective 3 >

Rewrite each linear equation as a function of x.

(a) $y = 3x - 4$

Because the equation is already solved for y, we simply replace y with $f(x)$.

$f(x) = 3x - 4$

(b) $2x - 3y = 6$

This equation must first be solved for y.

$-3y = -2x + 6$

$y = \dfrac{2}{3}x - 2$

We then rewrite the equation as

$f(x) = \dfrac{2}{3}x - 2$

 Check Yourself 4

Rewrite each equation as a function of x.

(a) $y = -2x + 5$ **(b)** $3x + 5y = 15$

The process of finding the graph of a linear function is identical to the process of finding the graph of a linear equation.

Example 5	Graphing a Linear Function

< Objective 4 >

Graph the function

$f(x) = 3x - 5$

We could use the slope and y-intercept to graph the line, or we can find three points (the third is a checkpoint) and draw the line through them. We will do the latter.

$f(0) = -5 \qquad f(1) = -2 \qquad f(2) = 1$

We use the three points $(0, -5)$, $(1, -2)$, and $(2, 1)$ to graph the line.

 Check Yourself 5

Graph the function

$f(x) = 5x - 3$

One benefit of having a function written in $f(x)$ form is that it makes it fairly easy to substitute values for x. In Example 5, we substituted the values 0, 1, and 2. Sometimes it is useful to substitute nonnumeric values for x.

Example 6	Substituting Nonnumeric Values for *x*

Let $f(x) = 2x + 3$. Evaluate f as indicated.

(a) $f(a)$

Substituting a for x in our equation, we see that

$$f(a) = 2a + 3$$

(b) $f(2 + h)$

Substituting $2 + h$ for x in our equation, we get

$$f(2 + h) = 2(2 + h) + 3$$

Distributing the 2 and then simplifying, we have

$$f(2 + h) = 4 + 2h + 3$$
$$= 2h + 7$$

Check Yourself 6

Let $f(x) = 4x - 2$. Evaluate f as indicated.

(a) $f(b)$ **(b)** $f(4 + h)$

Functions and function notation are used in many applications, as illustrated by Example 7.

Example 7	Applications of Functions

The profit function for stereos sold by S-Bar Electronics is given by

$$P(x) = 34x - 3,500$$

in which x is the number of stereos sold and $P(x)$ is the profit produced by selling x stereos.

NOTE

f is one name choice for a function. In applications, functions are often named in some way that relates to their usage.

(a) Determine the profit if S-Bar Electronics sells 75 stereos.

$$P(75) = 34(75) - 3,500$$
$$= 2,550 - 3,500$$
$$= -950$$

Because the profit is negative, we determine that S-Bar Electronics suffers a *loss* of $950 if it sells only 75 stereos.

(b) Determine the profit from the sale of 150 stereos.

$$P(150) = 34(150) - 3,500$$
$$= 5,100 - 3,500$$
$$= 1,600$$

S-Bar Electronics earns a profit of $1,600 when it sells 150 stereos.

(c) Find the break-even point (that is, the number of stereos that must be sold for the profit to equal zero).

We set $P(x) = 0$ and solve for x. That is, we are trying to find x so that the profit is zero.

$$0 = 34x - 3,500$$
$$3,500 = 34x$$
$$x = \frac{3,500}{34} \approx 103$$

The break-even point for the profit equation is 103 stereos.

NOTE

If they sell 103 stereos, they earn a profit of $2; however, if they sell 102 stereos, they lose $32. Since the number of stereos sold must be a whole number, we set the break-even point to 103.

Check Yourself 7

The manager at Sweet Eats determines that the profit produced by serving x dinners is given by the function

$$P(x) = 42x - 620$$

(a) Determine the profit if the restaurant serves 20 dinners.
(b) How many dinners must be served to reach the break-even point?

In the following example, we see an application from the field of electronics.

 Example 8   **Vocational Application**

A temperature sensor outputs a voltage for a given temperature in degrees Celsius. The output voltage is linearly related to the temperature. The function's rule for this sensor is

$$f(x) = 0.28x + 2.2$$

(a) Find $f(0)$ and interpret this result.

$$f(0) = 0.28(0) + 2.2 = 2.2$$

This says that when the temperature is 0°C, the voltage is 2.2 volts.

(b) Find $f(22)$ and interpret this result.

$$f(22) = 0.28(22) + 2.2 = 8.36$$

This says that when the temperature is 22°C, the voltage is 8.36 volts.

Check Yourself 8

The horsepower of an engine is a function of the torque (in foot-pounds) multiplied by the rpms x, multiplied by 2π, and then divided by 33,000. Given that the torque of an engine is 324 foot-pounds, the rule for the horsepower function is

$$h(x) = \frac{(2\pi)(324)x}{33,000}$$

Find and interpret $h(3,000)$.

Check Yourself ANSWERS

1. **(a)** 1; **(b)** 3; **(c)** -33 2. **(a)** -2; **(b)** 52; **(c)** -4 3. **(a)** $(0, -2)$;
(b) $(3, 52)$; **(c)** $\left(-\dfrac{1}{2}, -4\right)$ 4. **(a)** $f(x) = -2x + 5$; **(b)** $f(x) = -\dfrac{3}{5}x + 3$

5.

6. **(a)** $4b - 2$; **(b)** $4h + 14$ 7. **(a)** \$220; **(b)** 15   8. $h(3,000) \approx 185.07$;
At 3,000 rpm the horsepower is 185.

Reading Your Text

The following fill-in-the-blank exercises are designed to ensure that you understand some of the key vocabulary used in this section.

SECTION 7.5

(a) When we substitute values for the variable(s) in an expression, we _____ the expression.

(b) Rewriting an expression as some simpler equivalent expression is called _____ the expression.

(c) Function notation allows us to note the input value of x along with the _____ value of the function.

(d) In a "profit" application, the _____ point is the minimum number of items x that must be sold for the profit to be nonnegative.

Basic Skills | Advanced Skills | Vocational-Technical Applications | Calculator/Computer | Above and Beyond | Getting Ready

7.5 exercises

< Objectives 1–2 >

In exercises 1–10, evaluate each function for the value specified.

1. $f(x) = x^2 - x - 2$; find (a) $f(0)$, (b) $f(-2)$, and (c) $f(1)$.

2. $f(x) = x^2 - 7x + 10$; find (a) $f(0)$, (b) $f(5)$, and (c) $f(-2)$.

3. $f(x) = 3x^2 + x - 1$; find (a) $f(-2)$, (b) $f(0)$, and (c) $f(1)$.

4. $f(x) = -x^2 - x - 2$; find (a) $f(-1)$, (b) $f(0)$, and (c) $f(2)$.

5. $f(x) = x^3 - 2x^2 + 5x - 2$; find (a) $f(-3)$, (b) $f(0)$, and (c) $f(1)$.

6. $f(x) = -2x^3 + 5x^2 - x - 1$; find (a) $f(-1)$, (b) $f(0)$, and (c) $f(2)$.

7. $f(x) = -3x^3 + 2x^2 - 5x + 3$; find (a) $f(-2)$, (b) $f(0)$, and (c) $f(3)$.

 > Videos

8. $f(x) = -x^3 + 5x^2 - 7x - 8$; find (a) $f(-3)$, (b) $f(0)$, and (c) $f(2)$.

< Objective 3 >

In exercises 9–16, rewrite each equation as a function of x.

9. $y = -3x + 2$ **10.** $y = 5x + 7$

11. $3x + 2y = 6$ **12.** $4x + 3y = 12$

MathZone

Boost your grade at mathzone.com!
> Practice Problems
> NetTutor
> Self-Tests
> e-Professors
> Videos

Name _____

Section _____ Date _____

Answers

1. _____

2. _____

3. _____

4. _____

5. _____

6. _____

7. _____

8. _____

9. _____

10. _____

11. _____

12. _____

Answers

13. _____

14. _____

15. _____

16. _____

17. _____

18. _____

19. _____

20. _____

21. _____

22. _____

23. _____

13. $-2x + 6y = 9$

14. $-3x + 4y = 11$

15. $-5x - 8y = -9$ > Videos

16. $4x - 7y = -10$

< Objective 4 >

In exercises 17–20, graph the functions.

17. $f(x) = 3x + 7$

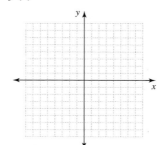

18. $f(x) = -2x - 5$

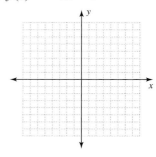

19. $f(x) = -2x + 7$ > Videos

20. $f(x) = -3x + 8$

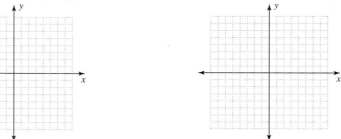

21. **BUSINESS AND FINANCE** The inventor of a new product believes that the cost of producing the product is given by the function

$$C(x) = 1.75x + 7,000$$

How much does it cost to produce 2,000 units of her invention?

22. **BUSINESS AND FINANCE** The inventor of a new product believes that the cost of producing the product is given by the function

$$C(x) = 3.25x + 9,000$$

How much does it cost to produce 5,000 units of his invention?

23. **BUSINESS AND FINANCE** If the inventor in exercise 21 charges $4 per unit, then her profit for producing and selling x units is given by the function

$$P(x) = 2.25x - 7,000$$

(a) What is her profit if she sells 2,000 units?

(b) What is her profit if she sells 5,000 units?

(c) What is the break-even point for sales?

24. **BUSINESS AND FINANCE** If the inventor in exercise 22 charges $8 per unit, then his profit for producing and selling x units is given by the function

$P(x) = 4.75x - 9,000$

(a) What is his profit if he sells 6,000 units?

(b) What is his profit if he sells 9,000 units?

(c) What is the break-even point for sales?

25. **SCIENCE AND MEDICINE** The height of a ball thrown in the air is given by the function

$h(t) = -16t^2 + 80t$

where t is the number of seconds that elapse after the ball is thrown.

(a) Find $h(2)$.

(b) Find $h(2.5)$.

(c) Find $h(5)$. What has happened here?

26. **SCIENCE AND MEDICINE** The height of a ball dropped from a roof 144 ft above the ground is given by the function

$h(t) = -16t^2 + 144$

where t is the number of seconds that elapse after the ball is dropped.

(a) Find $h(2)$.

(b) Find $h(2.5)$.

(c) Find $h(3)$. What has happened here?

Determine whether each of the following statements is **true** *or* **false.**

27. The equation for a vertical line could represent a function.

28. The equation for a horizontal line could represent a function.

Complete each of the following statements with **never, sometimes,** *or* **always.**

29. It is _____ possible to use 0 as an input value for a function.

30. The break-even point for a product _____ occurs at the minimum number of units necessary for the profit function to be nonnegative.

| Basic Skills | **Advanced Skills** | Vocational-Technical Applications | Calculator/Computer | Above and Beyond | Getting Ready |

In exercises 31–36, if $f(x) = 5x - 1$, find the following:

31. $f(a)$

32. $f(2r)$

> Videos

33. $f(x + 1)$

34. $f(a - 2)$

Answers

24. _____

25. _____

26. _____

27. _____

28. _____

29. _____

30. _____

31. _____

32. _____

33. _____

34. _____

Answers

35. _____

36. _____

37. _____

38. _____

39. _____

40. _____

41. _____

42. _____

43. _____

44. _____

45. _____

46. _____

47. _____

35. $f(x + h)$

36. $\dfrac{f(x + h) - f(x)}{h}$

In exercises 37–40, if $g(x) = -3x + 2$, find the following:

37. $g(m)$

38. $g(5n)$

39. $g(x + 2)$

40. $g(s - 1)$

In exercises 41–44, let $f(x) = 2x + 3$.

41. Find $f(1)$.

42. Find $f(3)$.

43. Form the ordered pairs $(1, f(1))$ and $(3, f(3))$.

44. Write the equation of the line passing through the points determined by the ordered pairs in exercise 43.

Basic Skills | Advanced Skills | **Vocational-Technical Applications** | Calculator/Computer | Above and Beyond | Getting Ready

45. CONSTRUCTION TECHNOLOGY The cost of building a house is $90 per square foot plus $12,000 for the foundation.

(a) Create a function to represent the cost of building a house and its relation to the number of square feet.

(b) What is the cost to build a house that has 1,800 square feet?

46. AUTOMOTIVE TECHNOLOGY A car engine burns 3.4 ounces of fuel per minute plus an additional 2 ounces during start-up.

(a) Write a function for the fuel consumption of a car as a function of the running time t, in minutes.

(b) How much fuel does a car consume if it runs for 20 minutes?

47. MECHANICAL ENGINEERING The time it takes a welder to fill an order is a function of 2 minutes for each piece plus 23 minutes for setup.

(a) Express this function, where p is the number of pieces.

(b) How long would it take to complete an order for 120 parts?

48. **ALLIED HEALTH** Dimercaprol (BAL) is used to treat arsenic poisoning in mammals. The recommended dose is 4 milligrams (mg) per kilogram (kg) of the animal's weight.

    (a) Write a function describing the relationship between the recommended dose, in milligrams, and the animal's weight, in kilograms.

    (b) How much BAL must be administered to a 5-kg cat?

49. **ALLIED HEALTH** Yohimbine is used to reverse the effects of xylazine in deer. The recommended dose is 0.125 milligram (mg) per kilogram (kg) of the deer's weight.

    (a) Write a function describing the relationship between the recommended dose, in milligrams, and the deer's weight, in kilograms.

    (b) How much yohimbine must be administered to a 15-kg fawn?

50. **ALLIED HEALTH** A brain tumor originally weighs 41 grams (g). Every day of chemotherapy treatment reduces the weight of the tumor by 0.83 g.

    (a) Write a function describing the relationship between the weight of the tumor, in grams, and the number of days spent in chemotherapy.

    (b) How much does the tumor weigh after 2 weeks of treatment?

51. Let $f(x) = 5x - 2$. Find (a) $f(4) - f(3)$; (b) $f(9) - f(8)$; (c) $f(12) - f(11)$. (d) How do the results of parts (a) through (c) compare to the slope of the line that is the graph of f?

52. Repeat exercise 51 with $f(x) = 7x + 1$.

53. Repeat exercise 51 with $f(x) = mx + b$.

54. Based on your work in exercises 51 through 53, write a paragraph discussing how slope relates to function notation.

Answers

Answers

48. _____

49. _____

50. _____

51. _____

52. _____

53. _____

54. _____

1. **(a)** -2; **(b)** 4; **(c)** -2 3. **(a)** 9; **(b)** -1; **(c)** 3
5. **(a)** -62; **(b)** -2; **(c)** 2 7. **(a)** 45; **(b)** 3; **(c)** -75

9. $f(x) = -3x + 2$ 11. $f(x) = -\dfrac{3}{2}x + 3$ 13. $f(x) = \dfrac{1}{3}x + \dfrac{3}{2}$

15. $f(x) = -\dfrac{5}{8}x + \dfrac{9}{8}$

17. $f(x) = 3x + 7$

19. $f(x) = -2x + 7$

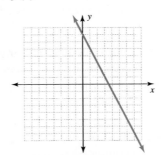

21. $10,500   **23.** **(a)** Loss: $2,500; **(b)** profit: $4,250; **(c)** 3,112 units
25. **(a)** 96; **(b)** 100; **(c)** 0; The ball hits the ground at 5 s.
27. False   **29.** sometimes   **31.** $5a - 1$ **33.** $5x + 4$
35. $5x + 5h - 1$ **37.** $-3m + 2$ **39.** $-3x - 4$ **41.** 5
43. $(1, 5), (3, 9)$ **45.** **(a)** $f(x) = 90x + 12,000$; **(b)** $174,000
47. **(a)** $f(p) = 2p + 23$; **(b)** 263 min   **49.** **(a)** $f(x) = 0.125x$; **(b)** 1.875 mg
51. **(a)** 5; **(b)** 5; **(c)** 5; **(d)** same   **53.** **(a)** m; **(b)** m; **(c)** m; **(d)** same

Definition/Procedure	Example	Reference

The Slope-Intercept Form

The slope-intercept form for the equation of a line is

$y = mx + b$

in which the line has slope m and y-intercept $(0, b)$.

For the equation $y = \dfrac{2}{3}x - 3$,

the slope m is $\dfrac{2}{3}$ and b, which

determines the y-intercept is -3.

Section 7.1

p. 663

Parallel and Perpendicular Lines

Two nonvertical lines are parallel if and only if they have the same slope, that is, when

$m_1 = m_2$

$y = 3x - 5$ and
$y = 3x + 2$ are parallel

Section 7.2

p. 682

Two lines are perpendicular if and only if their slopes are negative reciprocals, that is, when

$m_1 \cdot m_2 = -1$

$y = 5x + 2$ and $y = -\dfrac{1}{5}x - 3$

are perpendicular

p. 683

Continued

Definition/Procedure	Example	Reference

The Point-Slope Form

The equation of a line with slope m that passes through the point (x_1, y_1) is

$$y - y_1 = m(x - x_1)$$

The line with slope $\dfrac{1}{3}$ passing through (4, 3) has the equation

$$y - 3 = \frac{1}{3}(x - 4)$$

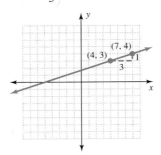

Section 7.3

p. 693

Graphing Linear Inequalities

The Graphing Steps

1. Replace the inequality symbol with an equal sign to form the equation of the boundary line of the graph.
2. Graph the boundary line. Use a dotted line if equality is not included ($<$ or $>$). Use a solid line if equality is included ($\leq$ or $\geq$).
3. Choose any convenient test point *not* on the line.
4. If the inequality is *true* at the checkpoint, shade the half-plane including the test point. If the inequality is *false* at the checkpoint, shade the half-plane that does not include the checkpoint.

To graph $x - 2y < 4$:
$x - 2y = 4$ is the boundary line. Using (0, 0) as the checkpoint, we have

$$(0) - 2(0) \overset{?}{<} 4$$
$$0 < 4 \qquad \text{(True)}$$

Shade the half-plane that includes (0, 0).

Section 7.4

p. 709

An Introduction to Functions

Given a function f, $f(c)$ designates the value of the function when the variable is equal to c.

$$f(x) = 2x^3 - x^2 + 1$$
$$f(-2) = 2(-2)^3 - (-2)^2 + 1$$
$$= 2(-8) - (4) + 1$$
$$= -19$$

Section 7.5

p. 724

This summary exercise set is provided to give you practice with each of the objectives of this chapter. Each exercise is keyed to the appropriate chapter section. When you are finished, you can check your answers to the odd-numbered exercises against those presented in the back of the text. If you have difficulty with any of these questions, go back and reread the examples from that section. Your instructor will give you guidelines on how best to use these exercises in your instructional setting.

7.1 *Find the slope and y-intercept of the line represented by each of the following equations.*

1. $y = 2x + 5$

2. $y = -4x - 3$

3. $y = -\dfrac{3}{4}x$

4. $y = \dfrac{2}{3}x + 3$

5. $2x + 3y = 6$

6. $5x - 2y = 10$

7. $y = -3$

8. $x = 2$

Write the equation of the line with the given slope and y-intercept. Then graph each line using the slope and y-intercept.

9. Slope = 2; *y*-intercept: $(0, 3)$

10. Slope $= \dfrac{3}{4}$; *y*-intercept: $(0, -2)$

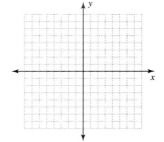

11. Slope: $-\dfrac{2}{3}$; *y*-intercept: $(0, 2)$

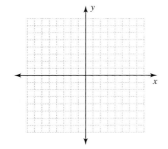

7.2 *Are the following pairs of lines* **parallel, perpendicular,** *or* **neither?**

12. L_1 through $(-3, -2)$ and $(1, 3)$
   L_2 through $(0, 3)$ and $(4, 8)$

13. L_1 through $(-4, 1)$ and $(2, -3)$
   L_2 through $(0, -3)$ and $(2, 0)$

14. L_1 with equation $x + 2y = 6$
   L_2 with equation $x + 3y = 9$

15. L_1 with equation $4x - 6y = 18$
   L_2 with equation $2x - 3y = 6$

7.3 *Write the equation of the line passing through each of the following points with the indicated slope. Give your results in slope-intercept form, where possible.*

16. $(0, -5)$, $m = \dfrac{2}{3}$

17. $(0, -3)$, $m = 0$

18. $(2, 3)$, $m = 3$

19. $(4, 3)$, m is undefined

20. $(3, -2)$, $m = \dfrac{5}{3}$

21. $(-2, -3)$, $m = 0$

22. $(-2, -4)$, $m = -\dfrac{5}{2}$

23. $(-3, 2)$, $m = -\dfrac{4}{3}$

24. $\left(\dfrac{2}{3}, -5\right)$, $m = 0$

25. $\left(-\dfrac{5}{2}, -1\right)$, m is undefined

Write the equation of the line L satisfying each of the following sets of geometric conditions.

26. L passes through $(-3, -1)$ and $(3, 3)$.

27. L passes through $(0, 4)$ and $(5, 3)$.

28. L has slope $\dfrac{3}{4}$ and y-intercept $(0, 3)$.

29. L passes through $(4, -3)$ with a slope of $-\dfrac{5}{4}$.

The Streeter/Hutchison Series in Mathematics   Beginning Algebra

30. L has y-intercept $(0, -4)$ and is parallel to the line with equation $3x - y = 6$.

31. L passes through $(3, -2)$ and is perpendicular to the line with equation $3x - 5y = 15$.

32. L passes through $(2, -1)$ and is perpendicular to the line with the equation $3x - 2y = 5$.

33. L passes through the point $(-5, -2)$ and is parallel to the line with the equation $4x - 3y = 9$.

7.4 Graph each of the following inequalities.

34. $x + y \leq 4$

35. $x - y > 5$

36. $2x + y < 6$

37. $2x - y \geq 6$

38. $x > 3$

39. $y \leq 2$

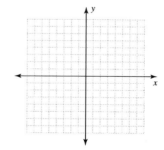

7.5 In exercises 40–45, evaluate $f(x)$ for the value specified.

40. $f(x) = x^2 - 3x + 5$; find (a) $f(0)$, (b) $f(-1)$, and (c) $f(1)$.

41. $f(x) = -2x^2 + x - 7$; find (a) $f(0)$, (b) $f(2)$, and (c) $f(-2)$.

42. $f(x) = x^3 - x^2 - 2x + 5$; find (a) $f(-1)$, (b) $f(0)$, and (c) $f(2)$.

43. $f(x) = -x^2 + 7x - 9$; find (a) $f(-3)$, (b) $f(0)$, and (c) $f(1)$.

44. $f(x) = 3x^2 - 5x + 1$; find (a) $f(-1)$, (b) $f(0)$, and (c) $f(2)$.

45. $f(x) = x^3 + 3x - 5$; find (a) $f(2)$, (b) $f(0)$, and (c) $f(1)$.

In exercises 46–49, rewrite each equation as a function of x.

46. $y = 4x + 7$

47. $y = -7x - 3$

48. $4x + 5y = 40$

49. $-3x - 2y = 12$

In exercises 50–55, graph the function.

50. $f(x) = 2x + 3$

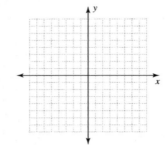

51. $f(x) = 3x - 6$

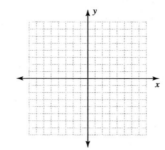

52. $f(x) = -5x + 6$

53. $f(x) = -x + 3$

54. $f(x) = -3x - 2$

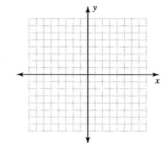

55. $f(x) = -2x + 6$

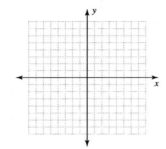

In exercises 56–61, evaluate each function as indicated.

56. $f(x) = 5x + 3$; find $f(2)$ and $f(0)$.

57. $f(x) = -3x + 5$; find $f(0)$ and $f(1)$.

58. $f(x) = 7x - 5$; find $f\left(\dfrac{5}{4}\right)$ and $f(-1)$.

59. $f(x) = -2x + 5$; find $f(0)$ and $f(-2)$.

60. $f(x) = -5x + 3$; find $f(a)$, $f(2b)$, and $f(x + 2)$.

61. $f(x) = 7x - 1$; find $f(a)$, $f(3b)$, and $f(x - 1)$.

The purpose of this self-test is to help you check your progress and to review for the next in-class exam. Allow yourself about an hour to take this test. At the end of that hour check your answers against those given in the back of the text. Section references accompany the answers. If you missed any questions, go back to those sections and reread the examples until you master the concepts.

Answers

Find the slope and y-intercept of the line represented by each of the following equations.

1. $4x - 5y = 10$

2. $y = -\dfrac{2}{3}x - 9$

Write the equation of the line with the given slope and y-intercept. Then graph each line.

3. Slope -3 and y-intercept $(0, 6)$

4. Slope 5 and y-intercept $(0, -3)$

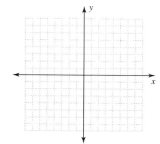

Determine whether the following pairs of lines are **parallel, perpendicular,** *or* **neither.**

5. L_1 through $(2, 5)$ and $(4, 9)$ L_2 through $(-7, 1)$ and $(-2, 11)$

6. L_1 with equation $y = 5x - 8$ L_2 with equation $5y + x = 3$

7. Find the equation of the line through $(-5, 8)$ and perpendicular to the line $4x + 2y = 8$.

Write the equation of the line L satisfying each of the following sets of geometric conditions.

8. L passes through the points $(-4, 3)$ and $(-1, 7)$.

9. L passes through $(-3, 7)$ and is perpendicular to $2x - 3y = 7$.

10. L has y-intercept $(0, 5)$ and is parallel to $4x + 6y = 12$.

11. L has y-intercept $(0, -8)$ and is parallel to the x-axis.

1. _____

2. _____

3. _____

4. _____

5. _____

6. _____

7. _____

8. _____

9. _____

10. _____

11. _____

Answers

Graph each of the following inequalities.

12. _____

13. _____

14. _____

15. _____

16. _____

17. _____

18. _____

12. $x + y < 3$

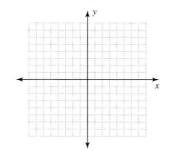

13. $3x + y \geq 9$

14. $x \leq 7$

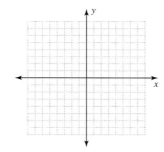

In each of the following, evaluate $f(x)$ for the value given.

15. $f(x) = x^2 - 4x - 5$; find $f(0)$ and $f(-2)$.

16. $f(x) = -x^3 + 5x - 3x^2 - 8$; find $f(-1)$ and $f(1)$.

17. $f(x) = -7x - 15$; find $f(0)$ and $f(-3)$.

18. $f(x) = 3x - 25$; find $f(a)$ and $f(x - 1)$.

Activity 7 ::
Graphing with the Internet

Each activity in this text is designed to either enhance your understanding of the topics of the preceding chapter, provide you with a mathematical extension of those topics, or both. The activities can be undertaken by one student, but they are better suited for a small group project. Occasionally, it is only through discussion that different facets of the activity become apparent. For material related to this activity, visit the text website at www.mhhe.com/streeter.

I. Find a Graphing Tutorial on the Internet

Search the Internet to find a website that allows you to create a graph from an equation. You may use one of the major search engines or go through an algebra tutorial website that you are familiar with.

II. Use the Graphing Tutorial

When you find a website that allows you to enter an equation or function to be graphed, enter the following function.

$y = 2x$ (*Hint:* You may need to use $f(x) = 2x$ instead of y.)

The "viewing window" or "range" describes the limits of the coordinate system (that is, the minimum and maximum values for the variables).

1. What range is shown?

2. Does the website allow you to change the range?

To your existing graph, add the graph of the equation $y = -3x + 5$.

3. Find the point at which the lines intersect. Does the website provide you with a way to do this or do you need to estimate the point?

4. Briefly describe how the plot distinguishes between your first and second equations.

III. Evaluate the Website

5. Describe any shortcomings to the graphing capability. What improvements might you recommend?

6. Describe other algebra tutorial content available on the website.

7. Consider some topic from your algebra course that you found difficult. Does the website provide tutorial information for this topic? How useful is the tutorial provided?

Name _____

Section _____ Date _____

The following questions are presented to help you review concepts from earlier chapters. This is meant as a review and not as a comprehensive exam. The answers are presented in the back of the text. Section references accompany the answers. If you have difficulty with any of these questions, be certain to at least read through the summary related to those sections.

Answers

1. _____

2. _____

3. _____

4. _____

5. _____

6. _____

7. _____

8. _____

9. _____

10. _____

11. _____

12. _____

13. _____

14. _____

15. _____

16. _____

Perform the indicated operation.

1. $3x^2y^2 - 5xy - 2x^2y^2 + 2xy$

2. $\dfrac{36m^5n^2}{27m^2n}$

3. $(x^2 - 3x + 5) - (x^2 - 2x - 4)$

4. $(5z^2 - 3z) - (2z^2 - 5)$

Multiply.

5. $(2x - 3)(x + 7)$

6. $(2a - 2b)(a + 4b)$

Divide.

7. $(x^2 + 3x + 2) \div (x - 3)$

8. $(x^4 - 2x) \div (x + 2)$

Solve each equation and check your results.

9. $5x - 2 = 2x - 6$

10. $3(x - 2) = 2(3x + 1) - 2$

Factor each polynomial completely.

11. $x^2 - x - 56$

12. $4x^3y - 2x^2y^2 + 8x^4y$

13. $8a^3 - 18ab^2$

14. $15x^2 - 21xy + 6y^2$

Find the slope of the line through the following pairs of points.

15. $(2, -4)$ and $(-3, -9)$

16. $(-1, 7)$ and $(3, -2)$

Perform the indicated operations.

17. $\dfrac{x^2 + 7x + 10}{x^2 + 5x} \cdot \dfrac{2x^2 - 7x + 6}{x^2 - 4}$

18. $\dfrac{2a^2 + 11a - 21}{a^2 - 49} \div (2a - 3)$

19. $\dfrac{5}{2m} + \dfrac{3}{m^2}$

20. $\dfrac{4}{x - 3} - \dfrac{2}{x}$

21. $\dfrac{3y}{y^2 + 5y + 4} + \dfrac{2y}{y^2 - 1}$

Solve the following equations for x.

22. $\dfrac{13}{4x} + \dfrac{3}{x^2} = \dfrac{5}{2x}$

23. $\dfrac{6}{x + 5} + 1 = \dfrac{3}{x - 5}$

Solve the following applications.

24. If the reciprocal of 4 times a number is subtracted from the reciprocal of the number, the result is $\dfrac{1}{12}$. What is the number?

25. Kyoko drove 280 mi to attend a business conference. In returning from the conference along a different route, the trip was only 240 mi, but traffic slowed her speed by 7 mi/h. If her driving time was the same both ways, what was her speed each way?

26. A laser printer can print 400 form letters in 30 min. At that rate, how long will it take the printer to complete a job requiring 1,680 letters?

Answers

17. _____

18. _____

19. _____

20. _____

21. _____

22. _____

23. _____

24. _____

25. _____

26. _____

Answers

27. _____

28. _____

29. _____

30. _____

27. Write the equation of the line perpendicular to the line $7x - y = 15$ with y-intercept of $(0, 2)$.

28. Write the equation of the line with slope of -5 and y-intercept $(0, 3)$.

29. Graph the inequality $4x - 2y \geq 8$.

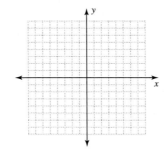

30. If $f(x) = x^2 + 3x$, find $f(-1)$.

INTRODUCTION

Most of the electricity in the United States is generated by the burning of fossil fuels (coal, oil, and gas); by nuclear fission; or by water-powered turbines in hydroelectric dams. About 65% of the electric power we use comes from burning fossil fuels. Because of this dependence on a nonrenewable resource and concern over pollution caused by burning fossil fuels, there has been some urgency in developing ways to utilize other power sources. Some of the most promising projects have been in solar- and wind-generated energy.

Alternative sources of energy are expensive compared to the traditional methods of generating electricity described above. As the price per kilowatt-hour (kWh) of electric power has increased (costs to residential users have increased by about $0.0028 per kWh per year since 1970), alternative energy sources look more promising. Additionally, the cost of manufacturing and installing banks of wind turbines in windy locations has declined.

When will the cost of generating electricity using wind power be equal to or less than the cost of using our traditional energy mix? Economists use systems of equations to make projections and then advise about the feasibility of investing in wind power plants for large cities.

CHAPTER

8

Systems of Linear Equations

CHAPTER 8 OUTLINE

Name _____

Section _____ Date _____

This pretest provides a preview of the types of exercises you will encounter in each section of this chapter. The answers for these exercises can be found in the back of the text. If you are working on your own, or ahead of the class, this pretest can help you find the sections in which you should focus more of your time.

Answers

1. _____

2. _____

3. _____

4. _____

5. _____

6. _____

7. _____

8. _____

9. _____

10. _____

11. _____

12. _____

8.1 Solve each of the following systems by graphing.

1. $-4x + y = 4$
$\quad 4x - y = 4$

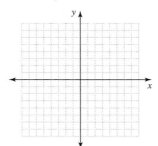

2. $2x + y = -2$
$\quad x - 3y = 6$

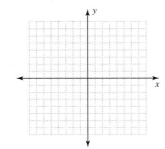

8.2–8.3 Solve each of the following systems.

3. $x - y = 9$
$\quad x + y = 3$

4. $3x - 4y = 2$
$\quad -6x + 8y = -4$

5. $x + y = 5$
$\quad x = y + 3$

6. $x - 2y = 5$
$\quad 3x + y = 8$

Solve the following problems. Be sure to show the equations used.

7. The sum of two numbers is 40, and their difference is 8. Find the two numbers.

8. A rope 30 m long is cut into two pieces so that one piece is 4 m longer than the other. How long is each piece?

9. Nila has 45 coins with a value of $8.40. If the coins are all dimes and quarters, how many of each coin does she have?

10. Jackson was able to travel 36 mi downstream in 4 h. In returning upstream, it took 6 h to make the trip. How fast can his canoe travel in still water? What was the rate of the river current?

8.4 Solve each of the following linear inequalities graphically.

11. $x + 2y < 4$
$\quad x - y < 5$

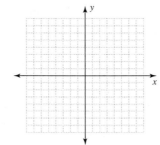

12. $2x - 6y \leq 12$
$\quad y \leq 3$

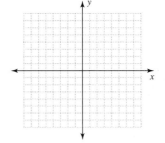

8.1

Systems of Linear Equations: Solving by Graphing

< 8.1 Objectives >

1 > Solve a consistent system of linear equations by graphing

2 > Solve an inconsistent system of linear equations by graphing

3 > Solve a dependent system of linear equations by graphing

From our work in Section 6.1, we know that an equation of the form $x + y = 3$ is a linear equation. Remember that its graph is a straight line. When we consider two equations together they form a **system of linear equations.** An example of such a system is

$$x + y = 3$$

$$3x - y = 5$$

A solution for a linear equation in two variables is any ordered pair that satisfies the equation. Often there is just one ordered pair that satisfies both equations of a system. It is called the **solution for the system.** For instance, there is one solution for the system above, (2, 1), because replacing x with 2 and y with 1, gives

$x + y = 3$	$3x - y = 5$
$(2) + (1) \overset{?}{=} 3$	$3 \cdot (2) - (1) \overset{?}{=} 5$
$3 = 3$	$6 - 1 \overset{?}{=} 5$
	$5 = 5$

NOTE

There is no other ordered pair that satisfies both equations.

Because both statements are true, the ordered pair (2, 1) satisfies both equations.

One approach to finding the solution for a system of linear equations is the **graphical method.** To use this, we graph the two lines on the same coordinate system. The coordinates of the point where the lines intersect is the solution for the system.

Example 1 | Solving by Graphing

< Objective 1 >

NOTE

Check that (5, 1) is the solution by substituting 5 for x and 1 for y in both equations.

Solve the system by graphing.

$$x + y = 6$$

$$x - y = 4$$

First, we determine solutions for the equations of our system. For $x + y = 6$, two solutions are (6, 0) and (0, 6). For $x - y = 4$, two solutions are (4, 0) and (0, −4). Using these intercepts, we graph the two equations. The lines intersect at the point (5, 1).

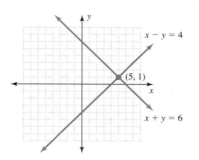

(5, 1) is the solution of the system.
It is the only point that lies on both lines.

Check Yourself 1

Solve the system by graphing.

$2x - y = 4$

$x + y = 5$

Example 2 shows how to graph a system when one of the equations represents a horizontal line.

Example 2	Solving by Graphing

Solve the system by graphing.

$3x + 2y = 6$

$y = 6$

For $3x + 2y = 6$, two solutions are (2, 0) and (0, 3). These represent the x- and y-intercepts of the graph of the equation. The equation $y = 6$ represents a horizontal line that crosses the y-axis at the point (0, 6). Using these intercepts, we graph the two equations. The lines will intersect at the point $(-2, 6)$. So this is the solution to our system.

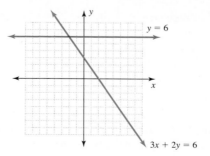

Check Yourself 2

Solve the system by graphing.

$$4x + 5y = 20$$
$$y = 8$$

The systems in Examples 1 and 2 both had exactly one solution. A system with one solution is called a **consistent system.** It is possible for a system of equations to have no solution. Such a system is called an **inconsistent system.** We present such a system here.

Example 3	**Solving an Inconsistent System**

< Objective 2 >

NOTE

In slope-intercept form, our equations are

$y = -2x + 2$

and

$y = -2x + 4$

Both lines have slope -2.

Solve the system by graphing.

$$2x + y = 2$$
$$2x + y = 4$$

We can graph the two lines as before. For $2x + y = 2$, two solutions are $(0, 2)$ and $(1, 0)$. For $2x + y = 4$, two solutions are $(0, 4)$ and $(2, 0)$. Using these intercepts, we graph the two equations.

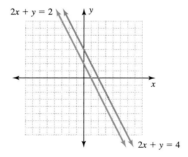

NOTE

When solving a system, we are searching for all ordered pairs that make all of the statements true. If we determine that there are no such pairs, we have successfully solved the system.

Notice that the slope for each of these lines is -2, but they have different y-intercepts. This means that the lines are parallel (they will never intersect). Because the lines have no points in common, there is no ordered pair that will satisfy both equations. The system has no solution. It is *inconsistent.*

Check Yourself 3

Solve the system by graphing.

$$x - 3y = 3$$
$$x - 3y = 6$$

There is one more possibility for linear systems, as Example 4 illustrates.

▶ **Example** 4	**Solving a Dependent System**

< Objective 3 >

NOTE

Multiplying the first equation by 2 results in the second equation.

Solve the system by graphing.

$$x - 2y = 4$$
$$2x - 4y = 8$$

Graphing as before and using the intercept method, we find

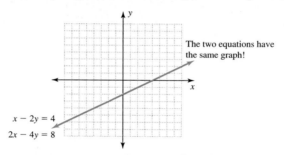

The two equations have the same graph!

$x - 2y = 4$
$2x - 4y = 8$

Because the graphs coincide, there are *infinitely many* solutions for this system. Every point on the graph of $x - 2y = 4$ is also on the graph of $2x - 4y = 8$, so any ordered pair satisfying $x - 2y = 4$ also satisfies $2x - 4y = 8$. This is called a *dependent* system, and any point on the line represents a solution.

Check Yourself 4

Solve the system by graphing.

$$x + y = 4$$
$$2x + 2y = 8$$

The following summarizes our work so far in this section.

Step by Step

To Solve a System of Equations by Graphing

Step 1   Graph both equations on the same coordinate system.

Step 2   Determine the solution to the system as follows.

a.   If the lines intersect at one point, the solution is the ordered pair corresponding to that point. This is called a **consistent system.**

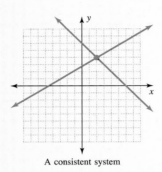

A consistent system

b.   If the lines are parallel, there are no solutions. This is called an **inconsistent system.**

> **NOTE**
>
> There is no ordered pair that lies on both lines.

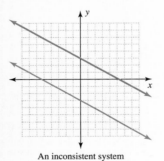

An inconsistent system

c.   If the two equations have the same graph, then the system has infinitely many solutions. This is called a **dependent system.**

> **NOTE**
>
> Any ordered pair that corresponds to a point on the line is a solution.

A dependent system

Step 3   Check the solution in both equations, if appropriate.

In the following example, we use the a graphical approach to solve an application from the field of medicine.

| Example 5 | Vocational Application |

A medical lab technician needs to determine how much 15% hydrochloric acid (HCl) solution (x) must be mixed with 5% HCl (y) to produce 50 milliliters of a 9% solution. To solve this problem, graph $x + y = 50$ and $15x + 5y = 450$ on the same set of axes, and determine the intersection point of the two lines.

The graphs of the two equations are shown here.

The intersection point appears to be (20, 30). Substituting into each equation verifies this. So, 20 mL of 15% HCl should be mixed with 30 mL of 5% HCl to obtain 50 mL of 9% HCl.

Check Yourself 5

A medical lab technician needs to determine how much 6 molar (M) copper sulfate ($CuSO_4$) solution (x) must be mixed with 2 M ($CuSO_4$) (y) to produce 200 milliliters of a 3 M solution. To solve this problem, graph $x + y = 200$ and $6x + 2y = 600$ on the same set of axes, and determine the intersection point of the two lines.

Check Yourself ANSWERS

1.

2.

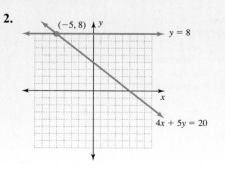

3. There is no solution. The lines are parallel, so the system is inconsistent.

$x - 3y = 3$

$x - 3y = 6$

4. There are infinitely many solutions.

$x + y = 4$
$2x + 2y = 8$

A dependent system

5.

$x + y = 200$

$6x + 2y = 600$

Solution: (50, 150); 50 mL of 6 molar copper sulfate should be mixed with 150 mL of 2 molar copper sulfate.

Reading Your Text

The following fill-in-the-blank exercises are designed to ensure that you understand some of the key vocabulary used in this section.

SECTION 8.1

(a) When we consider two linear equations together, they form a _____ of linear equations.

(b) An ordered pair that satisfies both equations of a system is called a _____ for the system.

(c) A system with exactly one solution is called a _____ system.

(d) If there are infinitely many solutions for a system, the system is called _____.

MathZone

Boost your grade at mathzone.com!

> Practice Problems
> NetTutor
> Self-Tests
> e-Professors
> Videos

Name _____

Section _____ Date _____

Answers

1. _____

2. _____

3. _____

4. _____

5. _____

6. _____

< Objectives 1–3 >

Solve each of the following systems by graphing.

1. $2x + 2y = 12$
  $x - y = 4$

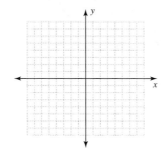

2. $x - y = 8$
  $x + y = 2$

3. $-x + y = 3$
  $x + y = 5$

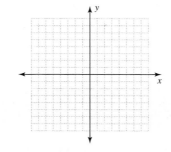

4. $x + y = 7$
  $-x + y = 3$

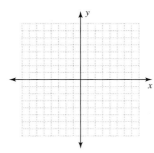

5. $x + 2y = 4$
  $x - y = 1$

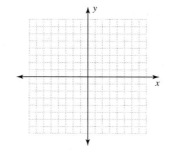

> Videos

6. $3x + y = 6$
  $x + y = 4$

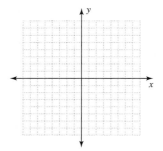

7. $2x + y = 8$
$2x - y = 0$

8. $x - 2y = -2$
$x + 2y = 6$

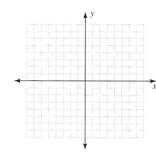

7. _____

8. _____

9. _____

10. _____

11. _____

12. _____

13. _____

14. _____

9. $x + 3y = 12$
$2x - 3y = 6$

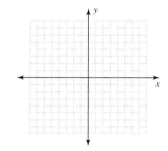

10. $2x - y = 4$
$2x - y = 6$

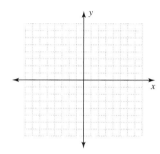

11. $3x + 2y = 12$
$y = 3$

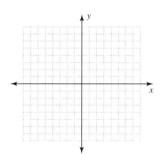

> Videos

12. $x - 2y = 8$
$3x - 2y = 12$

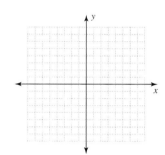

13. $x - y = 4$
$2x - 2y = 8$

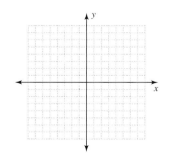

> Videos

14. $2x - y = 8$
$x = 2$

Answers

15. _____

16. _____

17. _____

18. _____

19. _____

20. _____

21. _____

22. _____

15. $x - 4y = -4$
$\quad\ \ x + 2y = \ \ 8$

16. $\quad x - 6y = 6$
$\quad -x + \ \ y = 4$

17. $3x - 2y = 6$
$\quad 2x - \ \ y = 5$

18. $4x + 3y = 12$
$\quad\ \ x + \ \ y = \ \ 2$

19. $3x - y = 3$
$\quad 3x - y = 6$

  Videos

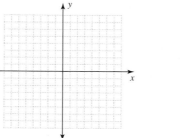

20. $3x - 6y = 9$
$\quad\ x - 2y = 3$

21. $\quad\quad 2y = \ \ 3$
$\quad x - 2y = -3$

22. $\quad x + \ \ y = -6$
$\quad -x + 2y = \ \ \ 6$

23. $x = 4$
$y = -6$

24. $x = -3$
$y = 5$

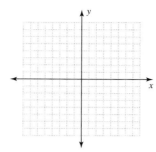

23. _____

24. _____

25. _____

26. _____

25. CONSTRUCTION The gambrel roof pictured here has several missing dimensions. These dimensions can be calculated using a system of equations.

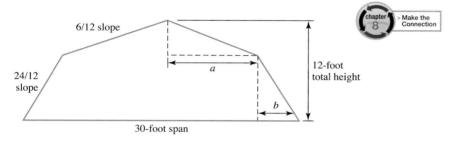

The system of equations is

$$2a + 2b = 30$$

$$\frac{24}{12}a + \frac{6}{12}b = 12$$

Solve this system of equations graphically.

26. CONSTRUCTION The beam shown in the figure is 15 feet long and has a load on each end.

To find the point where the beam balances, we use the following system of equations:

$$x + y = 15$$

$$80x = 120y$$

Solve this system of equations graphically.

Answers

27. _____

28. _____

29. _____

30. _____

31. _____

32. _____

33. _____

Determine whether each of the following statements is **true** *or* **false.**

27. A dependent system has an infinite number of solutions.

28. A linear system could have exactly two solutions.

Complete each of the following statements with **never, sometimes,** *or* **always.**

29. A linear system _____ has at least one solution.

30. If the graphs of two linear equations in a system have different slopes, the system _____ has exactly one solution.

| Basic Skills | **Advanced Skills** | Vocational-Technical Applications | Calculator/Computer | Above and Beyond | Getting Ready |

31. Find values for *m* and *b* in the following system so that the solution to the system is (1, 2).

$$mx + 3y = 8$$
$$-3x + 4y = b$$

32. Find values for *m* and *b* in the following system so that the solution to the system is (−3, 4).

$$5x + 7y = b$$
$$mx + y = 22$$

| Basic Skills | Advanced Skills | **Vocational-Technical Applications** | Calculator/Computer | Above and Beyond | Getting Ready |

 Calculator

33. **MECHANICAL ENGINEERING** At 2,100°C, a 60% aluminum oxide and 40% chromium oxide alloy separates into two different alloys. The first alloy is 78% Al_2O_3 and 22% Cr_2O_3, and the second is 50% Al_2O_3 and 50% Cr_2O_3. If the total amount of the alloy present is 7,000 grams, use a system of equations and solve by graphing to find out how many grams of each type the alloy separates into.

chapter 8 > Make the Connection

Use the system: $0.78x + 0.50y = 4,200$ and $0.22x + 0.50y = 2,800$

where *x* is the amount of the 78%/22% alloy and *y* is the amount of the 50%/50% alloy.

34. MECHANICAL ENGINEERING For a plating bath, 10,000 liters of 13% electrolyte solution is required. You have 8% and 16% solutions in stock.

Use the following system:
$$x + y = 10,000$$
$$0.08x + 0.16y = 1,300$$

chapter 8 > Make the Connection

where x is the amount of 8% solution and y is the amount of 16% solution.

Solve by graphing to determine how much of each type of solution to use.

35. MANUFACTURING A manufacturer has two machines that produce door handles. On Monday, machine A operates for 10 hours and machine B operates for 7 hours, and 290 door handles are produced. On Tuesday, machine A operates for 6 hours and machine B operates for 12 hours, and 330 door handles are produced.
    Use the following system:

$$10x + 7y = 290$$
$$6x + 12y = 330$$

chapter 8 > Make the Connection

where x is the number of handles produced by machine A in an hour, and y is the number of handles produced by machine B in an hour.

Solve the system graphically.

Basic Skills | Advanced Skills | Vocational-Technical Applications | Calculator/Computer | **Above and Beyond** | Getting Ready
▲

36. Complete the following statements in your own words:

"To solve an equation means to . . ."

"To solve a system of equations means to . . ."

37. A system of equations such as the one that follows is sometimes called a "2-by-2 system of linear equations."

$$3x + 4y = 1$$
$$x - 2y = 6$$

Explain this term.

Basic Skills | Advanced Skills | Vocational-Technical Applications | Calculator/Computer | Above and Beyond | **Getting Ready**
▲

Simplify each of the following expressions.

(a) $(2x + y) + (x - y)$ **(b)** $(x + y) + (-x + y)$
(c) $(3x + 2y) + (-3x - 3y)$ **(d)** $(x - 5y) + (2x + 5y)$
(e) $2(x + y) + (3x - 2y)$ **(f)** $2(2x - y) + (-4x - 3y)$
(g) $3(2x + y) + 2(-3x + y)$ **(h)** $3(2x - 4y) + 4(x + 3y)$

Answers

34. _____

35. _____

36. _____

37. _____

(a) _____

(b) _____

(c) _____

(d) _____

(e) _____

(f) _____

(g) _____

(h) _____

Answers

1. $\left.\begin{array}{r} 2x + 2y = 12 \\ x - y = 4 \end{array}\right\}$ $(5, 1)$

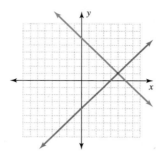

3. $\left.\begin{array}{r} -x + y = 3 \\ x + y = 5 \end{array}\right\}$ $(1, 4)$

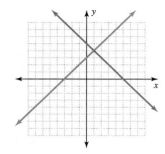

5. $\left.\begin{array}{r} x + 2y = 4 \\ x - y = 1 \end{array}\right\}$ $(2, 1)$

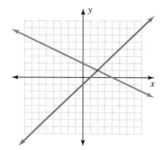

7. $\left.\begin{array}{r} 2x + y = 8 \\ 2x - y = 0 \end{array}\right\}$ $(2, 4)$

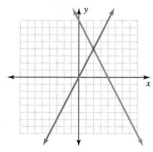

9. $\left.\begin{array}{r} x + 3y = 12 \\ 2x - 3y = 6 \end{array}\right\}$ $(6, 2)$

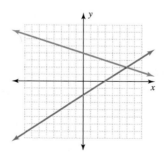

11. $\left.\begin{array}{r} 3x + 2y = 12 \\ y = 3 \end{array}\right\}$ $(2, 3)$

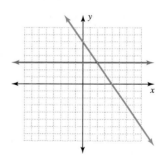

13. $\left.\begin{array}{r} x - y = 4 \\ 2x - 2y = 8 \end{array}\right\}$ Dependent

Infinitely many solutions

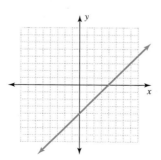

15. $\left.\begin{array}{r} x - 4y = -4 \\ x + 2y = 8 \end{array}\right\}$ $(4, 2)$

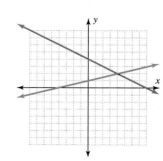

17. $3x - 2y = 6 \atop 2x - y = 5\}$ $(4, 3)$

19. $3x - y = 3 \atop 3x - y = 6\}$ Inconsistent

21. $2y = 3 \atop x - 2y = -3\}$ $\left(0, \dfrac{3}{2}\right)$

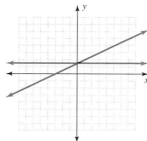

23. $x = 4 \atop y = -6\}$ $(4, -6)$

25. $a = 3, b = 12$ **27.** True   **29.** sometimes   **31.** $m = 2, b = 5$

33. $x = 2{,}500$ grams; $y = 4{,}500$ grams   **35.** A: 15 per hour; B: 20 per hour

37. Above and Beyond   **(a)** $3x$ **(b)** $2y$ **(c)** $-y$ **(d)** $3x$

(e) $5x$ **(f)** $-5y$ **(g)** $5y$ **(h)** $10x$

Systems of Linear Equations: Solving by the Addition Method

< 8.2 Objectives >

1 > Solve systems of linear equations using the addition method

2 > Solve applications of systems of linear equations

The graphical method of solving equations, shown in Section 8.1, has two definite disadvantages. First, it is time-consuming to graph each system that you want to solve. More importantly, the graphical method is not precise. For instance, look at the graph of the system

$$x - 2y = 4$$
$$3x + 2y = 6$$

which follows:

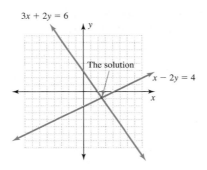

The exact solution for the system is $\left(\dfrac{5}{2}, -\dfrac{3}{4}\right)$, which is difficult to read from the graph.

Fortunately, there are algebraic methods that do not have this disadvantage and will allow you to find exact solutions for a system of equations.

One algebraic method of finding a solution is called the **addition method.**

| | **Example 1** | Solving a System by the Addition Method |

< Objective 1 >

Solve the system.

$$x + y = 8$$
$$x - y = 2$$

NOTES

This method uses the fact that if

$a = b$ and $c = d$

then

$a + c = b + d$

This is the **additive property** of equality. Note that by the additive property, if equals are added to equals, the resulting sums are equal.

This is also called **solution by elimination** for this reason.

Note that the coefficients of the y-terms are the *additive inverses* of one another (1 and -1) and that adding the two equations will "eliminate" the variable y. That addition step is shown here.

$$\begin{aligned} x + y &= 8 \\ x - y &= 2 \\ \hline 2x \quad\;\; &= 10 \\ x &= 5 \end{aligned}$$

By adding, we eliminate the variable y. The resulting equation contains *only* the variable x.

We now know that 5 is the x-coordinate of our solution. Substitute 5 for x into *either* of the original equations.

$$x + y = 8$$
$$(5) + y = 8$$
$$y = 3$$

So (5, 3) is the solution.

To check, replace x and y with these values in *both* of the original equations.

$$\begin{aligned} x + y &= 8 \\ (5) + (3) &\overset{?}{=} 8 \\ 8 &= 8 \quad \text{(True)} \end{aligned}$$

$$\begin{aligned} x - y &= 2 \\ (5) - (3) &\overset{?}{=} 2 \\ 2 &= 2 \quad \text{(True)} \end{aligned}$$

Because (5, 3) satisfies both equations, it is the solution.

Check Yourself 1

Solve the system by adding.

$$x - y = -2$$
$$x + y = \;\;\; 6$$

Example 2   **Solving a System by the Addition Method**

Solve the system.

$$-3x + 2y = \;\;\; 12$$
$$\;\;\; 3x - \;\; y = -9$$

In this case, adding will eliminate the x-terms.

NOTE

It does not matter which variable is eliminated. Choose the one that requires less work.

$$\begin{aligned} -3x + 2y &= \;\;\; 12 \\ 3x - \;\; y &= -9 \\ \hline y &= \;\;\; 3 \end{aligned}$$

Now substitute 3 for y in either equation. From the first equation

$$\begin{aligned} -3x + 2(3) &= \;\; 12 \\ -3x &= \;\;\; 6 \\ x &= -2 \end{aligned}$$

and $(-2, 3)$ is the solution.

Show that you get the same *x*-coordinate by substituting 3 for *y* in the second equation rather than in the first. Remember to check your solution in both equations.

Check Yourself 2

Solve the system by adding.

$$5x - 2y = 9$$
$$-5x + 3y = -11$$

Note that in both Examples 1 and 2 we found an equation in a single variable by adding. We could do this because the coefficients of one of the variables were opposites. This gave 0 as a coefficient for one of the variables after we added the two equations. In some systems, you will not be able to directly eliminate either variable by adding. However, an equivalent system can always be written by multiplying one or both of the equations by a nonzero constant so that the coefficients of *x* (or of *y*) are opposites. Example 3 illustrates this approach.

> **Example 3**   Solving a System by the Addition Method

NOTE

Remember that multiplying both sides of an equation by some nonzero number does not change the solutions. So even though we have "altered" the equations, they are equivalent and have the same solutions.

Solve the system.

$$2x + y = 13$$
$$3x + y = 18$$

Note that adding the equations in this form does not eliminate either variable. You would still have terms in *x* and in *y*. However, look at what happens if we multiply both sides of the second equation by -1 as the first step.

$$2x + y = 13 \quad\xrightarrow{}\quad 2x + y = 13$$
$$3x + y = 18 \quad\xrightarrow[\text{by } -1]{\text{Multiply}}\quad -3x - y = -18$$

Now we can add.

$$
\begin{array}{rcr}
2x + y &=& 13 \\
-3x - y &=& -18 \\
\hline
-x &=& -5 \\
x &=& 5
\end{array}
$$

Substitute 5 for *x* into either equation. We choose the first:

$$2(5) + y = 13$$
$$y = 3$$

(5, 3) is the solution. We leave it to the reader to check this solution.

Check Yourself 3

Solve the system by adding.

$$x - 2y = 9$$
$$x + 3y = -1$$

To summarize, multiplying both sides of one of the equations by a nonzero constant can yield an equivalent system in which the coefficients of the x-terms or the y-terms are opposites. This means that a variable can be eliminated by adding.

Example 4   Solving a System by the Addition Method

Solve the system.

$$x + 4y = 2$$
$$3x - 2y = -22$$

One approach is to multiply both sides of the second equation by 2. Do you see that the coefficients of the y-terms will then be opposites?

$$x + 4y = 2 \longrightarrow x + 4y = 2$$
$$3x - 2y = -22 \xrightarrow{\text{Multiply by 2}} 6x - 4y = -44$$

If we add the resulting equations, the variable y is eliminated and we can solve for x.

> **NOTE**
>
> Now the coefficients of the y-terms are opposites.

$$\begin{array}{r} x + 4y = 2 \\ 6x - 4y = -44 \\ \hline 7x = -42 \\ x = -6 \end{array}$$

Now substitute -6 for x in the first equation of this example to find y.

> **NOTE**
>
> We could substitute -6 for x in the second equation to find y.

$$(-6) + 4y = 2$$
$$4y = 8$$
$$y = 2$$

So $(-6, 2)$ is the solution.

Again you should check this result. As is often the case, there are several ways to solve the system. For example, what if we multiply both sides of our original equation by -3? The coefficients of the x-terms will then be opposites and adding will eliminate the variable x so that we can solve for y. Try that for yourself in the Check Yourself 4 exercise.

Check Yourself 4

Solve the system by eliminating x.

$$x + 4y = 2$$
$$3x - 2y = -22$$

It may be necessary to multiply each equation separately so that one of the variables will be eliminated when the equations are added. Example 5 illustrates this approach.

Example 5   **Solving a System by the Addition Method**

NOTE

Again, the addition method is often called the elimination method.

Solve the system.

$$4x + 3y = 11$$
$$3x - 2y = 4$$

Do you see that, if we want to have integer coefficients, multiplying in one equation does not help in this case? We have to multiply in both equations.

To eliminate x, we can multiply both sides of the first equation by 3 and both sides of the second equation by -4. The coefficients of the x-terms will then be opposites.

NOTE

The minus sign is used with the 4 so that the coefficients of the x-term are opposites.

$$4x + 3y = 11 \xrightarrow[\text{by 3}]{\text{Multiply}} 12x + 9y = 33$$
$$3x - 2y = 4 \xrightarrow[\text{by }-4]{\text{Multiply}} -12x + 8y = -16$$

Adding the resulting equations gives

$$17y = 17$$
$$y = 1$$

Now substituting 1 for y in the first equation, we have

NOTE

Check (2, 1) in both equations of the original system.

$$4x + 3(1) = 11$$
$$4x = 8$$
$$x = 2$$

and (2, 1) is the solution.

          Check Yourself 5

Solve the system by eliminating *y*.

$$4x + 3y = 11$$
$$3x - 2y = 4$$

The following summarizes the solution steps we have illustrated.

Step by Step

To Solve a System of Linear Equations by Adding

Step 1 If necessary, multiply both sides of one or both equations by nonzero numbers to form an equivalent system in which the coefficients of one of the variables are opposites.

Step 2 Add the equations of the new system.

Step 3 Solve the resulting equation for the remaining variable.

Step 4 Substitute the value found in step 3 into either of the original equations to find the value of the second variable.

Step 5 Check your solution in both of the original equations.

In Section 8.1 we saw that some systems had *infinitely* many solutions. Example 6 shows how this is indicated when we are using the addition method of solving equations.

Example 6   Solving a Dependent System

Solve the system.

$$x + 3y = -2$$
$$3x + 9y = -6$$

We multiply both sides of the first equation by -3.

NOTE

The lines coincide. That is the case whenever *adding eliminates both variables* and a true statement results.

$$x + 3y = -2 \xrightarrow[\text{by } -3]{\text{Multiply}} -3x - 9y = 6$$
$$3x + 9y = -6 \longrightarrow \quad 3x + 9y = -6$$
$$\overline{\qquad\qquad 0 = 0}$$

Adding, we see that both variables have been eliminated, and we have the true statement $0 = 0$.

Look at the graph of the system.

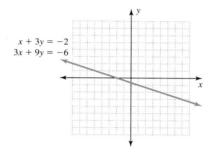

$x + 3y = -2$
$3x + 9y = -6$

As we see, the two equations have the *same* graph. This means that the system is *dependent,* and there are *infinitely many solutions.* Any (x, y) that satisfies $x + 3y = -2$ also satisfies $3x + 9y = -6$.

 Check Yourself 6

Solve the system by adding.

$$x - 2y = 3$$
$$-2x + 4y = -6$$

Earlier we encountered systems that had *no* solutions. Example 7 illustrates what happens when we try to solve such a system with the addition method.

 Example 7   Solving an Inconsistent System

Solve the system.

$$3x - y = 4$$
$$-6x + 2y = -5$$

NOTE

Be sure to multiply the 4 by 2.

We multiply both sides of the first equation by 2.

$$3x - y = 4 \xrightarrow[\text{by } 2]{\text{Multiply}} 6x - 2y = 8$$
$$-6x + 2y = -5 \longrightarrow -6x + 2y = -5$$
$$\overline{\qquad\qquad 0 = 3}$$

We now add the two equations.

Again both variables have been eliminated by addition. But this time we have the *false* statement $0 = 3$ because we tried to solve a system whose graph consists of two parallel lines. Because the two lines do not intersect, there is *no* solution for the system. It is *inconsistent*.

Check Yourself 7

Solve the system by adding.

$$5x + 15y = 20$$
$$x + 3y = 3$$

RECALL

In Chapter 2, all the unknowns in the problem had to be expressed in terms of a single variable.

In Chapter 2 we solved word problems by using equations in a single variable. Now you have the background to use two equations in two variables to solve word problems. The five steps for solving word problems stay the same (in fact, we give them again for reference in our first application example). Many students find that using two equations and two variables makes writing the necessary equation much easier, as Example 8 illustrates.

Example 8   Solving an Application with Two Linear Equations

< Objective 2 >

Ryan bought 8 pens and 7 pencils and paid a total of $14.80. Ashleigh purchased 2 pens and 10 pencils and paid $7. Find the cost for a single pen and a single pencil.

Thompson's Stationery
3784 Main St.
Thank you for
your business

8 pens
7 pencils

Total: 14.80

Step 1   You want to find the cost of a single pen and the cost of a single pencil.

Step 2   Let x be the cost of a pen and y be the cost of a pencil.

Step 3   Write the two necessary equations.

$$8x + 7y = 14.80$$
$$2x + 10y = 7.00$$

In the first equation, $8x$ is the total cost of the pens Ryan bought and $7y$ is the total cost of the pencils Ryan bought. The second equation is formed in a similar fashion.

Step 4   Solve the system formed in step 3. We multiply the second equation by -4. Then adding will eliminate the variable x.

$$8x + 7y = 14.80$$
$$-8x - 40y = -28.00$$

Now adding the equations, we have

$$-33y = -13.20$$
$$y = 0.40$$

Substituting 0.40 for y in the first equation, we have

$$8x + 7(0.40) = 14.80$$
$$8x + 2.80 = 14.80$$
$$8x = 12.00$$
$$x = 1.50$$

Step 5   From the results of step 4 we see that the pens are $1.50 each and the pencils are 40¢ each.

To check these solutions, replace x with 1.50 and y with 0.40 in the first equation.

$$8(1.50) + 7(0.40) \stackrel{?}{=} 14.80$$
$$12.00 + 2.80 \stackrel{?}{=} 14.80$$
$$14.80 = 14.80 \quad \text{(True)}$$

We leave it to you to check these values in the second equation.

Check Yourself 8

Alana bought three digital tapes and two compact disks on sale for $66. At the same sale, Chen bought three digital tapes and four compact disks for $96. Find the individual prices for a tape and a disk.

Example 9 shows how sketches can be helpful in setting up a problem.

Example 9   Using a Sketch to Help Solve an Application

An 18-ft board is cut into two pieces, one of which is 4 ft longer than the other. How long is each piece?

Step 1   You want to find the two lengths.

Step 2   Let x be the length of the longer piece and y the length of the shorter piece.

Step 3   Write the equations for the solution.

NOTE

Our second equation could also be written as

$x = y + 4$

$x + y = 18$ ⟵——— The total length is 18.
$x - y = \ \ 4$ ⟵——— The difference in
lengths is 4.

Step 4   To solve the system, add:

$$x + y = 18$$
$$\underline{x - y = \ \ 4}$$
$$2x \qquad = 22$$
$$\quad x = 11$$

Replace x with 11 in the first equation.

$$(11) + y = 18$$
$$\qquad y = \ \ 7$$

The longer piece has length 11 ft, and the shorter piece has length 7 ft.

Step 5   We leave it to you to check this result in the original problem.

Check Yourself 9

A 20-ft board is cut into two pieces, one of which is 6 ft longer than the other. How long is each piece?

Using two equations in two variables also helps in solving **mixture problems.**

 Example 10   Solving a Mixture Problem Involving Coins

Winnifred has collected \$4.50 in nickels and dimes. If she has 55 coins, how many of each kind of coin does she have?

NOTE

We choose appropriate variables—n for nickels, d for dimes.

RECALL

The value of a number of coins is the value per coin times the number of coins: $5n$, $10d$, etc.

Step 1   You want to find the number of nickels and the number of dimes.

Step 2   Let

n = number of nickels

d = number of dimes

Step 3   Write the equations for the solution.

$n + d = 55$ ⟵── There are 55 coins in all.

$5n + 10d = 450$

Value of   Value   Total value
nickels   of dimes   (in cents)

Step 4   We now have the system

$$n + d = 55$$
$$5n + 10d = 450$$

We choose to solve this system by addition. Multiply the first equation by -5. We then add the equations to eliminate the variable n.

$$-5n - 5d = -275$$
$$\underline{5n + 10d = 450}$$
$$5d = 175$$
$$d = 35$$

We now substitute 35 for d in the first equation.

$$n + (35) = 55$$
$$n = 20$$

There are 20 nickels and 35 dimes.

Step 5   We leave it to you to check this result. Just verify that the value of these coins is $4.50.

Check Yourself 10

Tickets for a play cost $8 or $6. If 350 tickets were sold in all and receipts were $2,500, how many tickets of each price were sold?

We can also solve mixture problems involving percentages with two equations in two unknowns. Look at Example 11.

 Example 11   Solving a Mixture Problem Involving Chemicals

There are two solutions in a chemistry lab: a 20% acid solution and a 60% acid solution. How many milliliters of each should be mixed to produce 200 mL of a 44% acid solution?

20%   60%   44%
x mL y mL   200 mL

Step 1 You need to know the amount of each solution to use.

Step 2 Let

x = amount of 20% acid solution

y = amount of 60% acid solution

Step 3 Note that a 20% acid solution is 20% acid and 80% water.

We can write equations from the total amount of the solution, here 200 mL, and from the amount of acid in that solution. Many students find a chart helpful in organizing the information at this point. Here, for example, we might have

	Amount of Solution	**% Acid**	**Amount of Acid**
x		0.20	$0.20x$
y		0.60	$0.60y$
Final solution	200	0.44	(0.44)(200)

Now we are ready to form our system.

$$x + y = 200$$

$$0.20x + 0.60y = 0.44(200)$$

Acid in 20%   Acid in 60%   Acid in
solution   solution   mixture

Step 4 If we multiply the second equation by 100 to clear it of decimals, we have

$$x + \quad y = \quad 200 \xrightarrow{\substack{\text{Multiply} \\ \text{by } -20}} -20x - 20y = -4,000$$

$$20x + 60y = 8,800 \xrightarrow{\hspace{2cm}} 20x + 60y = \quad 8,800$$

$$40y = \quad 4,800$$

$$y = \quad 120$$

Substituting 120 for y in the first equation, we have

$$x + (120) = 200$$

$$x = \quad 80$$

The amounts to be mixed are 80 mL (20% acid solution) and 120 mL (60% acid solution).

Step 5 You can check this solution by verifying that the amount of acid from the 20% solution added to the amount from the 60% solution is equal to the amount of acid in the mixture.

NOTE

The amount of acid is the amount of solution times the percentage of acid (as a decimal). That is the key to forming the third column of our table.

NOTE

The first equation is the total amount of the solution from the first column of our table.

NOTE

The second equation is the amount of acid from the third column of our table. The sum of the acid in the two solutions equals the acid in the mixture.

Check Yourself 11

You have a 30% alcohol solution and a 50% alcohol solution. How much of each solution should be combined to make 400 mL of a 45% alcohol solution?

A related kind of application involves interest. The key equation involves the *principal* (the amount invested), the annual *interest rate,* the *time* (in years) that the money is invested, and the amount of *interest* you receive.

$$I = P \cdot r \cdot t$$

Interest   Principal   Rate   Time

For 1 year we have

$$I = P \cdot r \qquad \text{if } t = 1$$

Example 12   Solving an Investment Application

Jeremy inherits $20,000 and invests part of the money in bonds with an interest rate of 11%. The remainder of the money is in savings at a 9% rate. What amount has he invested at each rate if he receives $2,040 in interest for 1 year?

Step 1   You want to find the amounts invested at 11% and at 9%.

Step 2   Let x = the amount invested at 11% and y = the amount invested at 9%. Once again you may find a chart helpful at this point.

	Principal	**Rate**	**Interest**
	x	11%	$0.11x$
	y	9%	$0.09y$
Totals	20,000		2,040

Step 3   Form the equations for the solution, using the first and third columns of the table.

$x + y = 20,000$ ⟵——— He has $20,000 invested in all.

$0.11x + 0.09y = 2,040$

The interest   The interest   The total
at 11%   at 9%   interest
(rate · principal)

NOTES

The amount invested at 11% could be represented by y and the amount invested at 9% by x.

The formula $I = P \cdot r$ (interest equals principal times rate) is the key to forming the third column of our table.

NOTE

The decimal forms of 11% and 9% are used in the equation.

Step 4   To solve the following system, use addition.

$$x + \quad y = 20{,}000$$
$$0.11x + 0.09y = \quad 2{,}040$$

To do this, multiply both sides of the first equation by -9. Multiplying both sides of the second equation by 100 clears decimals. Adding the resulting equations eliminates y.

$$-9x - 9y = -180{,}000$$
$$\underline{11x + 9y = \quad 204{,}000}$$
$$2x \qquad = \quad 24{,}000$$
$$x = \quad 12{,}000$$

Now, substitute 12,000 for x in the first equation and solve for y.

$$(12{,}000) + y = 20{,}000$$
$$y = \quad 8{,}000$$

Jeremy has $12,000 invested at 11% and $8,000 invested at 9%.

NOTE

Be sure to answer the question asked in the problem.

Step 5   To check, the interest at 11% is ($12,000)(0.11), or $1,320. The interest at 9% is ($8,000)(0.09), or $720. The total interest is $2,040, and the solution is verified.

 Check Yourself 12

> Jan has $2,000 more invested in a stock that pays 9% interest than in a savings account paying 8%. If her total interest for 1 year is $860, how much does she have invested at each rate?

In Chapters 2 and 5, we solved **motion problems;** they involve a distance traveled, the rate, and the time of travel. Example 13 shows the use of $d = r \cdot t$ in forming a system of equations to solve a motion problem.

Example 13	Solving a Motion Problem

RECALL

Distance, rate, and time of travel are related by the equation

$$d = r \cdot t$$

Distance Rate Time

A boat can travel 36 mi downstream in 2 h. Coming back upstream, the trip takes 3 h. Find the rate of the boat in still water and the rate of the current.

Step 1 You want to find the two rates (of the boat and the current).

Step 2 Let

x = rate of boat in still water

y = rate of current

Step 3 To write the equations, think about the following: What is the effect of the current? Suppose the boat's rate in still water is 10 mi/h and the current is 2 mi/h.

The current *increases* the rate *downstream* to 12 mi/h (10 + 2). The current *decreases* the rate *upstream* to 8 mi/h (10 − 2). So here the rate downstream is $x + y$ and the rate upstream is $x - y$. At this point a chart of information is helpful.

	Distance	Rate	Time
Downstream	36	$x + y$	2
Upstream	36	$x - y$	3

From the relationship $d = r \cdot t$ we use our table to write the system

$36 = 2(x + y)$ (From line 1 of our table)

$36 = 3(x - y)$ (From line 2 of our table)

Step 4 Removing the parentheses in the equations of step 3, we have

$2x + 2y = 36$

$3x - 3y = 36$

By either of our earlier methods, this system gives values of 15 for x and 3 for y. The rate in still water is 15 mi/h, and the rate of the current is 3 mi/h. We leave the check to you.

Check Yourself 13

A light plane flies 480 mi with the wind in 4 h. In returning against the wind, the trip takes 6 h. What is the rate of the plane in still air? What is the rate of the wind?

Check Yourself ANSWERS

1. (2, 4)   **2.** (1, −2)   **3.** (5, −2)   **4.** (−6, 2)   **5.** (2, 1)
6. There are infinitely many solutions. It is a dependent system.
7. There is no solution. The system is inconsistent.   **8.** Tape $12, disk $15
9. 7 ft, 13 ft   **10.** 150 $6 tickets, 200 $8 tickets
11. 100 mL (30%), 300 mL (50%)   **12.** $4,000 at 8%, $6,000 at 9%
13. Plane's rate in still air, 100 mi/h; wind's rate, 20 mi/h

Reading Your Text

The following fill-in-the-blank exercises are designed to ensure that you understand some of the key vocabulary used in this section.

SECTION 8.2

(a) When the coefficients of the y-terms are _____ of one another, adding the two equations eliminates the variable y.

(b) To check, replace x and y with values in *both* of the _____ equations.

(c) An equivalent system can always be obtained by _____ one or both of the equations by a nonzero constant so that the coefficients of x (or of y) are opposites.

(d) When a system is *dependent,* there are _____ solutions.

Basic Skills | Advanced Skills | Vocational-Technical Applications | Calculator/Computer | Above and Beyond | Getting Ready

< Objective 1 >

Solve each of the following systems by addition. If a unique solution does not exist, state whether the system is inconsistent or dependent.

1. $x + y = 6$
$x - y = 4$

2. $x - y = 8$
$x + y = 2$

3. $2x - y = 1$
$-2x + 3y = 5$

4. $x - 2y = 2$
$x + 2y = -14$

5. $x + 2y = -2$
$3x + 2y = -12$

6. $4x - 3y = 22$
$4x + 5y = 6$

7. $2x + y = 8$
$2x + y = 2$

8. $5x + 4y = 7$
$5x - 2y = 19$

9. $3x - 5y = 2$
$2x - 5y = -2$

10. $2x - y = 4$
$2x - y = 6$

11. $x + y = 3$
$3x - 2y = 4$
 > Videos

12. $x - y = -2$
$2x + 3y = 21$

13. $-5x + 2y = -3$
$x - 3y = -15$

14. $x + 5y = 10$
$-2x - 10y = -20$

15. $5x + 2y = 28$
$x - 4y = -23$

16. $7x + 2y = 17$
$x - 5y = 13$

17. $3x - 4y = 2$
$-6x + 8y = -4$

18. $-x + 5y = 19$
$4x + 3y = -7$

19. $3x - 2y = 12$
$5x - 3y = 21$

20. $-4x + 5y = -6$
$5x - 2y = 16$

21. $7x + 4y = 20$
$5x + 6y = 19$
 > Videos

22. $5x + 4y = 5$
$7x - 6y = 36$

23. $2x - 7y = 6$
$-4x + 3y = -12$

24. $3x + 2y = -18$
$7x - 6y = -42$

MathZone

Boost your grade at mathzone.com!
> Practice Problems
> NetTutor
> Self-Tests
> e-Professors
> Videos

Name _____

Section _____ Date _____

Answers

1. _____   2. _____

3. _____   4. _____

5. _____   6. _____

7. _____

8. _____   9. _____

10. _____

11. _____   12. _____

13. _____

14. _____

15. _____   16. _____

17. _____

18. _____   19. _____

20. _____   21. _____

22. _____   23. _____

24. _____

Answers

25. _____

26. _____

27. _____

28. _____

29. _____

30. _____

31. _____

32. _____

33. _____

34. _____

35. _____

36. _____

37. _____

38. _____

25. $5x - y = 20$
    $4x + 3y = 16$

26. $3x + y = -5$
    $5x - 4y = 20$

27. $3x + y = 1$
    $5x + y = 2$

28. $2x - y = 2$
    $2x + 5y = -1$

29. $5x - 2y = \dfrac{9}{5}$
    $3x + 4y = -1$

30. $2x + 3y = -\dfrac{1}{12}$
    $5x + 4y = \dfrac{2}{3}$

< Objective 2 >

Solve each of the following problems. Be sure to show the equations used for the solution.

31. **NUMBER PROBLEM** The sum of two numbers is 40. Their difference is 8. Find the two numbers.

32. **NUMBER PROBLEM** Eight eagle stamps and two raccoon stamps cost $2.80. Three eagle stamps and four raccoon stamps cost $2.35. Find the cost of each kind of stamp.

33. **NUMBER PROBLEM** Xavier bought five red delicious apples and four Granny Smith apples at a cost of $4.81. Dean bought one of each kind at a cost of $1.08. Find the cost for each kind of apple. > Videos

34. **NUMBER PROBLEM** Eight disks and five zip disks cost a total of $27.50. Two disks and four zip disks cost $16.50. Find the unit cost for each.

35. **CRAFTS** A 30-m rope is cut into two pieces so that one piece is 6 m longer than the other. How long is each piece?

36. **CRAFTS** An 18-ft board is cut into two pieces, one of which is twice as long as the other. How long is each piece?

37. **BUSINESS AND FINANCE** A coffee merchant has coffee beans that sell for $9 per pound and $12 per pound. The two types are to be mixed to create 100 lb of a mixture that will sell for $11.25 per pound. How much of each type of bean should be used in the mixture?

38. **BUSINESS AND FINANCE** Peanuts sell for $2 per pound and cashews sell for $5 per pound. How much of each type of nut is needed to create a 20-lb of a mixture that would sell for $2.75 per pound?

39. SCIENCE AND MEDICINE A chemist has a 25% and a 50% acid solution. How much of each solution should be used to form 200 mL of a 35% acid solution?

25% acid   50% acid

40. SCIENCE AND MEDICINE A pharmacist wishes to prepare 150 mL of a 20% alcohol solution. She has a 30% solution and a 15% solution in her stock. How much of each should be used in forming the desired mixture?

41. BUSINESS AND FINANCE Otis has a total of $12,000 invested in two accounts. One account pays 8% and the other 9%. If his interest for 1 year is $1,010, how much does he have invested at each rate?

42. BUSINESS AND FINANCE Amy invests a part of $8,000 in bonds paying 12% interest. The remainder is in a savings account at 8%. If she receives $840 in interest for 1 year, how much does she have invested at each rate?

> Videos

43. SCIENCE AND MEDICINE A light plane flies 450 mi with the wind in 3 h. Flying back against the wind, the plane takes 5 h to make the trip. What was the rate of the plane in still air? What was the rate of the wind?

44. SCIENCE AND MEDICINE An airliner made a trip of 1,800 mi in 3 h, flying east across the country with the jetstream directly behind it. The return trip, against the jetstream, took 4 h. Find the speed of the plane in still air and the speed of the jetstream.

Determine whether each of the following statements is **true** *or* **false.**

45. Multiplying both sides of an equation by a nonzero constant will change the solutions for that equation.

46. When you have found the value of one variable, you can substitute it into either of the original equations to find the value of the other variable.

Answers

39. _____

40. _____

41. _____

42. _____

43. _____

44. _____

45. _____

46. _____

Answers

47. _____

48. _____

49. _____

50. _____

51. _____

52. _____

53. _____

54. _____

Complete each of the following statements with **never, sometimes,** *or* **always.**

47. Both variables are _____ eliminated when the equations of a linear system are added.

48. It is _____ possible to use the addition method to solve a linear system.

Basic Skills | **Advanced Skills** | Vocational-Technical Applications | Calculator/Computer | Above and Beyond | Getting Ready
▲

Solve the following systems by adding. If a unique solution does not exist, state whether the system is inconsistent or dependent.

49. $\dfrac{x}{3} - \dfrac{y}{4} = -\dfrac{1}{2}$

$\dfrac{x}{2} - \dfrac{y}{5} = \dfrac{3}{10}$ ⊙ > Videos

50. $\dfrac{1}{3}x - \dfrac{1}{2}y = \dfrac{5}{6}$

$\dfrac{1}{2}x - \dfrac{2}{5}y = \dfrac{9}{10}$

51. $0.4x - 0.2y = 0.6$

$0.5x - 0.6y = 9.5$

52. $0.2x + 0.37y = 0.8$

$-0.6x + 1.4y = 2.62$

Basic Skills | Advanced Skills | **Vocational-Technical Applications** | Calculator/Computer | Above and Beyond | Getting Ready
▲

53. **CONSTRUCTION** In the vaulted split-level truss shown here, the dimensions x and y can be determined by the following system of equations:

$$\dfrac{3}{12}x + 8 = \dfrac{5}{12}y$$
$$x + 2 = y$$

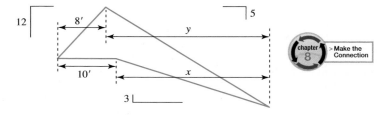

Solve this system using the addition method.

54. **MECHANICAL ENGINEERING** The forces (in pounds) on the arm of an industrial lift provide the equations

$1.5x + 2.4y = 90$ chapter 8 > Make the Connection

$2.7x - 1.8y = 30$

(Round your results to the nearest hundredth).

55. **BUSINESS AND FINANCE** Production for this week is up by 2,600 units from last week. If total production for the two weeks is 27,200 units, the production for each week can be given by

$$P_2 - 2,600 = P_1$$
$$P_1 + P_2 \quad = 27,200$$

Solve the system of equations using the addition method.

56. **MANUFACTURING** A manufacturer produces drive assemblies and relays. The drive assemblies sell for \$12 and the relays sell for \$3. On a given day, 118 items are produced with a total value of \$1,038. The production is described by the equations

$$12x + 3y = 1,038$$
$$x + y = 118$$

where x represents the number of drive assemblies and y represents the number of relays.

Solve this system using the addition method.

Basic Skills | Advanced Skills | Vocational-Technical Applications | Calculator/Computer | **Above and Beyond** | Getting Ready

Work with a partner to solve the following problem.

57. Your friend Valerie contacts you about going into business with her. She wants to start a small manufacturing business making and selling sweaters to specialty boutiques. She explains that the initial investment for each of you will be \$1,500 for a knitting machine. She has worked hard to come up with an estimate for expenses and thinks that they will be close to \$1,600 a month for overhead. She says that each sweater manufactured will cost \$28 to produce and that the sweaters will sell for at least \$70. She wants to know if you are willing to invest the money you have saved for college costs. You have faith in Valerie's ability to carry out her plan. But, you have worked hard to save this money. Use graphs and equations to help you decide whether this is a good opportunity. Think about whether you need more information from Valerie. Write a letter summarizing your thoughts.

Basic Skills | Advanced Skills | Vocational-Technical Applications | Calculator/Computer | Above and Beyond | **Getting Ready**

Solve each of the following equations.

(a) $2x + 3(x + 1) = 13$ (b) $3(y - 1) + 4y = 18$
(c) $x + 2(3x - 5) = 25$ (d) $3x - 2(x - 7) = 12$

Answers

55. _____

56. _____

57. _____

(a) _____

(b) _____

(c) _____

(d) _____

Answers

1. $(5, 1)$ **3.** $(2, 3)$ **5.** $\left(-5, \dfrac{3}{2}\right)$ **7.** Inconsistent system

9. $(4, 2)$ **11.** $(2, 1)$ **13.** $(3, 6)$ **15.** $\left(3, \dfrac{13}{2}\right)$

17. Dependent system   **19.** $(6, 3)$ **21.** $\left(2, \dfrac{3}{2}\right)$ **23.** $(3, 0)$

25. $(4, 0)$ **27.** $\left(\dfrac{1}{2}, -\dfrac{1}{2}\right)$ **29.** $\left(\dfrac{1}{5}, -\dfrac{2}{5}\right)$ **31.** 24, 16

33. Red delicious 49¢, Granny Smith 59¢   **35.** 18 m, 12 m

37. 25 lb at \$9, 75 lb at \$12   **39.** 120 mL of 25%, 80 mL of 50%

41. \$7,000 at 8%, \$5,000 at 9%   **43.** 120 mi/h, 30 mi/h

45. False   **47.** sometimes   **49.** $(3, 6)$ **51.** $(-11, -25)$

53. $x = 43$ ft; $y = 45$ ft   **55.** $P_1 = 12,300$; $P_2 = 14,900$

57. Above and Beyond   **(a)** 2   **(b)** 3   **(c)** 5   **(d)** -2

8.3

Systems of Linear Equations: Solving by Substitution

< 8.3 Objectives >

1 > Solve systems using the substitution method

2 > Choose an appropriate method for solving a system

3 > Solve applications of systems of equations

In Sections 8.1 and 8.2, we looked at graphing and addition as methods of solving linear systems. A third method is called the **substitution method.**

Example 1	Solving a System by Substitution

< Objective 1 >

Solve by substitution.

$$x + y = 12$$
$$y = 3x$$

Notice that the second equation says that y and $3x$ name the same quantity. So we may substitute $3x$ for y in the first equation. We then have

Replace y with $3x$ in the first equation.

$$\downarrow$$

$$x + 3x = 12$$
$$4x = 12$$
$$x = 3$$

We can now substitute 3 for x in either original equation to find the corresponding y-coordinate of the solution. We use the first equation:

$$(3) + y = 12$$
$$y = 9$$

So $(3, 9)$ is the solution.

This last step is identical to the one you saw in Section 8.2. As before, you can substitute the known coordinate value back into either of the original equations to find the value of the remaining variable. The check is also identical.

NOTE

The resulting equation contains only the variable x, so substitution is just another way of eliminating one of the variables from our system.

NOTE

The solution for a system is written as an ordered pair.

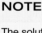 **Check Yourself 1**

Solve by substitution.

$$x - y = 9$$
$$y = 4x$$

The same technique can be readily used any time one of the equations is *already solved* for *x* or for *y*, as Example 2 illustrates.

Example 2   Solving a System by Substitution

Solve by substitution.

$$2x + 3y = 3$$
$$y = 2x - 7$$

Because the second equation tells us that *y* is $2x - 7$, we can replace *y* with $2x - 7$ in the first equation. This gives

NOTE

Now *y* is eliminated from the equation, and we can proceed to solve for *x*.

$$2x + 3\overbrace{(2x - 7)}^{y} = 3 \qquad \text{Distribute the factor 3.}$$
$$2x + 6x - 21 = 3$$
$$8x = 24$$
$$x = 3$$

We now know that 3 is the *x*-coordinate for the solution. So substituting 3 for *x* in the second equation, we have

$$y = 2(3) - 7$$
$$= 6 - 7$$
$$= -1$$

The solution is $(3, -1)$. Once again you should verify this result by letting $x = 3$ and $y = -1$ in the original system.

Check Yourself 2

Solve by substitution.

$$2x - 3y = 6$$
$$x = 4y - 2$$

As we have seen, the substitution method works very well when one of the given equations is already solved for *x* or *y*. It is also useful if you can readily solve for *x* or for *y* in one of the equations.

Example 3   Solving a System by Substitution

Solve by substitution.

$$x - 2y = 5$$
$$3x + y = 8$$

Neither equation is solved for a variable. That is easily handled in this case. Solving for *x* in the first equation, we have

$$x = 2y + 5$$

Now substitute $2y + 5$ for x in the second equation.

$$3(2y + 5) + y = \quad 8 \qquad \text{Distribute the factor 3.}$$
$$6y + 15 + y = \quad 8$$
$$7y = -7$$
$$y = -1$$

Substituting -1 for y in the second equation yields

$$3x + (-1) = 8$$
$$3x = 9$$
$$x = 3$$

So $(3, -1)$ is the solution. You should check this result by substituting 3 for x and -1 for y in the equations of the original system.

Check Yourself 3

Solve by substitution.

$$3x - \ y = 5$$
$$x + 4y = 6$$

In Example 3, we could have solved the second equation for y instead. The second equation becomes $y = -3x + 8$. Then substituting $-3x + 8$ for y in the first equation gives

$$x - 2(-3x + 8) = 5$$
$$x + 6x - 16 = 5$$
$$7x = 21$$
$$x = 3$$

Substituting 3 for x in the second equation yields

$$3(3) + y = \quad 8$$
$$y = -1$$

So, once again we see that the solution for the system is $(3, -1)$.

Inconsistent systems and dependent systems show up in a fashion similar to that which we saw in Section 8.2. Example 4 illustrates the approach to solving such systems.

Example 4 | Solving an Inconsistent or Dependent System

Solve the following systems by substitution.

(a) $4x - 2y = 6$
$$y = 2x - 3$$

NOTE

Be sure to change both signs in the parentheses.

From the second equation we substitute $2x - 3$ for y in the first equation.

$$4x - 2(2x - 3) = 6$$
$$4x - 4x + 6 = 6$$
$$6 = 6$$

Both variables have been eliminated, and we have the true statement $6 = 6$.

Recall from Section 8.2 that a true statement tells us that the graphs of the two equations are lines that coincide. We call this system dependent. There are an infinite number of solutions.

(b) $3x - 6y = 9$
$\qquad x = 2y + 2$

Substitute $2y + 2$ for x in the first equation.

$$3(2y + 2) - 6y = 9$$
$$6y + 6 - 6y = 9$$
$$6 = 9$$

This time we have a false statement.

This means that the system is *inconsistent* and that the graphs of the two equations are parallel lines. There is no solution.

Check Yourself 4

Indicate whether the given system is inconsistent (no solution) or dependent (an infinite number of solutions).

(a) $5x + 15y = 10$
$\qquad x = -3y + 1$

(b) $12x - 4y = 8$
$\qquad y = 3x - 2$

The following summarizes our work in this section.

Step by Step

To Solve a System of Linear Equations by Substitution

Step 1	Solve one of the given equations for x or y. If this is already done, go on to step 2.
Step 2	Substitute this expression for x or for y into the other equation.
Step 3	Solve the resulting equation for the remaining variable.
Step 4	Substitute the known value into either of the original equations to find the value of the second variable.
Step 5	Check your solution in both of the original equations.

You have now seen three different ways to solve systems of linear equations: by graphing, adding, and substitution. The natural question is, Which method should you use in a given situation?

Graphing is the least exact of the methods because solutions may have to be estimated.

The algebraic methods—addition and substitution—give exact solutions, and both always work for systems of linear equations. In fact, you may have noticed that several examples in this section could just as easily have been solved by adding (Example 3, for instance).

The choice of which algebraic method (substitution or addition) to use is yours and depends largely on the given system. Here are some guidelines to help you choose an appropriate method for solving a linear system.

Property

Choosing an Appropriate Method for Solving a System

1. If one of the equations is already solved for x (or for y), then substitution is the preferred method.
2. If the coefficients of x (or of y) in the two equations are the same, or opposites, then addition is the preferred method.
3. If solving for x (or for y) in either of the given equations results in fractional coefficients, then addition is the preferred method.

 Example 5   **Choosing an Appropriate Method for Solving a System**

< Objective 2 >

Select the most appropriate method for solving each of the following systems.

(a) $5x + 3y = 9$
  $2x - 7y = 8$

Addition is the most appropriate method because solving for a variable results in fractional coefficients.

(b) $7x + 26 = 8$
  $x = 3y - 5$

Substitution is the most appropriate method because the second equation is already solved for x.

(c) $8x - 9y = 11$
  $4x + 9y = 15$

Addition is the most appropriate method because the coefficients of y are opposites.

 Check Yourself 5

Select the most appropriate method for solving each of the following systems.

(a) $2x + 5y = 3$
  $8x - 5y = -13$

(b) $4x - 3y = 2$
  $y = 3x - 4$

(c) $3x - 5y = 2$
  $x = 3y - 2$

(d) $5x - 2y = 19$
  $4x + 6y = 38$

It is quite possible that, for a given system of equations, both algebraic methods will work equally well! We have solved the system in Example 3 (twice!) by substitution. In Example 6, we return to the same system and use addition.

Example 6	Solving a System by Addition

Solve by the addition method.

$$x - 2y = 5$$
$$3x + y = 8$$

We multiply the second equation by 2 in order to eliminate y.

$$x - 2y = 5$$
$$6x + 2y = 16$$

Adding, we get

$$7x = 21$$
$$x = 3$$

Substituting this value for x in the second equation, we have

$$3(3) + y = 8$$
$$y = -1$$

So, again we obtain the solution $(3, -1)$. When solving a system, then, we should choose the method that seems simplest.

Check Yourself 6

Solve the following system two ways: by substitution and by addition.

$$2x - 5y = 10$$
$$x - y = 8$$

Number problems, such as those presented in Chapter 2, are sometimes more easily solved by the methods presented in this section. Example 7 illustrates this approach.

Example 7	Solving a Number Problem by Substitution

< Objective 3 >

The sum of two numbers is 25. If the second number is 5 less than twice the first number, what are the two numbers?

RECALL

1. What do you want to find?
2. Assign variables. This time we use two letters, x and y.
3. Write equations for the solution. Here two equations are needed because we have introduced two variables.
4. Solve the system of equations.
5. Check the result.

Step 1   You want to find the two unknown numbers.

Step 2   Let $x =$ the first number and $y =$ the second number.

Step 3

$x + y = 25$

The sum   is 25.

$y = 2x - 5$

The second   is 5 less than
number   twice the first.

NOTE

We use the substitution method because the second equation is already solved for y.

Step 4

$$x + y = 25$$
$$y = 2x - 5$$

Substitute $2x - 5$ for y in the first equation.

$$x + (2x - 5) = 25$$
$$3x - 5 = 25$$
$$x = 10$$

From the first equation,

$$(10) + y = 25$$
$$y = 15$$

The two numbers are 10 and 15.

Step 5 The sum of the numbers is 25. The second number, 15, is 5 less than twice the first number, 10. The solution checks.

Check Yourself 7

The sum of two numbers is 28. The second number is 4 more than twice the first number. What are the numbers?

Sketches are always helpful in solving applications from geometry. Let's look at such an example.

 Example 8   Solving an Application from Geometry

The length of a rectangle is 3 m more than twice its width. If the perimeter of the rectangle is 42 m, find the dimensions of the rectangle.

Step 1 You want to find the dimensions (length and width) of the rectangle.

Step 2 Let L be the length of the rectangle and W the width. Now draw a sketch of the problem.

NOTE

We used x and y as our two variables in the previous examples. Use whatever letters you want. The process is the same, and sometimes it helps you remember what letter stands for what. Here $L = $ length and $W = $ width.

Step 3   Write the equations for the solution.

$L = \underbrace{2W + 3}$
3 more than twice
the width

$\underbrace{2L + 2W} = 42$
The perimeter

Step 4   Solve the system.

$$L = 2W + 3$$
$$2L + 2W = 42$$

From the first equation we can substitute $2W + 3$ for L in the second equation.

$$2(2W + 3) + 2W = 42$$
$$4W + 6 + 2W = 42$$
$$6W = 36$$
$$W = 6$$

Replace W with 6 to find L.

$$L = 2(6) + 3$$
$$= 12 + 3$$
$$= 15$$

The length is 15 m, the width is 6 m.

6 m

15 m

Step 5   Check these results. The length, 15 m, is 3 m more than twice the width, 6 m. The perimeter is $2L + 2W$, which should give us 42 m.

$$2(15) + 2(6) \overset{?}{=} 42$$
$$30 + 12 \overset{\checkmark}{=} 42 \qquad \text{True}$$

Check Yourself 8

The length of each of the two equal legs of an isosceles triangle is 5 in. less than the length of the base. If the perimeter of the triangle is 50 in., find the lengths of the legs and the base.

Our final example for this section comes from the field of mechanical engineering.

Example 9   Vocational Application

The antifreeze concentration in an industrial cooling system needs to be at 45%. If the system holds 32 gallons of coolant and is currently at a concentration of 30%, how much of the solution needs to be removed and replaced with pure antifreeze to bring the concentration up to the required level?

NOTE

We use substitution because one equation is already solved for a variable.

This can be solved with the system of equations

$$0.30x + 1y = 14.4$$
$$x + y = 32$$

where x is the amount of coolant at 30%, and y is the amount removed and replaced with pure antifreeze. Solve this system of equations by substitution.

Solving the first equation for y, we have

$$y = 14.4 - 0.30x$$

We substitute the expression $14.4 - 0.30x$ for y in the second equation:

$$x + (14.4 - 0.30x) = 32$$
$$x + 14.4 - 0.30x = 32$$
$$0.7x = 17.6$$
$$x = \frac{17.6}{0.7} \approx 25.14$$

So, $y \approx 6.86$.

This means that about 6.86 gallons of antifreeze currently in the system need to be removed and replaced with pure antifreeze.

Check Yourself 9

The antifreeze concentration in an industrial cooling system needs to be at 50%. If the system holds 36 gallons of coolant and is currently at a concentration of 35%, how much of the solution needs to be removed and replaced with pure antifreeze to bring the concentration up to the required level?

This can be solved with the system of equations

$$0.35 + 1y = 18$$
$$x + y = 36$$

where x is the amount of coolant at 35%, and y is the amount removed and replaced with pure antifreeze. Solve this system of equations by substitution.

Check Yourself ANSWERS

1. $(-3, -12)$ **2.** $(6, 2)$ **3.** $(2, 1)$
4. **(a)** Inconsistent system; **(b)** dependent system
5. **(a)** Addition; **(b)** substitution; **(c)** substitution; **(d)** addition   **6.** $(10, 2)$
7. The numbers are 8 and 20.   **8.** The legs have length 15 in.; the base is 20 in.   **9.** $x = 27.7$ gal; $y = 8.3$ gal

Reading Your Text

The following fill-in-the-blank exercises are designed to ensure that you understand some of the key vocabulary used in this section.

SECTION 8.3

(a) The _____ method works very well when one of the given equations is already solved for x or y.

(b) When a system is inconsistent, the graphs of the two equations are _____ lines.

(c) _____ is the least exact of the methods for solving a system of equations.

(d) If solving for x (or for y) in either of the given equations will result in fractional coefficients, then the _____ is the preferred method.

Basic Skills | Advanced Skills | Vocational-Technical Applications | Calculator/Computer | Above and Beyond | Getting Ready

8.3 exercises

< Objective 1 >

Solve each of the following systems by substitution.

1. $2x - y = 10$
$\quad\quad x = -2y$

2. $x + 3y = 10$
$\quad\quad 3x = y$

3. $3x + 2y = 12$
$\quad\quad y = 3x$

4. $4x - 3y = 24$
$\quad\quad y = -4x$

5. $x - y = 4$
$\quad\quad x = 2y - 2$

6. $x - y = 7$
$\quad\quad y = 2x - 12$

7. $2x + y = 7$
$\quad\quad y - x = -8$

8. $3x - y = -15$
$\quad\quad x = y - 7$

9. $3x + 4y = 9$
$\quad\quad y - 3x = 1$

10. $5x - 2y = -5$
$\quad\quad y - 5x = 3$

11. $3x - 18y = 4$
$\quad\quad x = 6y + 2$

12. $4x + 5y = 6$
$\quad\quad y = 2x - 10$

13. $5x - 3y = 6$
$\quad\quad y = 3x - 6$

14. $8x - 4y = 16$
$\quad\quad y = 2x - 4$

15. $x + 3y = 7$
$\quad\quad x - y = 3$

16. $2x - y = -4$
$\quad\quad x + y = -5$

17. $6x - 3y = 9$
$\quad\quad -2x + y = -3$

18. $5x - 6y = 21$
$\quad\quad x - 2y = 5$

19. $x - 7y = 3$
$\quad\quad 2x - 5y = 15$

20. $4x - 12y = 5$
$\quad\quad -x + 3y = -1$

< Objectives 1–2 >

Solve each of the following systems by using either addition or substitution. If a unique solution does not exist, state whether the system is dependent or inconsistent.

21. $2x + 3y = -6$
$\quad\quad x = 3y + 6$

22. $7x + 3y = 31$
$\quad\quad y = -2x + 9$

23. $2x - y = 1$
$\quad\quad -2x + 3y = 5$

24. $x + 3y = 12$
$\quad\quad 2x - 3y = 6$

MathZone

Boost your grade at mathzone.com!
> Practice Problems
> NetTutor
> Self-Tests
> e-Professors
> Videos

Name _____

Section _____ Date _____

Answers

1. _____   2. _____

3. _____   4. _____

5. _____   6. _____

7. _____   8. _____

9. _____   10. _____

11. _____   12. _____

13. _____

14. _____

15. _____   16. _____

17. _____

18. _____   19. _____

20. _____   21. _____

22. _____   23. _____

24. _____

Answers

25. _____

26. _____

27. _____

28. _____

29. _____

30. _____

31. _____

32. _____

33. _____

34. _____

35. _____

36. _____

37. _____

25. $6x + 2y = 4$
$\qquad y = -3x + 2$

26. $3x - 2y = 15$
$\qquad -x + 5y = -5$

27. $x + 2y = -2$
$\quad 3x + 2y = -12$

28. $10x + 2y = 7$
$\qquad y = -5x + 3$

29. $2x - 3y = 14$
$\quad 4x + 5y = -5$

30. $2x + 3y = 1$
$\quad 5x + 3y = 16$

< Objective 3 >

Solve each of the following problems. Be sure to show the equation used for the solution.

31. NUMBER PROBLEM The sum of two numbers is 100. The second is 3 times the first. Find the two numbers.

32. NUMBER PROBLEM The sum of two numbers is 70. The second is 10 more than 3 times the first. Find the numbers.

33. NUMBER PROBLEM The sum of two numbers is 56. The second is 4 less than twice the first. What are the two numbers?

34. NUMBER PROBLEM The difference of two numbers is 4. The larger is 8 less than twice the smaller. What are the two numbers?

35. NUMBER PROBLEM The difference of two numbers is 22. The larger is 2 more than 3 times the smaller. Find the two numbers.

36. NUMBER PROBLEM One number is 18 more than another, and the sum of the smaller number and twice the larger number is 45. Find the two numbers.

37. BUSINESS AND FINANCE Two packages together weigh 32 kilograms (kg). The smaller package weighs 6 kg less than the larger. How much does each package weigh?

38. **BUSINESS AND FINANCE** A washer-dryer combination costs $1,200. If the washer costs $220 more than the dryer, what does each appliance cost separately?

39. **SOCIAL SCIENCE** In a town election, the winning candidate had 220 more votes than the loser. If 810 votes were cast in all, how many votes did each candidate receive?

40. **BUSINESS AND FINANCE** An office desk and chair together cost $850. If the desk costs $50 less than twice as much as the chair, what did each cost?

41. **GEOMETRY** The length of a rectangle is 2 in. more than twice its width. If the perimeter of the rectangle is 34 in., find the dimensions of the rectangle.

42. **GEOMETRY** The perimeter of an isosceles triangle is 37 in. The length of each of the two equal legs is 6 in. less than 3 times the length of the base. Find the lengths of the three sides.

Determine whether each of the following statements is **true** *or* **false.**

43. Graphing is always the preferred method for solving a system.

44. The graphs of the equations are the same line in a dependent system.

Complete each of the following statements with **never, always,** *or* **sometimes.**

45. It is _____ possible to use the substitution method to solve a linear system.

46. The substitution method is _____ easier to use than the addition method.

Answers

38. _____

39. _____

40. _____

41. _____

42. _____

43. _____

44. _____

45. _____

46. _____

Answers

47. _____

48. _____

49. _____

50. _____

51. _____

52. _____

53. _____

Basic Skills | **Advanced Skills** | Vocational-Technical Applications | Calculator/Computer | Above and Beyond | Getting Ready
▲

Solve each system.

47. $\dfrac{1}{3}x + \dfrac{1}{2}y = 5$

$\dfrac{x}{4} - \dfrac{y}{5} = -2$

48. $\dfrac{5x}{2} - y = \dfrac{9}{10}$

$\dfrac{3x}{4} + \dfrac{5y}{6} = \dfrac{2}{3}$

49. $0.4x - 0.2y = 0.6$
$2.5x - 0.3y = 4.7$

50. $0.4x - 0.1y = 5$
$6.4x + 0.4y = 60$

Basic Skills | Advanced Skills | **Vocational-Technical Applications** | Calculator/Computer | Above and Beyond | Getting Ready
▲

51. **ALLIED HEALTH** A medical lab technician needs to determine how much 9% sulfuric acid (H_2SO_4) solution (x) must be mixed with 2% H_2SO_4 (y) to produce 75 milliliters of a 4% solution. From this information, the technician derives two linear equations:

$x + y = 75$
$9x + 2y = 300$

Solve this system of equations by substitution, and report how much of each type of solution is needed, to the nearest tenth of a milliliter.

chapter 8 > Make the Connection

52. **ALLIED HEALTH** A medical lab technician needs to determine how much 40% alcohol solution (x) must be mixed with 25% alcohol (y) to produce 800 milliliters of a 35% solution. From this information, the technician derives two linear equations:

$x + y = 800$
$40x + 25y = 28,000$

Solve this system of equations by substitution, and report how much of each type of solution is needed, to the nearest tenth of a milliliter.

chapter 8 > Make the Connection

53. **ALLIED HEALTH** A medical lab technician needs to determine how much 20% saline solution (x) must be mixed with 5% saline (y) to produce 100 milliliters of a 12% solution. From this information, the technician derives two linear equations:

$x + y = 100$
$20x + 5y = 1,200$

Solve this system of equations by substitution, and report how much of each type of solution is needed, to the nearest tenth of a milliliter.

chapter 8 > Make the Connection

54. **ALLIED HEALTH** A medical lab technician needs to determine how much 8.2 molar (M) calcium chloride ($CaCl_2$) solution (x) must be mixed with 3.5 M $CaCl_2$ (y) to produce 400 milliliters of a 5 M solution. From this information, the technician derives two linear equations:

$$x + \quad y = 400$$
$$8.2x + 3.5y = 2,000$$

Solve this system of equations by substitution, and report how much of each type of solution is needed, to the nearest tenth of a milliliter.

54. _____

55. _____

56. _____

57. _____

55. **CONSTRUCTION TECHNOLOGY** A 24-ft beam has a weight on each end. Find the point where a pivot can be placed so that the beam balances.

This system of equations can be used to find the balance point:

$$280x = 90y$$
$$x + y = 24$$

Solve this system of equations by substitution (round your results to the nearest hundredth).

56. **CONSTRUCTION TECHNOLOGY** A customer wants a dual-slope roof with one side having a 12/12 slope and the other side having a 5/12 slope. If the total span of the roof is to be 34 feet, find the length of the 5/12 span and the 12/12 span.

This truss is described by this system of equations:

$$\frac{12}{12}x = \frac{5}{12}y$$
$$x + y = 34$$

Solve this system of equations.

Basic Skills | Advanced Skills | Vocational-Technical Applications | Calculator/Computer | **Above and Beyond** | Getting Ready

57. You have a part-time job writing the *Consumer Concerns* column for your local newspaper. Your topic for this week is clothes dryers, and you are planning to compare the Helpmate and the Whirlgarb dryers, both readily available in stores in your area. The information you have is that the Helpmate dryer is listed at $520, and it costs 22.5¢ to dry an average-sized load at the utility rates in your city. The Whirlgarb dryer is listed at $735, and it costs 15.8¢ to run for each normal load. The maintenance costs for both dryers are about the same. Working with a partner, write a short article giving your readers helpful advice about these appliances. What should they consider when buying one of these clothes dryers?

Answers

(a) _____

(b) _____

(c) _____

(d) _____

(e) _____

(f) _____

Graph the solution sets for the following linear inequalities.

(a) $x + y > 8$

(b) $2x - y \leq 6$

(c) $3x + 4y \geq 12$

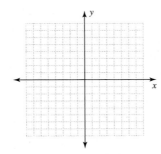

(d) $y > 2x$

(e) $y \leq -3$

(f) $x > 5$

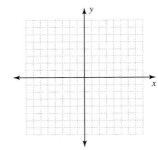

Answers

1. $(4, -2)$ **3.** $\left(\dfrac{4}{3}, 4\right)$ **5.** $(10, 6)$ **7.** $(5, -3)$ **9.** $\left(\dfrac{1}{3}, 2\right)$

11. No solution   **13.** $(3, 3)$ **15.** $(4, 1)$

17. Infinite number of solutions   **19.** (10, 1)   **21.** (0, −2)

23. (2, 3)   **25.** Dependent system   **27.** $\left(-5, \dfrac{3}{2}\right)$ **29.** $\left(\dfrac{5}{2}, -3\right)$

31. 25, 75   **33.** 20, 36   **35.** 32, 10   **37.** 13 kg, 19 kg
39. Winner 515, loser 295   **41.** Width 5 in., length 12 in.
43. False   **45.** always   **47.** (0, 10)   **49.** (2, 1)
51. 21.4 mL of 9%; 53.6 mL of 2%   **53.** 46.7 mL of 20%; 53.3 mL of 5%
55. $x = 5.84$ ft; $y = 18.16$ ft   **57.** Above and Beyond

(a) $x + y > 8$ **(b)** $2x - y \le 6$

      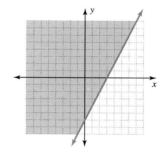

(c) $3x + 4y \ge 12$ **(d)** $y > 2x$

      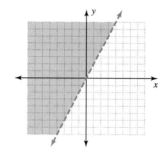

(e) $y \le -3$ **(f)** $x > 5$

      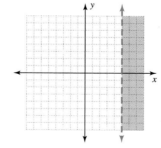

Systems of Linear Inequalities

< 8.4 Objectives >

1 > Graph a system of linear inequalities

2 > Solve an application of linear inequalities

NOTE

You might want to review graphing linear inequalities in Section 7.4 at this point.

Our previous work in this chapter dealt with finding the solution set of a system of linear equations. That solution set represented the points of intersection of the graphs of the equations in the system. In this section, we extend that idea to include systems of linear inequalities.

In this case, the solution set for each inequality is all ordered pairs that satisfy that inequality. *The graph of the solution set of a system of linear inequalities* is then the intersection of the graphs of the individual inequalities.

Example 1 | **Solving a System by Graphing**

< Objective 1 >

Solve the following system of linear inequalities by graphing.

$$x + y > 4$$
$$x - y < 2$$

We start by graphing each inequality separately. The boundary line is drawn, and using $(0, 0)$ as a test point, we see that we should shade the half-plane above the line in both graphs.

RECALL

The boundary line is dashed to indicate it is *not* included in the graph.

$x + y > 4$

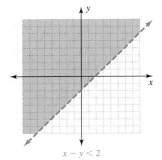

$x - y < 2$

NOTE

Points on the lines are not included in the solution set.

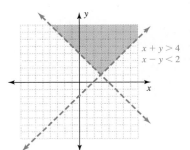

$x + y > 4$
$x - y < 2$

In practice, the graphs of the two inequalities are combined on the same set of axes, as shown in the graph at the left. This graph of the solution set of the original system is the intersection of the two original graphs.

The Streeter/Hutchison Series in Mathematics   Beginning Algebra

Check Yourself 1

Solve the following system of linear inequalities by graphing.

$$2x - y < 4$$
$$x + y < 3$$

In applications we often have restrictions on the possible values of x and y. We consider such restrictions in Example 2.

▶ **Example 2**   **Solving a System by Graphing**

Solve the following system of linear inequalities by graphing.

$$1 \leq x \leq 3$$
$$2 \leq y \leq 5$$

RECALL

The boundaries are solid because the symbol reads "less than or equal to."

The statement $1 \leq x \leq 3$ says that $1 \leq x$ *and* $x \leq 3$. We graph the vertical lines $x = 1$ and $x = 3$ as solid lines, and then shade points to the right of 1 and to the left of 3. Similarly, to graph $2 \leq y \leq 5$, we graph horizontal lines $y = 2$ and $y = 5$, and we shade between.

Putting this together, we have the following solution.

Check Yourself 2

Solve the following system of linear inequalities by graphing.

$$-2 \leq x \leq 3$$
$$-4 \leq y \leq -1$$

Most applications of systems of linear inequalities lead to bounded regions. This generally requires a system of three or more inequalities, as shown in Example 3.

Example 3	Solving a System by Graphing

Solve the following system of linear inequalities by graphing.

$$2x + y \leq 10$$
$$x + y \leq 7$$
$$x \geq 1$$
$$y \geq 2$$

On the same set of axes, we graph the boundary line of each of the inequalities. For example, the following graph shows the boundary line for $2x + y \leq 10$.

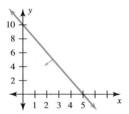

We then decide, since $(0, 0)$ is a solution, to shade below the line. Rather than shading now, we indicate the direction of solutions with an arrow.

Continuing in this fashion, we graph each boundary line on the same set of axes. The set of solutions is the intersection of the four indicated regions.

Check Yourself 3

Solve the following system of linear inequalities by graphing.

$$2x - y \leq 8$$
$$x + y \leq 7$$
$$x \geq 0$$
$$y \geq 0$$

Next, we look at an application of our work with systems of linear inequalities.

| Example 4 | Solving a Business Application |

< Objective 2 >

A manufacturer produces a standard model and a deluxe model of a 13-in. television set. The standard model requires 12 h of labor to produce, whereas the deluxe model requires 18 h. The labor available is limited to 360 h per week. Also, the plant capacity is limited to producing a total of 25 sets per week. Write a system of inequalities representing this situation. Then, draw a graph of the region representing the number of sets that can be produced, given these conditions.

We let x represent the number of standard-model sets produced and y the number of deluxe-model sets. Because the labor is limited to 360 h, we have

NOTE

The total labor is limited to (or less than or equal to) 360 h.

$$12x \quad + \quad 18y \le 360$$

12 h per   18 h per
standard set   deluxe set

The total production, here $x + y$ sets, is limited to 25, so we can write

$$x + y \le 25$$

For convenience in graphing, we divide both expressions in the first inequality by 6, to write the equivalent system:

NOTE

We have $x \ge 0$ and $y \ge 0$ because the number of sets produced cannot be negative.

$$2x + 3y \le 60$$
$$x + y \le 25$$
$$x \ge 0$$
$$y \ge 0$$

We now graph the system of inequalities as before. The shaded area represents all possibilities in terms of the number of sets that can be produced.

NOTE

The shaded area is called the *feasible region.* All points in the region meet the given conditions of the problem and represent possible production options.

Check Yourself 4

A manufacturer produces DVD players and compact disk players. The DVD players require 10 h of labor to produce and the disk players require 20 h. The labor hours available are limited to 300 h per week. Existing orders require that at least 10 DVD players and at least 5 disk players be produced per week.

Write a system of inequalities representing this situation. Then, draw a graph of the region representing the possible production options.

✓ Check Yourself ANSWERS

1. $2x - y < 4$
$x + y < 3$

2.

3. $2x - y \leq 8$
$x + y \leq 7$
$x \geq 0$
$y \geq 0$

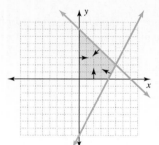

4. Let x be the number of DVD players and y be the number of CD players. The system is

$$10x + 20y \leq 300$$
$$x \geq 10$$
$$y \geq 5$$

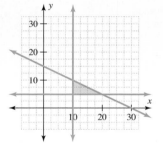

Reading Your Text ◀

The following fill-in-the-blank exercises are designed to ensure that you understand some of the key vocabulary used in this section.

SECTION 8.4

(a) The graph of the solution set of a system of linear inequalities is the _____ of the graphs of the individual inequalities.

(b) The boundary line is _____ to indicate that it is *not* part of the solution set.

(c) Most applications of systems of linear inequalities lead to _____ regions.

(d) The shaded area in an application problem is called the _____ region.

8.4 exercises

< Objective 1 >

Solve each of the following systems of linear inequalities graphically.

1. $x + 2y \leq 4$
$ x - y \geq 1$

2. $3x - y > 6$
$x + y < 6$

Name _____

Section _____ Date _____

Answers

1. _____

2. _____

3. _____

4. _____

5. _____

6. _____

3. $3x + y < 6$
$x + y > 4$

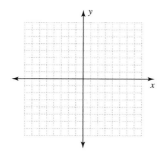

4. $2x + y \geq 8$
$x + y \geq 4$

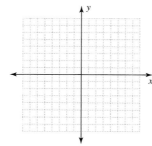

5. $x + 3y \leq 12$
$2x - 3y \leq 6$

 > Videos

6. $x - 2y > 8$
$3x - 2y > 12$

Answers

7. $3x + 2y \leq 12$
$x \geq 2$

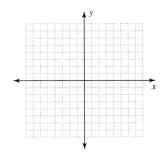

8. $2x + y \leq 6$
$y \geq 1$

9. $2x + y \leq 8$
$x > 1$
$y > 2$

 > Videos

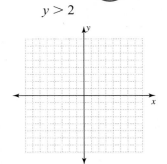

10. $3x - y \leq 6$
$x \geq 1$
$y \leq 3$

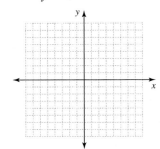

11. $x + 2y \leq 8$
$2 \leq x \leq 6$
$y \geq 0$

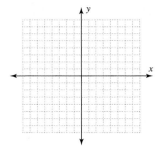

12. $x + y < 6$
$0 \leq y \leq 3$
$x \geq 1$

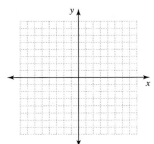

13. $3x + y \leq 6$
$x + y \leq 4$
$x \geq 0$
$y \geq 0$

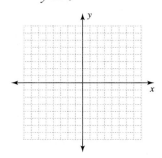

14. $x - 2y \geq -2$
$x + 2y \leq 6$
$x \geq 0$
$y \geq 0$

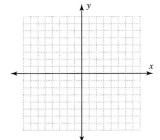

15. $4x + 3y \leq 12$
$x + 4y \leq 8$
$x \geq 0$
$y \geq 0$

16. $2x + y \leq 8$
$x + y \geq 3$
$x \geq 0$
$y \geq 0$

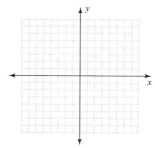

Answers

15. _____

16. _____

17. _____

18. _____

19. _____

17. $x - 4y \leq -4$
$x + 2y \leq 8$
$x \geq 2$

18. $x - 3y \geq -6$
$x + 2y \geq 4$
$x \leq 4$

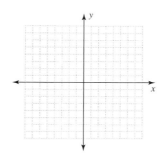

< Objective 2 >

Draw the appropriate graphs in each of the following.

19. BUSINESS AND FINANCE A manufacturer produces both two-slice and four-slice toasters. The two-slice toaster takes 6 h of labor to produce and the four-slice toaster 10 h. The labor available is limited to 300 h per week, and the total production capacity is 40 toasters per week. Write a system of inequalities representing this situation. Then, draw a graph of the feasible region given these conditions, in which x is the number of two-slice toasters and y is the number of four-slice toasters.

Answers

20. _____

21. _____

22. _____

23. _____

24. _____

25. _____

20. BUSINESS AND FINANCE A small firm produces both AM and AM/FM car radios. The AM radios take 15 h to produce, and the AM/FM radios take 20 h. The number of production hours is limited to 300 h per week. The plant's capacity is limited to a total of 18 radios per week, and existing orders require that at least 4 AM radios and at least 3 AM/FM radios be produced per week. Write a system of inequalities representing this situation. Then, draw a graph of the feasible region given these conditions, in which x is the number of AM radios and y the number of AM/FM radios.

Determine whether each of the following statements is **true** *or* **false.**

21. The boundary lines in a system of linear inequalities are always drawn as dashed lines.

22. If the boundary lines are drawn as dashed lines, points on the lines are included in the solution set.

Complete each of the following statements with **never, sometimes,** *or* **always.**

23. The graph of the solution set of a system of two linear inequalities _____ includes the origin.

24. The graph of the solution set of a system of two linear inequalities is _____ bounded.

| Basic Skills | **Advanced Skills** | Vocational-Technical Applications | Calculator/Computer | Above and Beyond | Getting Ready |

25. Write the system of inequalities whose graph is the shaded region.

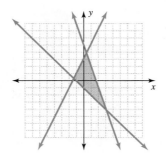

26. Write the system of inequalities whose graph is the shaded region.

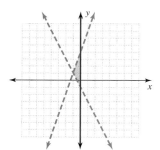

27. Describe a system of linear inequalities for which there is no solution.

Answers

Answers

26. _____

27. _____

1. $x + 2y \leq 4$
$\quad x - y \geq 1$

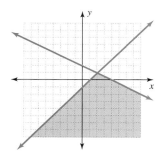

3. $3x + y < 6$
$\quad x + y > 4$

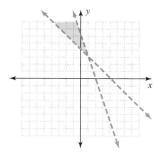

5. $x + 3y \leq 12$
$\quad 2x - 3y \leq 6$

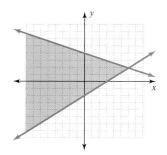

7. $3x + 2y \leq 12$
$\quad x \geq 2$

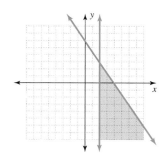

9. $2x + y \leq 8$
$x > 1$
$y > 2$

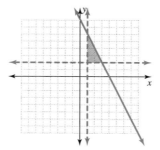

11. $x + 2y \leq 8$
$2 \leq x \leq 6$
$y \geq 0$

13. $3x + y \leq 6$
$x + y \leq 4$
$x \geq 0$
$y \geq 0$

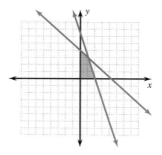

15. $4x + 3y \leq 12$
$x + 4y \leq 8$
$x \geq 0$
$y \geq 0$

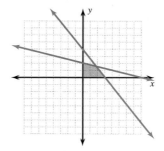

17. $x - 4y \leq -4$
$x + 2y \leq 8$
$x \geq 2$

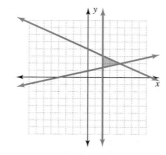

19. $6x + 10y \leq 300$
$x + y \leq 40$
$x \geq 0$
$y \geq 0$

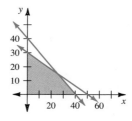

21. False   **23.** sometimes   **25.** $y \leq 2x + 3$
$y \leq -3x + 5$
$y \geq -x - 1$

27. Above and Beyond

Definition/Procedure	Example	Reference

Systems of Linear Equations: Solving by Graphing　　　　Section 8.1

A System of Equations

Two or more equations considered together.	$x + y = 4$ $2x - y = 5$	*p.* 751

Solution

A solution for a system of two equations in two unknowns is an ordered pair that satisfies each equation of the system.	$(x, y) = (3, 1)$	*p.* 751

Solving by Graphing

1. Graph both equations on the same coordinate system.
2. The system may have
   a. *One solution.* The lines intersect at one point (a consistent system). The solution is the ordered pair corresponding to that point.
   b. *No solution.* The lines are parallel (an inconsistent system).
   c. *Infinitely many solutions.* The two equations have the same graph (a dependent system). Any ordered pair corresponding to a point on the line is a solution.

p. 755

A consistent system

An inconsistent system

A dependent system

Continued

Definition/Procedure	Example	Reference

Systems of Linear Equations: Solving by the Addition Method

Solving by Adding

1. If necessary, multiply both sides of one or both equations by nonzero numbers to form an equivalent system in which the coefficients of one of the variables are opposites.
2. Add the equations of the new system.
3. Solve the resulting equation for the remaining variable.
4. Substitute the value found in step 3 into either of the original equations to find the value of the second variable.
5. Check your solution in both of the original equations.

Example:

$2x - y = 4$
$3x + 2y = 13$

Multiply the first equation by 2.

$4x - 2y = 8$
$3x + 2y = 13$

Add.

$7x = 21$
$x = 3$

In the original equation,

$2(3) - y = 4$
$y = 2$

$(3, 2)$ is the solution.

Reference: Section 8.2 — p. 770

Applying Systems of Equations

Often word problems can be solved by using two variables and two equations to represent the unknowns and the given relationships in the problem.

The Solution Steps

1. Read the problem carefully. Then reread it to decide what you are asked to find.
2. Assign variables to the unknown quantities.
3. Translate the problem to the language of algebra to form a system of equations.
4. Solve the system.
5. Verify your solution in the original problem.

Reference: p. 773

Systems of Linear Equations: Solving by Substitution

Solving by Substitution

1. Solve one of the given equations for x or for y. If this is already done, go on to step 2.
2. Substitute this expression for x or for y into the other equation.
3. Solve the resulting equation for the remaining variable.
4. Substitute the value found in step 3 into either of the original equations to find the value of the second variable.
5. Check your solution in both of the original equations.

Example:

$x - 2y = 3$
$2x + 3y = 13$

From the first equation,

$x = 2y + 3$

Substitute in the second equation:

$2(2y + 3) + 3y = 13$
$4y + 6 + 3y = 13$
$7y + 6 = 13$
$7y = 7$
$y = 1$

$x = 2(1) + 3$
$x = 5$

Reference: Section 8.3 — p. 790

Definition/Procedure	Example	Reference

Systems of Linear Inequalities

Section 8.4

A *system of linear inequalities* is two or more linear inequalities considered together. The *graph of the solution set* of a system of linear inequalities is the intersection of the graphs of the individual inequalities.

Solving Systems of Linear Inequalities Graphically

1. Graph each inequality, shading the appropriate half-plane, on the same set of coordinate axes.
2. The graph of the system is the intersection of the regions shaded in step 1.

To solve

$x + 2y \leq 8$

$x + \ y \leq 6$

$\qquad x \geq 0$

$\qquad y \geq 0$

graphically

p. 804

This summary exercise set is provided to give you practice with each of the objectives of this chapter. Each exercise is keyed to the appropriate chapter section. When you are finished, you can check your answers to the odd-numbered exercises against those presented in the back of the text. If you have difficulty with any of these questions, go back and reread the examples from that section. Your instructor will give you guidelines on how best to use these exercises in your instructional setting.

8.1 *Solve each of the following systems by graphing.*

1. $x + y = 6$
$\quad x - y = 2$

2. $x - y = 8$
$\quad 2x + y = 7$

3. $x + 2y = 4$
$\quad x + 2y = 6$

4. $2x - y = 8$
$\quad\quad y = 2$

5. $2x - 4y = 8$
$\quad x - 2y = 4$

6. $3x + 2y = 6$
$\quad 4x - \;\; y = 8$

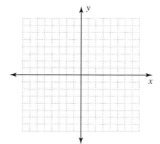

8.2 *Solve each of the following systems by addition. If a unique solution does not exist, state whether the system is inconsistent or dependent.*

7. $x + y = 8$
$\quad x - y = 2$

8. $-x - y = \;\; 4$
$\quad\quad x - y = -8$

9. $2x - 3y = 16$
$\quad 5x + 3y = 19$

10. $2x + y = 7$
$\quad 3x - y = 3$

11. $3x - 5y = 14$
$\quad 3x + 2y = \;\; 7$

12. $2x - 4y = 8$
$\quad\;\; x - 2y = 4$

13. $4x - 3y = -22$
$\quad 4x + 5y = -6$

14. $5x - 2y = 17$
$\quad 3x - 2y = \;\; 9$

15. $4x - 3y = 10$
$\quad 2x - 3y = \;\; 6$

16. $2x + 3y = -10$
$-2x + 5y = 10$

17. $3x + 2y = 3$
$6x + 4y = 5$

18. $3x - 2y = 23$
$x + 5y = -15$

19. $5x - 2y = -1$
$10x + 3y = 12$

20. $x - 3y = 9$
$5x - 15y = 45$

21. $2x - 3y = 18$
$5x - 6y = 42$

22. $3x + 7y = 1$
$4x - 5y = 30$

23. $5x - 4y = 12$
$3x + 5y = 22$

24. $6x + 5y = -6$
$9x - 2y = 10$

25. $4x - 3y = 7$
$-8x + 6y = -10$

26. $3x + 2y = 8$
$-x - 5y = -20$

27. $3x - 5y = -14$
$6x + 3y = -2$

8.3 *Solve each of the following systems by substitution. If a unique solution does not exist, state whether the system is inconsistent or dependent.*

28. $x + 2y = 10$
$y = 2x$

29. $x - y = 10$
$x = -4y$

30. $2x - y = 10$
$x = 3y$

31. $2x + 3y = 2$
$y = x - 6$

32. $4x + 2y = 4$
$y = 2 - 2x$

33. $x + 5y = 20$
$x = y + 2$

34. $6x + y = 2$
$y = 3x - 4$

35. $2x + 6y = 10$
$x = 6 - 3y$

36. $2x + y = 9$
$x - 3y = 22$

37. $x - 3y = 17$
$2x + y = 6$

38. $2x + 3y = 4$
$y = 2$

39. $4x - 5y = -2$
$x = -3$

40. $-6x + 3y = -4$
$y = -\dfrac{2}{3}$

41. $5x - 2y = -15$
$y = 2x + 6$

42. $3x + y = 15$
$x = 2y + 5$

Solve each of the following systems by either addition or substitution. If a unique solution does not exist, state whether the system is inconsistent or dependent.

43. $x - 4y = 0$
$\quad\ 4x + y = 34$

44. $2x + y = 2$
$\qquad\quad y = -x$

45. $3x - 3y = 30$
$\qquad\quad x = -2y - 8$

46. $5x + 4y = 40$
$\quad\ \ x + 2y = 11$

47. $x - 6y = -8$
$\quad 2x + 3y = 4$

48. $4x - 3y = 9$
$\quad 2x + y = 12$

49. $9x + y = 9$
$\quad\ x + 3y = 14$

50. $3x - 2y = 8$
$\quad -6x + 4y = -16$

51. $3x - 2y = 8$
$\quad 2x - 3y = 7$

Solve the following problems. Be sure to show the equations used.

52. NUMBER PROBLEM The sum of two numbers is 40. If their difference is 10, find the two numbers.

53. NUMBER PROBLEM The sum of two numbers is 17. If the larger number is 1 more than 3 times the smaller, what are the two numbers?

54. NUMBER PROBLEM The difference of two numbers is 8. The larger number is 2 less than twice the smaller. Find the numbers.

55. BUSINESS AND FINANCE Five writing tablets and three pencils cost $8.25. Two tablets and two pencils cost $3.50. Find the cost for each item.

56. CONSTRUCTION A cable 200 ft long is cut into two pieces so that one piece is 12 ft longer than the other. How long is each piece?

57. BUSINESS AND FINANCE An amplifier and a pair of speakers cost $925. If the amplifier costs $75 more than the speakers, what does each cost?

58. BUSINESS AND FINANCE A sofa and chair cost $850 as a set. If the sofa costs $100 more than twice as much as the chair, what is the cost of each?

59. GEOMETRY The length of a rectangle is 4 cm more than its width. If the perimeter of the rectangle is 64 cm, find the dimensions of the rectangle.

60. GEOMETRY The perimeter of an isosceles triangle is 29 in. The lengths of the two equal legs are 2 in. more than twice the length of the base. Find the lengths of the three sides.

61. NUMBER PROBLEM Darryl has 30 coins with a value of $5.50. If they are all nickels and quarters, how many of each kind of coin does he have?

62. BUSINESS AND FINANCE Tickets for a concert sold for $11 and $8. If 600 tickets were sold for one evening and the receipts were $5,550, how many of each kind of ticket were sold?

63. SCIENCE AND MEDICINE A laboratory has a 20% acid solution and a 50% acid solution. How much of each should be used to produce 600 mL of a 40% acid solution?

64. SCIENCE AND MEDICINE A service station wishes to mix 40 L of a 78% antifreeze solution. How many liters of a 75% solution and a 90% solution should be used in forming the mixture?

65. BUSINESS AND FINANCE Martha has $18,000 invested. Part of the money is invested in a bond that yields 11% interest. The remainder is in her savings account, which pays 7%. If she earns $1,660 in interest for 1 year, how much does she have invested at each rate?

66. SCIENCE AND MEDICINE A boat travels 24 mi upstream in 3 h. It then takes 3 h to go 36 mi downstream. Find the speed of the boat in still water and the speed of the current.

67. SCIENCE AND MEDICINE A plane flying with the wind makes a trip of 2,200 mi in 4 h. Returning against the wind, it can travel only 1,800 mi in 4 h. What is the plane's rate in still air? What is the wind speed?

8.4 Solve each system of linear inequalities.

68. $x - y < 7$
$x + y > 3$

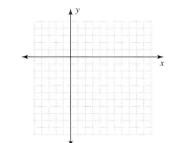

69. $x - 2y \leq -2$
$x + 2y \leq 6$

70. $x - 6y < 6$
$-x + y < 4$

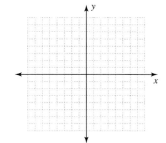

71. $2x + y \leq 8$
$x \geq 1$
$y \geq 0$

72. $2x + y \leq 6$
$x \geq 1$
$y \geq 0$

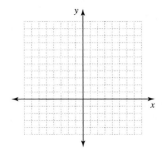

73. $4x + y \leq 8$
$x \geq 0$
$y \geq 2$

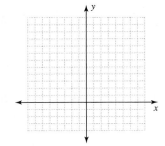

74. $4x + 2y \leq 8$
$x + y \leq 3$
$x \geq 0$
$y \geq 0$

75. $3x + y \leq 6$
$x + y \leq 4$
$x \geq 0$
$y \geq 0$

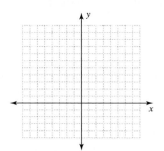

The purpose of this self-test is to help you check your progress and to review for the next in-class exam. Allow yourself about an hour to take this test. At the end of that hour check your answers against those given in the back of the text. Section references accompany the answers. If you missed any questions, go back to those sections and reread the examples until you master the concepts.

Answers

Solve each of the following systems by graphing. If a unique solution does not exist, state whether the system is inconsistent or dependent.

1. $x + y = 5$
$\quad x - y = 3$

2. $x + 2y = 8$
$\quad x - \ y = 2$

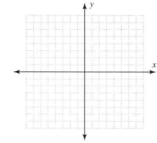

3. $x - 3y = 3$
$\quad x - 3y = 6$

4. $4x - \ y = \ \ 4$
$\quad x - 2y = -6$

1. _____

2. _____

3. _____

4. _____

5. _____

6. _____

7. _____

8. _____

9. _____

10. _____

11. _____

12. _____

Solve each of the following systems by addition. If a unique solution does not exist, state whether the system is inconsistent or dependent.

5. $x + y = 5$
$\quad x - y = 3$

6. $x + 2y = 8$
$\quad x - \ y = 2$

7. $\quad 3x + \ y = 6$
$\quad -3x + 2y = 3$

8. $3x + 2y = 11$
$\quad 5x + 2y = 15$

9. $3x - 6y = 12$
$\quad x - 2y = \ 4$

10. $4x + \ y = 2$
$\quad 8x - 3y = 9$

11. $2x - 5y = \ 2$
$\quad 3x + 4y = 26$

12. $\quad x + 3y = 6$
$\quad 3x + 9y = 9$

Answers

13. _____

14. _____

15. _____

16. _____

17. _____

18. _____

19. _____

20. _____

21. _____

22. _____

23. _____

24. _____

25. _____

26. _____

27. _____

28. _____

Solve each of the following systems by substitution. If a unique solution does not exist, state whether the system is inconsistent or dependent.

13. $x + y = 8$
$\quad\quad y = 3x$

14. $x - y = 9$
$\quad\quad x = -2y$

15. $2x - y = 10$
$\quad\quad x = y + 4$

16. $x - 3y = -7$
$\quad\quad y = x - 1$

17. $3x + y = -6$
$\quad\quad y = 2x + 9$

18. $4x + 2y = 8$
$\quad\quad y = 3 - 2x$

19. $5x + y = 10$
$\quad\quad x + 2y = -7$

20. $3x - 2y = 5$
$\quad\quad 2x + y = 8$

Solve each of the following problems. Be sure to show the equations used.

21. NUMBER PROBLEM The sum of two numbers is 30, and their difference is 6. Find the two numbers.

22. CONSTRUCTION A rope 50 m long is cut into two pieces so that one piece is 8 m longer than the other. How long is each piece?

23. GEOMETRY The length of a rectangle is 4 in. less than twice its width. If the perimeter of the rectangle is 64 in., what are the dimensions of the rectangle?

24. NUMBER PROBLEM Murray has 30 coins with a value of $5.70. If the coins are all dimes and quarters, how many of each coin does he have?

25. SCIENCE AND MEDICINE Jackson was able to travel 36 mi downstream in 2 h. In returning upstream, it took 3 h to make the trip. How fast can his boat travel in still water? What was the rate of the river current?

Solve each system of inequalities.

26. $x + y < 3$
$\quad\quad x - 2y < 6$

27. $4y + 3x \geq 12$
$\quad\quad x \geq 1$

28. $2y + x \leq 8$
$\quad\quad y + x \leq 6$
$\quad\quad x \geq 0$
$\quad\quad y \geq 0$

Activity 8 ::
Growth of Children—Fitting a Linear Model to Data

When you walk into a home where children have lived, you can often find a wall with notches showing the heights of the children at various points in their lives. Nearly every time you bring a child to see the doctor, the child's height and weight are recorded, regardless of the reason for the office visit.

The National Institutes of Health (NIH) through the Centers for Disease Control and Prevention (CDC) publishes and updates data detailing heights and weights of children for the populace as a whole and for what they consider to be healthy children. These data were most recently updated in November 2000.

In this activity, we explore the graphing of trend data and its predictive capability. We use a child's height and weight at various ages for our data.

I.

1. Locate the medical records of a child between 2–5 years old whom you know.

2. Create a table with three columns and seven rows. Label the table "Height and Weight of [name]; Year 2."

3. Label the first column "Age (months)," the second column, "Height (in.)," and the third column, "Weight (lb)."

4. In the first column, write the numbers (one to each row): 14, 16, 18, 20, 22, 24.

5. In the second column, list the child's height at each of the months listed in the first column.

6. Do likewise in the third column with the child's weight.

II.

1. Create a scatterplot of the data in your table.

2. Describe the trend of the data. Fit a line to the data and give the equation of the line.

3. Use the equation of the line to predict the child's height and weight at 10 months of age.

4. Use the equation of the line to predict the child's height and weight at 28 months of age.

5. Use the equation of the line to predict the child's height and weight at 240 months (20 years) of age.

6. According to the model, how tall will the child be when he or she is 50 years old (600 months)?

7. Discuss the accuracy of the predictions made in steps 3 to 6. What can you discern about using scatterplot data to make predictions in general?

Sample Data Set and Solutions

I.

Median Heights and Weights for Children: Year 2

Age (months)	Median Height for Girls (in.)	Median Weight for Boys (lb)
14	30	24
16	30.75	25
18	31.5	25.75
20	32.5	26.25
22	33	27.25
24	34	28

Source: Centers for Disease Control and Prevention.

II. 1.

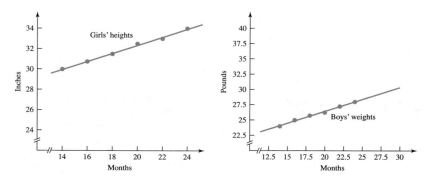

2. Both sets of data appear linear. The lines shown have the following equations:

Girls' heights: $y = 0.4x + 24.4$
Boys' weights: $y = 0.39x + 18.6$

3–6.

Predictions

	Age (months)			
	10	**28**	**240**	**600**
Girls' heights (in.)	28.4	35.6	120.4	264.4
Boys' weights (lb)	22.5	29.52	112.2	252.6

While the predictions for 10 and 28 months are reasonably accurate, 20-year-old women tend to be shorter than the 10 ft given by the prediction and 20-year-old men tend to be heavier than the predicted 112 lb. Likewise, women do not grow to be 22 ft tall by the age of 50.

7. We need to be careful that we stay close to the original data source when using data to predict. That is, we can reasonably predict a person's height at 10 months from the data of their second year of age, but we cannot extrapolate that to their height 50 years later.

The following exercises are presented to help you review concepts from earlier chapters. This is meant as a review and not as a comprehensive exam. The answers are presented in the back of the text. Section references accompany the answers. If you have difficulty with any of these exercises, be certain to at least read through the summary related to those sections.

Name _____

Section _____   Date _____

Answers

Perform each of the indicated operations.

1. $(5x^2 - 9x + 3) + (3x^2 + 2x - 7)$

2. Subtract $9w^2 + 5w$ from the sum of $8w^2 - 3w$ and $2w^2 - 4$.

3. $7xy(4x^2y - 2xy + 3xy^2)$

4. $(3s - 7)(5s + 4)$

5. $\dfrac{5x^3y - 10x^2y^2 + 15xy^2}{-5xy}$

6. $\dfrac{4x^2 + 6x - 4}{2x - 1}$

Solve the following equation for x.

7. $5 - 3(2x - 7) = 8 - 4x$

Factor each of the following polynomials completely.

8. $24a^3 - 16a^2$

9. $7m^2n - 21mn - 49mn^2$

10. $a^2 - 64b^2$

11. $5p^3 - 80pq^2$

12. $a^2 - 14a + 48$

13. $2w^3 - 8w^2 - 42w$

Solve each of the following equations.

14. $x^2 - 9x + 20 = 0$

15. $2x^2 - 32 = 0$

Solve the following applications.

16. Twice the square of a positive integer is 35 more than 9 times that integer. What is the integer?

1. _____

2. _____

3. _____

4. _____

5. _____

6. _____

7. _____

8. _____

9. _____

10. _____

11. _____

12. _____

13. _____

14. _____

15. _____

16. _____

Answers

17. _____

18. _____

19. _____

20. _____

21. _____

22. _____

23. _____

24. _____

25. _____

26. _____

17. The length of a rectangle is 2 in. more than 3 times its width. If the area of the rectangle is 85 in.^{2}, find the dimensions of the rectangle.

Write each fraction in simplest form.

18. $\dfrac{m^2 - 4m}{3m - 12}$

19. $\dfrac{a^2 - 49}{3a^2 + 22a + 7}$

Perform the indicated operations.

20. $\dfrac{3x^2 + 9x}{x^2 - 9} \cdot \dfrac{2x^2 - 9x + 9}{2x^3 - 3x^2}$

21. $\dfrac{4w^2 - 25}{2w^2 - 5w} \div (6w + 15)$

Graph each of the following equations.

22. $x - y = 5$

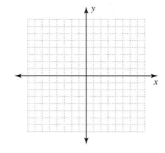

23. $y = \dfrac{2}{3}x + 3$

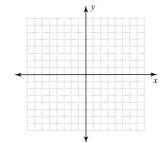

24. $2x - 5y = 10$

25. $y = -5$

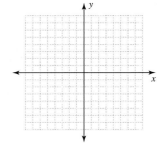

26. Find the slope of the line through the pair of points $(-2, -3)$ and $(5, 7)$.

Answers

27. Find the slope and *y*-intercept of the line described by the equation $5x - 3y = 15$.

28. Given the slope and *y*-intercept for the following line, write the equation of the line. Then graph the line.

Slope = 2; *y*-intercept: $(0, -5)$

Graph each of the following inequalities.

29. $x + 2y < 6$

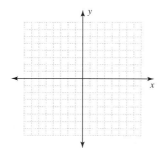

30. $3x - 4y \geq 12$

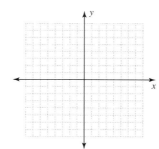

Solve the following system by graphing.

31. $3x + 2y = 6$
    $x + 2y = -2$

Answers

Beginning Algebra   The Streeter/Hutchison Series in Mathematics   © The McGraw-Hill Companies. All Rights Reserved.

32. _____

33. _____

34. _____

35. _____

36. _____

37. _____

38. _____

39. _____

40. _____

Solve each of the following systems. If a unique solution does not exist, state whether the system is inconsistent or dependent.

32. $5x + 2y = 30$
$\quad\ x - 4y = 17$

33. $2x - 6y = 8$
$\qquad\ x = 3y + 4$

34. $4x - 5y = 20$
$\quad\ 2x + 3y = 10$

35. $4x + 2y = 11$
$\quad\ 2x + \ y = \ 5$

36. $4x - 3y = 7$
$\quad\ 6x + 6y = 7$

Solve each of the following applications.

37. One number is 4 less than 5 times another. If the sum of the numbers is 26, what are the two numbers?

38. Cynthia bought five blank VHS tapes and four cassette tapes for $28.50. Charlie bought four VHS tapes and two cassette tapes for $21.00. Find the cost of each type of tape.

39. Receipts for a concert, attended by 450 people, were $27,750. If reserved-seat tickets were $70 and general-admission tickets were $40, how many of each type of ticket were sold?

40. Anthony invested part of his $12,000 inheritance in a bond paying 9% and the other part in a savings account paying 6%. If his interest from the two investments was $930 in the first year, how much did he have invested at each rate?

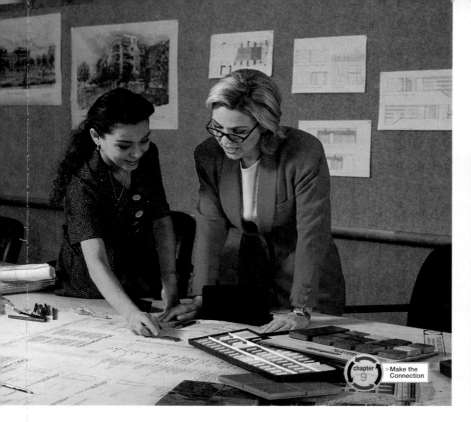

INTRODUCTION

Engineers and architects must include safety plans when designing public buildings. The Uniform Building Code states size and location requirements for exits: "If two exits are required in a building, they must be placed apart a distance not less than one-half the length of the maximum overall diagonal dimension of the building. . . ." Stated in algebraic terms, if the building is rectangular and if d is the distance between exits, l is the length of the building, and w is the width of the building, then

$$d \geq \frac{1}{2}\sqrt{l^2 + w^2}$$

For example, if a rectangular building is 50 ft by 40 ft, the diagonal dimension is $\sqrt{50^2 + 40^2}$, and the distance d between the exits must be equal to or more than half of this value. Thus,

$$d \geq \frac{1}{2}\sqrt{50^2 + 40^2}$$

In this case, the distance between the exits must be 32 ft or more.

The radical sign is often used in the measurement of distances and is based on the Pythagorean theorem, which describes the relationship between the sides of a right triangle. The algebraic expression above is an example of how algebra can make complicated statements easier to understand.

CHAPTER

Exponents and Radicals

CHAPTER 9 OUTLINE

Name _____

Section _____ Date _____

Answers

1. _____

2. _____

3. _____

4. _____ 5. _____

6. _____ 7. _____

8. _____ 9. _____

10. _____ 11. _____

12. _____ 13. _____

14. _____ 15. _____

16. _____ 17. _____

18. _____ 19. _____

20. _____ 21. _____

22. _____

23. _____

24. _____

25. _____

This pretest provides a preview of the types of exercises you will encounter in each section of this chapter. The answers for these exercises can be found in the back of the text. If you are working on your own, or ahead of the class, this pretest can help you identify the sections in which you should focus more of your time.

Evaluate, if possible.

9.1
1. $\sqrt{144}$ **2.** $\sqrt[3]{64}$ **3.** $\sqrt{-121}$ **4.** $-\sqrt[3]{-125}$

Simplify each of the following radical expressions.

9.2
5. $\sqrt{72}$ **6.** $\sqrt{27x^3}$ **7.** $\sqrt{\dfrac{9}{36}}$ **8.** $\sqrt{\dfrac{7}{25}}$

Simplify by combining like terms.

9.3
9. $3\sqrt{7} - 2\sqrt{7} + 6\sqrt{7}$ **10.** $\sqrt{5} + \sqrt{20}$

11. $3\sqrt{12} - \sqrt{27}$ **12.** $\sqrt{18} - \sqrt{98} + \sqrt{50}$

Simplify each of the following radical expressions.

9.4
13. $\sqrt{2x} \cdot \sqrt{5x}$ **14.** $\sqrt{2}(\sqrt{5} - 3\sqrt{2})$

15. $(\sqrt{3} + 2)(\sqrt{3} + 3)$ **16.** $\dfrac{\sqrt{5}}{\sqrt{3}}$ **17.** $\dfrac{44 - \sqrt{605}}{11}$

Find length x in each triangle. Express your answer to the nearest hundredth.

9.6
18.

19.

20.

21.

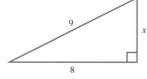

Solve the following word problem.

22. If the diagonal of a rectangle is 25 cm and the width of the rectangle is 16 cm, what is the length of the rectangle? Express your answer to the nearest hundredth of a centimeter.

Find the distance between the given two points. Express your answer to the nearest hundredth.

23. $(-4, 9)$ and $(-18, 9)$ **24.** $(-3, 4)$ and $(-6, -2)$

9.5 **25.** Solve $\sqrt{x - 3} = 9$ for x.

9.1

Roots and Radicals

< 9.1 Objectives >

1 > Use radical notation to represent roots

2 > Approximate a square root

3 > Evaluate cube and fourth roots

4 > Distinguish between rational and irrational numbers

In Chapter 3, we discussed the properties of exponents. Over the next four sections, we will work with a new notation that "reverses" the process of raising to a power.

We know that a statement such as

$$x^2 = 9$$

it is read as "x squared equals 9."

Here we are concerned with the relationship between the variable x and the number 9. We call that relationship the **square root** and say, equivalently, that "x is the square root of 9."

We know from experience that x must be 3 (because $3^2 = 9$) or -3 [because $(-3)^2 = 9$]. We see that 9 has two square roots, 3 and -3. In fact, every positive number has *two* square roots. In general, if $x^2 = a$, we call x a *square root of a*.

We are now ready for our new notation. The symbol $\sqrt{}$ is called a **radical sign.** We just saw that 3 is the positive square root of 9. We also call 3 the **principal square root** of 9 and write

$$\sqrt{9} = 3$$

to indicate that 3 is the principal square root of 9.

> **NOTE**
>
> The symbol $\sqrt{}$ first appeared in print in 1525. In Latin, "radix" means **root,** and this was contracted to a small *r*. The present symbol may have evolved from the manuscript form of that small *r*.

Definition

Square Root	$\sqrt{a}$ is the *principal* square root of a. It is the nonnegative number whose square is a.

Example 1 — **Finding Principal Square Roots**

< Objective 1 >

Find the following square roots.

(a) $\sqrt{49} = 7$ Because 7 is the positive number we must square to get 49.

(b) $\sqrt{\dfrac{4}{9}} = \dfrac{2}{3}$ Because $\dfrac{2}{3}$ is the positive number we must square to get $\dfrac{4}{9}$.

Check Yourself 1

Find the following square roots.

(a) $\sqrt{64}$ **(b)** $\sqrt{144}$ **(c)** $\sqrt{\dfrac{16}{25}}$

Beginning Algebra   The Streeter/Hutchison Series in Mathematics   © The McGraw-Hill Companies. All Rights Reserved.

NOTES

When you use the radical sign, you are referring to the *positive square root:*

$\sqrt{25} = 5$

$-\sqrt{x}$ is the negative square root of *x*.

Each positive number has two distinct square roots. For instance, 25 has square roots of 5 and -5 because

$5^2 = 25$ and $(-5)^2 = 25$

If you want to indicate the negative square root, you must use a minus sign in front of the radical.

$-\sqrt{25} = -5$

Example 2   **Finding Square Roots**

Find the following square roots.

(a) $\sqrt{100} = 10$ The principal root.

(b) $-\sqrt{100} = -10$ The negative square root.

(c) $-\sqrt{\dfrac{9}{16}} = -\dfrac{3}{4}$

> CAUTION

Be Careful! Do not confuse $-\sqrt{9}$ with $\sqrt{-9}$

The expression $-\sqrt{9}$ is -3, whereas $\sqrt{-9}$ is not a real number. Values such as $\sqrt{-9}$ will be discussed in subsequent mathematics courses.

Check Yourself 2

Find the following square roots.

(a) $\sqrt{16}$ **(b)** $-\sqrt{16}$ **(c)** $-\sqrt{\dfrac{16}{25}}$

Every number that we encounter in this text is a **real number.** The square roots of negative numbers are *not* real numbers. For instance, $\sqrt{-9}$ is *not* a real number because there is *no* real number x such that

$x^2 = -9$

Example 3 summarizes our discussion thus far.

Example 3   **Finding Square Roots**

Evaluate each of the following square roots.

(a) $\sqrt{36} = 6$ **(b)** $\sqrt{121} = 11$

(c) $-\sqrt{64} = -8$ **(d)** $\sqrt{-64}$ is not a real number.

(e) $\sqrt{0} = 0$ Because $0 \cdot 0 = 0.$

Check Yourself 3

Evaluate, if possible.

(a) $\sqrt{81}$ **(b)** $\sqrt{49}$ **(c)** $-\sqrt{49}$ **(d)** $\sqrt{-49}$

All calculators have square-root keys, but the only integers for which the calculator gives the exact value of the square root are perfect square integers. For all other positive integers, *a calculator gives only an approximation of the correct value.* In Example 4 we use a calculator to approximate square roots.

Example 4   **Approximating Square Roots**

< Objective 2 >

Use your calculator to approximate each square root to the nearest hundredth.

(a) $\sqrt{45} \approx 6.708203932 \approx 6.71$ **(b)** $\sqrt{8} \approx 2.83$

(c) $\sqrt{20} \approx 4.47$ **(d)** $\sqrt{273} \approx 16.52$

NOTE

The $\approx$ sign means "is approximately equal to."

Check Yourself 4

Use your calculator to approximate each square root to the nearest hundredth.

(a) $\sqrt{3}$ **(b)** $\sqrt{14}$ **(c)** $\sqrt{91}$ **(d)** $\sqrt{756}$

NOTE

$\sqrt[3]{8}$ is read "the cube root of 8."

As we mentioned earlier, finding the square root of a number is the reverse of squaring a number. We can extend that idea to work with other roots of numbers. For instance, the **cube root** of a number is the quantity we must cube (or raise to the third power) to get the original number. For example, the cube root of 8 is 2 because $2^3 = 8$, and we write

$$\sqrt[3]{8} = 2$$

The parts of a radical expression are summarized as follows.

Definition

Parts of a Radical Expression

Every radical expression contains three parts as shown here. The principal *n*th root of *a* is written as

Index

$$\sqrt[n]{a}$$

Radical
sign   Radicand

NOTES

The index for $\sqrt[3]{a}$ is 3.

The index of 2 for square roots is generally not written. We understand that $\sqrt{a}$ is the principal square root of a.

NOTE

The *even power* of a real number is always *positive* or *zero*.

NOTE

It would be helpful for your work here and in future mathematics classes to memorize these roots.

To illustrate, the *cube root* of 64 is written

$$\underset{\text{of 3}}{\text{Index}\longrightarrow}\ \sqrt[3]{64} = 4$$

because $4^3 = 64$. And

$$\underset{\text{of 4}}{\text{Index}\longrightarrow}\ \sqrt[4]{81} = 3$$

is the *fourth root* of 81 because $3^4 = 81$.

We can find roots of negative numbers as long as the index is *odd* (3, 5, etc.). For example,

$$\sqrt[3]{-64} = -4$$

because $(-4)^3 = -64$.

If the index is *even* (2, 4, etc.), roots of negative numbers are *not* real numbers. For example,

$$\sqrt[4]{-16}$$

is not a real number because there is no real number x such that $x^4 = -16$.

The following table shows the most common roots.

Square Roots		Cube Roots	Fourth Roots
$\sqrt{1} = 1$	$\sqrt{49} = 7$	$\sqrt[3]{1} = 1$	$\sqrt[4]{1} = 1$
$\sqrt{4} = 2$	$\sqrt{64} = 8$	$\sqrt[3]{8} = 2$	$\sqrt[4]{16} = 2$
$\sqrt{9} = 3$	$\sqrt{81} = 9$	$\sqrt[3]{27} = 3$	$\sqrt[4]{81} = 3$
$\sqrt{16} = 4$	$\sqrt{100} = 10$	$\sqrt[3]{64} = 4$	$\sqrt[4]{256} = 4$
$\sqrt{25} = 5$	$\sqrt{121} = 11$	$\sqrt[3]{125} = 5$	$\sqrt[4]{625} = 5$
$\sqrt{36} = 6$	$\sqrt{144} = 12$		

You can use this table in Example 5, which summarizes the discussion so far.

Example 5   Evaluating Roots

< Objective 3 >

NOTES

The cube root of a negative number is negative.

The fourth root of a negative number is not a real number.

Evaluate each of the following.

(a) $\sqrt[5]{32} = 2$ because $2^5 = 32$.

(b) $\sqrt[3]{-125} = -5$ because $(-5)^3 = -125$.

(c) $\sqrt[4]{-81}$ is not a real number.

Check Yourself 5

Evaluate, if possible.

(a) $\sqrt[3]{64}$ **(b)** $\sqrt[4]{16}$ **(c)** $\sqrt[4]{-256}$ **(d)** $\sqrt[3]{-8}$

The radical notation helps us to distinguish between two important types of numbers: rational numbers and irrational numbers.

A **rational number** can be represented by a fraction whose numerator and denominator are integers and whose denominator is not zero. The form of a rational number is

$$\frac{a}{b} \qquad a \text{ and } b \text{ are integers, } b \neq 0$$

Certain square roots are rational numbers also. For example,

$$\sqrt{4} \qquad \sqrt{25} \qquad \text{and} \qquad \sqrt{64}$$

represent the rational numbers 2, 5, and 8, respectively.

An **irrational number** is a number that *cannot* be written as the ratio of two integers. For example, the square root of any positive number that is not itself a perfect square is an irrational number. Because the radicands are *not* perfect squares, the expressions $\sqrt{2}$, $\sqrt{3}$, and $\sqrt{5}$ represent irrational numbers.

Example 6   Identifying Rational Numbers

< Objective 4 >

Which of the following numbers are rational and which are irrational?

$$\sqrt{\frac{2}{3}} \qquad \sqrt{\frac{4}{9}} \qquad \sqrt{7} \qquad \sqrt{16} \qquad \sqrt{25}$$

Here $\sqrt{7}$ and $\sqrt{\frac{2}{3}}$ are irrational numbers. The numbers $\sqrt{16}$ and $\sqrt{25}$ are rational because 16 and 25 are perfect squares. Also $\sqrt{\frac{4}{9}}$ is rational because $\sqrt{\frac{4}{9}} = \frac{2}{3}$.

Check Yourself 6

Determine whether each root is rational or irrational.

(a) $\sqrt{26}$ **(b)** $\sqrt{49}$ **(c)** $\sqrt{\frac{6}{7}}$

(d) $\sqrt{105}$ **(e)** $\sqrt{\frac{16}{9}}$

An important fact about the irrational numbers is that their decimal representations are always *nonterminating* and *nonrepeating*. We can therefore only approximate irrational numbers with a decimal that has been rounded. A calculator can be used to find roots. However, note that the values found for the irrational roots are only approximations. For instance, $\sqrt{2}$ is approximately 1.414 (to three decimal places), and we can write

$$\sqrt{2} \approx 1.414$$

With a calculator we find that

$$(1.414)^2 = 1.999396$$

NOTE

For this reason we refer to the number line as the **real number line**.

The set of all rational numbers and the set of all irrational numbers together form the set of *real numbers.* The real numbers represent every point that can be pictured on the number line. Some examples are shown.

The following diagram summarizes the relationships among various numeric sets.

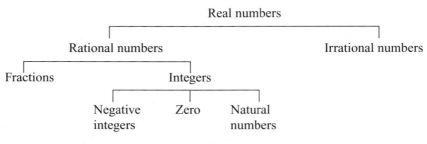

We conclude our work in this section by developing a general result that we will need later. Start by looking at two numerical examples.

$$\sqrt{2^2} = \sqrt{4} = 2$$
$$\sqrt{(-2)^2} = \sqrt{4} = 2 \qquad \text{because } (-2)^2 = 4$$

NOTE

This is because the principal square root of a number is always positive or zero.

Consider the value of $\sqrt{x^2}$ when x is positive or negative.

In the first equation when $x = 2$:   In the second equation when $x = -2$:

$$\sqrt{2^2} = 2 \qquad\qquad\qquad \sqrt{(-2)^2} \ne -2$$
$$\sqrt{(-2)^2} = -(-2) = 2$$

Comparing the results of these two equations, we see that $\sqrt{x^2}$ is x if x is positive (or 0) and $\sqrt{x^2}$ is $-x$ if x is negative. We can write

$$\sqrt{x^2} = \begin{cases} x & \text{when } x \ge 0 \\ -x & \text{when } x < 0 \end{cases}$$

From your earlier work with absolute values you will remember that

$$|x| = \begin{cases} x & \text{when } x \ge 0 \\ -x & \text{when } x < 0 \end{cases}$$

and we can summarize the discussion by writing

$$\sqrt{x^2} = |x| \qquad \text{for any real number } x$$

Example 7   **Evaluating Radical Expressions**

NOTE

Alternatively in part (b), we could write

$$\sqrt{(-4)^2} = \sqrt{16} = 4$$

Evaluate each of the following.

(a) $\sqrt{5^2} = 5$

(b) $\sqrt{(-4)^2} = |-4| = 4$

Check Yourself 7

Evaluate.

(a) $\sqrt{6^2}$ **(b)** $\sqrt{(-6)^2}$

Check Yourself ANSWERS

1. (a) 8; **(b)** 12; **(c)** $\dfrac{4}{5}$ **2. (a)** 4; **(b)** -4; **(c)** $-\dfrac{4}{5}$ **3. (a)** 9; **(b)** 7; **(c)** -7;
(d) not a real number   **4. (a)** ≈ 1.73; **(b)** ≈ 3.74; **(c)** ≈ 9.54; **(d)** ≈ 27.50
5. (a) 4; **(b)** 2; **(c)** not a real number; **(d)** -2 **6. (a)** Irrational;
(b) rational (because $\sqrt{49} = 7$); **(c)** irrational; **(d)** irrational;
(e) rational $\left(\text{because } \sqrt{\dfrac{16}{9}} = \dfrac{4}{3}\right)$ **7. (a)** 6; **(b)** 6

Reading Your Text

The following fill-in-the-blank exercises are designed to ensure that you understand some of the key vocabulary used in this section.

SECTION 9.1

(a) $\sqrt{a}$ is the positive (or _____) square root of a.

(b) To indicate a _____ square root, place a minus sign in front of the radical.

(c) A calculator gives only an _____ to the correct value for finding the square root of an integer that is not a perfect square.

(d) Every radical expression contains three parts, the radical sign, the index, and the _____.

9.1 exercises

Name _____

Section _____ Date _____

Answers

1. _____ 2. _____

3. _____ 4. _____

5. _____

6. _____

7. _____

8. _____

9. _____ 10. _____

11. _____

12. _____ 13. _____

14. _____ 15. _____

16. _____

17. _____

18. _____ 19. _____

20. _____

< Objectives 1–3 >

Evaluate, if possible.

1. $\sqrt{25}$

2. $\sqrt{121}$

3. $\sqrt{400}$

4. $\sqrt{64}$

5. $-\sqrt{144}$ > Videos

6. $\sqrt{-121}$

7. $\sqrt{-81}$

8. $-\sqrt{81}$

9. $\sqrt{\dfrac{36}{25}}$

10. $-\sqrt{\dfrac{1}{25}}$

11. $\sqrt{-\dfrac{4}{25}}$ > Videos

12. $\sqrt{\dfrac{4}{25}}$

13. $\sqrt{\dfrac{121}{100}}$

14. $\sqrt{\dfrac{256}{25}}$

15. $\sqrt[3]{-64}$ > Videos

16. $\sqrt[4]{-16}$

17. $\sqrt[4]{-81}$

18. $-\sqrt[3]{64}$

19. $-\sqrt[3]{27}$

20. $\sqrt[3]{-27}$

21. $\sqrt[4]{1{,}296}$

22. $\sqrt[3]{1{,}000}$

23. $\sqrt[3]{\dfrac{1}{27}}$

24. $\sqrt[3]{-\dfrac{8}{27}}$

< Objective 4 >

State whether each of the following roots is rational or irrational.

25. $\sqrt{21}$

26. $\sqrt{36}$

27. $\sqrt{100}$

28. $\sqrt{7}$

29. $\sqrt[3]{9}$

30. $\sqrt[3]{27}$

31. $\sqrt[4]{16}$

32. $\sqrt{\dfrac{4}{9}}$

33. $\sqrt{\dfrac{9}{15}}$

34. $\sqrt[3]{5}$

35. $\sqrt[3]{-27}$

36. $-\sqrt[4]{81}$ > Videos

37. GEOMETRY The area of a square is 32 ft². Find the length of a side to the nearest hundredth.

38. GEOMETRY The area of a square is 83 ft². Find the length of the side to the nearest hundredth.

39. GEOMETRY The area of a circle is 147 ft². Find the radius to the nearest hundredth.

40. GEOMETRY If the area of a circle is 72 cm², find the radius to the nearest hundredth.

41. SCIENCE AND MEDICINE The time in seconds that it takes for an object to fall from rest is given by $t = \dfrac{1}{4}\sqrt{s}$, in which s is the distance fallen (in feet). Find the time required for an object to fall to the ground from a building that is 800 ft high.

42. SCIENCE AND MEDICINE Find the time required for an object to fall to the ground from a building that is 1,400 ft high. (Use the formula given in exercise 41.)

Answers

21. _____

22. _____

23. _____

24. _____

25. _____

26. _____

27. _____

28. _____

29. _____

30. _____

31. _____

32. _____

33. _____

34. _____

35. _____

36. _____

37. _____

38. _____

39. _____ 40. _____

41. _____ 42. _____

SECTION 9.1   841

Answers

43. _____

44. _____

45. _____

46. _____

47. _____

48. _____

49. _____

50. _____

51. _____

52. _____

53. _____

54. _____

55. _____

56. _____

57. _____ 58. _____

59. _____ 60. _____

61. _____ 62. _____

63. _____ 64. _____

65. _____ 66. _____

67. _____ 68. _____

| Basic Skills | **Advanced Skills** | Vocational-Technical Applications | Calculator/Computer | Above and Beyond | Getting Ready |

Determine whether each of the following statements is **true** *or* **false.**

43. $\sqrt{16x^{16}} = 4x^4$

44. $\sqrt{(x-4)^2} = x - 4$

45. $\sqrt{16x^{-4}y^{-4}}$ is a real number

46. $\sqrt{x^2 + y^2} = x + y$

47. $\dfrac{\sqrt{x^2 - 25}}{x - 5} = \sqrt{x + 5}$

48. $\sqrt{2} + \sqrt{6} = \sqrt{8}$

< Objectives 1–3 >

For exercises 49–54, find the two expressions that are equivalent.

49. $\sqrt{-16}, -\sqrt{16}, -4$

50. $-\sqrt{25}, -5, \sqrt{-25}$

51. $\sqrt[3]{-125}, -\sqrt[3]{125}, |-5|$

52. $\sqrt[5]{-32}, -\sqrt[5]{32}, |-2|$

53. $\sqrt[4]{10{,}000}, 100, \sqrt[3]{1{,}000}$

54. $10^2, \sqrt{10{,}000}, \sqrt[3]{100{,}000}$

| Basic Skills | Advanced Skills | Vocational-Technical Applications | **Calculator/Computer** | Above and Beyond | Getting Ready |

< Objective 2 >

Use your calculator to approximate the square root to the nearest hundredth.

55. $\sqrt{11}$

56. $\sqrt{14}$

57. $\sqrt{7}$

58. $\sqrt{23}$

59. $\sqrt{65}$

60. $\sqrt{78}$

61. $\sqrt{\dfrac{2}{5}}$ > Videos

62. $\sqrt{\dfrac{4}{3}}$

63. $\sqrt{\dfrac{8}{9}}$

64. $\sqrt{\dfrac{7}{15}}$

65. $-\sqrt{18}$

66. $-\sqrt{31}$

67. $-\sqrt{27}$

68. $-\sqrt{65}$

In exercises 69–71, the area is given in square feet. Find the length of a side of the square. Round your answer to the nearest hundredth of a foot.

69.

10 ft²

70.

13 ft²

71.

17 ft²

Answers

69. _____

70. _____

71. _____

72. _____

73. _____

74. _____

75. _____

76. _____

77. _____

72. Suppose that a weight is attached to a string of length L, and the other end of the string is held fixed. If we pull the weight and then release it, allowing the weight to swing back and forth, we can observe the behavior of a simple pendulum. The period T is the time required for the weight to complete a full cycle, swinging forward and then back. The following formula may be used to describe the relationship between T and L.

$$T = 2\pi\sqrt{\frac{L}{g}}$$

If L is expressed in centimeters, then $g = 980$ cm/s². For each of the following string lengths, calculate the corresponding period. Round to the nearest tenth of a second.

(a) 30 cm   **(b)** 50 cm   **(c)** 70 cm   **(d)** 90 cm   **(e)** 110 cm

73. Is there any prime number whose square root is an integer? Explain your answer.

74. Determine the missing binomial in the following: $(\sqrt{3} - 2)(\quad) = -1$.

75. Try the following using your calculator.

(a) Choose a number greater than 1 and find its square root. Then find the square root of the result and continue in this manner, observing the successive square roots. Do these numbers seem to be approaching a certain value? If so, what?

(b) Choose a number greater than 0 but less than 1 and find its square root. Then find the square root of the result, and continue in this manner, observing successive square roots. Do these numbers seem to be approaching a certain value? If so, what?

76. (a) Can a number be equal to its own square root?

(b) Other than the number(s) found in part **(a)**, is a number always greater than its square root? Investigate.

77. Let a and b be positive numbers. If a is greater than b, is it always true that the square root of a is greater than the square root of b? Investigate.

Answers

(a) _____

(b) _____

(c) _____

(d) _____

(e) _____

(f) _____

(g) _____

(h) _____

Find each of the following products.

(a) $(4x^2)(2x)$ **(b)** $(9a^4)(5a)$ **(c)** $(16m^2)(3m)$ **(d)** $(8b^3)(2b)$

(e) $(27p^6)(3p)$ **(f)** $(81s^4)(s^3)$ **(g)** $(100y^4)(2y)$ **(h)** $(49m^6)(2m)$

Answers

1. 5   **3.** 20   **5.** -12 **7.** Not a real number   **9.** $\dfrac{6}{5}$

11. Not a real number   **13.** $\dfrac{11}{10}$ **15.** -4 **17.** Not a real number

19. -3 **21.** 6   **23.** $\dfrac{1}{3}$ **25.** Irrational   **27.** Rational

29. Irrational   **31.** Rational   **33.** Irrational   **35.** Rational

37. 5.66 ft   **39.** 6.84 ft   **41.** 7.07 s   **43.** False   **45.** True

47. False   **49.** $-\sqrt{16}, -4$ **51.** $\sqrt[3]{-125}, -\sqrt[3]{125}$

53. $\sqrt[4]{10{,}000}, \sqrt[3]{1{,}000}$ **55.** 3.32   **57.** 2.65   **59.** 8.06   **61.** 0.63

63. 0.94   **65.** -4.24 **67.** -5.20 **69.** 3.16 ft   **71.** 4.12 ft

73. No; Above and Beyond   **75.** Above and Beyond

77. Above and Beyond   **(a)** $8x^3$ **(b)** $45a^5$ **(c)** $48m^3$ **(d)** $16b^4$

(e) $81p^7$ **(f)** $81s^7$ **(g)** $200y^5$ **(h)** $98m^7$

9.2

Simplifying Radical Expressions

< 9.2 Objectives >

1 > Simplify expressions involving numeric radicals

2 > Simplify expressions involving algebraic radicals

In Section 9.1, we introduced radical notation. For most applications, we want to make sure that all radical expressions are in *simplest form*. To accomplish this, the following three conditions must be satisfied.

Property

Square Root Expressions in Simplest Form

An expression involving square roots is in *simplest form* if

1. There are no perfect-square factors in a radical.

2. No fraction appears inside a radical.

3. No radical appears in the denominator of a fraction.

For instance, considering condition 1,

$\sqrt{17}$ is in simplest form because 17 has *no* perfect-square factors

whereas

$\sqrt{12}$ is *not* in simplest form

because it does contain a perfect-square factor:

$$\sqrt{12} = \sqrt{4 \cdot 3}$$

A perfect square

To simplify radical expressions, we need to develop two important properties. First, look at the following expressions:

$$\sqrt{4 \cdot 9} = \sqrt{36} = 6$$
$$\sqrt{4} \cdot \sqrt{9} = 2 \cdot 3 = 6$$

Because this tells us that $\sqrt{4 \cdot 9} = \sqrt{4} \cdot \sqrt{9}$, the following general rule for radicals is suggested.

Property

Property 1 of Radicals

For any nonnegative real numbers a and b,

$$\sqrt{ab} = \sqrt{a} \cdot \sqrt{b}$$

In words, the square root of a product is the product of the square roots.

Example 1 illustrates how this property is applied in simplifying expressions when radicals are involved.

| **Example** 1 | **Simplifying Radical Expressions** |

< Objective 1 >

NOTE

Perfect-square factors are 1, 4, 9, 16, 25, 36, 49, 64, 81, 100, and so on.

NOTE

We removed the perfect-square factor from inside the radical, so the expression is in simplest form.

Simplify each expression.

(a) $\sqrt{12} = \sqrt{4 \cdot 3}$

A perfect square

$= \sqrt{4} \cdot \sqrt{3}$ Apply Property 1.

$= 2\sqrt{3}$

(b) $\sqrt{45} = \sqrt{9 \cdot 5}$

A perfect square

$= \sqrt{9} \cdot \sqrt{5}$

$= 3\sqrt{5}$

(c) $\sqrt{72} = \sqrt{36 \cdot 2}$

We look for the *largest* perfect-square factor, here 36. Then apply Property 1.

A perfect square

$= \sqrt{36} \cdot \sqrt{2}$

$= 6\sqrt{2}$

(d) $5\sqrt{18} = 5\sqrt{9 \cdot 2}$

A perfect square

$= 5 \cdot \sqrt{9} \cdot \sqrt{2} = 5 \cdot 3 \cdot \sqrt{2} = 15\sqrt{2}$

Be careful!   Even though

$\sqrt{a \cdot b} = \sqrt{a} \cdot \sqrt{b}$

the expression $\sqrt{a + b}$ is *not the same* as $\sqrt{a} + \sqrt{b}$

Let $a = 4$ and $b = 9$, and substitute.

$\sqrt{a + b} = \sqrt{4 + 9} = \sqrt{13}$

$\sqrt{a} + \sqrt{b} = \sqrt{4} + \sqrt{9} = 2 + 3 = 5$

Because $\sqrt{13} \neq 5$, we see that the expressions $\sqrt{a + b}$ and $\sqrt{a} + \sqrt{b}$ are not, in general, the same.

> CAUTION

The root of a product is equal to the product of the roots. The root of a sum **is not** equal to the sum of the roots.

Check Yourself 1

Simplify.

(a) $\sqrt{20}$ **(b)** $\sqrt{75}$ **(c)** $\sqrt{98}$ **(d)** $\sqrt{48}$

The process is the same if variables are involved in a radical expression. In our remaining work with radicals, we will assume that all variables represent positive real numbers.

| **Example** 2 | **Simplifying Algebraic Radical Expressions** |

< Objective 2 >

Simplify each of the following radicals.

(a) $\sqrt{x^3} = \sqrt{x^2 \cdot x}$

A perfect square

$= \sqrt{x^2} \cdot \sqrt{x}$

$= x\sqrt{x}$

NOTE

$\sqrt{x^2} = x$ (as long as x is not negative).

(b) $\sqrt{4b^3} = \sqrt{4 \cdot b^2 \cdot b}$

Perfect squares

$= \sqrt{4b^2} \cdot \sqrt{b}$

$= 2b\sqrt{b}$

RECALL

We want the perfect-square factor to have the largest possible even exponent, here 4. Keep in mind that $a^2 \cdot a^2 = a^4$

(c) $\sqrt{18a^5} = \sqrt{9 \cdot a^4 \cdot 2a}$

Perfect squares

$= \sqrt{9a^4} \cdot \sqrt{2a}$

$= 3a^2\sqrt{2a}$

 Check Yourself 2

Simplify.

(a) $\sqrt{9x^3}$ **(b)** $\sqrt{27m^3}$ **(c)** $\sqrt{50b^5}$

To develop a second property for radicals, look at the following expressions:

$$\sqrt{\frac{16}{4}} = \sqrt{4} = 2$$

$$\frac{\sqrt{16}}{\sqrt{4}} = \frac{4}{2} = 2$$

Because $\sqrt{\dfrac{16}{4}} = \dfrac{\sqrt{16}}{\sqrt{4}}$, a second general rule for radicals is suggested.

Property

Property 2 of Radicals

For any positive real numbers a and b,

$$\sqrt{\frac{a}{b}} = \frac{\sqrt{a}}{\sqrt{b}}$$

In words, the square root of a quotient is the quotient of the square roots.

This property is used in a fashion similar to Property 1 in simplifying radical expressions. Remember that our second condition for a radical expression to be in simplest form states that no fraction should appear inside a radical. Example 3 illustrates how expressions that violate that condition are simplified.

Example 3   **Simplifying Radical Expressions**

Write each expression in simplest form.

NOTE

Apply Property 2 to write the numerator and denominator as separate radicals.

(a) $\sqrt{\dfrac{9}{4}} = \dfrac{\sqrt{9}}{\sqrt{4}}$ Remove any perfect squares from the radical.

$= \dfrac{3}{2}$

(b) $\sqrt{\dfrac{2}{25}} = \dfrac{\sqrt{2}}{\sqrt{25}}$ Apply Property 2.

$= \dfrac{\sqrt{2}}{5}$

NOTE

Factor $8x^2$ as $4x^2 \cdot 2$.

(c) $\sqrt{\dfrac{8x^2}{9}} = \dfrac{\sqrt{8x^2}}{\sqrt{9}}$ Apply Property 2.

$= \dfrac{\sqrt{4x^2 \cdot 2}}{3}$

$= \dfrac{\sqrt{4x^2} \cdot \sqrt{2}}{3}$ Apply Property 1 in the numerator.

$= \dfrac{2x\sqrt{2}}{3}$

Check Yourself 3

Simplify.

(a) $\sqrt{\dfrac{25}{16}}$ (b) $\sqrt{\dfrac{7}{9}}$ (c) $\sqrt{\dfrac{12x^2}{49}}$

In the three expressions in Example 3, the denominator of the fraction appearing in the radical was a perfect square, and we were able to write each expression in simplest radical form by removing that perfect square from the radical.

If the denominator of the fraction in the radical is *not* a perfect square, we can still apply Property 2 of radicals. As we show in Example 4, the third condition for a radical to be in simplest form is then violated, and a new technique is necessary.

| Example 4 | Simplifying Radical Expressions |

NOTE

We begin by applying Property 2.

Write each expression in simplest form.

(a) $\sqrt{\dfrac{1}{3}} = \dfrac{\sqrt{1}}{\sqrt{3}} = \dfrac{1}{\sqrt{3}}$

Do you see that $\dfrac{1}{\sqrt{3}}$ is still not in simplest form because of the radical in the denominator? To solve this problem, we multiply the numerator and denominator by $\sqrt{3}$. Note that the denominator will become

$$\sqrt{3} \cdot \sqrt{3} = \sqrt{9} = 3$$

We then have

RECALL

We can do this because we are multiplying the fraction by $\dfrac{\sqrt{3}}{\sqrt{3}}$ or 1, which does not change its value.

$$\dfrac{1}{\sqrt{3}} = \dfrac{1 \cdot \sqrt{3}}{\sqrt{3} \cdot \sqrt{3}} = \dfrac{\sqrt{3}}{3}$$

The expression $\dfrac{\sqrt{3}}{3}$ is now in simplest form because all three of our conditions are satisfied.

NOTE

$\sqrt{2} \cdot \sqrt{5} = \sqrt{2 \cdot 5} = \sqrt{10}$
$\sqrt{5} \cdot \sqrt{5} = 5$

(b) $\begin{aligned}\sqrt{\dfrac{2}{5}} &= \dfrac{\sqrt{2}}{\sqrt{5}} \\ &= \dfrac{\sqrt{2} \cdot \sqrt{5}}{\sqrt{5} \cdot \sqrt{5}} \\ &= \dfrac{\sqrt{10}}{5}\end{aligned}$

and the expression is in simplest form because again our three conditions are satisfied.

NOTE

We multiply numerator and denominator by $\sqrt{7}$ to "clear" the denominator of the radical. This is also known as "rationalizing" the denominator.

(c) $\begin{aligned}\sqrt{\dfrac{3x}{7}} &= \dfrac{\sqrt{3x}}{\sqrt{7}} \\ &= \dfrac{\sqrt{3x} \cdot \sqrt{7}}{\sqrt{7} \cdot \sqrt{7}} \\ &= \dfrac{\sqrt{21x}}{7}\end{aligned}$

The expression is in simplest form.

Check Yourself 4

Simplify.

(a) $\sqrt{\dfrac{1}{2}}$ **(b)** $\sqrt{\dfrac{2}{3}}$ **(c)** $\sqrt{\dfrac{2y}{5}}$

In Example 5, we demonstrate that the properties of radicals introduced in this section apply to roots other than square roots.

Example 5   **Simplifying Radical Expressions**

Write each in simplest form.

(a) $\sqrt[3]{54} = \sqrt[3]{3^3 \cdot 2}$ Find the prime factors for the radicand. We
$= \sqrt[3]{3^3} \cdot \sqrt[3]{2}$ are particularly interested in factors raised
to powers that are multiples of the index, 3.
$= 3\sqrt[3]{2}$ The expression is now in simplest form.

(b) $\sqrt[3]{\dfrac{x^4}{8}} = \dfrac{\sqrt[3]{x^3 \cdot x}}{\sqrt[3]{8}}$

$= \dfrac{\sqrt[3]{x^3} \cdot \sqrt[3]{x}}{\sqrt[3]{2^3}}$ Find the prime factors for the radicand and
rewrite as the quotient of two radicals.

$= \dfrac{x\sqrt[3]{x}}{2}$ The expression is now in simplest form.

Check Yourself 5

Write each in simplest form.

(a) $\sqrt[3]{128}$ **(b)** $\sqrt[3]{\dfrac{a^3 b^5}{27}}$

Check Yourself ANSWERS

1. (a) $2\sqrt{5}$; **(b)** $5\sqrt{3}$; **(c)** $7\sqrt{2}$; **(d)** $4\sqrt{3}$ **2. (a)** $3x\sqrt{x}$; **(b)** $3m\sqrt{3m}$;
(c) $5b^2\sqrt{2b}$ **3. (a)** $\dfrac{5}{4}$; **(b)** $\dfrac{\sqrt{7}}{3}$; **(c)** $\dfrac{2x\sqrt{3}}{7}$ **4. (a)** $\dfrac{\sqrt{2}}{2}$; **(b)** $\dfrac{\sqrt{6}}{3}$;
(c) $\dfrac{\sqrt{10y}}{5}$ **5. (a)** $4\sqrt[3]{2}$; **(b)** $\dfrac{ab\sqrt[3]{b^2}}{3}$

Reading Your Text

The following fill-in-the-blank exercises are designed to ensure that you understand some of the key vocabulary used in this section.

SECTION 9.2

(a) In simplest form radical expressions have no perfect-square _____ in a radical.

(b) In simplest form radical expressions contain no _____ inside the radical.

(c) The square root of a product is the _____ of the square roots.

(d) For an expression involving square roots to be in simplest form, no radical appears in the _____ of a fraction.

Basic Skills | Advanced Skills | Vocational-Technical Applications | Calculator/Computer | Above and Beyond | Getting Ready

< Objectives 1–2 >

Use Property 1 to simplify each of the following radical expressions. Assume that all variables represent positive real numbers.

1. $\sqrt{48}$ > Videos

2. $\sqrt{50}$

3. $\sqrt{28}$

4. $\sqrt{108}$

5. $\sqrt{45}$

6. $\sqrt{80}$

7. $\sqrt{54}$

8. $\sqrt{180}$

9. $\sqrt{200}$

10. $\sqrt{96}$

11. $\sqrt{567}$

12. $\sqrt{300}$

13. $3\sqrt{12}$

14. $5\sqrt{24}$

15. $\sqrt{3x^2}$

16. $\sqrt{7a^2}$

17. $\sqrt{3y^4}$

18. $\sqrt{10x^6}$

19. $\sqrt{2r^3}$

20. $\sqrt{7a^7}$

21. $\sqrt{125b^2}$ > Videos

22. $\sqrt{98m^4}$

23. $\sqrt{24x^4}$

24. $\sqrt{72x^3}$

25. $\sqrt{54a^5}$

26. $\sqrt{200y^6}$

27. $\sqrt{x^3y^2}$

28. $\sqrt{a^2b^5}$

29. $\sqrt{a^6b^4c}$ > Videos

30. $\sqrt{x^3y^4z^2}$

MathZone

Boost your grade at mathzone.com!
> Practice Problems
> NetTutor
> Self-Tests
> e-Professors
> Videos

Name _____

Section _____ Date _____

Answers

1. _____   2. _____

3. _____   4. _____

5. _____   6. _____

7. _____   8. _____

9. _____   10. _____

11. _____   12. _____

13. _____   14. _____

15. _____   16. _____

17. _____   18. _____

19. _____   20. _____

21. _____   22. _____

23. _____   24. _____

25. _____   26. _____

27. _____   28. _____

29. _____   30. _____

Answers

31. _____

32. _____

33. _____

34. _____

35. _____

36. _____

37. _____

38. _____

39. _____

40. _____

41. _____

42. _____

43. _____

44. _____

45. _____

46. _____

47. _____

48. _____

Use Property 2 to simplify each of the following radical expressions.

31. $\sqrt{\dfrac{9}{16}}$

32. $\sqrt{\dfrac{49}{25}}$

33. $\sqrt{\dfrac{3}{4}}$

34. $\sqrt{\dfrac{7}{16}}$

35. $\sqrt{\dfrac{3}{49}}$ > Videos

36. $\sqrt{\dfrac{10}{49}}$

Determine whether each of the following statements is **true** *or* **false.**

37. A simplified radical might have a fraction inside the radical.

38. We cannot, in general, rewrite $\sqrt{a + b}$ as $\sqrt{a} + \sqrt{b}$.

Complete each of the following statements with **never, sometimes,** *or* **always.**

39. For positive real numbers a and b, it is _____ true that $\sqrt{ab} = \sqrt{a} \cdot \sqrt{b}$.

40. A radical in simplest form _____ has a radical in the denominator.

| Basic Skills | **Advanced Skills** | Vocational-Technical Applications | Calculator/Computer | Above and Beyond | Getting Ready |

Use the properties for radicals to simplify each of the following expressions. Assume that all variables represent positive real numbers.

41. $\sqrt{\dfrac{8a^2}{25}}$

42. $\sqrt{\dfrac{12y^2}{49}}$

43. $\sqrt{\dfrac{1}{5}}$

44. $\sqrt{\dfrac{1}{7}}$

45. $\sqrt{\dfrac{5}{2}}$

46. $\sqrt{\dfrac{5}{3}}$

47. $\sqrt{\dfrac{3a}{5}}$

48. $\sqrt{\dfrac{2x}{5}}$

49. $\sqrt{\dfrac{3x^4}{7}}$

50. $\sqrt{\dfrac{5m^2}{2}}$

51. $\sqrt{\dfrac{8s^3}{7}}$ Videos

52. $\sqrt{\dfrac{12x^3}{5}}$

Basic Skills | Advanced Skills | Vocational-Technical Applications | Calculator/Computer | **Above and Beyond** | Getting Ready

Decide whether each of the following is already written in simplest form. If it is not, explain what needs to be done.

53. $\sqrt{10mn}$

54. $\sqrt{18ab}$

55. $\sqrt{\dfrac{98x^2y}{7x}}$

56. $\dfrac{\sqrt{6xy}}{3x}$

57. Find the area and perimeter of this square:

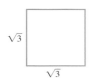

One of these measures, the area, is a rational number, and the other, the perimeter, is an irrational number. Explain how this happened. Will the area always be a rational number? Explain.

58. (a) Evaluate the three expressions $\dfrac{n^2 - 1}{2}, n, \dfrac{n^2 + 1}{2}$ using odd values of n: 1, 3, 5, 7, and so forth. Make a chart like the one that follows and complete it.

n	$a = \dfrac{n^2 - 1}{2}$	$b = n$	$c = \dfrac{n^2 + 1}{2}$	a^2	b^2	c^2
1						
3						
5						
7						
9						
11						
13						
15						

(b) Check for each of these sets of three numbers to see whether this statement is true: $\sqrt{a^2 + b^2} = \sqrt{c^2}$. For how many of your sets of three did this work? Sets of three numbers for which this statement is true are called *Pythagorean triples* because $a^2 + b^2 = c^2$. Can the radical equation be written as $\sqrt{a^2 + b^2} = a + b$? Explain your answer.

Answers

49. _____

50. _____

51. _____

52. _____

53. _____

54. _____

55. _____

56. _____

57. _____

58. _____

The Streeter/Hutchison Series in Mathematics   Beginning Algebra

Answers

(a) _____

(b) _____

(c) _____

(d) _____

(e) _____

(f) _____

(g) _____

(h) _____

Basic Skills | Advanced Skills | Vocational-Technical Applications | Calculator/Computer | Above and Beyond | **Getting Ready**
▲

Use the distributive property to combine the like terms in each of the following expressions.

(a) $5x + 6x$ **(b)** $8a - 3a$

(c) $10y - 12y$ **(d)** $7m + 10m$

(e) $9a + 7a - 12a$ **(f)** $5s - 8s + 4s$

(g) $12m + 3n - 6m$ **(h)** $8x + 5y - 4x$

Answers

1. $4\sqrt{3}$ **3.** $2\sqrt{7}$ **5.** $3\sqrt{5}$ **7.** $3\sqrt{6}$ **9.** $10\sqrt{2}$ **11.** $9\sqrt{7}$
13. $6\sqrt{3}$ **15.** $x\sqrt{3}$ **17.** $y^2\sqrt{3}$ **19.** $r\sqrt{2r}$ **21.** $5b\sqrt{5}$

23. $2x^2\sqrt{6}$ **25.** $3a^2\sqrt{6a}$ **27.** $xy\sqrt{x}$ **29.** $a^3b^2\sqrt{c}$ **31.** $\dfrac{3}{4}$

33. $\dfrac{\sqrt{3}}{2}$ **35.** $\dfrac{\sqrt{3}}{7}$ **37.** False   **39.** always   **41.** $\dfrac{2a\sqrt{2}}{5}$

43. $\dfrac{\sqrt{5}}{5}$ **45.** $\dfrac{\sqrt{10}}{2}$ **47.** $\dfrac{\sqrt{15a}}{5}$ **49.** $\dfrac{x^2\sqrt{21}}{7}$ **51.** $\dfrac{2s\sqrt{14s}}{7}$

53. Simplest form   **55.** Remove the perfect-square factors from the radical and simplify.   **57.** Above and Beyond   **(a)** $11x$ **(b)** $5a$ **(c)** $-2y$
(d) $17m$ **(e)** $4a$ **(f)** s **(g)** $6m + 3n$ **(h)** $4x + 5y$

9.3

Adding and Subtracting Radicals

< 9.3 Objectives >

1 > Add and subtract expressions involving numeric radicals

2 > Add and subtract expressions involving algebraic radicals

Two radicals that have the same index and the same radicand (the expression inside the radical) are called **like radicals.** For example,

$2\sqrt{3}$ and $5\sqrt{3}$ are like radicals.

$\sqrt{2}$ and $\sqrt{5}$ are not like radicals—they have different radicands.

$\sqrt{5}$ and $\sqrt[3]{5}$ are not like radicals—they have different indices (2 and 3, representing a square root and a cube root).

Like radicals can be added (or subtracted) in the same way as like terms. We apply the distributive property and then combine the coefficients:

$$2\sqrt{5} + 3\sqrt{5} = (2 + 3)\sqrt{5} = 5\sqrt{5}$$

NOTE

Indices is the plural of *index*.

Example 1   **Adding and Subtracting Like Numeric Radicals**

< Objective 1 >

NOTE

Apply the distributive property and then combine the coefficients.

Simplify each expression.

(a) $5\sqrt{2} + 3\sqrt{2} = (5 + 3)\sqrt{2} = 8\sqrt{2}$

(b) $7\sqrt{5} - 2\sqrt{5} = (7 - 2)\sqrt{5} = 5\sqrt{5}$

(c) $8\sqrt{7} - \sqrt{7} + 2\sqrt{7} = (8 - 1 + 2)\sqrt{7} = 9\sqrt{7}$

Check Yourself 1

Simplify.

(a) $2\sqrt{5} + 7\sqrt{5}$ **(b)** $9\sqrt{7} - \sqrt{7}$

(c) $5\sqrt{3} - 2\sqrt{3} + \sqrt{3}$

If a sum or difference involves terms that are *not* like radicals, we may be able to combine terms after simplifying the radicals according to our earlier rules.

> ▶ **Example 2**   **Adding and Subtracting Numeric Radicals**

Simplify each expression.

(a) $3\sqrt{2} + \sqrt{8}$

We do not have like radicals, but we can simplify $\sqrt{8}$. Remember that

$$\sqrt{8} = \sqrt{4 \cdot 2} = 2\sqrt{2}$$

so

$$3\sqrt{2} + \sqrt{8} = 3\sqrt{2} + \overset{\sqrt{8}}{2\sqrt{2}}$$
$$= (3 + 2)\sqrt{2} = 5\sqrt{2}$$

NOTE

The radicals can now be combined. Do you see why?

(b) $5\sqrt{3} - \sqrt{12} = 5\sqrt{3} - \sqrt{4 \cdot 3}$
$$= 5\sqrt{3} - \sqrt{4} \cdot \sqrt{3}$$
$$= 5\sqrt{3} - 2\sqrt{3}$$
$$= 3\sqrt{3}$$

 **Check Yourself 2**

Simplify.

(a) $\sqrt{2} + \sqrt{18}$ **(b)** $5\sqrt{3} - \sqrt{27}$

Example 3 illustrates the need to apply our earlier methods for adding fractions when working with radical expressions.

> ▶ **Example 3**   **Adding Radical Expressions**

Add $\dfrac{\sqrt{5}}{3} + \dfrac{2}{\sqrt{5}}$.

Our first step is to rationalize the denominator of the second fraction, to write the sum as

$$\frac{\sqrt{5}}{3} + \frac{2\sqrt{5}}{\sqrt{5} \cdot \sqrt{5}}$$

or

$$\frac{\sqrt{5}}{3} + \frac{2\sqrt{5}}{5}$$

The LCD of the fractions is 15 and rewriting each fraction with that denominator, we have

$$\frac{\sqrt{5} \cdot 5}{3 \cdot 5} + \frac{2\sqrt{5} \cdot 3}{5 \cdot 3} = \frac{5\sqrt{5} + 6\sqrt{5}}{15}$$
$$= \frac{11\sqrt{5}}{15}$$

Check Yourself 3

Subtract $\dfrac{3}{\sqrt{10}} - \dfrac{\sqrt{10}}{5}$.

If variables are involved in radical expressions, the process of combining terms proceeds in a fashion similar to that shown in Examples 1 and 2. Consider Example 4. We again assume that all variables represent positive real numbers.

 Example 4 | **Simplifying Expressions Involving Variables**

< Objective 2 >

NOTES

Because like radicals are involved, we apply the distributive property and combine terms as before.

Simplify the first term.

The radicals can now be combined.

Simplify each expression.

(a) $5\sqrt{3x} - 2\sqrt{3x} = 3\sqrt{3x}$

(b) $2\sqrt{3a^3} + 5a\sqrt{3a}$
$= 2\sqrt{a^2 \cdot 3a} + 5a\sqrt{3a}$
$= 2\sqrt{a^2} \cdot \sqrt{3a} + 5a\sqrt{3a}$
$= 2a\sqrt{3a} + 5a\sqrt{3a}$
$= 7a\sqrt{3a}$

Check Yourself 4

Simplify each expression.

(a) $2\sqrt{7y} + 3\sqrt{7y}$ **(b)** $\sqrt{20a^2} - a\sqrt{45}$

 Example 5 | **Adding or Subtracting Algebraic Radical Expressions**

Add or subtract as indicated.

(a) $2\sqrt{6} + \sqrt{\dfrac{2}{3}}$

We apply the quotient property to the *second term* and rationalize the denominator.

$$\sqrt{\frac{2}{3}} = \frac{\sqrt{2}}{\sqrt{3}} = \frac{\sqrt{2} \cdot \sqrt{3}}{\sqrt{3} \cdot \sqrt{3}} = \frac{\sqrt{6}}{3}$$

NOTES

Multiply by $\dfrac{\sqrt{3}}{\sqrt{3}}$, or 1.

$\dfrac{\sqrt{6}}{3}$ and $\dfrac{1}{3}\sqrt{6}$ are equivalent.

So

$$2\sqrt{6} + \sqrt{\frac{2}{3}} = 2\sqrt{6} + \frac{\sqrt{6}}{3}$$

$$= \left(2 + \frac{1}{3}\right)\sqrt{6} = \frac{7}{3}\sqrt{6}$$

(b) $\sqrt{20x} - \sqrt{\dfrac{x}{5}}$

Again we first simplify the two expressions. So

$$\sqrt{20x} - \sqrt{\dfrac{x}{5}} = 2\sqrt{5x} - \dfrac{\sqrt{x} \cdot \sqrt{5}}{\sqrt{5} \cdot \sqrt{5}}$$

$$= 2\sqrt{5x} - \dfrac{\sqrt{5x}}{5}$$

$$= \left(2 - \dfrac{1}{5}\right)\sqrt{5x} = \dfrac{9}{5}\sqrt{5x}$$

 Check Yourself 5

Add or subtract as indicated.

(a) $3\sqrt{7} + \sqrt{\dfrac{1}{7}}$ **(b)** $\sqrt{40x} - \sqrt{\dfrac{2x}{5}}$

The process of adding or subtracting like radicals is identical when working with cube roots or beyond. Example 6 demonstrates this idea.

▶ **Example 6** **Adding or Subtracting Cube Roots**

Perform the indicated operation.

(a) $2\sqrt[3]{7} + 3\sqrt[3]{7}$

$2\sqrt[3]{7} + 3\sqrt[3]{7} = 5\sqrt[3]{7}$ We can add the like radicals.

(b) $5x\sqrt[3]{2x} - \sqrt[3]{54x^4}$

$5x\sqrt[3]{2x} - \sqrt[3]{54x^4} = 5x\sqrt[3]{2x} - \sqrt[3]{2 \cdot 3^3 \cdot x \cdot x^3}$ Factor the radicand.

$= 5x\sqrt[3]{2x} - \sqrt[3]{3^3 x^3} \cdot \sqrt[3]{2x}$ Find the perfect-cube factors.

$= 5x\sqrt[3]{2x} - 3x \cdot \sqrt[3]{2x}$ Subtract the like terms.

$= 2x\sqrt[3]{2x}$

 Check Yourself 6

Perform the indicated operation.

(a) $5\sqrt[4]{10} - 3\sqrt[4]{10}$ **(b)** $3a^2\sqrt[3]{5a} + \sqrt[3]{40a^7}$

 Check Yourself ANSWERS

1. **(a)** $9\sqrt{5}$; **(b)** $8\sqrt{7}$; **(c)** $4\sqrt{3}$ **2.** **(a)** $4\sqrt{2}$; **(b)** $2\sqrt{3}$ **3.** $\dfrac{\sqrt{10}}{10}$

4. **(a)** $5\sqrt{7y}$; **(b)** $-a\sqrt{5}$ **5.** **(a)** $\dfrac{22}{7}\sqrt{7}$; **(b)** $\dfrac{9}{5}\sqrt{10x}$

6. **(a)** $2\sqrt[4]{10}$; **(b)** $5a^2\sqrt[3]{5a}$

Reading Your Text

The following fill-in-the-blank exercises are designed to ensure that you understand some of the key vocabulary used in this section.

SECTION 9.3

(a) Radicals that have the same index and the same _____ are called like radicals.

(b) Like radicals can be added by using the _____ property.

(c) In order to assure that they are real numbers, we assume that all radical variables are _____.

(d) Just as we refer to the root two as the square root, we refer to the root three as the _____ root.

Name _____

Section _____ Date _____

Answers

1. _____
2. _____
3. _____
4. _____
5. _____
6. _____

7. _____ 8. _____

9. _____ 10. _____

11. _____ 12. _____

13. _____ 14. _____

15. _____ 16. _____

17. _____ 18. _____

19. _____ 20. _____

21. _____ 22. _____

23. _____ 24. _____

25. _____ 26. _____

Basic Skills | Advanced Skills | Vocational-Technical Applications | Calculator/Computer | Above and Beyond | Getting Ready

< Objective 1 >

Simplify by combining like terms.

1. $3\sqrt{2} + 5\sqrt{2}$

2. $\sqrt{3} + 5\sqrt{3}$

3. $11\sqrt{7} - 4\sqrt{7}$

4. $7\sqrt{3} - 5\sqrt{2}$

5. $5\sqrt{7} + 3\sqrt{6}$

6. $3\sqrt{5} - 5\sqrt{5}$

7. $3\sqrt{5} - 7\sqrt{5}$

8. $2\sqrt{11} + 5\sqrt{11}$

9. $2\sqrt{3x} + 5\sqrt{3x}$

10. $7\sqrt{2a} - 3\sqrt{2a}$

11. $2\sqrt{3} + \sqrt{3} + 3\sqrt{3}$

12. $3\sqrt{5} + 2\sqrt{5} + \sqrt{5}$

13. $6\sqrt{11} - 5\sqrt{11} + 3\sqrt{11}$

14. $3\sqrt{10} - 2\sqrt{10} + \sqrt{10}$

> Videos

15. $2\sqrt{5x} + 5\sqrt{5x} - \sqrt{5x}$

16. $8\sqrt{3b} - 2\sqrt{3b} + \sqrt{3b}$

17. $2\sqrt{3} + \sqrt{12}$

18. $7\sqrt{3} + 2\sqrt{27}$

19. $\sqrt{20} - \sqrt{5}$

20. $\sqrt{98} - 3\sqrt{2}$

21. $2\sqrt{6} - \sqrt{54}$

22. $2\sqrt{3} - \sqrt{27}$

23. $\sqrt{72} + \sqrt{50}$ > Videos

24. $\sqrt{27} - \sqrt{12}$

25. $3\sqrt{12} - \sqrt{48}$

26. $5\sqrt{8} + 2\sqrt{18}$

27. $\sqrt{3} + \sqrt{\dfrac{1}{3}}$

28. $\sqrt{6} - \sqrt{\dfrac{1}{6}}$

29. $\sqrt{12} + \sqrt{27} - \sqrt{3}$

30. $\sqrt{50} + \sqrt{32} - \sqrt{8}$

31. $3\sqrt{24} - \sqrt{54} + \sqrt{6}$

32. $\sqrt{63} - 2\sqrt{28} + 5\sqrt{7}$

33. $\dfrac{\sqrt{12}}{3} - \dfrac{1}{\sqrt{3}}$

34. $\dfrac{\sqrt{20}}{5} + \dfrac{2}{\sqrt{5}}$

Determine whether each of the following statements is **true** *or* **false**.

35. Unlike radicals can sometimes be added or subtracted.

36. To add or subtract like radicals, we must use the distributive property.

| Basic Skills | **Advanced Skills** | Vocational-Technical Applications | Calculator/Computer | Above and Beyond | Getting Ready |

< Objective 2 >

Simplify by combining like terms.

37. $a\sqrt{27} - 2\sqrt{3a^2}$

38. $5\sqrt{2y^2} - 3y\sqrt{8}$

39. $5\sqrt{3x^3} + 2\sqrt{27x}$

40. $7\sqrt{2a^3} - \sqrt{8a}$

41. GEOMETRY Find the perimeter of the rectangle shown in the figure.

42. GEOMETRY Find the perimeter of the rectangle shown in the figure. Write your answer in radical form.

Answers

27. _____

28. _____

29. _____

30. _____

31. _____

32. _____

33. _____

34. _____

35. _____

36. _____

37. _____

38. _____

39. _____

40. _____

41. _____

42. _____

Answers

43. _____

44. _____

45. _____

46. _____

47. _____

48. _____

49. _____

50. _____

51. _____

52. _____

(a) _____

(b) _____

(c) _____

(d) _____

(e) _____

(f) _____

(g) _____

(h) _____

43. GEOMETRY Find the perimeter of the triangle shown in the figure. Write your answer in radical form.

44. GEOMETRY Find the perimeter of the triangle shown in the figure. Write your answer in radical form.

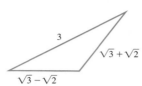

Basic Skills | Advanced Skills | Vocational-Technical Applications | **Calculator/Computer** | Above and Beyond | Getting Ready

Use a calculator to find a decimal approximation for each of the following. Round your answer to the nearest hundredth.

45. $\sqrt{3} - \sqrt{2}$

46. $\sqrt{7} + \sqrt{11}$

47. $\sqrt{5} + \sqrt{3}$

48. $\sqrt{17} - \sqrt{13}$

49. $4\sqrt{3} - 7\sqrt{5}$

50. $8\sqrt{2} + 3\sqrt{7}$

51. $5\sqrt{7} + 8\sqrt{13}$

52. $7\sqrt{2} - 4\sqrt{11}$

Basic Skills | Advanced Skills | Vocational-Technical Applications | Calculator/Computer | Above and Beyond | **Getting Ready**

Perform the indicated multiplication.

(a) $2(x + 5)$

(b) $3(a - 3)$

(c) $m(m - 8)$

(d) $y(y + 7)$

(e) $(w + 2)(w - 2)$

(f) $(x - 3)(x + 3)$

(g) $(x + y)(x + y)$

(h) $(b - 7)(b - 7)$

Answers

1. $8\sqrt{2}$ **3.** $7\sqrt{7}$ **5.** Cannot be simplified   **7.** $-4\sqrt{5}$
9. $7\sqrt{3x}$ **11.** $6\sqrt{3}$ **13.** $4\sqrt{11}$ **15.** $6\sqrt{5x}$ **17.** $4\sqrt{3}$
19. $\sqrt{5}$ **21.** $-\sqrt{6}$ **23.** $11\sqrt{2}$ **25.** $2\sqrt{3}$ **27.** $\dfrac{4}{3}\sqrt{3}$

29. $4\sqrt{3}$ **31.** $4\sqrt{6}$ **33.** $\dfrac{1}{3}\sqrt{3}$ **35.** True   **37.** $a\sqrt{3}$

39. $(5x+6)\sqrt{3x}$ **41.** 26   **43.** $2\sqrt{3}+3$ **45.** 0.32

47. 3.97   **49.** -8.72 **51.** 42.07   **(a)** $2x+10$ **(b)** $3a-9$ **(c)** m^2-8m
(d) y^2+7y **(e)** w^2-4 **(f)** x^2-9 **(g)** $x^2+2xy+y^2$ **(h)** $b^2-14b+49$

Multiplying and Dividing Radicals

< 9.4 Objectives >

1 > Multiply expressions involving radical expressions

2 > Divide expressions involving radical expressions

In Section 9.2 we stated the first property for radicals:

$\sqrt{ab} = \sqrt{a} \cdot \sqrt{b}$ when a and b are any nonnegative real numbers

That property has been used to simplify radical expressions up to this point. Suppose we want to find a product, such as $\sqrt{3} \cdot \sqrt{5}$.

    We can use our first radical rule in the opposite manner.

RECALL

The product of square roots is equal to the square root of the product of the radicands.

$\sqrt{a} \cdot \sqrt{b} = \sqrt{ab}$

so

$\sqrt{3} \cdot \sqrt{5} = \sqrt{3 \cdot 5} = \sqrt{15}$

    We may have to simplify after multiplying, as Example 1 illustrates.

Example 1   **Multiplying Radical Expressions**

< Objective 1 >

Multiply and then simplify each expression.

(a) $\sqrt{5} \cdot \sqrt{10} = \sqrt{5 \cdot 10} = \sqrt{50}$
$$= \sqrt{25 \cdot 2} = 5\sqrt{2}$$

(b) $\sqrt{12} \cdot \sqrt{6} = \sqrt{12 \cdot 6} = \sqrt{72}$
$$= \sqrt{36 \cdot 2} = \sqrt{36} \cdot \sqrt{2} = 6\sqrt{2}$$

An alternative approach would be to simplify $\sqrt{12}$ first.
$$\sqrt{12} \cdot \sqrt{6} = 2\sqrt{3}\,\sqrt{6} = 2\sqrt{18}$$
$$= 2\sqrt{9 \cdot 2} = 2\sqrt{9}\,\sqrt{2}$$
$$= 2 \cdot 3\sqrt{2} = 6\sqrt{2}$$

(c) $\sqrt{10x} \cdot \sqrt{2x} = \sqrt{20x^2} = \sqrt{4x^2 \cdot 5}$
$$= \sqrt{4x^2} \cdot \sqrt{5} = 2x\sqrt{5}$$

Check Yourself 1

Simplify.

(a) $\sqrt{3} \cdot \sqrt{6}$ (b) $\sqrt{3} \cdot \sqrt{18}$ (c) $\sqrt{8a} \cdot \sqrt{3a}$

If coefficients are involved in a product, we can use the commutative and associative properties to change the order and grouping of the factors. This is illustrated in Example 2.

Example 2   **Multiplying Radical Expressions**

NOTE

In practice, it is not necessary to show the intermediate steps.

Multiply.

$$(2\sqrt{5})(3\sqrt{6}) = (2 \cdot 3)(\sqrt{5} \cdot \sqrt{6})$$
$$= 6\sqrt{5 \cdot 6}$$
$$= 6\sqrt{30}$$

Check Yourself 2

Multiply $(3\sqrt{7})(5\sqrt{3})$.

The distributive property can also be applied in multiplying radical expressions. Consider the following.

Example 3   **Multiplying Radical Expressions**

Multiply.

(a) $\sqrt{3}(\sqrt{2} + \sqrt{3})$

$$= \sqrt{3} \cdot \sqrt{2} + \sqrt{3} \cdot \sqrt{3} \qquad \text{The distributive property}$$
$$= \sqrt{6} + 3 \qquad \text{Multiply the radicals.}$$

(b) $\sqrt{5}(2\sqrt{6} + 3\sqrt{3})$

$$= \sqrt{5} \cdot 2\sqrt{6} + \sqrt{5} \cdot 3\sqrt{3} \qquad \text{The distributive property}$$
$$= 2 \cdot \sqrt{5} \cdot \sqrt{6} + 3 \cdot \sqrt{5} \cdot \sqrt{3} \qquad \text{The commutative property}$$
$$= 2\sqrt{30} + 3\sqrt{15}$$

Check Yourself 3

Multiply.

(a) $\sqrt{5}(\sqrt{6} + \sqrt{5})$ (b) $\sqrt{3}(2\sqrt{5} + 3\sqrt{2})$

The FOIL pattern we used for multiplying binomials in Section 3.4 can also be applied in multiplying radical expressions. This is shown in Example 4.

Example 4 | Multiplying Radical Expressions

Multiply.

(a) $(\sqrt{3} + 2)(\sqrt{3} + 5)$

$$= \sqrt{3} \cdot \sqrt{3} + 5\sqrt{3} + 2\sqrt{3} + 2 \cdot 5$$
$$= 3 + 5\sqrt{3} + 2\sqrt{3} + 10 \qquad \text{Combine like terms.}$$
$$= 13 + 7\sqrt{3}$$

 > CAUTION

Be careful! This result *cannot* be further simplified: 13 and $7\sqrt{3}$ are *not* like terms.

(b) $(\sqrt{7} + 2)(\sqrt{7} - 2) = \sqrt{7} \cdot \sqrt{7} - 2\sqrt{7} + 2\sqrt{7} - 4$

$$= 7 - 4 = 3$$

(c) $(\sqrt{3} + 5)^2 = (\sqrt{3} + 5)(\sqrt{3} + 5)$

$$= \sqrt{3} \cdot \sqrt{3} + 5\sqrt{3} + 5\sqrt{3} + 5 \cdot 5$$
$$= 3 + 5\sqrt{3} + 5\sqrt{3} + 25$$
$$= 28 + 10\sqrt{3}$$

NOTE

You can use the pattern $(a + b)(a - b) = a^2 - b^2$, where $a = \sqrt{7}$ and $b = 2$, for the same result. $\sqrt{7} + 2$ and $\sqrt{7} - 2$ are called **conjugates** of each other. Note that their product is the rational number 3. The product of conjugates is *always rational*.

NOTE

The quotient of square roots is equal to the square root of the quotient of the radicands.

 Check Yourself 4

Multiply.

(a) $(\sqrt{5} + 3)(\sqrt{5} - 2)$ **(b)** $(\sqrt{3} + 4)(\sqrt{3} - 4)$ **(c)** $(\sqrt{2} - 3)^2$

We can also use our second property for radicals in the opposite manner.

$$\frac{\sqrt{a}}{\sqrt{b}} = \sqrt{\frac{a}{b}}$$

One use of this property to divide radical expressions is illustrated in Example 5.

Example 5 | Dividing Radical Expressions

< Objective 2 >

Simplify.

(a) $\dfrac{\sqrt{48}}{\sqrt{3}} = \sqrt{\dfrac{48}{3}} = \sqrt{16} = 4$

NOTE

The clue to recognizing when to use this approach is in noting that 48 is divisible by 3.

(b) $\dfrac{\sqrt{200}}{\sqrt{2}} = \sqrt{\dfrac{200}{2}} = \sqrt{100} = 10$

(c) $\dfrac{\sqrt{125x^2}}{\sqrt{5}} = \sqrt{\dfrac{125x^2}{5}} = \sqrt{25x^2} = 5x$

 Check Yourself 5

Simplify.

(a) $\dfrac{\sqrt{75}}{\sqrt{3}}$ **(b)** $\dfrac{\sqrt{81s^2}}{\sqrt{9}}$

There is one final quotient form that you may encounter in simplifying expressions, and it will be extremely important in our work with quadratic equations in Chapter 10. This form is shown in Example 6.

 | **Example 6** | Simplifying Radical Expressions

 > CAUTION

Be careful! Students are sometimes tempted to write

$$\frac{\cancel{3} + 6\sqrt{2}}{\cancel{3}} = 1 + 6\sqrt{2}$$

This is *not* correct. We must divide *both terms* of the numerator by the common factor.

Simplify the expression

$$\frac{3 + \sqrt{72}}{3}$$

First, we must simplify the radical in the numerator.

$$\frac{3 + \sqrt{72}}{3} = \frac{3 + \sqrt{36 \cdot 2}}{3} \qquad \text{Use Property 1 to simplify } \sqrt{72}.$$

$$= \frac{3 + \sqrt{36} \cdot \sqrt{2}}{3} = \frac{3 + 6\sqrt{2}}{3}$$

$$= \frac{3(1 + 2\sqrt{2})}{3} = 1 + 2\sqrt{2} \qquad \textit{Factor the numerator and then divide by the common factor 3.}$$

 Check Yourself 6

Simplify $\dfrac{15 + \sqrt{75}}{5}$.

In Section 9.2, we said that for an expression involving square roots to be in simplest form, no radical can appear in the denominator of a fraction. Changing a fraction that has a radical in the denominator to an equivalent fraction that has no radical in the denominator is called **rationalizing the denominator.** The primary reason for rationalizing denominators used to be so that a better arithmetic estimation could be made. With the availability of calculators, that is now rarely necessary. However, the process continues to be useful in higher-level mathematics. So that we can review this process, we present a straightforward example below.

 | **Example 7** | Rationalizing a Denominator

Rewrite each fraction so that there are no radicals in the denominator.

(a) $\dfrac{\sqrt{7}}{\sqrt{2}}$

$$\frac{\sqrt{7}}{\sqrt{2}} = \frac{\sqrt{7} \cdot \sqrt{2}}{\sqrt{2} \cdot \sqrt{2}}$$

We can always multiply the numerator and denominator of a fraction by the same value. Here, we multiply the numerator and denominator by $\sqrt{2}$. This will make the denominator a rational number.

$$\frac{\sqrt{7} \cdot \sqrt{2}}{\sqrt{2} \cdot \sqrt{2}} = \frac{\sqrt{14}}{2} \qquad \text{We now have an equivalent fraction with a rational number for its denominator.}$$

(b) $\dfrac{\sqrt{3x}}{\sqrt{54}}$

$$\dfrac{\sqrt{3x}}{\sqrt{54}} = \dfrac{\sqrt{3x}}{\sqrt{9 \cdot 6}} = \dfrac{\sqrt{3x}}{3\sqrt{6}}$$ First, simplify the denominator as much as possible.

$$\dfrac{\sqrt{3x}}{3\sqrt{6}} = \dfrac{\sqrt{3x} \cdot \sqrt{6}}{3\sqrt{6} \cdot \sqrt{6}} = \dfrac{\sqrt{18x}}{18}$$ Multiply numerator and denominator by $\sqrt{6}$.

$$\dfrac{\sqrt{18x}}{18} = \dfrac{\sqrt{9 \cdot 2x}}{18} = \dfrac{3\sqrt{2x}}{18}$$ Factor out the perfect square in the numerator.

$$= \dfrac{\sqrt{2x}}{6}$$ Simplify the fraction.

Check Yourself 7

Rewrite each fraction so that there are no radicals in the denominator.

(a) $\dfrac{\sqrt{17}}{\sqrt{3}}$ **(b)** $\dfrac{\sqrt{3a}}{\sqrt{12}}$

Check Yourself ANSWERS

1. (a) $3\sqrt{2}$; **(b)** $3\sqrt{6}$; **(c)** $2a\sqrt{6}$ **2.** $15\sqrt{21}$ **3. (a)** $\sqrt{30} + 5$;
(b) $2\sqrt{15} + 3\sqrt{6}$ **4. (a)** $-1 + \sqrt{5}$; **(b)** -13; **(c)** $11 - 6\sqrt{2}$
5. (a) 5; **(b)** $3s$ **6.** $3 + \sqrt{3}$ **7. (a)** $\dfrac{\sqrt{51}}{3}$; **(b)** $\dfrac{\sqrt{a}}{2}$

Reading Your Text

The following fill-in-the-blank exercises are designed to ensure that you understand some of the key vocabulary used in this section.

SECTION 9.4

(a) The product of two square roots is equal to the _____ of their product.

(b) The _____ pattern can be used to multiply two binomials.

(c) 13 and $7\sqrt{7}$ are not _____ terms.

(d) The product of conjugates is always _____ .

< Objective 1 >

Perform the indicated multiplication. Then simplify each radical expression.

1. $\sqrt{7} \cdot \sqrt{5}$

2. $\sqrt{3} \cdot \sqrt{7}$

3. $\sqrt{5} \cdot \sqrt{11}$

4. $\sqrt{13} \cdot \sqrt{5}$

5. $\sqrt{3} \cdot \sqrt{10m}$

6. $\sqrt{7a} \cdot \sqrt{13}$

7. $\sqrt{2x} \cdot \sqrt{15}$

8. $\sqrt{15} \cdot \sqrt{2b}$

9. $\sqrt{3} \cdot \sqrt{7} \cdot \sqrt{2}$

10. $\sqrt{5} \cdot \sqrt{7} \cdot \sqrt{3}$

11. $\sqrt{3} \cdot \sqrt{12}$

12. $\sqrt{8} \cdot \sqrt{8}$

13. $\sqrt{10x} \cdot \sqrt{10x}$

14. $\sqrt{5a} \cdot \sqrt{15a}$

15. $\sqrt{27} \cdot \sqrt{2}$

16. $\sqrt{8} \cdot \sqrt{10}$

17. $\sqrt{2x} \cdot \sqrt{6x}$

18. $\sqrt{3a} \cdot \sqrt{15a}$

19. $3\sqrt{2x} \cdot \sqrt{6x}$

20. $3\sqrt{2a} \cdot \sqrt{5a}$

21. $(3a\sqrt{3a})(5\sqrt{7a})$ > Videos

22. $(2x\sqrt{5x})(3\sqrt{11x})$

23. $(3\sqrt{5})(2\sqrt{10})$

24. $(4\sqrt{3})(3\sqrt{6})$

25. $\sqrt{3}(\sqrt{2} + \sqrt{3})$

26. $\sqrt{3}(\sqrt{5} - \sqrt{3})$

27. $\sqrt{3}(2\sqrt{5} - 3\sqrt{3})$ > Videos

28. $\sqrt{7}(2\sqrt{3} + 3\sqrt{7})$

29. $(\sqrt{3} + 5)(\sqrt{3} + 3)$

30. $(\sqrt{3} - 5)(\sqrt{3} + 2)$

31. $(\sqrt{5} - 1)(\sqrt{5} + 3)$ > Videos

32. $(\sqrt{2} + 3)(\sqrt{2} - 7)$

33. $(\sqrt{5} - 2)(\sqrt{5} + 2)$

34. $(\sqrt{7} + 5)(\sqrt{7} - 5)$

MathZone

Boost your grade at mathzone.com!

> Practice Problems
> NetTutor
> Self-Tests
> e-Professors
> Videos

Name _____

Section _____ Date _____

Answers

1. _____ 2. _____

3. _____ 4. _____

5. _____ 6. _____

7. _____ 8. _____

9. _____ 10. _____

11. _____ 12. _____

13. _____ 14. _____

15. _____ 16. _____

17. _____ 18. _____

19. _____ 20. _____

21. _____ 22. _____

23. _____ 24. _____

25. _____ 26. _____

27. _____ 28. _____

29. _____ 30. _____

31. _____ 32. _____

33. _____ 34. _____

Answers

35. _____

36. _____

37. _____

38. _____

39. _____

40. _____

41. _____

42. _____

43. _____

44. _____

45. _____

46. _____

47. _____

48. _____

49. _____

50. _____

51. _____

52. _____

53. _____

54. _____

55. _____

56. _____

35. $(\sqrt{7} + 3)(\sqrt{7} - 3)$

36. $(\sqrt{11} - 3)(\sqrt{11} + 3)$

37. $(\sqrt{x} + 3)(\sqrt{x} - 3)$

38. $(\sqrt{a} - 4)(\sqrt{a} + 4)$

39. $(\sqrt{3} + 2)^2$

40. $(\sqrt{5} - 3)^2$

41. $(\sqrt{y} - 5)^2$ > Videos

42. $(\sqrt{x} + 4)^2$

Determine whether each of the following statements is **true** *or* **false.**

43. The square root of a quotient is equal to the quotient of the square roots.

44. The square root of a sum is equal to the sum of the square roots.

Complete each of the following statements with **never, sometimes,** *or* **always.**

45. The product of conjugates is _____ rational.

46. The product of two irrational numbers is _____ rational.

| Basic Skills | **Advanced Skills** | Vocational-Technical Applications | Calculator/Computer | Above and Beyond | Getting Ready |

< Objective 2 >

Perform the indicated division. Rationalize the denominator if necessary. Then simplify each radical expression.

47. $\dfrac{\sqrt{98}}{\sqrt{2}}$ > Videos

48. $\dfrac{\sqrt{108}}{\sqrt{3}}$

49. $\dfrac{\sqrt{72a^2}}{\sqrt{2}}$

50. $\dfrac{\sqrt{48m^2}}{\sqrt{3}}$

51. $\dfrac{4 + \sqrt{48}}{4}$

52. $\dfrac{12 + \sqrt{108}}{6}$

53. $\dfrac{5 + \sqrt{175}}{5}$

54. $\dfrac{18 + \sqrt{567}}{9}$

55. $\dfrac{-8 - \sqrt{512}}{4}$ > Videos

56. $\dfrac{-9 - \sqrt{108}}{3}$

57. $\dfrac{6 + \sqrt{18}}{3}$

58. $\dfrac{6 - \sqrt{20}}{2}$

59. $\dfrac{15 - \sqrt{75}}{5}$

60. $\dfrac{8 + \sqrt{48}}{4}$

61. **GEOMETRY** Find the area of the rectangle shown.

$\sqrt{3}$

$\sqrt{11}$

62. **GEOMETRY** Find the area of the rectangle shown.

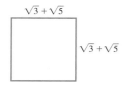

$\sqrt{3} + \sqrt{5}$

$\sqrt{3} + \sqrt{5}$

| Basic Skills | Advanced Skills | Vocational-Technical Applications | Calculator/Computer | **Above and Beyond** | Getting Ready |

---

63. Complete this statement: "$\sqrt{2} \cdot \sqrt{5} = \sqrt{10}$ because . . ."

64. Explain why $2\sqrt{3} + 5\sqrt{3} = 7\sqrt{3}$ but $7\sqrt{3} + 3\sqrt{5} \neq 10\sqrt{8}$.

65. When you look out over an unobstructed landscape or seascape, the distance to the visible horizon depends on your height above the ground. The equation

$$d = \sqrt{\dfrac{3}{2}h}$$

is a good estimate of this, in which d = distance to horizon in miles and h = height of viewer above the ground. Work with a partner to make a chart of distances to the horizon given different elevations. Use the actual heights of tall buildings or prominent landmarks in your area. The local library should have a list of these. Be sure to consider the view to the horizon you get when flying in a plane. What would your elevation have to be to see from one side of your city or town to the other? From one side of your state or county to the other?

Answers

57. _____

58. _____

59. _____

60. _____

61. _____

62. _____

63. _____

64. _____

65. _____

Answers

Basic Skills | Advanced Skills | Vocational-Technical Applications | Calculator/Computer | Above and Beyond | **Getting Ready**
▲

Solve the following equations.

(a) $x^2 = 16$ (b) $x^2 = 25$ (c) $2x^2 = 72$

(d) $5x^2 = 125$ (e) $(x - 1)^2 = 9$ (f) $(x + 2)^2 = 16$

Answers

1. $\sqrt{35}$ **3.** $\sqrt{55}$ **5.** $\sqrt{30m}$ **7.** $\sqrt{30x}$ **9.** $\sqrt{42}$ **11.** 6

13. $10x$ **15.** $3\sqrt{6}$ **17.** $2x\sqrt{3}$ **19.** $6x\sqrt{3}$ **21.** $15a^2\sqrt{21}$

23. $30\sqrt{2}$ **25.** $\sqrt{6} + 3$ **27.** $2\sqrt{15} - 9$ **29.** $18 + 8\sqrt{3}$

31. $2 + 2\sqrt{5}$ **33.** 1 **35.** -2 **37.** $x - 9$ **39.** $7 + 4\sqrt{3}$

41. $y - 10\sqrt{y} + 25$ **43.** True   **45.** always   **47.** 7 **49.** $6a$

51. $1 + \sqrt{3}$ **53.** $1 + \sqrt{7}$ **55.** $-2 - 4\sqrt{2}$ **57.** $2 + \sqrt{2}$

59. $3 - \sqrt{3}$ **61.** $\sqrt{33}$ **63.** Above and Beyond

65. Above and Beyond   (a) $-4, 4$ (b) $-5, 5$ (c) $-6, 6$

(d) $-5, 5$ (e) $-2, 4$ (f) $-6, 2$

9.5

Solving Radical Equations

< 9.5 Objective >

1 > Solve an equation containing a radical expression

NOTE

$$x^2 = 1$$
$$x^2 - 1 = 0$$
$$(x + 1)(x - 1) = 0$$
So the solutions are 1 and -1.

In this section, we establish some procedures for solving equations involving radical expressions. The basic technique involves raising both sides of an equation to some power. However, doing so requires some caution.

For example, begin with the equation $x = 1$. Squaring both sides gives $x^2 = 1$, which has two solutions, 1 and -1. Clearly -1 is not a solution to the original equation. We refer to -1 as an *extraneous solution*.

We must be aware of the possibility of extraneous solutions any time we raise both sides of an equation to any *even power*. Having said that, we are now prepared to introduce the power property of equality.

Property

The Power Property of Equality	Given any two expressions a and b and any positive integer n,
	If $a = b$, then $a^n = b^n$.

Although we never lose a solution when applying the power property, we often find an extraneous one as a result of raising both sides of the equation to some power. Because of this, it is very important that you *check all solutions.*

Example 1 | Solving a Radical Equation

< Objective 1 >

NOTE

$(\sqrt{x + 2})^2 = x + 2$
This is why squaring both sides of the equation removes the radical.

Solve $\sqrt{x + 2} = 3$.

Squaring each side, we have

$$(\sqrt{x + 2})^2 = 3^2$$
$$x + 2 = 9$$
$$x = 7$$

Substituting 7 into the original equation, we find

$$\sqrt{(7) + 2} \stackrel{?}{=} 3$$
$$\sqrt{9} \stackrel{?}{=} 3$$
$$3 = 3$$

Because 7 is the only value that makes this a true statement, the solution for the equation is 7.

 Check Yourself 1

Solve the equation $\sqrt{x-5} = 4$.

▶ **Example 2**   **Solving a Radical Equation**

Beginning Algebra   The Streeter/Hutchison Series in Mathematics   © The McGraw-Hill Companies. All Rights Reserved.

NOTES

Applying the power property will remove the radical only if that radical is isolated on one side of the equation.

Notice that on the right $(-1)^2 = 1$.

Solve $\sqrt{4x+5} + 1 = 0$.

First isolate the radical on the left side:

$$\sqrt{4x+5} = -1$$

Then, square both sides,

$$(\sqrt{4x+5})^2 = (-1)^2$$
$$4x + 5 = 1$$

solve for x to get

$$x = -1$$

NOTE

2 is never equal to 0, so -1 is *not* a solution for the original equation.

Now check the solution by substituting -1 for x in the original equation:

$$\sqrt{4(-1)+5} + 1 \overset{?}{=} 0$$
$$\sqrt{1} + 1 \overset{?}{=} 0$$

and $2 \neq 0$

Because -1 is an extraneous solution, there is *no solution* to the original equation.

 Check Yourself 2

Solve $\sqrt{3x-2} + 2 = 0$.

Next, consider an example in which the procedure involves squaring a binomial.

▶ **Example 3**   **Solving a Radical Equation**

Solve $\sqrt{x+3} = x + 1$.

We can square each side, as before.

$$(\sqrt{x+3})^2 = (x+1)^2$$
$$x + 3 = x^2 + 2x + 1$$

Simplifying this gives us the quadratic equation

$$x^2 + x - 2 = 0$$

NOTE

Verify this for yourself by substituting 1 and then -2 for x in the original equation.

> C A U T I O N

Factoring, we have

$$(x - 1)(x + 2) = 0$$

which gives us the possible solutions

$$x = 1 \qquad \text{or} \qquad x = -2$$

Now we check for extraneous solutions and find that $x = 1$ is a valid solution, but that $x = -2$ does not yield a true statement.

Be careful!   Sometimes (as in this example), one side of the equation contains a binomial. In that case, we must remember the middle term when we square the binomial. The square of a binomial *is always a trinomial*.

    Check Yourself 3

Solve $\sqrt{x - 5} = x - 7$.

It is not always the case that one of the solutions is extraneous. We may have zero, one, or two valid solutions when we generate a quadratic from a radical equation.

In Example 4 we see a case in which both of the solutions derived satisfy the equation.

Example 4   **Solving a Radical Equation**

Solve $\sqrt{7x + 1} - 1 = 2x$.

First, *we must isolate the term involving the radical.*

$$\sqrt{7x + 1} = 2x + 1$$

We can now square both sides of the equation.

$$7x + 1 = 4x^2 + 4x + 1$$

Now we write the quadratic equation in standard form.

$$4x^2 - 3x = 0$$

Factoring, we have

$$x(4x - 3) = 0$$

which yields two possible solutions

$$x = 0 \qquad \text{or} \qquad x = \frac{3}{4}$$

Checking the solutions by substitution, we find that both values for x give true statements, as follows.

Letting x be 0, we have

$$\sqrt{7(0) + 1} - 1 \stackrel{?}{=} 2(0)$$
$$\sqrt{1} - 1 \stackrel{?}{=} 0$$

or $0 = 0$ A true statement

Letting x be $\frac{3}{4}$, we have

$$\sqrt{7\left(\frac{3}{4}\right) + 1} - 1 \stackrel{?}{=} 2\left(\frac{3}{4}\right)$$

$$\sqrt{\frac{25}{4}} - 1 \stackrel{?}{=} \frac{3}{2}$$

$$\frac{5}{2} - 1 \stackrel{?}{=} \frac{3}{2}$$

$$\frac{3}{2} = \frac{3}{2} \qquad \text{Again a true statement. The solutions are 0 and } \frac{3}{4}.$$

Check Yourself 4

Solve $\sqrt{5x + 1} - 1 = 3x$.

Many real-world applications of the material in this section require that a formula be solved for one of the variables. We look at an example of this process next.

    Example 5   Solving a Radical Equation for a Given Variable

Solve the equation $h = \sqrt{pq}$ for the variable p.
    Solving for the variable p requires that p be isolated on one side of the equation. We need to eliminate the radical in order to isolate the p.

$$h^2 = (\sqrt{pq})^2$$

$$h^2 = pq$$

$$\frac{h^2}{q} = p \qquad \text{or} \qquad p = \frac{h^2}{q}$$

Check Yourself 5

Solve the equation $2a = \sqrt{3ab}$ for the variable b.

We summarize our work in this section in the following algorithm for solving equations involving radicals.

Step by Step

Solving Equations Involving Radicals	Step 1	Isolate a radical on one side of the equation.
	Step 2	Raise each side of the equation to the smallest power that eliminates the isolated radical.
	Step 3	If any radicals remain in the equation derived in step 2, return to step 1 and continue until no radical remains.
	Step 4	Solve the resulting equation to determine any possible solutions.
	Step 5	Check all solutions to determine whether extraneous solutions resulted from step 2.

Check Yourself ANSWERS

1. 21   **2.** No solution   **3.** 9   **4.** $0, -\dfrac{1}{9}$ **5.** $\dfrac{4a}{3}$

Reading Your Text

The following fill-in-the-blank exercises are designed to ensure that you understand some of the key vocabulary used in this section.

SECTION 9.5

(a) The basic technique for solving a radical equation is to raise both _____ of an equation to some power.

(b) We must be aware of the possibility of extraneous solutions any time we raise both sides of an equation to any _____ power.

(c) We never _____ a solution when using the power property of equality.

(d) Squaring a binomial always results in _____ terms.

Name _____

Section _____ Date _____

Answers

1. _____   2. _____

3. _____   4. _____

5. _____   6. _____

7. _____   8. _____

9. _____   10. _____

11. _____   12. _____

13. _____

14. _____

15. _____   16. _____

17. _____   18. _____

19. _____

20. _____

21. _____

22. _____

< Objective 1 >

Solve each of the equations. Be sure to check your solutions.

1. $\sqrt{x} = 2$

2. $\sqrt{x} - 3 = 0$

3. $2\sqrt{y} - 1 = 0$

4. $3\sqrt{2z} = 9$

5. $\sqrt{m + 5} = 3$ > Videos

6. $\sqrt{y + 7} = 5$

7. $\sqrt{x - 3} = 4$

8. $\sqrt{x + 4} = 3$

9. $\sqrt{2x - 1} = 3$

10. $\sqrt{3x + 1} = 4$

11. $\sqrt{2x + 4} - 4 = 0$

12. $\sqrt{3x + 3} - 6 = 0$

13. $\sqrt{3x - 2} + 2 = 0$

14. $\sqrt{4x + 1} + 3 = 0$

15. $x + \sqrt{x - 1} = 7$

16. $x - \sqrt{x + 2} = 10$

17. $x + \sqrt{2x - 5} = 10$ > Videos

18. $2x - \sqrt{3x - 2} = 8$

19. $\sqrt{2x - 3} + 1 = 3$

20. $\sqrt{3x + 1} - 2 = -1$

21. $2\sqrt{3z + 2} - 1 = 5$

22. $3\sqrt{4q - 1} - 2 = 7$

23. $\sqrt{15 - 2x} = x$

24. $\sqrt{48 - 2y} = y$

25. $\sqrt{x + 5} = x - 1$

26. $\sqrt{2x - 1} = x - 8$

27. $\sqrt{3m - 2} + m = 10$

28. $\sqrt{2x + 1} + x = 7$

29. $\sqrt{t + 9} + 3 = t$ > Videos

30. $\sqrt{2y + 7} + 4 = y$

Complete each of the following statements with **never, sometimes,** *or* **always.**

31. When applying the power property of equality, you _____ lose a solution.

32. If you square both sides of an equation, you _____ find an extraneous solution.

33. The square of a binomial is _____ a trinomial.

34. If we generate a quadratic equation from a radical, we _____ find more than one solution.

Basic Skills | **Advanced Skills** | Vocational-Technical Applications | Calculator/Computer | Above and Beyond | Getting Ready
▲

Solve for the indicated variable.

35. $h = \sqrt{pq}$ for q

36. $c = \sqrt{a^2 + b^2}$ for a

37. $v = \sqrt{2gR}$ for R

38. $v = \sqrt{2gR}$ for g

39. $r = \sqrt{\dfrac{S}{2\pi}}$ for S > Videos

40. $r = \sqrt{\dfrac{3V}{4\pi}}$ for V

Answers

23. _____   24. _____

25. _____   26. _____

27. _____   28. _____

29. _____   30. _____

31. _____

32. _____

33. _____

34. _____

35. _____

36. _____

37. _____

38. _____

39. _____

40. _____

Answers

(a) _____

(b) _____

(c) _____

(d) _____

(e) _____

(f) _____

Evaluate the following. Round your answer to the nearest thousandth.

(a) $\sqrt{16}$ (b) $\sqrt{49}$

(c) $\sqrt{121}$ (d) $\sqrt{12}$

(e) $\sqrt{27}$ (f) $\sqrt{98}$

Answers

1. 4   **3.** $\dfrac{1}{4}$ **5.** 4   **7.** 19   **9.** 5   **11.** 6   **13.** No solution

15. 5   **17.** 7   **19.** $\dfrac{7}{2}$ **21.** $\dfrac{7}{3}$ **23.** 3   **25.** 4   **27.** 6

29. 7   **31.** never   **33.** always   **35.** $q = \dfrac{h^2}{p}$ **37.** $R = \dfrac{v^2}{2g}$

39. $S = 2\pi r^2$ (a) 4   (b) 7   (c) 11   (d) 3.464
(e) 5.196   (f) 9.899

9.6

Applications of the Pythagorean Theorem

< 9.6 Objectives >

1 > Apply the Pythagorean theorem in solving problems

2 > Use the distance formula to find the distance between two points

Perhaps the most famous theorem in all of mathematics is the **Pythagorean theorem.** The theorem was named for the Greek mathematician Pythagoras, born in 572 B.C.E. Pythagoras was the founder of the Greek society the Pythagoreans. Although the theorem bears Pythagoras's name, his own work on this theorem is uncertain because the Pythagoreans credited all new discoveries to their founder.

Property

| The Pythagorean Theorem | For every right triangle, the square of the length of the hypotenuse is equal to the sum of the squares of the lengths of the legs. |

c is the hypotenuse.

a and b are the legs.

$$c^2 = a^2 + b^2$$

▶ **Example 1**   **Verifying the Pythagorean Theorem**

< Objective 1 >

Verify the Pythagorean theorem for the given triangles.

(a) For the following right triangle,

$$5^2 \overset{?}{=} 3^2 + 4^2$$
$$25 \overset{?}{=} 9 + 16$$
$$25 = 25$$

(b) For the following right triangle,

$$13^2 \overset{?}{=} 12^2 + 5^2$$

$$169 \overset{?}{=} 144 + 25$$

$$169 = 169$$

Check Yourself 1

Verify the Pythagorean theorem for the right triangle shown.

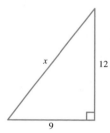

The Pythagorean theorem can be used to find the length of one side of a right triangle when the lengths of the two other sides are known.

 Example 2   **Solving for the Length of the Hypotenuse**

Find length x.

NOTE

x is longer than the given sides because it is the hypotenuse.

$$x^2 = 9^2 + 12^2$$
$$= 81 + 144$$
$$= 225$$

so

$$x = 15 \qquad \text{or} \qquad \underbrace{x = -15}$$

We reject this solution because a length must be positive.

Check Yourself 2

Find length x.

Sometimes, one or more of the lengths of the sides may be represented by an irrational number.

Example 3   Solving for the Length of the Leg

 > Calculator

Find length x. Use a calculator to approximate x to the nearest tenth.

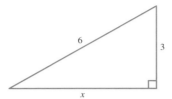

$$3^2 + x^2 = 6^2$$
$$9 + x^2 = 36$$
$$x^2 = 27$$
$$x = \pm\sqrt{27}$$

NOTE

You can approximate $3\sqrt{3}$ (or $\sqrt{27}$) with a calculator.

but distance cannot be negative, so

$$x = \sqrt{27}$$

So x is approximately 5.2.

Check Yourself 3

Find length x and approximate it to the nearest tenth.

The Pythagorean theorem can be applied to solve a variety of geometric problems.

Example 4   Solving for the Length of the Diagonal

 > Calculator

Find, to the nearest tenth, the length of the diagonal of a rectangle that is 8 cm long and 5 cm wide. Let x be the unknown length of the diagonal:

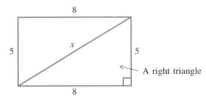

A right triangle

NOTES

Always draw and label a sketch showing the information from a problem when geometric figures are involved.

Again, distance cannot be negative, so we eliminate $x = -\sqrt{89}$.

So

$$x^2 = 5^2 + 8^2$$
$$= 25 + 64$$
$$= 89$$
$$x = \sqrt{89}$$

Thus

$$x \approx 9.4 \text{ cm}$$

Check Yourself 4

The diagonal of a rectangle is 12 in. and its width is 6 in. Find its length to the nearest tenth.

The application in Example 5 also makes use of the Pythagorean theorem.

▶   **Example** 5   Solving an Application

NOTE

Always check to see if your final answer is reasonable.

How long must a guy wire be to reach from the top of a 30-ft pole to a point on the ground 20 ft from the base of the pole?

Again, be sure to draw a sketch of the problem.

$$x^2 = 20^2 + 30^2$$
$$= 400 + 900$$
$$= 1,300$$
$$x = \sqrt{1,300}$$
$$\approx 36 \text{ ft}$$

Check Yourself 5

A 16-ft ladder leans against a wall with its base 4 ft from the wall. How far above the floor is the top of the ladder?

NOTE

Distances are always positive, so we use only the principal square root.

To find the distance between any two points in the plane, we use a formula derived from the Pythagorean theorem. First, we need an alternate form of the Pythagorean theorem.

Property

The Pythagorean Theorem

Given a right triangle in which c is the length of the hypotenuse, we have the equation

$$c^2 = a^2 + b^2$$

We can rewrite the formula as

$$c = \sqrt{a^2 + b^2}$$

We use this form of the Pythagorean theorem in Example 6.

Example 6   **Finding the Distance Between Two Points**

< Objective 2 >

Find the distance from (2, 3) to (5, 7).

The distance can be seen as the hypotenuse of a right triangle.

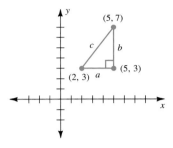

The lengths of the two legs can be found by finding the difference of the two x-coordinates and the difference of the two y-coordinates. So

$$a = 5 - 2 = 3 \qquad \text{and} \qquad b = 7 - 3 = 4$$

The distance c can then be found using the formula

$$c = \sqrt{a^2 + b^2}$$

or, in this case,

$$c = \sqrt{3^2 + 4^2}$$
$$c = \sqrt{9 + 16}$$
$$= \sqrt{25}$$
$$= 5$$

The distance is 5 units.

Check Yourself 6

Find the distance between (0, 2) and (5, 14).

If we call our points (x_1, y_1) and (x_2, y_2), we can state the **distance formula.**

Property

Distance Formula

The distance between points (x_1, y_1) and (x_2, y_2) can be found using the formula

$$d = \sqrt{(x_2 - x_1)^2 + (y_2 - y_1)^2}$$

Example 7   Finding the Distance Between Two Points

Find the distance between $(-2, 5)$ and $(2, -3)$. Simplify the radical answer.
   Using the formula,

$$d = \sqrt{[2 - (-2)]^2 + [(-3) - 5]^2}$$

$$= \sqrt{(4)^2 + (-8)^2}$$

$$= \sqrt{16 + 64}$$

$$= \sqrt{80}$$

$$= 4\sqrt{5}$$

NOTE

$\sqrt{80} = \sqrt{16 \cdot 5} = 4\sqrt{5}$

Check Yourself 7

Find the distance between (2, 5) and (−5, 2).

In Example 7, you were asked to find the distance between $(-2, 5)$ and $(2, -3)$.

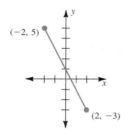

To form a right triangle, we include the point $(-2, -3)$.

Note that the lengths of the two sides of the right triangle are 4 and 8. By the Pythagorean theorem, the hypotenuse must have length $\sqrt{4^2 + 8^2} = \sqrt{80} = 4\sqrt{5}$. The distance formula is an application of the Pythagorean theorem.
   You can use the square-root key on a calculator to approximate the length of a diagonal line. This is particularly useful in checking to see if an object is square or rectangular.

Example 8   Approximating Length with a Calculator

> Calculator

Approximate the length of the diagonal of the given rectangle. The diagonal forms the hypotenuse of a triangle with legs 12.2 in. and 15.7 in. The length of the diagonal is $\sqrt{12.2^2 + 15.7^2} = \sqrt{395.33} \approx 19.88$ in. Use your calculator to confirm the approximation.

15.7 in.

12.2 in.

Check Yourself 8

Approximate the length of the diagonal of the rectangle to the nearest tenth.

13.7 in.

19.7 in.

There are many real-world applications of the Pythagorean theorem. The following is one that occurs regularly.

| Example 9 | An Application of the Pythagorean Theorem |

A 12-foot ladder is placed against an outside wall so that the bottom is 5 feet from the wall. How high on the wall does the ladder reach?

The ladder is the hypotenuse of a right triangle with one leg measuring 5 feet. Using the Pythagorean theorem yields

$$h^2 + 5^2 = 12^2$$
$$h^2 = 12^2 - 5^2 = 144 - 25$$
$$h = \sqrt{144 - 25}$$
$$h = \sqrt{119}$$
$$h \approx 10.9$$

The ladder rests approximately 10.9 feet up the wall.

Check Yourself 9

A 15-foot ladder is placed against an outside wall so that the bottom is 7 feet from the wall. How high on the wall will the ladder reach (to the nearest tenth of a foot)?

Check Yourself ANSWERS

1. $10^2 \overset{?}{=} 8^2 + 6^2$; $100 \overset{?}{=} 64 + 36$; $100 = 100$ **2.** 13   **3.** $2\sqrt{7}$; or approximately 5.3   **4.** The length is approximately 10.4 in.
5. The height is approximately 15.5 ft.   **6.** 13 units   **7.** $\sqrt{58}$
8. $\approx$24.0 in.   **9.** 13.3 feet

Reading Your Text

The following fill-in-the-blank exercises are designed to ensure that you understand some of the key vocabulary used in this section.

SECTION 9.6

(a) Pythagoras was a _____ mathematician.

(b) For every right triangle, the square of the length of the _____ is equal to the sum of the squares of the lengths of the legs.

(c) When solving an equation for a length, we reject solutions that are _____.

(d) You can use the square-root key on a calculator to get an _____ of a square root.

< Objective 1 >

Find the length x in each triangle. Express your answer in simplified radical form.

1.

2.

3.

> Videos

4.

5.

6.

In exercises 7–12, express your answer to the nearest thousandth.

7. GEOMETRY Find the diagonal of a rectangle with a length of 10 cm and a width of 7 cm.

> Videos

8. GEOMETRY Find the diagonal of a rectangle with 5 in. width and 7 in. length.

9. GEOMETRY Find the width of a rectangle whose diagonal is 12 ft and whose length is 10 ft.

10. GEOMETRY Find the length of a rectangle whose diagonal is 9 in. and whose width is 6 in.

11. CONSTRUCTION How long must a guy wire be to run from the top of a 20-ft pole to a point on the ground 8 ft from the base of the pole?

Name _____

Section _____ Date _____

Answers

1. _____

2. _____

3. _____

4. _____

5. _____

6. _____

7. _____

8. _____

9. _____

10. _____

11. _____

Answers

12. _____

13. _____

14. _____

15. _____

16. _____

17. _____

18. _____

12. Construction The base of a 15-ft ladder is 5 ft away from a wall. How high from the floor is the top of the ladder?

Find the altitude of each triangle.

13.

14.

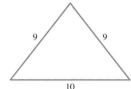

15. Construction A homeowner wishes to insulate her attic with fiberglass insulation to conserve energy. The insulation comes in 40-cm-wide rolls that are cut to fit between the rafters in the attic. If the roof is 6 m from peak to eave and the attic space is 2 m high at the peak, how long does each of the pieces of insulation need to be? Round to the nearest tenth.

16. Construction For the home described in exercise 15, if the roof is 7 m from peak to eave and the attic space is 3 m high at the peak, how long does each of the pieces of insulation need to be? Round to the nearest tenth.

17. Manufacturing A solar collector and its stand are in the shape of a right triangle. The collector is 5.00 m long, the upright leg is 3.00 m long, and the base leg is 4.00 m long. Because of inefficiencies due to the collector's position, it needs to be raised by 0.50 m on the upright leg. How long will the new base leg be? Round to the nearest tenth.

18. Manufacturing A solar collector and its stand are in the shape of a right triangle. The collector is 5.00 m long, the upright leg is 2.00 m long, and the base leg is 4.58 m long. Because of inefficiencies due to the collector's position, it needs to be lowered by 0.50 m on the upright leg. How long will the new base leg be? Round to the nearest tenth.

Find the distance between each pair of points.

19. $(2, 0)$ and $(-4, 0)$

20. $(-3, 0)$ and $(4, 0)$

21. $(0, -2)$ and $(0, -9)$

22. $(0, 8)$ and $(0, -4)$

23. $(2, 5)$ and $(5, 2)$

24. $(3, 3)$ and $(5, 7)$

25. $(5, 1)$ and $(3, 8)$

26. $(2, 9)$ and $(7, 4)$

27. $(-2, 8)$ and $(1, 5)$ > Videos

28. $(2, 6)$ and $(-3, 4)$

29. $(6, -1)$ and $(2, 2)$

30. $(2, -8)$ and $(1, 0)$

31. $(-1, -1)$ and $(2, 5)$

32. $(-2, -2)$ and $(3, 3)$

33. $(-2, 9)$ and $(-3, 3)$

34. $(4, -1)$ and $(0, -5)$

35. $(-1, -4)$ and $(-3, 5)$

36. $(-2, 3)$ and $(-7, -1)$

37. $(-2, -4)$ and $(-4, 1)$

38. $(-1, -1)$ and $(4, -2)$

39. $(-4, -2)$ and $(-1, -5)$

40. $(-2, -2)$ and $(-4, -4)$

41. $(-2, 0)$ and $(-4, -1)$ > Videos

42. $(-5, -2)$ and $(-7, -1)$

*Determine whether each of the following statements is **true** or **false**.*

43. It is possible to have a right triangle that has two equal sides.

44. In a right triangle, there could be an angle whose measure is more than $180°$.

*Complete each of the following statements with **never, sometimes,** or **always**.*

45. The hypotenuse is _____ the longest side of a right triangle.

46. The distance between two points is _____ positive.

| Basic Skills | **Advanced Skills** | Vocational-Technical Applications | Calculator/Computer | Above and Beyond | Getting Ready |

Use the distance formula to show that each set of points describes an isosceles triangle (a triangle with two sides of equal length).

47. $(-3, 0)$, $(2, 3)$, and $(1, -1)$

48. $(-2, 4)$, $(2, 7)$, and $(5, 3)$

Answers

19. _____ 20. _____

21. _____ 22. _____

23. _____ 24. _____

25. _____ 26. _____

27. _____ 28. _____

29. _____ 30. _____

31. _____ 32. _____

33. _____ 34. _____

35. _____ 36. _____

37. _____ 38. _____

39. _____ 40. _____

41. _____ 42. _____

43. _____ 44. _____

45. _____

46. _____

47. _____

48. _____

Answers

49. _____

50. _____

51. _____

52. _____

53. _____

54. _____

55. _____

49. GEOMETRY The length of one leg of a right triangle is 3 in. more than the other. If the length of the hypotenuse is 15 in., what are the lengths of the two legs?

50. GEOMETRY The length of a rectangle is 1 cm longer than its width. If the diagonal of the rectangle is 5 cm, what are the dimensions (the length and width) of the rectangle?

Use the Pythagorean theorem to determine the length of each line segment. Where appropriate, round to the nearest hundredth.

51.

> Videos

52.

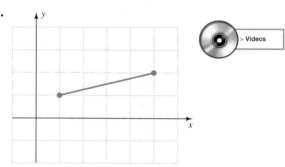

53.

For each figure, use the slope concept and the Pythagorean theorem to show that the figure is a square. (Recall that a square must have four right angles and four equal sides.) Then give the area of the square to the nearest hundredth.

54.

55.

56.

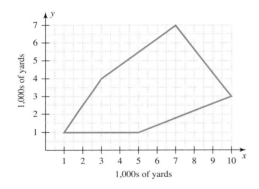

Answers

56. _____

57. _____

58. _____

59. _____

60. _____

61. _____

Basic Skills | Advanced Skills | **Vocational-Technical Applications** | Calculator/Computer | Above and Beyond | Getting Ready

57. CONSTRUCTION TECHNOLOGY The diagram for a jetport is shown here.

Use the distance formula to find the total length of fence around the jetport. (Note that each tick mark is 1,000 yards.)

Basic Skills | Advanced Skills | Vocational-Technical Applications | Calculator/Computer | **Above and Beyond** | Getting Ready

58. Suppose that two legs of a right triangle each have length 5. Find the length of the hypotenuse. Express your answer as a simplified radical.

59. Repeat exercise 58 using legs of length 6.

60. Repeat exercise 58 using legs of length 7.

61. Generalize the results of exercises 58–60. That is, if the two legs of a right triangle each have length x, what do you think the length of the hypotenuse will be?

Answers

62.

_____

62. BUSINESS AND FINANCE Your architectural firm just received this memo.

To: Algebra Expert Architecture, Inc.
From: Microbeans Coffee Company, Inc.
Re: Design for On-Site Day Care Facility
Date: Aug. 10, 2005

We are requesting that you submit a design for a nursery for preschool children. We are planning to provide free on-site day care for the workers at our corporate headquarters.

    The nursery should be large enough to serve the needs of 20 preschoolers. There will be three child care workers in this facility. We want the nursery to be 3,000 ft^2 in area. It needs a playroom, a small kitchen and eating space, and bathroom facilities. There should be some space to store toys and books, many of which should be accessible to children. The company plans to put this facility on the first floor on an outside wall so the children can go outside to play without disturbing workers. You are free to add to this design as you see fit.

    Please send us your design drawn to a scale of 1 ft to 0.25 in., with precise measurements and descriptions. We would like to receive this design 1 week from today. Please give us some estimate of the cost of this renovation to our building.

chapter 9 > Make the Connection

Submit a design, keeping in mind that the design has to conform to strict design specifications for buildings designated as nurseries, including:

1. Number of exits: Two exits for the first 7 people and one exit for every additional 7 people.

2. Width of exits: The total width of exits in inches shall not be less than the total occupant load served by an exit multiplied by 0.3 for stairs and 0.2 for other exits. No exit shall be less than 3 ft wide and 6 ft 8 in. high.

3. Arrangements of exits: If two exits are required, they shall be placed a distance apart equal to but not less than one-half the length of the maximum overall diagonal dimension of the building or area to be served measured in a straight line between exits. Where three or more exits are required, two shall be placed as above and the additional exits arranged a reasonable distance apart.

4. Distance to exits: Maximum distance to travel from any point to an exterior door shall not exceed 100 ft.

Answers

1. 15   **3.** 15   **5.** $2\sqrt{6}$ **7.** ≈ 12.207 cm   **9.** ≈ 6.633 ft
11. ≈ 21.541 ft   **13.** 4   **15.** $8\sqrt{2} \approx 11.3$ m   **17.** ≈ 3.6 m
19. 6   **21.** 7   **23.** $3\sqrt{2}$ **25.** $\sqrt{53}$ **27.** $3\sqrt{2}$ **29.** 5
31. $3\sqrt{5}$ **33.** $\sqrt{37}$ **35.** $\sqrt{85}$ **37.** $\sqrt{29}$ **39.** $3\sqrt{2}$ **41.** $\sqrt{5}$
43. True   **45.** always   **47.** Sides have length $\sqrt{34}$, $\sqrt{17}$, and $\sqrt{17}$
49. 9 in., 12 in.   **51.** 4.12   **53.** 5   **55.** 13   **57.** 22,991 yards
59. $6\sqrt{2}$ **61.** $x\sqrt{2}$

Definition/Procedure	Example	Reference
Roots and Radicals		Section 9.1
Square Roots		
$\sqrt{x}$ is the principal (or positive) square root of x. It is the nonnegative number we must square to get x. $-\sqrt{x}$ is the negative square root of x. The square root of a negative number is not a real number.	$\sqrt{49} = 7$ $-\sqrt{49} = -7$ $\sqrt{-49}$ is not a real number.	*p. 833*
Other Roots $\sqrt[3]{x}$ is the cube root of x. $\sqrt[4]{x}$ is the fourth root of x.	$\sqrt[3]{64} = 4$ because $4^3 = 64$. $\sqrt[4]{81} = 3$ because $3^4 = 81$.	*pp. 835–836*
Simplifying Radical Expressions		Section 9.2
An expression involving square roots is in *simplest form* if **1.** There are no perfect-square factors in a radical. **2.** No fraction appears inside a radical. **3.** No radical appears in the denominator.		*p. 845*
To simplify a radical expression, use the following properties. The square root of a product is the product of the square roots. $\sqrt{ab} = \sqrt{a} \cdot \sqrt{b}$	$\begin{aligned} \sqrt{40} &= \sqrt{4 \cdot 10} \\ &= \sqrt{4} \cdot \sqrt{10} \\ &= 2\sqrt{10} \\ \sqrt{12x^3} &= \sqrt{4x^2 \cdot 3x} \\ &= \sqrt{4x^2} \cdot \sqrt{3x} \\ &= 2x \cdot \sqrt{3x} \end{aligned}$	*p. 845*
The square root of a quotient is the quotient of the square roots. $\sqrt{\dfrac{a}{b}} = \dfrac{\sqrt{a}}{\sqrt{b}}$	$\sqrt{\dfrac{5}{16}} = \dfrac{\sqrt{5}}{\sqrt{16}} = \dfrac{\sqrt{5}}{4}$ $\sqrt{\dfrac{2y}{3}} = \dfrac{\sqrt{2y}}{\sqrt{3}} = \dfrac{\sqrt{2y} \cdot \sqrt{3}}{\sqrt{3} \cdot \sqrt{3}}$ $\quad = \dfrac{\sqrt{6y}}{\sqrt{9}} = \dfrac{\sqrt{6y}}{3}$	*p. 847*
Adding and Subtracting Radicals		Section 9.3
Like radicals have the same index and the same radicand (the expression inside the radical). Like radicals can be added (or subtracted) in the same way as like terms. Apply the distributive law and combine the coefficients.	$3\sqrt{5}$ and $2\sqrt{5}$ are like radicals. $\begin{aligned} 2\sqrt{3} + 3\sqrt{3} &= (2+3)\sqrt{3} \\ &= 5\sqrt{3} \\ 5\sqrt{7} - 2\sqrt{7} &= (5-2)\sqrt{7} \\ &= 3\sqrt{7} \end{aligned}$	*p. 855*

Continued

Definition/Procedure	Example	Reference

Certain expressions can be combined after one or more of the terms involving radicals are simplified.

$$\sqrt{12} + \sqrt{3} = 2\sqrt{3} + \sqrt{3}$$
$$= (2 + 1)\sqrt{3}$$
$$= 3\sqrt{3}$$

p. 855

Multiplying and Dividing Radicals

Section 9.4

Multiplying

To multiply radical expressions, use the first property of radicals in the following way:

$$\sqrt{a} \cdot \sqrt{b} = \sqrt{ab}$$

$$\sqrt{6} \cdot \sqrt{15} = \sqrt{6 \cdot 15} = \sqrt{90}$$
$$= \sqrt{9 \cdot 10}$$
$$= 3\sqrt{10}$$

p. 864

The distributive property also applies when multiplying radical expressions.

$$\sqrt{5}(\sqrt{3} + 2\sqrt{5})$$
$$= \sqrt{5} \cdot \sqrt{3} + \sqrt{5} \cdot 2\sqrt{5}$$
$$= \sqrt{15} + 10$$

p. 865

The FOIL pattern allows us to find the product of binomial radical expressions.

$$(\sqrt{5} + 2)(\sqrt{5} - 1)$$
$$= \sqrt{5} \cdot \sqrt{5} - \sqrt{5} + 2\sqrt{5} - 2$$
$$= 3 + \sqrt{5}$$
$$(\sqrt{10} + 3)(\sqrt{10} - 3)$$
$$= 10 - 9 = 1$$

p. 865

Dividing

To divide radical expressions, use the second property of radicals in the following way:

$$\frac{\sqrt{a}}{\sqrt{b}} = \sqrt{\frac{a}{b}}$$

$$\frac{\sqrt{50}}{\sqrt{2}} = \sqrt{\frac{50}{2}}$$
$$= \sqrt{25}$$
$$= 5$$

p. 866

Solving Radical Equations

Section 9.5

Power Property of Equality

To solve an equation involving radicals, apply the power property of equality:

Given any two expressions a and b and any positive integer n,

if $a = b$, then $a^n = b^n$.

p. 873

Definition/Procedure	Example	Reference

Solving Equations Involving Radicals

1. Isolate a radical on one side of the equation.
2. Raise each side of the equation to the smallest power that will eliminate the radical.
3. If any radicals remain in the equation derived in step 2, return to step 1 and continue the process.
4. Solve the resulting equation to determine any possible solutions.
5. Check all solutions to determine whether extraneous solutions may have resulted from step 2.

Solve

$$\sqrt{2x - 3} + x = 9$$
$$\sqrt{2x - 3} = -x + 9$$
$$2x - 3 = x^2 - 18x + 81$$
$$0 = x^2 - 20x + 84$$
$$0 = (x - 6)(x - 14)$$
$$x = 6 \quad \text{or} \quad x = 14$$

By substitution, 6 is the only valid solution.

p. 876

Applications of the Pythagorean Theorem

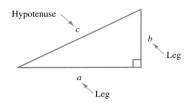

In words, for every right triangle, the square of the length of the hypotenuse is equal to the sum of the squares of the lengths of the legs.

$$c^2 = a^2 + b^2$$

Find length x:

$$x^2 = 10^2 + 6^2$$
$$= 100 + 36$$
$$= 136$$
$$x = \sqrt{136} \quad \text{or} \quad 2\sqrt{34}$$

Section 9.6

p. 881

This summary exercise set is provided to give you practice with each of the objectives of this chapter. Each exercise is keyed to the appropriate chapter section. When you are finished, you can check your answers to the odd-numbered exercises against those presented in the back of the text. If you have difficulty with any of these questions, go back and reread the examples from that section. Your instructor will give you guidelines on how best to use these exercises in your instructional setting.

9.1 *Evaluate if possible.*

1. $\sqrt{81}$

2. $-\sqrt{49}$

3. $\sqrt{-49}$

4. $\sqrt[3]{64}$

5. $\sqrt[3]{-64}$

6. $\sqrt[4]{81}$

7. $\sqrt[4]{-81}$

9.2 *Simplify each of the following radical expressions. Assume that all variables represent positive real numbers.*

8. $\sqrt{50}$

9. $\sqrt{45}$

10. $\sqrt{7a^3}$

11. $\sqrt{20x^4}$

12. $\sqrt{49m^5}$

13. $\sqrt{200b^3}$

14. $\sqrt{147r^3s^2}$

15. $\sqrt{108a^2b^5}$

16. $\sqrt{\dfrac{10}{81}}$

17. $\sqrt{\dfrac{18x^2}{25}}$

18. $\sqrt{\dfrac{12m^5}{49}}$

19. $\sqrt{\dfrac{3}{7}}$

20. $\sqrt{\dfrac{3a}{2}}$

21. $\sqrt{\dfrac{8x^2}{7}}$

9.3 *Simplify by combining like terms.*

22. $\sqrt{3} + 4\sqrt{3}$

23. $9\sqrt{5} - 3\sqrt{5}$

24. $3\sqrt{2} + 2\sqrt{3}$

25. $3\sqrt{3a} - \sqrt{3a}$

26. $7\sqrt{6} - 2\sqrt{6} + \sqrt{6}$

27. $5\sqrt{3} + \sqrt{12}$

28. $3\sqrt{18} - 5\sqrt{2}$

29. $\sqrt{32} - \sqrt{18}$

30. $\sqrt{27} - \sqrt{3} + 2\sqrt{12}$

31. $\sqrt{8} + 2\sqrt{27} - \sqrt{75}$

32. $x\sqrt{18} - 3\sqrt{8x^2}$

9.4 *Simplify each radical expression.*

33. $\sqrt{6} \cdot \sqrt{5}$

34. $\sqrt{3} \cdot \sqrt{6}$

35. $\sqrt{3x} \cdot \sqrt{2}$

36. $\sqrt{2} \cdot \sqrt{8} \cdot \sqrt{3}$

37. $\sqrt{5a} \cdot \sqrt{10a}$

38. $\sqrt{2}(\sqrt{3} + \sqrt{5})$

39. $\sqrt{7}(2\sqrt{3} - 3\sqrt{7})$

40. $(\sqrt{3} + 5)(\sqrt{3} - 3)$

41. $(\sqrt{15} - 3)(\sqrt{15} + 3)$

42. $(\sqrt{2} + 3)^2$

43. $\dfrac{\sqrt{7x^3}}{\sqrt{3}}$

44. $\dfrac{18 - \sqrt{20}}{2}$

9.5 *Solve each of the following equations. Be sure to check your solutions.*

45. $\sqrt{x - 5} = 4$

46. $\sqrt{3x - 2} + 2 = 5$

47. $\sqrt{y + 7} = y - 5$

48. $\sqrt{2x - 1} + x = 8$

9.6 *Find length x in each triangle. Express your answer in simplified radical form.*

49.

50.

51.

52.

53.

54.

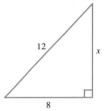

Solve each of the following word problems. Approximate your answer to one decimal place where necessary.

55. Find the diagonal of a rectangle whose length is 12 in. and whose width is 9 in.

56. Find the length of a rectangle whose diagonal has a length of 10 cm and whose width is 5 cm.

57. How long must a guy wire be to run from the top of an 18-ft pole to a point on level ground 16 ft away from the base of the pole?

58. The length of one leg of a right triangle is 2 in. more than the length of the other. If the length of the hypotenuse of the triangle is 10 in., what are the lengths of the two legs?

Find the distance between each pair of points.

59. $(-3, 2)$ and $(-7, 2)$

60. $(2, 0)$ and $(5, 9)$

61. $(-2, 7)$ and $(-5, -1)$

62. $(5, -1)$ and $(-2, 3)$

63. $(-3, 4)$ and $(-2, -5)$

64. $(6, 4)$ and $(-3, 5)$

Beginning Algebra

The Streeter/Hutchison Series in Mathematics

© The McGraw-Hill Companies. All Rights Reserved.

Name _____

Section _____ Date _____

Answers

1. _____

2. _____

3. _____

4. _____

5. _____

6. _____

7. _____

8. _____

9. _____

10. _____

11. _____

12. _____

13. _____

14. _____

15. _____

16. _____

17. _____

18. _____

900

The purpose of this self-test is to help you check your progress and to review for the next in-class exam. Allow yourself about an hour to take this test. At the end of that hour check your answers against those given in the back of the text. Section references accompany the answers. If you missed any questions, go back to those sections and reread the examples until you master the concepts.

Evaluate if possible.

1. $\sqrt{121}$

2. $\sqrt[3]{27}$

3. $\sqrt{-144}$

4. $-\sqrt[3]{-64}$

Simplify each of the following radical expressions.

5. $\sqrt{75}$

6. $\sqrt{24a^3}$

7. $\sqrt{\dfrac{16}{25}}$

8. $\sqrt{\dfrac{5}{9}}$

Simplify by combining like terms.

9. $2\sqrt{10} - 3\sqrt{10} + 5\sqrt{10}$

10. $3\sqrt{8} - \sqrt{18}$

11. $2\sqrt{50} - \sqrt{8} - \sqrt{50}$

12. $\sqrt{20} + \sqrt{45} - \sqrt{5}$

Simplify each of the following radical expressions.

13. $\sqrt{3x} \cdot \sqrt{6x}$

14. $(\sqrt{5} + 3)(\sqrt{5} + 2)$

15. $\dfrac{\sqrt{7}}{\sqrt{2}}$

16. $\dfrac{14 + 3\sqrt{98}}{7}$

Solve the following equations for x.

17. $\sqrt{x - 2} = 9$

18. $\sqrt{3x + 4} + x = 8$

Find length x in each triangle. Write the answer in simplified radical form.

19.

20.

21.

22.

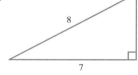

23. If the diagonal of a rectangle is 12 cm and the width of the rectangle is 7 cm, what is the length of the rectangle? Round to the nearest thousandth.

Find the distance between the two points.

24. $(-3, 7)$ and $(-12, 7)$

25. $(-2, 5)$ and $(-9, -1)$

Answers

19. _____

20. _____

21. _____

22. _____

23. _____

24. _____

25. _____

Activity 9 ::
The Swing of the Pendulum

The action of a pendulum seems simple. Scientists have studied the characteristics of a swinging pendulum and found them to be quite useful. In 1851 in Paris, Jean Foucault (pronounced "Foo-koh") used a pendulum to clearly demonstrate the rotation of the earth about its own axis.

A pendulum can be as simple as a string or cord with a weight fastened to one end. The other end is fixed, and the weight is allowed to swing. We define the **period** of a pendulum to be the amount of time required for the pendulum to make one complete swing (back and forth). The question we pose is: How does the *period* of a pendulum relate to the *length* of the pendulum?

For this activity, you need a piece of string that is approximately 1 m long. Fasten a weight (such as a small hexagonal nut) to one end, and then place clear marks on the string every 10 cm up to 70 cm, measured from the center of the weight.

1. Working with one or two partners, hold the string at the mark that is 10 cm from the weight. Pull the weight to the side with your other hand and let it swing freely. To estimate the period, let the weight swing through 30 periods, record the time in the given table, and then divide by 30. Round your result to the nearest hundredth of a second and record it. (*Note:* If you are unable to perform the experiment and collect your own data, you can use the sample data collected in this manner and presented at the end of this activity.)

   Repeat the described procedure for each length indicated in the given table.

Length of string, cm	10	20	30	40	50	60	70
Time for 30 periods, s							
Time for 1 period, s							

2. Let L represent the length of the pendulum and T represent the time period that results from swinging that pendulum. Fill out the table:

L	10	20	30	40	50	60	70
T							

3. Which variable, L or T, is viewed here as the independent variable?

4. On graph paper, draw horizontal and vertical axes, but plan to graph the data points in the first quadrant only. Explain why this is so.

5. With the independent variable marked on the horizontal axis, scale the axes appropriately, keeping an eye on your data.

6. Plot your data points. Should you connect them with a smooth curve?

7. What period *T* would correspond to a string length of 0? Include this point on your graph.

8. Use your graph to predict the period for a string length of 80 cm.

9. Verify your prediction by measuring the period when the string is held at 80 cm (as described in step 1). How close did your experimental estimate come to the prediction made in step 8?

You have created a graph showing *T* as a function of *L*. The shape of the graph may not be familiar to you yet. The shape of your pendulum graph fits that of a square-root function.

Sample Data

Length of string, cm	10	20	30	40	50	60	70
Time for 30 periods, s	19	27	33	38	42	46	49

Name _____

Section _____ Date _____

Answers

The following exercises are presented to help you review concepts from earlier chapters. This is meant as a review and not as a comprehensive exam. The answers are presented in the back of the text. Section references accompany the answers. If you have difficulty with any of these exercises, be certain to at least read through the summary related to those sections.

Simplify the following expressions.

1. $8x^2y^3 - 5x^3y - 5x^2y^3 + 3x^3y$

1. _____

2. $(4x^2 - 2x + 7) - (-3x^2 + 4x - 5)$

2. _____

3. _____

Evaluate each expression when $x = 2$, $y = -1$, and $z = -4$.

3. $2xyz^2 - 4x^2y^2z$ **4.** $-2xyz + 2x^2y^2$

4. _____

5. _____

Solve the following equations for x.

6. _____

5. $-3x - 2(4 - 6x) = 10$

7. _____

6. $5x - 3(4 - 2x) = 6(2x - 3)$

8. _____

7. Solve the inequality $3x - 11 < 5x - 19$.

9. _____

10. _____

Perform the indicated operations.

11. _____

8. $2x^2y(3x^2 - 5x + 19)$ **9.** $(5x + 3y)(4x - 7y)$

12. _____

Factor each of the following completely.

13. _____

10. $36xy - 27x^3y^2$ **11.** $8x^2 - 26x + 15$

Perform the indicated operations.

12. $\dfrac{2}{3x + 21} - \dfrac{3}{5x + 35}$

13. $\dfrac{x^2 - x - 6}{x^2 - x - 20} \div \dfrac{x^2 + x - 2}{x^2 + 3x - 4}$

Graph each of the following:

14. $4x + 5y = 20$

15. $5x - 4y \geq 20$

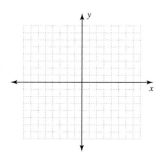

16. Find the slope of the line through the points $(2, 9)$ and $(-1, -6)$.

17. Given that the slope of a line is $-\dfrac{3}{2}$ and the y-intercept is $(0, 5)$, write the equation of the line.

Solve each of the following systems. If a unique solution does not exist, state whether the system is inconsistent or dependent.

18. $4x - 5y = 20$
$2x + 3y = 10$

19. $4x + 7y = 24$
$8x + 14y = 12$

Solve the following application. Be sure to show the system of equations used for your solution.

20. Amir was able to travel 80 mi downstream in 5 h. Returning upstream, he took 8 h to make the trip. How fast can he travel in still water, and what was the rate of the current?

Evaluate each root, if possible.

21. $\sqrt{144}$

22. $-\sqrt{144}$

23. $\sqrt{-144}$

24. $\sqrt[3]{-27}$

Answers

14. _____

15. _____

16. _____

17. _____

18. _____

19. _____

20. _____

21. _____

22. _____

23. _____

24. _____

Answers

25. _____

26. _____

27. _____

28. _____

29. _____

30. _____

Simplify each of the following radical expressions.

25. $a\sqrt{20} - 2\sqrt{45a^2}$

26. $\dfrac{\sqrt{8x^3}}{\sqrt{3}}$

27. $\dfrac{12 - \sqrt{72}}{3}$

28. $\sqrt{98x^2}$

29. $\sqrt{150m^3n^2}$

30. $\sqrt{\dfrac{12a^2}{25}}$

CHAPTER

10

INTRODUCTION

Large cities often commission fireworks artists to choreograph elaborate displays on holidays. Such displays look like beautiful paintings in the sky, in which the fireworks seem to dance to well-known popular and classical music. The displays are feats of engineering and very accurate timing. Suppose the designer wants a second set of rockets of a certain color and shape to be released after the first set reaches a specific height and explodes. He or she must know the strength of the initial liftoff and use a quadratic equation to determine the proper time for setting off the second round.

The equation $h = -16t^2 + 100t$ gives the height in feet t seconds after the rockets are shot into the air if the initial velocity is 100 feet per second. Using this equation, the designer knows how high the rocket will ascend and when it will begin to fall. The designer can time the next round to achieve the desired effect. Displays that involve large banks of fireworks in shows that last up to an hour are programmed using computers, but quadratic equations are at the heart of the mechanism that creates the beautiful effects.

Quadratic Equations

CHAPTER 10 OUTLINE

Name _____

Section _____ Date _____

This pretest will provide a preview of the types of exercises you will encounter in each section of this chapter. The answers for these exercises can be found in the back of the text. If you are working on your own, or ahead of the class, this pretest can help you find the sections in which you should focus more of your time.

Answers

1. _____

2. _____

3. _____

4. _____

5. _____

6. _____

7. _____

8. _____

9. _____

10. _____

11. _____

12. _____

13. _____

14. _____

15. _____

Solve each of the equations for x.

10.1   **1.** $x^2 = 17$ **2.** $x^2 - 12 = 0$

3. $(x - 1)^2 = 5$ **4.** $9x^2 = 14$

Solve each of the equations by completing the square.

10.2   **5.** $x^2 - 3x - 10 = 0$ **6.** $x^2 - 5x + 2 = 0$

7. $x^2 - 4x - 4 = 0$ **8.** $2x^2 - 4x - 7 = 0$

Use the quadratic formula to solve each of the equations.

10.3   **9.** $x^2 + 5x - 14 = 0$ **10.** $x^2 - 10x + 25 = 0$

11. $x^2 + 3x = 5$ **12.** $2x^2 = 3x + 3$

13. $3x = 4 + \dfrac{2}{x}$ **14.** $(x - 1)(x + 4) = 3$

Graph each quadratic equation after completing the given table of values.

10.4   **15.** $y = x^2 + 2$

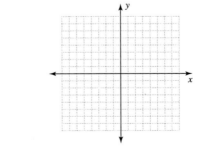

x	y
-2	
-1	
0	
1	
2	

16. $y = x^2 + 4x$

x	y
−4	
−3	
−2	
−1	
0	

17. $y = x^2 + x - 6$

x	y
−2	
−1	
0	
1	
2	

18. $y = -x^2 + 3$

x	y
−2	
−1	
0	
1	
2	

More on Quadratic Equations

< 10.1 Objectives >

1 > Solve equations of the form $ax^2 = k$

2 > Solve equations of the form $(x - h)^2 = k$

We now have more tools for solving quadratic equations. In Sections 10.1–10.3 we will be using the ideas from Chapter 9 to extend our solution techniques.

In Section 4.6 we identified all equations of the form

$$ax^2 + bx + c = 0$$

as quadratic equations in standard form. In that section, we discussed solving these equations whenever the quadratic expression was factorable. In this chapter, we want to extend our equation-solving techniques so that we can find solutions for all such quadratic equations.

First, we review the factoring method of solution that we introduced in Chapter 4.

Example 1   **Solving Quadratic Equations by Factoring**

Solve each quadratic equation by factoring.

(a) $x^2 = -7x - 12$

First, we write the equation in standard form.

NOTE

Add $7x$ and 12 to both sides of the equation. The quadratic expression must be *set equal to* 0.

$$x^2 + 7x + 12 = 0$$

Once the equation is in standard form, we can factor the quadratic member.

$$(x + 3)(x + 4) = 0$$

Finally, using the zero-product principle, we solve the equations $x + 3 = 0$ and $x + 4 = 0$ as follows:

$$x = -3 \qquad \text{or} \qquad x = -4$$

(b) $x^2 = 16$

NOTE

Here we factor the quadratic member of the equation as a difference of squares.

Again, we write the equation in standard form.

$$x^2 - 16 = 0$$

Factoring, we have

$$(x + 4)(x - 4) = 0$$

The Streeter/Hutchison Series in Mathematics   Beginning Algebra

Finally, the solutions are

$$x = -4 \qquad \text{or} \qquad x = 4$$

Check Yourself 1

Solve each of the following quadratic equations.

(a) $x^2 - 4x = 45$ **(b)** $w^2 = 25$

Certain quadratic equations can be solved by other methods, such as the square-root method. Return to the equation in part **(b)** of Example 1.

Beginning with

$$x^2 = 16$$

we can take the square root of each side, to write

$$\sqrt{x^2} = \sqrt{16}$$

From Section 9.1, we know that this is equivalent to

$$\sqrt{x^2} = 4$$

or

$$|x| = 4$$

Values for x of 4 or -4 will both satisfy this last equation, and so we have the two solutions

$$x = 4 \qquad \text{or} \qquad x = -4$$

We usually write the solutions as

$$x = \pm 4$$

Two more equations solved by this method are shown in Example 2.

RECALL

By definition

$$\sqrt{x^2} = |x|$$

NOTE

$x = \pm 4$ is simply a convenient "shorthand" for indicating the two solutions, and we generally will go directly to this form.

Example 2   **Solving Equations by the Square-Root Method**

< Objective 1 >

Solve each of the following equations by the square-root method.

(a) $x^2 = 9$

By taking the square root of each side, we have

$$\sqrt{x^2} = \sqrt{9}$$
$$|x| = 3$$
$$x = \pm 3$$

(b) $x^2 = 5$

Again, we take the square root of each side to write our two solutions as

$$\sqrt{x^2} = \sqrt{5}$$
$$|x| = \sqrt{5}$$
$$x = \pm\sqrt{5}$$

Check Yourself 2

Solve.

(a) $x^2 = 100$ (b) $t^2 = 15$

You may have to add or subtract on both sides of the equation to write an equation in the form of those in Example 2, as Example 3 illustrates.

 Example 3   **Solving Equations by the Square-Root Method**

Solve $x^2 - 8 = 0$.
   First, add 8 to both sides of the equation. We have

$$x^2 = 8$$

Now take the square root of both sides.

RECALL

$\sqrt{8} = \sqrt{4 \cdot 2}$
   $= \sqrt{4} \cdot \sqrt{2}$
   $= 2\sqrt{2}$

$$x = \pm\sqrt{8}$$

Normally, the solutions are written in simplest form. In this case we have

$$x = \pm 2\sqrt{2}$$

Check Yourself 3

Solve.

(a) $x^2 - 18 = 0$ (b) $x^2 + 1 = 7$

NOTE

In the form $ax^2 = k$
a is the coefficient of x^2 and k is some number.

To solve a quadratic equation of the form $ax^2 = k$, divide both sides of the equation by a as the first step. This is shown in Example 4.

 Example 4   **Solving Equations by the Square-Root Method**

Solve $4x^2 = 3$.
   Divide both sides of the equation by 4.

$$x^2 = \frac{3}{4}$$

Now take the square root of both sides.

RECALL

$\sqrt{\dfrac{3}{4}} = \dfrac{\sqrt{3}}{\sqrt{4}}$
   $= \dfrac{\sqrt{3}}{2}$

$$x = \pm\sqrt{\frac{3}{4}}$$

Again write your result in simplest form, so

$$x = \pm\frac{\sqrt{3}}{2}$$

Check Yourself 4

Solve $9x^2 = 5$.

Equations of the form $(x - h)^2 = k$ can also be solved by taking the square root of both sides. Consider Example 5.

Example 5   Solving Equations by the Square-Root Method

< Objective 2 >

Solve $(x - 1)^2 = 6$.

Again, take the square root of both sides of the equation.

$$x - 1 = \pm\sqrt{6}$$

Now add 1 to both sides of the equation to isolate x.

$$x = 1 \pm \sqrt{6}$$

Check Yourself 5

Solve $(x + 2)^2 = 12$.

Equations of the form $a(x - h)^2 = k$ can also be solved if each side of the equation is divided by a first, as shown in Example 6.

Example 6   Solving Equations by the Square-Root Method

Solve $3(x - 2)^2 = 5$.

$$(x - 2)^2 = \frac{5}{3}$$

NOTE

$$\sqrt{\frac{5}{3}} = \frac{\sqrt{5}}{\sqrt{3}} \cdot \frac{\sqrt{3}}{\sqrt{3}} = \frac{\sqrt{15}}{3}$$

$$x - 2 = \pm\sqrt{\frac{5}{3}} = \frac{\pm\sqrt{15}}{3}$$

$$x = 2 \pm \frac{\sqrt{15}}{3}$$

$$x = \frac{6}{3} \pm \frac{\sqrt{15}}{3}$$

$$x = \frac{6 \pm \sqrt{15}}{3}$$

Check Yourself 6

Solve $5(x + 3)^2 = 2$.

What about an equation such as the following?

$$x^2 + 5 = 0$$

If we apply the methods of Examples 4–6, we first subtract 5 from both sides, to write

$$x^2 = -5$$

Taking the square root of both sides gives

$$x = \pm\sqrt{-5}$$

RECALL

The Pythagorean theorem was introduced in Section 9.6.

But we know there are no square roots of -5 in the real numbers, so this equation has *no real-number solutions*. You might work with this type of equation in your next algebra course.

Many applied problem situations can be modeled with quadratic equations. Such problems may arise in geometry, construction, physics, and economics, to name but a few.

The following example relies on the use of the Pythagorean theorem, studied previously.

▷   **Example 7**   **Vocational Application**

How long must a guy wire be to reach from the top of a 30-ft pole to a point on the ground 20 ft from the base of the pole? Round to the nearest tenth of a foot.

We start by drawing a sketch of the problem.

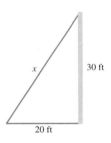

Using the Pythagorean theorem, we write

$$x^2 = 20^2 + 30^2$$
$$x^2 = 400 + 900$$
$$x^2 = 1{,}300$$

This is a quadratic equation that can be solved using the square-root method.

$$x = \pm\sqrt{1{,}300}$$

NOTE

Always check to see if your final answer is reasonable.

Since x must be positive, we reject $-\sqrt{1{,}300}$ and keep $\sqrt{1{,}300}$. To the nearest tenth of a foot, we have $x = 36.1$ ft.

        Check Yourself 7

A 16-ft ladder leans against a wall with its base 4 ft from the wall. How far above the floor, to the nearest tenth of a foot, is the top of the ladder?

Check Yourself ANSWERS

1. (a) $-5, 9$; (b) $-5, 5$ 2. (a) ± 10; (b) $\pm\sqrt{15}$

3. (a) $\pm 3\sqrt{2}$; (b) $\pm\sqrt{6}$ 4. $\pm\dfrac{\sqrt{5}}{3}$ 5. $-2 \pm 2\sqrt{3}$

6. $\dfrac{-15 \pm \sqrt{10}}{5}$ 7. 15.5 ft

Reading Your Text

The following fill-in-the-blank exercises are designed to ensure that you understand some of the key vocabulary used in this section.

SECTION 10.1

(a) Equations of the form $ax^2 + bx + c = 0$ are called _____ equations in standard form.

(b) To solve an equation of the form $ax^2 = k$, _____ both sides of the equation by a as the first step.

(c) Equations of the form $(x - h)^2 = k$ can be solved by taking the _____ of both sides.

(d) The equation $x^2 + 5 = 0$ has no _____ solutions.

10.1 exercises

MathZone

Boost your grade at mathzone.com!

> Practice Problems
> NetTutor
> Self-Tests
> e-Professors
> Videos

Name _____

Section _____ Date _____

Answers

1. _____	2. _____
3. _____	4. _____
5. _____	6. _____
7. _____	8. _____
9. _____	10. _____
11. _____	12. _____
13. _____	14. _____
15. _____	16. _____
17. _____	18. _____
19. _____	20. _____
21. _____	22. _____
23. _____	24. _____
25. _____	26. _____
27. _____	28. _____

< Objective 1 >

Solve each of the equations for x.

1. $x^2 = 5$ **2.** $x^2 = 15$

3. $x^2 = 33$ **4.** $x^2 = 43$

5. $x^2 - 7 = 0$ **6.** $x^2 - 13 = 0$

7. $x^2 - 20 = 0$ > Videos   **8.** $x^2 = 28$

9. $x^2 = 40$ **10.** $x^2 - 54 = 0$

11. $x^2 + 3 = 12$ **12.** $x^2 - 7 = 18$

13. $x^2 + 5 = 8$ **14.** $x^2 - 4 = 17$

15. $x^2 - 2 = 16$ **16.** $x^2 + 6 = 30$

17. $9x^2 = 25$ **18.** $16x^2 = 9$

19. $49x^2 = 11$ **20.** $16x^2 = 3$

21. $4x^2 = 7$ > Videos   **22.** $25x^2 = 13$

< Objective 2 >

Solve each equation for x.

23. $(x - 1)^2 = 5$ **24.** $(x - 3)^2 = 10$

25. $(x + 1)^2 = 12$ **26.** $(x + 2)^2 = 32$

27. $(x - 3)^2 = 24$ **28.** $(x - 5)^2 = 27$

29. $(x + 5)^2 = 25$ > Videos

30. $(x + 2)^2 = 16$

31. $3(x - 5)^2 = 7$

32. $2(x - 5)^2 = 3$

33. $4(x + 5)^2 = 9$ > Videos

34. $16(x + 2)^2 = 25$

35. $-2(x + 2)^2 = -6$

36. $-5(x + 4)^2 = -10$

37. $-4(x - 2)^2 = -5$

38. $-9(x - 2)^2 = -11$

Solve each of the following applications. Give all answers to the nearest thousandth.

39. CONSTRUCTION How long must a guy wire be to run from the top of a 20-ft pole to a point on the ground 8 ft from the base of the pole?

40. CONSTRUCTION How long must a guy wire be to run from the top of a 16-ft pole to a point on the ground 6 ft from the base of the pole?

41. CONSTRUCTION The base of a 15-ft ladder is 5 ft away from a wall. How far above the floor is the top of the ladder?

42. CONSTRUCTION The base of an 18-ft ladder is 4 ft away from a wall. How far above the floor is the top of the ladder?

43. GEOMETRY One leg of a right triangle is twice the length of the other. The hypotenuse is 8 m long. Find the length of each leg.

*Determine whether each of the following statements is **true** or **false**.*

44. Not all quadratic expressions are factorable.

45. Some quadratic equations have no real-number solutions.

*Complete each of the following statements with **never, sometimes,** or **always**.*

46. An equation of the form $(x - h)^2 = k$, where k is positive, _____ has two distinct solutions.

47. An equation of the form $x^2 = k$ _____ has real-number solutions.

Answers

29. _____

30. _____

31. _____

32. _____

33. _____

34. _____

35. _____

36. _____

37. _____

38. _____

39. _____

40. _____

41. _____

42. _____

43. _____

44. _____

45. _____

46. _____

47. _____

Answers

48.

49.

50.

51.

52.

53.

54.

55.

56.

57.

58.

Solve each equation for x.

48. $(2x + 11)^2 - 9 = 0$

49. $(3x + 14)^2 - 25 = 0$

50. $x^2 - 2x + 1 = 7$
(*Hint:* Factor the left-hand side.)

51. $x^2 + 4x + 4 = 7$
(*Hint:* Factor the left-hand side.)

 > Videos

52. **NUMBER PROBLEM** The square of a number decreased by 2 is equal to the negative of the number. Find the numbers.

53. **NUMBER PROBLEM** The square of 2 more than a number is 64. Find the numbers.

54. **NUMBER PROBLEM** The square of the sum of a number and 5 is 36. Find the numbers.

55. **ALLIED HEALTH** A toxic chemical is introduced into a protozoan culture. The number of deaths per hour N is given by the equation

$$N = 363 - 3t^2$$

where t is the number of hours after the chemical's introduction. How long will it take before the protozoa stop dying?

56. **ALLIED HEALTH** One technique of controlling cancer is to use radiation therapy. After such a treatment, the total number of cancerous cells N, in thousands, can be estimated by the equation

$$N = 121 - 4t^2$$

where t is the number of days of treatment. How many days of treatment are required to kill all the cancer cells?

57. **ALLIED HEALTH** One technique of controlling cancer is to use radiation therapy. After such a treatment, the total number of cancerous cells N, in thousands, can be estimated by the equation

$$N = 169 - 4t^2$$

where t is the number of days of treatment. How many days of treatment are required to kill all the cancer cells?

58. **MANUFACTURING TECHNOLOGY** The volume V of structural lumber (in cubic feet) that can be harvested from a coniferous tree is given by the formula

$$V = \frac{1}{350}D^2H - \frac{1}{5}$$

where H is the height of the tree (in feet) and D is the diameter (in inches) of the tree at the base. Calculate the necessary base diameter for a 48-foot-tall tree if we desire a volume of 70 cubic feet.

59. **MECHANICAL ENGINEERING** The deflection d (in inches) of a beam loaded with a single concentrated load is described by the equation

$$d = \frac{x^2 - 64}{200}$$

Find the location x (in feet) if the deflection is equal to 0.085 inch.

Basic Skills | Advanced Skills | Vocational-Technical Applications | Calculator/Computer | **Above and Beyond** | Getting Ready

▲

60. In this section, you solved quadratic equations by "extracting roots," taking the square root of both sides after writing one side as the square of a binomial. But what if the algebraic expression cannot be written this way? Work with another student to decide what needs to be added to each of the following expressions to make it a "perfect-square trinomial." Label the dimensions of the squares and the area of each section.

(a)

$$x^2 + 6x + \underline{\quad} = (x + ?)^2$$

(b)
$$n^2 + 10n + \underline{\quad} = (n + ?)^2$$

(c)
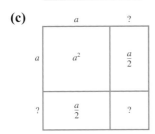
$$a^2 + a + \underline{\quad} = (a + ?)^2$$

(d)
$$x^2 - 12x + \underline{\quad} = (x - ?)^2$$

(e) $x^2 + 20x + \underline{\quad} = (x + ?)^2$ (f) $n^2 - 16n + \underline{\quad} = (n - ?)^2$

Answers

(a) _____

(b) _____

(c) _____

(d) _____

(e) _____

(f) _____

(g) _____

(h) _____

Multiply each of the following expressions.

(a) $(x + 1)^2$ **(b)** $(x + 5)^2$

(c) $(x - 2)^2$ **(d)** $(x - 7)^2$

(e) $(x + 4)^2$ **(f)** $(x - 3)^2$

(g) $(2x + 5)^2$ **(h)** $(2x - 1)^2$

Answers

1. $\pm\sqrt{5}$ **3.** $\pm\sqrt{33}$ **5.** $\pm\sqrt{7}$ **7.** $\pm 2\sqrt{5}$ **9.** $\pm 2\sqrt{10}$

11. ± 3 **13.** $\pm\sqrt{3}$ **15.** $\pm 3\sqrt{2}$ **17.** $\pm\dfrac{5}{3}$ **19.** $\pm\dfrac{\sqrt{11}}{7}$

21. $\pm\dfrac{\sqrt{7}}{2}$ **23.** $1 \pm \sqrt{5}$ **25.** $-1 \pm 2\sqrt{3}$ **27.** $3 \pm 2\sqrt{6}$

29. $-10, 0$ **31.** $\dfrac{15 \pm \sqrt{21}}{3}$ **33.** $-\dfrac{13}{2}, -\dfrac{7}{2}$ **35.** $-2 \pm \sqrt{3}$

37. $\dfrac{4 \pm \sqrt{5}}{2}$ **39.** 21.541 ft   **41.** 14.142 ft   **43.** 3.578 m, 7.155 m

45. True   **47.** sometimes   **49.** $-3, -\dfrac{19}{3}$ **51.** $-2 \pm \sqrt{7}$

53. $6, -10$ **55.** 11 hours   **57.** 6.5 days

59. -9 ft and $+9$ ft from the center

(a) $x^2 + 2x + 1$ (b) $x^2 + 10x + 25$ (c) $x^2 - 4x + 4$

(d) $x^2 - 14x + 49$ (e) $x^2 + 8x + 16$ (f) $x^2 - 6x + 9$

(g) $4x^2 + 20x + 25$ (h) $4x^2 - 4x + 1$

10.2

Completing the Square

< 10.2 Objectives >

1 > Complete the square for a trinomial expression

2 > Solve a quadratic equation by completing the square

We can solve a quadratic equation such as

$$x^2 - 2x + 1 = 5$$

very easily if we notice that the expression on the left is a perfect-square trinomial. Factoring, we have

$$(x - 1)^2 = 5$$

so

$$x - 1 = \pm\sqrt{5} \qquad \text{or} \qquad x = 1 \pm \sqrt{5}$$

The solutions for the original equation are then $1 + \sqrt{5}$ and $1 - \sqrt{5}$.

In fact, *every* quadratic equation can be written with a perfect-square trinomial on the left. That is the basis for the **completing-the-square method** for solving quadratic equations.

First, let's look at two perfect-square trinomials.

$$x^2 + 6x + \ 9 = (x + 3)^2$$
$$x^2 - 8x + 16 = (x - 4)^2$$

There is an important relationship between the coefficient of the middle term (the x-term) and the constant.

In the first equation,

$$\left(\frac{1}{2} \cdot 6\right)^2 = 3^2 = 9$$

The x-coefficient   The constant

In the second equation,

$$\left[\frac{1}{2}(-8)\right]^2 = (-4)^2 = 16$$

The x-coefficient   The constant

It is always true that, in a perfect-square trinomial with a coefficient of 1 for x^2, the square of one-half of the x-coefficient is equal to the constant term.

> **NOTE**
>
> Here we have used the square-root method, once the factoring was done.

(▶) **Example 1**   **Completing the Square**

< Objective 1 >

(a) Find the term that should be added to $x^2 + 4x$ so that the expression is a perfect-square trinomial.

NOTE

The coefficient of x^2 must be 1 before the added term is found.

To complete the square of $x^2 + 4x$, add the square of one-half of **4** (the x-coefficient).

$$x^2 + 4x + \left(\frac{1}{2} \cdot 4\right)^2 \qquad \text{or} \qquad x^2 + 4x + 2^2 \qquad \text{or} \qquad x^2 + 4x + 4$$

The trinomial $x^2 + 4x + 4$ is a perfect square because

$$x^2 + 4x + 4 = (x + 2)^2$$

(b) Find the term that should be added to $x^2 - 10x$ so that the expression is a perfect-square trinomial.

To complete the square of $x^2 - 10x$, add the square of one-half of -10 (the x-coefficient).

$$x^2 - 10x + \left[\frac{1}{2}(-10)\right]^2 \qquad \text{or} \qquad x^2 - 10x + (-5)^2 \qquad \text{or} \qquad x^2 - 10x + 25$$

Check for yourself, by factoring, that this is a perfect-square trinomial.

Check Yourself 1

Complete the square and factor.

(a) $x^2 + 2x$ (b) $x^2 - 12x$

We can now use the process of Example 1 along with the solution methods of Section 10.1 to solve a quadratic equation.

Example 2   **Solving a Quadratic Equation by Completing the Square**

< Objective 2 >

Solve $x^2 + 4x - 2 = 0$ by completing the square.

$$x^2 + 4x = 2$$

We find the term needed to complete the square by squaring one-half of the x-coefficient.

$$\left(\frac{1}{2} \cdot 4\right)^2 = 2^2 = 4$$

NOTE

This *completes the square* on the left.

We now add 4 to both sides of the equation.

$$x^2 + 4x + 4 = 2 + 4$$

Now factor on the left and simplify on the right.

$$(x + 2)^2 = 6$$

Now solving as before, we have

$$x + 2 = \pm\sqrt{6}$$
$$x = -2 \pm \sqrt{6}$$

Check Yourself 2

Solve by completing the square.

$$x^2 + 6x - 4 = 0$$

To complete the square in this manner, the coefficient of x^2 must be 1. Example 3 illustrates the solution process when the coefficient of x^2 is not equal to 1.

Example 3   **Solving a Quadratic Equation by Completing the Square**

Solve $2x^2 - 4x - 5 = 0$ by completing the square.

$$2x^2 - 4x - 5 = 0$$ Add 5 to both sides.

$$2x^2 - 4x = 5$$ Because the coefficient of x^2 is not 1 (here it is 2), divide every term by 2. This will make the new leading coefficient equal to 1.

$$x^2 - 2x = \frac{5}{2}$$

$$x^2 - 2x + 1 = \frac{5}{2} + 1$$ Complete the square and solve as before.

$$(x - 1)^2 = \frac{7}{2}$$

$$x - 1 = \pm\sqrt{\frac{7}{2}}$$

$$x - 1 = \pm\frac{\sqrt{14}}{2}$$ Simplify the radical on the right.

$$x = 1 \pm \frac{\sqrt{14}}{2}$$

or

$$x = \frac{2 \pm \sqrt{14}}{2}$$

NOTE

$$\sqrt{\frac{7}{2}} = \frac{\sqrt{7}}{\sqrt{2}} \cdot \frac{\sqrt{2}}{\sqrt{2}}$$

$$= \frac{\sqrt{14}}{2}$$

NOTE

We have combined the terms on the right with the common denominator of 2.

Check Yourself 3

Solve by completing the square.

$$3x^2 - 6x + 2 = 0$$

The completing-the-square method is easiest to use when the coefficient of x is even. If it is odd, the method still works, but it will definitely involve fractions. Consider Example 4.

| Example 4 | Solving a Quadratic Equation by Completing the Square |

Solve $2x^2 - 6x - 9 = 0$ by completing the square.

$2x^2 - 6x - 9 = 0$ Add 9 to both sides.

$2x^2 - 6x = 9$ Divide every term by 2.

$x^2 - 3x = \dfrac{9}{2}$

Be careful here. $\left[\dfrac{1}{2}(-3)\right]^2 = \left(-\dfrac{3}{2}\right)^2 = \dfrac{9}{4}$. So add $\dfrac{9}{4}$ to both sides.

$x^2 - 3x + \dfrac{9}{4} = \dfrac{9}{2} + \dfrac{9}{4}$

$\left(x - \dfrac{3}{2}\right)^2 = \dfrac{27}{4}$

To factor the trinomial, remember how we got $\dfrac{9}{4}$: by squaring $-\dfrac{3}{2}$.

$x - \dfrac{3}{2} = \pm\sqrt{\dfrac{27}{4}}$

$x - \dfrac{3}{2} = \pm\dfrac{\sqrt{27}}{2}$

$x - \dfrac{3}{2} = \pm\dfrac{3\sqrt{3}}{2}$

$x = \dfrac{3}{2} \pm \dfrac{3\sqrt{3}}{2}$

or

$x = \dfrac{3 \pm 3\sqrt{3}}{2}$

NOTE

We simplify $\sqrt{27}$ as follows:
$\sqrt{27} = \sqrt{9 \cdot 3} = \sqrt{9} \cdot \sqrt{3}$
$= 3\sqrt{3}$

Check Yourself 4

Solve by completing the square.

$2x^2 + 10x - 3 = 0$

Example 5 illustrates a geometric application using the Pythagorean theorem.

| Example 5 | Solving a Geometry Application |

The length of one leg of a right triangle is 2 cm more than the other. If the length of the hypotenuse is 6 cm, what are the lengths of the two legs? Round to the nearest tenth of a foot.

Draw a sketch of the problem, labeling the known and unknown lengths. Here, if one leg is represented by x, the other must be represented by $x + 2$.

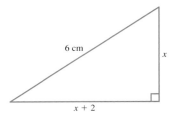

Use the Pythagorean theorem to form an equation.

$$x^2 + (x + 2)^2 = 6^2$$
$$x^2 + x^2 + 4x + 4 = 36$$
$$2x^2 + 4x - 32 = 0 \qquad \text{Divide both sides by 2.}$$
$$x^2 + 2x - 16 = 0$$

This equation can be solved by completing the square.

$$x^2 + 2x = 16$$
$$x^2 + 2x + 1 = 16 + 1$$
$$(x + 1)^2 = 17$$
$$x + 1 = \pm\sqrt{17}$$
$$x = -1 \pm \sqrt{17} \approx 3.1 \quad \text{or} \quad -5.1$$

We generally reject the negative solution in a geometric problem.

If $x \approx 3.1$, then $x + 2 \approx 5.1$. The lengths of the legs are approximately 3.1 and 5.1 cm.

Check Yourself 5

The length of one leg of a right triangle is 1 in. more than the other. If the length of the hypotenuse is 3 in., what are the lengths of the legs? Round to the nearest thousandth of an inch.

We close this section with an application that arises in the field of economics. Related to the production of a certain item, there are two important equations: a **supply** equation and a **demand** equation. Each is related to the price of the item. The **equilibrium** price is the price for which supply equals demand.

Example 6 — Solving a Business Application

The demand equation for a certain computer chip is given by

$$D = -4p + 50$$

The supply equation is predicted to be

$$S = -p^2 + 20p - 6$$

Find the equilibrium price.

We wish to know what price p results in equal supply and demand, so we write

$$-p^2 + 20p - 6 = -4p + 50$$
$$-56 = p^2 - 24p$$
or $p^2 - 24p = -56$

Then, completing the square,

$$p^2 - 24p + 144 = -56 + 144$$
$$(p - 12)^2 = 88$$
$$p - 12 = \pm\sqrt{88}$$
$$p = 12 \pm \sqrt{88}$$
$$p = 21.38 \text{ or } 2.62$$

You should confirm that when $p = 2.62$, supply and demand are positive.

RECALL

We can simplify
$12 \pm \sqrt{88} = 12 \pm 2\sqrt{22}$
but if we are finding the price using a calculator, there is no need to simplify the radical.

Here we must be careful: Supply and demand must be positive. When $p = 21.38$, they are negative, so the equilibrium price is $2.62.

Check Yourself 6

The demand equation for a certain item is given by

$D = -2p + 32$

The supply equation is predicted to be

$S = -p^2 + 18p - 5$

Find the equilibrium price.

We summarize the steps used to solve a quadratic equation by completing the square.

Step by Step

Solving a Quadratic Equation by Completing the Square		
	Step 1	Write the equation in the form $ax^2 + bx = k$ so that the variable terms are on the left side and the constant is on the right side.
	Step 2	If the coefficient of x^2 is not 1, divide both sides of the equation by that coefficient.
	Step 3	Add the square of one-half the coefficient of x to both sides of the equation.
	Step 4	The left side of the equation is now a perfect-square trinomial. Factor and solve as before.

Check Yourself ANSWERS

1. **(a)** $x^2 + 2x + 1 = (x + 1)^2$; **(b)** $x^2 - 12x + 36 = (x - 6)^2$

2. $-3 \pm \sqrt{13}$ 3. $\dfrac{3 \pm \sqrt{3}}{3}$ 4. $\dfrac{-5 \pm \sqrt{31}}{2}$ 5. 1.562 in., 2.562 in.

6. $2.06

Reading Your Text

The following fill-in-the-blank exercises are designed to ensure that you understand some of the key vocabulary used in this section.

SECTION 10.2

(a) Every quadratic equation can be written with a _____ trinomial on the left.

(b) In a perfect-square trinomial with a coefficient of 1 for x^2, the square of one-half of the x-coefficient is equal to the _____ term.

(c) We find the term needed to _____ the square by squaring one-half of the x-coefficient.

(d) The completing-the-square method is easiest to use when the coefficient of x is an _____ number.

Basic Skills | Advanced Skills | Vocational-Technical Applications | Calculator/Computer | Above and Beyond | Getting Ready

Name _____

Section _____ Date _____

Answers

1. _____   2. _____

3. _____   4. _____

5. _____   6. _____

7. _____   8. _____

9. _____   10. _____

11. _____   12. _____

13. _____   14. _____

15. _____

16. _____

17. _____

18. _____

19. _____

20. _____

21. _____   22. _____

< Objective 1 >

Determine whether each of the following trinomials is a perfect square.

1. $x^2 - 14x + 49$

2. $x^2 + 9x + 16$

3. $x^2 - 18x - 81$ > Videos

4. $x^2 + 10x + 25$

5. $x^2 - 18x + 81$

6. $x^2 - 24x + 48$

Find the constant term that should be added to make each of the following expressions a perfect-square trinomial.

7. $x^2 + 6x$

8. $x^2 - 8x$

9. $x^2 - 10x$

10. $x^2 + 5x$

11. $x^2 + 9x$ > Videos

12. $x^2 - 20x$

< Objective 2 >

Solve each of the following quadratic equations by completing the square.

13. $x^2 + 4x - 12 = 0$

14. $x^2 - 6x + 8 = 0$

15. $x^2 - 2x - 5 = 0$ > Videos

16. $x^2 + 4x - 7 = 0$

17. $x^2 + 3x - 27 = 0$

18. $x^2 + 5x - 3 = 0$

19. $x^2 + 6x - 1 = 0$

20. $x^2 + 4x - 4 = 0$

21. $x^2 - 5x + 6 = 0$

22. $x^2 - 6x - 3 = 0$

23. $x^2 + 6x - 5 = 0$

24. $x^2 - 2x = 1$

25. $x^2 = 9x + 5$ ⊙ > Videos

26. $x^2 = 4 - 7x$

Solve each of the following applications. Give all answers to the nearest thousandth.

27. GEOMETRY The length of one leg of a right triangle is 4 in. more than the other. If the length of the hypotenuse is 8 in., what are the lengths of the two legs?

28. GEOMETRY The length of a rectangle is 2 cm longer than its width. If the diagonal of the rectangle is 4 cm, what are the dimensions (the length and the width) of the rectangle?

29. GEOMETRY The width of a rectangle is 6 ft less than its length. If the area of the rectangle is 75 ft², what are the dimensions of the rectangle?

30. GEOMETRY The length of a rectangle is 8 cm more than its width. If the area of the rectangle is 90 cm², find the dimensions.

31. SCIENCE AND MEDICINE The equation

$$h = -16t^2 - 32t + 320$$

gives the height of a ball, thrown downward from the top of a 320-ft building with an initial velocity of 32 ft/s. Find the time it takes for the ball to reach a height of 160 ft.

(chapter 10) > Make the Connection

32. SCIENCE AND MEDICINE The equation

$$h = -16t^2 - 32t + 320$$

gives the height of a ball, thrown downward from the top of a 320-ft building with an initial velocity of 32 ft/s. Find the time it takes for the ball to reach a height of 64 ft.

(chapter 10) > Make the Connection

33. BUSINESS AND FINANCE The demand equation for a certain type of printer is given by

$$D = -200p + 35,000$$

The supply equation is predicted to be

$$S = -p^2 + 400p - 20,000$$

Find the equilibrium price.

Answers

23. _____

24. _____

25. _____

26. _____

27. _____

28. _____

29. _____

30. _____

31. _____

32. _____

33. _____

Answers

34. _____

35. _____

36. _____

37. _____

38. _____

39. _____

40. _____

41. _____

42. _____

43. _____

44. _____

45. _____

46. _____

34. BUSINESS AND FINANCE The demand equation for a certain type of printer is given by

$$D = -80p + 7{,}000$$

The supply equation is predicted to be

$$S = -p^2 + 220p - 8{,}000$$

Find the equilibrium price.

Determine whether each of the following statements is **true** *or* **false.**

35. To complete the square when the coefficient of x^2 is not 1, one of the first steps is to divide both sides of the equation by that coefficient.

36. If the coefficient of x is odd, the completing-the-square method will not work.

Complete each of the following statements with **never, sometimes,** *or* **always.**

37. When completing the square for $x^2 + bx$, the number added is _____ negative.

38. A quadratic equation can _____ be solved by completing the square.

| Basic Skills | **Advanced Skills** | Vocational-Technical Applications | Calculator/Computer | Above and Beyond | Getting Ready |

Solve each of the following quadratic equations by completing the square.

39. $2x^2 - 6x + 1 = 0$

40. $2x^2 + 10x + 11 = 0$

41. $2x^2 - 4x + 1 = 0$

42. $2x^2 - 8x + 5 = 0$

43. $4x^2 - 2x - 1 = 0$ > Videos

44. $3x^2 - x - 2 = 0$

Solve the following problems.

45. NUMBER PROBLEM If the square of 3 more than a number is 9, find the number(s).

46. NUMBER PROBLEM If the square of 2 less than an integer is 16, find the number(s).

47. Business and Finance The revenue for selling x units of a product is given by

$$R = x\left(25 - \frac{1}{2}x\right)$$

Find the number of units sold if the revenue is $294.50.

48. Number Problem Find two consecutive positive integers such that the sum of their squares is 85.

Basic Skills | Advanced Skills | **Vocational-Technical Applications** | Calculator/Computer | Above and Beyond | Getting Ready
▲

49. Allied Health An experimental drug is being tested on a bacteria colony. It is found that t days after the colony is treated, the number N of bacteria per cubic centimeter is given by the equation

$$N = -20t^2 - 120t + 1,000$$

In how many days will the colony be reduced to 200 bacteria?

50. Mechanical Engineering The rotational moment in a shaft is given by the formula

$$M = -30x + 2x^2$$

Find the value of x when the moment is equal to 152.

51. Manufacturing Technology Suppose that the cost C, in dollars, of producing x chairs is given by

$$C = 2x^2 - 40x + 2,400$$

How many chairs can be produced for $4,650?

52. Manufacturing Technology Suppose that the profit P, in dollars, of producing and selling x appliances is given by

$$P = -3x^2 + 240x - 1,800$$

How many appliances must be produced and sold to achieve a profit of $2,325?

53. Manufacturing Technology A small manufacturer's weekly profit P, in dollars, is given by

$$P = -3x^2 + 270x$$

Find the number of items x that must be produced to realize a profit of $5,775.

54. Manufacturing Technology The demand equation for a certain computer chip is given by

$$D = -5p + 62$$

The supply equation is predicted to be

$$S = -p^2 + 23p - 11$$

Find the equilibrium price.

Answers

47. _____

48. _____

49. _____

50. _____

51. _____

52. _____

53. _____

54. _____

Answers

(a) _____

(b) _____

(c) _____

(d) _____

(e) _____

(f) _____

Evaluate the expression $b^2 - 4ac$ for each set of values.

(a) $a = 1, b = 1, c = -3$ **(b)** $a = 1, b = -1, c = -1$

(c) $a = 1, b = -8, c = -3$ **(d)** $a = 1, b = -2, c = -1$

(e) $a = -2, b = 4, c = -2$ **(f)** $a = 2, b = -3, c = 4$

Answers

1. Yes   **3.** No   **5.** Yes   **7.** 9   **9.** 25   **11.** $\dfrac{81}{4}$ **13.** $-6, 2$

15. $1 \pm \sqrt{6}$ **17.** $\dfrac{-3 \pm 3\sqrt{13}}{2}$ **19.** $-3 \pm \sqrt{10}$ **21.** $2, 3$

23. $-3 \pm \sqrt{14}$ **25.** $\dfrac{9 \pm \sqrt{101}}{2}$ **27.** 3.292 in., 7.292 in.

29. 6.165 ft by 12.165 ft   **31.** 2.317 s   **33.** \$112.92   **35.** True

37. never   **39.** $\dfrac{3 \pm \sqrt{7}}{2}$ **41.** $\dfrac{2 \pm \sqrt{2}}{2}$ **43.** $\dfrac{1 \pm \sqrt{5}}{4}$

45. $-6, 0$ **47.** 19, 31   **49.** 4 days   **51.** 45 chairs

53. 35 or 55 items   **(a)** 13   **(b)** 5   **(c)** 76   **(d)** 8

(e) 0   **(f)** -23

10.3

The Quadratic Formula

< 10.3 Objectives >

1 > Write a quadratic equation in standard form

2 > Use the quadratic formula to solve a quadratic equation

We are now ready to derive and use the **quadratic formula,** which will allow us to solve all quadratic equations. We derive the formula by using the method of completing the square.

To use the quadratic formula, the quadratic equation you want to solve must be in *standard form.* That form is

$$ax^2 + bx + c = 0 \qquad \text{in which } a \neq 0$$

| | **Example** 1 | **Writing Quadratic Equations in Standard Form** |

< Objective 1 >

Write each equation in standard form.

(a) $2x^2 - 5x + 3 = 0$

The equation is already in standard form.

$$a = 2 \qquad b = -5 \qquad \text{and} \qquad c = 3$$

(b) $5x^2 + 3x = 5$

The equation is *not* in standard form. Rewrite it by adding -5 to both sides.

$$5x^2 + 3x - 5 = 0 \qquad \text{Standard form}$$

$$a = 5 \qquad b = 3 \qquad \text{and} \qquad c = -5$$

Check Yourself 1

Rewrite each quadratic equation in standard form.

(a) $x^2 - 3x = 5$ **(b)** $3x^2 = 7 - 2x$

Once a quadratic equation is written in standard form, we will be able to find any solutions to the equation. Remember that a solution is a value for x that will make the equation true.

What follows is the derivation of the quadratic formula, which can be used to solve quadratic equations.

Step by Step

Deriving the Quadratic Formula

NOTE

This is the completing-the-square step that makes the left-hand side a perfect square.

Let $ax^2 + bx + c = 0$, in which $a \neq 0$.

$ax^2 + bx = -c$	Subtract c from both sides.
$x^2 + \dfrac{b}{a}x = -\dfrac{c}{a}$	Divide both sides by a.
$x^2 + \dfrac{b}{a}x + \dfrac{b^2}{4a^2} = \dfrac{b^2}{4a^2} - \dfrac{c}{a}$	Add $\dfrac{b^2}{4a^2}$ to both sides.
$\left(x + \dfrac{b}{2a}\right)^2 = \dfrac{b^2 - 4ac}{4a^2}$	Factor on the left and add the fractions on the right.
$x + \dfrac{b}{2a} = \pm\sqrt{\dfrac{b^2 - 4ac}{4a^2}}$	Take the square root of both sides.
$x + \dfrac{b}{2a} = \pm\dfrac{\sqrt{b^2 - 4ac}}{2a}$	Simplify the radical on the right.
$x = -\dfrac{b}{2a} \pm \dfrac{\sqrt{b^2 - 4ac}}{2a}$	Add $-\dfrac{b}{2a}$ to both sides.
$x = \dfrac{-b \pm \sqrt{b^2 - 4ac}}{2a}$	Use the common denominator, $2a$.

Property

The Quadratic Formula

If $ax^2 + bx + c = 0$, then

$$x = \frac{-b \pm \sqrt{b^2 - 4ac}}{2a}$$

Let's use the quadratic formula to solve some equations.

Example 2   Using the Quadratic Formula to Solve an Equation

Use the quadratic formula to solve $x^2 - 5x + 4 = 0$.

The equation is in standard form, so first identify a, b, and c.

$$x^2 - 5x + 4 = 0$$

$a = 1 \quad b = -5 \quad c = 4$

We now substitute the values for a, b, and c into the formula.

NOTES

The leading coefficient is 1, so $a = 1$.

Simplify the expression.

$$x = \frac{-b \pm \sqrt{b^2 - 4ac}}{2a}$$

$$= \frac{-(-5) \pm \sqrt{(-5)^2 - 4(1)(4)}}{2(1)}$$

$$= \frac{5 \pm \sqrt{25 - 16}}{2}$$

$$= \frac{5 \pm \sqrt{9}}{2}$$

$$= \frac{5 \pm 3}{2}$$

NOTE

These results could also have been found by factoring the original equation. You should check that for yourself.

Now,

$$x = \frac{5 + 3}{2} \quad \text{or} \quad x = \frac{5 - 3}{2}$$
$$= 4 \qquad\qquad = 1$$

The solutions are 4 and 1.

 Check Yourself 2

Use the quadratic formula to solve $x^2 - 2x - 8 = 0$. Check your result by factoring.

The main use of the quadratic formula is to solve equations that *cannot* be factored.

 Example 3   **Using the Quadratic Formula to Solve an Equation**

Use the quadratic formula to solve $2x^2 = x + 4$.

First, the equation *must be written* in standard form to find a, b, and c.

$$2x^2 - x - 4 = 0$$

$a = 2 \qquad b = -1 \qquad c = -4$

NOTE

Substitute the values for a, b, and c into the formula.

$$x = \frac{-b \pm \sqrt{b^2 - 4ac}}{2a}$$
$$= \frac{-(-1) \pm \sqrt{(-1)^2 - 4(2)(-4)}}{2(2)}$$
$$= \frac{1 \pm \sqrt{1 + 32}}{4}$$
$$= \frac{1 \pm \sqrt{33}}{4}$$

 Check Yourself 3

Use the quadratic formula to solve $3x^2 = 3x + 4$.

 Example 4   **Using the Quadratic Formula to Solve an Equation**

Use the quadratic formula to solve $x^2 - 2x = 4$.

In standard form, the equation is

$$x^2 - 2x - 4 = 0$$

$a = 1 \qquad b = -2 \qquad c = -4$

NOTE

Again substitute the values into the quadratic formula.

$$x = \frac{-(-2) \pm \sqrt{(-2)^2 - 4(1)(-4)}}{2(1)}$$
$$= \frac{2 \pm \sqrt{20}}{2}$$

NOTE

20 has a perfect-square factor. So

$\sqrt{20} = \sqrt{4 \cdot 5}$

$\qquad = 2\sqrt{5}$

Unless you are finding a decimal approximation, you should always write your solution in simplest form.

$$x = \frac{2 \pm 2\sqrt{5}}{2}$$

$$\quad = \frac{2(1 \pm \sqrt{5})}{2} \qquad \text{Now factor the numerator and divide by the common factor 2.}$$

$$\quad = 1 \pm \sqrt{5}$$

Check Yourself 4

Use the quadratic formula to solve $3x^2 = 2x + 4$.

Sometimes equations have common factors. Factoring first simplifies these equations, making them easier to solve. This is illustrated in Example 5.

Example 5   **Using the Quadratic Formula to Solve an Equation**

Use the quadratic formula to solve $3x^2 - 6x - 3 = 0$.

Because the equation is in standard form, we could use

$$a = 3 \qquad b = -6 \qquad \text{and} \qquad c = -3$$

in the quadratic formula. There is, however, a better approach.

Note the common factor of 3 in the quadratic member of the original equation. Factoring, we have

$$3(x^2 - 2x - 1) = 0$$

and dividing both sides of the equation by 3 gives

$$x^2 - 2x - 1 = 0$$

Now let $a = 1$, $b = -2$, and $c = -1$. Then

NOTE

The advantage to this approach is that these values will require much less simplification after we substitute into the quadratic formula.

$$x = \frac{-(-2) \pm \sqrt{(-2)^2 - 4(1)(-1)}}{2 \cdot 1}$$

$$\quad = \frac{2 \pm \sqrt{8}}{2}$$

$$\quad = \frac{2 \pm 2\sqrt{2}}{2}$$

$$\quad = \frac{2(1 \pm \sqrt{2})}{2}$$

$$\quad = 1 \pm \sqrt{2}$$

Check Yourself 5

Use the quadratic formula to solve $4x^2 - 20x = 12$.

In applications that lead to quadratic equations, you may want to find approximate values for the solutions.

 Example 6   **Using the Quadratic Formula to Solve an Equation**

Use the quadratic formula to solve $x^2 - 5x + 5 = 0$ and write your solutions in approximate decimal form.

Substituting $a = 1$, $b = -5$, and $c = 5$ gives

$$x = \frac{-(-5) \pm \sqrt{(-5)^2 - 4(1)(5)}}{2(1)}$$

$$= \frac{5 \pm \sqrt{5}}{2}$$

Use your calculator to find $\sqrt{5} \approx 2.236$, so

$$x \approx \frac{5 + 2.236}{2} \qquad \text{or} \qquad x \approx \frac{5 - 2.236}{2}$$

$$= \frac{7.236}{2} \qquad\qquad\qquad = \frac{2.764}{2}$$

$$= 3.618 \qquad\qquad\qquad = 1.382$$

 Check Yourself 6

Use the quadratic formula to solve $x^2 - 3x - 5 = 0$ and approximate the solutions in decimal form to the nearest thousandth.

You may be wondering whether the quadratic formula can be used to solve all quadratic equations. It can, but not all quadratic equations will have real-number solutions, as Example 7 shows.

 Example 7   **Using the Quadratic Formula to Solve an Equation**

NOTE

Make sure the quadratic equation is in standard form. $x^2 - 3x = -5$ is equivalent to $x^2 - 3x + 5 = 0$.

Use the quadratic formula to solve $x^2 - 3x = -5$.

Substituting $a = 1$, $b = -3$, and $c = 5$, we have

$$x = \frac{-(-3) \pm \sqrt{(-3)^2 - 4(1)(5)}}{2(1)}$$

$$= \frac{3 \pm \sqrt{-11}}{2}$$

In this case, there are no real-number solutions because the radicand is negative.

Check Yourself 7

Use the quadratic formula to solve $x^2 - 3x = -3$.

The part of the quadratic formula, $b^2 - 4ac$, that is under the radical is called the **discriminant.** By computing this quantity, we can determine what type of solutions we would find if we were to solve completely. For example, for the quadratic equation in the previous example, $x^2 - 3x = -5$, we found that $a = 1$, $b = -3$, and $c = 5$. The value of the discriminant is

$$b^2 - 4ac = (-3)^2 - 4(1)(5) = 9 - 20 = -11$$

Since the discriminant is negative, we know immediately that there are no real-number solutions.

The following information can be obtained simply by evaluating the discriminant.

1. If the discriminant is negative, there are no real-number solutions.
2. If the discriminant is zero, there is exactly one real-number solution. In fact, the solution will be a rational number.
3. If the discriminant is positive and is a perfect square, there are two rational-number solutions.
4. And, if the discriminant is positive but is not a perfect square, there are two irrational-number solutions.

Example 8   **Using the Discriminant to Determine the Type of Solutions**

For each equation, evaluate the discriminant, determine how many real solutions there are, and classify them.

(a) $2x^2 - 5x + 1 = 0$

First evaluate the discriminant:

$$b^2 - 4ac = (-5)^2 - 4(2)(1) = 25 - 8 = 17$$

Since 17 is positive and is not a perfect square, there are two irrational solutions.

(b) $10x - x^2 = 25$

We begin by writing this in standard form:

$$x^2 - 10x + 25 = 0$$

Then

$$b^2 - 4ac = (-10)^2 - 4(1)(25) = 100 - 100 = 0$$

Since the discriminant is zero, there is exactly one rational solution.

(c) $3x^2 - x + 2 = 0$

We have

$$b^2 - 4ac = (-1)^2 - 4(3)(2) = 1 - 24 = -23$$

There are no real-number solutions.

Check Yourself 8

For each equation, evaluate the discriminant, determine how many real solutions there are, and classify them.

(a) $8x - 16 = x^2$ **(b)** $2x^2 + 3x = 20$ **(c)** $x^2 + 5x + 7 = 0$

We summarize the steps used for solving equations by using the quadratic formula.

Step by Step

Solving Equations by using the Quadratic Formula

Step 1   Rewrite the equation in standard form.

$$ax^2 + bx + c = 0$$

Step 2   If a common factor exists, divide both sides of the equation by that common factor.

Step 3   Identify the coefficients a, b, and c.

Step 4   Substitute values for a, b, and c into the formula

$$x = \frac{-b \pm \sqrt{b^2 - 4ac}}{2a}$$

Step 5   Simplify the right side of the expression formed in step 4 to write the solutions for the original equation.

Often, applied problems will lead to quadratic equations that must be solved by the methods of Section 10.2 or this section.

Example 9   **Solving a Construction Application**

A rectangular garden is to be surrounded by a walkway of constant width. The garden's dimensions are 5 m by 8 m. The total area, garden plus walkway, is to be 100 m². What must be the width of the walkway? Round to the nearest hundredth of a meter.

First we sketch the situation.

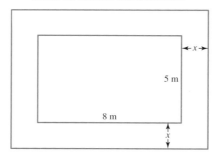

The area of the entire figure may be represented by

$$\underbrace{(8 + 2x)}_{\text{Length}}\underbrace{(5 + 2x)}_{\text{Width}} \qquad \text{(Do you see why?)}$$

Since the area is to be 100 m², we write

$$(8 + 2x)(5 + 2x) = 100$$

$$40 + 16x + 10x + 4x^2 = 100$$

$$4x^2 + 26x - 60 = 0$$

$$2x^2 + 13x - 30 = 0$$

$$x = \frac{-13 \pm \sqrt{13^2 - 4(2)(-30)}}{2(2)}$$

$$= \frac{-13 \pm \sqrt{409}}{4}$$

$$x \approx 1.8059 \quad \text{or} \quad \approx -8.3059$$

Since x must be positive, we have (rounded) $x = 1.81$ m.

Check Yourself 9

A rectangular swimming pool is 16 ft by 40 ft. A tarp to go over the pool also covers a strip of equal width surrounding the pool. If the area of the tarp is 1100 ft^2, how wide is the covered strip around the pool? Round to the nearest tenth of a foot.

NOTE

We are neglecting air resistance in this discussion.

When a ball is thrown directly upward, or if some sort of projectile is fired upward, the height of the object is approximated by a quadratic function of time. This concept is illustrated in Example 10.

Example 10   Solving a Science Application

Suppose that a ball is thrown upward with an initial velocity of 80 ft/s. (The initial velocity is the speed with which the ball leaves the thrower's hand. The speed will then decrease as the ball rises.) If the ball is released at a height of 5 ft, the height equation may be written as follows:

$$h = -16t^2 + 80t + 5$$

When, to the nearest hundredth of a second, will the ball be at a height of 93 ft?
    Since the desired height is 93 ft, we write

$$93 = -16t^2 + 80t + 5$$

We must solve for t:

$$93 = -16t^2 + 80t + 5 \qquad \text{We need a 0 on one side.}$$
$$16t^2 - 80t + 88 = 0 \qquad \text{Divide through by 8.}$$
$$2t^2 - 10t + 11 = 0 \qquad \text{Use the quadratic formula.}$$

$$t = \frac{10 \pm \sqrt{(-10)^2 - 4(2)(11)}}{2(2)}$$

NOTE

Use a calculator to check decimal results such as these.

$$= \frac{10 \pm \sqrt{12}}{4} \approx 1.63 \quad \text{or} \quad 3.37$$

The ball reaches a height of 93 ft twice: first, on the way up, at 1.63 s, and second, on the way down, at 3.37 s.

Check Yourself 10

Using the same height equation provided in Example 10, determine, to the nearest hundredth of a second, when the ball will be at a height of 77 ft.

In Example 11, we look at a similar application. It highlights the importance of proper interpretation of results.

Example 11 — Solving a Science Application

A projectile is fired upward from a platform in such a way that the object will miss the platform (fortunately!) on the way down. If the height of the platform is 20 ft, and the initial velocity is 200 ft/s, the height is given by

$$h = -16t^2 + 200t + 20$$

To the nearest tenth of a second, when will the projectile hit the ground?

The projectile will hit the ground when the height is 0, so we write

$$0 = -16t^2 + 200t + 20 \qquad \text{Divide through by } -4.$$
$$0 = 4t^2 - 50t - 5$$
$$t = \frac{50 \pm \sqrt{(-50)^2 - 4(4)(-5)}}{2(4)}$$
$$= \frac{50 \pm \sqrt{2{,}580}}{8} \approx 12.599 \quad \text{or} \quad -0.099$$

The projectile will hit at about 12.6 s. We reject the negative solution, as it would indicate a time *before* the projectile was fired.

Check Yourself 11

A projectile fired upward from a platform follows the height equation

$$h = -16t^2 + 160t + 40$$

When will it hit the ground? Round to the nearest tenth of a second.

Check Yourself ANSWERS

1. **(a)** $x^2 - 3x - 5 = 0$; **(b)** $3x^2 + 2x - 7 = 0$

2. $x = 4, -2$ 3. $x = \dfrac{3 \pm \sqrt{57}}{6}$

4. $x = \dfrac{1 \pm \sqrt{13}}{3}$

5. $x = \dfrac{5 \pm \sqrt{37}}{2}$ 6. $x \approx 4.193$ or -1.193

7. $\dfrac{3 \pm \sqrt{-3}}{2}$, no real solutions

8. **(a)** 1 rational solution; **(b)** 2 rational solutions; **(c)** no real solutions

9. 3.6 ft   10. 1.18 s, 3.82 s   11. 10.2 s

Reading Your Text

The following fill-in-the-blank exercises are designed to ensure that you understand some of the key vocabulary used in this section.

SECTION 10.3

(a) To use the quadratic formula, the quadratic equation you want to solve must be in _____ form.

(b) The main use of the quadratic formula is to solve equations that cannot be _____.

(c) The part of the quadratic formula that is under the radical is called the _____.

(d) If the discriminant is positive and is a perfect square, there are two _____ number solutions.

< Objectives 1–2 >

Use the quadratic formula to solve each of the following quadratic equations.

1. $x^2 + 9x + 20 = 0$ > Videos

2. $x^2 - 9x + 14 = 0$

3. $x^2 - 4x + 3 = 0$

4. $x^2 - 13x + 22 = 0$

5. $3x^2 + 2x - 1 = 0$

6. $x^2 - 8x + 16 = 0$

7. $x^2 + 5x = -4$

8. $4x^2 + 5x = 6$

9. $x^2 = 6x - 9$ > Videos

10. $2x^2 - 5x = 3$

11. $2x^2 - 3x - 7 = 0$

12. $x^2 - 5x + 2 = 0$

13. $x^2 + 2x - 4 = 0$

14. $x^2 - 4x + 2 = 0$

15. $2x^2 - 3x = 3$

16. $3x^2 - 2x + 1 = 0$

17. $3x^2 - 2x = 6$ > Videos

18. $4x^2 = 4x + 5$

19. $3x^2 + 3x + 2 = 0$

20. $2x^2 - 3x = 6$

MathZone

Boost your grade at mathzone.com!

> Practice Problems
> NetTutor
> Self-Tests
> e-Professors
> Videos

Name _____

Section _____ Date _____

Answers

1. _____ 2. _____

3. _____ 4. _____

5. _____ 6. _____

7. _____ 8. _____

9. _____

10. _____

11. _____

12. _____

13. _____

14. _____

15. _____

16. _____

17. _____

18. _____

19. _____

20. _____

Answers

21.

22.

23.

24.

25.

26.

27.

28.

29.

30.

31.

32.

33.

34.

35.

36.

37.

21. $5x^2 = 8x - 2$

22. $5x^2 - 2 = 2x$

23. $2x^2 - 9 = 4x$

24. $3x^2 - 6x = 2$

In exercises 25–32, evaluate the discriminant, determine how many real solutions there are, and classify them.

25. $2x^2 - 5x - 3 = 0$

26. $x^2 = 8x - 16$

27. $5x^2 - 3x + 2 = 0$ > Videos

28. $4x^2 - 7x + 2 = 0$

29. $3x^2 + 7x + 4 = 0$

30. $x^2 + 3x + 4 = 0$

31. $2x^2 - 2 = 5x$

32. $6x^2 + 11x - 35 = 0$

33. GEOMETRY One leg of a right triangle is 1 in. shorter than the other leg. The hypotenuse is 4 in. longer than the shorter side. Find the length of each side.

34. GEOMETRY The hypotenuse of a given right triangle is 6 cm longer than the shorter leg. The length of the shorter leg is 2 cm less than that of the longer leg. Find the lengths of the three sides.

35. CONSTRUCTION A rectangular field is 300 ft by 500 ft. A roadway of width x ft is to be built just inside the field. What is the widest the roadway can be and still leave 100,000 ft^2 in the region?

36. CONSTRUCTION A rectangular garden is to be surrounded by a walkway of constant width. The garden's dimensions are 20 ft by 28 ft. The total area, garden plus walkway, is to be 1,100 ft^2. What must be the width of the walkway?

37. SCIENCE AND MEDICINE The equation

$$h = -16t^2 + 112t$$

gives the height of an arrow, shot upward from the ground with an initial velocity of 112 ft/s, where t is the time after the arrow leaves the ground. Find the time it takes for the arrow to reach a height of 120 ft.

chapter 10 > Make the Connection

38. SCIENCE AND MEDICINE The equation

$$h = -16t^2 + 112t$$

gives the height of an arrow, shot upward from the ground with an initial velocity of 112 ft/s, where t is the time after the arrow leaves the ground. Find the time it takes for the arrow to reach a height of 180 ft.

39. SCIENCE AND MEDICINE If a ball is thrown vertically upward from a height of 6 ft, with an initial velocity of 64 ft/s, its height h after t s is given by

$$h = -16t^2 + 64t + 6$$

How long does it take the ball to return to the ground?

40. SCIENCE AND MEDICINE If a ball is thrown vertically upward from a height of 5 ft, with an initial velocity of 96 ft/s, its height h after t s is given by

$$h = -16t^2 + 96t + 5$$

How long does it take the ball to return to the ground?

Determine whether each of the following equations is **true** *or* **false.**

41. The solutions of a quadratic equation are always rational.

42. A ball thrown vertically might reach a specified height two times.

Complete each of the following statements with **never, sometimes,** *or* **always.**

43. The quadratic formula can _____ be used to solve a quadratic equation.

44. If the value of $b^2 - 4ac$ is negative, the equation $ax^2 + bx + c = 0$ _____ has real-number solutions.

Basic Skills | **Advanced Skills** | Vocational-Technical Applications | Calculator/Computer | Above and Beyond | Getting Ready

45. $3x - 5 = \dfrac{1}{x}$ **46.** $x + 3 = \dfrac{1}{x}$

47. $(x - 2)(x + 1) = 3$ **48.** $(x - 3)(x + 2) = 5$

Answers

38. _____

39. _____

40. _____

41. _____

42. _____

43. _____

44. _____

45. _____

46. _____

47. _____

48. _____

Answers

49. _____

50. _____

51. _____

52. _____

53. _____

54. _____

55. _____

56. _____

57. _____

58. _____

59. _____

60. _____

61. _____

62. _____

Solve the following equations. (Hint: First multiply both sides of the given equation by an expression that will eliminate fractions.)

49. $\dfrac{3}{x} + \dfrac{5}{x^2} = 9$

50. $\dfrac{8}{x} - \dfrac{3}{x^2} = -6$

51. $\dfrac{x}{x+1} + \dfrac{10x}{x^2+4x+3} = \dfrac{15}{x+3}$

52. $x - \dfrac{9x}{x-2} = \dfrac{-10}{x-2}$

Solve the following quadratic equations by factoring or by any of the techniques of this chapter.

53. $(x-1)^2 = 7$

54. $(2x+3)^2 = 5$

55. $2x^2 - 8x + 3 = 0$

56. $x^2 + 2x - 1 = 0$

57. $x^2 - 9x - 4 = 6$

58. $5x^2 + 10x + 2 = 2$

59. $4x^2 - 8x + 3 = 5$ ⊙ > Videos

60. $x^2 + 4x = 21$

Basic Skills ⏐ Advanced Skills ⏐ **Vocational-Technical Applications** ⏐ Calculator/Computer ⏐ Above and Beyond ⏐ Getting Ready
▲

61. **ALLIED HEALTH** The concentration C, in nanograms per milliliter (ng/mL), of digoxin, a medication prescribed for congestive heart failure, is given by the equation

$$C = -0.0015t^2 + 0.0845t + 0.7170$$

where t is the number of hours since the drug was taken orally. For a drug to have a beneficial effect, its concentration in the bloodstream must exceed a certain value, the minimum therapeutic level, which for digoxin is 0.8 ng/mL.

(a) How long will it take for the drug to start having an effect?

(b) How long does it take for the drug to stop having an effect?

62. **ALLIED HEALTH** The concentration C, in micrograms per milliliter (mcg/mL), of phenobarbital, an anticonvulsant medication, is given by the equation

$$C = -1.35t^2 + 10.81t + 7.38$$

where t is the number of hours since the drug was taken orally. For a drug to have a beneficial effect, its concentration in the bloodstream must exceed a certain value, the minimum therapeutic level, which for phenobarbital is 10 mcg/mL.

(a) How long will it take for the drug to start having an effect?

(b) How long does it take for the drug to stop having an effect?

63. **ALLIED HEALTH** One technique of controlling cancer is to use radiation therapy. After such a treatment, the total number of cancerous cells N, in thousands, can be estimated by the equation

$$N = -3t^2 + 6t + 140$$

where t is the number of days of treatment. How many days of treatment are required to kill all the cancer cells?

64. **MANUFACTURING TECHNOLOGY** The Scribner log rule is used to calculate the volume V (in cubic feet) of a 16-foot log given the diameter (in inches) of the smaller end (inside the bark). The formula for the log rule is

$$V = 0.79D^2 - 2D - 4$$

Find the diameter of the small end of a log that has a volume of 272 cubic feet.

Basic Skills | Advanced Skills | Vocational-Technical Applications | Calculator/Computer | **Above and Beyond** | Getting Ready

65. Work with a partner to decide all values of b in the following equations that will give one or more real-number solutions.

(a) $3x^2 + bx - 3 = 0$
(b) $5x^2 + bx + 1 = 0$
(c) $-3x^2 + bx - 3 = 0$
(d) Write a rule for judging whether an equation has solutions by looking at it in standard form.

66. Which method of solving a quadratic equation seems simplest to you? Which method should you try first?

67. Complete this statement: "You can tell an equation is quadratic and not linear by . . ."

Basic Skills | Advanced Skills | Vocational-Technical Applications | Calculator/Computer | Above and Beyond | **Getting Ready**

Evaluate each of the given expressions for the value of the variable given.

(a) $x^2 + 3x - 5; x = 3$ **(b)** $x^2 - 3x - 5; x = -2$
(c) $3x^2 + 4x - 6; x = -2$ **(d)** $-2x^2 - 5x + 3; x = 4$
(e) $-5x^2 - 5x + 6; x = -1$ **(f)** $\frac{2}{3}x^2 - \frac{1}{3}x + 5; x = 6$

Answers

63. _____
64. _____
65. _____
66. _____
67. _____
(a) _____
(b) _____
(c) _____
(d) _____
(e) _____
(f) _____

Answers

1. $-4, -5$ **3.** $3, 1$ **5.** $-1, \dfrac{1}{3}$ **7.** $-4, -1$ **9.** 3

11. $\dfrac{3 \pm \sqrt{65}}{4}$ **13.** $-1 \pm \sqrt{5}$ **15.** $\dfrac{3 \pm \sqrt{33}}{4}$ **17.** $\dfrac{1 \pm \sqrt{19}}{3}$

19. No real-number solutions   **21.** $\dfrac{4 \pm \sqrt{6}}{5}$ **23.** $\dfrac{2 \pm \sqrt{22}}{2}$

25. 49; 2 rational   **27.** -31; no real solutions   **29.** 1; 2 rational

31. 41; 2 irrational   **33.** 7.899 in., 8.899 in., 11.899 in.

35. 34.168 ft   **37.** 1.321 or 5.679 s   **39.** 4.092 s   **41.** False

43. always   **45.** $\dfrac{5 \pm \sqrt{37}}{6}$ **47.** $\dfrac{1 \pm \sqrt{21}}{2}$ **49.** $\dfrac{1 \pm \sqrt{21}}{6}$

51. 5 **53.** $1 \pm \sqrt{7}$ **55.** $\dfrac{4 \pm \sqrt{10}}{2}$ **57.** $-1, 10$

59. $\dfrac{2 \pm \sqrt{6}}{2}$ **61. (a)** 1 hour; **(b)** 55.3 hours   **63.** 8 days

65. Above and Beyond   **67.** Above and Beyond   **(a)** 13   **(b)** 5
(c) -2 **(d)** -49 **(e)** 6   **(f)** 27

10.4

Graphing Quadratic Equations

< 10.4 Objectives >

1 > Graph a quadratic equation by plotting points

2 > Find the axis of symmetry of a parabola

3 > Find the *x*-intercepts of a parabola

In Section 6.3 you learned to graph first-degree (or linear) equations. Similar methods allow you to graph quadratic equations of the form

$$y = ax^2 + bx + c \qquad a \neq 0$$

The first thing to notice is that the graph of an equation in this form is not a straight line. The graph is always a curve called a **parabola.**

Here are some examples:

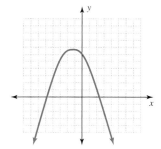

To graph quadratic equations, start by finding solutions for the equation. We begin by completing a table of values. Choose any convenient values for *x*. Then use the given equation to compute the corresponding values for *y*. Example 1 illustrates this process.

949

Example 1   Completing a Table of Values

< Objective 1 >

Complete the ordered pairs to form solutions for the equation $y = x^2$. Then show these results in a table of values.

$(-2, \ ), (-1, \ ), (0, \ ), (1, \ ), (2, \ )$

For example, to complete the pair $(-2, \ )$, substitute -2 for x in the given equation.

$y = (-2)^2 = 4$

So $(-2, 4)$ is a solution.

Substituting the other values for x in this manner gives the following table of values for $y = x^2$:

x	y
-2	4
-1	1
0	0
1	1
2	4

Check Yourself 1

Complete the ordered pairs to form solutions for $y = x^2 + 2$ and form a table of values.

$(-2, \ ), (-1, \ ), (0, \ ), (1, \ ), (2, \ )$

We can now plot points in the Cartesian coordinate system that correspond to solutions to the equation.

Example 2   Plotting Some Solution Points

Plot the points from the table of values corresponding to $y = x^2$ from Example 1.

x	y
-2	4
-1	1
0	0
1	1
2	4

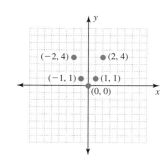

Notice that the y-axis acts as a mirror. Do you see that any point graphed in quadrant I will be "reflected" in quadrant II?

Check Yourself 2

Plot the points from the table of values formed in Check Yourself 1.

The graph of the equation can be drawn by joining the points with a smooth curve.

| | Example 3 | Completing the Graph of the Solution Set |

RECALL

The graph must be a parabola.

NOTE

Notice that a parabola **does not** come to a point.

Draw the graph of $y = x^2$.

We can now draw a smooth curve between the points found in Example 2 to form the graph of $y = x^2$.

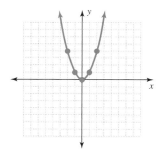

Check Yourself 3

Draw a smooth curve between the points plotted in Check Yourself 2.

You can use any convenient values for x in forming your table of values. You should use as many pairs as are necessary to get the correct shape of the graph (a parabola).

| **Example** 4 | **Graphing the Solution Set** |

Graph $y = x^2 - 2x$. Use integer values of x from -1 to 3.

First, determine solutions for the equation. For instance, if $x = -1$,

$$y = (-1)^2 - 2(-1)$$
$$= 1 + 2$$
$$= 3$$

and $(-1, 3)$ is a solution for the given equation.

Substituting the other values for x, we can form the following table of values. We then plot the corresponding points and draw a smooth curve to form our graph.

The graph of $y = x^2 - 2x$.

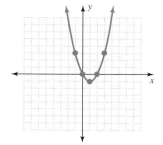

x	y
-1	3
0	0
1	-1
2	0
3	3

Check Yourself 4

Graph $y = x^2 + 4x$. Use integer values of x from -4 to 0.

Choosing values for x is also a valid method of graphing a quadratic equation that contains three terms.

Example 5   Graphing the Solution Set

Graph $y = x^2 - x - 2$. Use integer values of x from -2 to 3. We show the computation for two of the solutions.

$$\text{If } x = -2:$$
$$y = (-2)^2 - (-2) - 2$$
$$= 4 + 2 - 2$$
$$= 4$$

$$\text{If } x = 3:$$
$$y = (3)^2 - (3) - 2$$
$$= 9 - 3 - 2$$
$$= 4$$

You should substitute the remaining values for x into the given equation to verify the other solutions shown in the following table of values.

The graph of $y = x^2 - x - 2$.

x	y
-2	4
-1	0
0	-2
1	-2
2	0
3	4

Check Yourself 5

Graph $y = x^2 - 4x + 3$. Use integer values of x from -1 to 4.

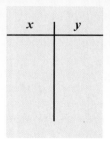

x	y

In Example 6, the graph looks significantly different from previous graphs.

Example 6   Graphing the Solution Set

Graph $y = -x^2 + 3$. Use integer values from -2 to 2.
Again we show two computations.

NOTE

$-(-2)^2 = -4$

$$\text{If } x = -2:$$
$$y = -(-2)^2 + 3$$
$$= -4 + 3$$
$$= -1$$

$$\text{If } x = 1:$$
$$y = -(1)^2 + 3$$
$$= -1 + 3$$
$$= 2$$

Verify the remainder of the solutions shown in the following table of values for yourself.

The graph of $y = -x^2 + 3$.

x	y
−2	−1
−1	2
0	3
1	2
2	−1

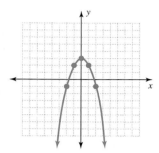

There is an important difference between this graph and the others we have seen. This time the parabola opens downward! Can you guess why? The answer is in the coefficient of the x^2-term.

If the coefficient of x^2 is *positive*, the parabola opens *upward*.

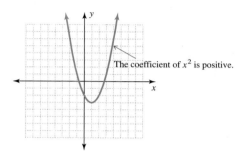

The coefficient of x^2 is positive.

If the coefficient of x^2 is *negative*, the parabola opens *downward*.

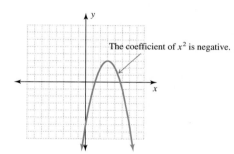

The coefficient of x^2 is negative.

Check Yourself 6

Graph $y = -x^2 - 2x$. Use integer values from −3 to 1.

x	y

There are two other terms we would like to introduce before closing this section on graphing quadratic equations. As you may have noticed, all the parabolas that we graphed are symmetric about a vertical line. This is called the **axis of symmetry** for the parabola.

The point at which the parabola intersects that vertical line (this will be the lowest—or the highest—point on the parabola) is called the **vertex.**

RECALL

$x = h$ (for any number h) is the equation of a vertical line.

The axis of symmetry

The vertex

As we construct a table of values, we want to include x-values that will span the parabola's turning point: the vertex. For a quadratic equation $y = ax^2 + bx + c$, the following is an equation for the axis of symmetry:

$$x = \frac{-b}{2a}$$

Once we have this, we can choose x-values on either side of the axis of symmetry to help us build a table. Look at Example 7.

Example 7	**Using the Axis of Symmetry to Create a Table and Graph**

< Objective 2 >

Given $y = -x^2 - 8x - 12$, complete the following.

(a) Write the equation for the axis of symmetry.

Since $a = -1$ and $b = -8$, we have

$$x = \frac{-b}{2a} = \frac{-(-8)}{2(-1)} = \frac{8}{-2} = -4$$

Thus the vertical line $x = -4$ is the axis of symmetry.

(b) Use the equation to create a table of values for the quadratic equation.

Finding values on both sides of -4, we make the following table:

x	y
-6	0
-5	3
-4	4
-3	3
-2	0

(c) Sketch the graph. Show the axis of symmetry as a dotted vertical line.

Beginning Algebra   The Streeter/Hutchison Series in Mathematics

NOTE

Here again the parabola opens downward.

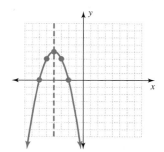

Check Yourself 7

For $y = x^2 - 7x + 4$, find the axis of symmetry, create a table of values, and sketch the graph.

The x-intercepts of a parabola are the points where the parabola touches the x-axis. We can locate these directly without actually graphing the parabola. The key is to set y equal to 0 and then solve for x.

Example 8   **Finding the x-Intercepts of a Parabola**

< Objective 3 >

Find the x-intercepts for each equation.

(a) $y = x^2 - 5x + 2$

We set $y = 0$, and solve.

$0 = x^2 - 5x + 2$ Use the quadratic formula.

NOTE

If decimal approximations are desired, we have $(0.44, 0)$ and $(4.56, 0)$.

$$x = \frac{5 \pm \sqrt{(-5)^2 - 4(2)}}{2} = \frac{5 \pm \sqrt{17}}{2}$$

There are two x-intercepts: $\left(\dfrac{5 - \sqrt{17}}{2}, 0\right)$ and $\left(\dfrac{5 + \sqrt{17}}{2}, 0\right)$.

(b) $y = x^2 + 6x + 9$

$0 = x^2 + 6x + 9$

$0 = (x + 3)(x + 3)$

$x = -3$

There is only one x-intercept: $(-3, 0)$. The parabola touches (but does not cross) the x-axis at this point.

(c) $y = x^2 - 4x + 5$

$0 = x^2 - 4x + 5$

$$x = \frac{4 \pm \sqrt{(-4)^2 - 4(5)}}{2}$$

$$= \frac{4 \pm \sqrt{-4}}{2}$$

Because these values are not real numbers, there are no *x*-intercepts. The parabola does not touch the *x*-axis.

Check Yourself 8

Find the *x*-intercepts for each equation.

(a) $y = x^2 + 3x - 5$ **(b)** $y = x^2 - 10x + 25$

RECALL

For the equation
$y = ax^2 + bx + c$
the axis of symmetry has equation
$x = \dfrac{-b}{2a}$.

In many applications involving quadratic graphs, the point on the graph of a parabola that is often desired is the vertex. As we saw earlier, the vertex must lie on the axis of symmetry. This means that, if we determine the equation of the axis of symmetry, we know the *x*-coordinate of the vertex. If we then substitute this *x*-value into the equation to determine the corresponding *y*-value, we will have the vertex.

 Example 9   **A Vocational Application**

An experimental drug is being tested on a bacteria colony. It is found that *t* days after the colony is treated, the number *N* of bacteria per cubic centimeter is given by the equation

$$N = 15t^2 - 100t + 700$$

NOTE

On a parabola, the vertex represents either a minimum point (if $a > 0$) or a maximum point (if $a < 0$).

(a) When will the bacteria colony be at a minimum?

We know that the graph of this equation is a parabola, and that the graph opens up. (Why?) So the vertex represents a minimum point for the graph.

To find the vertex, we first find the equation of the axis of symmetry.

$$t = \frac{-b}{2a} = \frac{-(-100)}{2(15)} = \frac{100}{30} = \frac{10}{3}$$

This means the bacteria colony will be at a minimum after $\dfrac{10}{3}$, or $3\dfrac{1}{3}$, days.

(b) What is that minimum number?

We now find *N* that corresponds to this *t*-value.

$$N = 15\left(\frac{10}{3}\right)^2 - 100\left(\frac{10}{3}\right) + 700$$

NOTE

The vertex of the parabola is
$\left(3\dfrac{1}{3}, 533\dfrac{1}{3}\right)$.

Using a calculator, we obtain $N = 533\dfrac{1}{3}$.

So the minimum number of bacteria is approximately 533.

Check Yourself 9

The number of people who are sick *t* days after the outbreak of a flu epidemic is given by the equation

$$P = -t^2 + 120t + 20$$

(a) On what day will the maximum number of people be sick?
(b) How many people will be sick on that day?

Check Yourself ANSWERS

1.

x	y
−2	6
−1	3
0	2
1	3
2	6

2.

3. $y = x^2 + 2$

4. $y = x^2 + 4x$

x	y
−4	0
−3	−3
−2	−4
−1	−3
0	0

5. $y = x^2 - 4x + 3$

x	y
−1	8
0	3
1	0
2	−1
3	0
4	3

6. $y = -x^2 - 2x$

x	y
−3	−3
−2	0
−1	1
0	0
1	−3

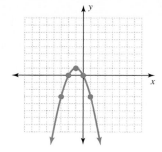

7. $y = x^2 - 7x + 4$; axis: $x = \dfrac{7}{2}$

x	y
1	−2
2	−6
3	−8
4	−8
5	−6
6	−2

8. (a) $\left(\dfrac{-3 - \sqrt{29}}{2}, 0\right)$ and $\left(\dfrac{-3 + \sqrt{29}}{2}, 0\right)$; **(b)** $(5, 0)$

9. (a) Day 60; **(b)** 3,620 people

Reading Your Text

The following fill-in-the-blank exercises are designed to ensure that you understand some of the key vocabulary used in this section.

SECTION 10.4

(a) The graph of $y = ax^2 + bx + c$, in which $a \neq 0$, is always a curve called a _____.

(b) If the coefficient of x^2 is negative, the parabola opens _____.

(c) All the parabolas that we graphed are symmetric about a vertical line, which is called the _____.

(d) The point at which a parabola intersects its axis of symmetry is called the _____.

10.4 exercises

Name _____

Section _____ Date _____

Answers

1. _____

2. _____

3. _____

4. _____

< Objective 1 >

Graph each of the following quadratic equations after completing the given table of values.

1. $y = x^2 + 1$

x	y
−2	
−1	
0	
1	
2	

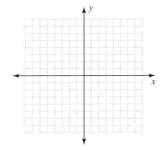

2. $y = x^2 - 2$

x	y
−2	
−1	
0	
1	
2	

3. $y = x^2 - 4$

x	y
−2	
−1	
0	
1	
2	

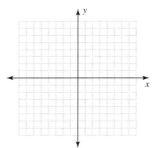

4. $y = x^2 + 3$

x	y
−2	
−1	
0	
1	
2	

5. $y = x^2 - 4x$

x	y
0	
1	
2	
3	
4	

5. _____

6. _____

7. _____

8. _____

6. $y = x^2 + 2x$

x	y
-3	
-2	
-1	
0	
1	

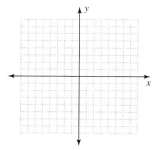

7. $y = x^2 + x$

x	y
-2	
-1	
0	
1	
2	

8. $y = x^2 - 3x$

x	y
-1	
0	
1	
2	
3	

Answers

9. _____

10. _____

11. _____

12. _____

9. $y = x^2 - 2x - 3$

x	y
-1	
0	
1	
2	
3	

10. $y = x^2 - 5x + 6$

x	y
0	
1	
2	
3	
4	

11. $y = x^2 - x - 6$

x	y
-1	
0	
1	
2	
3	

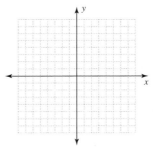

12. $y = x^2 + 3x - 4$

x	y
-4	
-3	
-2	
-1	
0	

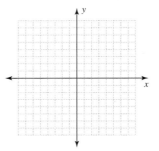

13. $y = -x^2 + 2$

x	y
-2	
-1	
0	
1	
2	

Answers

13. _____

14. _____

15. _____

16. _____

14. $y = -x^2 - 2$

x	y
-2	
-1	
0	
1	
2	

15. $y = -x^2 - 4x$ > Videos

x	y
-4	
-3	
-2	
-1	
0	

16. $y = -x^2 + 2x$

x	y
-1	
0	
1	
2	
3	

Answers

17. _____

18. _____

19. _____

20. _____

21. _____

22. _____

23. _____

24. _____

25. _____

26. _____

27. _____

28. _____

(a) _____

(b) _____

(c) _____

(d) _____

(e) _____

(f) _____

(g) _____

(h) _____

Complete each of the following statements with **never, sometimes,** *or* **always.**

17. The vertex of a parabola is _____ located on the axis of symmetry.

18. The vertex of a parabola is _____ the highest point on the graph.

19. The graph of $y = ax^2 + bx + c$ _____ intersects the x-axis.

20. The graph of $y = ax^2 + bx + c$ _____ has more than two x-intercepts.

21. The graph of $y = ax^2 + bx + c$ _____ intersects the y-axis.

22. The graph of $y = ax^2 + bx + c$ _____ intersects the y-axis more than once.

| Basic Skills | **Advanced Skills** | Vocational-Technical Applications | Calculator/Computer | Above and Beyond | Getting Ready |

Match each graph with the correct equation on the left.

23.

24.

25.

26.

27.

28.

29.

30.

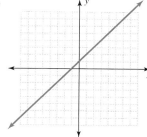

Answers

29. _____

30. _____

31. _____

32. _____

33. _____

< Objective 2 >

For each equation in exercises 31–34, identify the axis of symmetry, create a suitable table of values, and sketch the graph (including the axis of symmetry).

31. $y = x^2 + 4x$

x	y

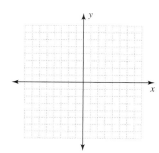

32. $y = x^2 - 5x + 3$

x	y

33. $y = -x^2 - 3x + 3$ > Videos

x	y

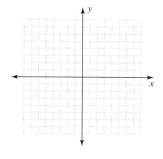

Answers

34. _____

35. _____

36. _____

37. _____

38. _____

39. _____

40. _____

41. _____

42. _____

43. _____

44. _____

45. _____

46. _____

47. _____

48. _____

49. _____

50. _____

34. $y = -x^2 + 6x - 2$

< Objective 3 >

For each equation in exercises 35–40, find the x-intercepts.

35. $y = x^2 - 2x - 8$ **36.** $y = x^2 + x - 6$

37. $y = x^2 + 3x - 6$ **38.** $y = x^2 - 5x - 4$

39. $y = x^2 - 2x + 1$ **40.** $y = x^2 - x + 2$

For each equation in exercises 41–48, find the vertex.

41. $y = x^2 - 4x$ **42.** $y = x^2 + 5x$

43. $y = x^2 + 3x - 4$ **44.** $y = x^2 - 2x - 3$

45. $y = -x^2 + 2x$ **46.** $y = -x^2 - 3x + 3$

47. $y = 2x^2 - 12x + 5$ **48.** $y = -2x^2 - 10x + 7$

| Basic Skills | Advanced Skills | **Vocational-Technical Applications** | Calculator/Computer | Above and Beyond | Getting Ready |

49. ALLIED HEALTH The concentration C, in nanograms per milliliter (ng/mL), of digoxin, a medication prescribed for congestive heart failure, is given by the equation

$$C = -0.0015t^2 + 0.0845t + 0.7170$$

where t is the number of hours since the drug was taken orally.

(a) When will the drug's concentration be at a maximum?

(b) Determine the drug's maximum concentration in the bloodstream.

50. ALLIED HEALTH A patient's body temperature T, in degrees Fahrenheit, t hours after taking acetaminophen, a commonly prescribed analgesic, can be approximated by the equation

$$T = 0.4t^2 - 2.6t + 103$$

(a) When will the patient's temperature be at a minimum?

(b) Determine the patient's minimum temperature.

Answers

1. $y = x^2 + 1$

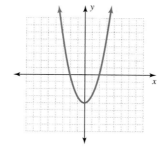

3. $y = x^2 - 4$

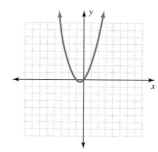

5. $y = x^2 - 4x$

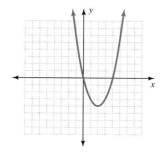

7. $y = x^2 + x$

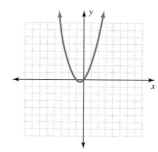

9. $y = x^2 - 2x - 3$

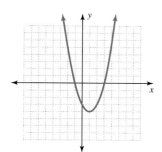

11. $y = x^2 - x - 6$

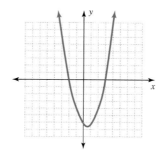

13. $y = -x^2 + 2$

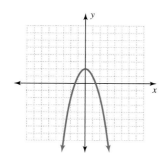

15. $y = -x^2 - 4x$

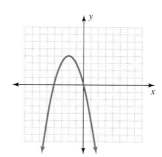

17. always   **19.** sometimes   **21.** always
23. f   **25.** a   **27.** b   **29.** e

31. Axis: $x = -2$

x	y
-4	0
-3	-3
-2	-4
-1	-3
0	0

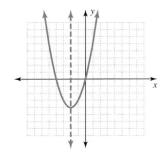

33. Axis: $x = -\dfrac{3}{2}$

x	y
-4	-1
-3	3
-2	5
-1	5
0	3
1	-1

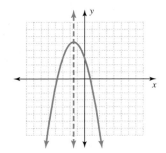

35. $(-2, 0), (4, 0)$ **37.** $\left(\dfrac{-3 - \sqrt{33}}{2}, 0\right), \left(\dfrac{-3 + \sqrt{33}}{2}, 0\right)$

39. $(1, 0)$ **41.** $(2, -4)$ **43.** $\left(-1\dfrac{1}{2}, -6\dfrac{1}{4}\right)$ **45.** $(1, 1)$

47. $(3, -13)$ **49. (a)** 28.2 hours; **(b)** 1.91 ng/mL

Definition/Procedure	Example	Reference

More on Quadratic Equations

Solving Quadratic Equations by Factoring

1. Add or subtract the necessary terms on both sides of the equation so that the equation is in standard form (set equal to 0).

2. Factor the quadratic expression.

3. Set each factor equal to 0.

4. Solve the resulting equations to find the solutions.

5. Check each solution by substituting in the original equation.

Section 10.1

To solve:

$$x^2 + 7x = 30$$
$$x^2 + 7x - 30 = 0$$
$$(x + 10)(x - 3) = 0$$
$$x + 10 = 0 \quad \text{or} \quad x - 3 = 0$$

$x = -10$ and $x = 3$ are solutions.

p. 910

Square-Root Property

If $x^2 = k$, then $x = \sqrt{k}$ or $x = -\sqrt{k}$.

If $(x - h)^2 = k$, then $x = h \pm \sqrt{k}$.

To solve:

$$(x - 3)^2 = 5$$
$$x - 3 = \pm\sqrt{5}$$
$$x = 3 \pm \sqrt{5}$$

p. 912

Completing the Square

Completing the Square

To solve a quadratic equation by completing the square:

1. Write the equation in the form

$ax^2 + bx = k$

so that the variable terms are on the left side and the constant is on the right side.

2. If the leading coefficient of x^2 is not 1, divide both sides by that coefficient.

3. Add the square of one-half the coefficient of x to both sides of the equation.

4. The left side of the equation is now a perfect-square trinomial. Factor and solve as before.

Section 10.2

To solve:

$$2x^2 + 2x - 1 = 0$$
$$2x^2 + 2x = 1$$
$$x^2 + x = \frac{1}{2}$$
$$x^2 + x + \left(\frac{1}{2}\right)^2 = \frac{1}{2} + \left(\frac{1}{2}\right)^2$$
$$\left(x + \frac{1}{2}\right)^2 = \frac{3}{4}$$
$$x + \frac{1}{2} = \pm\sqrt{\frac{3}{4}} = \pm\frac{\sqrt{3}}{2}$$
$$x = \frac{-1 \pm \sqrt{3}}{2}$$

p. 926

Continued

Definition/Procedure	Example	Reference

The Quadratic Formula

Section 10.3

The Quadratic Formula

To solve an equation by formula:

1. Rewrite the equation in standard form.

   $ax^2 + bx + c = 0$

2. If a common factor exists, divide both sides of the equation by that common factor.

3. Identify the coefficients a, b, and c.

4. Substitute the values for a, b, and c into the quadratic formula.

   $$x = \frac{-b \pm \sqrt{b^2 - 4ac}}{2a}$$

5. Simplify the right side of the expression formed in step 4 to write the solutions for the original equation.

To solve:

$x^2 - 2x = 4$

Write the equation as

$$x^2 - 2x - 4 = 0$$
$$a = 1 \qquad b = -2 \qquad c = -4$$

$$x = \frac{-(-2) \pm \sqrt{(-2)^2 - 4(1)(-4)}}{2(1)}$$

$$= \frac{2 \pm \sqrt{20}}{2}$$

$$= \frac{2 \pm 2\sqrt{5}}{2} = \frac{2(1 \pm \sqrt{5})}{2}$$

$$= 1 \pm \sqrt{5}$$

p. 939

Graphing Quadratic Equations

Section 10.4

To graph equations of the form

$y = ax^2 + bx + c$

1. Find the axis of symmetry using

   $$x = \frac{-b}{2a}$$

2. Form a table of values by choosing x-values that span the axis of symmetry.

3. Plot the points from the table of values.

4. Draw a smooth curve between the points.

The graph of a quadratic equation will always be a parabola. The parabola opens upward if a, the coefficient of the x^2-term, is positive.

$y = x^2 - 4x$

$$x = \frac{-(-4)}{2(1)} = \frac{4}{2} = 2$$

Choose x-values to the right and left of 2.

x	y
-1	5
0	0
1	-3
2	-4
3	-3
4	0
5	5

p. 949

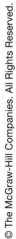

Definition/Procedure	Example	Reference
The x-intercepts of a parabola are the points where the parabola intersects the x-axis. They may be found by setting y equal to 0, and then solving for x.	To find the x-intercepts of $$y = x^2 - 4x - 6$$ set $y = 0$ and solve. $$0 = x^2 - 4x - 6$$ $$x = \frac{4 \pm \sqrt{(-4)^2 - 4(-6)}}{2}$$ $$= \frac{4 \pm \sqrt{40}}{2} = \frac{4 \pm 2\sqrt{10}}{2}$$ $$= 2 \pm \sqrt{10}$$ The x-intercepts are $(2 - \sqrt{10}, 0)$ and $(2 + \sqrt{10}, 0)$.	*p.* 956

The Streeter/Hutchison Series in Mathematics   Beginning Algebra

This summary exercise set is provided to give you practice with each of the objectives of this chapter. Each exercise is keyed to the appropriate chapter section. When you are finished, you can check your answers to the odd-numbered exercises against those presented in the back of the text. If you have difficulty with any of these questions, go back and reread the examples from that section. Your instructor will give you guidelines on how best to use these exercises in your instructional setting.

10.1 *Solve each of the following equations for x by the square-root method.*

1. $x^2 = 10$

2. $x^2 = 48$

3. $x^2 - 20 = 0$

4. $x^2 + 2 = 8$

5. $(x - 1)^2 = 5$

6. $(x + 2)^2 = 8$

7. $(x + 3)^2 = 5$

8. $64x^2 = 25$

9. $4x^2 = 27$

10. $9x^2 = 20$

11. $25x^2 = 7$

12. $7x^2 = 3$

10.2 *Solve each of the following equations by completing the square.*

13. $x^2 - 3x - 10 = 0$

14. $x^2 - 8x + 15 = 0$

15. $x^2 - 5x + 2 = 0$

16. $x^2 - 2x - 2 = 0$

17. $x^2 - 4x - 4 = 0$

18. $x^2 + 3x = 7$

19. $x^2 - 4x = -2$

20. $x^2 + 3x = 5$

21. $x^2 - x = 7$

22. $2x^2 + 6x = 12$

23. $2x^2 - 4x - 7 = 0$

24. $3x^2 + 5x + 1 = 0$

10.3 *Solve each of the following equations using the quadratic formula.*

25. $x^2 - 5x - 14 = 0$

26. $x^2 - 8x + 16 = 0$

27. $x^2 + 5x - 3 = 0$

28. $x^2 - 7x - 1 = 0$

29. $x^2 - 6x + 1 = 0$

30. $x^2 - 3x + 5 = 0$

31. $3x^2 - 4x = 2$

32. $2x - 3 = \dfrac{3}{x}$

33. $(x - 1)(x + 4) = 3$

34. $x^2 - 5x + 7 = 5$

35. $2x^2 - 8x = 12$

36. $5x^2 = 15 - 15x$

Solve by factoring or by any of the methods of this chapter.

37. $5x^2 = 3x$

38. $(2x - 3)(x + 5) = -11$

39. $(x - 1)^2 = 10$

40. $2x^2 = 7$

41. $2x^2 = 5x + 4$

42. $2x^2 - 4x = 30$

43. $2x^2 = 5x + 7$

44. $3x^2 - 4x = 2$

45. $3x^2 + 6x - 15 = 0$

46. $x^2 - 3x = 2(x + 5)$

47. $x - 2 = \dfrac{2}{x}$

48. $3x + 1 = \dfrac{5}{x}$

10.4 *Graph each quadratic equation after completing the table of values.*

49. $y = x^2 + 3$

50. $y = x^2 - 2$

51. $y = x^2 - 3x$

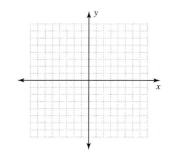

x	y
−1	
0	
1	
2	
3	

52. $y = x^2 + 4x$

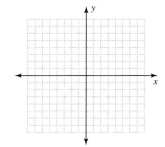

x	y
−4	
−3	
−2	
−1	
0	

53. $y = x^2 - x - 2$

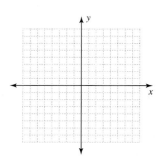

x	y
−1	
0	
1	
2	
3	

54. $y = x^2 - 4x + 3$

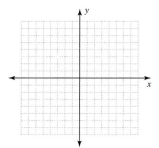

x	y
0	
1	
2	
3	
4	

55. $y = x^2 + 2x - 3$

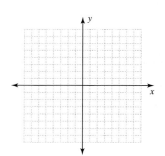

x	y
−3	
−2	
−1	
0	
1	

56. $y = 2x^2$

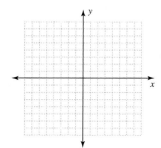

x	y
−2	
−1	
0	
1	
2	

57. $y = 2x^2 - 3$

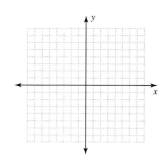

x	y
−2	
−1	
0	
1	
2	

58. $y = -x^2 + 3$

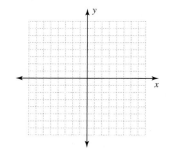

x	y
−2	
−1	
0	
1	
2	

59. $y = -x^2 - 2$

x	y
-2	
-1	
0	
1	
2	

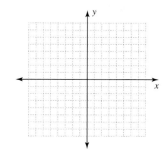

60. $y = -x^2 + 4x$

x	y
0	
1	
2	
3	
4	

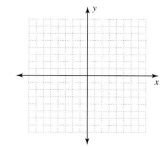

For each equation, identify the axis of symmetry, create a suitable table of values, and sketch the graph.

61. $y = x^2 + 6x + 3$

x	y
-5	
-4	
-3	
-2	
-1	

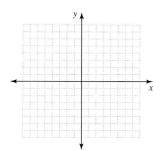

62. $y = -x^2 - 4x + 3$

x	y
-4	
-3	
-2	
-1	
0	

63. $y = -x^2 + 5x + 2$

x	y
0	
1	
2	
3	
4	
5	

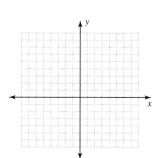

64. $y = x^2 - 3x + 4$

x	y
-1	
0	
1	
2	
3	
4	

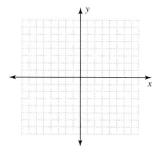

For each equation, find the x-intercepts of the graph.

65. $y = x^2 + 2x - 15$

66. $y = x^2 - 3x - 4$

67. $y = x^2 + 6x + 3$

68. $y = x^2 - 3x + 4$

69. $y = -x^2 + 2x - 3$

70. $y = -x^2 - 4x + 1$

The Streeter/Hutchison Series in Mathematics   Beginning Algebra

Solve each of the following applications. Round results to the nearest hundredth.

71. **GEOMETRY** The longer leg of a right triangle is 3 cm less than twice the shorter leg. The hypotenuse is 18 cm long. Find the length of each leg.

72. **CONSTRUCTION** A rectangular garden is to be surrounded by a walkway of constant width. The garden's dimensions are 15 ft by 24 ft. The total area, garden plus walkway, is to be 750 ft^2. What must be the width of the walkway?

73. **SCIENCE AND MEDICINE** If a ball is thrown vertically upward from a height of 5 ft, with an initial velocity of 80 ft/s, its height h after t seconds is given by $h = -16t^2 + 80t + 5$. Find the time it takes for the ball to reach a height of 75 ft.

74. **SCIENCE AND MEDICINE** If a ball is thrown vertically upward from a height of 5 ft, with an initial velocity of 80 ft/s, its height h after t seconds is given by $h = -16t^2 + 80t + 5$. How long does it take the ball to return to the ground?

75. **BUSINESS AND FINANCE** Suppose that the cost C, in dollars, of producing x items is given by $C = 3x^2 - 50x + 1,800$. How many items can be produced for $13,000?

76. **BUSINESS AND FINANCE** The demand equation for a certain computer chip is given by

$D = -2p + 25$

The supply equation is predicted to be

$S = -p^2 + 16p + 5$

Find the equilibrium price.

The purpose of this self-test is to help you check your progress and to review for the next in-class exam. Allow yourself about an hour to take this test. At the end of that hour check your answers against those given in the back of the text. Section references accompany the answers. If you missed any questions, go back to those sections and reread the examples until you master the concepts.

Name _____

Section _____ Date _____

Answers

Solve each of the following equations for x.

1. $x^2 = 15$

2. $x^2 - 8 = 0$

3. $(x - 1)^2 = 7$

4. $9x^2 = 10$

Solve each of the following equations by completing the square.

5. $x^2 - 2x - 8 = 0$

6. $x^2 + 3x - 1 = 0$

7. $x^2 + 2x - 5 = 0$

8. $2x^2 - 5x + 1 = 0$

Solve each of the following equations by using the quadratic formula.

9. $x^2 - 2x - 3 = 0$

10. $x^2 - 6x + 9 = 0$

11. $x^2 - 5x = 2$

12. $2x^2 = 2x + 5$

13. $2x - 1 = \dfrac{4}{x}$

14. $(x - 1)(x + 3) = 2$

Graph each quadratic equation after completing the given table of values.

15. $y = x^2 + 4$

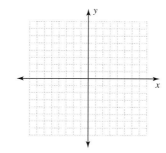

x	y
-2	
-1	
0	
1	
2	

1. _____

2. _____

3. _____

4. _____

5. _____

6. _____

7. _____

8. _____

9. _____

10. _____

11. _____

12. _____

13. _____

14. _____

15. _____

Answers

16. _____

17. _____

18. _____

19. _____

20. _____

16. $y = x^2 - 2x$

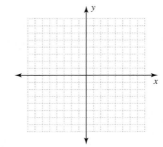

x	y
−1	
0	
1	
2	
3	

17. $y = x^2 + x - 2$

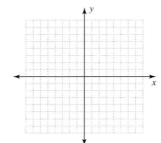

x	y
−2	
−1	
0	
1	
2	

Give the axis of symmetry and the x-intercepts.

18. $y = x^2 - 4x - 21$

Solve the following applications. Round results to the nearest hundredth.

19. GEOMETRY If the length of a rectangle is 4 ft less than 3 times its width and the area of the rectangle is 96 ft^2, what are the dimensions of the rectangle?

20. SCIENCE AND MEDICINE If a pebble is dropped toward a pond from the top of a 180-ft building, its height h after t seconds is given by $h = -16t^2 + 180$. How long does it take the pebble to pass through a height of 30 ft?

Activity 10 ::
The Gravity Model

Each activity in this text is designed to either enhance your understanding of the topics of the preceding chapter, provide you with a mathematical extension of those topics, or both. The activities can be undertaken by one student, but they are better suited for a small group project. Occasionally it is only through discussion that different facets of the activity become apparent. For material related to this activity, visit the text website at www.mhhe.com/streeter.

This activity requires two to three people and an open area outside. You will also need a ball (preferably a baseball or some other small, heavy ball), a stopwatch, and a tape measure.

1. Designate one person as the "ball thrower."

2. The ball thrower should throw the ball straight up into the air, as high as possible. While this is happening, another group member should take note of exactly where the ball thrower releases the ball. Be careful that no one gets hit by the ball as it comes back down.

3. Use the tape measure to determine the height above the ground that the thrower released the ball. Convert your measurement to feet, using decimals to indicate parts of a foot. For example, 3 inches is 0.25 ft. Record this as the "initial height."

4. Repeat steps 2 and 3 to ensure that you have the correct initial height.

5. The thrower should now throw the ball straight up, as high as possible. The person with the stopwatch should time the ball until it lands.

6. Repeat step 5 twice more, recording the time.

7. Take the average (mean) of your three recorded times. We will use this number as the "hang time" of the ball for the remainder of this activity.

According to Isaac Newton (1642–1727), the height of an object with initial velocity v_0 and initial height s_0 is given by

$$s = -16t^2 + v_0 t + s_0$$

in which the height s is measured in feet and t represents the time, in seconds.

The initial velocity v_0 is positive when the object is thrown upward, and it is negative when the object is thrown downward.

For example, the height of a ball thrown upward from a height of 4 ft with an initial velocity of 50 ft/s is given by

$$s = -16t^2 + 50t + 4$$

To determine the height of the ball 2 seconds after release, we evaluate the polynomial for $t = 2$.

$$= -16(2)^2 + 50(2) + 4$$
$$= 40$$

The ball is 40 feet high after 2 seconds.

Take note that the height of the ball when it lands is zero.

8. Substitute the time found in step 7 for t, your initial height for s_0, and 0 for s in the height equation. Solve the resulting equation for the initial velocity, v_0.

9. Now write your height equation using the initial velocity found in step 8 along with the initial height. Your equation should have two variables, s and t.

10. What is the height of the ball after 1 second?

11. The maximum height of the ball will be attained when $t = \dfrac{v_0}{32}$. Determine this time and find the maximum height of the thrown ball.

12. If a ball is dropped from a height of 256 feet, how long will it take to hit the ground? (*Hint:* If a ball is dropped, its initial velocity is 0 ft/s.)

The following exercises are presented to help you review concepts from earlier chapters. This is meant as a review and not as a comprehensive exam. The answers are presented in the back of the text. Section references accompany the answers. If you have difficulty with any of these exercises, be certain to at least read through the summary related to those sections.

Name _____

Section _____ Date _____

Simplify the following expressions.

1. $6x^2y - 4xy^2 + 5x^2y - 2xy^2$

2. $(3x^2 + 2x - 5) - (2x^2 - 3x + 2)$

Evaluate each expression when $x = 2$, $y = -3$, and $z = 4$.

3. $4x^2y - 3z^2y^2$

4. $-3x^2y^2z^2 - 2xyz$

5. Solve for x: $4x - 2(3x - 5) = 8$.

6. Solve the inequality $4x + 15 > 2x + 19$.

Perform the indicated operations.

7. $3xy(2x^2 - x + 5)$

8. $(2x + 5)(3x - 2)$

9. $(3x + 4y)(3x - 4y)$

Factor each of the following completely.

10. $16x^2y^2 - 8xy^3$

11. $8x^2 - 2x - 15$

12. $25x^2 - 16y^2$

Perform the indicated operations.

13. $\dfrac{7}{4x + 8} - \dfrac{5}{7x + 14}$

14. $\dfrac{5x + 5}{x - 2} \cdot \dfrac{x^2 - 4x + 4}{x^2 - 1}$

15. $\dfrac{3x^2 + 8x - 3}{15x^2} + \dfrac{3x - 1}{5x^2}$

Answers

1. _____

2. _____

3. _____

4. _____

5. _____

6. _____

7. _____

8. _____

9. _____

10. _____

11. _____

12. _____

13. _____

14. _____

15. _____

Answers

16. _____

17. _____

18. _____

19. _____

20. _____

21. _____

22. _____

23. _____

24. _____

25. _____

Graph the following equations.

16. $3x - 2y = 6$

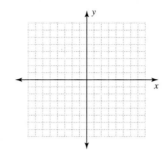

17. $y = 4x - 5$

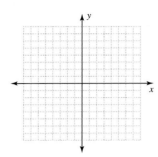

18. Find the slope of the line through the points $(2, 9)$ and $(-1, -6)$.

19. Given that the slope of a line is 2 and the y-intercept is $(0, -5)$, write the equation of the line.

20. Graph the inequality $x + 2y < 6$.

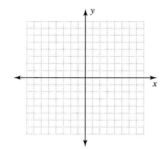

Solve each of the following systems. If a unique solution does not exist, state whether the system is inconsistent or dependent.

21. $2x - 3y = 6$
    $x - 3y = 2$

22. $2x + y = 4$
    $y = 2x - 8$

23. $5x + 2y = 8$
    $x - 4y = 17$

24. $2x - 6y = 8$
    $x = 3y + 4$

Solve each of the following applications. Be sure to show the system of equations used for your solution.

25. One number is 4 less than 5 times another. If the sum of the numbers is 26, what are the two numbers?

The Streeter/Hutchison Series in Mathematics   Beginning Algebra

26. Receipts for a concert attended by 450 people were $2,775. If reserved-seat tickets were $7 and general admission tickets were $4, how many of each type of ticket were sold?

27. A chemist has a 30% acid solution and a 60% solution already prepared. How much of each of the two solutions should be mixed to form 300 mL of a 50% solution?

Evaluate each root, if possible.

28. $\sqrt{169}$

29. $-\sqrt{169}$

30. $\sqrt{-169}$

31. $\sqrt[3]{-64}$

Simplify each of the following radical expressions.

32. $\sqrt{12} + 3\sqrt{27} - \sqrt{75}$

33. $3\sqrt{2a} \cdot 5\sqrt{6a}$

34. $(\sqrt{2} - 5)(\sqrt{2} + 3)$

35. $\dfrac{8 - \sqrt{32}}{4}$

Solve each of the following equations.

36. $x^2 - 72 = 0$

37. $x^2 + 6x - 3 = 0$

38. $2x^2 - 3x = 2(x + 1)$

Graph each of the following quadratic equations.

39. $y = x^2 - 2$

40. $y = x^2 - 4x$

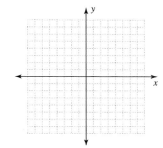

Answers

26. _____

27. _____

28. _____

29. _____

30. _____

31. _____

32. _____

33. _____

34. _____

35. _____

36. _____

37. _____

38. _____

39. _____

40. _____

Answers

Beginning Algebra   The Streeter/Hutchison Series in Mathematics   © The McGraw-Hill Companies. All Rights Reserved.

41. _____

42. _____

Solve the following applications.

41. The equation $h = -16t^2 - 64t + 250$ gives the height of a ball, thrown downward from the top of a 250-ft building with an initial velocity of 64 ft/s. Find the time it takes for the ball to reach a height of 100 ft. Round your answer to the nearest thousandth.

42. The demand equation for a certain type of printer is given by

$$D = -120p + 16,000$$

The supply equation is predicted to be

$$S = -p^2 + 260p - 9,000$$

Find the equilibrium price.

Evaluate the given expressions.

1. $|-25| - |-11|$

2. $16 + (-22)$

3. $(-41) - (-15)$

4. $(-5)(-3)(-7)$

5. $\dfrac{3(-2) - 8}{-7 - (-4)(3)}$

6. $6 - 2^3 \cdot 5$

Evaluate the expressions for the given values of the variables.

7. $b^2 - 4ac$ for $a = -3$, $b = -4$, and $c = 2$

8. $-x^2 - 7x - 3$ for $x = -2$

9. Write the expression $9 \cdot p \cdot p \cdot p \cdot q \cdot q \cdot q \cdot q \cdot q$ in exponential form.

Simplify the expressions using the properties of exponents. Write all answers using positive exponents only.

10. $z^{-11}z^5$

11. $(5c^8d^7)^2$

12. $\dfrac{4x^8y^5z^3}{2x^6y^9z^7}$

Perform the indicated operations. Write each answer in simplified form.

13. $2x(x + 3) + 5$

14. $(6x^2 - 3x - 20) - 2(4x^2 - 16x + 11)$

15. $(7x - 9)(4x + 5)$

16. $(3x + 4y)(3x - 4y)$

17. $(5x - 2y)^2$

18. $x(x - 3y) - 2y(y + 6x)$

Solve the given equations.

19. $5x - 9 = -7x - 3$

20. $2x - 3(x - 2) = 8$

21. $\dfrac{5 - x}{-2} = 3x$

22. Solve the inequality $4x - 3 > 6x - 2$.

Name _____

Section _____ Date _____

Answers

1. _____ 2. _____

3. _____ 4. _____

5. _____ 6. _____

7. _____ 8. _____

9. _____

10. _____

11. _____

12. _____

13. _____

14. _____

15. _____

16. _____

17. _____

18. _____

19. _____

20. _____

21. _____

22. _____

Answers

Beginning Algebra · The Streeter/Hutchison Series in Mathematics

23. _____

24. _____

25. _____

26. _____

27. _____

28. _____

29. _____

30. _____

31. _____

32. _____

33. _____

34. _____

35. _____

36. _____

37. _____

38. _____

Factor each expression completely.

23. $10x^2 - 490$

24. $x^2 - 12x + 36$

25. $4p^2 - p - 18$

26. $3xy + 3xz - 5y - 5z$

Simplify the given expression.

27. $\dfrac{2y - 36}{3y^2 - 54y}$

28. $\dfrac{4z}{z + 8} + \dfrac{32}{z + 8}$

29. $\dfrac{6a}{a^2 - 9} - \dfrac{5a}{a^2 + a - 6}$

30. $\dfrac{y^2 + y - 2}{y + 5} \cdot \dfrac{3y + 3}{9y - 9}$

31. $\dfrac{x^2 - 3x - 10}{3x} \div \dfrac{5x - 25}{15x^2}$

Solve the given equation by the indicated method.

32. $3x^2 + 5x - 2 = 0$ by factoring

33. $x^2 + x - 3 = 0$ by using the quadratic formula

34. $x^2 + 6x = 5$ by completing the square

35. Solve the equation $\dfrac{1}{x - 1} + \dfrac{x + 1}{x^2 + 2x - 3} = \dfrac{1}{x + 3}$

36. Find the slope of the line through the points $(2, -3)$ and $(5, 9)$.

37. Find the slope of the line whose equation is $3x - 4y = 12$.

38. Find the equation of the line that passes through the point $(4, -2)$ and is parallel to the line $2x + y = 6$.

39. Graph the line whose equation is $4x + 5y = 20$.

In exercises 40–43, perform the indicated operations and simplify the result.

40. $-\sqrt{\dfrac{64}{25}}$

41. $\sqrt{18x^5y^6}$

42. $3\sqrt{20} - 2\sqrt{125}$

43. $(\sqrt{5} + 2)(\sqrt{5} - 8)$

44. Solve the system of equations

$$2x + 3y = 4$$
$$4x - 2y = 8$$

45. Determine the equation of the line in the following graph.

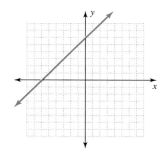

46. The length of a rectangle is 2 cm more than 3 times the width. The perimeter is 44 cm. Find the dimensions of the rectangle.

47. Find the equation of the line that passes through the points $(-1, -3)$ and $(2, 6)$.

Answers

39. _____

40. _____

41. _____

42. _____

43. _____

44. _____

45. _____

46. _____

47. _____

Answers

48. _____

49. _____

50. _____

48. One number is 3 more than 6 times another. If the sum of the numbers is 38, find the two numbers.

49. A store marks up items to make a 30% profit. If an item sells for $3.25, what does it cost before the markup?

50. Graph the equation $y = x^2 - 3x + 2$.

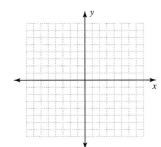

Answers to Pretests, Reading Your Text, Summary Exercises, Self-Tests, and Cumulative Reviews

Pretest Chapter 0

1. 1, 2, 3, 6, 7, 14, 21, 42
2. Prime: 2, 3, 7, 17, 23; composite: 6, 9, 18, 21
3. (a) $2 \times 2 \times 3 \times 5$; (b) $2 \times 5 \times 5 \times 7$ **4.** (a) 4; (b) 6
5. (a) 20; (b) 180 **6.** $\frac{5}{4}$ **7.** $\frac{3}{2}$ **8.** $\frac{19}{12}$ **9.** $\frac{7}{18}$
10. 3.767 **11.** 22.8404 **12.** $12\frac{3}{5}$ **13.** $\frac{3}{4}$ **14.** 2.435
15. 6 **16.** 8 **17.** 13 **18.** 2 **19.** (a) $\frac{23}{100}$; (b) 0.23
20. $3\frac{1}{2}\%$
21.

22. $-4, -2, -1, 0, 1, 5$ **23.** Max: 7; min: -5 **24.** 5
25. 6 **26.** 6 **27.** 6 **28.** 6 **29.** 16 **30.** -23

Reading Your Text

Section 0.1 (a) multiplication; (b) prime; (c) relatively; (d) multiple
Section 0.2 (a) proper; (b) equivalent; (c) LCD; (d) fraction
Section 0.3 (a) decimal; (b) repeating; (c) percents; (d) left
Section 0.4 (a) addition; (b) exponent; (c) grouping; (d) Parentheses
Section 0.5 (a) negative; (b) minus; (c) integers; (d) ascending

Summary Exercises for Chapter 0

1. 1, 2, 4, 13, 26, 52 **3.** 1, 2, 4, 19, 38, 76
5. Prime: 2, 5, 7, 11, 17, 23, 43; composite: 14, 21, 27, 39
7. $2^2 \cdot 3 \cdot 5 \cdot 7$ **9.** $2^2 \cdot 3^2 \cdot 5$ **11.** 1 **13.** 60 **15.** 16
17. 36 **19.** $\frac{6}{22}, \frac{9}{33}, \frac{12}{44}$ **21.** $\frac{3}{8}$ **23.** $\frac{1}{9}$ **25.** $\frac{2}{3}$
27. $\frac{13}{9}$ **29.** $\frac{7}{18}$ **31.** 11.523 **33.** 19.96696
35. $9\frac{37}{60}$ **37.** $2\frac{19}{24}$ **39.** $2\frac{1}{8}$ **41.** 8.222 **43.** $\frac{1}{50}$
45. $\frac{3}{8}$ **47.** $2\frac{1}{3}$ **49.** 0.75 **51.** 0.0625 **53.** 0.006
55. 6% **57.** 700% **59.** 0.5% **61.** 40%
63. $266\frac{2}{3}\%$ **65.** 3 **67.** 80 **69.** 41 **71.** 20
73. 324 **75.** 11 **77.** 4 **79.** 15
81.

83. $-8, -7, -3, 0, 1, 2, 3, 7, 8$ **85.** Max: 8; min: -9
87. 63 **89.** 9 **91.** -9 **93.** 20 **95.** 6 **97.** 14

Self-Test for Chapter 0

[0.1] **1.** Prime: 5, 13, 17, 31; composite: 9, 22, 27, 45
2. $2 \cdot 2 \cdot 2 \cdot 3 \cdot 11$ **3.** 12 **4.** 8 **5.** 108 **6.** 36

[0.2–0.3] **7.** $\frac{2}{7}$ **8.** $\frac{3}{4}$ **9.** $\frac{19}{12}$ **10.** $\frac{2}{21}$ **11.** 7.375
12. 3.884 **13.** 22.9635 **14.** 79.91 **15.** $3\frac{3}{7}$ **16.** $7\frac{11}{12}$
17. $1\frac{11}{18}$ **18.** 7.35 [0.3] **19.** $\frac{7}{100}$ **20.** $\frac{18}{25}$ **21.** 0.42
22. 0.06 **23.** 1.6 **24.** 3% **25.** 4.2% **26.** 40%
27. 62.5% [0.4] **28.** 4^4 **29.** 9^5 **30.** 3 **31.** 65
32. 144 **33.** 7 **34.** 8 **35.** 7
[0.5] **36.**

37. $-6, -3, -2, 0, 2, 4, 5$ **38.** Max: 6; min: -5 **39.** 7
40. 7 **41.** 11 **42.** 11 **43.** -19 **44.** -40 **45.** 19

Pretest Chapter 1

1. Commutative property of multiplication
2. Distributive property
3. Associative property of addition **4.** -10 **5.** -1
6. -5 **7.** -1 **8.** -3 **9.** -19 **10.** 12
11. 0 **12.** 21 **13.** 14 **14.** $\frac{1}{2}$ **15.** 7 **16.** $x - 8$
17. $\frac{w}{17x}$ **18.** Not an expression **19.** Expression
20. -3 **21.** 55 **22.** -7 **23.** $8w^2t$
24. $-a^2 + 4a + 3$ **25.** $12xy^2$

Reading Your Text

Section 1.1 (a) calendar; (b) properties; (c) associative; (d) area
Section 1.2 (a) negative; (b) absolute; (c) Zero; (d) opposite
Section 1.3 (a) negative; (b) positive; (c) one; (d) zero
Section 1.4 (a) variables; (b) difference; (c) multiplication;
    (d) letter
Section 1.5 (a) variable; (b) fraction; (c) operations; (d) principal
Section 1.6 (a) term; (b) number; (c) like; (d) distributive
Section 1.7 (a) add; (b) base; (c) multiply; (d) subtract

Summary Exercises for Chapter 1

1. Associative property of addition
3. Associative property of multiplication **5.** $72 = 72$
7. $20 = 20$ **9.** $80 = 80$ **11.** $3 \cdot 7 + 3 \cdot 4$
13. $\frac{1}{2} \cdot 5 + \frac{1}{2} \cdot 8$ **15.** -11 **17.** 0 **19.** -18
21. -4 **23.** -5 **25.** 17 **27.** 0 **29.** 3
31. 157 **33.** -96 **35.** 4.24 **37.** 2.048
39. 29.75 **41.** -70 **43.** 45 **45.** 0 **47.** $-\frac{3}{2}$
49. 5 **51.** 9 **53.** -4 **55.** -1 **57.** -7
59. $y + 5$ **61.** $8a$ **63.** $5mn$ **65.** $17x + 3$
67. Yes **69.** No **71.** 3 **73.** 80 **75.** 41 **77.** 20
79. 324 **81.** 11 **83.** -7 **85.** 15 **87.** 19
89. 25 **91.** 1 **93.** -3 **95.** $4a^3, -3a^2$
97. $5m^2, -4m^2, m^2$ **99.** $12c$ **101.** $2a$ **103.** $3xy$

105. $19a + b$ **107.** $3x^3 + 9x^2$ **109.** $10a^3$ **111.** x^7
113. x **115.** $2p^2$ **117.** $5m^5n^2$ **119.** $8p^2q^2$ **121.** $20x^7$
123. $24x^5y^5$ **125.** $12x^6y$ **127.** $25 - x$ **129.** $2x + 5$
131. $6n - 7$ **133.** $0.1x + 0.25q$

Self-Test for Chapter 1

[1.1] **1.** Commutative property of multiplication
2. Distributive property   **3.** Associative property of addition
4. 21 **5.** $20x + 12$ **6.** Not an expression   **7.** Expression
[1.2] **8.** -13 **9.** -3 **10.** -21 **11.** 1 **12.** -6
13. -21 **14.** 9 **15.** 0 **[1.3]** **16.** -40 **17.** 63
18. -27 **19.** -24 **[1.4]** **20.** $a - 5$ **21.** $6m$
22. $4(m + n)$ **23.** $\dfrac{a + b}{3}$ **[1.5]** **24.** -25 **25.** 3
26. -5 **27.** Undefined   **28.** 17 **29.** 65 **30.** 144
31. -9 **32.** -4 **[1.6]** **33.** $13a$ **34.** $19x + 5y$
35. $8a^2$ **[1.7]** **36.** a^{14} **37.** $8x^7y^3$ **38.** $3x^6$ **39.** $4ab^3$
40. x^9 **41.** $2x - 8$ **42.** $2w + 4$

Pretest Chapter 2

1. No   **2.** Yes   **3.** 8 **4.** -12 **5.** 7 **6.** 35
7. -2 **8.** 2 **9.** $W = \dfrac{P - 2L}{2}$ or $W = \dfrac{P}{2} - L$
10. $y = \dfrac{5x - 14}{3}$ or $y = \dfrac{5}{3}x - \dfrac{14}{3}$ **11.** $4x + 5 = 17$
12. $4(y + 6) = 10y + 6$ **13.** 6 **14.** $15, 17$
15. $4\text{ cm} \times 13\text{ cm}$ **16.** 540 **17.** $13,125$ **18.** 15%
19. $x \le 15$ **20.** $x \le -1$

Reading Your Text

Section 2.1 **(a)** fresh; **(b)** equation; **(c)** true; **(d)** solutions
Section 2.2 **(a)** equivalent; **(b)** multiplication; **(c)** dividing;
          (d) reciprocal
Section 2.3 **(a)** isolate; **(b)** addition; **(c)** original; **(d)** denominators
Section 2.4 **(a)** Formulas; **(b)** solving; **(c)** read; **(d)** equal
Section 2.5 **(a)** draw; **(b)** Mixture; **(c)** amount; **(d)** identifying
Section 2.6 **(a)** inequality; **(b)** smaller; **(c)** solution; **(d)** negative

Summary Exercises for Chapter 2

1. Yes   **3.** Yes   **5.** No   **7.** 2 **9.** -7 **11.** 5
13. 1 **15.** -7 **17.** 7 **19.** -4 **21.** 32 **23.** 27
25. 3 **27.** -2 **29.** $\dfrac{7}{2}$ **31.** 18 **33.** 6 **35.** $\dfrac{2}{5}$
37. 6 **39.** 6 **41.** 4 **43.** 5 **45.** $\dfrac{1}{2}$ **47.** $\dfrac{P - 2W}{2}$ or
$\dfrac{P}{2} - W$ **49.** $\dfrac{2A}{b}$ **51.** $mq + p$ **53.** 8 **55.** $17, 19, 21$
57. Susan: 7 yr; Larry: 9 yr; Nathan: 14 yr   **59.** 22%
61. $18,200$ **63.** $2,800$ **65.** 6.5%
67. Before: $3,150$; after: $3,276$ **69.** 500 s   **71.** 114.50
73. $x > -5$

75. $x \ge -3$

77. $x \le -15$

79. $x \ge 3$

81. $x \le 7$

83. $x > -6$

Self-Test for Chapter 2

[2.1] **1.** No   **2.** Yes   **[2.1–2.3]** **3.** 11 **4.** 12 **5.** 7
6. 7 **7.** -12 **8.** 25 **9.** 3 **10.** 4 **11.** $-\dfrac{2}{3}$
12. $-\dfrac{9}{4}$ **[2.4]** **13.** $\dfrac{C}{2\pi}$ **14.** $\dfrac{3V}{B}$ **15.** $\dfrac{6 - 3x}{2}$
[2.6] **16.** $x \le 14$

17. $x < -4$

18. $x \ge \dfrac{4}{3}$

19. $x > -1$

[2.4–2.5] **20.** 7 **21.** $21, 22, 23$
22. Juwan: 6; Jan: 12; Rick: 17   **23.** 10 in., 21 in.   **24.** 5%
25. $35,000$

Cumulative Review Chapters 0–2

[1.2–1.3] **1.** 4 **2.** -12 **3.** 8 **4.** 3 **5.** -18
6. 44 **7.** -5 **8.** 10 **9.** 0 **10.** Undefined
[1.5] **11.** 20 **12.** -11 **13.** 27 **14.** -28 **15.** -4
16. 2 **[1.6–1.7]** **17.** $3x^2y$ **18.** $6x^4 - 10x^3y$
19. $x - 2y + 3$ **20.** $12x^2 + 3x$ **[2.1–2.3]** **21.** 5
22. -24 **23.** $\dfrac{5}{4}$ **24.** $-\dfrac{2}{5}$ **25.** 5
[2.4] **26.** $\dfrac{I}{Pt}$ **27.** $\dfrac{2A}{b}$ **28.** $\dfrac{c - ax}{b}$
[2.6] **29.** $x < 3$

30. $x \le -\dfrac{3}{2}$

31. $x > 4$

32. $x \ge \dfrac{4}{3}$

[2.4–2.5] **33.** 13 **34.** $42, 43$ **35.** 7 **36.** 420
37. 5 cm, 17 cm   **38.** 8 in., 13 in., 16 in.   **39.** 2.5%
40. 7.5%

Pretest Chapter 3

1. x^{12} **2.** $8x^5y^7$ **3.** $3x^3y$ **4.** $4x^6y^8$ **5.** x^{16}
6. $\dfrac{2y^3}{x^5}$ **7.** Binomial   **8.** Trinomial   **9.** $2x^2 - 2x - 2$
10. $9x^2 - 11x + 6$ **11.** $12x^3y^3 - 6x^2y^2 + 21x^2y^4$
12. $6x^2 - 11x - 10$ **13.** $x^2 - 4y^2$ **14.** $16m^2 + 40m + 25$
15. $3x^3 - 14x^2y + 17xy^2 - 6y^3$ **16.** $9x^3 - 30x^2y + 25xy^2$
17. $4y - 5x^2y^3$ **18.** $x + 2$ **19.** $x - 3$
20. $3x - 5 - \dfrac{5}{x + 4}$

The Streeter/Hutchison Series in Mathematics   ·   Beginning Algebra

Reading Your Text

Section 3.1 **(a)** multiplication; **(b)** factor; **(c)** term; **(d)** coefficient
Section 3.2 **(a)** zero; **(b)** negative; **(c)** light; **(d)** Light
Section 3.3 **(a)** sum; **(b)** first; **(c)** place; **(d)** terms
Section 3.4 **(a)** exponents; **(b)** first; **(c)** outer; **(d)** three
Section 3.5 **(a)** exponents; **(b)** simplify; **(c)** three; **(d)** zero

Summary Exercises for Chapter 3

1. x^7 **3.** x **5.** $2p^2$ **7.** $5m^5n^2$ **9.** $8p^2q^2$ **11.** $4a^2b^2$
13. $72x^{12}y^8$ **15.** x **17.** $27y^{12}$ **19.** -4 **21.** -19
23. Binomial   **25.** Trinomial   **27.** Binomial   **29.** $9x, 1$
31. $x + 5, 1$ **33.** $7x^6 + 9x^4 - 3x, 6$ **35.** 1   **37.** 1
39. $\dfrac{1}{3^3}$ **41.** $\dfrac{4}{x^4}$ **43.** $\dfrac{1}{m^2}$ **45.** $\dfrac{x^5}{y^5}$ **47.** $\dfrac{1}{a^{18}}$
49. 5.1×10^4 cps   **51.** 3.22×10^9 **53.** 2×10^{17}
55. $21a^2 - 2a$ **57.** $5y^3 + 4y$ **59.** $5x^2 + 3x + 10$
61. $x - 2$ **63.** $-9w^2 - 10w$ **65.** $9b^2 + 8b - 2$
67. $2x^2 - 2x - 9$ **69.** $5a^5$ **71.** $54p^5$ **73.** $15x - 40$
75. $-10r^3s^2 + 25r^2s^2$ **77.** $x^2 + 9x + 20$ **79.** $a^2 - 49b^2$
81. $a^2 + 7ab + 12b^2$ **83.** $6x^2 - 19xy + 15y^2$
85. $y^3 - y + 6$ **87.** $x^3 - 8$ **89.** $2x^3 - 2x^2 - 60x$
91. $x^2 + 14x + 49$ **93.** $4w^2 - 20w + 25$
95. $a^2 + 14ab + 49b^2$ **97.** $x^2 - 25$ **99.** $4m^2 - 9$
101. $25r^2 - 4s^2$ **103.** $2x^3 - 20x^2 + 50x$
105. $3a^3$ **107.** $3a - 2$ **109.** $-3rs + 6r^2$ **111.** $x - 5$
113. $x - 3 + \dfrac{2}{x - 5}$ **115.** $x^2 + 2x - 1 + \dfrac{-4}{6x + 2}$
117. $x^2 + x + 2 + \dfrac{1}{x + 2}$

Self-Test for Chapter 3

[3.1] 1. a^{14} **2.** $15x^3y^7$ **3.** $2x^3$ **4.** $4ab^3$ **5.** $27x^6y^3$
6. $\dfrac{4w^4}{9t^6}$ **7.** $16x^{18}y^{17}$ **8.** 6   **9.** Binomial   **10.** Trinomial
11. $8x^4 - 3x^2 - 7; 8, -3, -7; 4$ **[3.2] 12.** 1   **13.** 6
14. $\dfrac{1}{y^5}$ **15.** $\dfrac{3}{b^7}$ **16.** $\dfrac{1}{y^4}$ **17.** $\dfrac{1}{p^{10}}$
[3.3] 18. $10x^2 - 12x - 7$ **19.** $7a^3 + 11a^2 - 3a$
20. $3x^2 + 11x - 12$ **21.** $b^2 - 7b - 5$ **22.** $7a^2 - 10a$
23. $4x^2 + 5x - 6$ **24.** $2x^2 - 7x + 5$
[3.4] 25. $15a^3b^2 - 10a^2b^2 + 20a^2b^3$ **26.** $3x^2 + x - 14$
27. $2x^3 + 7x^2y - xy^2 - 2y^3$ **28.** $8x^2 - 14xy - 15y^2$
29. $12x^3 + 11x^2y - 5xy^2$ **30.** $9m^2 + 12mn + 4n^2$
31. $a^2 - 49b^2$ **[3.5] 32.** $2x^2 - 3y$ **33.** $4c^2 - 6 + 9cd$
34. $x - 6$ **35.** $x + 2 + \dfrac{10}{2x - 3}$ **36.** $2x^2 - 3x + 2 + \dfrac{7}{3x + 1}$
37. $x^2 - 4x + 5 + \dfrac{-4}{x - 1}$ **[3.2] 38.** 1.68×10^{20}
39. 3.12×10^{-10} **40.** 5.2×10^{19}

Cumulative Review Chapters 0–3

[1.2–1.3] 1. 17   **2.** 6   **3.** 150   **4.** 4   **[1.5] 5.** 55
6. $-\dfrac{26}{21}$ **[3.1] 7.** $9x^{16}$ **8.** $\dfrac{x^{10}}{y^6}$ **9.** $8x^9y^3$ **[3.2] 10.** 7
11. 1   **12.** $\dfrac{1}{x^4}$ **13.** $\dfrac{3}{x^2}$ **14.** $\dfrac{1}{x^4}$ **15.** $\dfrac{1}{x^3y^3}$
[1.6] 16. $4x^5y$ **[3.3] 17.** $x^2 + 7x$ **18.** $2x - 2y$
[3.4] 19. $x^2 - 2x - 15$ **20.** $x^2 + 2xy + y^2$

21. $9x^2 - 24xy + 16y^2$ **[3.5] 22.** $x + 4$
[3.4] 23. $x^3 - xy^2$ **[2.3] 24.** -2 **25.** -2 **26.** 84
27. 1   **[2.4] 28.** $B = 2A - b$ **[2.6] 29.** $x \geq -2$
30. $x < -22$ **[2.5] 31.** Sam: \$510; Larry: \$250   **32.** 37, 39
33. \$2,120   **34.** \$645

Pretest Chapter 4

1. $5(3c + 7)$ **2.** $4q^3(2q - 5)$ **3.** $6(x^2 - 2x + 4)$
4. $7cd(c^2d - 3 + 2d^2)$ **5.** $(b - 3)(b + 5)$
6. $(x + 4)(x + 6)$ **7.** $(x - 9)(x - 5)$ **8.** $(a + 3b)(a + 4b)$
9. $(3y - 4)(y + 3)$ **10.** $(5w + 3)(w + 4)$
11. $(3x + 7y)(2x - 3y)$ **12.** $x(2x + 3)(x - 5)$
13. $(b + 7)(b - 7)$ **14.** $(6p + q)(6p - q)$ **15.** $(3x - 2y)^2$
16. $3x(3y - 4x)(3y + 4x)$ **17.** 4, 7   **18.** $-2, 7$
19. $\dfrac{3}{5}, -2$ **20.** 0, 2

Reading Your Text

Section 4.1 **(a)** greatest common; **(b)** coefficients; **(c)** factors; **(d)** same
Section 4.2 **(a)** multiplication; **(b)** middle; **(c)** positive; **(d)** factors
Section 4.3 **(a)** plus; **(b)** opposite; **(c)** GCF; **(d)** algorithm
Section 4.4 **(a)** multiples; **(b)** factored; **(c)** greatest common; **(d)** sum
Section 4.5 **(a)** GCF; **(b)** binomial; **(c)** completely; **(d)** grouping
Section 4.6 **(a)** quadratic; **(b)** zero-product; **(c)** zero; **(d)** repeated

Summary Exercises for Chapter 4

1. $6(3a + 4)$ **3.** $8s^2(3t - 2)$ **5.** $7s^2(5s - 4)$
7. $9m^2n(2n - 3 + 2n^2)$ **9.** $8ab(a + 3 - 2b)$
11. $(x + y)(2x - y)$ **13.** $(x + 4)(x + 5)$ **15.** $(a - 4)(a + 3)$
17. $(x + 6)(x + 6)$ **19.** $(b - 7c)(b + 3c)$
21. $m(m + 7)(m - 5)$ **23.** $3y(y - 7)(y - 9)$
25. $(3x + 5)(x + 1)$ **27.** $(2b - 3)(b - 3)$
29. $(5x - 3)(2x - 1)$ **31.** $(3y - 5z)(3y + 4z)$
33. $4x(2x + 1)(x - 5)$ **35.** $3x(2x - 3)(x + 1)$
37. $(p + 7)(p - 7)$ **39.** $(m + 3n)(m - 3n)$
41. $(5 + z)(5 - z)$ **43.** $(5a + 6b)(5a - 6b)$
45. $3w(w + 2z)(w - 2z)$ **47.** $2(m + 6n^2)(m - 6n^2)$
49. $(x + 4)^2$ **51.** $(2x + 3)^2$ **53.** $x(4x + 5)^2$
55. $(x - 4)(x + 5)$ **57.** $(3x + 2)(2x - 5)$
59. $x(2x + 3)(3x - 2)$ **61.** $1, -\dfrac{3}{2}$ **63.** 0, 10   **65.** $-3, 5$
67. $2, \dfrac{5}{4}$ **69.** 0, 3   **71.** $-4, 4$

Self-Test for Chapter 4

[4.1] 1. $6(2b + 3)$ **2.** $3p^2(3p - 4)$ **3.** $5(x^2 - 2x + 4)$
4. $6ab(a - 3 + 2b)$ **[4.4] 5.** $(a - 5)(a - 5)$
6. $(8m + n)(8m - n)$ **7.** $(7x + 4y)(7x - 4y)$
8. $2b(4a + 5b)(4a - 5b)$ **[4.2] 9.** $(a - 7)(a + 2)$
10. $(b + 3)(b + 5)$ **11.** $(x - 4)(x - 7)$
12. $(y + 10z)(y + 2z)$ **13.** $(x + 2)(x - 5)$
14. $(2x - 3)(3x + 1)$ **[4.5] 15.** $(2x - 1)(x + 8)$
16. $(3w + 7)(w + 1)$ **17.** $(4x - 3y)(2x + y)$
18. $3x(2x + 5)(x - 2)$ **[4.6] 19.** 3, 5   **20.** $-1, 4$
21. $-1, \dfrac{2}{3}$ **22.** 0, 3   **23.** 0, 4   **24.** $-3, 8$ **25.** 7

Cumulative Review Chapters 0–4

[1.2–1.3] 1. 17 **2.** -2 **[3.3] 3.** $9x^2 - x - 5$
4. $-4a^2 - 2a - 5$ **5.** $6b^2 + 8b - 3$
[3.4] 6. $15r^3s^2 - 12r^2s^2 + 18r^2s^3$ **7.** $6a^3 - 5a^2b + 3ab^2 - b^3$
[3.5] 8. $-y^2 + 3xy - 2x^2$ **9.** $3a + 2$ **10.** $x^2 - 2x + \dfrac{5}{2x + 4}$

[2.3] 11. $x = -2$ **[2.6] 12.** $x \le \dfrac{33}{5}$ **[2.4] 13.** $t = \dfrac{2S - na}{n}$
[3.1] 14. x^{17} **15.** $6x^5y^7$ **16.** $9x^4y^6$ **17.** $4xy^2$
18. $108x^8$ **[4.5] 19.** $12w^4(3w - 4)$ **20.** $5xy(x - 3 + 2y)$
21. $(5x + 3y)^2$ **22.** $4p(p + 6q)(p - 6q)$ **23.** $(a + 3)(a + 1)$
24. $2w(w^2 - 2w - 12)$ **25.** $(3x + 2y)(x + 3y)$ **[4.6] 26.** 3, 4
27. $-4, 4$ **28.** $\dfrac{2}{3}, -1$ **29.** 6 **30.** 5 in. by 21 in.

Pretest Chapter 5

1. $-\dfrac{3b^6}{5a^2}$ **2.** $\dfrac{x + 4}{2}$ **3.** $\dfrac{x - 1}{2x}$ **4.** $\dfrac{13a}{6}$ **5.** 5
6. $x + 6$ **7.** $\dfrac{5w - 6}{2w^2}$ **8.** $\dfrac{3b + 3}{b(b - 3)}$ **9.** $-\dfrac{11}{6(x - 1)}$
10. $\dfrac{10}{x - 5}$ **11.** $\dfrac{a}{2b^2}$ **12.** $\dfrac{x + 3}{2x}$ **13.** $2b^2$ **14.** $\dfrac{x + y}{4x}$
15. $\dfrac{3x}{2}$ **16.** $\dfrac{y}{2y + x}$ **17.** 3 **18.** $-2, 5$ **19.** 40
20. 6 **21.** 5 **22.** 7 **23.** 5, 20
24. Going: 48 mi/h; returning: 40 mi/h **25.** 15 ft, 40 ft

Reading Your Text

Section 5.1 (a) equivalent; **(b)** nonzero; **(c)** simplest;
(d) negative
Section 5.2 (a) multiplying; **(b)** simplify; **(c)** invert; **(d)** zero
Section 5.3 (a) like; **(b)** simplest; **(c)** parentheses; **(d)** factor
Section 5.4 (a) rational; **(b)** common; **(c)** simplest; **(d)** opposite
Section 5.5 (a) invert; **(b)** nonzero; **(c)** fraction; **(d)** LCD
Section 5.6 (a) expressions; **(b)** LCD; **(c)** excluded;
(d) original
Section 5.7 (a) original; **(b)** time; **(c)** rate; **(d)** variable

Summary Exercises for Chapter 5

1. $\dfrac{2}{3a}$ **3.** $\dfrac{w^2 - 25}{2w - 8}$ **5.** $\dfrac{-m - 1}{m + 3}$ **7.** $\dfrac{2}{3x}$ **9.** $\dfrac{x}{2}$
11. $\dfrac{2}{3p}$ **13.** $\dfrac{2x - 3}{x}$ **15.** $\dfrac{a + b}{4a}$ **17.** $\dfrac{x}{3}$ **19.** $\dfrac{11}{x + 2}$
21. $\dfrac{2r - s}{r}$ **23.** 2 **25.** $\dfrac{7x}{6}$ **27.** $\dfrac{5m - 6}{2m^2}$
29. $\dfrac{3x + 3}{x(x - 3)}$ **31.** $\dfrac{3w - 5}{(w - 5)(w - 3)}$ **33.** $-\dfrac{11}{6(x - 1)}$
35. $\dfrac{5a}{(a + 4)(a - 1)}$ **37.** $\dfrac{2}{3x}$ **39.** $\dfrac{y + x}{y - x}$ **41.** $\dfrac{n - m}{n + m}$
43. $\dfrac{a + 3}{a - 3}$ **45.** None **47.** $-1, 2$ **49.** $-1, -2$
51. 7 **53.** 40 **55.** 6 **57.** 7 **59.** 8 **61.** 4, 8
63. 4, 12 **65.** 48 mi/h, 40 mi/h **67.** 120 mi/h
69. 150 mL

Self-Test for Chapter 5

[5.1] 1. $-\dfrac{3x^4}{4y^2}$ **2.** $\dfrac{4}{a}$ **3.** $\dfrac{x + 1}{x - 2}$ **[5.3] 4.** a **5.** 2
6. 5 **[5.4] 7.** $\dfrac{17x}{15}$ **8.** $\dfrac{3s - 2}{s^2}$ **9.** $\dfrac{4x + 17}{(x - 2)(x + 3)}$
10. $\dfrac{15}{w - 5}$ **[5.2] 11.** $\dfrac{4p^2}{7q}$ **12.** $\dfrac{2}{x - 1}$ **13.** $\dfrac{3}{4y}$ **14.** $\dfrac{3}{m}$
[5.5] 15. $\dfrac{2}{3x}$ **16.** $\dfrac{n}{2n + m}$ **[5.6] 17.** 4 **18.** $-3, 3$
[5.7] 19. 36 **20.** 2, 6 **21.** 6 **22.** 4 **23.** 4, 12
24. 50 mi/h, 45 mi/h **25.** 20 ft, 35 ft

Cumulative Review Chapters 0–5

[1.6–1.7] 1. $-2xy$ **2.** $\dfrac{4a^2}{3}$ **3.** $2x^2 - 5x + 6$
4. $3a^2 + 6a + 1$ **[0.4] 5.** 31 **[0.5] 6.** 1
[3.4] 7. $2x^2 - xy - 6y^2$ **8.** $x^2 + 11x + 28$
[3.5] 9. $2x - 1 + \dfrac{1}{x + 2}$ **10.** $x + 1 - \dfrac{4}{x - 1}$ **[2.3] 11.** 4
12. -2 **[4.5] 13.** $(x - 7)(x + 2)$ **14.** $3mn(m - 2n + 3)$
15. $(a + 3b)(a - 3b)$ **16.** $2x(x - 6)(x - 8)$ **[2.4] 17.** 7
[5.7] 18. 4 **19.** 264 ft/s **[4.6] 20.** 5 in. by 7 in.
[5.1] 21. $\dfrac{m}{3}$ **22.** $\dfrac{a - 7}{3a + 1}$ **[5.4] 23.** $\dfrac{8r + 3}{6r^2}$
24. $\dfrac{x + 33}{3(x - 3)(x + 3)}$ **[5.2] 25.** $\dfrac{3}{x}$ **26.** $\dfrac{1}{3w}$
[5.5] 27. $\dfrac{x - 1}{2x + 1}$ **28.** $\dfrac{n}{3n + m}$ **[5.6] 29.** $\dfrac{6}{5}$ **30.** $-\dfrac{9}{2}, 7$

Pretest Chapter 6

1. $(15, 3), (18, 6)$ **2.** $(0, 3), (2, 0)$
3. $(1, 3), (0, 5), (-3, 11)$ **4.** Answers will vary.
5. $A(2, 3), B(0, -5), C(-3, 5)$
6.

7.

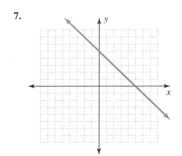

— The Streeter/Hutchison Series in Mathematics — Beginning Algebra

8.

9.

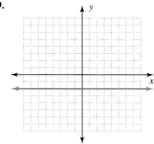

10. 1   **11.** −1   **12.** 8

13.

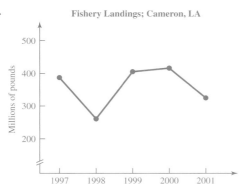

Fishery Landings; Cameron, LA

14. Predicted: answers will vary; actual: 350 million pounds
15. $1,000

Reading Your Text

Section 6.1 (a) solution; (b) expressions; (c) pair; (d) ordered-pair
Section 6.2 (a) origin; (b) quadrants; (c) negative; (d) zero
Section 6.3 (a) linear; (b) vertical; (c) horizontal; (d) intercept
Section 6.4 (a) slope; (b) run; (c) undefined; (d) variation
Section 6.5 (a) table; (b) circle; (c) legend; (d) line

Summary Exercises for Chapter 6

1. Yes   **3.** Yes   **5.** No   **7.** $(6, 0), (3, -3), (0, -6)$
9. $(3, 0), (-3, 4), (0, 2)$ **11.** $(4, 4), (0, 8), (8, 0), (6, 2)$
13. $(3, 0), (6, -2), (9, -4), (-3, 4)$ **15.** $(0, 10), (2, 8),$
$(4, 6), (6, 4)$ **17.** $(0, -2), (3, 0), (6, 2), (9, 4)$ **19.** $(4, 6)$

21. $(-1, -5)$ **23.–25.**

27.

29.

31.

33.

35.

37.

39.

41.

43.

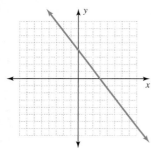

45. $y = -\dfrac{3}{2}x + 3$

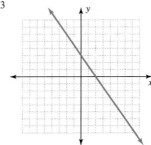

47. -3 **49.** $\dfrac{2}{3}$ **51.** $-\dfrac{5}{2}$ **53.** Undefined

55. -2 **57.** $-\dfrac{2}{3}$

59.

61.

63. $\dfrac{5}{3}$

65.

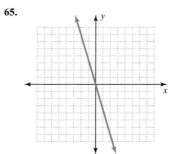

67. 41,378   **69.** 52%   **71.** 23%   **73.** 7,029,990
75. 40%   **77.** 500%

79.

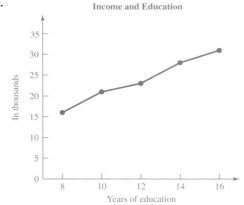

Income and Education

Self-Test for Chapter 6

[6.1] 1. $(3, 6), (9, 0)$ **2.** $(4, 0), (5, 4)$ **3.** $(3, 3), (6, 2), (9, 1)$
4. $(3, 0), (0, 4), \left(\dfrac{3}{4}, 3\right)$ **5.** Answers will vary.
6. Answers will vary. **[6.2] 7.** $(4, 2)$ **8.** $(-4, 6)$
9. $(0, -7)$

10–12.

[6.3] 13.

14.

15.

16.

17.

18.

[6.4] 19. 1 **20.** $\dfrac{3}{4}$ **21.** Undefined **22.** 0 **23.** -7
24.

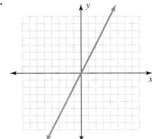

25. 5   **[6.5] 26.** 15%   **27.** $17,200,000   **28.** 2,500
29. Decreased by 2,000   **30.** 1980 to 1985; 2,000

Cumulative Review Chapters 0–6

[1.2–1.3] 1. 3   **2.** 5   **3.** 37   **4.** −53   **5.** 69
6. −120   **7.** −5   **8.** 3   **[1.5] 9.** 108   **10.** 3

11. 69   **12.** 9   **[2.3] 13.** $-\dfrac{4}{3}$ **14.** $-\dfrac{8}{3}$

15. 10   **16.** 1   **17.** $C = \dfrac{5}{9}(F - 32)$ **[2.6] 18.** $x < 4$

19. $x \le 3$ **[3.1] 20.** $\dfrac{1}{x^4 y^6}$ **21.** $\dfrac{y^5}{x}$ **22.** 1

[3.3] 23. $5x^2 - 10$ **24.** $-7a^2 + 7a + 2$ **25.** 19   **26.** 26
[3.4] 27. $6x^2 + 2xy - 20y^2$ **28.** $6x^3 - 3x^2 - 45x$
29. $4a^2 - 49b^2$ **[4.5] 30.** $4pn^2(3p + 5 - 4n)$
31. $(y - 3)(y^2 - 5)$ **32.** $b(3a + 7)(3a - 7)$
33. $2(3x + 2)(x - 1)$ **34.** $(2a + 3b)(3a - b)$

[4.6] 35. $-3, 11$ **36.** $\dfrac{4}{5}, \dfrac{2}{7}$ **[5.1] 37.** $\dfrac{-5a^3}{3b^2}$

38. $\dfrac{w - 2}{w + 3}$ **[5.3] 39.** $\dfrac{a + 5}{a(a - 5)}$ **40.** $\dfrac{10}{w - 5}$

[5.2] 41. $\dfrac{3x^3}{4}$ **42.** $\dfrac{m - 4}{4m}$ **[5.6] 43.** 6   **44.** 5

[6.3] 45.

46.

47.

[6.4] 48. 2

49.

50. 30   **[2.5] 51.** Width: 5 in.; length: 7 in.   **52.** 41, 43, 45
[2.5] 53. $200

Pretest Chapter 7

1. Slope: -2; y-intercept: $(0, 3)$ **2.** Slope: $\dfrac{5}{4}$; y-intercept: $(0, -5)$

3. $y = 5x - 2$

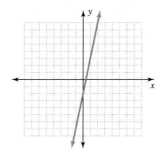

4. $y = -4x + 6$

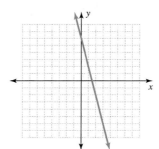

5. Parallel   **6.** Perpendicular   **7.** $y = -4x - 12$
8. $y = -\dfrac{5}{3}x + \dfrac{7}{3}$ **9.** $y = \dfrac{3}{5}x - 3$

10.

11.

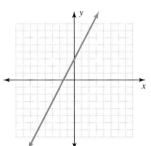

12. 13; −3

Reading Your Text

Section 7.1 **(a)** slope; **(b)** Fixed; **(c)** variable; **(d)** marginal
Section 7.2 **(a)** parallel; **(b)** intersection; **(c)** perpendicular;
(d) Horizontal
Section 7.3 **(a)** point-slope; **(b)** horizontal; **(c)** vertical;
(d) reciprocals
Section 7.4 **(a)** half-plane; **(b)** dotted; **(c)** origin; **(d)** feasible
Section 7.5 **(a)** evaluate; **(b)** simplifying; **(c)** output; **(d)** break-even

Summary Exercises for Chapter 7

1. Slope: 2; y-intercept: $(0, 5)$ **3.** Slope: $-\frac{3}{4}$; y-intercept: $(0, 0)$

5. Slope: $-\frac{2}{3}$; y-intercept: $(0, 2)$ **7.** Slope: 0; y-intercept: $(0, -3)$

9. $y = 2x + 3$

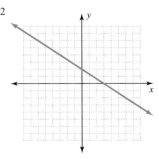

11. $y = -\frac{2}{3}x + 2$

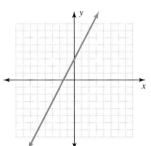

13. Perpendicular   **15.** Parallel   **17.** $y = -3$

19. $x = 4$ **21.** $y = -3$ **23.** $y = -\frac{4}{3}x - 2$

25. $x = -\frac{5}{2}$ **27.** $y = -\frac{1}{5}x + 4$ **29.** $y = -\frac{5}{4}x + 2$

31. $y = -\frac{5}{3}x + 3$ **33.** $y = \frac{4}{3}x + \frac{14}{3}$

35.

37.

39.

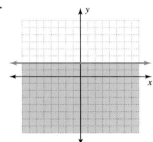

41. **(a)** −7; **(b)** −13; **(c)** −17   **43.** **(a)** −39; **(b)** −9; **(c)** −3
45. **(a)** 9; **(b)** −5; **(c)** −1   **47.** $f(x) = -7x - 3$

49. $f(x) = -\frac{3}{2}x - 6$

51.

53.

55.

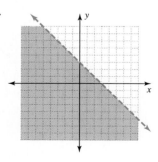

57. 5, 2   **59.** 5, 9   **61.** $7a - 1, 21b - 1, 7x - 8$

Self-Test for Chapter 7

[7.1] 1. Slope: $\dfrac{4}{5}$; y-intercept: $(0, -2)$ **2.** Slope: $-\dfrac{2}{3}$;
y-intercept: $(0, -9)$

3. $y = -3x + 6$

4. $y = 5x - 3$

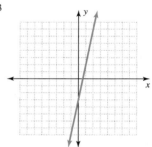

[7.2] 5. Parallel   **6.** Perpendicular   **7.** $y = \dfrac{1}{2}x + \dfrac{21}{2}$

[7.3] 8. $y = \dfrac{4}{3}x + \dfrac{25}{3}$ **9.** $y = -\dfrac{3}{2}x + \dfrac{5}{2}$

10. $y = -\dfrac{2}{3}x + 5$ **11.** $y = -8$

[7.4] 12.

13.

14.

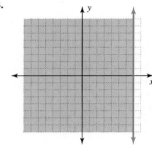

[7.5] 15. $-5; 7$ **16.** $-15; -7$ **17.** $-15; 6$
18. $3a - 25; 3x - 28$

Cumulative Review Chapters 0–7

1. [3.3] $x^2y^2 - 3xy$ **2.** $\dfrac{4m^3n}{3}$ **3.** $-x + 9$

4. $3z^2 - 3z + 5$ **[3.4] 5.** $2x^2 + 11x - 21$

6. $2a^2 + 6ab - 8b^2$ **[3.5] 7.** $x + 6 + \dfrac{20}{x - 3}$

8. $x^3 - 2x^2 + 4x - 10 + \dfrac{20}{x - 2}$ **[2.3] 9.** $-\dfrac{4}{3}$

10. -2 **[4.5] 11.** $(x - 8)(x + 7)$ **12.** $2x^2y(2x - y + 4x^2)$

13. $2a(2a + 3b)(2a - 3b)$ **14.** $3(5x - 2y)(x - y)$

[6.4] 15. 1 **16.** $-\dfrac{9}{4}$ **[5.2–5.4] 17.** $\dfrac{2x - 3}{x}$ **18.** $\dfrac{1}{a - 7}$

19. $\dfrac{5m + 6}{2m^2}$ **20.** $\dfrac{2x + 6}{x(x - 3)}$ **21.** $\dfrac{5y}{(y + 4)(y - 1)}$

[5.6] 22. -4 **23.** $7, -10$ **[5.7] 24.** 9
25. Going: 49 mi/h; returning: 42 mi/h   **26.** 126 min

[7.2] 27. $y = -\dfrac{1}{7}x + 2$ **[7.1] 28.** $y = -5x + 3$

[7.4] 29.

[7.5] 30. -2

Pretest Chapter 8

1. Inconsistent

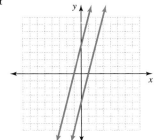

2. $(0, -2)$ **3.** $(6, -3)$

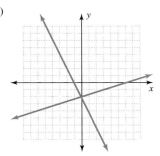

4. Dependent system   **5.** $(4, 1)$ **6.** $(3, -1)$ **7.** 16, 24
8. 13 m, 17 m   **9.** 19 dimes, 26 quarters
10. Canoe: 7.5 mi/h; current: 1.5 mi/h
11.

12.

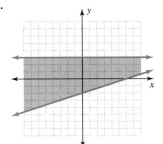

Reading Your Text

Section 8.1 **(a)** system; **(b)** solution; **(c)** consistent; **(d)** dependent
Section 8.2 **(a)** opposites; **(b)** original; **(c)** multiplying;
            (d) infinitely many
Section 8.3 **(a)** substitution; **(b)** parallel; **(c)** Graphing;
            (d) addition
Section 8.4 **(a)** intersection; **(b)** dashed; **(c)** bounded; **(d)** feasible

Summary Exercises for Chapter 8

1. $(4, 2)$

3. No solution

5. Infinite number of solutions   **7.** $(5, 3)$ **9.** $(5, -2)$

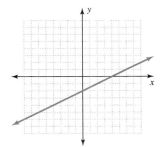

11. $(3, -1)$ **13.** $(-4, 2)$ **15.** $\left(2, -\dfrac{2}{3}\right)$

17. Inconsistent system   **19.** $\left(\dfrac{3}{5}, 2\right)$ **21.** $(6, -2)$

23. $(4, 2)$ **25.** Inconsistent system   **27.** $\left(-\dfrac{4}{3}, 2\right)$

29. $(8, -2)$ **31.** $(4, -2)$ **33.** $(5, 3)$

35. Inconsistent system   **37.** $(5, -4)$ **39.** $(-3, -2)$

41. $(-3, 0)$ **43.** $(8, 2)$ **45.** $(4, -6)$ **47.** $\left(0, \dfrac{4}{3}\right)$

49. $\left(\dfrac{1}{2}, \dfrac{9}{2}\right)$ **51.** $(2, -1)$ **53.** 4, 13

55. Tablet $1.50, pencil $0.25   **57.** Speakers $425, amplifier $500
59. Width 14 cm, length 18 cm   **61.** 10 nickels, 20 quarters
63. 200 mL of 20%, 400 mL of 50%   **65.** $10,000 at 11%,
$8,000 at 7%   **67.** Plane 500 mi/h, wind 50 mi/h

69.

71.

73.

75.

Self-Test for Chapter 8

[8.1] 1. (4, 1)

2. (4, 2)

3. Inconsistent system

4. (2, 4)

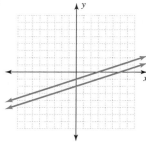

[8.2] 5. (4, 1)   **6.** (4, 2)   **7.** (1, 3)   **8.** $\left(2, \dfrac{5}{2}\right)$

9. Dependent system   **10.** $\left(\dfrac{3}{4}, -1\right)$ **11.** (6, 2)

12. Inconsistent system   **[8.3] 13.** (2, 6)   **14.** (6, −3)
15. (6, 2)   **16.** (5, 4)   **17.** (−3, 3)
18. Inconsistent system   **19.** (3, −5)   **20.** (3, 2)
21. 12, 18   **22.** 21 m, 29 m   **23.** Width 12 in., length 20 in.
24. 12 dimes, 18 quarters   **25.** Boat 15 mi/h, current 3 mi/h

[8.4] 26.

27.

28.

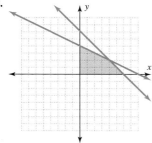

Cumulative Review Chapters 0–8

[3.3] 1. $8x^2 - 7x - 4$ **2.** $w^2 - 8w - 4$

[3.4] 3. $28x^3y^2 - 14x^2y^2 + 21x^2y^3$ **4.** $15s^2 - 23s - 28$

[3.5] 5. $-x^2 + 2xy - 3y$ **6.** $2x + 4$

[2.3] 7. 9   **[4.1–4.4] 8.** $8a^2(3a - 2)$ **9.** $7mn\,(m - 3 - 7n)$

10. $(a + 8b)(a - 8b)$ **11.** $5p(p + 4q)(p - 4q)$

12. $(a - 6)(a - 8)$ **13.** $2w(w - 7)(w + 3)$

[4.6] 14. 4, 5   **15.** $-4, 4$ **16.** 7   **17.** 5 in. by 17 in.

[5.1] 18. $\dfrac{m}{3}$ **19.** $\dfrac{a - 7}{3a + 1}$ **[5.2] 20.** $\dfrac{3}{x}$ **21.** $\dfrac{1}{3w}$

[7.1] 22.

23.

24.

25.

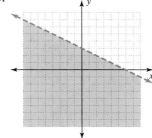

[6.4] 26. $\dfrac{10}{7}$ **[7.1] 27.** Slope: $\dfrac{5}{3}$; y-intercept: $(0, -5)$

28. $y = 2x - 5$

[7.4] 29.

30.

[8.1] 31. $(4, -3)$

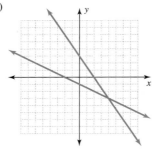

[8.2–8.3] 32. $\left(7, -\dfrac{5}{2}\right)$ **33.** Dependent system   **34.** $(5, 0)$

35. Inconsistent system   **36.** $\left(\dfrac{3}{2}, -\dfrac{1}{3}\right)$ **37.** 5, 21

38. VHS \$4.50, cassette \$1.50   **39.** 325 at \$70 and 125 at \$40
40. \$5,000 at 6%, \$7,000 at 9%

Pretest Chapter 9

1. 12   **2.** 4   **3.** Not a real number   **4.** 5   **5.** $6\sqrt{2}$

6. $3x\sqrt{3x}$ **7.** $\dfrac{1}{2}$ **8.** $\dfrac{\sqrt{7}}{5}$ **9.** $7\sqrt{7}$ **10.** $3\sqrt{5}$

11. $3\sqrt{3}$ **12.** $\sqrt{2}$ **13.** $x\sqrt{10}$ **14.** $\sqrt{10} - 6$

15. $9 + 5\sqrt{3}$ **16.** $\dfrac{\sqrt{15}}{3}$ **17.** $4 - \sqrt{5}$ **18.** 15

19. 24   **20.** 8.49   **21.** 4.12   **22.** 19.21 cm   **23.** 14
24. 6.71   **25.** 84

Reading Your Text

Section 9.1 (a) principal; **(b)** negative; **(c)** approximation;
    (d) radicand
Section 9.2 (a) factors; **(b)** fractions; **(c)** product; **(d)** denominator
Section 9.3 (a) radicand; **(b)** distributive; **(c)** positive; **(d)** cube
Section 9.4 (a) square root; **(b)** FOIL; **(c)** like; **(d)** rational
Section 9.5 (a) sides; **(b)** even; **(c)** lose; **(d)** three
Section 9.6 (a) Greek; **(b)** hypotenuse; **(c)** negative;
    (d) approximate

Summary Exercises for Chapter 9

1. 9   **3.** Not a real number   **5.** -4 **7.** Not a real number
9. $3\sqrt{5}$ **11.** $2x^2\sqrt{5}$ **13.** $10b\sqrt{2b}$ **15.** $6ab^2\sqrt{3b}$
17. $\dfrac{3x\sqrt{2}}{5}$ **19.** $\dfrac{\sqrt{21}}{7}$ **21.** $\dfrac{2x\sqrt{14}}{7}$ **23.** $6\sqrt{5}$
25. $2\sqrt{3a}$ **27.** $7\sqrt{3}$ **29.** $\sqrt{2}$ **31.** $2\sqrt{2} + \sqrt{3}$
33. $\sqrt{30}$ **35.** $\sqrt{6x}$ **37.** $5a\sqrt{2}$ **39.** $2\sqrt{21} - 21$
41. 6   **43.** $\dfrac{x\sqrt{21x}}{3}$ **45.** 21   **47.** 9   **49.** 10   **51.** 15
53. $5\sqrt{5}$ **55.** 15 in.   **57.** 24.1 ft   **59.** 4   **61.** $\sqrt{73}$
63. $\sqrt{82}$

Self-Test for Chapter 9

[9.1] 1. 11   **2.** 3   **3.** Not a real number   **4.** 4
[9.2] 5. $5\sqrt{3}$ **6.** $2a\sqrt{6a}$ **7.** $\dfrac{4}{5}$ **8.** $\dfrac{\sqrt{5}}{3}$

[9.3] 9. $4\sqrt{10}$ **10.** $3\sqrt{2}$ **11.** $3\sqrt{2}$ **12.** $4\sqrt{5}$
[9.4] 13. $3x\sqrt{2}$ **14.** $11 + 5\sqrt{5}$ **15.** $\dfrac{\sqrt{14}}{2}$
16. $2 + 3\sqrt{2}$ **[9.5] 17.** 83   **18.** 4   **[9.6] 19.** 20
20. 12   **21.** $3\sqrt{5}$ **22.** $\sqrt{15}$ **23.** 9.747 cm   **24.** 9
25. $\sqrt{85}$

Cumulative Review Chapters 0–9

[1.6] 1. $3x^2y^3 - 2x^3y$ **[3.3] 2.** $7x^2 - 6x + 12$ **[3.1] 3.** 0
4. -8 **[2.3] 5.** 2   **6.** 6   **[2.6] 7.** $x > 4$
[3.4] 8. $6x^4y - 10x^3y + 38x^2y$ **9.** $20x^2 - 23xy - 21y^2$
[4.5] 10. $9xy(4 - 3x^2y)$ **11.** $(4x - 3)(2x - 5)$
[5.4] 12. $\dfrac{1}{15(x + 7)}$ **[5.2] 13.** $\dfrac{x - 3}{x - 5}$

[6.3] 14.

[7.4] 15.

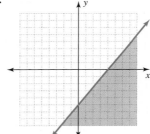

[6.4] 16. 5   **[7.3] 17.** $y = -\dfrac{3}{2}x + 5$ **[8.2] 18.** $(5, 0)$
19. Inconsistent system   **20.** Boat: 13 mi/h; current: 3 mi/h
[9.1] 21. 12   **22.** -12 **23.** Not a real number   **24.** -3
[9.2–9.4] 25. $-4a\sqrt{5}$ **26.** $\dfrac{2x\sqrt{6x}}{3}$ **27.** $4 - 2\sqrt{2}$
28. $7x\sqrt{2}$ **29.** $5mn\sqrt{6m}$ **30.** $\dfrac{2a\sqrt{3}}{5}$

Pretest Chapter 10

1. $\pm\sqrt{17}$ **2.** $\pm 2\sqrt{3}$ **3.** $1 \pm \sqrt{5}$ **4.** $\pm\dfrac{\sqrt{14}}{3}$
5. $-2, 5$ **6.** $\dfrac{5 \pm \sqrt{17}}{2}$ **7.** $2 \pm 2\sqrt{2}$
8. $\dfrac{2 \pm 3\sqrt{2}}{2}$ **9.** $-7, 2$ **10.** 5   **11.** $\dfrac{-3 \pm \sqrt{29}}{2}$
12. $\dfrac{3 \pm \sqrt{33}}{4}$ **13.** $\dfrac{2 \pm \sqrt{10}}{3}$ **14.** $\dfrac{-3 \pm \sqrt{37}}{2}$

15.

16.

17.

18.

Reading Your Text

Section 10.1 **(a)** quadratic; **(b)** divide; **(c)** square root; **(d)** real-number

Section 10.2 **(a)** perfect-square; **(b)** constant; **(c)** complete; **(d)** even

Section 10.3 **(a)** standard; **(b)** factored; **(c)** discriminant; **(d)** rational

Section 10.4 **(a)** parabola; **(b)** downward; **(c)** axis of symmetry; **(d)** vertex

Summary Exercises for Chapter 10

1. $\pm\sqrt{10}$ **3.** $\pm2\sqrt{5}$ **5.** $1\pm\sqrt{5}$ **7.** $-3\pm\sqrt{5}$

9. $\pm\dfrac{3\sqrt{3}}{2}$ **11.** $\pm\dfrac{\sqrt{7}}{5}$ **13.** $-2,5$ **15.** $\dfrac{5\pm\sqrt{17}}{2}$

17. $2\pm2\sqrt{2}$ **19.** $2\pm\sqrt{2}$ **21.** $\dfrac{1\pm\sqrt{29}}{2}$ **23.** $\dfrac{2\pm3\sqrt{2}}{2}$

25. $-2,7$ **27.** $\dfrac{-5\pm\sqrt{37}}{2}$ **29.** $3\pm2\sqrt{2}$ **31.** $\dfrac{2\pm\sqrt{10}}{3}$

33. $\dfrac{-3\pm\sqrt{37}}{2}$ **35.** $2\pm\sqrt{10}$ **37.** $0,\dfrac{3}{5}$ **39.** $1\pm\sqrt{10}$

41. $\dfrac{5\pm\sqrt{57}}{4}$ **43.** $-1,\dfrac{7}{2}$ **45.** $-1\pm\sqrt{6}$ **47.** $1\pm\sqrt{3}$

49.

51.

53.

55.

57.

59.

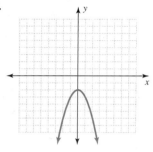

61. Axis: $x = -3$

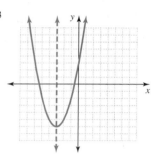

63. Axis: $x = \dfrac{5}{2}$

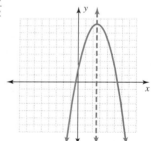

65. $(-5, 0), (3, 0)$ **67.** $(-3 - \sqrt{6}, 0), (-3 + \sqrt{6}, 0)$
69. None   **71.** 9.23 cm, 15.46 cm   **73.** 1.13 s, 3.87 s
75. 70 items

Self-Test for Chapter 10

[10.1] **1.** $\pm\sqrt{15}$ **2.** $\pm2\sqrt{2}$ **3.** $1 \pm \sqrt{7}$ **4.** $\pm\dfrac{\sqrt{10}}{3}$

[10.2] **5.** $4, -2$ **6.** $\dfrac{-3 \pm \sqrt{13}}{2}$ **7.** $-1 \pm \sqrt{6}$

8. $\dfrac{5 \pm \sqrt{17}}{4}$ [10.3] **9.** $-1, 3$ **10.** 3   **11.** $\dfrac{5 \pm \sqrt{33}}{2}$

12. $\dfrac{1 \pm \sqrt{11}}{2}$ **13.** $\dfrac{1 \pm \sqrt{33}}{4}$ **14.** $-1 \pm \sqrt{6}$
[10.4] **15.**

16.

17.

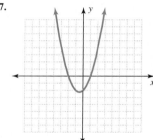

18. Axis: $x = 2$; $(-3, 0)$ and $(7, 0)$ **19.** 6.36 ft by 15.08 ft
20. 3.06 s

Cumulative Review Chapters 0–10

[1.6] **1.** $11x^2y - 6xy^2$ **2.** $x^2 + 5x - 7$ [1.5] **3.** -480
4. $-1{,}680$ [2.3] **5.** 1   [2.6] **6.** $x > 2$
[3.4] **7.** $6x^3y - 3x^2y + 15xy$ **8.** $6x^2 + 11x - 10$
9. $9x^2 - 16y^2$ [4.1–4.4] **10.** $8xy^2(2x - y)$
11. $(2x - 3)(4x + 5)$ **12.** $(5x - 4y)(5x + 4y)$

[5.2, 5.4] **13.** $\dfrac{29}{28(x + 2)}$ **14.** $\dfrac{5(x - 2)}{x - 1}$

15. $\dfrac{(3x - 1)(x + 6)}{15x^2}$

[6.4] **16.**

17.

[6.5] 18. 5   **[7.1] 19.** $y = 2x - 5$

[7.4] 20.

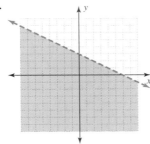

[8.2–8.3] 21. $\left(4, \dfrac{2}{3}\right)$ **22.** $(3, -2)$ **23.** $\left(3, -\dfrac{7}{2}\right)$

24. Dependent system   **25.** 5, 21

26. 325 reserved seats, 125 general admissions

27. 100 mL of 30%, 200 mL of 60%   **[9.1] 28.** 13   **29.** -13

30. Not a real number   **31.** -4 **[9.2–9.4] 32.** $6\sqrt{3}$

33. $30a\sqrt{3}$ **34.** $-13 - 2\sqrt{2}$ **35.** $2 - \sqrt{2}$

[10.1–10.3] 36. $\pm 6\sqrt{2}$ **37.** $-3 \pm 2\sqrt{3}$ **38.** $\dfrac{5 \pm \sqrt{41}}{4}$

[10.4] 39.

40.

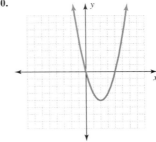

41. 1.657 s   **42.** $84.64

Graph Paper

Use these grids for graphing. Make as many copies of this page as you need. You can access an electronic version of these grids through the book's website. **Please visit www.mhhe.com/baratto**. Click on the cover for this book, and then click on the Graph Paper link in the Information Center.

The Streeter/Hutchison Series in Mathematics